THIRD EDITION

Basic Mathematics for College Students

Alan S. Tussy

Citrus College

R. David Gustafson

Rock Valley College

THOMSON

BROOKS/COLE

Australia • Canada • Mexico • Singapore • Spain • United Kingdom • United States

THOMSON

BROOKS/COLE

Basic Mathematics for College Students, Third Edition
Alan S. Tussy and R. David Gustafson

Executive Editor: Jennifer Laugier

Development Editor: Kirsten Markson

Assistant Editor: Rebecca Subity

Technology Project Manager: Sarah Woicicki

Marketing Manager: Greta Kleinert

Marketing Assistant: Jessica Bothwell

Advertising Project Manager: Bryan Vann

Project Manager, Editorial Production: Hal Humphrey

Art Director: Vernon Boes

Print/Media Buyer: Barbara Britton

Permissions Editor: Joohee Lee

Production Service: Helen Walden

Text Designer: Diane Beasley

Art Editor: Helen Walden

Photo Researcher: Helen Walden

Copy Editor: Carol Reitz

Illustrator: Lori Heckelman/LHI Technical Illustration

Cover Designer: Cheryl Carrington

Cover Image: Pete Atkinson/Getty Images

Cover Printer: Quebecor World/Dubuque

Compositor: G & S Book Services

Printer: Quebecor World/Dubuque

Study Skills Workshop photos from Getty Images

For more information about our products, contact us at:
Thomson Learning Academic Resource Center
1-800-423-0563

For permission to use material from this text or product, submit a request online at **http://www.thomsonrights.com**.

Any additional questions about permissions can be submitted by email to **thomsonrights@thomson.com**.

Library of Congress Control Number: 2004118198
Student Edition: ISBN 0-534-42223-3
Annotated Instructor's Edition: ISBN 0-534-42224-1

Thomson Higher Education
10 Davis Drive
Belmont, CA 94002
USA

Asia
Thomson Learning
5 Shenton Way #01-01
UIC Building
Singapore 068808

Australia/New Zealand
Thomson Learning
102 Dodds Street
Southbank, Victoria 3006
Australia

Canada
Nelson
1120 Birchmount Road
Toronto, Ontario M1K 5G4
Canada

Europe/Middle East/Africa
Thomson Learning
High Holborn House
50/51 Bedford Row
London WC1R 4LR
United Kingdom

Latin America
Thomson Learning
Seneca, 53
Colonia Polanco
11560 Mexico D.F.
Mexico

Spain/Portugal
Paraninfo
Calle Magallanes, 25
28015 Madrid, Spain

CONTENTS

6   Ratio, Proportion, and Measurement   327

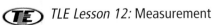
7   Descriptive Statistics   387

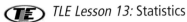
8   An Introduction to Algebra   419

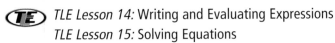

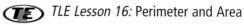

9   An Introduction to Geometry   491

Appendix I
Polynomials   A-1

Appendix II
Inductive and Deductive Reasoning   A-15

Appendix III
Roots and Powers   A-22

Appendix IV
Answers to Selected Exercises   A-23
(Appears in the Student Edition only)

PREFACE

For the Instructor

The purpose of this textbook is to teach students how to read, write, and think mathematically using the language of mathematics. It is written for students studying basic mathematics for the first time and for those who need a review of the basics. *Basic Mathematics for College Students,* Third Edition, employs a variety of instructional methods that reflect the recommendations of NCTM and AMATYC. You will find extensive opportunities for skill practice and the well-defined pedagogy that are hallmarks of a traditional approach to teaching mathematics. You will also find that we emphasize the reasoning, modeling, and communication skills that are a part of today's mathematics reform movement.

The third edition retains the basic philosophy of the second edition. However, we have made several improvements as a direct result of the comments and suggestions we received from instructors and students. Our goal has been to make the book more enjoyable to read, easier to understand, and more relevant.

New to This Edition

New features make this the most student relevant and engaging edition of this popular text.

New Chapter Openers with TLE Labs: TLE (The Learning Equation) is interactive courseware that uses a guided inquiry approach to teaching developmental math concepts. Each chapter opener integrates a TLE Lab with a real-world application, which offers an opportunity to seamlessly integrate TLE lessons into your course. For more information on TLE, instructors can see the Preview at the beginning of their book.

Check Your Knowledge: New pretests at the beginning of each chapter are helpful in gauging a student's knowledge base for the upcoming chapter. Instructors can assign the pretest to see how well students are prepared for the material in the chapter, and customize subsequent lessons based on the needs of their students. Students can also take the pretest themselves as a warm-up for the chapter and to structure review. Answers to the pretests appear at the back of the book.

Study Skills Workshop: This complete mini-course in mathematics study skills helps students and instructors tackle the problem of lack of preparedness and inadequate study habits in an organized series of lessons. Each chapter opens with a one-page Study Skills Workshop sequenced to address relevant study skills issues as the students move through the course. For example, students learn how to use a calendar to schedule study times in the first lesson, best practices for study groups around the time of a midterm exam, and how to effectively study for a final in one of the last lessons. This helpful reference can be used in the classroom or assigned as homework.

Think It Through: Each chapter contains one or two of these features, which make the connection between mathematics and student life. These problems are student-relevant and require mathematics skills from the chapter to be applied to a real-life situation. Topics include tuition costs, statistics about college life, job opportunities, and many more topics directly connected to the student experience.

A colorful new design visually organizes information on the page.

Proven Features of *Basic Mathematics*

- The authors' proven five-step problem-solving strategy teaches students to analyze the problem, form an equation, solve the equation, state the conclusion, and check the result. In a step-by-step manner, this approach clarifies the thought process and mathematical skills necessary to solve a wide variety of problems. As a result, students' confidence is increased and their problem-solving abilities are strengthened.

- STUDY SETS are found at the end of every section and feature a unique organization, tailored to improve students' ability to read, write, and communicate mathematical ideas, thereby approaching topics from a variety of perspectives. Each comprehensive STUDY SET is divided into seven parts: VOCABULARY, CONCEPTS, NOTATION, PRACTICE, APPLICATIONS, WRITING, and REVIEW.

  - VOCABULARY, NOTATION, and WRITING problems help students improve their ability to read, write, and communicate mathematical ideas.

  - The CONCEPT problems section in the STUDY SETS reinforces major ideas through exploration and fosters independent thinking and the ability to interpret graphs and data.

  - PRACTICE problems in the STUDY SETS provide the necessary drill for mastery while the APPLICATIONS provide opportunities for students to deal with real-life situations. Each STUDY SET concludes with a REVIEW section that consists of problems randomly selected from previous sections.

- SELF CHECK problems, adjacent to most worked examples, reinforce concepts and build confidence. The answer to each Self Check follows the problem to give students instant feedback.

- The KEY CONCEPT section is a one-page review, found at the end of each chapter, which revisits the importance of the role the concept plays in the overall picture.

- REAL-LIFE APPLICATIONS are presented from a number of disciplines, including science, business, economics, manufacturing, entertainment, history, art, music, and mathematics.

- CALCULATOR SNAPSHOT sections introduce keystrokes and show how scientific calculators can be used to solve application problems, for instructors who wish to integrate calculators into their course.

- CUMULATIVE REVIEW EXERCISES at the end of every chapter except Chapter 1 help students retain what they have learned in prior chapters.

For detailed information on the ancillary resources available for this text, instructors can see the Preview section at the beginning of their book.

Acknowledgments

We are grateful to the following people who reviewed this manuscript and the other manuscripts in the paperback series at various stages of development. They all had valuable suggestions that have been incorporated into the text.

The following people reviewed the first and second editions:

Linda Beattie
Western New Mexico University

Julia Brown
Atlantic Community College

Linda Clay
Albuquerque TVI

John Coburn
Saint Louis Community College–
Florissant Valley

Sally Copeland
Johnson County Community College

Ben Cornelius
Oregon Institute of Technology

James Edmondson
Santa Barbara Community College

David L. Fama
Germanna Community College

Barbara Gentry
Parkland College

Laurie Hoecherl
Kishwaukee College

Judith Jones
Valencia Community College

Therese Jones
Amarillo College

Joanne Juedes
University of Wisconsin–Marathon
County

Dennis Kimzey
Rogue Community College

Sally Lesik
Holyoke Community College

Elizabeth Morrison
Valencia Community College

Jan Alicia Nettler
Holyoke Community College

Scott Perkins
Lake–Sumter Community College

Angela Peterson
Portland Community College

J. Doug Richey
Northeast Texas Community College

Angelo Segalla
Orange Coast College

June Strohm
Pennsylvania State Community College–
DuBois

Rita Sturgeon
San Bernardino Valley College

Jo Anne Temple
Texas Technical University

Sharon Testone
Onondaga Community College

Marilyn Treder
Rochester Community College

Thomas Vanden Eynden
Thomas More College

The following people reviewed the books in this series in preparation for the third edition:

Cedric E. Atkins
Mott Community College

William D. Barcus
SUNY, Stony Brook

Kathy Bernunzio
Portland Community College

Girish Budhwar
United Tribes Technical College

Sharon Camner
Pierce College–Fort Steilacoom

Robin Carter
Citrus College

Ann Corbeil
Massasoit Community College

Carolyn Detmer
Seminole Community College

Maggie Flint
Northeast State Technical Community
College

Charles Ford
Shasta College

Michael Heeren
Hamilton College

Monica C. Kurth
Scott Community College

Sandra Lofstock
St. Petersberg College–Tarpon Springs
Center

Marge Palaniuk
United Tribes Technical College

Jane Pinnow
University of Wisconsin–Parkside

Eric Sims
Art Institute of Dallas

Annette Squires
Palomar College

Lee Ann Spahr
Durham Technical Community College

John Strasser
Scottsdale Community College

Stuart Swain
University of Maine at Machias

Celeste M. Teluk
D'Youville College

Sven Trenholm
Herkeimer County Community College

Stephen Whittle
Augusta State University

Mary Lou Wogan
Klamath Community College

Without the talents and dedication of the editorial, marketing, and production staff of Brooks/Cole, this revision of *Basic Mathematics* could not have been so well accomplished. We express our sincere appreciation for the hard work of Bob Pirtle, Jennifer Laugier, Helen Walden, Lori Heckleman, Vernon Boes, Diane Beasley, Sarah Woicicki, Greta Kleinert, Jessica Bothwell, Bryan Vann, Kirsten Markson, Rebecca Subity, Hal Humphrey, Jolene Rhodes, Christine Davis, Diane Koenig, and G & S Typesetters for their help in creating the book. Special thanks to David Casey of Citrus College for his hard work on the pretests and to Sheila Pisa for writing the excellent Study Skills Workshops.

Alan S. Tussy
R. David Gustafson

For the Student

Success in Mathematics

To be successful in mathematics, you need to know how to study it. The following checklist will help you develop your own personal strategy to study and learn the material. The suggestions below require some time and self-discipline on your part, but it will be worth the effort. This will help you get the most out of the course.

As you read each of the following statements, place a check mark in the box if you can truthfully answer Yes. If you can't answer Yes, think of what you might do to make the suggestion part of your personal study plan. You should go over this checklist several times during the semester to be sure you are following it.

Preparing for the Class

❑ I have made a commitment to myself to give this course my best effort.
❑ I have the proper materials: a pencil with an eraser, paper, a notebook, a ruler, a calculator, and a calendar or day planner.
❑ I am willing to spend a minimum of two hours doing homework for every hour of class.
❑ I will try to work on this subject every day.
❑ I have a copy of the class syllabus. I understand the requirements of the course and how I will be graded.
❑ I have scheduled a free hour after the class to give me time to review my notes and begin the homework assignment.

Class Participation

❑ I know my instructor's name.
❑ I will regularly attend the class sessions and be on time.
❑ When I am absent, I will find out what the class studied, get a copy of any notes or handouts, and make up the work that was assigned when I was gone.

❏ I will sit where I can hear the instructor and see the board.

❏ I will pay attention in class and take careful notes.

❏ I will ask the instructor questions when I don't understand the material.

❏ When tests, quizzes, or homework papers are passed back and discussed in class, I will write down the correct solutions for the problems I missed so that I can learn from my mistakes.

Study Sessions

❏ I will find a comfortable and quiet place to study.

❏ I realize that reading a math book is different from reading a newspaper or a novel. Quite often, it will take more than one reading to understand the material.

❏ After studying an example in the textbook, I will work the accompanying Self Check.

❏ I will begin the homework assignment only after reading the assigned section.

❏ I will try to use the mathematical vocabulary mentioned in the book and used by my instructor when I am writing or talking about the topics studied in this course.

❏ I will look for opportunities to explain the material to others.

❏ I will check all my answers to the problems with those provided in the back of the book (or with the *Student Solutions Manual*) and resolve any differences.

❏ My homework will be organized and neat. My solutions will show all the necessary steps.

❏ I will work some review problems every day.

❏ After completing the homework assignment, I will read the next section to prepare for the coming class session.

❏ I will keep a notebook containing my class notes, homework papers, quizzes, tests, and any handouts — all in order by date.

Special Help

❏ I know my instructor's office hours and am willing to go in to ask for help.

❏ I have formed a study group with classmates that meets regularly to discuss the material and work on problems.

❏ When I need additional explanation of a topic, I use the tutorial videos and the interactive CD, as well as the Web site.

❏ I make use of extra tutorial assistance that my school offers for mathematics courses.

❏ I have purchased the *Student Solutions Manual* that accompanies this text, and I use it.

To follow each of these suggestions will take time. It takes a lot of practice to learn mathematics, just as with any other skill.

No doubt, you will sometimes become frustrated along the way. This is natural. When it occurs, take a break and come back to the material after you have had time to clear your thoughts. Keep in mind that the skills and discipline you learn in this course will help make for a brighter future. Good luck!

iLrn Tutorial Quick Start Guide

iLrn Can Help You Succeed in Math

iLrn™ is an online program that facilitates math learning by providing resources and practice to help you succeed in your math course. Your instructor chose to use iLrn because it provides online opportunities for learning (Explanations found by clicking

Read Book), practice (Exercises), and evaluating (Quizzes). It also gives you a way to keep track of your own progress and manage your assignments.

The mathematical notation in iLrn is the same as that you see in your textbooks, in class, and when using other math tools like a graphing calculator. iLrn can also help you run calculations, plot graphs, enter expressions, and grasp difficult concepts. You will encounter various problem types as you work through iLrn, all of which are designed to strengthen your skills and engage you in learning in different ways.

▌ Logging in to 1Pass

Registering with the PIN Code on the 1Pass Card *Situation:* Your instructor has not given you a PIN code for an online course, but you have a textbook with a 1Pass PIN code. With 1Pass, you have one simple PIN code access to all media resources associated with your textbook. Please refer to your 1Pass card for a complete list of those resources.

Initial Log-in

To access your web gateway through 1Pass:

1. Check the outside of your textbook to see if there is an additional 1Pass card.
2. Take this card (and the additional 1Pass card if appropriate) and go to http://1pass.thomson.com.
3. Type in your 1Pass access code (or codes).
4. Follow the directions on the screen to set up your personal username and password.
5. Click through to launch your personal portal.
6. Access the media resources associated with your text . . . all the resources are just one click away.
7. Record your username and password for future visits and be sure to use the same username for all Thomson Learning resources.

For tech support, contact us at 1 (800) 423-0563.

> You will be asked to enter a valid e-mail address and password. Save your password in a safe place. You will need them to log in the next time you use 1Pass. Only your e-mail address and password will allow you to reenter 1Pass.

Subsequent Log-in

1. Go to **http://1pass.thomson.com.**
2. Type your e-mail address and password (see boxed information above) in the "Existing Users" box; then click on **Login.**

▌ Navigating through Your iLrn Tutorial

To navigate between chapters and sections, use the drop-down menu below the top navigation bar. This will give you access to the study activities available for each section.

The view of a tutorial in iLrn looks like this.

Math Toolbar

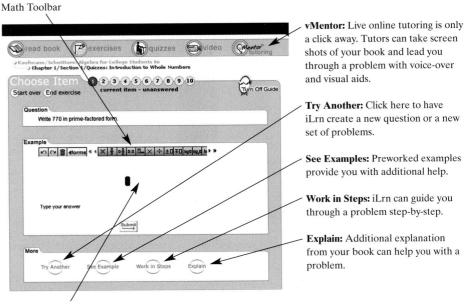

vMentor: Live online tutoring is only a click away. Tutors can take screen shots of your book and lead you through a problem with voice-over and visual aids.

Try Another: Click here to have iLrn create a new question or a new set of problems.

See Examples: Preworked examples provide you with additional help.

Work in Steps: iLrn can guide you through a problem step-by-step.

Explain: Additional explanation from your book can help you with a problem.

Type your answer here.

Online Tutoring with vMentor

Access to iLrn also means access to online tutors and support through vMentor™, which provides live homework help and tutorials. To access vMentor while you are working in the Exercises or Tutorial areas in iLrn, click on the **vMentor Tutoring** button at the top right of the navigation bar above the problem or exercise.

Next, click on the **vMentor** button; you will be taken to a Web page that lists the steps for entering a vMentor classroom. If you are a first-time user of vMentor, you might need to download Java software before entering the class for the first class. You can either take an Orientation Session or log in to a vClass from the links at the bottom of the opening screen.

All vMentor Tutoring is done through a vClass, an Internet-based virtual classroom that features two-way audio, a shared whiteboard, chat, messaging, and experienced tutors.

You can access vMentor Sunday through Thursday, as follows:

5 p.m. to 9 p.m. Pacific Time

6 p.m. to 10 p.m. Mountain Time

7 p.m. to 11 p.m. Central Time

8 p.m. to midnight Eastern Time

If you need additional help using vMentor, you can access the Participant Quick Reference Guide at this Web site: **http://www.elluminate.com/support/docs/Elive_Participant_Quick_Reference_Guide_6.0.pdf.**

Interact with TLE Online Labs

Use TLE Online Labs to explore and reinforce key concepts introduced in this text. These electronic labs give you access to additional instruction and practice problems, so you can explore each concept interactively, at your own pace. Not only will you be better prepared, but you will perform better in the class overall.

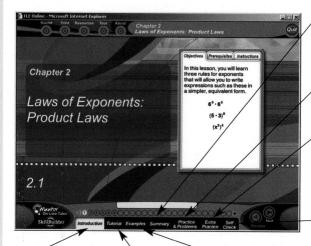

Summary: The Summary revisits the problem presented in the Introduction and encourages you to apply the mathematics you learned in the Tutorial and Examples.

Practice & Problems: Practice & Problems presents up to 25 questions organized in four or five categories.

Extra Practice: Extra practice presents questions like those in the Examples. After each question you have the option to try again, see the answer, see a sample solution, or try another question of the same type.

Self-Check: Self-Check presents up to 10 dynamically generated questions. To complete a lesson you must obtain the minimum standard (about 70%).

Introduction: Each lesson opens with objectives and prerequisites and provides brief instructions on using TLE.

Tutorial: The Tutorial provides the main instruction for the lesson. Hint and Success Tips teach strategies that can be used to solve the problem.

Examples: The examples expand on what you learned in the Tutorial. A hidden picture is progressively revealed as you complete each example.

APPLICATIONS INDEX

Examples that are applications are shown with boldface page numbers.

Exercises that are applications are shown with lightface page numbers.

Whole Numbers

Getty Images

Office managers play an important role in many businesses. They oversee the day-to-day activities of a company, making sure that the business runs smoothly and efficiently. To be an effective office manager, one needs excellent organizational, planning, and communication skills. Strong mathematical skills are also necessary to perform such job responsibilities as scheduling meetings, managing payroll and budgets, and designing office workspace layouts.

To learn more about the use of mathematics in the business world, visit The Learning Equation on the Internet at http://tle.brookscole.com. (The log-in instructions are in the Preface.) For Chapter 1, the online lessons are:

- *TLE* Lesson 1: Whole Numbers
- *TLE* Lesson 2: Prime Factors and Exponents
- *TLE* Lesson 3: Order of Operations

Check Your Knowledge

1. The set of _____ numbers is {1, 2, 3, 4, 5, . . .}, and the set of _____ numbers is {0, 1, 2, 3, 4, 5, . . .}.

2. The distance around a rectangle is called its _____.

3. The property that guarantees that we can add two numbers in either order is called the _____ property of addition. The property that allows us to group numbers in an addition in any way we wish is called the _____ property of addition.

4. Numbers that are to be multiplied are called _____. The result of a multiplication problem is called a _____. The answer to a division problem is called the _____.

5. A _____ number is a whole number, greater than 1, that has only 1 and itself as factors.

6. Write 3,737 in expanded notation.

7. Round 186,250 to the nearest hundred.

Refer to the data in the table.

Day	1	2	3	4	5
Temperature (Celsius)	13	8	12	5	7

8. Use the data to make a bar graph.

9. Use the data to make a line graph.

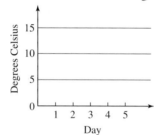

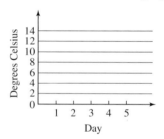

10. Place one of the symbols < or > in the blank to make a true statement. 27 ☐ 19

11. Add: 3,742
        +1,379

12. Subtract 289 from 347.

13. In 2003, *People* magazine had a paid circulation of 3,603,115. By what amount did this exceed *The National Enquirer*, with a paid circulation of 1,541,618?

14. Multiply: 432
            × 57

15. Divide: $79\overline{)4,537}$.

16. Find the perimeter and area of a rectangle that is 13 ft wide and 19 ft long.

17. Find the prime factorization of 950.

Evaluate each expression.

18. $3 + 4 - 2$

19. $3 + 4 \cdot 25$

20. $\dfrac{(4^3 - 2) + 7}{5(2 + 4) - 7}$

21. $3 \cdot 7 - 2\,[10 - 3(5 - 2)]$

22. Julia scored 95, 85, 73, 62, and 0 on five math quizzes. Find her mean (average) quiz score.

Study Skills Workshop

GET ORGANIZED!

Students who have had difficulty learning math in the past may think that their problem is not being born with the ability to "do math." This isn't true! Learning math is a skill and, much like learning to play a musical instrument, takes daily, organized practice. It is also a sequential process; what you learn one day will be used again as a foundation for a new concept. Thus, it is especially important to be prepared and willing to begin work in your math class from the first day. Below are some strategies to get you off to a good start.

Attend Class. Attending class every meeting is one of the most important things you can do to succeed. Your instructor not only explains material and gives examples to support your text, but may also discuss topics that do not show up in your book, or may make changes in homework assignments or test dates. Getting to know at least a few of your classmates is also important to your success. Find a classmate or two whom you can depend on for information, who can help with homework, or with whom you can form a study group.

Make a Calendar. Because daily practice is so important in learning math, it is a good idea to set up a calendar that lists all of your time commitments. You may wonder how much time is appropriate to budget for your classes. A general rule of thumb is to allot 2 hours outside of class for every lecture hour. That means if your class meets for 3 hours per week, plan on 6 hours per week for homework and study. Remember that this is for each class. On your calendar, write in the times for your classes, the time that you need to spend on homework, and other regular time commitments (like work, social obligations, church activities, etc.).

Gather Needed Materials. All math classes require textbooks, notebooks, pencils (with big erasers!) and usually as much scrap paper as you can gather. A good source of scrap paper is often a computer lab on your campus. To be sure you have everything you need, check with your instructor. Have your materials by your second class meeting and bring them to every class meeting thereafter. Additional materials that may be of use outside of class are the online tutorial program iLrn (www.iLrn.com) and the Video Skillbuilder CD-ROM that is packaged with your textbook.

What Does Your Instructor Expect From You? Your instructor's syllabus is documentation of his or her expectations. Often your instructor will detail in the syllabus how your grade is determined, when office hours are held, and where you can get help outside of class. Read the syllabus thoroughly and make sure you understand all that is required. If something is not clear to you, contact the instructor as soon as possible.

ASSIGNMENT

1. Download a calendar online at series.brookscole.com/tussypaperback, or make your own calendar with class times and study times for each course you are taking as well as times for work and other essential activities (e.g. church activities, social obligations, time with children, etc.). You may also want to schedule additional time to study a week before a test. Also include time for physical exercise and rest — this is important to reduce the effects of stress that school brings.

2. Download and print out the Course Information Sheet online at series.brookscole.com/tussypaperback, or make a list of the following items:
   a. Instructor's name, office location, office hours, phone number, email address
   b. Test dates, if scheduled
   c. What work determines your course grade and how grades are calculated

3. Write down the name, phone number, and email address of at least two classmates.

4. Does your school have tutorial services or a math lab/learning center? Where are they located? What are their hours of operation?

In this chapter, we will use the operations of addition, subtraction, multiplication, and division to solve problems that involve whole numbers.

1.1 An Introduction to the Whole Numbers

- Sets of numbers • Place value • Expanded notation • Graphing on the number line
- Ordering of the whole numbers • Rounding whole numbers • Tables and graphs

In this section, we discuss natural numbers and whole numbers. These numbers are used to answer questions such as How many?, How fast?, How heavy?, and How far?

- The movie *Titanic* won 11 Academy Awards.
- The fastest roller coaster in the world is the *Top Thrill Dragster* at Cedar Point, Sandusky, Ohio. It reaches speeds of up to 120 mph.
- The Statue of Liberty weighs 225 tons.
- The driving distance between New York City and Los Angeles is 2,786 miles.

▌ Sets of numbers

A **set** is a collection of objects. Two basic sets in mathematics are the natural numbers (the numbers that we count with) and the whole numbers. When writing a set, we use **braces { }** to enclose its **members** (or **elements**).

The set of natural numbers

$\{1, 2, 3, 4, 5, 6, 7, 8, 9, 10, 11, 12, \ldots\}$

The set of whole numbers

$\{0, 1, 2, 3, 4, 5, 6, 7, 8, 9, 10, 11, 12, \ldots\}$

The three dots at the end of the lists above indicate that these sets continue on forever. There is no largest natural number or whole number.

Since every natural number is also a whole number, we say that the set of natural numbers is a **subset** of the set of whole numbers. However, not all whole numbers are natural numbers, because 0 is a whole number but not a natural number.

▌ Place value

When we express a whole number with a *numeral* containing the *digits* 0, 1, 2, 3, 4, 5, 6, 7, 8, 9, we say that we have written the number in **standard notation.** The position of a digit in a numeral determines its value. In the numeral 325, the 5 is in the *ones column,* the 2 is in the *tens column,* and the 3 is in the *hundreds column.*

3 2 5

Hundreds column ⌐↑↑⌐ Ones column

Tens column

To make numerals easy to read, we use commas to separate their digits into groups of three, called **periods.** Each period has a name, such as *ones, thousands,* *millions,* and so on. The following table shows the place value of each digit in the numeral 345,576,402,897,415, which is read as

three hundred forty-five trillion, five hundred seventy-six billion, four hundred two million, eight hundred ninety-seven thousand, four hundred fifteen

345 trillion			576 billion			402 million			897 thousand			4 hundred fifteen		
3	4	5	5	7	6	4	0	2	8	9	7	4	1	5
Trillions			Billions			Millions			Thousands			Ones		
Hundreds	Tens	Ones	Hundreds	Tens	Ones	Hundreds	Tens	Ones	Hundreds	Tens	Ones	Hundreds	Tens	Ones

As we move to the left in this table, the place value of each column is 10 times greater than the column to its right. This is why we call our number system a *base-10 number system.*

EXAMPLE 1 **TV news.** In 2003, there were 73,365,880 basic cable subscribers in the United States. Which digit in 73,365,880 tells the number of hundreds?

Solution In 73,365,880, the hundreds column is the third column from the right. The digit 8 tells the number of hundreds.

Self Check 1
In 2003, there were 158,722,000 cellular telephone subscribers in the United States. Which digit in 158,722,000 tells the number of ten thousands?

Answer 2

▧ Expanded notation

In the numeral 6,352, the digit 6 is in the thousands column, 3 is in the hundreds column, 5 is in the tens column, and 2 is in the ones (or units) column. The meaning of 6,352 becomes clear when we write it in **expanded notation.**

6 thousands + 3 hundreds + 5 tens + 2 ones

We read the numeral 6,352 as "six thousand, three hundred fifty-two."

EXAMPLE 2 Write each number in expanded notation: **a.** 63,427 and **b.** 1,251,609.

Solution
a. 6 ten thousands + 3 thousands + 4 hundreds + 2 tens + 7 ones

We read this number as "sixty-three thousand, four hundred twenty-seven."

b. 1 million + 2 hundred thousands + 5 ten thousands + 1 thousand + 6 hundreds + 0 tens + 9 ones

Since 0 tens is zero, the expanded notation can also be written as

1 million + 2 hundred thousands + 5 ten thousands + 1 thousand + 6 hundreds + 9 ones

We read this number as "one million, two hundred fifty-one thousand, six hundred nine."

Self Check 2
Write 808,413 in expanded notation.

Answer 8 hundred thousands + 8 thousands + 4 hundreds + 1 ten + 3 ones. Read as "eight hundred eight thousand, four hundred thirteen."

EXAMPLE 3 Write twenty-three thousand forty in standard notation.

Solution In expanded notation, the number is written as

2 ten thousands + 3 thousands + 4 tens   There are 0 hundreds and 0 ones.

In standard notation, this is written as 23,040.

▮ Graphing on the number line

Whole numbers can be illustrated by drawing points on the **number line,** a line that is used to represent numbers graphically. Like a ruler, the number line is straight and has uniform markings. (See Figure 1-1). To construct the number line, we begin on the left with a point on the line representing the number 0. This point is called the **origin.** We then proceed to the right, drawing equally spaced marks and labeling them with whole numbers that increase progressively in size. The arrowhead at the right indicates that the number line continues forever.

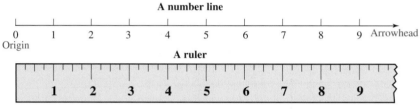

FIGURE 1-1

Using a process known as **graphing,** we can represent a single number or a set of numbers on a number line. *The graph of a number* is the point on the number line that corresponds to that number. *To graph a number* means to locate its position on the number line and highlight it with a heavy dot. Figure 1-2 shows the graphs of 5 and 8.

FIGURE 1-2

▮ Ordering of the whole numbers

As we move to the right on the number line, the numbers get larger. Because 8 lies to the right of 5, we say that 8 is greater than 5. The **inequality symbol** > ("is greater than") can be used to write this fact.

8 > 5   Read as "8 is greater than 5."

Since 8 > 5, it is also true that 5 < 8. (Read as "5 is less than 8.")

! COMMENT To distinguish between these two inequality symbols, remember that they always point to the smaller of the two numbers involved.

8 > 5   5 < 8
⌐ Points to the ⌐
smaller number

EXAMPLE 4 Place an < or an > symbol in the box to make a true statement: **a.** 3 ☐ 7 and **b.** 18 ☐ 16.

Solution
a. Since 3 is to the left of 7 on the number line, 3 < 7.

b. Since 18 is to the right of 16 on the number line, 18 > 16.

Self Check 4
Place an < or an > symbol in the box to make a true statement:
a. 12 ☐ 4 **b.** 7 ☐ 10

Answers **a.** >, **b.** <

▌ Rounding whole numbers

When we don't need exact results, we often round numbers. For example, when a teacher with 36 students orders 40 textbooks, he has rounded the actual number to the *nearest ten,* because 36 is closer to 40 than it is to 30.

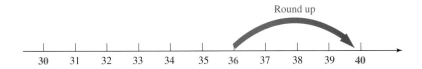

When a geologist says that the height of Alaska's Mount McKinley is "about 20,300 feet," she has rounded to the *nearest hundred,* because its actual height of 20,320 feet is closer to 20,300 than it is to 20,400.

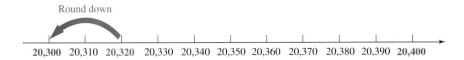

To round a whole number, we follow an established set of rules. To round a number to the nearest ten, for example, we locate the **rounding digit** in the tens column. If the **test digit** to the right of that column (the digit in the ones column) is 5 or greater, we *round up* by increasing the tens digit by 1 and placing a 0 in the ones column. If the test digit is less than 5, we *round down* by leaving the tens digit unchanged and placing a 0 in the ones column.

EXAMPLE 5 Round each number to the nearest ten: **a.** 3,764 and **b.** 12,087.

Solution
a. We find the rounding digit in the tens column, which is 6.

      ⌐ Rounding digit
     ↓
3,764
    ↑
    └ Test digit

We then look at the test digit to the right of 6, the 4 in the ones column. Since 4 < 5, we round down by leaving the 6 unchanged and replacing the test digit with 0. The rounded answer is 3,760.

Self Check 5
Round each number to the nearest ten:
a. 35,642
b. 3,756

b. We find the rounding digit in the tens column, which is 8.

$\quad\quad\quad\quad\underset{\displaystyle 12{,}0\underset{\uparrow}{8}7}{\overset{\displaystyle \overset{\llcorner\text{Rounding digit}}{\downarrow}}{}}$

┌─ Rounding digit
↓
12,087
   ↑
   └─ Test digit

We then look at the test digit to the right of 8, the 7 in the ones column. Because $7 > 5$, we round up by adding 1 to 8 and replacing the test digit with 0. The rounded answer is 12,090.

Answers a. 35,640, **b.** 3,760

A similar method is used to round numbers to the nearest hundred, the nearest thousand, the nearest ten thousand, and so on.

Rounding a whole number

1. To round a number to a certain place, locate the rounding digit in that place.
2. Look at the test digit to the right of the rounding digit.
3. If the test digit is 5 or greater, round up by adding 1 to the rounding digit and changing all of the digits to the right of the rounding digit to 0.

   If the test digit is less than 5, round down by keeping the rounding digit and changing all of the digits to the right of the rounding digit to 0.

Self Check 6
Round 365,283 to the nearest hundred.

EXAMPLE 6   Round 7,960 to the nearest hundred.

Solution   First, we find the rounding digit in the hundreds column, which is 9.

┌─ Rounding digit
↓
7,960
  ↑
  └─ Test digit

We then look at the 6 to the right of 9. Because $6 > 5$, we round up and increase 9 in the hundreds column by 1. Since the 9 in the hundreds column represents 900, increasing 9 by 1 represents increasing 900 to 1,000. Thus, we replace the 9 with a 0 and add 1 to the 7 in the thousands column. Finally, we replace the two rightmost digits with 0's. The rounded answer is 8,000.

Answer 365,300

Self Check 7
Round the elevation of Denver **a.** to the nearest hundred feet and **b.** to the nearest thousand feet.

EXAMPLE 7   U.S. cities.   In 2003, Denver was the nation's 26th largest city. Round the 2003 population of Denver given in Figure 1-3 **a.** to the nearest thousand   and **b.** to the nearest ten thousand.

Denver
CITY LIMIT
Pop. 557, 478   Elev. 5,280

FIGURE 1-3

Solution
a. The rounding digit in the thousands column is 7.
   The test digit, 4, is less than 5, so we round down.
   To the nearest thousand, Denver's population in 2003 was 557,000.

b. The rounding digit in the ten thousands column is 5. The test digit, 7, is greater than 5, so we round up. To the nearest ten thousand, Denver's population in 2003 was 560,000.

Answers a. 5,300 ft, **b.** 5,000 ft

Re-entry Students

"A re-entry student is considered one who is the age of 25 or older, or those students that have had a break in their academic work for 5 years or more. Nationally, this group of students is growing at an astounding rate."
Student Life and Leadership Department, University Union, Cal Poly University, San Luis Obispo

Some common concerns expressed by adult students considering returning to school are listed below in Column I. Match each concern to an encouraging reply in Column II.

Column I	Column II
1. I'm too old to learn.	**a.** Many students qualify for some type of financial aid.
2. I don't have the time.	
3. I didn't do well in school the first time around. I don't think a college would accept me.	**b.** Taking even a single class puts you one step closer to your educational goal.
	c. There's no evidence that older students can't learn as well as younger ones.
4. I'm afraid I won't fit in.	**d.** More than 41% of the students in college are older than 25.
5. I don't have the money to pay for college.	**e.** Typically, community colleges and career schools have an open admissions policy.

Adapted from *Common Concerns for Adult Students,* Minnesota Higher Education Services Office

▍ Tables and graphs

The table in Figure 1-4(a) is an example of the use of whole numbers. It shows the number of women elected to the United States House of Representatives in the congressional elections held every two years from 1996 to 2004.

Year	Number of women elected
1996	54
1998	56
2000	59
2002	60
2004	65

Source: Center for American Women and Politics

(a)

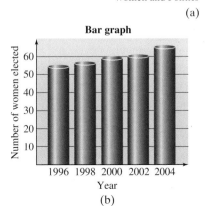

(b)

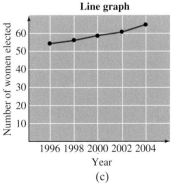

(c)

FIGURE 1-4

In Figure 1-4(b) on the previous page, the election results are presented in a **bar graph.** The horizontal scale is labeled "Year" and scaled in units of 2 years. The vertical scale is labeled "Number of women elected" and scaled in units of 10. The bar directly over each year extends to a height indicating the number of women elected to Congress that year.

Another way to present the information in the table is with a **line graph.** Instead of using a bar to denote the number of women elected, we use a dot drawn at the correct height. After drawing data points for 1996, 1998, 2000, 2002, and 2004, we connect the points with line segments to create the line graph in Figure 1-4(c).

Section 1.1 STUDY SET

VOCABULARY   *Fill in the blanks.*

1. A _____ is a collection of objects.
2. The set of _____ numbers is {1, 2, 3, 4, 5, . . .}, and the set of _____ numbers is {0, 1, 2, 3, 4, 5, . . .}.
3. When 297 is written as 2 hundreds + 9 tens + 7 ones, it is written in _____ notation.
4. If we _____ 627 to the nearest ten, we get 630.
5. Using a process known as graphing, we can represent whole numbers as points on a _____ line.
6. The symbols > and < are _____ symbols.

CONCEPTS   *Consider the numeral 57,634.*

7. What digit is in the tens column?
8. What digit is in the thousands column?
9. What digit is in the hundreds column?
10. What digit is in the ten thousands column?
11. What set of numbers is obtained when 0 is combined with the natural numbers?
12. Place the numbers 25, 17, 37, 15, 45 in order from smallest to largest.
13. Graph: 1, 3, 5, and 7.

14. Graph: 0, 2, 4, 6, and 8.

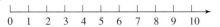

15. Graph the whole numbers less than 6.

16. Graph the whole numbers between 2 and 8.

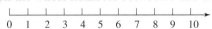

Place an > or an < symbol in the box to make a true statement.

17. 47 ⬜ 41
18. 53 ⬜ 67
19. 309 ⬜ 300
20. 841 ⬜ 814

21. 2,052 ⬜ 2,502
22. 999 ⬜ 998
23. Since 4 < 7, it is also true that 7 ⬜ 4.
24. Since 9 > 0, it is also true that 0 ⬜ 9.

NOTATION   *Fill in the blanks.*

25. The symbols { }, called _____, are used when writing a set.
26. The symbol > means ___ _____ _____, and the symbol < means ___ _____ _____.

PRACTICE   *Write each number in expanded notation and then write it in words.*

27. 245
28. 508
29. 3,609
30. 3,960
31. 32,500
32. 73,009
33. 104,401
34. 570,003

Write each number in standard notation.

35. 4 hundreds + 2 tens + 5 ones
36. 7 hundreds + 7 tens + 7 ones
37. 2 thousands + 7 hundreds + 3 tens + 6 ones
38. 7 billions + 3 hundreds + 5 tens
39. Four hundred fifty-six
40. Three thousand seven hundred thirty-seven
41. Twenty-seven thousand five hundred ninety-eight
42. Seven million, four hundred fifty-two thousand, eight hundred sixty

43. Nine thousand one hundred thirteen

44. Nine hundred thirty

45. Ten million, seven hundred thousand, five hundred six

46. Eighty-six thousand four hundred twelve

Round 79,593 to the nearest . . .

47. ten   **48.** hundred

49. thousand   **50.** ten thousand

Round 5,925,830 to the nearest . . .

51. thousand   **52.** ten thousand

53. hundred thousand   **54.** million

Round $419,161 to the nearest . . .

55. $10   **56.** $100

57. $1,000   **58.** $10,000

▌ APPLICATIONS

59. GAME SHOWS On *The Price is Right* television show, the winning contestant is the person who comes closest to (without going over) the price of the item up for bid. Which contestant shown below will win if they are bidding on a bedroom set that has a suggested retail price of $4,745?

Donna	Tyronne	Aisha	Coby
$4,995	$4,550	$4,551	$4,200

60. PRESIDENTS The following list shows the ten youngest U.S. presidents and their ages (in years/days) when they took office. Construct a two-column table that presents the data in order, beginning with the youngest president.

C. Arthur   50 yr/350 days   U. Grant   46 yr/236 days

G. Cleveland   47 yr/351 days   J. Kennedy   43 yr/236 days

W. Clinton   46 yr/154 days   F. Pierce   48 yr/101 days

M. Filmore   50 yr/184 days   J. Polk   49 yr/122 days

J. Garfield   49 yr/105 days   T. Roosevelt   42 yr/322 days

61. MISSIONS TO MARS The United States, Russia, Europe, and Japan have launched Mars space probes. The graph in the next column shows the success rate of the missions, by decade.

   a. What decade had the greatest number of successful or partially successful missions? How many?

   b. What decade had the greatest number of unsuccessful missions? How many?

   c. Which decade had the greatest number of missions? How many?

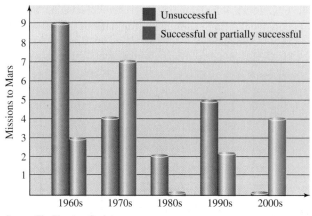

Source: The Planetary Society

62. BANKING The illustration shows the number of banks that were closed or taken over by federal agencies during the years 1935–1995.

   a. During what two time spans was there an upsurge in bank failures?

   b. In what year were there the most bank failures? Estimate the number of banks that failed that year.

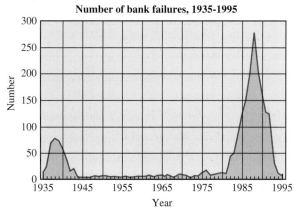

Source: FDIC Division of Research and Statistics

63. ENERGY RESERVES Construct a bar graph using the data in the table.

NATURAL GAS RESERVES, 2003 (IN TRILLION CUBIC FEET)	
United States	187
Venezuela	148
Canada	60
Argentina	27
Mexico	9

Source: *Oil and Gas Journal*

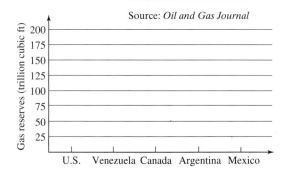

64. ENERGY RESERVES Refer to Exercise 63, and construct a line graph using the data in the table on the previous page.

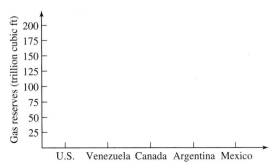

65. COFFEE Construct a line graph using the data in the table.

STARBUCKS LOCATIONS	
Year	Number
1997	1,412
1998	1,886
1999	2,135
2000	3,501
2001	4,709
2002	5,886
2003	7,225
2004	8,337

Source: Starbucks Company

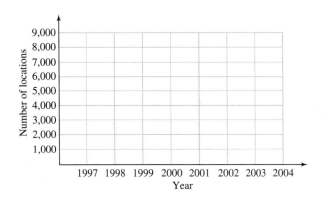

66. COFFEE Construct a bar graph using the data in the table above.

67. Complete each check by writing the amount in words on the proper line.

a.

No. 201	March 9, 20 05
Payable to Davis Chevrolet	$ 15,601.00
	DOLLARS
45-365-02	Don Smith

b.

No. 7890	Aug. 12, 20 05
Payable to Dr. Anderson	$ 3,433.00
	DOLLARS
45-828-02	Juan Decito

68. ANNOUNCEMENTS One style used when printing formal invitations and announcements is to write all numbers in words. Use this style to write each of the following phrases.

a. This diploma awarded this 27th day of June, 2005.

b. The suggested contribution for the fundraiser is $850 a plate, or an entire table may be purchased for $5,250.

69. EDITING Edit this excerpt from a history text by circling all numbers written in words and rewriting them using digits.

> Abraham Lincoln was elected with a total of one million, eight hundred sixty-five thousand, five hundred ninety-three votes — four hundred eighty-two thousand, eight hundred eighty more than the runner-up, Stephen Douglas. He was assassinated after having served a total of one thousand five hundred three days in office. Lincoln's Gettysburg Address, a mere two hundred sixty-nine words long, was delivered at the battle site where forty-three thousand four hundred forty-nine casualties occurred.

70. READING METERS The amount of electricity used in a household is measured in kilowatt-hours (kwh). Determine the reading on the meter shown below. (When the pointer is between two numbers, read the lower number.)

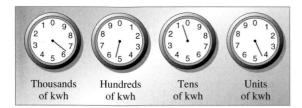

| Thousands of kwh | Hundreds of kwh | Tens of kwh | Units of kwh |

71. SPEED OF LIGHT The speed of light in a vacuum is 299,792,458 meters per second. Round this number

   a. to the nearest hundred thousand meters per second.

   b. to the nearest million meters per second.

72. CLOUDS Draw a vertical number line scaled from 0 to 40,000 feet, in units of 5,000 feet. Graph each cloud type given in the table in the next column at the proper altitude.

Cloud type	Altitude (ft)
Altocumulus	21,000
Cirrocumulus	37,000
Cirrus	38,000
Cumulonimbus	15,000
Cumulus	8,000
Stratocumulus	9,000
Stratus	4,000

WRITING

73. Explain why the natural numbers are called the counting numbers.

74. Explain how you would round 687 to the nearest ten.

75. The houses in a new subdivision are priced "in the low 130s." What does this mean?

76. A million is a thousand thousands. Explain why this is so.

77. Many television infomercials offer the viewer creative ways to make a six-figure income. What is a six-figure income? What is the smallest and what is the largest six-figure income?

78. What whole number is associated with each of the following words?

duo	decade	a grand	four score
dozen	trio	century	a pair

1.2 Adding Whole Numbers

- Properties of addition • Adding whole numbers with more than one digit
- The perimeter of a rectangle and a square • Adding whole numbers using a calculator

By mastering addition of whole numbers, we can solve many problems. For example, to find the distance around a rectangle, we need to add the lengths of the rectangle's four sides. To prepare an annual budget, we need to add separate line items. To find the cost of a shirt and a pair of jeans, we must add their costs.

Properties of addition

Adding whole numbers corresponds to combining sets of objects. For example, if a set of 4 objects is combined with a set of 5 objects, we get a total of 9 objects.

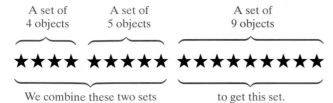

A set of 4 objects   A set of 5 objects   A set of 9 objects

We combine these two sets   to get this set.

This corresponds to the addition fact

$4 + 5 = 9$ Read as "4 plus 5 equals 9."

In this addition, 4 and 5 are called **addends** or **terms,** and 9 is called the **sum.** This addition is often shown in a vertical format.

$$
\begin{array}{l}
4 \leftarrow \text{Addend} \\
\underline{+5} \leftarrow \text{Addend} \\
9 \leftarrow \text{Sum}
\end{array}
$$

If we combine the sets in the opposite order, we will get the same sum.

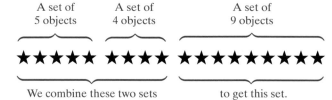

| A set of 5 objects | A set of 4 objects | A set of 9 objects |

We combine these two sets   to get this set.

This corresponds to the addition fact

$$
5 + 4 = 9 \qquad \text{or} \qquad \begin{array}{r} 5 \\ +4 \\ \hline 9 \end{array} \qquad \text{Read as "5 plus 4 equals 9."}
$$

These examples illustrate that two whole numbers can be added in either order to get the same sum. This property is called the **commutative property of addition.**

Commutative property of addition

The order in which whole numbers are added does not change the sum.
For example,

$$
6 + 5 = 5 + 6
$$

Table 1-1 summarizes the basic addition facts.

- To find the sum of 6 and 8 using the table, we find the intersection of the row starting with **6** and the column headed by **8.** The sum is 14.
- To find the sum of 8 and 6, we find the intersection of the row starting with **8** and the column headed by **6.** Once again, the sum is 14.

·	0	1	2	3	4	5	6	7	8	9
0	0	0	0	0	0	0	0	0	0	0
1	0	1	2	3	4	5	6	7	8	9
2	0	2	**4**	6	8	10	12	14	16	18
3	0	3	6	**9**	12	15	18	21	24	27
4	0	4	8	12	**16**	20	24	28	32	36
5	0	5	10	15	20	**25**	30	35	40	45
6	0	6	12	18	24	30	**36**	42	48	54
7	0	7	14	21	28	35	42	**49**	56	63
8	0	8	16	24	32	40	48	56	**64**	72
9	0	9	18	27	36	45	54	63	72	**81**

TABLE 1-1

We note that the answers in the table above the diagonal line in bold print are identical to the answers below the diagonal line. This illustrates that addition is commutative.

EXAMPLE 1   Find each sum: **a.** $3 + 6$,   **b.** $5 + 7$,   **c.** $8 + 9$,   and   **d.** $9 + 5$.

Solution

a. $3 + 6 = 9$ **b.** $5 + 7 = 12$

c. $8 + 9 = 17$ **d.** $9 + 5 = 14$

Self Check 1
Find each sum: **a.** $6 + 7$ and
b. $7 + 4$.

Answers **a.** 13,   **b.** 11

To find the sum of three whole numbers, we add two of them and then add the sum to the third number. In the following examples, we add $3 + 4 + 7$ in two ways. We will use the grouping symbols (   ), called **parentheses,** to show this. It is standard practice to perform the operations within the parentheses first.

Method 1: Group 3 and 4

$(\mathbf{3 + 4}) + 7 = \mathbf{7} + 7$ Because of the parentheses, add 3 and 4 first to get 7.

$ = 14$ Then add 7 and 7 to get 14.

Method 2: Group 4 and 7

$3 + (\mathbf{4 + 7}) = 3 + \mathbf{11}$ Because of the parentheses, add 4 and 7 first to get 11.

$ = 14$ Then add 3 and 11 to get 14.

Either way, the sum is 14. It does not matter how we group or associate numbers in addition. This property is called the **associative property of addition.**

> **Associative property of addition**
>
> The way in which whole numbers are grouped does not affect their sum. For example,
>
> $$(2 + 5) + 4 = 2 + (5 + 4)$$

EXAMPLE 2   Find each sum: **a.** $(5 + 7) + 8$ and   **b.** $5 + (7 + 8)$.

Solution

a. $(\mathbf{5 + 7}) + 8 = \mathbf{12} + 8$ Perform the addition within the parentheses first: $5 + 7 = 12$.

$ = 20$ Perform the addition.

b. $5 + (\mathbf{7 + 8}) = 5 + \mathbf{15}$ Perform the addition within the parentheses first: $7 + 8 = 15$.

$ = 20$ Perform the addition.

Self Check 2
Find each sum: **a.** $4 + (6 + 3)$
and   **b.** $(4 + 6) + 3$.

Answers **a.** 13,   **b.** 13

Whenever we add 0 to a whole number, the number is unchanged. This property is called the **addition property of 0.**

> **Addition property of 0**
>
> The sum of any whole number and 0 is that whole number. For example,
>
> $$3 + 0 = 3, \qquad 5 + 0 = 5, \qquad \text{and} \qquad 0 + 9 = 9$$

Self Check 3
Find each sum: **a.** 7 + (3 + 0)
and   **b.** (8 + 5) + 0.

Answers **a.** 10,   **b.** 13

EXAMPLE 3   Find each sum: **a.** (5 + 0) + 3   and   **b.** 0 + (3 + 9).

Solution
a. **(5 + 0)** + 3 = **5** + 3   Perform the addition within the parentheses first: 5 + 0 = 5.
                      = 8   Perform the addition.

b. 0 + **(3 + 9)** = 0 + **12**   Perform the addition within the parentheses first: 3 + 9 = 12.
                      = 12   Perform the addition.

■ Adding whole numbers with more than one digit

We can add whole numbers greater than 10 using a vertical format. We simply add the numbers in each corresponding column. If an addition of the numbers in any one column exceeds 9, we must carry.

Self Check 4
Add: 131 + 232 + 221 + 312.

Answer 896

EXAMPLE 4   Add: 421 + 123 + 245.

Solution   We write the digits in a vertical format, with the ones digits in a column, the tens digits in a column, and the hundreds digits in a column. We start at the right and add the ones digits, then the tens digits, and finally the hundreds digits.

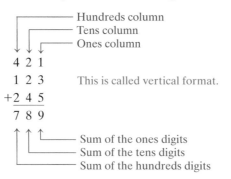

This is called vertical format.

The sum is 789.

Self Check 5
Add: 35 + 47.

Answer 82

EXAMPLE 5   Add: 27 + 15.

Solution   We write the digits in a vertical format, with the ones digits in a column and the tens digits in a column. We begin by adding the digits in the ones column: 7 + 5 = 12. Because 12 = 1 ten + 2 ones, we write 2 in the ones column of the answer and carry 1 to the tens column.

$$\begin{array}{r} \overset{1}{2}\,7 \\ +\underline{1\,5} \\ 2 \end{array}$$ Add the digits in the ones column: 7 + 5 = 12. Carry 1 to the tens column.

Then we add the digits in the tens column.

$$\begin{array}{r} \overset{1}{2}\,7 \\ +\underline{1\,5} \\ 4\,2 \end{array}$$ Add the digits in the tens column: 1 + 2 + 1 = 4. Place the result of 4 in the tens column of the answer.

The sum is 42.

EXAMPLE 6   Add: 9,835 + 692 + 7,275.

Solution   We write the digits in a vertical format with their corresponding digits aligned. Then we add the numbers, one column at a time.

$$
\begin{array}{r}
\overset{1}{9}\,8\,3\,5 \\
6\,9\,2 \\
+7\,2\,7\,5 \\
\hline
2
\end{array}
$$
   Add the digits in the ones column: 5 + 2 + 5 = 12. Write 2 in the ones column of the answer and carry 1 to the tens column.

$$
\begin{array}{r}
\overset{2\,1}{9}\,8\,3\,5 \\
6\,9\,2 \\
+7\,2\,7\,5 \\
\hline
0\,2
\end{array}
$$
   Add the digits in the tens column: 1 + 3 + 9 + 7 = 20. Write 0 in the tens column of the answer and carry 2 to the hundreds column.

$$
\begin{array}{r}
\overset{1\,2\,1}{9}\,8\,3\,5 \\
6\,9\,2 \\
+7\,2\,7\,5 \\
\hline
8\,0\,2
\end{array}
$$
   Add the digits in the hundreds column: 2 + 8 + 6 + 2 = 18. Write 8 in the hundreds column of the answer and carry 1 to the thousands column.

$$
\begin{array}{r}
\overset{1\,2\,1}{9}\,8\,3\,5 \\
6\,9\,2 \\
+\ \ 7\,2\,7\,5 \\
\hline
1\,7\,8\,0\,2
\end{array}
$$
   Add the digits in the thousands column: 1 + 9 + 7 = 17. Write 7 in the thousands column of the answer and write 1 in the ten thousands column.

The sum is 17,802.

To check whether the result in Example 6 is reasonable, we can round the addends and **estimate** the answer: 9,835 is a little less than 10,000, 692 is a little less than 700, and 7,275 is a little greater than 7,000. We can estimate the answer to be about 10,000 + 700 + 7,000 = 17,700. Therefore, the result 17,802 seems reasonable. (We will study estimation in more detail later in this chapter.)

Words such as *increase, gain, credit, up, forward, rises, in the future,* and *to the right* are used to indicate addition.

EXAMPLE 7   **Calculating temperatures.**   At noon, the temperature in Helena, Montana, was 31°. By 1:00 P.M., the temperature had increased 5°, and by 2:00 P.M., it had risen another 7°. Find the temperature at 2:00 P.M.

Solution   To the temperature at noon, we add the two increases.

31 + 5 + 7

The two additions are done working from left to right.

$$31 + 5 + 7 = 36 + 7$$
$$= 43$$

The temperature at 2:00 P.M. was 43°.

Self Check 8

By 1700, the populations of the four colonies were New Hampshire 5,000, New York 19,100, Massachusetts 55,900, and Virginia 58,600. Find the total population.

EXAMPLE 8   History.   The populations of four American colonies in 1630 are shown in Figure 1-5. Find the total population.

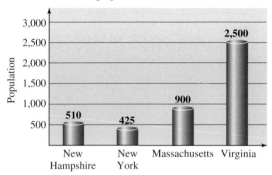

FIGURE 1-5

Solution   The word *total* indicates that we must add the populations of the colonies.

$$
\begin{array}{r}
\overset{2}{5}10 \\
425 \\
900 \\
+2{,}500 \\
\hline
4{,}335
\end{array}
$$

Align the numerals vertically. Add the digits, one column at a time, working from right to left.

The total population was 4,335.

Answer 138,600

The perimeter of a rectangle and a square

Figure 1-6(a) is an example of a four-sided figure called a **rectangle.** Either of the longer sides of a rectangle is called its **length** and either of the shorter sides is called its **width.** Together, the length and width are called the **dimensions** of the rectangle. For any rectangle, opposite sides have the same measure.

When all four of the sides of a rectangle have the same measure, we call the rectangle a **square.** An example of a square is shown in Figure 1-6(b).

A rectangle

Length

Width   Width

Length

(a)

A square

Side

Side   Side

Side

(b)

FIGURE 1-6

The distance around a rectangle or a square is called its **perimeter.** To find the perimeter of a rectangle, we add the lengths of its four sides.

| The perimeter of a rectangle | = | length | + | length | + | width | + | width |

To find the perimeter of a square, we add the lengths of its four sides.

| The perimeter of a square | = | side | + | side | + | side | + | side |

EXAMPLE 9 **Perimeters.** Find the perimeter of the dollar bill shown in Figure 1-7.

Solution To find the perimeter of the rectangular-shaped bill, we add the lengths of its four sides.

$$\begin{array}{r} \scriptstyle 2\,2 \\ 156 \\ 156 \\ 65 \\ +\ 65 \\ \hline 442 \end{array}$$

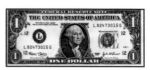

Width = 65 mm

Length = 156 mm

mm stands for millimeters

FIGURE 1-7

The perimeter is 442 mm.

To see whether this result is reasonable, we estimate the answer. Because the rectangle is about 150 mm by 70 mm, its perimeter is approximately 150 + 150 + 70 + 70, or 440 mm. An answer of 442 mm is reasonable.

Self Check 9
A Monopoly game board is a square, with sides 19 inches long. Find the perimeter of the board.

Answer 76 in.

Adding whole numbers using a calculator

Vehicle production — CALCULATOR SNAPSHOT

In 2003, Japan produced 8,487,065 new passenger cars and 1,665,612 new trucks. To find the total number of cars and trucks that Japan produced that year, we must add: 8,487,065 + 1,665,612. We can perform this addition using a calculator by entering

8487065 [+] 1665612 [=] 10152677

The total number of cars and trucks produced by Japan in 2003 was 10,152,677.

Section 1.2 STUDY SET

VOCABULARY *Fill in the blanks.*

1. Numbers that are to be added are called ________.

2. When two numbers are added, the result is called a ______.

3. The figure on the left is an example of a ________. The figure on the right is an example of a ________.

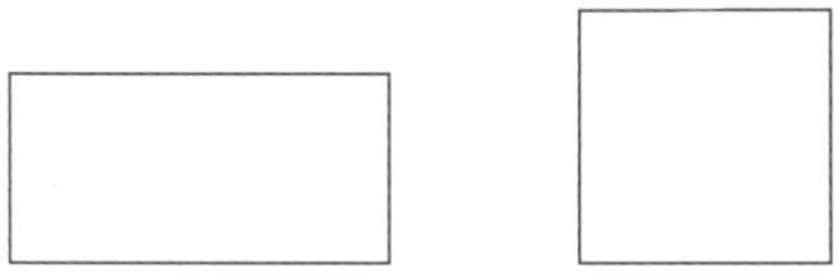

4. Label the *length* and the *width* of the rectangle.

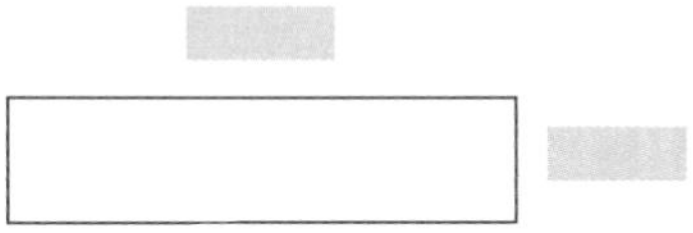

5. Together, the length and width of a rectangle are called its ________.

6. When all the sides of a rectangle have the same measure, we call the rectangle a ________.

7. The property that guarantees that we can add two numbers in either order is called the ________ property of addition.

8. The property that allows us to group numbers in an addition in any way we wish is called the ________ property of addition.

9. The distance around a rectangle is called its ________.

10. To check whether the result of an addition is reasonable, we can round the addends and ________ the answer.

CONCEPTS   *What property of addition is shown?*

11. $3 + 4 = 4 + 3$

12. $(3 + 4) + 5 = 3 + (4 + 5)$

13. $7 + (8 + 2) = (7 + 8) + 2$

14. $(8 + 5) + 1 = 1 + (8 + 5)$

15. $(3 + 5) + 2 = (5 + 3) + 2$

16. $(6 + 5) + 3 = 6 + (5 + 3)$

17. Fill in the blank: Any number added to ▦ stays the same.

18. In evaluating $(12 + 8) + 5$, which addition should be performed first?

NOTATION   *Fill in the blanks.*

19. The symbols (   ) are called _____.

20. The plus sign + means _____.

Express the following facts in words.

21. $33 + 12 = 45$

22. $28 + 22 = 50$

Complete each solution.

23. $(36 + 11) + 5 = $ ▦ $+ 5$
$$= 52$$

24. $12 + (15 + 2) = 12 + $ ▦
$$= 29$$

PRACTICE   *Perform each addition.*

25. $25 + 13$

26. $47 + 12$

27. $156 + 305$

28. $647 + 38$

29. $(95 + 16) + 39$

30. $832 + (97 + 27)$

31. $25 + (321 + 17)$

32. $(4,231 + 213) + 5,234$

33. $\begin{array}{r} 76 \\ +45 \\ \hline \end{array}$ **34.** $\begin{array}{r} 87 \\ +56 \\ \hline \end{array}$

35. $\begin{array}{r} 93 \\ +47 \\ \hline \end{array}$ **36.** $\begin{array}{r} 59 \\ +65 \\ \hline \end{array}$

37. $\begin{array}{r} 632 \\ +347 \\ \hline \end{array}$ **38.** $\begin{array}{r} 423 \\ +570 \\ \hline \end{array}$

39. $\begin{array}{r} 1,372 \\ + \ 613 \\ \hline \end{array}$ **40.** $\begin{array}{r} 2,477 \\ + \ 693 \\ \hline \end{array}$

41. $\begin{array}{r} 6,427 \\ +3,573 \\ \hline \end{array}$ **42.** $\begin{array}{r} 3,567 \\ +8,778 \\ \hline \end{array}$

43. $\begin{array}{r} 8,539 \\ +7,368 \\ \hline \end{array}$ **44.** $\begin{array}{r} 5,799 \\ +6,879 \\ \hline \end{array}$

45. $\begin{array}{r} 1,246 \\ 578 \\ + \ \ 37 \\ \hline \end{array}$ **46.** $\begin{array}{r} 4,689 \\ 3,422 \\ + \ \ 26 \\ \hline \end{array}$

47. $\begin{array}{r} 3,156 \\ 1,578 \\ + \ 578 \\ \hline \end{array}$ **48.** $\begin{array}{r} 2,379 \\ 4,779 \\ +2,339 \\ \hline \end{array}$

Find the perimeter of each rectangle or square.

49.
32 feet (ft) · 12 ft

50.
127 meters (m) · 91 m

51.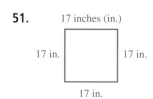
17 inches (in.) · 17 in. · 17 in. · 17 in.

52.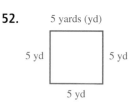
5 yards (yd) · 5 yd · 5 yd · 5 yd

APPLICATIONS

53. DIMENSIONS OF A HOUSE Find the length of the house shown in the blueprint.

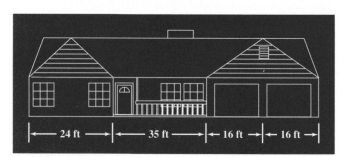

24 ft · 35 ft · 16 ft · 16 ft

54. GRADUATES During the first year of this century, 18,549 boys and 25,182 girls graduated from American high schools. Find the total number of graduates.

55. COOKIES In 2002–2003, the two top-selling brands of cookies were Oreos, with sales of \$449,772,768, and Chips Ahoy, with sales of \$317,724,064. What was their combined sales?

56. SPACE TRAVEL Astronaut Walter Schirra's first space flight orbited the Earth 6 times and lasted 9 hours. His second flight orbited the Earth 16 times and lasted 26 hours. How long was Schirra in space?

57. IMPORTS The table below shows the number of new passenger cars imported into the United States from various countries in 2003. Find the total number of cars the U.S. imported from these countries.

Country	Number of passenger cars
Canada	1,811,892
Germany	561,482
Japan	1,770,355
Mexico	680,214
South Korea	692,863
Sweden	119,773
United Kingdom	207,158

Source: Bureau of the Census, Foreign Trade Division

58. INVESTMENTS A student owned the following shares of stock: Microsoft, 250; GE, 175; and Verizon, 312. How many shares did the student own?

59. TAX DEDUCTIONS For tax purposes, a woman kept the mileage records shown in the table. Find the total number of miles that she drove.

Month	Miles driven
January	2,345
February	1,712
March	1,778
April	445
May	1,003
June	2,774

60. BUDGETS A department head in a company prepared an annual budget with the line items shown. Find the projected number of dollars to be spent.

Line item	Amount
Equipment	$17,242
Utilities	5,443
Travel	2,775
Supplies	10,553
Development	3,225
Maintenance	1,075

61. CANDY The graph below shows U.S. candy sales in 2003 during four holiday periods. Find the sum of these seasonal candy sales.

Valentine's Day — $1,040,000,000
Easter — $1,810,000,000
Halloween — $1,993,000,000
Winter Holidays — $1,390,000,000

Source: National Confectioners Association

62. BANKING A savings account contains $3,712. How much will be in the account after deposits of $4,673, $3,237, and $7,635?

63. FLAGS To decorate a city flag, yellow fringe is to be sewn around its outside edges, as shown. The fringe is sold by the inch. How many inches of fringe must be purchased to complete the project?

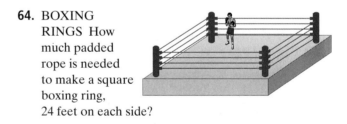

34 in.

64 in.

64. BOXING RINGS How much padded rope is needed to make a square boxing ring, 24 feet on each side?

■ **WRITING**

65. Explain why the operation of addition is commutative.

66. Explain why the operation of addition is associative.

■ **REVIEW** *Write each numeral in expanded notation.*

67. 3,125

68. 60,037

Round 6,354,784 to the indicated place.

69. Nearest ten

70. Nearest hundred

71. Nearest ten thousand

72. Nearest hundred thousand

1.3 Subtracting Whole Numbers

- Subtracting whole numbers   • Subtracting whole numbers with more than one digit
- Checking subtractions   • Subtracting whole numbers using a calculator   • Mixed operations

When we have mastered subtraction of whole numbers, we can solve more problems from science, business, and other fields. For example, to find the difference between two temperatures, we subtract them. To find the profit of a business, we subtract expenses from income. To find the sale price of a car, we subtract the discount from the regular price.

Subtracting whole numbers

The subtraction of whole numbers determines how many objects remain after some objects are removed from a set. For example, if we start with a set of 9 objects and take away 4 objects, we are left with a set of 5 objects.

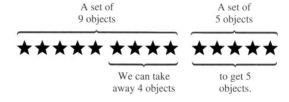

This corresponds to the subtraction fact

$$9 - 4 = 5 \quad \text{Read as "9 minus 4 equals 5."}$$

In this subtraction, 9 is called the **minuend,** 4 is called the **subtrahend,** and 5 is called the **difference.** Subtraction is often shown in a vertical format.

$$
\begin{array}{r}
9 \leftarrow \text{Minuend} \\
-4 \leftarrow \text{Subtrahend} \\
\hline
5 \leftarrow \text{Difference}
\end{array}
$$

With whole numbers, we cannot subtract in the opposite order and find the difference $4 - 9$, because we cannot take 9 objects away from 4 objects. Since the subtraction of whole numbers cannot be done in either order, the operation of subtraction is not commutative.

Subtraction is not associative either, because if we group numbers in different ways, we get different results. For example,

$$(9 - 5) - 1 = 4 - 1 \quad \text{but} \quad 9 - (5 - 1) = 9 - 4$$
$$= 3 \qquad\qquad\qquad\qquad = 5$$

EXAMPLE 1   Find each difference: **a.** $9 - 3$,   **b.** $8 - 5$,   **c.** $9 - (6 - 3)$,   and **d.** $(9 - 6) - 3$.

Solution

a. $9 - 3 = 6$ **b.** $8 - 5 = 3$

c. $9 - (6 - 3) = 9 - 3$ Perform the subtraction within parentheses first: $6 - 3 = 3$.
$\qquad\qquad\quad = 6$ Perform the subtraction.

d. $(9 - 6) - 3 = 3 - 3$ Perform the subtraction within parentheses first: $9 - 6 = 3$.
$\qquad\qquad\quad = 0$ Perform the subtraction.

Subtracting whole numbers with more than one digit

We can subtract whole numbers greater than 10 using a vertical format. We simply subtract the numbers in each corresponding column. When we must subtract a larger digit from a smaller digit, we will need to *borrow*.

EXAMPLE 2 Subtract 27 from 59.

Self Check 2
Subtract 32 from 68.

Solution The number to be subtracted is 27. When we translate to math symbols, we must reverse the order in which 27 and 59 appear in the sentence.

To perform the subtraction, we can write the digits in a vertical format, with the ones digits in a column and the tens digits in a column. We then subtract the ones digits and subtract the tens digits.

Subtract 27 from 59.

$$59 - 27$$

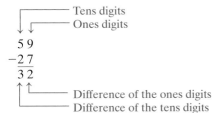

Tens digits
Ones digits

$$\begin{array}{r} 5\,9 \\ -2\,7 \\ \hline 3\,2 \end{array}$$

Difference of the ones digits
Difference of the tens digits

The difference is 32.

Answer 36

EXAMPLE 3 Subtract 15 from 32.

Self Check 3
Subtract 36 from 63.

Solution This sentence translates to: $32 - 15$. We write the digits in a vertical format, with the digits aligned in their corresponding columns.

$$\begin{array}{r} 3\,2 \\ -1\,5 \end{array}$$

Since 5 cannot be subtracted from 2, we borrow 1 ten from the tens column.

$$\begin{array}{r} \overset{2}{\cancel{3}}\ \overset{12}{\cancel{2}} \\ -1\ \ 5 \\ \hline 7 \end{array}$$
To subtract in the ones column, borrow 1 ten from the tens column and add the 10 to the ones column of the minuend, giving 12. Then subtract: $12 - 5 = 7$.

$$\begin{array}{r} \overset{2}{\cancel{3}}\ \overset{12}{\cancel{2}} \\ -1\ \ 5 \\ \hline 1\ \ 7 \end{array}$$
Subtract in the tens column.

Thus, $32 - 15 = 17$.

Answer 27

EXAMPLE 4 Find: $2{,}021 - 576$.

Self Check 4
Find: $2{,}021 - 1{,}445$.

Solution We write the digits in a vertical format, with their corresponding digits aligned.

$$\begin{array}{r} 2\,0\,2\,1 \\ -\ \ 5\,7\,6 \end{array}$$

To subtract in the ones column, we borrow 1 ten from the tens column and add it to the ones column to get 11 in the ones column. We then subtract: $11 - 6 = 5$.

$$
\begin{array}{r}
2\ 0\ \overset{1}{\cancel{2}}\ \overset{11}{\cancel{1}} \\
-\ \ \ 5\ 7\ 6 \\
\hline
5
\end{array}
$$

Since we can't subtract 7 from 1 in the tens column, we must borrow. Because there is a 0 in the hundreds column of the minuend, we must borrow from the thousands column.

$$
\begin{array}{r}
1\ \overset{9}{\cancel{\overset{11}{\cancel{10}}}}\ \overset{11}{\cancel{1}}\ 11 \\
2\ 0\ 2\ \cancel{1} \\
-\ \ \ 5\ 7\ 6 \\
\hline
4\ 5
\end{array}
$$

To subtract in the tens column, borrow 1 thousand from the thousands column and write it as 10 hundreds in the hundreds column. Borrow 1 hundred from the hundreds column and think of it as 10 tens. Add these 10 tens to the 1 ten that is left in the tens column to get 11 tens. Then subtract: $11 - 7 = 4$.

$$
\begin{array}{r}
1\ \overset{9}{\cancel{10}}\ \overset{11}{\cancel{1}}\ 11 \\
2\ 0\ 2\ 1 \\
-\ \ 5\ 7\ 6 \\
\hline
4\ 4\ 5
\end{array}
$$

Subtract in the hundreds column: $9 - 5 = 4$.

$$
\begin{array}{r}
1\ \overset{9}{\cancel{10}}\ \overset{11}{\cancel{1}}\ 11 \\
2\ 0\ 2\ 1 \\
-\ \ 5\ 7\ 6 \\
\hline
1\ 4\ 4\ 5
\end{array}
$$

Subtract in the thousands column: $1 - 0 = 1$.

The difference is 1,445.

To see that the result in Example 4 is reasonable, we can round the minuend and subtrahend and estimate the answer: 2,021 is a little more than 2,000, and 576 is a little less than 600. We can estimate the answer to be about $2,000 - 600 = 1,400$. Therefore, the result 1,445 seems reasonable.

Answer 576

◾ Checking subtractions

We have introduced the operation of subtraction as a take-away process.

$$
9 - 4 = 5 \qquad \text{or} \qquad
\begin{array}{r}
9 \\
-\ 4 \\
\hline
5
\end{array}
\qquad \text{means} \qquad \text{"9 take away 4 equals 5."}
$$

We can also think of subtraction as the opposite of addition.

$$
9 - 4 = 5 \qquad \text{or} \qquad
\begin{array}{r}
9 \\
-\ 4 \\
\hline
5
\end{array}
\qquad \text{means} \qquad
\begin{array}{l}
\text{"What must I add to 4 to get 9?"} \\
\text{The answer is 5.}
\end{array}
$$

This provides a way to check subtractions. If a subtraction is done correctly, the sum of the difference and the subtrahend will always equal the minuend. For example, to check the subtraction in Example 4, we add the difference (1,445) to the subtrahend (576) and show that the sum is the minuend (2,021).

$$
\begin{array}{rl}
2021 & \leftarrow \text{Minuend} \\
-\ \ 576 & \leftarrow \text{Subtrahend} \\
\hline
1445 & \leftarrow \text{Difference}
\end{array}
\qquad
\textbf{\textit{Check:}}
\quad
\begin{array}{r}
\overset{1\ 1\ 1}{1445} \\
+\ \ 576 \\
\hline
2021
\end{array}
$$

EXAMPLE 5   Check the subtraction:

$$
\begin{array}{r}
3682 \\
-1954 \\
\hline
1728
\end{array}
$$

Solution   We verify that $1,728 + 1,954 = 3,682$.

$$
\begin{array}{r}
1728 \\
+1954 \\
\hline
3682
\end{array}
$$

The answer checks.

Self Check 5
Determine whether the subtraction is correct by adding:

$$
\begin{array}{r}
9,784 \\
-4,792 \\
\hline
4,892
\end{array}
$$

Answer incorrect

Words such as *minus, decrease, loss, debit, down, backward, fall, reduce, in the past,* and *to the left* indicate subtraction.

EXAMPLE 6   **Vehicle crashes.**   In 2000, the number of motor vehicle traffic crashes in the United States was 6,394,000. That number declined in 2001, dropping by 71,000. In 2002, it fell by an additional 7,000. How many motor vehicle traffic crashes were there in 2002?

Solution   The words *dropping* and *fell* indicate subtraction. We can show the calculations necessary to solve this example in a single expression, as shown below. The two subtractions are done working from left to right.

$$6,394,000 - 71,000 - 7,000 = \mathbf{6,323,000} - 7,000$$
$$= 6,316,000$$

In 2002, there were 6,316,000 motor vehicle traffic crashes in the United States.

Self Check 6
According to the Recording Industry Association of America, in 2001, manufacturers shipped 881,900,000 CDs. That number declined in 2002, dropping by 78,600,000. In 2003, it fell by an additional 57,400,000. How many CDs were shipped in 2003?

Answer 745,900,000

Subtracting whole numbers using a calculator

To answer questions about *how much more* or *how many more*, subtraction can be used.

High school sports	**CALCULATOR SNAPSHOT**

In 2004, the number of boys that participated in high school sports was 4,038,253 and the number of girls that participated was 2,865,299. To determine how many more boys participated than girls, we must subtract: $4,038,253 - 2,865,299$. We can perform this subtraction using a calculator by entering

4038253 $\boxed{-}$ 2865299 $\boxed{=}$ $\boxed{1172954}$

In 2004, we see that 1,172,954 more boys than girls participated in high school sports.

Mixed operations

Additions and subtractions often appear in the same problem. It is important to read the problem carefully, locate the useful information, and organize it correctly.

Self Check 7
One share of ABC Corporation stock cost $75. The price fell $7 per share. However, it recovered and rose $13 per share. What is its current price?

Answer $81

EXAMPLE 7   Bus passengers. Twenty-seven people were riding a bus on Route 47. At the Seventh Street stop, 16 riders got off the bus and 5 got on. How many riders were left on the bus?

Solution   The route and street number are not important. The phrase *got off the bus* indicates subtraction, and the phrase *got on* indicates addition. The number of riders on the bus can be found by calculating $27 - 16 + 5$. Working from left to right, we have

$$27 - 16 + 5 = 11 + 5$$
$$= 16$$

There were 16 riders left on the bus.

! COMMENT When making the calculation in Example 7, we must perform the subtraction first. If the addition is done first, we obtain an incorrect answer of 6. For expressions containing addition and subtraction, perform them *as they occur from left to right.*

$$27 - 16 + 5 = 27 - 21$$
$$= 6$$

Section 1.3   STUDY SET

VOCABULARY   *Fill in the blanks.*

1. The result of a subtraction is called a _____.
2. In a subtraction, the _____ is always subtracted from the minuend.

CONCEPTS   *Fill in the blanks.*

3. Subtraction is the opposite of _____.
4. The subtraction $7 - 3 = 4$ is related to the addition  + ▨ = ▨.
5. When an expression contains both additions and subtractions, we perform the operations as they occur from _____ to _____.
6. The operation of _____ can be used to check the result of a subtraction.

NOTATION   *Fill in the blanks.*

7. The minus symbol − means _____. The plus symbol + means _____.
8. Express the fact in words:
   $$28 - 22 = 6$$

9. Translate the following sentence to math symbols: Subtract 30 from 83.
10. Complete the solution:

    $$36 - 11 + 5 = \boxed{} + 5$$
    $$= \boxed{}$$

PRACTICE   *Check each subtraction with an addition.*

11.   29
      −17
      ―――
       12

12.   325
      −237
      ―――
       78

13.   453
      −327
      ―――
       136

14.   2,698
      −1,569
      ―――
       1,129

Perform each subtraction.

15. $17 - 14$
16. $42 - 31$
17. $39 - 14$
18. $45 - 32$
19. $174 - 71$
20. $257 - 155$
21. $416 - 357$
22. $787 - 696$

23.   367
      −343

24.   224
      −122

25.   423
      −305

26.   330
      −270

27.	1,537 − 579	**28.**	2,470 − 863
29.	4,267 −2,578	**30.**	7,356 −3,578
31.	17,246 − 6,789	**32.**	34,510 −27,593
33.	15,700 −15,397	**34.**	35,021 −23,999
35.	67,305 −23,746	**36.**	93,001 −35,002
37.	29,307 −10,008	**38.**	40,012 −19,045

39. Subtract 199 from 301.

40. Subtract 78 from 2,047.

41. Subtract 249 from 50,009.

42. Subtract 198 from 20,020.

Perform the operations.

43. 633 − 598 − 30

44. 600 − 497 − 60

45. 852 − 695 + 40

46. 397 − 348 + 65

47. 120 + 30 − 40

48. 600 + 99 − 54

■ APPLICATIONS

49. GARDENING Beverly planted 27 seedlings, but only 21 of them survived. How many plants died?

50. RETAILING Phil sold 42 toasters on Tuesday, 7 more than he sold on Monday. How many did he sell on Monday?

51. TRANSPORTATION For a 17-mile trip, Wanda paid the taxi driver $23. If $5 was a tip, how much was the fare?

52. MAGAZINES *TV Guide* had a recent annual circulation of 16,969,260. By what amount did this exceed the circulation of *Reader's Digest*, with 16,566,650 readers?

53. DOW JONES AVERAGE How much did the Dow rise on the day described by the graph?

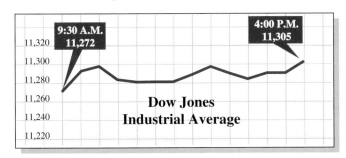

54. JEWELRY MAKING Gold melts at about 1,947° F. The melting point of silver is 183° F lower. What is the melting point of silver?

55. BANKING A savings account contained $370. After a deposit of $40 and a withdrawal of $197, how much was left in the account?

56. TRAVEL A student wants to make a 2,221-mile trip in three days. If she drives 751 miles on the first day and 875 miles on the second day, how far must she travel on the third day?

57. MAGAZINES The monthly circulation of *Motor Trend* magazine in 2001 was 1,272,053. In 2002, the monthly circulation grew by 11,207. In 2003, the monthly circulation decreased by 20,230. What was the monthly circulation in 2003?

58. LENGTH OF A MOTOR Find the length of the motor on the machine shown in the blueprint.

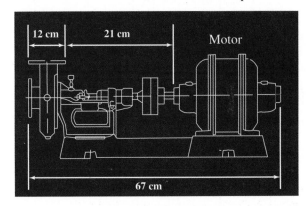

Refer to the following table. To use this salary schedule, note that a third-year teacher with 15 units of coursework beyond a Bachelor's Degree (B.D.) makes $30,887 per year (Step 3/Column 2).

59. How much more will a teacher who is now at Step 2/ Column 2 make next year, when she has gained 1 year of teaching experience?

60. How much more will a teacher who is now at Step 4/ Column 2 make next year if he takes enough coursework to move to Column 3?

TEACHERS' SALARY SCHEDULE ABC UNIFIED SCHOOL DISTRICT			
Years teaching	Column 1: B.D.	Column 2: B.D. + 15	Column 3: B.D. + 30
Step 1	$26,785	$28,243	$29,701
Step 2	$28,107	$29,565	$31,023
Step 3	$29,429	$30,887	$32,345
Step 4	$30,751	$32,209	$33,667
Step 5	$32,073	$33,531	$34,989

61. DALMATIANS See the graph on the next page. How many fewer Dalmatians were registered in 2000 as compared to the year when they were at their height of popularity?

62. DALMATIANS See the graph below. Between which two years was the drop in registrations the greatest? What was that drop?

Number of new Dalmatians registered with the American Kennel Club

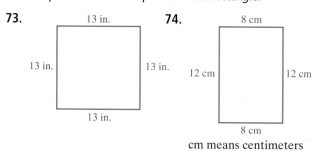

21,603 30,225 38,927 42,816 42,621 36,714 32,972 22,726 9,722 4,652 3,084

'90 '91 '92 '93 '94 '95 '96 '97 '98 '99 '00
Year

▌ WRITING

63. Explain why the operation of subtraction is not commutative.

64. List five words that indicate subtraction.

65. Explain how addition can be used to check subtraction.

66. The symbol ≠ means is not equal to. Show that

$$(20 - 8) - 5 \neq 20 - (8 - 5)$$

Explain why we have agreed always to perform subtractions from left to right.

▌ REVIEW   *Round 5,370,645 to the indicated place.*

67. Nearest ten

68. Nearest hundred

69. Nearest thousand

70. Nearest ten thousand

71. Nearest hundred thousand

72. Nearest million

Find the perimeter of the square and the rectangle.

73.

13 in.

13 in.   13 in.

13 in.

74.

8 cm

12 cm   12 cm

8 cm

cm means centimeters

Perform each addition.

75.   345
          672
        +513

76.   813
          487
        +654

1.4 Multiplying Whole Numbers

- Properties of multiplication • Multiplying whole numbers by numbers with one digit
- Multiplying by powers of 10
- Multiplying whole numbers by numbers with more than one digit
- Finding the area of a rectangle • Multiplying whole numbers using a calculator

Our problem-solving tools become even more effective when we learn how to multiply whole numbers. This operation enables us to find the areas of geometric figures and to solve business problems. For example, to find the area of a rectangle, we need to multiply its length by its width. To figure a paycheck, we need to multiply the number of hours worked by the hourly rate of pay.

▌ Properties of multiplication

We can use three symbols to indicate multiplication.

Symbols used for multiplication		
Symbol		**Example**
×	times symbol	4×5 or 12×234
·	raised dot	$4 \cdot 5$ or $117 \cdot 225$
( )	parentheses	$(4)(5)$ or $4(5)$ or $4(5)$

Multiplication is repeated addition. For example, 4×5, $4 \cdot 5$, and $4(5)$ all mean the sum of four 5's.

$$4 \times 5 = 4 \cdot 5 = 4(5) = \overbrace{5 + 5 + 5 + 5}^{\text{The sum of four 5's}} = 20$$

In the above multiplication, the result of 20 is called the **product.** The numbers that were multiplied (4 and 5) are called **factors.**

$$\underset{\underset{4}{\downarrow}}{\text{Factor}} \ \cdot \ \underset{\underset{5}{\downarrow}}{\text{Factor}} \ = \ \underset{\underset{20}{\downarrow}}{\text{Product}}$$

Multiplication is often shown in a vertical format.

$$
\begin{array}{r}
5 \leftarrow \text{Factor} \\
\times\ 4 \leftarrow \text{Factor} \\
\hline
20 \leftarrow \text{Product}
\end{array}
$$

The multiplications 5×4, $5 \cdot 4$, and $5(4)$ all mean the sum of five 4's.

$$5 \times 4 = 5 \cdot 4 = 5(4) = \overbrace{4 + 4 + 4 + 4 + 4}^{\text{The sum of five 4's}} = 20$$

We have seen that $4 \cdot 5 = 20$ and $5 \cdot 4 = 20$. The results are the same. These examples illustrate that the order in which we multiply two numbers does not affect the result. This property is called the **commutative property of multiplication.**

Commutative property of multiplication

The order in which whole numbers are multiplied does not affect their product. For example,

$$4 \times 6 = 6 \times 4, \qquad 7 \cdot 5 = 5 \cdot 7, \qquad \text{and} \qquad 7(8) = 8(7)$$

Table 1-2 on the next page summarizes the basic multiplication facts.

- To find the product of 6 and 8 using the table, we find the intersection of the row starting with 6 and the column headed by 8. The product is 48.
- To find the product of 8 and 6, we find the intersection of the row starting with 8 and the column headed by 6. Once again, the product is 48.

You should memorize this table if you have not already done so.

From the table, we see that whenever we multiply a number by 0, the product is 0. For example,

$$0 \cdot 5 = 0, \qquad 0 \cdot 8 = 0, \qquad \text{and} \qquad 9 \cdot 0 = 0$$

We also see that whenever we multiply a number by 1, the number remains the same. For example,

$$3 \cdot 1 = 3, \qquad 7 \cdot 1 = 7, \qquad \text{and} \qquad 1 \cdot 9 = 9$$

These examples suggest the multiplication properties of 0 and 1.

·	0	1	2	3	4	5	6	7	8	9
0	**0**	0	0	0	0	0	0	0	0	0
1	0	**1**	2	3	4	5	6	7	8	9
2	0	2	**4**	6	8	10	12	14	16	18
3	0	3	6	**9**	12	15	18	21	24	27
4	0	4	8	12	**16**	20	24	28	32	36
5	0	5	10	15	20	**25**	30	35	40	45
6	0	6	12	18	24	30	**36**	42	48	54
7	0	7	14	21	28	35	42	**49**	56	63
8	0	8	16	24	32	40	48	56	**64**	72
9	0	9	18	27	36	45	54	63	72	**81**

TABLE 1-2

We note that the answers in the table above the diagonal line in bold print are identical to the answers below the diagonal line. This illustrates that multiplication is commutative.

> **Multiplication properties of 0 and 1**
>
> The product of any whole number and 0 is 0.
>
> The product of any whole number and 1 is that whole number.

Application problems that involve repeated addition are often more easily solved using multiplication.

Self Check 1

At a rate of $8 per hour, how much will a school bus driver earn if she works from 8:00 A.M. until noon?

EXAMPLE 1   Computing wages.
Raul worked an 8-hour day at an hourly rate of $9. How much money did he earn?

Solution   For each of the 8 hours, Raul earned $9. His total pay for the day is the sum of eight 9's: $9 + 9 + 9 + 9 + 9 + 9 + 9 + 9$. This repeated addition can be calculated by multiplication.

$$\text{Total wages} = 8 \cdot 9$$
$$= 72 \qquad \text{See the multiplication table.}$$

Answer $32

Raul earned $72.

To multiply three numbers, we first multiply two of them and then multiply that result by the third number. In the following examples, we multiply $3 \cdot 2 \cdot 4$ in two ways. The parentheses show us which multiplication to do first.

Method 1: Group 3 · 2

$(\mathbf{3 \cdot 2}) \cdot 4 = \mathbf{6 \cdot 4}$ Multiply 3 and 2 to get 6.
$\qquad\qquad\quad = 24$ Then multiply 6 and 4 to get 24.

Method 2: Group 2 · 4

$3 \cdot (\mathbf{2 \cdot 4}) = 3 \cdot \mathbf{8}$ Multiply 2 and 4 to get 8.
$\qquad\qquad\quad = 24$ Then multiply 3 and 8 to get 24.

The answers are the same. This illustrates that changing the grouping when multiplying numbers does not affect the result. This property is called the **associative property of multiplication.**

> **Associative property of multiplication**
> The way in which whole numbers are grouped does not affect their product. For example,
>
> $(4 \cdot 3) \cdot 2 = 4 \cdot (3 \cdot 2)$

Multiplying whole numbers by numbers with one digit

To find the product $8 \cdot 47$, it is inconvenient to add eight 47's. Instead, we find the product using a multiplication process.

$$\begin{array}{r} 4\,7 \\ \times \quad 8 \\ \hline \end{array}$$ Write the factors in a column, with the corresponding digits aligned vertically.

$$\begin{array}{r} \overset{5}{4}\,7 \\ \times \quad 8 \\ \hline 6 \end{array}$$ Multiply 7 by 8. The product is 56. Place 6 in the ones column of the answer, and carry 5 to the tens column.

$$\begin{array}{r} \overset{5}{4}\,7 \\ \times \quad 8 \\ \hline 3\,7\,6 \end{array}$$ Multiply 4 by 8. The product is 32. To the 32, add the carried 5 to get 37. Place the 7 in the tens column of the answer and the 3 in the hundreds column.

The product is 376.

EXAMPLE 2 A car can travel 32 miles on 1 gallon of gasoline. How many miles can the car travel on 3 gallons of gasoline?

Solution Each of the 3 gallons of gasoline enables the car to go 32 miles. The total distance the car can travel is the sum of three 32's: $32 + 32 + 32$. This can be calculated by multiplication.

$$3 \cdot 32 = 96$$

On 3 gallons of gasoline, the car can travel 96 miles.

Self Check 2
How far can the car travel on 8 gallons of gasoline?

Answer 256 mi

Multiplying by powers of 10

The numbers 1, 10, 100, 1,000, 10,000, and so on, are called **powers of 10,** because each one is ten times the previous one. Table 1-3 shows several products involving 10, 100, 1,000 and 10,000.

$1 \cdot 10 = 10$	$1 \cdot 100 = 100$	$1 \cdot 1{,}000 = 1{,}000$	$1 \cdot 10{,}000 = 10{,}000$
$2 \cdot 10 = 20$	$2 \cdot 100 = 200$	$2 \cdot 1{,}000 = 2{,}000$	$2 \cdot 10{,}000 = 20{,}000$
$3 \cdot 10 = 30$	$3 \cdot 100 = 300$	$3 \cdot 1{,}000 = 3{,}000$	$3 \cdot 10{,}000 = 30{,}000$
$4 \cdot 10 = 40$	$4 \cdot 100 = 400$	$4 \cdot 1{,}000 = 4{,}000$	$4 \cdot 10{,}000 = 40{,}000$
$5 \cdot 10 = 50$	$5 \cdot 100 = 500$	$5 \cdot 1{,}000 = 5{,}000$	$5 \cdot 10{,}000 = 50{,}000$

TABLE 1-3

The results in the table show that every product is a whole number ending with one or more zeros. This fact makes it easy to multiply a number by a power of 10. For example,

To multiply 14 by 10, we simply attach one zero to the right of 14: $10 \cdot 14 = 140$.

To multiply 14 by 100, we simply attach two zeros to the right of 14: $100 \cdot 14 = 1,400$.

To multiply 14 by 1,000, we simply attach three zeros to the right of 14: $1,000 \cdot 14 = 14,000$.

To multiply 14 by 10,000, we simply attach four zeros to the right of 14: $10,000 \cdot 14 = 140,000$.

These examples suggest the following rule.

Multiplying by powers of 10

To find the product of a whole number and a power of 10:

1. Determine how many zeros are in the power of 10.

2. Attach that number of zeros to the right side of the whole number.

Self Check 3
Find each product:
a. $25 \cdot 100$

b. $875(1,000)$

Answers **a.** 2,500,   **b.** 875,000

EXAMPLE 3   Find each product: **a.** $63 \cdot 10$,   **b.** $45 \cdot 100$,   and **c.** $912(10,000)$.

Solution

a. $63 \cdot 10 = 630$ Since 10 has one zero, attach one 0 after 63.

b. $45 \cdot 100 = 4,500$ Since 100 has two zeros, attach two zeros after 45.

c. $912(10,000) = 9,120,000$ Since 10,000 has four zeros, attach four zeros after 912.

We can handle zeros in a similar way in multiplications that do not involve powers of 10. For example, to find the product of 43 and 500, we can write 500 as $5 \cdot 100$ and find the product $43 \cdot 5 \cdot 100$.

$$43 \cdot 5 \cdot 100 = 215 \cdot 100 \quad \text{Find the product of 43 and 5.}$$
$$= 21,500 \quad \text{Since 100 has two zeros, attach two zeros after 215.}$$

Self Check 4
Find each product:
a. $15 \cdot 300$

b. $315 \cdot 2,000$

Answers **a.** 4,500,   **b.** 630,000

EXAMPLE 4   Find each product: **a.** $67 \cdot 20$ and   **b.** $45 \cdot 40,000$.

Solution

a. $67 \cdot \mathbf{20} = 67 \cdot \mathbf{2} \cdot \mathbf{10}$ Write 20 as $2 \cdot 10$.

$\qquad = 134 \cdot 10$ Multiply: $67 \cdot 2 = 134$.

$\qquad = 1,340$ Since 10 has one zero, attach one zero after 134.

b. $45 \cdot \mathbf{40,000} = 45 \cdot \mathbf{4} \cdot \mathbf{10,000}$ Write 40,000 as $4 \cdot 10,000$.

$\qquad = 180 \cdot 10,000$ Multiply: $45 \cdot 4 = 180$.

$\qquad = 1,800,000$ Since 10,000 has four zeros, attach four zeros after 180.

▍Multiplying whole numbers by numbers with more than one digit

To find the product 23 · 435, we use the multiplication process. Because 23 = 20 + 3, we multiply 435 by 20 and by 3, and add the products.

$$
\begin{array}{r}
4\,3\,5 \\
\times\ \ 2\,3 \\
\hline
\end{array}
$$
Write the factors in a column, with the corresponding digits aligned vertically.

We begin by multiplying 435 by 3.

$$
\begin{array}{r}
\overset{1}{} \\
4\,3\,5 \\
\times\ \ 2\,3 \\
\hline
5
\end{array}
$$
Multiply 5 by 3. The product is 15. Place 5 in the ones column and carry 1 to the tens column.

$$
\begin{array}{r}
\overset{1}{}\overset{1}{} \\
4\,3\,5 \\
\times\ \ 2\,3 \\
\hline
0\,5
\end{array}
$$
Multiply 3 by 3. The product is 9. To the 9, add the carried 1 to get 10. Place the 0 in the tens column and carry the 1 to the hundreds column.

$$
\begin{array}{r}
\overset{1}{}\overset{1}{} \\
4\,3\,5 \\
\times\ \ 2\,3 \\
\hline
1\,3\,0\,5
\end{array}
$$
Multiply 4 by 3. The product is 12. Add the 12 to the carried 1 to get 13. Write 13.

We continue by multiplying 435 by 2 tens, or 20.

$$
\begin{array}{r}
\overset{1}{} \\
4\,3\,5 \\
\times\ \ 2\,3 \\
\hline
1\,3\,0\,5 \\
0
\end{array}
$$
Multiply 5 by 2. The product is 10. Write 0 in the tens column and carry 1.

$$
\begin{array}{r}
\overset{1}{} \\
4\,3\,5 \\
\times\ \ 2\,3 \\
\hline
1\,3\,0\,5 \\
7\,0
\end{array}
$$
Multiply 3 by 2. The product is 6. Add 6 to the carried 1 to get 7. Write the 7. There is no carry.

$$
\begin{array}{r}
\overset{1}{} \\
4\,3\,5 \\
\times\ \ 2\,3 \\
\hline
1\,3\,0\,5 \\
8\,7\,0
\end{array}
$$
Multiply 4 by 2. The product is 8. There is no carry to add. Write the 8.

$$
\begin{array}{r}
4\,3\,5 \\
\times\ \ 2\,3 \\
\hline
1\,3\,0\,5 \\
8\,7\,0 \\
\hline
1\,0\,0\,0\,5
\end{array}
$$
Draw another line beneath the two completed rows. Add the numbers in the two rows. This sum gives the product of 435 and 23.

Thus, 23 · 435 = 10,005.

To check whether the result in the previous example is reasonable, we can round off and estimate the answer: 435 is a little less than 450, and 23 is a little more than 20. We can estimate the answer to be about 450 · 20 = 900. An answer of 10,005 is reasonable.

Self Check 5
Find the product: 706(350).

EXAMPLE 5   Find each product: **a.** 253 · 406   and   **b.** 2,007(1,006).

Solution

a.   253
         ×406
         1518   Multiply: 6 · 253 = 1,518. Align the 8 in the ones column.
         000   Multiply: 0 · 253 = 0. Align the rightmost 0 in the tens column.
         1012   Multiply: 4 · 253 = 1,012. Align the 2 in the hundreds column.
        102718   Add column by column.

The product is 102,718.

b.   2007
         ×1006
        12042   Multiply: 6 · 2,007 = 12,042. Align the 2 in the ones column.
        0000   Multiply: 0 · 2,007 = 0. Align the rightmost 0 in the tens column.
        0000   Multiply: 0 · 2,007 = 0. Align the rightmost 0 in the hundreds column.
        2007   Multiply: 1 · 2,007 = 2,007. Align the 7 in the thousands column.
       2019042   Add column by column.

The product is 2,019,042.

Answer 247,100

We can use multiplication to count objects arranged in rectangular patterns. For example, the following display on the left below shows a rectangular array consisting of 5 rows of 7 stars. The product 5 · 7, or 35, indicates the total number of stars.

Because multiplication is commutative, the array on the right, consisting of 7 rows of 5 stars, contains the same number of stars.

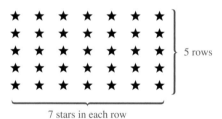

Self Check 6
On a color monitor, each of the pixels can be red, green, or blue. How many colored pixels does the computer control?

EXAMPLE 6   Computer science. To draw graphics on a computer screen, a computer controls each *pixel* (one dot on the screen). See Figure 1-8. If a computer-graphics image is 800 pixels wide and 600 pixels high, how many pixels does the computer control?

Solution   The graphics image is a rectangular array of pixels. Each of its 600 rows consists of 800 pixels. The total number of pixels is the product of 600 and 800.

          600 · 800 = 480,000   This could also be
                                 written as 600(800).

Answer 1,440,000

The computer controls 480,000 pixels.

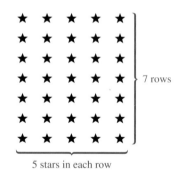

FIGURE 1-8

Finding the area of a rectangle

One important application of multiplication is finding the area of a rectangle. The **area of a rectangle** is the measure of the amount of surface it encloses. Area is measured in square units, such as square inches (denoted as in.^{2}) or square centimeters (denoted as cm^2). See Figure 1-9.

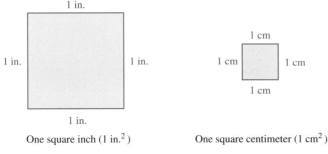

One square inch (1 in.^{2})   One square centimeter (1 cm^2)

FIGURE 1-9

The rectangle in Figure 1-10 has a length of 5 centimeters and a width of 3 centimeters. Since each small square covers an area of one square centimeter, each small square measures 1 cm^2. The small squares form a rectangular pattern, with 3 rows of 5 squares.

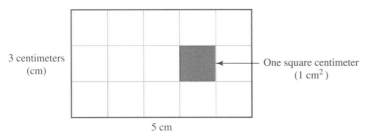

FIGURE 1-10

Because there are 5 · 3, or 15, small squares, the area of the rectangle is 15 cm^2. This suggests that the area of any rectangle is the product of its length and its width.

$$\text{Area of a rectangle} = \text{length} \cdot \text{width}$$

By using the letter *l* to represent the length and the letter *w* to represent the width, we can write this formula in simpler form.

Area of a rectangle

The area, *A*, of a rectangle is the product of the rectangle's length, *l*, and its width, *w*.

$$\text{Area} = (\text{length})(\text{width}) \qquad \text{or} \qquad A = l \cdot w$$

The formula $A = l \cdot w$ can be written more simply as $A = lw$.

EXAMPLE 7   Gift wrap. When completely unrolled, a long sheet of gift wrapping paper has the dimensions shown in Figure 1-11. How many square feet of gift wrap are on the roll?

3 ft

12 ft

FIGURE 1-11

Self Check 7
Find the area of a 9-inch-by-12-inch sheet of paper.

Solution   To find the number of square feet of paper, we need to find the area of the rectangle shown in the figure.

$A = l \cdot w$ This is the formula for the area of a rectangle.

$A = 12 \cdot 3$ Replace l with 12 and w with 3.

$= 36$ Perform the multiplication.

Answer 108 in.²

There are 36 square feet (ft²) of wrapping paper on the roll.

! COMMENT Remember that the perimeter of a rectangle is the distance around it. The area of a rectangle is a measure of the surface it encloses.

▮ Multiplying whole numbers using a calculator

CALCULATOR SNAPSHOT   **Calculating production**

The labor force of an electronics firm works two 8-hour shifts each day and manufactures 53 television sets each hour. To find how many sets will be manufactured in 5 days, we must find the following product:

2 shifts per day 8 hours per shift
↓ ↓
$2 \cdot 8 \cdot 53 \cdot 5$
↑ ↑
53 sets per hour 5 days

We can use a calculator to evaluate $2 \cdot 8 \cdot 53 \cdot 5$ by entering these numbers and pressing these keys.

2 ☒ 8 ☒ 53 ☒ 5 ▣ | 4240 |

4,240 television sets will be manufactured.

Section 1.4 STUDY SET

▮ **VOCABULARY**   *Fill in the blanks.*

1. Multiplication is repeated _____.

2. Numbers that are to be multiplied are called _____.

3. The result of a multiplication problem is called a _____.

4. The statement $6 \cdot 7 = 7 \cdot 6$ illustrates the _____ property of multiplication.

5. The statement $(5 \cdot 4) \cdot 7 = 5 \cdot (4 \cdot 7)$ illustrates the _____ property of multiplication.

6. If a square measures 1 inch on each side, its area is 1 _____ _____.

7. The _____ of a rectangle is a measure of the amount of surface it encloses.

8. The numbers 1, 10, 100, 1,000, and 10,000 are called _____ of 10.

▮ **CONCEPTS**

9. Write each repeated addition as a multiplication.
   a. $8 + 8 + 8 + 8$

   b. $15 + 15 + 15 + 15 + 15 + 15 + 15$

10. Using the numbers 2, 3, and 4, write a statement that illustrates the associative property of multiplication.

11. Using the numbers 5 and 8, write a statement that illustrates the commutative property of multiplication.

12. How do we find the amount of surface enclosed by a rectangle?

13. How many zeros do you attach to the right of 25 if you multiply 25 by 1,000?

14. Find a multiplication statement that finds the number of red squares.

15. Determine whether the concept of *perimeter* or that of *area* should be applied to find each of the following.

    a. The amount of floor space to carpet

    b. The amount of clear glass to be tinted

    c. The amount of lace needed to trim the sides of a handkerchief

16. Perform each multiplication.

    a. $1 \cdot 25$ **b.** $62(1)$

    c. $10 \cdot 0$ **d.** $0(4)$

▮ NOTATION

17. Write three symbols that are used for multiplication.

18. What does ft^2 mean?

▮ PRACTICE   *Perform each multiplication.*

19. $4 \cdot 7$ **20.** $7 \cdot 9$

21. $8 \cdot 7$ **22.** $9 \cdot 6$

23. $12 \cdot 7$ **24.** $15 \cdot 8$

25. 14×7 **26.** $19 \cdot 9$

27.   213   **28.**   863
     $\times \ \ 7$ $\times \ \ 9$

29.   779   **30.**   596
     $\times \ \ 8$ $\times \ \ 6$

Perform each multiplication without using pencil and paper or a calculator.

31. $37 \cdot 100$ **32.** $63 \cdot 1,000$

33. $75 \cdot 10$ **34.** $88 \cdot 10,000$

35. $107(10,000)$ **36.** $323(100)$

37. $512 \cdot 1,000$ **38.** $673 \cdot 10$

39. $56 \cdot 200$ **40.** $83 \cdot 3,000$

41. $315 \cdot 20$ **42.** $222 \cdot 5,000$

43. $275(30)$ **44.** $110(3,000)$

45. $202 \cdot 500$ **46.** $304 \cdot 300$

Perform each multiplication.

47. $27 \cdot 12$ **48.** $35 \cdot 17$

49. $67(87)$ **50.** $58(86)$

51.   99   **52.**   73
    $\times 77$ $\times 59$

53.   20   **54.**   78
    $\times 53$ $\times 20$

55.   112   **56.**   232
    $\times \ 23$ $\times \ 53$

57.   207   **58.**   768
    $\times \ 97$ $\times \ 70$

59.   516   **60.**   357
    $\times 302$ $\times 205$

61.   679   **62.**   889
    $\times 602$ $\times 507$

63. 5,619   **64.** 1,376
    $\times 3,020$ $\times 2,003$

65. 2,978   **66.** 2,003
    $\times 3,004$ $\times 5,030$

67. $25 \cdot 32 \cdot 3$ **68.** $37 \cdot 45 \cdot 7$

69. $42 \cdot 64 \cdot 17$ **70.** $17 \cdot 49 \cdot 63$

Find the area of each rectangle or square.

71.

72.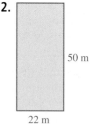

73.

74.

▮ APPLICATIONS

75. FIGURING WAGES Jennifer worked 12 hours at $11 per hour. How much did she earn?

76. FIGURING WAGES Manuel worked 23 hours at $9 per hour. How much did he earn?

77. NUTRITION There are 17 grams of fat in one Krispy Kreme chocolate-iced, custard-filled donut. How many total grams of fat are there in one dozen of those donuts?

78. RENTING APARTMENTS Mia owns an apartment building with 18 units. Each unit generates a monthly income of $450. Find her total monthly income.

79. WORD PROCESSING A student used the *Insert Table* options shown when typing a report. How many entries will the table hold?

Insert Table	☒
Number of Columns: 8	OK
Number of Rows: 9	Cancel

80. FILLING PRESCRIPTIONS How many tablets should a pharmacist put in the container shown in the illustration?

Ramirez
Pharmacy
No. 2173   11/04
**Take 2 tablets
3 times a day
for 14 days**
Expires: 11/05

81. ATTENDING CONCERTS A jazz quartet gave two concerts in each of 37 cities. Approximately 1,700 fans attended each concert. How many people heard the group?

82. CEREAL A cereal maker advertises "Two cups of raisins in every box." Find the number of cups of raisins in a case of 36 boxes of cereal.

83. ORANGE JUICE It takes 13 oranges to make one can of orange juice. Find the number of oranges used to make a case of 24 cans.

84. ROOM CAPACITY A college lecture hall has 17 rows of 33 seats. A sign on the wall reads, "Occupancy by more than 570 persons is prohibited." If the seats are filled and there is one instructor, is the college breaking the rule?

85. ELEVATORS There are 14 people in an elevator with a capacity of 2,000 pounds. If the average weight of a person on the elevator is 150 pounds, is the elevator overloaded?

86. CHANGING UNITS There are 12 inches in 1 foot. How many inches are in 80 feet?

87. MUSTANGS Mileage figures for a 2005 Ford Mustang GT are shown in the table. For city driving, how far can it travel on a tank of gas?

Fuel tank capacity	16 gal
Fuel economy (miles per gallon)	18 city/23 hwy

88. MUSTANGS How far can the Ford Mustang GT in Exercise 87 travel on a tank of gas during highway driving?

89. WRAPPING PRESENTS When completely unrolled, a long sheet of wrapping paper has the dimensions shown. How many square feet of gift wrap are on the roll?

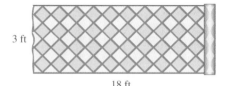

3 ft

18 ft

90. BIRDS How many times do a hummingbird's wings beat each minute?

65 wingbeats
per second

91. AREA OF WYOMING The state of Wyoming is approximately rectangular-shaped, with dimensions 360 miles long and 270 miles wide. Find its perimeter and its area.

92. COMPARING ROOMS Which has the greater area, a rectangular room that is 14 feet by 17 feet or a square room that is 16 feet on each side? Which has the greater perimeter?

93. CHESSBOARDS A chessboard consists of 8 rows, with 8 squares in each row. How many squares are there on a chessboard?

94. COMPARING MATTRESSES A queen-size mattress measures 60 inches by 80 inches, and a full-size mattress measures 54 inches by 75 inches. How much more sleeping surface is there on a queen-size mattress?

95. GARDENING A rectangular garden is 27 feet long and 19 feet wide. A path in the garden uses 125 square feet of space. How many square feet are left for planting?

96. POSTER BOARDS A rectangular-shaped poster board has dimensions of 24 inches by 36 inches. Find its area.

■ WRITING

97. Explain the difference between 1 foot and 1 square foot.

98. When two numbers are multiplied, the result is 0. What conclusion can be drawn about the numbers?

■ **REVIEW**

99. Consider the number 372,856. What digit is in the hundreds column?

100. Round 45,995 to the nearest thousand.

101. Add: 357, 39, and 476.

102. DISCOUNT SHOPPING A radio, originally priced at $97, has been marked down to $75. By how many dollars has the radio been discounted?

1.5 Dividing Whole Numbers

- Properties of division • Division and zero • Short division • Long division
- Tests for divisibility • Dividing whole numbers that end with zeros
- Dividing whole numbers using a calculator

We can solve more problems once we have learned to divide whole numbers. Mastering division will enable us to calculate how many miles a car can travel on one gallon of gasoline or to calculate the average cost of doing business.

■ Properties of division

If $12 is distributed equally among 4 people, we can divide to see that each person receives $3.

$$\frac{3}{4\overline{)12}}$$

Three symbols can be used to indicate division.

Symbols used for division			
Symbol		**Example**	
$\div$	division symbol	$12 \div 4$	or $1{,}242 \div 23$
$\overline{)}$	long division	$4\overline{)12}$	or $23\overline{)1{,}242}$
——	fraction bar	$\dfrac{12}{4}$	or $\dfrac{1{,}242}{23}$

In a division, the number being divided is called the **dividend** and the number we are dividing by is called the **divisor.** The answer is called a **quotient.**

$$\text{Dividend} \div \text{divisor} = \text{quotient} \qquad \text{Divisor}\overline{)\text{dividend}}^{\text{quotient}} \qquad \frac{\text{Dividend}}{\text{Divisor}} = \text{quotient}$$

EXAMPLE 1 In each part, identify the dividend, the divisor, and the quotient:

a. $12 \div 4 = 3$, **b.** $6\overline{)54}^{\,9}$, and **c.** $\dfrac{72}{8} = 9$.

Solution

a. Dividend $\rightarrow 12 \div 4 = 3 \leftarrow$ Quotient
             ↑
          Divisor

Self Check 1
Identify the dividend, the divisor, and the quotient in the statement: "36 divided by 9 equals 4."

$$9 \leftarrow \text{Quotient}$$

b. $\text{Divisor} \rightarrow 6\overline{)54} \leftarrow \text{Dividend}$

Answers Dividend is 36, divisor is 9, quotient is 4.

c. $\dfrac{\text{Dividend} \rightarrow 72}{\text{Divisor} \rightarrow 8} = 9 \leftarrow \text{Quotient}$

Division can be thought of as repeated subtraction. To divide 12 by 4, we ask, "How many 4's can be subtracted from 12?" Since exactly three 4's can be subtracted from 12 to get 0, we know that $12 \div 4 = 3$.

$$\overbrace{12 - 4 - 4 - 4}^{\text{Three 4's}} = 0$$

Division is also related to multiplication. For example, the division $\frac{12}{3}$ asks the question "What must I multiply 3 by to get 12?" Since the answer is 4,

$$\frac{12}{3} = 4 \quad \text{has the related multiplication statements:} \quad \mathbf{3 \cdot 4} = 12 \quad \text{or} \quad \mathbf{4 \cdot 3} = 12$$

Similarly,

$$\frac{20}{5} = 4 \quad \text{has the related multiplication statements:} \quad \mathbf{5 \cdot 4} = 20 \quad \text{or} \quad \mathbf{4 \cdot 5} = 20$$

Self Check 2
Write a related multiplication statement for $\dfrac{48}{6} = 8$.

EXAMPLE 2   For each division, write a related multiplication statement:

a. $16 \div 8 = 2$ and   **b.** $6\overline{)24}^{\,4}$.

Solution
a. $16 \div 8 = 2$ has the related multiplication statements $\mathbf{8 \cdot 2} = 16$ or $\mathbf{2 \cdot 8} = 16$.

Answer $6 \cdot 8 = 48$ or $8 \cdot 6 = 48$

b. $6\overline{)24}^{\,4}$ has the related multiplication statements $\mathbf{6 \cdot 4} = 24$ or $\mathbf{4 \cdot 6} = 24$.

▌ Division and zero

We now consider three types of division that involve zero. In the first case, we examine a division of zero; in the second, a division by zero; in the third, a division of zero by zero.

Division Statement	Related Multiplication Statement	Result
$\dfrac{0}{2} = ?$	$2 \cdot ? = 0$ ↑ This must be 0 if the product is to be 0.	$\dfrac{0}{2} = \mathbf{0}$
$\dfrac{2}{0} = ?$	$0 \cdot ? = 2$ ↑ There is no number that gives 2 when multiplied by 0.	There is no quotient.
$\dfrac{0}{0} = ?$	$0 \cdot ? = 0$ ↑ Any number times 0 is 0.	Any number can be the quotient.

We see that $\frac{0}{2} = 0$. Since $\frac{2}{0}$ does not have a quotient, we say that division of 2 by 0 is *undefined.* Since $\frac{0}{0}$ can be any number, we say that $\frac{0}{0}$ is *undetermined.*

Division with zero

1. The quotient of 0 divided by any nonzero number is 0. For example, $\frac{0}{17} = 0$.

2. The division of any nonzero number by 0 is undefined. For example, $\frac{17}{0}$ is undefined.

3. $\frac{0}{0}$ is undetermined.

The example $\frac{12}{1} = 12$ illustrates that *any number divided by 1 is the number itself.* The example $\frac{12}{12} = 1$ illustrates that *any nonzero number divided by itself is 1.*

Properties of division

Any number divided by 1 is that number. For example, $\frac{14}{1} = 14$.

Any nonzero number divided by itself is equal to 1. For example, $\frac{14}{14} = 1$.

▌ Short division

We can use a process called **short division** to divide a number by a one-digit divisor.

EXAMPLE 3   Divide 246 by 6.

Solution   First, we write the division in the form $6\overline{)246}$ and proceed as follows: Since 6 will not divide 2, we divide 24 by 6.

$$\overset{4}{6\overline{)246}} \qquad \frac{24}{6} = 4.\ \text{Write the 4 in the tens column above the division symbol.}$$

Then we divide 6 by 6.

$$\overset{41}{6\overline{)246}} \qquad \frac{6}{6} = 1.\ \text{Write the 1 in the ones column above the division symbol.}$$

We can check by verifying that $6 \cdot 41 = 246$.

$$\begin{array}{r} 41 \\ \times\ \ 6 \\ \hline 246 \end{array}$$

Self Check 3
Divide: $4\overline{)328}$.

Answer 82

When a division is not exact, we say that there is a **remainder.** For example, $\frac{15}{5} = 3$, because $5 \cdot 3 = 15$. However, $\frac{16}{5} = 3$ with 1 left over, because $5 \cdot 3 + 1 = 16$. In the next example, there is a remainder.

EXAMPLE 4   Divide 168 by 5.

Solution   First, we write the division in the form $5\overline{)168}$ and proceed as follows: Since 5 will not divide 1, we divide 16 by 5.

$$\overset{3}{5\overline{)16^18}} \qquad \frac{16}{5} = 3\ \text{with 1 left over. Put that 1 in front of the 8.}$$

Self Check 4
Divide: $7\overline{)169}$.

Then we divide 18 by 5.

$$\overset{3\ 3}{5\overline{)16^18}} \qquad \frac{18}{5} = 3 \text{ with 3 left over. This 3 is called the remainder.}$$

We write an R (for remainder) and the remainder of 3 to the right of the quotient.

$$\overset{3\ 3\,R\,3}{5\overline{)16^18}}$$

Answer   24 R 1

The result is a quotient of 33 with a remainder of 3. We can check our work by verifying that $5 \cdot 33 + 3 = 168$.

■ Long division

We can use **long division** to perform divisions when the divisor has more than one digit. For example, to divide 832 by 23, we proceed as follows:

Quotient →

Divisor → $23\overline{)832}$ Place the divisor and the dividend as indicated. The quotient will appear above the division symbol.

↑
Dividend

We find the quotient using the following division process:

$$\overset{4}{2\ 3\overline{)8\ 3\ 2}}$$
Ask: "How many times will 23 divide 83?" Because an estimate is 4, place 4 in the tens column of the quotient.

$$\begin{array}{r} 4 \\ 2\,3\overline{)8\ 3\ 2} \\ 9\ 2 \end{array}$$
Multiply $23 \cdot 4$ and place the answer, 92, under the 83. Because 92 is *larger* than 83, our estimate of 4 for the tens column of the quotient was too large.

$$\begin{array}{r} 3 \\ 2\,3\overline{)8\ 3\ 2} \\ 6\ 9\downarrow \\ \hline 1\ 4\ 2 \end{array}$$
Revise the estimate of the quotient to be 3. Multiply $23 \cdot 3$ to get 69, place 69 under the 83, draw a line, and subtract.

Bring down the 2 in the ones column.

$$\begin{array}{r} 3\ 7 \\ 2\,3\overline{)8\ 3\ 2} \\ 6\ 9 \\ \hline 1\ 4\ 2 \\ 1\ 6\ 1 \end{array}$$
Ask: "How many times will 23 divide 142?" The answer is approximately 7. Place 7 in the ones column of the quotient. Multiply $23 \cdot 7$ to get 161. Place 161 under 142. Because 161 is *larger* than 142, the estimate of 7 is too large.

$$\begin{array}{r} 3\ 6 \\ 2\,3\overline{)8\ 3\ 2} \\ 6\ 9 \\ \hline 1\ 4\ 2 \\ 1\ 3\ 8 \\ \hline 4 \end{array}$$
Revise the estimate of the quotient to be 6. Multiply: $23 \cdot 6 = 138$.

Place 138 under 142 and subtract.

The quotient is 36, and the remainder is 4. We can write this result as 36 R 4.

To check the result of a division, we multiply the divisor by the quotient and then add the remainder. The result should be the dividend.

Check: Quotient · divisor + remainder = dividend

$$36 \quad \cdot \quad 23 \quad + \quad 4 \quad = \quad 832$$
$$828 + 4 = 832$$
$$832 = 832$$

Application problems that involve forming equal-sized groups can often be solved using division.

EXAMPLE 5 Managing a soup kitchen.

A soup kitchen plans to feed 1,990 people. Because of space limitations, only 165 people can be served at one time. How many seatings will be necessary to feed everyone? How many will be served at the last seating?

Solution The 1,990 people can be fed 165 at a time. To find the number of seatings, we must divide.

$$
\begin{array}{r}
12 \\
165\overline{)1{,}990} \\
1\ 65\downarrow \\
\hline
340 \\
330 \\
\hline
10
\end{array}
$$

The quotient is 12, and the remainder is 10. Thirteen seatings will be needed: 12 full-capacity seatings and one partial seating to serve the remaining 10 people.

Self Check 5
Each gram of fat in a meal provides 9 calories. A fast-food meal contains 243 calories from fat. How many grams of fat does the meal contain?

Answer 27

■ Tests for divisibility

One number is *divisible* by another number if, when we divide them, the remainder is 0. There are tests to help us decide whether one number is divisible by another.

> **Tests for divisibility**
>
> A number is divisible by
> 2 if its last digit is divisible by 2.
> 3 if the sum of its digits is divisible by 3.
> 4 if the number formed by its last two digits is divisible by 4.
> 5 if its last digit is 0 or 5.
> 6 if it is divisible by 2 and 3.
> 9 if the sum of its digits is divisible by 9.
> 10 if its last digit is 0.

Tests for divisibility by other numbers are more complicated. To test a specific number for divisibility by a number other than 2, 3, 4, 5, 6, 9, or 10, perform a long division. If the division leaves no remainder, the specific number is divisible by the other number.

EXAMPLE 6 Is 534,840 divisible by **a.** 2, **b.** 3, **c.** 4 **d.** 5, **e.** 6, **f.** 9, and **g.** 10?

Solution
a. It is divisible by 2, because its last digit 0 is divisible by 2.

b. It is divisible by 3, because the sum of its digits is divisible by 3.

$$5 + 3 + 4 + 8 + 4 + 0 = 24 \quad \text{and} \quad \frac{24}{3} = 8$$

c. It is divisible by 4, because the number formed by its last two digits is divisible by 4.

$$\frac{40}{4} = 10$$

d. It is divisible by 5, because its last digit is 0 or 5.

e. It is divisible by 6, because it is divisible by 2 and 3.

Self Check 6
Is 73,311,435 divisible by

a. 2, **b.** 3, **c.** 5, **d.** 6, **e.** 9, and **f.** 10?

f. It is not divisible by 9, because the sum of its digits is not divisible by 9.

$$\frac{24}{9} = 2 \text{ with 6 left over}$$

g. It is divisible by 10, because its last digit is 0.

◼ Dividing whole numbers that end with zeros

There is a shortcut for dividing a dividend by a divisor when both end with zeros. We simply drop the ending zeros in the divisor and drop the same number of ending zeros in the dividend.

Self Check 7
Divide: 12,000 ÷ 1,500.

EXAMPLE 7 Perform each division: **a.** 80 ÷ 10, **b.** 47,000 ÷ 100, and **c.** 350)‾9,800.

Solution

There is one zero in the divisor.
↓
a. 80 ÷ 10 = 8 ÷ 1 = 8
  ↑ ↑
Drop one zero from the dividend and the divisor.

There are two zeros in the divisor.
↓
b. 47,000 ÷ 100 = 470 ÷ 1 = 470
   ↑   ↑
Drop two zeros from the dividend and the divisor.

c. To perform the division

$$350\overline{)9,800}$$

we drop one zero from the divisor and the dividend and perform the division $35\overline{)980}$.

$$\begin{array}{r} 28 \\ 35\overline{)980} \\ \underline{70}\!\downarrow \\ 280 \\ \underline{280} \\ 0 \end{array}$$

Answer 8

◼ Dividing whole numbers using a calculator

CALCULATOR SNAPSHOT   **Retailing**

A salesperson sold a number of calculators for $17 each, and her total sales were $1,819. To find the number of calculators she sold, we must divide the total sales by the cost of each calculator. We can use a calculator to evaluate 1,819 ÷ 17 by entering these numbers and pressing these keys.

1819 ÷ 17 = | 107 |

The salesperson sold 107 calculators.

Section 1.5 STUDY SET

VOCABULARY   *Fill in the blanks.*

1. The answer to a division problem is called the _____. In a division, the _____ is divided by the divisor.

2. If a division is not exact, the leftover part is called a _____.

3. One number is _____ by another number if, when we divide them, the remainder is 0.

4. To perform divisions when the divisor has more than one digit, we use the process of _____ division.

CONCEPTS   *Fill in the blanks.*

5. Zero divided by any nonzero number is ▨.

6. Any nonzero number divided by 0 is _____.

7. A nonzero number cannot be divided by ▨.

8. A number is divisible by ▨ if its last digit is divisible by 2.

9. A number is divisible by 3 if the _____ of its digits is divisible by 3.

10. A number is divisible by 4 if the number formed by its last _____ digits is divisible by 4.

11. A number is divisible by 5 if its last digit is ▨ or ▨.

12. A number is divisible by 6 if it is divisible by ▨ and ▨.

13. A number is divisible by 9 if the _____ of its digits is divisible by 9.

14. A number is divisible by ▨ if its last digit is 0.

15. $\dfrac{25}{25} = $ ▨

16. $\dfrac{25}{1} = $ ▨

17. $\dfrac{8}{0}$ is _____.

18. $\dfrac{0}{0}$ is _____.

19. In the division $3{,}800 \div 100$, _____ zeros should be dropped from the dividend and the divisor.

20. The division $63{,}000 \div 300$ is equivalent to the division ▨ $\div$ 3.

NOTATION

21. Write three symbols that can be used for division.

22. In a division, 35 R 4 means "a quotient of 35 and a _____ of 4."

PRACTICE   *Perform each division.*

23. $40 \div 5$

24. $40 \div 8$

25. $54 \div 9$

26. $72 \div 8$

27. $63 \div 7$

28. $48 \div 8$

29. $\dfrac{56}{8}$

30. $\dfrac{81}{9}$

Write a related multiplication statement for each division.

31. $24 \div 3 = 8$

32. $32 \div 4 = 8$

33. $7\overline{)56}$ (quotient 8)

34. $9\overline{)72}$ (quotient 8)

Perform each division.

35. $6\overline{)432}$

36. $8\overline{)368}$

37. $7\overline{)245}$

38. $9\overline{)513}$

39. $5\overline{)1{,}185}$

40. $7\overline{)2{,}415}$

41. $\dfrac{4{,}936}{8}$

42. $\dfrac{25{,}950}{6}$

43. $75 \div 15$

44. $315 \div 35$

45. $42 \div 14$

46. $65 \div 13$

47. $132 \div 11$

48. $132 \div 12$

49. $\dfrac{221}{17}$

50. $\dfrac{221}{13}$

51. $13\overline{)949}$

52. $73\overline{)949}$

53. $33\overline{)1{,}353}$

54. $41\overline{)1{,}353}$

55. $39\overline{)7{,}995}$

56. $71\overline{)7{,}313}$

57. $\dfrac{6{,}090}{29}$

58. $\dfrac{7{,}410}{13}$

59. $31\overline{)273}$

60. $25\overline{)290}$

61. $37\overline{)743}$

62. $79\overline{)931}$

63. $42\overline{)1{,}273}$

64. $83\overline{)3{,}280}$

65. $57\overline{)1{,}795}$

66. $99\overline{)9{,}876}$

Determine whether 149,850 is divisible by each number given.

67. 2

68. 3

69. 4

70. 5

71. 6

72. 9

73. 10

74. 8

Perform each division.

75. $35{,}000 \div 100$

76. $4{,}700 \div 100$

77. $89{,}000 \div 1{,}000$

78. $930{,}000 \div 1{,}000$

79. $2{,}480 \div 20$

80. $3{,}120 \div 30$

81. $\dfrac{64{,}000}{400}$

82. $\dfrac{125{,}000}{5{,}000}$

APPLICATIONS

83. DISTRIBUTING MILK Juan's first-grade class received 73 half-pint cartons of milk to distribute evenly to his 23 students. How many cartons were left over?

84. WAITING TABLES A waitress earned $96 in tips. If each customer tipped $3, how many customers did she serve?

85. DRAINING POOLS A 950,000-gallon pool empties in 16 hours. How many gallons are drained each hour?

86. RUNNING Brian runs 7 miles each day. In how many days will Brian run 371 miles?

87. LIFT SYSTEMS If the bus weighs 58,000 pounds, how much weight is on each jack?

88. MILEAGE A touring rock group travels in a bus that has a range of 700 miles on one tank (140 gallons) of gasoline. How far can the bus travel on one gallon of gas?

89. DOUGHNUTS How many dozen doughnuts must be ordered for a meeting if 156 people are expected to attend, and each person will be served one doughnut?

90. HAULING DIRT A 15-cubic-yard dump truck must haul 405 cubic yards of dirt to a construction site. How many trips must the truck make?

91. VOLLEYBALL A total of 216 girls tried out for a city volleyball league. How many girls should be put on each team roster if the following requirements must be met?

All the teams are to have the same number of players.

A reasonable number of players on a team is 7 to 10.

For scheduling purposes, there must be an even number of teams (2, 4, 6, 8, and so on).

92. WINDSCREENS A farmer intends to plant pine trees 12 feet apart so as to form a windscreen for her crops. How many trees should she buy if the length of the field is 744 feet?

93. PRICE OF A TEXTBOOK A store received $17,520 on the sale of 240 algebra textbooks. What was the cost of each book?

94. WATER DISCHARGE The Susquehanna River discharges 1,719,000 cubic feet of water into Chesapeake Bay in 45 seconds. How many cubic feet of water is discharged each second?

95. U.S. STATES To find the **population density** of a state, we can divide its population by its land area (in square miles). The result is the number of people per square mile. Use the data in the table to approximate the population density for each state.

State	2003 Population*	Land area* (square miles)
Arizona	5,500,000	110,000
Indiana	6,120,000	36,000
Rhode Island	1,100,000	1,000
South Carolina	4,200,000	30,000

*approximation
Source: Bureau of the Census

96. SCIENCE Light travels at approximately 11,160,000 miles in one minute. How far does it travel in one second?

WRITING

97. Explain why the division of two numbers is not commutative.

98. Explain why division of 0 is possible, but division by 0 is impossible.

REVIEW

99. What digit is in the thousands column of the number 372,856?

100. Round 45,995 to the nearest hundred.

101. Subtract 987 from 1,010.

102. DISCOUNTING CARS A car, originally priced at $17,550, is being sold for $13,970. By how many dollars has the price been decreased?

Estimation

We have used **estimation** as a means of checking the reasonableness of an answer. We now take a more in-depth look at the process of estimating.

Estimation is used to find an *approximate* answer to a problem. Estimates can be helpful in two ways. First, they serve as an accuracy check that can detect major computational errors. If an answer does not seem reasonable when compared to the estimate, the original problem should be reworked. Second, some situations call for only an approximate answer rather than the exact answer.

There are several ways to estimate, but there is one theme to all the methods: The numbers in the problem are simplified so that the computation can be made easily and quickly. The first method we discuss is called **front-end rounding.** Each number is rounded to its largest place value, so that all but the first digit of each number is zero.

EXAMPLE 1   Estimating sums, differences, and products.

a. Estimate the sum: $3,714 + 2,489 + 781 + 5,500 + 303$.

   Solution   Use front-end rounding.

$$
\begin{array}{rcl}
3,714 & \longrightarrow & 4,000 \\
2,489 & \longrightarrow & 2,000 \\
781 & \longrightarrow & 800 \\
5,500 & \longrightarrow & 6,000 \\
+\ 303 & \longrightarrow & +\ 300 \\
\hline
 & & 13,100
\end{array}
$$

Each number is rounded to its largest place value. All but the first digit is zero.

The estimate is 13,100.

If we compute $3,714 + 2,489 + 781 + 5,500 + 303$, the sum is 12,787. We can see that our estimate is close; it's just 313 more than 12,787. This example illustrates the tradeoff when using estimation: The calculations are easier to perform and they take less time, but the answers are not exact.

b. Estimate the difference: $46,721 - 13,208$.

   Solution   Use front-end rounding.

$$
\begin{array}{rcl}
46,721 & \longrightarrow & 50,000 \\
-13,208 & \longrightarrow & -10,000 \\
\hline
 & & 40,000
\end{array}
$$

Only the first digit is nonzero.

The estimate is 40,000.

c. Estimate the product: $334 \cdot 59$.

   Solution   Use front-end rounding.

$$
\begin{array}{rcl}
334 & \longrightarrow & 300 \\
\times\ 59 & \longrightarrow & \times\ 60 \\
\hline
 & & 18,000
\end{array}
$$

334 rounds to 300, and 59 rounds to 60.

The estimate is 18,000.

Self Check 1

a. Estimate the sum:

$$
\begin{array}{r}
6,780 \\
3,278 \\
566 \\
4,230 \\
+1,923 \\
\hline
\end{array}
$$

b. Estimate the difference:

$$
\begin{array}{r}
89,070 \\
-15,331 \\
\hline
\end{array}
$$

c. Estimate the product:

$$
\begin{array}{r}
707 \\
\times 251 \\
\hline
\end{array}
$$

Answers **a.** 16,600, **b.** 70,000, **c.** 210,000

To estimate quotients, we use a method that approximates both the dividend and the divisor so that they divide easily. With this method, some insight and intuition are needed. There is one rule of thumb for this method: If possible, round both numbers up or both numbers down.

Self Check 2
Estimate: 33,642 ÷ 42.

EXAMPLE 2   **Estimating quotients.**   Estimate the quotient: 170,715 ÷ 57.

Solution   Both numbers are rounded up. The division can then be done in your head.

$$\underbrace{170{,}715}_{} \div \underbrace{57}_{} \qquad 180{,}000 \div 60 = 3{,}000$$

The dividend is approximately

The divisor is approximately

Answer 800

The estimate is 3,000.

STUDY SET   *Use front-end rounding to find an estimate to check the reasonableness of each answer. Write yes if it appears reasonable and no if it does not.*

1.
```
  25,405
  11,222
   8,909
   1,076
  14,595
 +33,999
 ───────
  73,206
```

2.
```
  568,334
 − 31,225
 ────────
  497,109
```

3.
```
    451
 ×   73
 ──────
  39,923
```

4.
```
    616
 ×   98
 ──────
  60,368
```

Use estimation to check the reasonableness of each answer.

5. 57,238 ÷ 28 = 200

6. $322\overline{)13{,}202}$ (quotient 41)

Use an estimation procedure to answer each problem.

7. CAMPAIGNING The number of miles flown each day by a politician on a campaign swing are shown below. Estimate the number of miles she flew during this time.

Day 1	3,546 miles
Day 2	567
Day 3	1,203
Day 4	342
Day 5	2,699

8. SHOPPING MALLS The total sales income for a downtown mall in its first three years in operation are shown here.

2001	$5,234,301
2002	$2,898,655
2003	$6,343,433

Estimate the difference in income for 2002 and 2003 as compared to the first year, 2001.

9. GOLF COURSES Estimate the number of bags of grass seed needed to plant a fairway whose area is 86,625 square feet if the seed in each bag covers 2,850 square feet.

10. CENSUS Estimate the total population of the ten largest counties in the United States as of 2003.

LARGEST COUNTIES, BY POPULATION	
1. Los Angeles, CA	9,871,506
2. Cook, IL	5,351,552
3. Harris, TX	3,596,086
4. Maricopa, AZ	3,389,260
5. Orange, CA	2,957,766
6. San Diego, CA	2,930,886
7. Kings, NY	2,472,523
8. Miami-Dade, FL	2,341,167
9. Dallas, TX	2,284,096
10. Queens, NY	2,225,486

Source: Bureau of the Census

11. CURRENCY Estimate the number of $5 bills in circulation as of June 30, 2004, if the total value of the currency was $9,373,288,075.

12. CORPORATIONS In 2003, General Motors Corporation had sales of $185,524,000,000. Approximately how many times larger was this than the 2003 sales of IBM, which were $89,000,000,000?

1.6 Prime Factors and Exponents

- Factoring whole numbers • Even and odd whole numbers • Prime numbers
- Composite numbers • Finding prime factorizations with the tree method
- Exponents • Finding prime factorizations with the division method

In this section, we learn how to represent whole numbers in alternative forms. The procedures used to find these forms involve multiplication and division. We then discuss exponents, a way to represent repeated multiplication.

■ Factoring whole numbers

The statement $3 \cdot 2 = 6$ has two parts: the numbers that are being multiplied, and the answer. The numbers that are being multiplied are *factors,* and the answer is the *product.* We say that 3 and 2 are factors of 6.

Factors

Numbers that are multiplied together are called **factors.**

EXAMPLE 1   Find the factors of 12.

Solution   We need to find the possible ways that we can multiply two whole numbers to get a product of 12.

$$1 \cdot 12 = 12, \qquad 2 \cdot 6 = 12, \qquad \text{and} \qquad 3 \cdot 4 = 12$$

In order, from least to greatest, the factors of 12 are 1, 2, 3, 4, 6, and 12.

Self Check 1
Find the factors of 20.

Answer 1, 2, 4, 5, 10, and 20

Example 1 shows that 1, 2, 3, 4, 6, and 12 are the factors of 12. This observation was established by using multiplication facts. Each of these factors is related to 12 by division as well. Each of them divides 12, leaving a remainder of 0. Because of this fact, we say that 12 is divisible by each of its factors. When a division ends with a remainder of 0, we say that the division comes out even or that one of the numbers divides the other *exactly.*

Divisibility

One number is divisible by another if, when dividing them, we get a remainder of 0.

When we say that 3 is a factor of 6, we are using the word *factor* as a noun. The word *factor* is also used as a verb.

Factoring a whole number

To **factor** a whole number means to express it as the product of other whole numbers.

Self Check 2
Factor 18 using **a.** two factors
and **b.** three factors.

EXAMPLE 2   Factor 40 using **a.** two factors   and **b.** three factors.

Solution
a. There are several possibilities.

$$40 = 1 \cdot 40, \qquad 40 = 2 \cdot 20, \qquad 40 = 4 \cdot 10, \qquad \text{and} \qquad 40 = 5 \cdot 8$$

b. Again, there are several possibilities. Two of them are

$$40 = 5 \cdot 4 \cdot 2 \qquad \text{and} \qquad 40 = 2 \cdot 2 \cdot 10$$

Answers **a.** $1 \cdot 18, 2 \cdot 9, 3 \cdot 6,$
b. $2 \cdot 3 \cdot 3$

Even and odd whole numbers

> **Even and odd whole numbers**
> If a whole number is divisible by 2, it is called an **even** number.
> If a whole number is not divisible by 2, it is called an **odd** number.

The even whole numbers are the numbers

$$0, 2, 4, 6, 8, 10, 12, 14, 16, 18, \ldots$$

The odd whole numbers are the numbers

$$1, 3, 5, 7, 9, 11, 13, 15, 17, 19, \ldots$$

There are infinitely many even and infinitely many odd whole numbers.

Prime numbers

Self Check 3
Find the factors of 23.

EXAMPLE 3   Find the factors of 17.

Solution

$$1 \cdot 17 = 17$$

The only factors of 17 are 1 and 17.

Answer 1 and 23

In Example 3 and its Self Check, we saw that the only factors of 17 are 1 and 17, and the only factors of 23 are 1 and 23. Numbers that have only two factors, 1 and the number itself, are called **prime numbers.**

> **Prime numbers**
> A **prime number** is a whole number greater than 1 that has only 1 and itself as factors.

The prime numbers are the numbers

$$2, 3, 5, 7, 11, 13, 17, 19, 23, 29, 31, \ldots$$

The dots at the end of the list indicate that there are infinitely many prime numbers.

Note that the only even prime number is 2. Any other even whole number is divisible by 2, and thus has 2 as a factor, in addition to 1 and itself. Also note that not all odd whole numbers are prime numbers. For example, since 15 has factors of 1, 3, 5, and 15, it is not a prime number.

■ Composite numbers

The set of whole numbers contains many prime numbers. It also contains many numbers that are not prime.

> **Composite numbers**
>
> The **composite numbers** are whole numbers greater than 1 that are not prime.

The composite numbers are the numbers

$$4, 6, 8, 9, 10, 12, 14, 15, 16, 18, \ldots$$

The three dots at the end of the list indicate that there are infinitely many composite numbers.

EXAMPLE 4

a. Is 37 a prime number?   **b.** Is 45 a prime number?

Solution

a. Since 37 is a whole number greater than 1 and its only factors are 1 and 37, it is prime.

b. The factors of 45 are 1, 3, 5, 9, 15, and 45. Since there are factors other than 1 and 45, 45 is not prime. It is a composite number.

Self Check 4

a. Is 39 a prime number?

b. Is 57 a prime number?

Answers **a.** no, **b.** no

! COMMENT The numbers 0 and 1 are neither prime nor composite, because neither is a whole number greater than 1.

■ Finding prime factorizations with the tree method

Every composite number can be formed by multiplying a specific combination of prime numbers. The process of finding that combination is called **prime factorization.**

> **Prime factorization**
>
> To find the **prime factorization** of a whole number means to write it as the product of only prime numbers.

One method for finding the prime factorization of a number is called the **tree method.** We use the tree method in the diagrams below to find the prime factorization of 90 in two ways.

1. Factor 90 as $9 \cdot 10$.

2. Factor 9 and 10.

3. The process is complete when only prime numbers appear.

1. Factor 90 as $6 \cdot 15$.

2. Factor 6 and 15.

3. The process is complete when only prime numbers appear.

In either case, the prime factors are $2 \cdot 3 \cdot 3 \cdot 5$. Thus, the prime-factored form of 90 is $2 \cdot 3 \cdot 3 \cdot 5$. As we have seen, it does not matter how we factor 90. We will always get the same set of prime factors. No other combination of prime factors will multiply together and produce 90. This example illustrates an important fact about composite numbers.

> **Fundamental theorem of arithmetic**
> Any composite number has exactly one set of prime factors.

Self Check 5
Use a factor tree to find the prime factorization of 120.

EXAMPLE 5   Use a factor tree to find the prime factorization of 210.

Solution

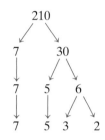

Factor 210 as $7 \cdot 30$.

Bring down the 7. Factor 30 as $5 \cdot 6$.

Bring down the 7 and the 5. Factor 6 as $3 \cdot 2$.

The prime factorization of 210 is $7 \cdot 5 \cdot 3 \cdot 2$. Writing the prime factors in order, from least to greatest, we have $210 = 2 \cdot 3 \cdot 5 \cdot 7$.

Answer $2 \cdot 2 \cdot 2 \cdot 3 \cdot 5$

Exponents

In the Self Check of Example 5, we saw that the prime factorization of 120 is $2 \cdot 2 \cdot 2 \cdot 3 \cdot 5$. Because this factorization has three factors of 2, we call 2 a *repeated factor*. To express a repeated factor, we can use an **exponent.**

> **Exponent and base**
> An **exponent** is used to indicate repeated multiplication. It tells how many times the **base** is used as a factor.

$$\underbrace{2 \cdot 2 \cdot 2}_{\text{Repeated factors}} = \overset{\text{The exponent is 3.}}{\underset{\text{The base is 2.}}{2^3}}$$ Read 2^3 as "2 to the third power" or "2 cubed."

The prime factorization of 120 can be written in a more compact form using exponents: $2 \cdot 2 \cdot 2 \cdot 3 \cdot 5 = 2^3 \cdot 3 \cdot 5$.

In the **exponential expression** 2^3, the number 2 is the base and 3 is the exponent. The expression is called a **power of 2.**

Self Check 6
Use exponents to write each prime factorization:

a. $3 \cdot 3 \cdot 7$ **b.** $5(5)(7)(7)$

c. $2 \cdot 2 \cdot 2 \cdot 3 \cdot 3 \cdot 5$

EXAMPLE 6   Use exponents to write each prime factorization: **a.** $5 \cdot 5 \cdot 5$, **b.** $7 \cdot 7 \cdot 11$,   and   **c.** $2(2)(2)(2)(3)(3)(3)$.

Solution

a. $5 \cdot 5 \cdot 5 = 5^3$ 5 is used as a factor 3 times.

b. $7 \cdot 7 \cdot 11 = 7^2 \cdot 11$ 7 is used as a factor 2 times.

c. $2(2)(2)(2)(3)(3)(3) = 2^4(3^3)$ 2 is used as a factor 4 times, and 3 is used as a factor 3 times.

Answers **a.** $3^2 \cdot 7$,   **b.** $5^2(7^2)$,
c. $2^3 \cdot 3^2 \cdot 5$

EXAMPLE 7   Find the value of each expression.

Solution

a. $7^2 = 7 \cdot 7 = 49$ Read 7^2 as "7 to the second power" or "7 squared."
                         Write the base 7 as a factor two times.

b. $2^5 = 2 \cdot 2 \cdot 2 \cdot 2 \cdot 2 = 32$ Read 2^5 as "2 to the fifth power."
                         Write the base 2 as a factor five times.

c. $10^4 = 10 \cdot 10 \cdot 10 \cdot 10 = 10,000$ Read 10^4 as "10 to the fourth power."
                         Write the base 10 as a factor four times.

d. $6^1 = 6$ Read 6^1 as "6 to the first power."
                         Write the base 6 once.

Self Check 7
Which of the numbers 3^5, 4^4, and 5^3 is the largest?

Answer $4^4 = 256$

❗ COMMENT Note that 2^5 means $2 \cdot 2 \cdot 2 \cdot 2 \cdot 2$. It does not mean $2 \cdot 5$. That is, $2^5 = 32$ and $2 \cdot 5 = 10$.

EXAMPLE 8   The prime factorization of a number is $2^3 \cdot 3^4 \cdot 5$. What is the number?

Solution   To find the number, we find the value of each power and then do the multiplication.

$$2^3 \cdot 3^4 \cdot 5 = 8 \cdot 81 \cdot 5 \qquad 2^3 = 8 \text{ and } 3^4 = 81.$$
$$= 648 \cdot 5 \qquad \text{Perform the multiplications, working from left to right.}$$
$$= 3,240$$

The number is 3,240.

Self Check 8
The prime factorization of a number is $3 \cdot 5^2 \cdot 7$. What is the number?

Answer 525

Bacterial growth

At the end of one hour, a culture contains two bacteria. Suppose the number of bacteria doubles every hour thereafter. Use exponents to determine how many bacteria the culture will contain after 24 hours.

    We can use Table 1-4 to help model the situation. From the table, we see a pattern developing: The number of bacteria in the culture after 24 hours will be 2^{24}. We can evaluate this exponential expression using the exponential key $\boxed{y^x}$ on a scientific calculator ($\boxed{x^y}$ on some models).

    To find the value of 2^{24}, we enter these numbers and press these keys.

$$2 \;\boxed{y^x}\; 24 \;\boxed{=}$$

Time	Number of bacteria
1 hr	$2 = 2^1$
2 hr	$4 = 2^2$
3 hr	$8 = 2^3$
4 hr	$16 = 2^4$
24 hr	$? = 2^{24}$

TABLE 1-4

$$\boxed{16777216}$$

Since $2^{24} = 16,777,216$, there will be 16,777,216 bacteria after 24 hours.

▮ Finding prime factorizations with the division method

We can also find the prime factorization of a whole number by division. For example, to find the prime factorization of 363, we begin the division method by choosing the *smallest* prime number that will divide the given number exactly. We continue this "inverted division" process until the result of the division is a prime number.

Step 1: The prime number 2 doesn't divide 363 exactly, but 3 does. The result is 121, which is not prime. We continue the division process.

$$3\overline{|363}$$
$$121$$

Step 2: Next, we choose the smallest prime number that will divide 121. The primes 2, 3, 5, and 7 don't divide 121 exactly, but 11 does. The result is 11, which is prime. We are done.

$$3\overline{|363}$$
$$11\overline{|121}$$
$$\mathbf{11}$$
$$363 = 3 \cdot \mathbf{11} \cdot \mathbf{11}$$

Using exponents, we can write the prime factorization of 363 as $3 \cdot 11^2$.

Self Check 9
Use the division method to find the prime factorization of 108. Use exponents to express the result.

Answer $2^2 \cdot 3^3$

EXAMPLE 9   Use the division method to find the prime factorization of 100. Use exponents to express the result.

Solution

2 divides 100 exactly. The result is 50, which is not prime. ⟶ $2\overline{|100}$
2 divides 50 exactly. The result is 25, which is not prime. ⟶ $2\overline{|50}$
5 divides 25 exactly. The result is 5, which is prime. We are done. ⟶ $5\overline{|25}$
5

The prime factorization of 100 is $2 \cdot 2 \cdot 5 \cdot 5$ or $2^2 \cdot 5^2$.

❗ COMMENT In Example 9, it would be incorrect to begin the division process with

$$10\overline{|100}$$

because 10 is not a prime number.

Section 1.6 STUDY SET

▮ **VOCABULARY**   *Fill in the blanks.*

1. Numbers that are multiplied together are called _____.

2. One number is _____ by another if the remainder is 0 when they are divided. When a division ends with a remainder of 0, we say that one of the numbers divides the other _____.

3. To _____ a whole number means to express it as the product of other whole numbers.

4. A _____ number is a whole number greater than 1 that has only 1 and itself as factors.

5. Whole numbers greater than 1 that are not prime numbers are called _____ numbers.

6. An _____ whole number is exactly divisible by 2. An _____ whole number is not exactly divisible by 2.

7. To prime factor a number means to write it as a product of only _____ numbers.

8. An _____ is used to represent repeated multiplication.

9. In the exponential expression 6^4, 6 is called the _____, and 4 is called the _____.

10. Another way to say "5 to the second power" is 5 _____. Another way to say "7 to the third power" is 7 _____.

CONCEPTS

11. Write 27 as the product of two factors.

12. Write 30 as the product of three factors.

13. The complete list of the factors of a whole number is given. What is the number?

   a. 2, 4, 22, 44, 11, 1

   b. 20, 1, 25, 100, 2, 4, 5, 50, 10

14. a. Find the factors of 24.

   b. Find the prime factorization of 24.

15. Find the factors of each number.

   a. 11

   b. 23

   c. 37

   d. From the results obtained in parts a–c, what can be said about 11, 23, and 37?

16. Suppose a number is divisible by 10. Is 10 a factor of the number?

17. If 4 is a factor of a whole number, will 4 divide the number exactly?

18. Give examples of whole numbers that have 11 as a factor.

The prime factorization of a whole number is given. Find the number.

19. $2 \cdot 3 \cdot 3 \cdot 5$ 20. $3^3 \cdot 2$

21. $11^2 \cdot 5$ 22. $2 \cdot 2 \cdot 2 \cdot 7$

23. Can we change the order of the base and the exponent in an exponential expression and obtain the same result? In other words, does $3^2 = 2^3$?

24. Find the prime factors of 20 and 35. What prime factor do they have in common?

25. Find the prime factors of 20 and 50. What prime factors do they have in common?

26. Find the prime factors of 30 and 165. What prime factors do they have in common?

27. Find the prime factors of 30 and 242. What prime factor do they have in common?

28. Find 1^2, 1^3, and 1^4. From the results, what can be said about any power of 1?

29. Finish the process of prime factoring 150. Compare the results.

30. Find three whole numbers less than 10 that would fit at the top of this tree diagram.

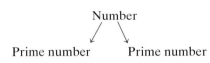

31. Complete the table.

Product of the factors of 12	Sum of the factors of 12
$1 \cdot 12$	
$2 \cdot 6$	
$3 \cdot 4$	

32. Consider 1, 4, 9, 16, 25, 36, 49, 64, 81, 100. Of the numbers listed, which is the *largest* factor of
   a. 18   b. 24   c. 50

33. When using the division method to find the prime factorization of an even number, what is an obvious choice with which to start the division process?

34. When using the division method to find the prime factorization of a number ending in 5, what is an obvious choice with which to start the division process?

NOTATION *Write the repeated multiplication represented by each expression.*

35. 7^3 36. 8^4

37. 3^5 38. 4^6

39. $5^2(11)$ 40. $2^3 \cdot 3^2$

Simplify each expression.

41. 10^1 42. 2^1

Use exponents to write each expression in simpler form.

43. $2 \cdot 2 \cdot 2 \cdot 2 \cdot 2$ 44. $3 \cdot 3 \cdot 3 \cdot 3 \cdot 3 \cdot 3$

45. $5(5)(5)(5)$ 46. $9(9)(9)$

47. $4(4)(5)(5)$ 48. $12 \cdot 12 \cdot 12 \cdot 16$

PRACTICE *Find the factors of each whole number.*

49. 10   50. 6

51. 40   52. 75

53. 18   **54.** 32

55. 44   **56.** 65

57. 77   **58.** 81

59. 100   **60.** 441

Find the prime factorization of each number.

61. 39   **62.** 20

63. 99   **64.** 105

65. 162   **66.** 400

67. 220   **68.** 126

69. 64   **70.** 243

71. 147   **72.** 98

Evaluate each exponential expression.

73. 3^4 **74.** 5^3

75. 2^5 **76.** 10^5

77. 12^2 **78.** 7^3

79. 8^4 **80.** 9^5

81. $3^2(2^3)$ **82.** $3^3(4^2)$

83. $2^3 \cdot 3^3 \cdot 4^2$ **84.** $3^2 \cdot 4^3 \cdot 5^2$

85. 🖩 234^3 **86.** 🖩 51^4

87. 🖩 $23^2 \cdot 13^3$ **88.** 🖩 $12^3 \cdot 15^2$

APPLICATIONS

89. PERFECT NUMBERS A whole number is called a **perfect number** when the sum of its factors that are less than the number equals the number. For example, 6 is a perfect number, because $1 + 2 + 3 = 6$. Find the factors of 28. Then use addition to show that 28 is also a perfect number.

90. CRYPTOGRAPHY Information is often transmitted in code. Many codes involve writing products of large primes, because they are difficult to factor. To see how difficult, try finding two prime factors of 7,663. (*Hint:* Both primes are greater than 70.)

91. LIGHT The illustration shows that the light energy that passes through the first unit of area, 1 yard away from the bulb, spreads out as it travels away from the source. How much area does that energy cover 2 yards, 3 yards, and 4 yards from the bulb? Express each answer using exponents.

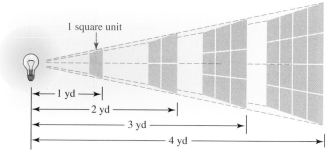

92. CELL DIVISION After one hour, a cell has divided to form another cell. In another hour, these two cells have divided so that four cells exist. In another hour, these four cells divide so that eight exist.

    a. How many cells exist at the end of the fourth hour?

    b. The number of cells that exist after each division can be found using an exponential expression. What is the base?

    c. Use a calculator to find the number of cells after 12 hours.

WRITING

93. Explain how to test a number to see whether it is prime.

94. Explain how to test a number to see whether it is even.

95. Explain the difference between the *factors* of a number and the *prime factorization* of the number.

96. Explain why it would be incorrect to say that the area of the square shown in the illustration is 25^2 ft. How should we express its area?

5 ft

5 ft

REVIEW

97. Round 230,999 to the nearest thousand.

98. Write the set of whole numbers.

99. What is $0 \div 15$?

100. Multiply: $15 \cdot (6 \cdot 9)$.

101. What is the formula for the area of a rectangle?

102. MARCHING BANDS When a university band lines up in eight rows of 15 musicians, there are five musicians left over. How many band members are there?

1.7 Order of Operations

- Order of operations • Evaluating expressions with no grouping symbols
- Evaluating expressions with grouping symbols • The arithmetic mean (average)

Punctuation marks, such as commas, quotations, and periods, serve purpose when writing compositions. They determine the way in which sentences are to be read and interpreted. To read and interpret mathematical expressions correctly, we must use an agreed-upon set of priority rules for the *order of operations.*

Order of operations

Suppose you are asked to contact a friend if you see a certain type of watch for sale while you are traveling in Europe. While in Switzerland, you spot the watch and send the following E-mail message.

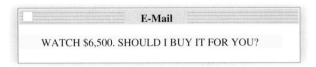

The next day, you get this response from your friend.

Something is wrong. One statement says to buy the watch at any price. The other says not to buy it, because it's too expensive. The placement of the exclamation point makes us read these statements differently, resulting in different meanings.

When we read a mathematical statement, the same kind of confusion is possible. For example, we consider

$$3 + 2 \cdot 5$$

The above expression contains two operations: addition and multiplication. We can *evaluate* it (find its value) in two ways. We can do the addition first and then do the multiplication. Or we can do the multiplication first and then do the addition. However, we get different results.

$3 + 2 \cdot 5 = 5 \cdot 5$	Add first: $3 + 2 = 5$.		$3 + 2 \cdot 5 = 3 + 10$	Multiply first: $2 \cdot 5 = 10$
$= 25$	Multiply 5 and 5.		$= 13$	Add 3 and 10.

Different results

If we don't establish an order of operations, the expression $3 + 2 \cdot 5$ will have two different answers. To avoid this possibility, we evaluate expressions in the following order.

> **Order of operations**
>
> Perform all calculations within parentheses and other grouping symbols in the following order, working from the innermost pair to the outermost pair.
>
> **1.** Evaluate all powers. That is, evaluate any expression with exponents.
>
> **2.** Perform all multiplications and divisions as they occur from left to right.
>
> **3.** Perform all additions and subtractions as they occur from left to right.
>
> When all grouping symbols have been removed, repeat steps 1–3 to complete the calculation.
>
> If a fraction bar is present, evaluate the expression above the bar (called the **numerator**) and the expression below the bar (called the **denominator**) separately. Then perform the division indicated by the fraction bar, if possible.

To evaluate $3 + 2 \cdot 5$ correctly, we must apply the rules for the order of operations. Since there are no parentheses and there are no exponents, we perform the multiplication first and then do the addition.

Ignore the addition for now ↓ ↓ and perform the multiplication first.

$$3 + 2 \cdot 5 = 3 + 10$$
$$= 13 \qquad \text{Now perform the addition.}$$

Using the rules for the order of operations, we see that the correct answer is 13.

▌Evaluating expressions with no grouping symbols

Self Check 1
Evaluate: $4 \cdot 3^3 - 6$.

EXAMPLE 1   Evaluate: $2 \cdot 4^2 - 8$.

Solution   Since the expression does not contain any grouping symbols, we begin with step 2 of the rules for the order of operations.

$$2 \cdot 4^2 - 8 = 2 \cdot 16 - 8 \qquad \text{Evaluate the power: } 4^2 = 16.$$
$$= 32 - 8 \qquad \text{Perform the multiplication: } 2 \cdot 16 = 32.$$
$$= 24 \qquad \text{Perform the subtraction.}$$

Answer 102

Self Check 2
Evaluate: $10 - 2 \cdot 3 + 24$.

EXAMPLE 2   Evaluate: $8 - 3 \cdot 2 + 16$.

Solution   Since the expression does not contain grouping symbols and since there are no powers to find, we look for multiplications or divisions to perform.

$$8 - 3 \cdot 2 + 16 = 8 - 6 + 16 \qquad \text{Perform the multiplication: } 3 \cdot 2 = 6.$$
$$= 2 + 16 \qquad \text{Working from left to right, do the subtraction: } 8 - 6 = 2.$$
$$= 18 \qquad \text{Perform the addition.}$$

Answer 28

❗ COMMENT Some students incorrectly think that additions are always done before subtractions. As Example 2 shows, this is not true. Working from left to right, we do the additions or subtractions *in the order in which they occur*. The same is true for multiplications and divisions.

EXAMPLE 3 Evaluate: $192 \div 6 - 5(3)2$.

Solution Although this expression contains parentheses, there are no calculations to perform within them. Since there are no powers, we perform multiplications and divisions as they are encountered from left to right.

$\mathbf{192 \div 6} - 5(3)2$	$= \mathbf{32} - 5(3)2$	Working from left to right, do the division: $192 \div 6 = 32$.
	$= 32 - 15(2)$	Working from left to right, do the multiplication: $5(3) = 15$.
	$= 32 - 30$	Perform the multiplication: $15(2) = 30$.
	$= 2$	Perform the subtraction.

Self Check 3
Evaluate: $36 \div 9 + 4(2)3$.

Answer 28

EXAMPLE 4 **Phone bills.** Figure 1-12 shows the rates for international telephone calls charged by a 10-10 long-distance company. A businesswoman calls Germany for 20 minutes, South Korea for 5 minutes, and Mexico City for 35 minutes. What is the total cost of the calls?

All rates are per minute.	
Canada	*10¢*
Germany	*23¢*
Jamaica	*68¢*
Mexico City	*42¢*
South Korea	*29¢*

FIGURE 1-12

Solution We can find the cost of a call (in cents) by multiplying the rate charged per minute by the length of the call (in minutes). To find the total cost, we add the costs of the three calls.

The cost of the call to Germany		The cost of the call to South Korea		The cost of the call to Mexico City
↓		↓		↓
23(20)	+	29(5)	+	42(35)

To evaluate this expression, we apply the rules for the order of operations.

$23(20) + 29(5) + 42(35)$	$= 460 + 145 + 1{,}470$	Perform the multiplications.
	$= 2{,}075$	Perform the additions.

The total cost of the calls is 2,075 cents, or \$20.75.

Evaluating expressions with grouping symbols

Grouping symbols determine the order in which an expression is to be evaluated. Examples of grouping symbols are parentheses ( ), brackets [ ], and the fraction bar ——.

In the next example, we have two similar-looking expressions. However, because of the parentheses, we evaluate them in a different order.

EXAMPLE 5 Evaluate each expression: **a.** $12 - 3 + 5$ and **b.** $12 - (3 + 5)$.

Solution

a. We perform the additions and subtractions as they occur, from left to right.

$\mathbf{12 - 3} + 5$	$= \mathbf{9} + 5$	Perform the subtraction: $12 - 3 = 9$.
	$= 14$	Perform the addition.

Self Check 5
Evaluate each expression:
a. $20 - 7 + 6$
b. $20 - (7 + 6)$

b. This expression contains parentheses. We must do the calculation within the parentheses first.

$$12 - (3 + 5) = 12 - 8 \qquad \text{Perform the addition: } 3 + 5 = 8.$$
$$= 4 \qquad \text{Perform the subtraction.}$$

Answers **a.** 19,   **b.** 7

Self Check 6
Evaluate: $(1 + 3)^4$.

EXAMPLE 6   Evaluate: $(2 + 6)^3$.

Solution   We begin by doing the calculation within the parentheses.

$$(2 + 6)^3 = 8^3 \qquad \text{Perform the addition.}$$
$$= 512 \qquad \text{Evaluate the exponential expression: } 8^3 = 8 \cdot 8 \cdot 8 = 512.$$

Answer 256

Self Check 7
Evaluate: $50 - 4(12 - 5 \cdot 2)$.

EXAMPLE 7   Evaluate: $5 + 2(13 - 5 \cdot 2)$.

Solution   This expression contains grouping symbols. We apply the rules for the order of operations within the parentheses first, to evaluate $13 - 5 \cdot 2$.

$$5 + 2(13 - 5 \cdot 2) = 5 + 2(13 - 10) \qquad \begin{array}{l}\text{Perform the multiplication within the}\\ \text{parentheses.}\end{array}$$
$$= 5 + 2(3) \qquad \begin{array}{l}\text{Perform the subtraction within the}\\ \text{parentheses.}\end{array}$$
$$= 5 + 6 \qquad \text{Perform the multiplication: } 2(3) = 6.$$
$$= 11 \qquad \text{Perform the addition.}$$

Answer 42

Sometimes an expression contains two or more sets of grouping symbols. Since it can be confusing to read an expression such as $16 + 2(14 - 3(5 - 2))$, we often use brackets in place of the second pair of parentheses.

$$16 + 2[14 - 3(5 - 2)]$$

If an expression contains more than one pair of grouping symbols, we always begin by working within the innermost pair and then work to the outermost pair.

Innermost parentheses
$$16 + 2[14 - 3(5 - 2)]$$
Outermost brackets

Self Check 8
Evaluate:
$140 - 7[4 + 3(6 - 2)]$.

EXAMPLE 8   Evaluate: $16 + 6[14 - 3(5 - 2)]$.

Solution

$$16 + 6[14 - 3(5 - 2)] = 16 + 6[14 - 3(3)] \qquad \begin{array}{l}\text{Perform the subtraction within the}\\ \text{parentheses.}\end{array}$$
$$= 16 + 6(14 - 9) \qquad \begin{array}{l}\text{Perform the multiplication within}\\ \text{the brackets. Since only one set of}\\ \text{grouping symbols is now needed,}\\ \text{write } 14 - 9 \text{ within parentheses.}\end{array}$$

$$= 16 + 6(5) \qquad \text{Perform the subtraction within the parentheses.}$$

$$= 16 + 30 \qquad \text{Perform the multiplication: } 6(5) = 30.$$

$$= 46 \qquad \text{Perform the addition.}$$

Answer 28

EXAMPLE 9   Evaluate: $\dfrac{2(13) - 2}{3(2^3)}$.

Self Check 9
Evaluate: $\dfrac{3(14) - 6}{2(3^2)}$.

Solution   A fraction bar is a grouping symbol. We evaluate the numerator and denominator separately and then perform the indicated division.

$$\dfrac{2(13) - 2}{3(2^3)} = \dfrac{26 - 2}{3(8)} \qquad \begin{array}{l}\text{In the numerator, perform the multiplication.} \\ \text{In the denominator, perform the calculation within the} \\ \text{parentheses.}\end{array}$$

$$= \dfrac{24}{24} \qquad \begin{array}{l}\text{In the numerator, perform the subtraction.} \\ \text{In the denominator, perform the multiplication.}\end{array}$$

$$= 1 \qquad \text{Perform the division.}$$

Answer 2

■ The arithmetic mean (average)

The **arithmetic mean,** or **average,** of several numbers is a value around which the numbers are grouped. It gives you an indication of the "center" of the set of numbers. When finding the mean of a set of numbers, we usually need to apply the rules for the order of operations.

> **Finding an arithmetic mean**
>
> To find the mean of a set of scores, divide the sum of the scores by the number of scores.

EXAMPLE 10 Basketball. In 1998, the Lady Vols of the University of Tennessee won the women's basketball championship, capping a perfect 39-0 season. Find their average margin of victory in their last four tournament games shown below.

Regional	Regional final	Semifinal	Championship
Beat Rutgers by 32 points	Beat North Carolina by 6 points	Beat Arkansas by 28 points	Beat Louisiana Tech by 18 points

Solution   To find the average margin of victory, add the margins of victory and divide by 4.

$$\text{Average} = \dfrac{32 + 6 + 28 + 18}{4}$$

$$= \dfrac{84}{4}$$

$$= 21$$

Their average margin of victory was 21 points.

Self Check 10
Syracuse University won the 2003 NCAA men's basketball championship. Find their average margin of victory in their six tournament games, which they won by 11, 12, 1, 16, 11, and 3 points.

Answer 9 points

CALCULATOR SNAPSHOT   Order of operations and parentheses

Scientific calculators have the rules for order of operations built in. Even so, some evaluations require the use of a left parenthesis key $\boxed{(}$ and a right parenthesis key $\boxed{)}$. For example, to evaluate $\frac{240}{20-15}$, we enter these numbers and press these keys.

$$240 \boxed{\div} \boxed{(} 20 \boxed{-} 15 \boxed{)} \boxed{=} \qquad \boxed{48}$$

THINK IT THROUGH   Preparing For Class

"Only about 13% of full-time students spend more than 25 hours a week preparing for class, the approximate number that faculty members say is needed to do well in college." The National Survey of Student Engagement Annual Report 2003

The National Survey of Student Engagement 2003 Annual Report Committee questioned thousands of full-time college students about their weekly activities. Use the given clues to determine the results of the survey shown below.

Full-time Student Time Usage per Week

Activity	Time per week
Preparing for class	14 hours
Working on-campus or off-campus	Four hours less than the time spent preparing for class
Participating in co-curricular activities	Half as many hours as the time spent working on-campus or off-campus
Relaxing and socializing	Two hours more than twice the time spent participating in co-curricular activities
Providing care for dependents	Three hours less than one-half of the time spent relaxing and socializing
Commuting to class	One hour more than the time spent providing care for dependents

Section 1.7   STUDY SET

VOCABULARY   *Fill in the blanks.*

1. The grouping symbols ( ) are called _____, and the symbols [ ] are called _____.

2. The expression above a fraction bar is called the _____. The expression below a fraction bar is called the _____.

3. To _____ the expression $2 + 5 \cdot 4$ means to find its value.

4. To find the _____ of several values, we add the values and divide by the number of values.

CONCEPTS

5. Consider $5(2)^2 - 1$. How many operations need to be performed to evaluate the expression? List them in the order in which they should be performed.

6. Consider $15 - 3 + (5 \cdot 2)^3$. How many operations need to be performed to evaluate this expression? List them in the order in which they should be performed.

7. Consider $\dfrac{5 + 5(7)}{2 + (8 - 4)}$. In the numerator, what operation should be done first? In the denominator, what operation should be done first?

8. In the expression $\dfrac{3 - 5(2)}{5(2) + 4}$, the bar is a grouping symbol. What does it separate?

9. Explain the difference between $2 \cdot 3^2$ and $(2 \cdot 3)^2$.

10. Use brackets to write $2(12 - (5 + 4))$ in clearer form.

NOTATION *Complete each solution to evaluate the expression.*

11. $28 - 5(2)^2 = 28 - 5(\;\;)$
$$= 28 - \blacksquare$$
$$= \blacksquare$$

12. $2 + (5 + 6 \cdot 2) = 2 + (5 + \blacksquare)$
$$= 2 + \blacksquare$$
$$= \blacksquare$$

13. $[4(2 + 7)] - 6 = [4(\;\;)] - 6$
$$= \blacksquare - 6$$
$$= \blacksquare$$

14. $\dfrac{5(3) + 12}{9 - 6} = \dfrac{\blacksquare + 12}{\blacksquare}$
$$= \dfrac{\blacksquare}{3}$$
$$= \blacksquare$$

PRACTICE *Evaluate each expression.*

15. $7 + 4 \cdot 5$
16. $10 - 2 \cdot 2$
17. $2 + 3(0)$
18. $5(0) + 8$
19. $20 - 10 + 5$
20. $80 - 5 + 4$
21. $25 \div 5 \cdot 5$
22. $6 \div 2 \cdot 3$
23. $7(5) - 5(6)$
24. $4 \cdot 2 + 2 \cdot 4$
25. $4^2 + 3^2$
26. $12^2 - 5^2$
27. $2 \cdot 3^2$
28. $3^3 \cdot 5$
29. $3 + 2 \cdot 3^4 \cdot 5$
30. $3 \cdot 2^3 \cdot 4 - 12$
31. $5 \cdot 10^3 + 2 \cdot 10^2 + 3 \cdot 10^1 + 9$
32. $8 \cdot 10^3 + 0 \cdot 10^2 + 7 \cdot 10^1 + 4$
33. $3(2)^2 - 4(2) + 12$
34. $5(1)^3 + (1)^2 + 2(1) - 6$

35. $(8 - 6)^2 + (4 - 3)^2$
36. $(2 + 1)^2 + (3 + 2)^2$
37. $60 - \left(6 + \dfrac{40}{8}\right)$
38. $7 + \left(5^3 - \dfrac{200}{2}\right)$
39. $6 + 2(5 + 4)$
40. $3(5 + 1) + 7$
41. $3 + 5(6 - 4)$
42. $7(9 - 2) - 1$
43. $(7 - 4)^2 + 1$
44. $(9 - 5)^3 + 8$
45. $6^3 - (10 + 8)$
46. $5^2 - (9 + 3)$
47. $50 - 2(4)^2$
48. $30 + 2(3)^3$
49. $16^2 - 4(2)(5)$
50. $8^2 - 4(3)(1)$
51. $39 - 5(6) + 9 - 1$
52. $15 - 3(2) - 4 + 3$
53. $(18 - 12)^3 - 5^2$
54. $(9 - 2)^2 - 3^3$
55. $2(10 - 3^2) + 1$
56. $1 + 3(18 - 4^2)$
57. $6 + \dfrac{25}{5} + 6(3)$
58. $15 - \dfrac{24}{6} + 8 \cdot 2$
59. $3\left(\dfrac{18}{3}\right) - 2(2)$
60. $2\left(\dfrac{12}{3}\right) + 3(5)$
61. $(2 \cdot 6 - 4)^2$
62. $2(6 - 4)^2$
63. $4[50 - (3^3 - 5^2)]$
64. $6[15 + (5 \cdot 2^2)]$
65. $80 - 2[12 - (5 + 4)]$
66. $15 + 5[12 - (2^2 + 4)]$
67. $2[100 - (5 + 4)] - 45$
68. $8[6(6) - 6^2] + 4(5)$
69. $\dfrac{10 + 5}{6 - 1}$
70. $\dfrac{18 + 12}{2(3)}$
71. $\dfrac{5^2 + 17}{6 - 2^2}$
72. $\dfrac{3^2 - 2^2}{(3 - 2)^2}$
73. $\dfrac{(3 + 5)^2 + 2}{2(8 - 5)}$
74. $\dfrac{25 - (2 \cdot 3 - 1)}{2 \cdot 9 - 8}$
75. $\dfrac{(5 - 3)^2 + 2}{4^2 - (8 + 2)}$
76. $\dfrac{(4^3 - 2) + 7}{5(2 + 4) - 7}$
77. $12{,}985 - (1{,}800 + 689)$
78. $\dfrac{897 - 655}{88 - 77}$
79. $3{,}245 - 25(16 - 12)^2$
80. $\dfrac{24^2 - 4^2}{22 + 58}$

APPLICATIONS *Write an expression to solve each problem and evaluate it.*

81. BUYING GROCERIES At the supermarket, Carlos has 2 cases of soda, 4 bags of potato chips, and 2 cans of dip in his cart. Each case of soda costs $6, each bag of chips costs $2, and each can of dip costs $1. Find the total cost of the groceries.

82. JUDGING The scores received by a junior diver are as follows:

5	2	4	6	3	4

The formula for computing the overall score for the dive is as follows:

1. Throw out the lowest score.
2. Throw out the highest score.
3. Divide the sum of the remaining scores by 4.

Find the diver's score.

83. BANKING When a customer deposits cash, a teller must complete a currency count on the back of the deposit slip. In the illustration, what is the total amount of cash being deposited?

Currency count, for financial use only			
24	x 1's		
—	x 2's		
6	x 5's		
10	x 10's		
12	x 20's		
2	x 50's		
1	x 100's		
	TOTAL $		

84. WRAPPING GIFTS How much ribbon is needed to wrap the package shown if 15 inches of ribbon are needed to make the bow?

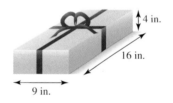

4 in.
16 in.
9 in.

85. SCRABBLE Illustration (a) shows part of the game board before and Illustration (b) shows it after the words *brick* and *aphid* were played. Determine the scoring for each word. (The number on each tile gives the point value of the letter.)

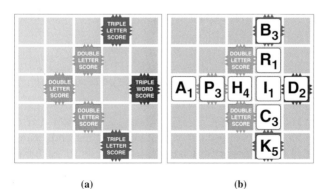

(a)   (b)

86. THE GETTYSBURG ADDRESS Here is an excerpt from Abraham Lincoln's Gettysburg Address:

> Fourscore and seven years ago, our fathers brought forth on this continent a new nation, conceived in liberty, and dedicated to the proposition that all men are created equal.

Lincoln's comments refer to the year 1776, when the United States declared its independence. If a score is 20 years, in what year did Lincoln deliver the Gettysburg Address?

87. CLIMATE One December week, the temperatures in Honolulu, Hawaii, were 75°, 80°, 83°, 80°, 77°, 72°, and 86°. Find the week's average (mean) temperature.

88. GRADES In a psychology class, a student had test scores of 94, 85, 81, 77, and 89. He also overslept, missed the final exam, and received a 0 on it. What was his test average in the class?

89. NATURAL NUMBERS What is the average (mean) of the first nine natural numbers?

90. ENERGY USAGE See the illustration. Find the average number of therms of natural gas used per month.

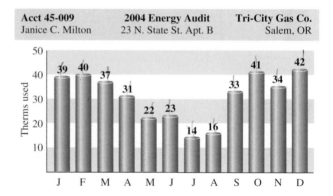

Acct 45-009	2004 Energy Audit	Tri-City Gas Co.
Janice C. Milton	23 N. State St. Apt. B	Salem, OR

91. FAST FOODS The table shows the sandwiches Subway advertises as its low-fat menu. What is the average (mean) number of calories for the group of sandwiches?

6-inch subs	Calories	Fat (g)
Veggie Delite	237	3
Turkey Breast	289	4
Turkey Breast & Ham	295	5
Ham	302	5
Roast Beef	303	5
Subway Club	312	5
Roasted Chicken Breast	348	6

92. TV RATINGS The list below shows the number of people watching *Who Wants to Be a Millionaire?* on five weeknights in November of 1999. How large was the average audience?

Monday	26,800,000
Tuesday	24,900,000
Wednesday	22,900,000
Thursday	25,900,000
Friday	21,900,000

93. SUM-PRODUCT NUMBERS

a. Evaluate the expression below, which is the sum of the digits of 135 times the product of the digits of 135.

$$(1 + 3 + 5)(1 \cdot 3 \cdot 5)$$

b. Write an expression representing the sum of the digits of 144 times the product of the digits of 144. Then evaluate the expression.

94. PRIME NUMBERS Show that 87 is the sum of the squares of the first four prime numbers.

WRITING

95. Explain why rules for the order of operations are necessary.

96. Explain the difference between the steps used to evaluate $5 \cdot 2^3$ and $(5 \cdot 2)^3$.

97. Explain the process of finding the mean of a large group of numbers. What does an average tell you?

98. What does it mean when we say to do all additions and subtractions *as they occur from left to right*?

REVIEW *Perform the operations.*

99.
$$\begin{array}{r} 4{,}029 \\ +3{,}271 \\ \hline \end{array}$$

100.
$$\begin{array}{r} 4{,}263 \\ -3{,}764 \\ \hline \end{array}$$

101.
$$\begin{array}{r} 417 \\ \times\ 23 \\ \hline \end{array}$$

102. $82\overline{)50{,}430}$

Order of Operations

When asked to evaluate a numerical expression, you must perform the operations in the expression in the proper order. One of the major objectives of this course is that you be able to apply the rules for the order of operations.

Evaluate the expression $2 + 3 \cdot 5$ in two ways.

Multiply first	Add first
$2 + 3 \cdot 5$	$2 + 3 \cdot 5$

1. Why are the rules for the order of operations necessary? Which method is correct?
2. Fill in the blanks to complete the rules for the order of operations.

Perform all calculations within parentheses and grouping symbols in the following order, working from the innermost pair to the outermost pair.

1. Evaluate all _____. That is, evaluate any expressions with _____.
2. Perform all _____ and divisions as they occur from _____ to right.
3. Perform all additions and _____ as they occur from left to _____.

To apply the rules for the order of operations, we must identify the operations involved in an expression. In the proper order, list the operations that must be performed to evaluate each of the following expressions.

3. $10 + 4 - 3^2$

4. $\dfrac{180}{6} - (4)3$

5. $2(3) - 12 \div 6 \cdot 3$

6. $2(3)^3(4) + 6$

After identifying the operations, we must perform them in the proper order. Evaluate each expression.

7. $10 + 4 - 3^2$

8. $\dfrac{180}{6} - (4)3$

9. $2(3) - 12 \div 6 \cdot 3$

10. $2(3)^3(4) + 6$

When expressions involve grouping symbols, we perform the operations inside them first. Evaluate each of the following expressions.

11. $2(4 + 3 \cdot 2)^2 + 6$

12. $1 + 3[6 - (1 + 5)]$

ACCENT ON TEAMWORK

SECTION 1.1
PLACE VALUE   Have each student in your group bring a calculator to class so that you can examine several different models. For each model, determine the largest number (if there is one) that can be entered on the display of the calculator. Then press the appropriate calculator keys to add 1 to that number. What does the display show?

LARGE NUMBERS   Bill Gates, founder of Microsoft Corporation, is said to be a billionaire. How many millions make one billion?

SECTION 1.2
PERIMETERS   Find the perimeter of each figure.

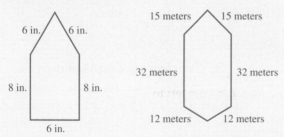

SECTION 1.3
SUBTRACTION   Explain how subtraction is related to addition. Use this idea to perform the following subtractions.

a. 9
  −5

b. 27
  −13

c. 25
  −18

SECTION 1.4
MAXIMUM AREA   A gardener has 80 feet of fencing to enclose a rectangular garden. Find the length and width that will enclose the greatest area.

SECTION 1.5
DIVISIBILITY TESTS   A number is divisible by 8 if the number formed by the last three digits is divisible by 8. Determine whether each of the following numbers is divisible by 8.

a. 1,216   **b.** 2,496

c. 4,160   **d.** 3,078

e. 16,928   **f.** 27,926

SECTION 1.6
COMMON FACTORS   The prime factorizations of 36 and 126 are shown below. The prime factors that are common to 36 and 126 (highlighted in color) are 2, 3, and 3.

$$36 = 2 \cdot 2 \cdot 3 \cdot 3$$
$$126 = 2 \cdot 3 \cdot 3 \cdot 7$$

Find the common prime factors for each of the following pairs of numbers.

a. 25, 45   **b.** 24, 60

c. 18, 45   **d.** 40, 112

e. 180, 210   **f.** 242, 198

SECTION 1.7
ORDER OF OPERATIONS   Consider the expression

$$5 + 8 \cdot 2^3 - 3 \cdot 2$$

Insert a set of parentheses somewhere in the expression so that, when it is evaluated, you obtain

a. 63   **b.** 132

c. 21   **d.** 127

CHAPTER REVIEW

An Introduction to the Whole Numbers

CONCEPT

A *set* is a collection of objects. The set of *natural numbers* is

$\{1, 2, 3, 4, 5, \ldots\}$

The set of *whole numbers* is

$\{0, 1, 2, 3, 4, 5, \ldots\}$

Whole numbers are often used in tables, bar graphs, and line graphs.

REVIEW EXERCISES

Consider the set $\{0, 2, \frac{3}{2}, 5, 7.2, 9\}$.

1. List each natural number in the set.

2. List each whole number in the set.

Consider the data in the table, listing the number of building permits issued in the city of Springsville for the period 2001–2004.

Year	2001	2002	2003	2004
Building permits	12	15	10	7

3. Construct a bar graph of the data.

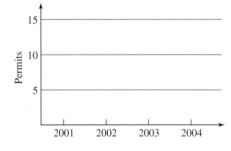

4. Construct a line graph of the data.

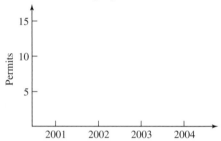

The digits in a *whole number* have *place value*.

Consider the number 2,365,720.

5. Which digit is in the ten thousands column?

6. Which digit is in the hundreds column?

A whole number is written in *expanded notation* when its digits are written with their place values.

Write each number in expanded notation.

7. 570,302

8. 37,309,054

Write each number in standard notation.

9. 3 thousands + 2 hundreds + 7 ones

10. Twenty-three million, two hundred fifty-three thousand, four hundred twelve

---

The symbol $<$ means "is less than." The symbol $>$ means "is greater than."

Place an $<$ or an $>$ symbol between the numerals to make a true statement.

11. 9 [ ] 7

12. 301 [ ] 310

To give approximate answers, we often use *rounded numbers*.

Round 2,507,348 to the specified place.

13. Nearest hundred

14. Nearest ten thousand

15. Nearest ten

16. Nearest hundred thousand

Adding Whole Numbers

Adding whole numbers corresponds to combining sets of objects. Do additions within parentheses first.

Find each sum.

17. $7 + 6$ **18.** $6 + 7$

19. $4 + (7 + 3)$ **20.** $(4 + 7) + 3$

21. $5 + (6 + 9)$ **22.** $(9 + 3) + 6$

Commutative property of addition: The order in which whole numbers are added does not affect their sum.

Perform each addition.

23. $135 + 213$ **24.** $4,447 + 7,478$

25. $\begin{array}{r} 236 \\ +282 \end{array}$ **26.** $\begin{array}{r} 5,345 \\ +\ \ 655 \end{array}$

Associative property of addition: The way whole number addends are grouped does not affect their sum.

27. What property of addition is shown?

   a. $19 + 6 = 6 + 19$ **b.** $101 + (99 + 57) = (101 + 99) + 57$

The *perimeter* of a rectangle is the distance around it.

28. Find the perimeter of the rectangle.

731 ft

642 ft

29. AIRPORTS The world's three busiest airports in 2003 are listed below. Find the total number of passengers passing through those airports.

Airport	Total passengers
Atlanta, Hartsfield	76,086,792
Chicago, O'Hare	69,354,154
London, Heathrow	63,468,620

Source: Airports Council International World Headquarters

30. What is the sum of three thousand seven hundred six and ten thousand nine hundred fifty-five?

Subtracting Whole Numbers

Subtracting whole numbers tells how many objects remain when some are removed from a set.

Perform each subtraction.

31. $8 - 5$ **32.** $9 - (7 - 2)$

33. Subtract 218 from 235.   **34.** $5,231 - 5,177$

35. $\begin{array}{r} 343 \\ -269 \end{array}$ **36.** $\begin{array}{r} 7,800 \\ -5,725 \end{array}$

37. TRAVEL A direct flight to San Francisco costs $237. A flight with one stop in Reno costs $192. How much can be saved by taking the inexpensive flight?

38. SAVINGS ACCOUNTS A savings account contains $931. If the owner deposits $271 and makes withdrawals of $37 and $380, find the final balance.

39. FARMING In a shipment of 350 animals, 124 were hogs, 79 were sheep, and the rest were cattle. Find the number of cattle in the shipment.

40. LAND AREA Use the data in the table to determine how much larger the land area of Russia is compared to that of Canada.

Country	Land area (square miles)
Russia	6,592,735
Canada	3,855,081

Source: *Time Almanac, 2005*

SECTION 1.4   *Multiplying Whole Numbers*

Multiplication is repeated addition. For example,

The sum of four 6's

$$4 \cdot 6 = 6 + 6 + 6 + 6$$
$$= 24$$

The result, 24, is called the *product*, and the 4 and 6 are called *factors*.

Commutative property of multiplication: The order in which whole numbers are multiplied does not affect their product.

Associative property of multiplication: The way whole number factors are grouped does not affect their product.

The *area A of a rectangle* is the product of its length *l* and its width *w*.

$$A = l \cdot w$$

Perform each multiplication.

41. $8 \cdot 7$ **42.** $7(8)$

43. $8 \cdot 0$ **44.** $7 \cdot 1$

45. $(5 \cdot 7) \cdot 6$ **46.** $5 \cdot (7 \cdot 6)$

Perform each multiplication.

47. $157 \cdot 21$ **48.** $3,723(48)$

49. $\begin{array}{r} 356 \\ \times\ 89 \\ \hline \end{array}$ **50.** $\begin{array}{r} 5,624 \\ \times\ \ 81 \\ \hline \end{array}$

51. MARCHING BANDS For a football halftime performance, the members of a marching band were assembled in a rectangular 22-row and 15-column formation. How many members are in the marching band?

52. What property of multiplication is shown?

     a. $2 \cdot (5 \cdot 7) = (2 \cdot 5) \cdot 7$ **b.** $100(50) = 50(100)$

Find the area of the rectangle and the square.

53.
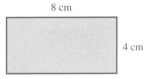
8 cm
4 cm

54.
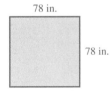
78 in.
78 in.

55. PAYCHECKS Sarah worked 12 hours at $9 per hour, and Santiago worked 14 hours at $8 per hour. Who earned more money?

56. SHOPPING There are 12 eggs in one dozen, and 12 dozen in one gross. How many eggs are in a shipment of 100 gross?

Dividing Whole Numbers

Division is an operation that determines how many times a number (the *divisor*) is contained in another number (the *dividend*). *Remember that you can never divide by 0.*

Perform each division, if possible.

57. $357 \div 17$

58. $1,443 \div 39$

59. $21\overline{)405}$

60. $54\overline{)1,269}$

61. $\dfrac{81}{27}$

62. $\dfrac{595}{35}$

63. $\dfrac{0}{10}$

64. $\dfrac{10}{0}$

65. TREATS If 745 candies are divided equally among 45 children, how many will each child receive? How many candies will be left over?

66. PURCHASING A county received a $850,000 grant to purchase some new police patrol cars. If a fully equipped patrol car costs $25,000, how many can the county purchase with the grant money?

Prime Factors and Exponents

Numbers that are multiplied together are called *factors*.

A *prime number* is a whole number greater than 1 that has only 1 and itself as factors. Whole numbers greater than 1 that are not prime are called *composite numbers*.

Whole numbers that are divisible by 2 are *even* numbers. Whole numbers that are not divisible by 2 are *odd* numbers.

The *prime factorization* of a whole number is the product of its prime factors.

An *exponent* is used to indicate repeated multiplication. In the expression 6^3, 6 is the *base* and 3 is the exponent.

Find all of the factors of each number.

67. 18

68. 25

Identify each number as a prime, composite, or neither.

69. 31

70. 100

71. 1

72. 0

73. 125

74. 47

Identify each number as an even or odd number.

75. 171

76. 214

77. 0

78. 1

Find the prime factorization of each number.

79. 42

80. 75

Write each expression using exponents.

81. $6 \cdot 6 \cdot 6 \cdot 6$

82. $5(5)(5)(13)(13)$

Evaluate each expression.

83. 5^3

84. 11^2

85. $2^3 \cdot 5^2$

86. $2^2 \cdot 3^3 \cdot 5^2$

SECTION 1.7 — *Order of Operations*

Perform mathematical operations in the following order:

Perform all calculations within parentheses and grouping symbols in the following order, working from the innermost pair to the outermost pair.

1. Evaluate all powers. That is, evaluate any expressions with exponents.

2. Perform all multiplications and divisions in order from left to right.

3. Perform all additions and subtractions in order from left to right.

The *arithmetic mean* (average) is a value around which number values are grouped.

Evaluate each expression.

87. $13 + 12 \cdot 3$

88. $35 - 15 + 3$

89. $(13 + 12) \cdot 3$

90. $(35 - 15) \div 5$

91. $8 \cdot 5 - 4 \div 2 \cdot 4$

92. $8 \cdot (5 - 4 \div 2)^2$

93. $2 + 3(10 - 4 \cdot 2)$

94. $4(20 - 5 \cdot 3 + 2) - 4$

95. $\dfrac{4(6) - 6}{2(3^2)}$

96. $\dfrac{12 + 3 \cdot 7}{5^2 - 14}$

97. $7 + 3[10 - 3(4 - 2)]$

98. $5 + 2[(15 - 3 \cdot 4) - 2]$

Find the arithmetic mean (average) of each set of scores.

99.

Test	1	2	3	4
Score	80	74	66	88

100.

Test	1	2	3	4	5
Score	73	77	81	0	69

1. List the whole numbers less than 5.

2. Write "five thousand two hundred sixty-six" in expanded notation.

3. Write "7 thousands + 5 hundreds + 7 ones" in standard notation.

4. Round 34,752,341 to the nearest million.

Refer to the data in the table.

Lot number	1	2	3	4
Defective bolts	7	10	5	15

5. Use the data to make a bar graph.

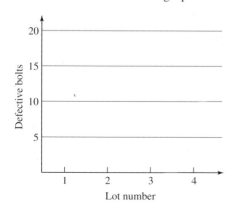

6. Use the data to make a line graph.

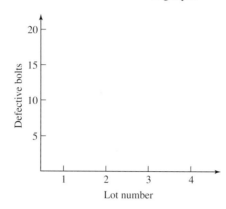

Place one of the symbols < or > between the numbers to make each statement true.

7. 15 ⬚ 10

8. 1,247 ⬚ 1,427

Perform each operation.

9. Add: 327 + 435.

10. Subtract 287 from 535.

11. Add: 4,521
     +3,579

12. Subtract: 4,521
         −3,579

13. A rectangle is 327 inches wide and 757 inches long. Find its perimeter.

14. STOCKS On Tuesday, a share of KBJ Company was selling at $73. The price rose $12 on Wednesday and fell $9 on Thursday. Find its price on Thursday.

Perform each operation.

15. Multiply: 53
   $\times\ 8$

16. Multiply: 367
   $\times\ 73$

17. Divide: $63\overline{)4{,}536}$.

18. Divide: $73\overline{)8{,}379}$.

19. Find the perimeter and the area of the square.

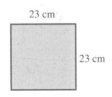

23 cm

23 cm

20. What property is illustrated by each statement?

   a. $18 \cdot (9 \cdot 40) = (18 \cdot 9) \cdot 40$

   b. $23{,}999 + 1 = 1 + 23{,}999$

21. Perform each operation, if possible.

   a. $15 \cdot 0$ **b.** $\dfrac{0}{15}$

22. P.E. CLASSES In a physical education class, the students stand in a rectangular formation of 8 rows and 12 columns when the instructor takes attendance. How many students are in the class?

23. Find the prime factorization of 1,260.

Evaluate each expression.

24. $3(4^2) - 2^2$.

25. $9 + 4 \cdot 5$.

26. $10 + 2[12 - 2(6 - 4)]$.

27. $\dfrac{3^3 - 2(6 - 5)^2}{33 - 9 + 1}$

28. GRADES A student scored 73, 52, and 70 on three exams and received 0 on two missed exams. Find his mean (average) score.

29. What information does the arithmetic mean (average) give us about a set of values?

30. Explain the difference between what the perimeter and the area of a rectangle measure.

31. What are parentheses, and how have they been used in this chapter?

The Integers

Getty Images

There are few things more breathtaking than a star-filled sky on a clear night. Stars come in different colors, sizes, shapes, and ages. You have probably noticed that some stars are bright and others are faint. Astronomers have developed a scale to classify the relative brightness of stars. The most brilliant stars (including the sun) are assigned negative number magnitudes while positive number magnitudes are assigned to the dimmer stars (and planets).

To learn more about positive and negative numbers, visit The Learning Equation on the Internet at http://tle.brookscole.com. (The log-in instructions are in the Preface.) For Chapter 2, the online lessons are:

• *TLE* Lesson 4: An Introduction to Integers
• *TLE* Lesson 5: Adding Integers
• *TLE* Lesson 6: Subtracting Integers

Check Your Knowledge

1. The _____ value of a number is the distance between the number and zero on a number line.

2. When 0 is added to a number, the number remains the same. We call 0 the additive _____.

3. Two numbers that are the same distance from 0 on the number line, but on opposite sides of it, are called _____.

4. The product of two integers with _____ signs is negative.

5. Insert one of the symbols $>$ or $<$ in the blank to make the statement true: $-15 \boxed{} -16$.

6. Find the mean (average) of the temperatures shown in the table on the left.

7. Add:

    a. $-27 + 13$

    b. $12 + (-12)$

    c. $(-2) + (-2) + (-2) + (-2)$

    d. $(-4 + 7) + [1 + (-6) + 4]$

8. Subtract:

    a. $7 - 13$ b. $3 - (-3)$ c. $0 - 5 - 7$

9. Find each product.

    a. $3(-3)$ b. $-5(-20)$ c. $(-3)(-2)(-4)$

10. Write the related multiplication statement for $\dfrac{-18}{3} = -6$.

11. Find each quotient, if possible.

    a. $\dfrac{36}{-9}$ b. $\dfrac{-900}{-30}$ c. $\dfrac{-3}{2 + 1 - 3}$

12. Evaluate each expression.

    a. $-(-7)$ b. $-|-7|$ c. $3|-6 + 2|$

13. Find each power:

    a. -3^2 b. $(-3)^2$ c. $(3 - 4)^2$

14. Evaluate:

    a. $24 \div 3 \cdot 2$ b. $6 + \dfrac{12}{-4} - 4^2$

    c. $-5 - 2[7 - (-3)(-2)^3]$ d. $\dfrac{-3^3 + (-4)(-6)}{-3(3 - 5)^2 + 9}$

15. The price of a certain computer dropped from $620 to $500 in six months. How much did the price drop per month?

16. On one lie detector test, a burglar scored -19, which indicates deception. However, on a second test, he scored -4, which is inconclusive. Find the difference in the scores.

Daily high temperatures

Sunday	1
Monday	7
Tuesday	-3
Wednesday	1
Thursday	0
Friday	-1
Saturday	2

Study Skills Workshop

ATTITUDES, PAST EXPERIENCE, AND LEARNING STYLES

Your personality will affect the way you learn math; some traits will work toward your benefit and some may not. The key is to maximize your strengths, minimize weaknesses, and find learning methods that will best suit your personal style.

Attitudes. How do you feel about math? How do you feel about taking control of your learning environment? What do you think of your chance for success in this class? If you respond to these questions with answers like: "Ugh! I'll never learn math!" or "My math teachers just never taught me well," you may be setting yourself up for failure. Try to alter your attitude slightly. Constructive thinking and positive self-talk may bring good results. Instead of telling yourself "I'm just stupid when it comes to math," try saying "Math is not my strongest subject, but I will work hard at it and learn new things this time." Know that in truly learning any new concept there are bound to be frustrations, but they can be overcome with support, strategy, and hard work.

Past Experience. Good or bad experiences in previous math classes may affect the way you view math now. Many students who have had negative math experiences in the past may feel anxiety and stress in the present. If you have high levels of stress or feel helpless when dealing with numbers, you may have math anxiety. These counterproductive feelings can be overcome with extra preparation, support services, relaxation techniques, and even hypnotherapy.

Learning Styles. What type of learner are you? The answer to this question will help you determine how you study, how you do your homework, and maybe even where you choose to sit in class. For example, visual-verbal learners learn best by reading and writing, so a good strategy for them in studying is to rewrite notes and examples. However, audio learners learn best by listening, so making audiotapes of important concepts may be their best study strategy. Kinesthetic learners like to move and do things with their hands, so incorporating the use of games or puzzles (often called manipulatives) in studying may be helpful.

ASSIGNMENT

1. Take a learning skills inventory test such as that found online at http://www
.metamath.com/multiple/multiple_choice_questions.cgi to determine what type of
learner you are. Once you have determined your learning style, write a plan on how
will you use this information to help you succeed in your class.
2. Describe your past experiences in math courses. Have they generally been good or
bad? Why? If you feel you have math anxiety, visit your school's counseling office
and schedule an appointment to find ways of dealing with this problem.

In this chapter, we introduce the concept of negative number and explore an extension of the set of whole numbers, called the integers.

2.1 An Introduction to the Integers

- The integers
- Extending the number line
- More on inequality
- Absolute value
- The opposite of a number
- The − symbol

We have seen that whole numbers can be used to describe many situations that arise in everyday life. For example, we can use whole numbers to express temperatures above zero, the balance in a checking account, or an altitude above sea level. However, we cannot use whole numbers to express temperatures below zero, the balance in a checking account that is overdrawn, or how far an object is below sea level. (See Figure 2-1.)

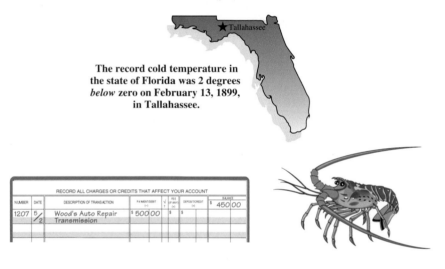

The record cold temperature in the state of Florida was 2 degrees *below* zero on February 13, 1899, in Tallahassee.

A check for $500 was written when there was only $450 in the account. The checking account is *overdrawn*.

The American lobster is found off the East Coast of North America at depths as much as 600 feet *below* sea level.

FIGURE 2-1

In this section, we see how negative numbers can be used to describe these three situations as well as many others.

The integers

To describe a temperature of 2 degrees (2°) above zero, a balance of $50, or 600 feet above sea level, we can use numbers called **positive numbers.** All positive numbers are greater than 0, and we can write them using a **positive sign** + or no sign at all.

In words	In symbols	Read as
2 degrees above zero	+2 or 2	positive two
A balance of $50	+50 or 50	positive fifty
600 feet above sea level	+600 or 600	positive six hundred

To describe a temperature of 2 degrees below zero, $50 overdrawn, or 600 feet below sea level, we need to use negative numbers. **Negative numbers** are numbers less than 0, and they are written using a **negative sign** −.

In words	In symbols	Read as
2 degrees below zero	-2	negative two
$50 overdrawn	-50	negative fifty
600 feet below sea level	-600	negative six hundred

> **Positive and negative numbers**
> **Positive numbers** are greater than 0. **Negative numbers** are less than 0.

! COMMENT Zero is neither positive nor negative.

We often call positive and negative numbers **signed numbers.** The first three of the following signed numbers are positive, and the last three are negative.

$$+12, \quad +26, \quad 515, \quad -12, \quad -26, \quad \text{and} \quad -515$$

The collection of positive whole numbers, the negatives of the whole numbers, and 0 is called the set of **integers** (read as "in-ti-jers").

> **The set of integers**
> $$\{\dots, -5, -4, -3, -2, -1, 0, 1, 2, 3, 4, 5, \dots\}$$

Since every natural number is an integer, we say that the set of natural numbers is a subset of the integers. See Figure 2-2. Since every whole number is an integer, we say that the set of whole numbers is a **subset** of the integers.

The set of natural numbers

The set of integers → $\{\dots, -7, -6, -5, -4, -3, -2, -1, 0, 1, 2, 3, 4, 5, 6, 7, \dots\}$

The set of whole numbers

FIGURE 2-2

! COMMENT Note that the negative integers and 0 are not natural numbers. Also note that negative integers are not whole numbers.

◼ Extending the number line

In Section 1.1, we introduced the number line. We can use an extension of the number line to learn about negative numbers.

Negative numbers can be represented on a number line by extending the line to the left. Beginning at the origin (the 0 point), we move to the left, marking equally spaced points as shown in Figure 2-3. As we move to the right on the number line, the values of the numbers increase. As we move to the left, the values of the numbers decrease.

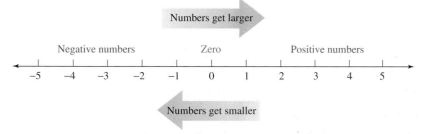

FIGURE 2-3

The thermometer shown in Figure 2-4(a) is an example of a vertical number line. It is scaled in degrees and shows a temperature of $-10°$. The time line shown in Figure 2-4(b) is an example of a horizontal number line. It is scaled in increments of 500 years.

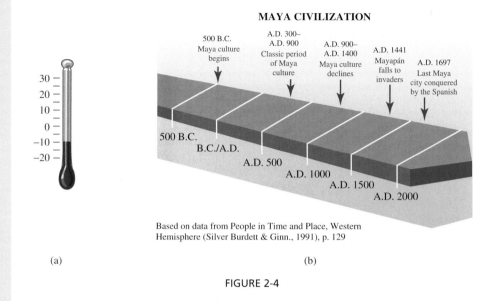

Based on data from People in Time and Place, Western Hemisphere (Silver Burdett & Ginn., 1991), p. 129

(a)   (b)

FIGURE 2-4

Self Check 1
On the number line, graph -4, -2, 1, and 3.

Answer

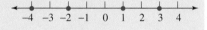

EXAMPLE 1   On the number line, graph -3, -1, 2, and 4.

Solution   To graph each integer, we locate its position on the number line and draw a dot.

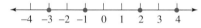

By extending the number line to include negative numbers, we can represent more situations using bar graphs and line graphs. For example, the bar graph shown in Figure 2-5 illustrates the annual profits *and losses* of Toys R Us over a nine-year period. Note that the profit in 2004 was $88 million and that the loss in 1999 was $132 million.

Toys R Us Net Income

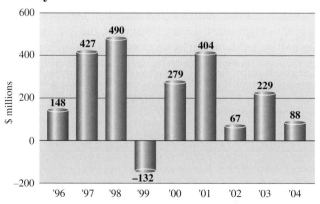

Source: Morningstar.com

FIGURE 2-5

THINK IT THROUGH

Credit Card Debt

"The most dangerous pitfall for many college students is the overuse of credit cards. Many banks do their best to entice new card holders with low or zero-interest cards." Gary Schatsky, certified financial planner

Which numbers on the credit card statement below are actually debts and, therefore, could be represented using negative numbers?

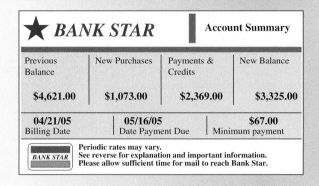

★ BANK STAR			Account Summary
Previous Balance	New Purchases	Payments & Credits	New Balance
$4,621.00	$1,073.00	$2,369.00	$3,325.00
04/21/05 Billing Date	05/16/05 Date Payment Due		$67.00 Minimum payment

BANK STAR — Periodic rates may vary. See reverse for explanation and important information. Please allow sufficient time for mail to reach Bank Star.

More on inequality

Recall that the symbol < means "is less than" and that > means "is greater than." Figure 2-6 shows the graph of the integers −2 and 1. Since −2 is to the left of 1 on the number line, −2 < 1. Since −2 < 1, it is also true that 1 > −2.

FIGURE 2-6

EXAMPLE 2 Use one of the symbols > or < to make each statement true:
a. 4 ___ −5   and   **b.** −4 ___ −2.

Solution
a. Since 4 is to the right of −5 on the number line, 4 > −5.
b. Since −4 is to the left of −2 on the number line, −4 < −2.

Self Check 2
Use one of the symbols > or < to make each statement true:
a. 6 ___ −6
b. −6 ___ −5

Answers **a.** >, **b.** <

Three other commonly used inequality symbols are the "is not equal to" symbol ≠, the "is less than or equal to" symbol ≤, and the "is greater than or equal to" symbol ≥.

$5 \neq 2$ Read as "5 is not equal to 2."

$6 \leq 10$ Read as "6 is less than or equal to 10." This statement is true, because 6 < 10.

$12 \leq 12$ Read as "12 is less than or equal to 12." This statement is true, because 12 = 12.

$17 \geq 15$ Read as "17 is greater than or equal to 15." This statement is true, because 17 > 15.

$20 \geq 20$ Read as "20 is greater than or equal to 20." This statement is true, because 20 = 20.

Self Check 3
Use an inequality symbol to
write "30 is less than or equal
to 35."

EXAMPLE 3   Use an inequality symbol to write   **a.** "8 is not equal to 5"   and
b. "50 is greater than or equal to 40."

Solution
a. $8 \neq 5$

Answer $30 \leq 35$

b. $50 \geq 40$

▮ Absolute value

Using the number line, we can see that the numbers 3 and -3 are both a distance of
3 units away from 0, as shown in Figure 2-7.

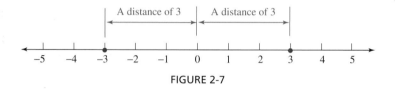

FIGURE 2-7

The **absolute value** of a number gives the distance between the number and 0 on
the number line. To indicate absolute value, the number is inserted between two verti-
cal bars, called the **absolute value symbol.** For example, we can write $|-3| = 3$. This is
read as "The absolute value of negative 3 is 3," and it tells us that the distance between
-3 and 0 on the number line is 3 units. From Figure 2-7, we also see that $|3| = 3$.

> **Absolute value**
>
> The **absolute value** of a number is the distance on the number line between the
> number and 0.

! COMMENT Absolute value expresses distance. The absolute value of a number
is always positive or 0. It is never negative.

Self Check 4
Evaluate each expression:
a. $|-9|$
b. $|4|$

EXAMPLE 4   Evaluate each expression: **a.** $|8|$,   **b.** $|-5|$,   and   **c.** $|0|$.

Solution
a. On the number line, the distance between 8 and 0 is 8. Therefore,

$$|8| = 8$$

b. On the number line, the distance between -5 and 0 is 5. Therefore,

$$|-5| = 5$$

c. On the number line, the distance between 0 and 0 is 0. Therefore,

$$|0| = 0$$

Answers **a.** 9,   **b.** 4

■ The opposite of a number

> **Opposites or negatives**
>
> Two numbers that are the same distance from 0 on the number line, but on opposite sides of it, are called **opposites** or **negatives.**

Figure 2-8 shows that for each natural number on the number line, there is a corresponding natural number, called its *opposite,* to the left of 0. For example, we see that 3 and −3 are opposites, as are −5 and 5. Note that 0 is its own opposite.

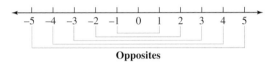

Opposites

FIGURE 2-8

To write the opposite of a number, a − symbol is used. For example, the opposite of 5 is −5 (read as "negative 5"). Parentheses are needed to express the opposite of a negative number. The opposite of −5 is written as −(−5). Since 5 and −5 are the same distance from 0, the opposite of −5 is 5. Therefore, −(−5) = 5. This leads to the following conclusion.

> **The double negative rule**
>
> The opposite of the negative of a number is that number.

Number	Opposite	
57	−57	Read as "negative fifty-seven."
−8	−(−8) = 8	Read as "the opposite of negative eight." Apply the double negative rule.
0	−0 = 0	The opposite of 0 is 0.

The concept of opposite can also be applied to an absolute value. For example, the opposite of the absolute value of −8 can be written as −|−8|. Think of this as a two-step process. Find the absolute value first, and then attach a − to that result.

First, find the absolute value.

$$-|-8| = -8$$

Then attach a − sign.

EXAMPLE 5 Simplify each expression: **a.** −(−44) and **b.** −|−225|.

Solution

a. −(−44) means the opposite of −44. Since the opposite of −44 is 44, we write

$$-(-44) = 44$$

b. −|−225| means the opposite of |−225|. Since |−225| = 225, and the opposite of 225 is −225, we write

$$-|-225| = -225$$

Self Check 5

Simplify each expression:
a. −(−1) and **b.** −|−99|.

Answers **a.** 1, **b.** −99

■ The − symbol

The − symbol is used to indicate a negative number, the opposite of a number, and the operation of subtraction. The key to interpreting the − symbol correctly is to examine the context in which it is used.

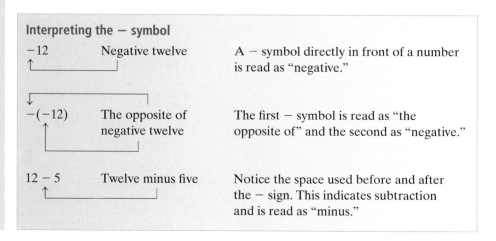

Interpreting the − symbol

−12	Negative twelve	A − symbol directly in front of a number is read as "negative."
−(−12)	The opposite of negative twelve	The first − symbol is read as "the opposite of" and the second as "negative."
12 − 5	Twelve minus five	Notice the space used before and after the − sign. This indicates subtraction and is read as "minus."

Section 2.1   STUDY SET

■ VOCABULARY   *Fill in the blanks.*

1. Numbers can be represented by points equally spaced on the number _____.

2. The point on the number line representing 0 is called the _____.

3. To _____ a number means to locate it on the number line and highlight it with a dot.

4. The graph of a number is the point on the number _____ that represents that number.

5. The symbols > and < are called _____ symbols.

6. _____ numbers are less than 0.

7. The _____ _____ of a number is the distance between the number and 0 on the number line.

8. Two numbers that are the same distance from 0 on the number line, but on opposite sides of it, are called _____.

9. {. . . , −3, −2, −1, 0, 1, 2, 3 . . .} is called the set of _____.

10. The double negative rule states that the opposite of the _____ of a number is that number.

■ CONCEPTS

11. Refer to each graph and use an inequality symbol > or < to make a true statement.

 a.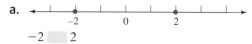

     −2 [ ] 2

 b.

     0 [ ] −1

 c.

     −1 [ ] 0 and 1 [ ] 0

12. Determine what is wrong with each number line.

 a.

 b.

 c.

 d.

13. Does every number on the number line have an opposite?

14. Is the absolute value of a number always positive?

15. Which of the following contains a minus sign:
$15 - 8$, $-(-15)$, or -15?

16. Is there a number that is both greater than 10 and less than 10 at the same time?

17. Express the fact $12 < 15$ using the $>$ symbol.

18. Express the fact $5 > 4$ using the $<$ symbol.

19. Represent each of these situations using a signed number.

   a. $225 overdrawn

   b. 10 seconds before liftoff

   c. 3 degrees below normal

   d. A deficit of $12,000

   e. A racehorse finished 2 lengths behind the leader.

20. Represent each of these situations using a signed number, and then describe its opposite in words.

   a. A trade surplus of $3 million

   b. A bacteria count 70 more than the standard

   c. A profit of $67

   d. A business $1 million in the "black"

   e. 20 units over their quota

21. If a number is less than 0, what type of number must it be?

22. If a number is greater than 0, what type of number must it be?

23. On the number line, what number is 3 units to the right of -7?

24. On the number line, what number is 4 units to the left of 2?

25. Name two numbers on the number line that are a distance of 5 away from -3.

26. Name two numbers on the number line that are a distance of 4 away from 3.

27. Which number is closer to -3 on the number line: 2 or -7?

28. Which number is farther from 1 on the number line: -5 or 8?

29. Give examples of the $-$ symbol used in three different ways.

30. What is the opposite of 0?

NOTATION

31. Translate each phrase to mathematical symbols.

   a. The opposite of negative eight

   b. The absolute value of negative eight

   c. Eight minus eight

   d. The opposite of the absolute value of negative eight

32. Write the set of integers.

PRACTICE   *Simplify each expression.*

33. $|9|$

34. $|12|$

35. $|-8|$

36. $|-1|$

37. $|-14|$

38. $|-85|$

39. $-|20|$

40. $-|110|$

41. $-|-6|$

42. $|0|$

43. $|203|$

44. $-|-11|$

45. -0

46. $-|0|$

47. $-(-11)$

48. $-(-1)$

49. $-(-4)$

50. $-(-9)$

51. $-(-12)$

52. $-(-25)$

Graph each set of numbers on the number line.

53. $\{-3, 0, 3, 4, -1\}$

54. $\{-4, -1, 2, 5, 1\}$

55. The opposite of -3, the opposite of 5, and the absolute value of -2

56. The absolute value of 3, the opposite of 3, and the number that is 1 less than -3

Insert one of the symbols $>$, $<$, or $=$ in the blank to make a true statement.

57. $-5 \quad 5$

58. $0 \quad -1$

59. $-12 \quad -6$

60. $-6 \quad -7$

61. $-10 \quad -11$

62. $-11 \quad -20$

63. $|-2| \quad 0$

64. $|-30| \quad -40$

Insert one of the symbols ≥ or ≤ in the blank to make a true statement.

65. −14 ⬛ −15   **66.** −77 ⬛ −76

67. 210 ⬛ 210   **68.** 37 ⬛ 37

69. −1,255 ⬛ −(−1,254) **70.** 0 ⬛ −3

71. −|−3| ⬛ 4   **72.** −|−163| ⬛ −150

▮ APPLICATIONS

73. FLIGHT OF A BALL A boy throws a ball from the top of a building, as shown. At the instant he does this, his friend starts a stopwatch and keeps track of the time as the ball rises to a peak and then falls to the ground. Use the vertical number line to complete the table by finding the position of the ball at each specified time.

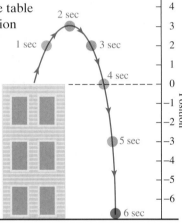

Time	Position of ball
1 sec	
2 sec	
3 sec	
4 sec	
5 sec	
6 sec	

74. SHOOTING GALLERIES At an amusement park, a shooting gallery contains moving ducks. The path of one duck is shown, along with the time it takes the duck to reach certain positions on the gallery wall. Complete the table using the horizontal number line in the illustration.

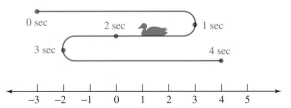

Time	Position of duck
0 sec	
1 sec	
2 sec	
3 sec	
4 sec	

75. TECHNOLOGY The readout from a testing device is shown. It is important to know the height of each of the three peaks and the depth of each of the three valleys. Use the vertical number line to find these numbers.

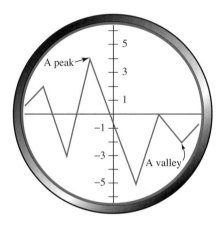

76. FLOODING A week of daily reports listing the height of a river in comparison to flood stage is given in the table. Complete the bar graph in the illustration.

Flood stage report	
Sun.	2 ft below
Mon.	3 ft over
Tue.	4 ft over
Wed.	2 ft over
Thu.	1 ft below
Fri.	3 ft below
Sat.	4 ft below

77. GOLF In golf, *par* is the standard number of strokes considered necessary on a given hole. A score of −2 indicates that a golfer used 2 strokes less than par. A score of +2 means 2 more strokes than par were used. In the illustration, each golf ball represents the score of a professional golfer on the 16th hole of a certain course.

   a. What score was shot most often on this hole?

   b. What was the best score on this hole?

   c. Explain why this hole appears to be too easy for a professional golfer.

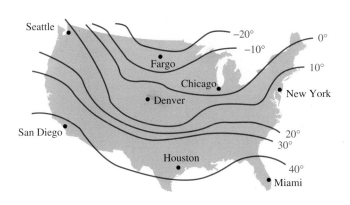

78. PAYCHECKS Examine the items listed on the following paycheck stub. Then write two columns on your paper — one headed "positive" and the other "negative." List each item under the appropriate heading.

Tom Dryden Dec. 04	Christmas bonus	$100
Gross pay $2,000	**Reductions**	
Overtime $300	Retirement	$200
Deductions	**Taxes**	
Union dues $30	Federal withholding	$160
U.S. Bonds $100	State withholding	$35

79. WEATHER MAPS The illustration shows the predicted Fahrenheit temperatures for a day in mid-January.

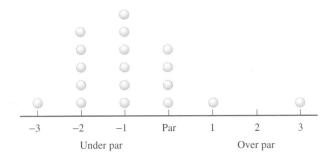

a. What is the temperature range for the region including Fargo, North Dakota?

b. According to the prediction, what is the warmest it should get in Houston?

c. According to this prediction, what is the coldest it should get in Seattle?

80. PROFITS/LOSSES The graph in the illustration shows the net income of Apple Computer Inc. for the years 2000–2004.

   a. In what year did the company suffer a loss? Estimate each loss.

   b. In what year did Apple have the greatest profit? Estimate it.

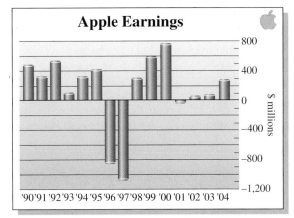

Source: Morningstar.com

81. HISTORY Number lines can be used to display historical data. Some important world events are shown on the time line in the illustration.

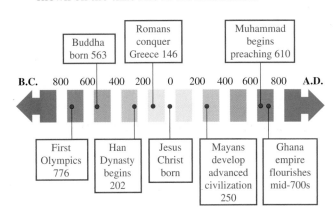

a. What basic unit is used to scale this time line?

b. What can be thought of as positive numbers?

c. What can be thought of as negative numbers?

d. What important event distinguishes the positive from the negative numbers?

82. ASTRONOMY Astronomers use a type of number line called the *apparent magnitude scale* to denote the brightness of objects in the sky. The brighter an object appears to an observer on Earth, the more negative is its apparent magnitude. Graph each of the following on the scale in the illustration.

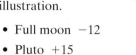

- Full moon −12
- Pluto +15
- Sirius (a bright star) −2
- Sun −26
- Venus −4
- Visual limit of binoculars +10
- Visual limit of large telescope +20
- Visual limit of naked eye +6

83. LINE GRAPHS Each thermometer in the illustration gives the daily high temperature in degrees Fahrenheit. Plot each daily high temperature on the grid and then construct a line graph.

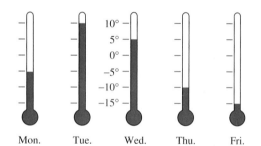

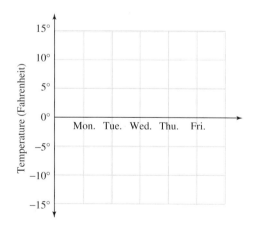

84. GARDENING The illustration in the next column shows the depths at which the bottoms of various types of flower bulbs should be planted. (The symbol ″ represents inches.)

a. At what depth should a tulip bulb be planted?

b. How much deeper are hyacinth bulbs planted than gladiolus bulbs?

c. Which bulb must be planted the deepest? How deep?

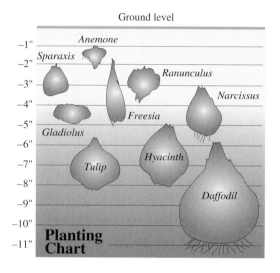

WRITING

85. Explain the concept of the opposite of a number.

86. What real-life situation do you think gave rise to the concept of a negative number?

87. Explain why the absolute value of a number is never negative.

88. Give an example of the use of the number line that you have seen in another course.

89. DIVING Divers use the terms *positive buoyancy*, *neutral buoyancy*, and *negative buoyancy* as shown. What do you think each of these terms means?

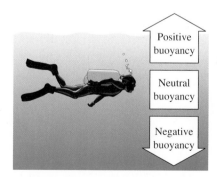

90. NEW ORLEANS The city of New Orleans, Louisiana, lies largely below sea level. Find out why the city is not under water.

REVIEW

91. Round 23,456 to the nearest hundred.

92. Evaluate: $19 - 2 \cdot 3$.

93. Subtract 2,081 from 2,842.

94. Divide 345 by 15.

95. Give the name of the property illustrated here:

$$(13 \cdot 2) \cdot 5 = 13 \cdot (2 \cdot 5)$$

96. Write four times five using three different notations.

2.2 Adding Integers

- Adding two integers with the same sign • Adding two integers with different signs
- The addition property of zero • The additive inverse of a number

A dramatic change in temperature occurred in 1943 in Spearfish, South Dakota. On January 22, at 7:30 A.M., the temperature was $-4°F$. In just two minutes, the temperature rose 49 degrees! To calculate the temperature at 7:32 A.M., we need to add 49 to -4.

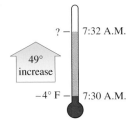

$$-4 + 49$$

To perform this addition, we must know how to add positive and negative integers. In this section, we develop rules to help us make such calculations.

Adding two integers with the same sign

$4 + 3$
both positive

To explain addition of signed numbers, we can use the number line. (See Figure 2-9.) To compute $4 + 3$, we begin at the **origin** (the point labeled 0) and draw an arrow 4 units long, pointing to the right. This represents positive 4. From that point, we draw an arrow 3 units long, pointing to the right, to represent positive 3. The second arrow points to the answer. Therefore, $4 + 3 = 7$.

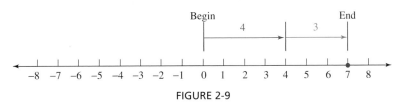

FIGURE 2-9

To check our work, let's think of the problem in terms of money. If you had $4 and earned $3 more, you would have a total of $7.

$-4 + (-3)$
both negative

To compute $-4 + (-3)$, we begin at the origin and draw an arrow 4 units long, pointing to the left. (See Figure 2-10.) This represents -4. From there, we draw an arrow 3 units long, pointing to the left, to represent -3. The second arrow points to the answer: $-4 + (-3) = -7$.

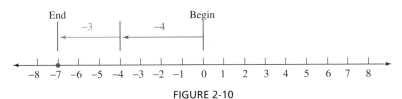

FIGURE 2-10

Let's think of this problem in terms of money. If you had a debt of $4 (negative 4) and then took on $3 more debt (negative 3), you would be in debt $7 (negative 7).

Here are some observations about the process of adding two numbers that have the same sign, using the number line.

- Both arrows point in the same direction and "build" on each other.
- The answer has the same sign as the two numbers being added.

$$4 \; + \; 3 \; = \; 7 \qquad\qquad -4 \; + \; (-3) \; = \; -7$$

Positive + positive = positive answer   Negative + negative = negative answer

These observations suggest the following rule.

> **Adding two integers with the same sign**
>
> To add two integers with the same sign, add their absolute values and attach their common sign to the sum. If both integers are positive, their sum is positive. If both integers are negative, their sum is negative.

! COMMENT When writing additions that involve signed numbers, write negative numbers within parentheses to separate the negative sign $-$ from the plus sign $+$.

$$9 + (-4) \qquad 9 + -4 \qquad \text{and} \qquad -9 + (-4) \qquad -9 + -4$$

EXAMPLE 1   Find the sum: $-9 + (-4)$.

Solution
Step 1: To add two integers with the same sign, we first add the absolute values of each of the integers. Since $|-9| = 9$ and $|-4| = 4$, we begin by adding 9 and 4.

$$9 + 4 = 13$$

Step 2: We then attach the common sign (which is negative) to this result. Therefore,

$$-9 + (-4) = -13$$

⌐— Make the answer negative.

After some practice, you will be able to do this kind of problem in your head. It will not be necessary to show all the steps as we have done here.

Self Check 2
Find the sum:
a. $7 + 5$

b. $-300 + (-100)$

EXAMPLE 2   Find the sum: **a.** $6 + 4$ and   **b.** $-80 + (-60)$.

Solution
a. Since both integers are positive, the answer is positive.

$$6 + 4 = 10$$

b. Since both integers are negative, the answer is negative.

$$-80 + (-60) = -140$$

Answers **a.** 12,   **b.** -400

■ Adding two integers with different signs

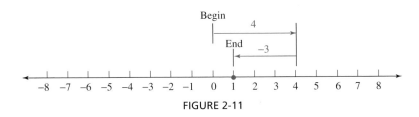

To compute 4 + (−3), we start at the origin and draw an arrow 4 units long, pointing to the right. (See Figure 2-11.) This represents positive 4. From there, we draw an arrow 3 units long, pointing to the left, to represent −3. The second arrow points to the answer: 4 + (−3) = 1.

FIGURE 2-11

In terms of money, if you had $4 (positive 4) and then took on a debt of $3 (negative 3), you would have $1 (positive 1).

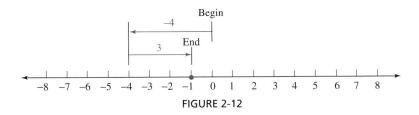

The problem −4 + 3 can be illustrated by drawing an arrow 4 units long from the origin, pointing to the left. (See Figure 2-12.) This represents −4. From there, we draw an arrow 3 units long, pointing to the right, to represent positive 3. The second arrow points to the answer: −4 + 3 = −1.

FIGURE 2-12

This problem can be thought of as owing $4 (negative 4) and then paying back $3 (positive 3). You will still owe $1 (negative 1).

The last two examples lead us to some observations about adding two integers with different signs, using the number line.

- The arrows representing the integers point in opposite directions.
- The longer of the two arrows determines the sign of the answer.

These observations suggest the following rule.

Adding two integers with different signs

To add two integers with different signs, subtract their absolute values, the smaller from the larger. Then attach to that result the sign of the integer with the larger absolute value.

EXAMPLE 3   Find the sum: $5 + (-7)$.

Solution

Step 1: To add two integers with different signs, we first subtract the smaller absolute value from the larger absolute value. Since $|5|$, which is 5, is smaller than $|-7|$, which is 7, we begin by subtracting 5 from 7.

$$7 - 5 = 2$$

Step 2: -7 has the larger absolute value, so we attach a negative sign to the result from step 1. Therefore,

$$5 + (-7) = -2$$

— Make the answer negative.

Self Check 4
Find the sum:
a. $-2 + 7$
b. $6 + (-9)$

EXAMPLE 4   Find the sum: **a.** $-8 + 5$ and   **b.** $11 + (-5)$.

Solution

a. Since -8 has the larger absolute value, the answer is negative.

$-8 + 5 = -3$ Because the signs of the numbers are different, subtract their absolute values, 5 from 8, to get 3. Attach a negative sign to that result.

b. Since 11 has the larger absolute value, the answer is positive.

$11 + (-5) = 6$ Subtract the absolute values, 5 from 11, to get 6. The answer is positive.

Answers a. 5,   **b.** -3

THINK IT THROUGH   Cash Flow

"College can be trial by fire — a test of how to cope with pressure, freedom, distractions, and a flood of credit card offers. It's easy to get into a cycle of overspending and unnecessary debt as a student." Planning for College, Wells Fargo Bank

If your income is less than your expenses, you have a *negative* cash flow. A negative cash flow can be a red flag that you should increase your income and/or reduce your expenses. Which of the following activities can increase income and which can decrease expenses?

- Buy generic or store-brand items.
- Get training and/or more education.
- Use your student ID to get discounts at stores, events, etc.
- Work more hours.
- Turn a hobby or skill into a money-making business.
- Tutor young students.
- Stop expensive habits, like smoking, buying snacks every day, etc.
- Attend free activities and free or discounted days at local attractions.
- Sell rarely used items, like an old CD player.
- Compare the prices of at least three products or at three stores before buying.

Based on the *Building Financial Skills* by National Endowment for Financial Education.

EXAMPLE 5 **Temperature change.** At the beginning of this section, we learned that at 7:30 A.M. on January 22, 1943, in Spearfish, South Dakota, the temperature was −4°. The temperature then rose 49 degrees in just two minutes. What was the temperature at 7:32 A.M.?

Solution The phrase *temperature rose 49 degrees* indicates addition. We need to add 49 to −4.

$$-4 + 49 = 45$$ Subtract the smaller absolute value, 4, from the larger absolute value, 49. The sum is positive.

At 7:32 A.M., the temperature was 45°F.

EXAMPLE 6 Add: $-3 + 5 + (-12) + 2$.

Solution This expression contains four integers. We add them, working from left to right.

$$\begin{aligned}
-3 + 5 + (-12) + 2 &= 2 + (-12) + 2 &&\text{Add: } -3 + 5 = 2.\\
&= -10 + 2 &&\text{Add: } 2 + (-12) = -10.\\
&= -8
\end{aligned}$$

Self Check 6
Add: $-12 + 8 + (-6) + 1$.

Answer −9

An alternative approach to problems like Example 6 is to add all the positive numbers, add all the negative numbers, and then add those results.

EXAMPLE 7 Find the sum: $-3 + 5 + (-12) + 2$.

Solution We can use the commutative property of addition to reorder the numbers and use the associative property of addition to group the positives together and the negatives together.

$$\begin{aligned}
-3 + 5 + (-12) + 2 &= \mathbf{5 + 2} + (\mathbf{-3}) + (\mathbf{-12}) &&\text{Reorder the numbers.}\\
&= (\mathbf{5 + 2}) + [(\mathbf{-3}) + (\mathbf{-12})] &&\text{Group the positives.}\\
& &&\text{Group the negatives.}
\end{aligned}$$

We perform the operations inside the grouping symbols first.

$$\begin{aligned}
(\mathbf{5 + 2}) + [(\mathbf{-3}) + (\mathbf{-12})] &= \mathbf{7} + (\mathbf{-15}) &&\text{Add the positives. Add the negatives.}\\
&= -8 &&\text{Add the numbers with different signs.}
\end{aligned}$$

Self Check 7
Find the sum:

$$-12 + 8 + (-6) + 1$$

Answer −9

EXAMPLE 8 Evaluate: $[-1 + (-5)] + (-7 + 5)$.

Solution By the rules for the order of operations, we must perform the operations within the grouping symbols first.

$$\begin{aligned}
[\mathbf{-1 + (-5)}] + (\mathbf{-7 + 5}) &= \mathbf{-6} + (\mathbf{-2}) &&\text{Perform the addition within the brackets.}\\
& &&\text{Perform the addition within parentheses.}\\
&= -8 &&\text{Add the numbers with the same sign.}
\end{aligned}$$

Self Check 8
Evaluate:

$$(-6 + -8) + [10 + (-17)].$$

Answer −21

    Entering negative numbers

Nigeria is the United States' second largest trading partner in Africa. To calculate the 2002 U.S. trade balance with Nigeria, we add the $1,057,700,000 worth of exports to Nigeria (considered positive) to the $5,945,300,000 worth of imports from Nigeria (considered negative). We can use a scientific calculator to perform the addition: 1,057,700,000 + (−5,945,300,000).

- We do not have to do anything special to enter a positive number. When we key in 1,057,700,000, a positive number is entered.
- To enter −5,945,300,000, we press the change-of-sign key $\boxed{+/-}$ *after entering* 5,945,300,000. Note that the change-of-sign key is different from the subtraction key $\boxed{-}$.

1057700000 $\boxed{+}$ 5945300000 $\boxed{+/-}$ $\boxed{=}$ $\boxed{\text{-4887600000}}$

In 2002, the United States had a trade balance of −$4,887,600,000 with Nigeria. Because the result is negative, it is called a *trade deficit.*

The addition property of zero

When 0 is added to a number, the number remains the same. For example, 5 + 0 = 5, and 0 + (−4) = −4. Because of this, we call 0 the **additive identity.**

> **Addition property of 0**
>
> The sum of any number and 0 is that number.

The additive inverse of a number

A second fact concerning 0 and the operation of addition can be demonstrated by considering the sum of a number and its opposite. To illustrate this, we use the number line in Figure 2-13 to add 6 and its opposite, −6. We see that 6 + (−6) = 0.

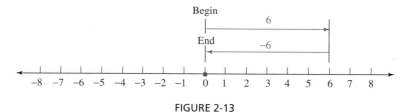

FIGURE 2-13

If the sum of two numbers is 0, the numbers are said to be **additive inverses** of each other. Since 6 + (−6) = 0, we say that 6 and −6 are additive inverses.

We can now classify a pair of numbers such as 6 and −6 in three ways: as opposites, negatives, or additive inverses.

> **The additive inverse of a number**
>
> Two numbers are said to be **additive inverses** if their sum is 0.

EXAMPLE 9   What is the additive inverse of −3? Justify your result.

Solution   The additive inverse of −3 is its opposite, 3. To justify the result, we add and show that the sum is 0.

$$-3 + 3 = 0$$

Self Check 9
What is the additive inverse of 12? Justify your result.

Answer −12; 12 + (−12) = 0

Section 2.2 STUDY SET

VOCABULARY   *Fill in the blanks.*

1. When 0 is added to a number, the number remains the same. We call 0 the additive _____.

2. Since −5 + 5 = 0, we say that 5 is the additive _____ of −5. We can also say that 5 and −5 are _____.

CONCEPTS   *Find each answer using the number line.*

3. −3 + 6
4. −3 + (−2)
5. −5 + 3
6. −1 + (−3)

7. **a.** Is the sum of two positive integers always positive?

   b. Is the sum of two negative integers always negative?

8. **a.** What is the sum of a number and its additive inverse?

   b. What is the sum of a number and its opposite?

9. Find each absolute value.

   a. |−7| **b.** |10|

10. If the sum of two numbers is 0, what can be said about the numbers?

Fill in the blanks.

11. To add two integers with unlike signs, _____ their absolute values, the smaller from the larger. Then attach to that result the sign of the number with the _____ absolute value.

12. To add two integers with like signs, add their _____ values and attach their common _____ to the sum.

NOTATION   *Complete each solution to evaluate the expression.*

13. −16 + (−2) + (−1) =  + (−1)

    =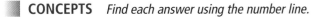

14. −8 + (−2) + 6 = ⬚ + 6

    = ⬚

15. (−3 + 8) + (−3) = ⬚ + (−3)

    = ⬚

16. −5 + [2 + (−9)] = −5 + (⬚)

    = ⬚

17. Explain why the expression −6 + −5 is not written correctly. How should it be written?

18. What mathematical symbol is suggested when the word *sum* is used?

PRACTICE   *Find the additive inverse of each number.*

19. −11   20. 9
21. −23   22. −43
23. 0   24. 1
25. 99   26. 250

Find each sum.

27. −6 + (−3)   28. −2 + (−3)
29. −5 + (−5)   30. −8 + (−8)
31. −6 + 7   32. −2 + 4
33. −15 + 8   34. −18 + 10
35. 20 + (−40)   36. 25 + (−10)
37. 30 + (−15)   38. 8 + (−20)
39. −1 + 9   40. −2 + 7
41. −7 + 9   42. −3 + 6
43. 5 + (−15)   44. 16 + (−26)
45. 24 + (−15)   46. −4 + 14
47. 35 + (−27)   48. 46 + (−73)
49. 24 + (−45)   50. −65 + 31

Evaluate each expression.

51. $-2 + 6 + (-1)$

52. $4 + (-3) + (-2)$

53. $-9 + 1 + (-2)$

54. $5 + 4 + (-6)$

55. $6 + (-4) + (-13) + 7$

56. $8 + (-5) + (-10) + 6$

57. $9 + (-3) + 5 + (-4)$

58. $-3 + 7 + 1 + (-4)$

59. Find the sum of -6, -7, and -8.

60. Find the sum of -11, -12, and -13.

Find each sum.

61. $-7 + 0$ **62.** $6 + 0$

63. $9 + 0$ **64.** $0 + (-15)$

65. $-4 + 4$ **66.** $18 + (-18)$

67. $2 + (-2)$ **68.** $-10 + 10$

69. What number must be added to -5 to obtain 0?

70. What number must be added to 8 to obtain 0?

Evaluate each expression.

71. $2 + (-10 + 8)$

72. $(-9 + 12) + (-4)$

73. $(-4 + 8) + (-11 + 4)$

74. $(-12 + 6) + (-6 + 8)$

75. $[-3 + (-4)] + (-5 + 2)$

76. $[9 + (-10)] + (-7 + 9)$

77. $[6 + (-4)] + [8 + (-11)]$

78. $[5 + (-8)] + [9 + (-15)]$

79. $-2 + [-8 + (-7)]$

80. $-8 + [-5 + (-2)]$

81. $789 + (-9,135)$

82. $2,701 + (-4,089)$

83. $-675 + (-456) + 99$

84. $-9,750 + (-780) + 2,345$

▮ **APPLICATIONS**   *Use signed numbers to help answer each question.*

85. G FORCES As a fighter pilot dives and loops, different forces are exerted on the body, just like the forces you experience when riding on a roller coaster. Some of the forces, called G's, are positive and some are negative. The force of gravity, 1G, is constant. Complete the diagram in the next column.

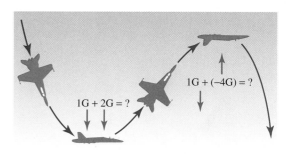

1G + 2G = ?

1G + (−4G) = ?

86. CHEMISTRY The first several steps of a chemistry lab experiment are listed here. The experiment begins with a compound that is stored at $-40°F$.

Step 1: Raise the temperature of the compound 200°.

Step 2: Add sulfur and then raise the temperature 10°.

Step 3: Add 10 milliliters of water, stir, and raise the temperature 25°.

What is the resulting temperature of the mixture after step 3?

87. CASH FLOW The maintenance costs, utilities, and taxes on a duplex are $900 per month. The owner of the apartments receives monthly rental payments of $450 and $380. Does this investment produce a positive cash flow each month?

88. JOGGING A businessman's lunchtime workout includes jogging up ten stories of stairs in his high-rise office building. If he starts on the fourth level below ground in the underground parking garage, on what story of the building will he finish his workout?

89. HEALTH Find the point total for the six risk factors (in blue) on the medical questionnaire below. Then use the table at the bottom of the form to determine the risk of contracting heart disease for the man whose responses are shown.

Age		Total Cholesterol	
Age	Points	Reading	Points
35	−4	280	3
Cholesterol		**Blood Pressure**	
HDL	Points	Systolic/Diastolic	Points
62	−3	124/100	3
Diabetic		**Smoker**	
	Points		Points
Yes	4	Yes	2

10-Year Heart Disease Risk			
Total Points	**Risk**	Total Points	**Risk**
−2 or less	1%	5	4%
−1 to 1	2%	6	6%
2 to 3	3%	7	6%
4	4%	8	7%

Source: National Heart, Lung, and Blood Institute

90. SPREADSHEETS Monthly rain totals for four counties are listed in the spreadsheet shown below. The −1 entered in cell B1 means that the rain total for Suffolk County for a certain month was 1 inch below average. We can analyze this data by asking the computer to perform various operations.

  a. To ask the computer to add the numbers in cells C1, C2, C3, and C4, we type SUM(C1:C4). Find this sum.

  b. Find SUM(B4:F4).

	A	B	C	D	E	F
1	Suffolk	−1	−1	0	+1	+1
2	Marin	0	−2	+1	+1	−1
3	Logan	−1	+1	+2	+1	+1
4	Tipton	−2	−2	+1	−1	−3

91. ATOMS An atom is composed of protons, neutrons, and electrons. A proton has a positive charge (represented by +1), a neutron has no charge, and an electron has a negative charge (−1). Two simple models of atoms are shown. What is the net charge of each atom?

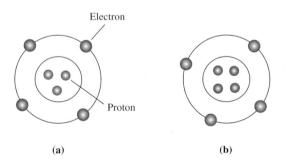

(a)   (b)

92. POLITICAL POLLS Six months before a general election, the incumbent senator found himself trailing the challenger by 18 points. To overtake his opponent, the campaign staff decided to use a four-part strategy. Each part of this plan is shown below, with the anticipated point gain.

  1. Intense TV ad blitz   +10
  2. Ask for union endorsement   +2
  3. Voter mailing   +3
  4. Get-out-the-vote campaign   +1

With these gains, will the incumbent overtake the challenger on election day?

93. FLOODING After a heavy rainstorm, a river that had been 4 feet under flood stage rose 11 feet in a 48-hour period. Find the height of the river after the storm in comparison to flood stage.

94. MILITARY SCIENCE During a battle, an army retreated 1,500 meters, regrouped, and advanced 3,500 meters. The next day, it had to retreat 1,250 meters. Find the army's net gain.

95. AIRLINES The graph below shows the annual net income for Delta Air Lines during the years 2000–2003. Estimate the company's total net income over this span of four years.

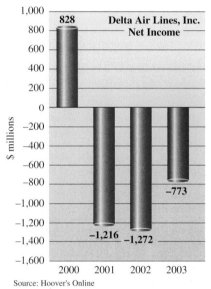

Source: Hoover's Online

96. ACCOUNTING On a financial balance sheet, debts (considered negative numbers) are denoted within parentheses. Assets (considered positive numbers) are written without parentheses. What is the 2004 fund balance for the preschool whose financial records are shown?

Community Care Preschool Balance Sheet, June 2004	
Fund balances	
Classroom supplies	$ 5,889
Emergency needs	927
Holiday program	(2,928)
Insurance	1,645
Janitorial	(894)
Licensing	715
Maintenance	(6,321)
BALANCE	?

WRITING

97. Is the sum of a positive and a negative number always positive? Explain why or why not.

98. How do you explain the fact that when asked to *add* −4 and 8, we must actually *subtract* to obtain the result?

99. Why is the sum of two negative numbers a negative number?

100. Write an application problem that will require adding −50 and −60.

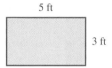 **REVIEW**

101. Find the area of the rectangle shown below.

5 ft

3 ft

102. A car with a tank that holds 15 gallons of gasoline goes 25 miles on 1 gallon. How far can it go on a full tank?

103. Find the perimeter of the rectangle in Exercise 101.

104. What property is illustrated by the statement
$5 \cdot 15 = 15 \cdot 5$?

105. Prime factor 125. Use exponents to express the result.

106. Perform the division: $\dfrac{144}{12}$.

2.3 Subtracting Integers

- Adding the opposite • Order of operations • Applications of subtraction

In this section, we study another way to think about subtraction. This new procedure is helpful when subtraction problems involve negative numbers.

Adding the opposite

The subtraction problem $6 - 4$ can be thought of as taking away 4 from 6. We can use the number line to illustrate this. (See Figure 2-14.) Beginning at the origin, we draw an arrow of length 6 units in the positive direction. From that point, we move back 4 units to the left. The answer, called the **difference,** is 2.

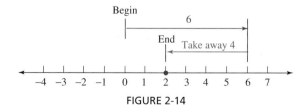

FIGURE 2-14

The work shown in Figure 2-14 looks like the illustration for the *addition* problem $6 + (-4) = 2$, shown in Figure 2-15.

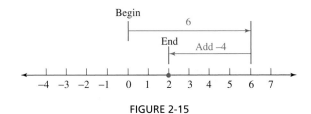

FIGURE 2-15

In the first problem, $6 - 4$, we subtracted 4 from 6. In the second, $6 + (-4)$, we added -4 (which is the opposite of 4) to 6. In each case, the result was 2.

Subtracting   Adding the opposite of 4
    ↓   ↓
$6 - 4 = 2$ $6 + (-4) = 2$

The same result

This observation helps to justify the following rule for subtraction.

Rule for subtraction

To subtract two integers, add the opposite of the second integer to the first integer.

In words, this rule says that *subtraction is the same as adding the opposite of the number to be subtracted.*

You won't need to use this rule for every subtraction problem. For example, $6 - 4$ is obviously 2; it does not need to be rewritten as adding the opposite. But for more complicated problems such as $-6 - 4$ or $3 - (-5)$, where the result is not obvious, the subtraction rule will be quite helpful.

EXAMPLE 1   Subtract: $-6 - 4$.

Solution   The number to be subtracted is 4. Applying the subtraction rule, we write

$-6 - 4 = -6 + (-4)$ Write the subtraction as an addition of the opposite of 4, which is -4. Write -4 within parentheses.

$\qquad\quad = -10$ To add -6 and -4, apply the rule for adding two negative numbers.

To check the result, we add the difference, -10, and the subtrahend, 4. We should obtain the minuend, -6.

$-10 + 4 = -6$

The answer, -10, checks.

Self Check 1

Find $-2 - 3$ and check the result.

Answer -5

EXAMPLE 2   Subtract: $3 - (-5)$.

Solution   The number being subtracted is -5.

$3 - (-5) = 3 + 5$ Write the subtraction as an addition of the opposite of -5, which is 5.

$\qquad\quad = 8$ Perform the addition.

Self Check 2

Subtract: $3 - (-2)$.

Answer 5

Self Check 3
Subtract −8 from −3.

EXAMPLE 3   Subtract −3 from −8.

Solution   The number being subtracted is −3, so we write it after −8.

$$-8 - (-3) = -8 + 3 \quad \text{Add the opposite of } -3, \text{ which is } 3.$$
$$= -5 \qquad \text{Perform the addition.}$$

Answer 5

Remember that any subtraction problem can be rewritten as an equivalent addition. We just add the opposite of the number that is to be subtracted.

Subtraction can be written as addition . . .

$$4 - 8 = 4 + (-8) = -4$$
$$4 - (-8) = 4 + 8 = 12$$
$$-4 - 8 = -4 + (-8) = -12$$
$$-4 - (-8) = -4 + 8 = 4$$

of the opposite of the
number to be subtracted.

▌ Order of operations

Expressions can contain repeated subtraction or subtraction in combination with grouping symbols. To work these problems, we apply the rules for the order of operations, listed on page 58.

Self Check 4
Evaluate: −3 − 5 − (−1).

EXAMPLE 4   Evaluate: −1 − (−2) − 10.

Solution   This problem involves two subtractions. We work from left to right, rewriting each subtraction as an addition of the opposite.

$$-1 - (-2) - 10 = -1 + 2 + (-10) \quad \text{Add the opposite of } -2, \text{ which is } 2. \text{ Add the}$$
$$\text{opposite of } 10, \text{ which is } -10. \text{ Write } -10 \text{ in}$$
$$\text{parentheses.}$$
$$= 1 + (-10) \qquad \text{Work from left to right. Add } -1 + 2.$$
$$= -9 \qquad \text{Perform the addition.}$$

Answer −7

Self Check 5
Evaluate: −2 − (−6 − 5).

EXAMPLE 5   Evaluate: −8 − (−2 − 2).

Solution   We must perform the subtraction within the parentheses first.

$$-8 - (-2 - 2) = -8 - [-2 + (-2)] \quad \text{Add the opposite of } 2, \text{ which is } -2.$$
$$\text{Since } -2 \text{ must be written within}$$
$$\text{parentheses, we write } -2 + (-2) \text{ within}$$
$$\text{brackets.}$$
$$= -8 - (-4) \qquad \text{Add } -2 \text{ and } -2. \text{ Since only one set of}$$
$$\text{grouping symbols is now needed, we}$$
$$\text{write } -4 \text{ within parentheses.}$$
$$= -8 + 4 \qquad \text{Add the opposite of } -4, \text{ which is } 4.$$
$$= -4 \qquad \text{Perform the addition.}$$

Answer 9

▮ Applications of subtraction

Things are constantly changing in our daily lives. The temperature, the amount of money we have in the bank, and our ages are examples. In mathematics, the operation of subtraction is used to measure change.

❗ **COMMENT** In general, to find the change in a quantity, we subtract the earlier value from the later value.

EXAMPLE 6 **Change of water level.**
On Monday, the water level in a city storage tank was 6 feet above normal. By Friday, the level had fallen to a mark 4 feet below normal. Find the change in the water level from Monday to Friday. (See Figure 2-16.)

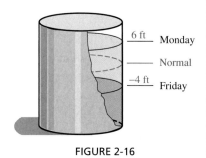

6 ft — Monday

— Normal

−4 ft — Friday

FIGURE 2-16

Solution We use subtraction to find the amount of change. The water levels of 4 feet below normal (the later value) and 6 feet above normal (the earlier value) can be represented by −4 and 6, respectively.

The water level Friday	minus	the water level Monday	is	the change in the water level.

$-4 - 6 = -4 + (-6)$ Add the opposite of 6, which is −6.

$ = -10$ Perform the addition. The negative result indicates that the water level fell.

The water level fell 10 feet from Monday to Friday.

In the next example, the number line serves as a mathematical model of a real-life situation. You will see how the operation of subtraction can be used to find the distance between two points on the number line.

EXAMPLE 7 **Artillery.** In a practice session, an artillery group fired two rounds at a target. The first landed 65 yards short of the target, and the second landed 50 yards past it. (See Figure 2-17.) How far apart were the two impact points?

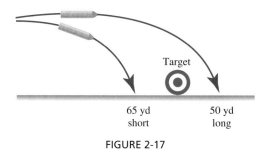

Target

65 yd
short

50 yd
long

FIGURE 2-17

Solution We can use the number line to model this situation. The target is the origin. The words *short of the target* indicate a negative number, and the words *past it* indicate a positive number. Therefore, we graph the impact points at −65 and 50 in Figure 2-18.

$$\text{Short} \qquad\qquad \text{Target} \qquad\qquad \text{Long}$$
$$-65 \qquad\qquad 0 \qquad\qquad 50$$

FIGURE 2-18

The phrase *how far apart* tells us to subtract.

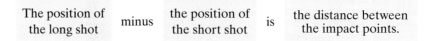

| The position of the long shot | minus | the position of the short shot | is | the distance between the impact points. |

$$50 - (-65) = 50 + 65 \qquad \text{Add the opposite of } -65.$$
$$ = 115 \qquad \text{Perform the addition.}$$

The impact points are 115 yards apart.

CALCULATOR SNAPSHOT

Subtraction with negative numbers

The world's highest peak is Mount Everest in the Himalayas. The greatest ocean depth yet measured lies in the Mariana Trench near the island of Guam in the western Pacific. (See Figure 2-19) To find the range between the highest peak and the greatest depth, we must subtract.

$$29,035 - (-36,025)$$

To perform this subtraction using a scientific calculator, we enter these numbers and press these keys.

29035 $\boxed{-}$ 36025 $\boxed{+/-}$ $\boxed{=}$ $\boxed{\text{65060}}$

The range is 65,060 feet between the highest peak and the lowest depth. We could have applied the subtraction rule to write $29,035 - (-36,025)$ as $29,035 + 36,025$ before using the calculator.

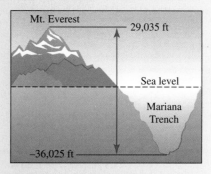

FIGURE 2-19

Section 2.3 STUDY SET

VOCABULARY *Fill in the blanks.*

1. The answer to a subtraction problem is called the
   _____.

2. Two numbers that are the same distance from 0 on
   the number line, but on opposite sides of it, are
   called _____.

CONCEPTS *Fill in the blanks.*

3. Subtraction is the same as _____ the _____ of
   the number to be subtracted.

4. Subtracting 3 is the same as adding ____.

5. Subtracting -6 is the same as adding ____.

6. The opposite of -8 is ____.

7. $9 - 5 = 9 \;\boxed{}\; (-5)$

8. **a.** $2 - 7 = 2 + \boxed{}$

   b. $2 - (-7) = 2 + \boxed{}$

   c. $-2 - 7 = -2 + \boxed{}$

   d. $-2 - (-7) = -2 + \boxed{}$

9. After using parentheses as grouping symbols, if
   another set of grouping symbols is needed, we use
   _____.

10. We can find the _____ in a quantity by subtracting
    the earlier value from the later value.

11. Write this problem using mathematical symbols:
    negative eight minus negative four.

12. Write this problem using mathematical symbols:
    negative eight subtracted from negative four.

13. Find the distance between -4 and 3 on the number
    line.

14. Find the distance between -10 and 1 on the number
    line.

15. Is subtracting 3 from 8 the same as subtracting 8
    from 3? Explain.

16. Evaluate each expression.

    a. $-2 - 0$ **b.** $0 - (-2)$

NOTATION *Complete each solution to evaluate each
expression.*

17. $1 - 3 - (-2) = 1 + (\boxed{}) + 2$
    $$= -2 + \boxed{}$$
    $$= \boxed{}$$

18. $-6 + 5 - (-5) = -6 + 5 + \boxed{}$
    $$= \boxed{} + 5$$
    $$= \boxed{}$$

19. $(-8 - 2) - (-6) = [-8 + (\boxed{})] - (-6)$
    $$= \boxed{} - (-6)$$
    $$= -10 + \boxed{}$$
    $$= \boxed{}$$

20. $-5 - (-1 - 4) = -5 - [-1 + (\boxed{})]$
    $$= -5 - (\boxed{})$$
    $$= -5 + \boxed{}$$
    $$= \boxed{}$$

PRACTICE *Find each difference.*

21. $8 - (-1)$ 22. $3 - (-8)$
23. $-4 - 9$ 24. $-7 - 6$
25. $-5 - 5$ 26. $-7 - 7$
27. $-5 - (-4)$ 28. $-9 - (-1)$
29. $-1 - (-1)$ 30. $-4 - (-3)$
31. $-2 - (-10)$ 32. $-6 - (-12)$
33. $0 - (-5)$ 34. $0 - 8$
35. $0 - 4$ 36. $0 - (-6)$
37. $-2 - 2$ 38. $-3 - 3$
39. $-10 - 10$ 40. $4 - 4$
41. $9 - 9$ 42. $4 - (-4)$
43. $-3 - (-3)$ 44. $-5 - (-5)$

Evaluate each expression.

45. $-4 - (-4) - 15$ 46. $-3 - (-3) - 10$
47. $-3 - 3 - 3$ 48. $-1 - 1 - 1$
49. $5 - 9 - (-7)$ 50. $6 - 8 - (-4)$
51. $10 - 9 - (-8)$ 52. $16 - 14 - (-9)$
53. $-1 - (-3) - 4$ 54. $-2 - 4 - (-1)$
55. $-5 - 8 - (-3)$ 56. $-6 - 5 - (-1)$
57. $(-6 - 5) - 3$ 58. $(-2 - 1) - 5$
59. $(6 - 4) - (1 - 2)$ 60. $(5 - 3) - (4 - 6)$
61. $-9 - (6 - 7)$ 62. $-3 - (6 - 12)$
63. $-8 - [4 - (-6)]$ 64. $-1 - [5 - (-2)]$
65. $[-4 + (-8)] - (-6)$ 66. $[-5 + (-4)] - (-2)$

67. Subtract −3 from 7.   **68.** Subtract 8 from −2.

69. Subtract −6 from −10.   **70.** Subtract −4 from −9.

Use a calculator to perform each subtraction.

71. −1,557 − 890

72. −345 − (−789)

73. 20,007 − (−496)

74. −979 − (−44,879)

75. −162 − (−789) − 2,303

76. −787 − 1,654 − (−232)

APPLICATIONS   *Use signed numbers to help answer each question.*

77. SCUBA DIVING A diver jumps from his boat into the water and descends 50 feet. He pauses to check his equipment and then descends an additional 70 feet. Use a signed number to represent the diver's final depth.

78. TEMPERATURE CHANGE Rashawn flew from his New York home to Hawaii for a week of vacation. He left blizzard conditions and a temperature of −6°, and stepped off the airplane into 85° weather. What temperature change did he experience?

79. READING PROGRAMS In a state reading test administered at the start of a school year, an elementary school's performance was 23 points below the county average. The principal immediately began a special tutorial program. At the end of the school year, retesting showed the students to be only 7 points below the average. How many points did the school's reading score improve over the year?

80. SUBMARINES A submarine was traveling 2,000 feet below the ocean's surface when the radar system warned of an impending collision with another sub. The captain ordered the navigator to dive an additional 200 feet and then level off. Find the depth of the submarine after the dive.

81. AMPERAGE During normal operation, the ammeter on a car reads +5. If the headlights, which draw a current of 7 amps, and the radio, which draws a current of 6 amps, are both turned on, what number will the ammeter register?

82. GIN RUMMY After a losing round, a card player must subtract the value of each of the cards left in his hand from his previous point total of 21. If face cards are counted as 10 points, what is his new score?

83. GEOGRAPHY Death Valley, California, is the lowest land point in the United States, at 282 feet below sea level. The lowest land point on the Earth is the Dead Sea, which is 1,348 feet below sea level. How much lower is the Dead Sea than Death Valley?

84. LIE DETECTOR TESTS On one lie detector test, a burglar scored −18, which indicates deception. However, on a second test, he scored −1, which is inconclusive. Find the difference in the scores.

85. FOOTBALL A college football team records the outcome of each of its plays during a game on a stat sheet. Find the net gain (or loss) after the 3rd play.

Down	Play	Result
1st	Run	Lost 1 yd
2nd	Pass — sack!	Lost 6 yd
Penalty	Delay of game	Lost 5 yd
3rd	Pass	Gained 8 yd
4th	Punt	—

86. ACCOUNTING Complete the balance sheet below. Then determine the overall financial condition of the company by subtracting the total liabilities from the total assets.

W a l k e r C o r p o r a t i o n				
Balance Sheet 2001				
Assets				
Cash	$11	1	0	9
Supplies	7	8	6	2
Land	67	5	4	3
Total assets	$			
Liabilities				
Accounts payable	$79	0	3	7
Income taxes	20	1	8	1
Total liabilities	$			

87. DIVING A diver jumps from a platform. After she hits the water, her momentum takes her to the bottom of the pool.

   a. Use the number line and signed numbers to model this situation. Show the top of the platform, the water level as 0, and the bottom of the pool.

      ⟵————————————⟶

   b. Find the total length of the dive from the top of the platform to the bottom of the pool.

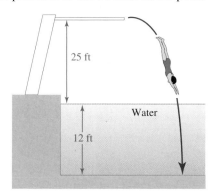

88. TEMPERATURE EXTREMES The highest and lowest temperatures ever recorded in several cities are shown below. List the cities in order, from the largest to smallest range in temperature extremes.

Extreme temperatures		
City	**Highest**	**Lowest**
Atlantic City, NJ	106	−11
Barrow, AK	79	−56
Kansas City, MO	109	−23
Norfolk, VA	104	−3
Portland, ME	103	−39

89. CHECKING ACCOUNTS Michael has $1,303 in his checking account. Can he pay his car insurance premium of $676, his utility bills of $121, and his rent of $750 without having to make another deposit? Explain.

90. HISTORY Two of the greatest Greek mathematicians were Archimedes (287–212 B.C.) and Pythagoras (569–500 B.C.). How many years apart were they born?

▮ WRITING

91. Explain what is meant when we say that subtraction is the same as addition of the opposite.

92. Give an example showing that it is possible to subtract something from nothing.

93. Explain how to check the result: $-7 - 4 = -11$.

94. Explain why students don't need to change every subtraction they encounter to an addition of the opposite. Give some examples.

▮ REVIEW

95. Round 5,989 to the nearest ten.

96. Round 5,999 to the nearest hundred.

97. List the factors of 20.

98. It takes 13 oranges to make one can of orange juice. Find the number of oranges used to make 12 cans.

99. Evaluate: $12^2 - (5 - 4)^2$.

100. What property does the following illustrate?

$$15 + 12 = 12 + 15$$

2.4 Multiplying Integers

- Multiplying two positive integers • Multiplying a positive and a negative integer
- Multiplying a negative and a positive integer • Multiplying by zero
- Multiplying two negative integers • Powers of integers

We now turn our attention to multiplication of integers. When we multiply two nonzero integers, the first factor can be positive or negative. The same is true for the second factor. This means that there are four possible combinations to consider.

Positive · positive	Positive · negative
Negative · positive	Negative · negative

In this section, we discuss these four combinations and use our observations to establish rules for multiplying two integers.

Multiplying two positive integers

4(3)
like signs
both positive

We begin by considering the product of two positive integers, 4(3). Since both factors are positive, we say that they have *like* signs. In Chapter 1, we learned that multiplication is repeated addition. Therefore, 4(3) represents the sum of four 3's.

$4(3) = 3 + 3 + 3 + 3$ Multiplication is repeated addition. Write 3 as an addend four times.

$4(3) = 12$ The result is 12, which is a positive number.

This result suggests that *the product of two positive integers is positive.*

Multiplying a positive and a negative integer

4(−3)
unlike signs
one positive, one negative

Next, we consider 4(−3). This is the product of a positive and a negative integer. The signs of these factors are *unlike*. According to the definition of multiplication, 4(−3) means that we are to add −3 four times.

$4(-3) = (-3) + (-3) + (-3) + (-3)$ Use the definition of multiplication. Write −3 as an addend four times.

$4(-3) =\ \ \ \ (-6) + (-3) + (-3)$ Work from left to right. Apply the rule for adding two negative numbers.

$4(-3) =\ \ \ \ (-9) + (-3)$ Work from left to right. Apply the rule for adding two negative numbers.

$4(-3) =\ \ \ \ \ \ \ \ -12$ Perform the addition.

This result is −12, which suggests that *the product of a positive integer and a negative integer is negative.*

Multiplying a negative and a positive integer

−3(4)
unlike signs
one negative, one positive

To develop a rule for multiplying a negative and a positive integer, we consider −3(4). Notice that the factors have *unlike* signs. Because of the commutative property of multiplication, the answer to −3(4) will be the same as the answer to 4(−3). We know that 4(−3) = −12 from the previous discussion, so −3(4) = −12. This suggests that *the product of a negative integer and a positive integer is negative.*

Putting the results of the last two cases together leads us to the rule for multiplying two integers with unlike signs.

Multiplying two integers with unlike signs

To multiply a positive integer and a negative integer, or a negative integer and a positive integer, multiply their absolute values. Then make the answer negative.

EXAMPLE 1   Find each product: **a.** $7(-5)$,   **b.** $20(-8)$,   and   **c.** $-8 \cdot 5$.

Solution   To multiply integers with unlike signs, we multiply their absolute values and make the product negative.

a. $7(-5) = -35$ Multiply the absolute values, 7 and 5, to get 35. Then make the answer negative.

b. $20(-8) = -160$ Multiply the absolute values, 20 and 8, to get 160. Then make the answer negative.

c. $-8 \cdot 5 = -40$ Multiply the absolute values, 8 and 5, to get 40. Then make the answer negative.

Self Check 1
Find each product:
a. $2(-6)$
b. $30(-2)$
c. $-15 \cdot 2$

Answers **a.** -12, **b.** -60, **c.** -30

! COMMENT   When writing multiplication involving signed numbers, do not write a negative sign – next to a raised dot · (the multiplication symbol). Instead, use parentheses to show the multiplication.

$$6(-2) \qquad 6 \cdot -2 \qquad \text{and} \qquad -6(-2) \qquad -6 \cdot -2$$

▮ Multiplying by zero

Before we can develop a rule for multiplying two negative integers, we need to examine multiplication by 0. If $4(3)$ means that we are to find the sum of four 3's, then $0(-3)$ means that we are to find the sum of zero -3's. Obviously, the sum is 0. Thus, $0(-3) = 0$.

The commutative property of multiplication guarantees that we can change the order of the factors in the multiplication problem without affecting the result.

$$(-3)(0) = 0(-3) \quad = \quad 0$$
$$\uparrow \qquad\qquad \uparrow \qquad\qquad \uparrow$$

Change the order   The result is
of the factors.   still 0.

We see that the order in which we write the factors 0 and -3 doesn't matter — their product is 0. This example suggests that the product of any number and 0 is 0.

> **Multiplying by 0**
>
> The product of any integer and 0 is 0.

EXAMPLE 2   Find $-12 \cdot 0$.

Solution   Since the product of any number and 0 is 0, we have

$$-12 \cdot 0 = 0$$

Self Check 2
Find $0(-56)$.

Answer 0

▮ Multiplying two negative integers

$-3(-4)$
like signs
both negative

To develop a rule for multiplying two negative integers, we consider the pattern displayed on the next page. There, we multiply -4 by a series of factors that decrease by 1. After determining each product, we graph each product on the number line (Figure 2-20). See if you can determine the answers to the last three multiplication problems by examining the pattern of answers leading up to them.

This factor decreases
by 1 as you read down   Look for a
the column.   ↓   pattern here.
                             ↓

$$4(-4) = -16$$
$$3(-4) = -12$$
$$2(-4) = -8$$
$$1(-4) = -4$$
$$0(-4) = 0$$
$$-1(-4) = ?$$
$$-2(-4) = ?$$
$$-3(-4) = ?$$

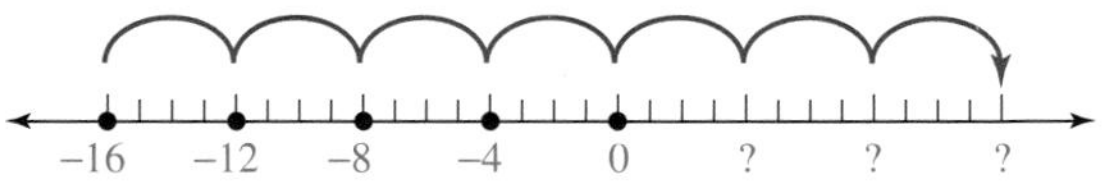

FIGURE 2-20

From the pattern, we see that

$$-1(-4) = 4$$
$$-2(-4) = 8$$
$$-3(-4) = 12$$
    ↑   ↑   ↑

For two negative factors,   the product is a positive.

These results suggest that *the product of two negative integers is positive.* Earlier in this section, we saw that the product of two positive integers is also positive. This leads to the following conclusion.

Multiplying two integers with like signs

To multiply two positive integers, or two negative integers, multiply their absolute values. The answer is positive.

Self Check 3
Find each product:
a. $-9(-7)$
b. $-12(-2)$

Answers **a.** 63, **b.** 24

EXAMPLE 3   Find each product: **a.** $-5(-9)$ and **b.** $-8(-10)$.

Solution   To multiply two negative integers, we multiply their absolute values and make the result positive.

a. $-5(-9) = 45$ Multiply the absolute values, 5 and 9, to get 45. The answer is positive.

b. $-8(-10) = 80$ Multiply the absolute values, 8 and 10, to get 80. The answer is positive.

We now summarize the rules for multiplying two integers.

Multiplying two integers

To multiply two integers, multiply their absolute values.

1. The product of two integers with *like* signs is positive.
2. The product of two integers with *unlike* signs is negative.

We can use a calculator to multiply signed numbers.

Multiplication with negative numbers   CALCULATOR SNAPSHOT

At Thanksgiving time, a large supermarket chain offered customers a free turkey with every grocery purchase of $100 or more. Each turkey cost the store $8, and 10,976 people took advantage of the offer. Since each of the 10,976 turkeys given away represented a loss of $8 (which can be expressed as -8 dollars), the company lost $10,976(-8)$ dollars. To find this product, we enter these numbers and press these keys on a scientific calculator.

$$10976 \boxed{\times} 8 \boxed{+/-} \boxed{=} \qquad\qquad \boxed{-87808}$$

The negative result indicates that with this promotion, the supermarket chain gave away $87,808 in turkeys.

EXAMPLE 4 Multiply: **a.** $-3(-2)(-6)(-5)$ and **b.** $-2(-4)(-5)$.

Solution

a. $-3(-2)(-6)(-5) = 6(-6)(-5)$ Work from left to right: $-3(-2) = 6$.

$\qquad\qquad\qquad\qquad = -36(-5)$ Work from left to right: $6(-6) = -36$.

$\qquad\qquad\qquad\qquad = 180$

b. $-2(-4)(-5) = 8(-5)$ Work from left to right: $-2(-4) = 8$.

$\qquad\qquad\qquad = -40$

> **Self Check 4**
> Multiply: **a.** $-1(-2)(-5)$ and **b.** $-2(-7)(-1)(-2)$.
>
> Answers **a.** -10, **b.** 28

Example 4, part a, illustrates that a product is positive when there is an even number of negative factors. Part b illustrates that a product is negative when there is an odd number of negative factors.

▌ Powers of integers

Recall that exponential expressions are used to represent repeated multiplication. For example, 2 to the third power, or 2^3, is a shorthand way of writing $2 \cdot 2 \cdot 2$. In this expression, 3 is the exponent and the base is positive 2. In the next example, we evaluate exponential expressions with bases that are negative numbers.

EXAMPLE 5 Find each power: **a.** $(-2)^4$ and **b.** $(-5)^3$.

Solution

a. $(-2)^4 = (-2)(-2)(-2)(-2)$ Write -2 as a factor 4 times.

$\qquad\quad = 4(-2)(-2)$ Work from left to right. Multiply -2 and -2 to get 4.

$\qquad\quad = -8(-2)$ Work from left to right. Multiply 4 and -2 to get -8.

$\qquad\quad = 16$ Perform the multiplication.

b. $(-5)^3 = (-5)(-5)(-5)$ Write -5 as a factor 3 times.

$\qquad\quad = 25(-5)$ Work from left to right. Multiply -5 and -5 to get 25.

$\qquad\quad = -125$ Perform the multiplication.

> **Self Check 5**
> Find each power:
> **a.** $(-3)^4$
> **b.** $(-4)^3$
>
> Answers **a.** 81, **b.** -64

In Example 5, part a, -2 was raised to an even power, and the answer was positive. In part b, another negative number, -5, was raised to an odd power, and the answer was negative. These results suggest a general rule.

> **Even and odd powers of a negative integer**
>
> When a negative integer is raised to an even power, the result is positive.
>
> When a negative integer is raised to an odd power, the result is negative.

Self Check 6
Find the power: $(-1)^8$.

EXAMPLE 6   Find the power: $(-1)^5$.

Solution   We have a negative integer raised to an odd power. The result will be negative.

$$(-1)^5 = (-1)(-1)(-1)(-1)(-1)$$
$$= -1$$

Answer 1

! COMMENT Although the expressions -3^2 and $(-3)^2$ look similar, they are not the same. In -3^2, the base is 3 and the exponent 2. The $-$ sign in front of 3^2 means the opposite of 3^2. In $(-3)^2$, the base is -3 and the exponent is 2. When we evaluate them, it becomes clear that they are not equivalent.

-3^2 represents *the opposite of* 3^2		$(-3)^2$ represents $(-3)(-3)$	
$-3^2 = -(3 \cdot 3)$	Write 3 as a factor 2 times.	$(-3)^2 = (-3)(-3)$	Write -3 as a factor 2 times.
$= -9$	Multiply within the parentheses first.	$= 9$	The product of two negative numbers is positive.

Notice that the results are different.

Self Check 7
Evaluate: **a.** -4^2 and
b. $(-4)^2$.

EXAMPLE 7   Evaluate: **a.** -2^2 and   **b.** $(-2)^2$.

Solution

a. $-2^2 = -(2 \cdot 2)$ Since 2 is the base, write 2 as a factor two times.

$= -4$ Perform the multiplication within the parentheses.

b. $(-2)^2 = (-2)(-2)$ The base is -2. Write it as a factor twice.

$= 4$ The signs are like, so the product is positive.

Answers a. -16,   **b.** 16

CALCULATOR SNAPSHOT   **Raising a negative number to a power**

Negative numbers can be raised to a power using a scientific calculator. We use the change-of-sign key $\boxed{+/-}$ and the power key $\boxed{y^x}$ (on the some calculators, $\boxed{x^y}$). For example, to evaluate $(-5)^6$, we enter these numbers and press these keys.

$5 \boxed{+/-} \boxed{y^x} 6 \boxed{=}$ $\boxed{15625}$

The result is 15,625.

Some scientific calculators require parentheses when entering a negative base raised to a power.

Section 2.4   STUDY SET

VOCABULARY   *Fill in the blanks.*

1. In the multiplication $-5(-4)$, the integers -5 and -4, which are being multiplied, are called _____. The answer, 20, is called the _____.

2. The definition of multiplication tells us that $3(-4)$ represents repeated _____ $-4 + (-4) + (-4)$.

3. In the expression -3^5,   is the base and 5 is the _____.

4. In the expression $(-3)^5$,   is the base and   is the exponent.

CONCEPTS   *Fill in the blanks.*

5. The product of two integers with _____ signs is negative.

6. The product of two integers with like signs is _____.

7. The _____ property of multiplication implies that $-2(-3) = -3(-2)$.

8. The product of 0 and any number is .

9. Find $-1(9)$. In general, what is the result when we multiply a positive number by -1?

10. Find $-1(-9)$. In general, what is the result when we multiply a negative number by -1?

11. When we multiply two integers, there are four possible combinations of signs. List each of them.

12. When we multiply two integers, there are four possible combinations of signs. How can they be grouped into two categories?

13. If each of the following powers were evaluated, what would be the *sign* of the result?
   a. $(-5)^{13}$ **b.** $(-3)^{20}$

14. A student claimed, "A positive and a negative is negative." What is wrong with this statement?

15. Find each absolute value.
   a. $|-3|$ **b.** $|12|$
   c. $|-5|$ **d.** $|9|$
   e. $|10|$ **f.** $|-25|$

16. Find each product and then graph it on a number line. What is the distance between every two products?

$$2(-2), \quad 1(-2), \quad 0(-2), \quad -1(-2), \quad -2(-2)$$

17. a. Complete the table.

Problem	Number of negative factors	Answer
$-2(-2)$		
$-2(-2)(-2)(-2)$		
$-2(-2)(-2)(-2)(-2)(-2)$		

   b. The answers entered in the table help to justify the following rule: The product of an _____ number of negative integers is positive.

18. a. Complete the table.

Problem	Number of negative factors	Answer
$-2(-2)(-2)$		
$-2(-2)(-2)(-2)(-2)$		
$-2(-2)(-2)(-2)(-2)(-2)(-2)$		

   b. The answers entered in the table help to justify the following rule: The product of an _____ number of negative integers is negative.

NOTATION   *Complete each solution to evaluate the expression.*

19. $\begin{aligned} -3(-2)(-4) &= (-4) \\ &= \end{aligned}$

20. $\begin{aligned} (-3)^4 &= (-3)(-3)(-3) \\ &= (-3)(-3) \\ &= (-3) \\ &= \end{aligned}$

21. Explain why the expression below is not written correctly. How should it be written?
$$-6 \cdot -5$$

22. Translate to mathematical symbols.
   a. the product of negative three and negative two
   b. negative five, squared
   c. the opposite of five squared

PRACTICE   *Find each product.*

23. $-9(-6)$ **24.** $-5(-5)$

25. $-3 \cdot 5$ **26.** $-6 \cdot 4$

27. $12(-3)$ **28.** $11(-4)$
29. $(-8)(-7)$ **30.** $(-9)(-3)$
31. $(-2)10$ **32.** $(-3)8$
33. $-40 \cdot 3$ **34.** $-50 \cdot 2$
35. $-8(0)$ **36.** $0(-27)$
37. $-1(-6)$ **38.** $-1(-8)$
39. $-7(-1)$ **40.** $-5(-1)$
41. $1(-23)$ **42.** $-35(1)$

Evaluate each expression.

43. $-6(-4)(-2)$ **44.** $-3(-2)(-3)$
45. $5(-2)(-4)$ **46.** $3(-3)(3)$
47. $2(3)(-5)$ **48.** $6(2)(-2)$
49. $6(-5)(2)$ **50.** $4(-2)(2)$
51. $(-1)(-1)(-1)$ **52.** $(-1)(-1)(-1)(-1)$
53. $-2(-3)(3)(-1)$ **54.** $5(-2)(3)(-1)$
55. $3(-4)(0)$ **56.** $-7(-9)(0)$
57. $-2(0)(-10)$ **58.** $-6(0)(-12)$

59. Find the product of -6 and the opposite of 10.

60. Find the product of the opposite of 9 and the opposite of 8.

Find each power.

61. $(-4)^2$ **62.** $(-6)^2$
63. $(-5)^3$ **64.** $(-6)^3$
65. $(-2)^3$ **66.** $(-4)^3$
67. $(-9)^2$ **68.** $(-10)^2$
69. $(-1)^5$ **70.** $(-1)^6$
71. $(-1)^8$ **72.** $(-1)^9$

Evaluate each expression.

73. $(-7)^2$ and -7^2
74. $(-5)^2$ and -5^2
75. -12^2 and $(-12)^2$
76. -11^2 and $(-11)^2$

Use a calculator to evaluate each expression.

77. $-76(787)$ **78.** $407(-32)$
79. $(-81)^4$ **80.** $(-6)^5$
81. $(-32)(-12)(-67)$ **82.** $(-56)(-9)(-23)$
83. $(-25)^4$ **84.** $(-41)^5$

APPLICATIONS   *Use signed numbers to help solve each problem.*

85. DIETING After giving a patient a physical exam, a physician felt that the patient should begin a diet. Two options were discussed.

	Plan #1	Plan #2
Length	10 weeks	14 weeks
Daily exercise	1 hr	30 min
Weight loss per week	3 lb	2 lb

a. Find the expected weight loss from each diet plan. Express each answer as a signed number.

b. With which plan should the patient expect to lose more weight? Explain why the patient might not choose it.

86. INVENTORIES A spreadsheet is used to record inventory losses at a warehouse. The items, their cost, and the number missing are listed in the table.

	A	B	C	D
1	Item	Cost	Number of units	$ losses
2	CD	$5	-11	
3	TV	$200	-2	
4	Radio	$20	-4	

a. What instruction should be given to find the total losses for each type of item? Find each of those losses and fill in column D.

b. What instruction should be given to find the *total* inventory losses for the warehouse? Find this number.

87. MAGNIFICATION Using an electronic testing device, a mechanic can check the emissions of a car. The results of the test are displayed on a screen.

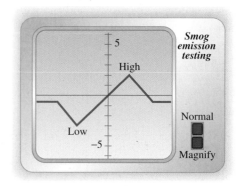

a. Find the high and low values for this test as shown on the screen.

b. By switching a setting on the monitor, the picture on the screen can be magnified. What would be the new high and new low if every value were doubled?

88. LIGHT Sunlight is a mixture of all colors. When sunlight passes through water, the water absorbs different colors at different rates, as shown.

a. Use a signed number to represent the depth to which red light penetrates water.

b. Green light penetrates 4 times deeper than red light. How deep is this?

c. Blue light penetrates 3 times deeper than orange light. How deep is this?

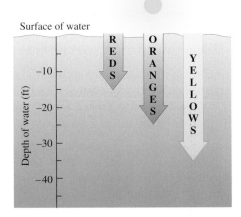

89. TEMPERATURE CHANGE A farmer, worried about his fruit trees suffering frost damage, calls the weather service for temperature information. He is told that temperatures will be decreasing approximately 4° every hour for the next five hours. What signed number represents the total change in temperature expected over the next five hours?

90. DEPRECIATION For each of the last four years, a businesswoman has filed a $200 depreciation allowance on her income tax return, for an office computer system. What signed number represents the total amount of depreciation written off over the four-year period?

91. EROSION A levee protects a town in a low-lying area from flooding. According to geologists, the banks of the levee are eroding at a rate of 2 feet per year. If something isn't done to correct the problem, what signed number indicates how much of the levee will erode during the next decade?

92. DECK SUPPORTS After a winter storm, a homeowner has an engineering firm inspect his damaged deck. Their report concludes that the original pilings were not anchored deep enough, by a factor of 3. What signed number represents the depth to which the pilings should have been sunk?

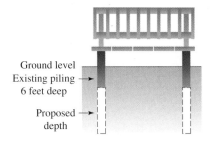

93. WOMEN'S NATIONAL BASKETBALL ASSOCIATION The average attendance for the WNBA Houston Comets is 8,110 a game. Suppose the team gives a sports bag, costing $3, to everyone attending a game. What signed number expresses the financial loss from this promotional giveaway?

94. HEALTH CARE A health care provider for a company estimates that 75 hours per week are lost by employees suffering from stress-related or preventable illness. In a 52-week year, how many hours are lost? Use a signed number to answer.

▌ WRITING

95. If a product contains an even number of negative factors, how do we know that the result will be positive?

96. Explain why the product of a positive number and a negative number is negative, using $5(-3)$ as an example.

97. Explain why the result is the opposite of the original number when a number is multiplied by -1.

98. Can you think of any number that yields a nonzero result when it is multiplied by 0? Explain your response.

▌ REVIEW

99. The prime factorization of a number is $3^2 \cdot 5$. What is the number?

100. Round 10,345 to the nearest hundred.

101. The enrollment at a college went from 10,200 to 12,300 in one year. What was the increase in enrollment?

102. Find the perimeter of a square with sides 6 yards long.

103. What does the symbol $<$ mean?

104. List the first ten prime numbers.

2.5 Dividing Integers

- The relationship between multiplication and division
- Rules for dividing integers • Division and zero

In this section, we develop rules for division of integers, just as we did for multiplication of integers. We also consider two types of division involving 0.

The relationship between multiplication and division

Every division fact containing three numbers has a related multiplication fact involving the same three numbers. For example,

$$\frac{6}{3} = 2 \quad \text{because} \quad 3(2) = 6 \quad \text{Remember that in the division statement, 6 is the } \textit{dividend, } \text{3 is the } \textit{divisor, } \text{and 2 is the } \textit{quotient.}$$

$$\frac{20}{5} = 4 \quad \text{because} \quad 5(4) = 20$$

Rules for dividing integers

We now use the relationship between multiplication and division to help develop rules for dividing integers. There are four cases to consider.

Case 1: In the first case, a positive integer is divided by a positive integer. From years of experience, we already know that the result is positive. Therefore, *the quotient of two positive integers is positive.*

Case 2: Next, we consider the quotient of two negative integers. As an example, consider the division $\frac{-12}{-2} = ?$. We can do this division by examining its related multiplication statement, $-2(?) = -12$. Our objective is to find the number that should replace the question mark. To do this, we use the rules for multiplying integers, introduced in the previous section.

Multiplication statement   **Division statement**

$-2(?) = -12$ $\dfrac{-12}{-2} = ?$

This must be *positive* 6 if the product is to be *negative* 12.   So the quotient is *positive* 6.

Therefore, $\frac{-12}{-2} = 6$. From this example, we can see that *the quotient of two negative integers is positive.*

Case 3: The third case we examine is the quotient of a positive integer and a negative integer. Let's consider $\frac{12}{-2} = ?$. Its equivalent multiplication statement is $-2(?) = 12$.

Multiplication statement   **Division statement**

$-2(?) = 12$ $\dfrac{12}{-2} = ?$

This must be -6 if the product is to be *positive* 12.   So the quotient is -6.

Therefore, $\frac{12}{-2} = -6$. This result shows that *the quotient of a positive integer and a negative integer is negative.*

Case 4: Finally, to find the quotient of a negative integer and a positive integer, let's consider $\frac{-12}{2} = ?$. Its equivalent multiplication statement is $2(?) = -12$.

Multiplication statement	**Division statement**
$2(?) = -12$	$\dfrac{-12}{2} = ?$

    └— This must be   So the quotient —┘
        -6 if the product   is -6.
        is to be -12.

Therefore, $\frac{-12}{2} = -6$. From this example, we can see that *the quotient of a negative integer and a positive integer is negative.*

We now summarize the results from the previous discussion.

Dividing two integers

To divide two integers, divide their absolute values.

1. The quotient of two integers with *like* signs is positive.

2. The quotient of two integers with *unlike* signs is negative.

The rules for dividing integers are similar to those for multiplying integers.

EXAMPLE 1   Find each quotient: **a.** $\dfrac{-35}{7}$ and   **b.** $\dfrac{20}{-5}$.

Solution   To divide integers with unlike signs, we find the quotient of their absolute values and make the quotient negative.

a. $\dfrac{-35}{7} = -5$ Divide the absolute values, 35 by 7, to get 5. The quotient is negative.

To check the result, we multiply the divisor, 7, and the quotient, -5. We should obtain the dividend, -35.

$$7(-5) = -35$$

The answer, -5, checks.

b. $\dfrac{20}{-5} = -4$ Divide the absolute values, 20 by 5, to get 4. The quotient is negative.

Self Check 1
Find each quotient and check the result:

a. $\dfrac{-45}{5}$

b. $\dfrac{60}{-20}$

Answers **a.** -9, **b.** -3

EXAMPLE 2   Divide: $\dfrac{-12}{-3}$.

Solution   The integers have like signs. The quotient will be positive.

$\dfrac{-12}{-3} = 4$ Divide the absolute values, 12 by 3, to get 4. The quotient is positive.

Self Check 2

Divide: $\dfrac{-21}{-3}$.

Answer 7

EXAMPLE 3   Price reductions.   Over the course of a year, a retailer reduced the price of a television set by an equal amount each month, because it was not selling. By the end of the year, the cost was $132 less than at the beginning of the year. How much did the price fall each month?

Solution   We label the drop in price of $132 for the year as -132. It occurred in 12 equal reductions. This indicates division.

$$\frac{-132}{12} = -11$$ The quotient of a negative number and a positive number is negative.

The drop in price each month was $11.

▮ Division and zero

To review the concept of division of 0, we look at $\frac{0}{2} = ?$. The equivalent multiplication statement is $2(?) = 0$.

Multiplication statement	**Division statement**
$2(?) = 0$	$\frac{0}{2} = ?$

└─ This must be   So the quotient ─┘
0 if the product   is 0.
is to be 0.

Therefore, $\frac{0}{2} = 0$. This example suggests that *the quotient of 0 divided by any nonzero number is 0.*

To review division by 0, let's look at $\frac{2}{0} = ?$. The equivalent multiplication statement is $0(?) = 2$.

Multiplication statement	**Division statement**
$0(?) = 2$	$\frac{2}{0} = ?$

└─ There is no number   There is no ─┘
that gives 2 when   quotient.
multiplied by 0.

Therefore, $\frac{2}{0}$ does not have an answer. We say that division by 0 is undefined. This example suggests that *the quotient of any number divided by 0 is undefined.*

Division with 0

1. If 0 is divided by any nonzero number, the quotient is 0.

2. Division by 0 is undefined.

Self Check 4

Find $\dfrac{0}{-4}$.

Answer 0

EXAMPLE 4   Find $\dfrac{-4}{0}$, if possible.

Solution   Since $\dfrac{-4}{0}$ is division by 0, the division is undefined.

Division with negative numbers   CALCULATOR SNAPSHOT

The Bureau of Labor Statistics estimated that the United States lost 146,400 jobs in the manufacturing sector of the economy in 2003. Because the jobs were lost, we write this as $-146,400$. To find the average number of manufacturing jobs lost each month, we divide: $\frac{-146,400}{12}$. To perform this division using a scientific calculator, we enter these numbers and press these keys.

146400 $\boxed{+/-}$ $\boxed{\div}$ 12 $\boxed{=}$ $\boxed{-12200}$

The average number of manufacturing jobs lost each month in 2003 was 12,200.

Section 2.5 STUDY SET

VOCABULARY   *Fill in the blanks.*

1. In $\dfrac{-27}{3} = -9$, the number -9 is called the _____, and the number 3 is the _____.

2. Division by 0 is _____. Division _____ 0 by a nonzero number is 0.

3. The _____ _____ of a number is the distance between it and 0 on the number line.

4. $\{\ldots, -4, -3, -2, -1, 0, 1, 2, 3, 4, \ldots\}$ is the set of _____.

5. The quotient of two negative integers is _____.

6. The quotient of a negative integer and a positive integer is _____.

CONCEPTS

7. Write the related multiplication statement for $\frac{-25}{5} = -5$.

8. Write the related multiplication statement for $\frac{0}{-15} = 0$.

9. Show that there is no answer for $\frac{-6}{0}$ by writing the related multiplication statement.

10. Find the value of $\frac{0}{5}$.

11. Write a related division statement for $5(-4) = -20$.

12. How do the rules for multiplying integers compare with the rules for dividing integers?

13. Determine whether each statement is always true, sometimes true, or never true.

    a. The product of a positive integer and a negative integer is negative.

    b. The sum of a positive integer and a negative integer is negative.

    c. The quotient of a positive integer and a negative integer is negative.

14. Determine whether each statement is always true, sometimes true, or never true.

    a. The product of two negative integers is positive.

    b. The sum of two negative integers is negative.

    c. The quotient of two negative integers is negative.

PRACTICE   *Find each quotient, if possible.*

15. $\dfrac{-14}{2}$ 16. $\dfrac{-10}{5}$

17. $\dfrac{-8}{-4}$ 18. $\dfrac{-12}{-3}$

19. $\dfrac{-25}{-5}$ 20. $\dfrac{-36}{-12}$

21. $\dfrac{-45}{-15}$ 22. $\dfrac{-81}{-9}$

23. $\dfrac{40}{-2}$ 24. $\dfrac{35}{-7}$

25. $\dfrac{50}{-25}$ 26. $\dfrac{80}{-40}$

27. $\dfrac{0}{-16}$ **28.** $\dfrac{0}{-6}$

29. $\dfrac{-6}{0}$ **30.** $\dfrac{-8}{0}$

31. $\dfrac{-5}{1}$ **32.** $\dfrac{-9}{1}$

33. $-5 \div (-5)$ **34.** $-11 \div (-11)$

35. $\dfrac{-9}{9}$ **36.** $\dfrac{-15}{15}$

37. $\dfrac{-10}{-1}$ **38.** $\dfrac{-12}{-1}$

39. $\dfrac{-100}{25}$ **40.** $\dfrac{-100}{50}$

41. $\dfrac{75}{-25}$ **42.** $\dfrac{300}{-100}$

43. $\dfrac{-500}{-100}$ **44.** $\dfrac{-60}{-30}$

45. $\dfrac{-200}{50}$ **46.** $\dfrac{-500}{100}$

47. Find the quotient of -45 and 9.

48. Find the quotient of -36 and -4.

49. Divide 8 by -2.

50. Divide -16 by -8.

 Use a calculator to perform each division.

51. $\dfrac{-13,550}{25}$ **52.** $\dfrac{-3,876}{-19}$

53. $\dfrac{272}{-17}$ **54.** $\dfrac{-6,776}{-77}$

▊ **APPLICATIONS**   *Use signed numbers to help solve each problem.*

55. TEMPERATURE DROP During a five-hour period, the temperature steadily dropped as shown. What was the average change in the temperature per hour over this five-hour time span?

60°

40°

56. PRICE DROPS Over a three-month period, the price of a DVR steadily fell as shown. What was the average monthly change in the price of the DVR over this period?

Was $300
NOW
$240

57. SUBMARINE DIVES In a series of three equal dives, a submarine is programmed to reach a depth of 3,030 feet below the ocean surface. What signed number describes how deep each of the three dives will be?

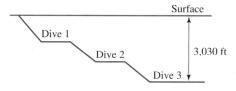

Surface

Dive 1

Dive 2

Dive 3

3,030 ft

58. GRAND CANYON A mule train is to travel from a stable on the rim of the Grand Canyon to a camp on the canyon floor, approximately 5,000 feet below the rim. If the guide wants the mules to be rested after every 1,000 feet of descent, how many stops will be made on the trip?

59. BASEBALL TRADES At the midway point of the season, a baseball team finds itself 12 games behind the league leader. Team management decides to trade for a talented hitter, in hopes of making up at least half of the deficit in the standings by the end of the year. Where in the league standings does management expect to finish at season's end?

60. BUDGET DEFICITS A politician proposed a two-year plan for cutting a county's $20-million budget deficit, as shown. If this plan is put into effect, what will be the change in the county's financial status in two years?

	Plan	Prediction
1st year	Raise taxes, drop subsidy programs	Will cut deficit in half
2nd year	Search out waste and fraud	Will cut remaining deficit in half

61. MARKDOWNS The owner of a clothing store decides to reduce the price on a line of jeans that are not selling. She feels she can afford to lose $300 of projected income on these pants. By how much can she mark down each of the 20 pairs of jeans?

62. WATER RESERVOIRS Over a week's time, engineers at a city water reservoir released enough water to lower the water level 35 feet. On average, how much did the water level change each day during this period?

63. PAY CUTS In a cost-cutting effort, a business decides to lower expenditures on salaries by $9,135,000. To do this, all of the 5,250 employees will have their salaries reduced by an equal dollar amount. How big a pay cut will each employee experience?

64. STOCK MARKET On Monday, the value of Maria's 255 shares of stock was at an all-time high. By Friday, the value had fallen $4,335. What was her per-share loss that week?

WRITING

65. Explain why the quotient of two negative numbers is positive.

66. Think of a real-life situation that could be represented by $\frac{0}{4}$. Explain why the answer would be 0.

67. Using a specific example, explain how multiplication can be used as a check for division.

68. Explain what it means when we say that division by 0 is undefined.

REVIEW

69. Evaluate: $3\left(\dfrac{18}{3}\right)^2 - 2(2)$.

70. List the set of whole numbers.

71. Find the prime factorization of 210.

72. The statement $(4 + 8) + 10 = 4 + (8 + 10)$ illustrates what property?

73. Is the following true? $17 \geq 17$

74. Does $8 - 2 = 2 - 8$?

75. Evaluate: 3^4.

76. Sharif has scores of 55, 70, 80, and 75 on four mathematics tests. What is his mean (average) score?

2.6 Order of Operations and Estimation

- Order of operations • Absolute value • Estimation

In this section, we evaluate expressions involving more than one operation. To do this, we apply the rules for the order of operations and the rules for working with integers. We also continue the discussion of estimating an answer. Estimation can be used when you need a quick indication of the size of the actual answer to a calculation.

Order of operations

In Section 1.7, we introduced the following rules for the order of operations: an agreed-upon sequence of steps for completing the operations of arithmetic.

> **Order of operations**
>
> Perform all calculations within parentheses and other grouping symbols in the following order, working from innermost pair to outermost pair.
>
> **1.** Evaluate all powers. That is, evaluate all expressions with exponents.
>
> **2.** Perform all multiplications and divisions as they occur from left to right.
>
> **3.** Perform all additions and subtractions as they occur from left to right.
>
> When all grouping symbols have been removed, repeat steps 1–3 to complete the calculation.
>
> If a fraction bar is present, evaluate the expression above the bar (the *numerator*) and the expression below the bar (the *denominator*) separately. Then do the division indicated by the fraction bar, if possible.

EXAMPLE 1   Evaluate: $-4(-3)^2 - (-2)$.

Self Check 1
Evaluate: $-5(-2)^2 - (-6)$.

Solution   This expression contains the operations of multiplication, raising to a power, and subtraction. The rules for the order of operations tell us to find the power first.

$$-4(\mathbf{-3})^2 - (-2) = -4(\mathbf{9}) - (-2) \qquad \text{Evaluate the exponential expression: } (-3)^2 = 9.$$

$$= -36 - (-2) \qquad \text{Perform the multiplication: } -4(9) = -36.$$

$$= -36 + 2 \qquad \text{To perform the subtraction, add the opposite of } -2.$$

$$= -34 \qquad \text{Perform the addition.}$$

Answer −14

Self Check 2

Evaluate: $4(2) + (-4)(-3)(-2)$.

Answer −16

EXAMPLE 2   Evaluate: $2(3) + (-5)(-3)(-2)$.

Solution   This expression contains the operations of multiplication and addition. By the rules for the order of operations, we perform the multiplications first.

$$2(3) + (-5)(-3)(-2) = 6 + (-30) \qquad \text{Working from left to right, perform the multiplications.}$$

$$= -24 \qquad \text{Perform the addition.}$$

Self Check 3

Evaluate: $45 \div (-5)3$.

Answer −27

EXAMPLE 3   Evaluate: $40 \div (-4)5$.

Solution   This expression contains the operations of division and multiplication. We perform the divisions and multiplications as they occur from left to right.

$$40 \div (\mathbf{-4})5 = \mathbf{-10} \cdot 5 \qquad \text{Perform the division first: } 40 \div (-4) = -10.$$

$$= -50 \qquad \text{Perform the multiplication.}$$

Self Check 4

Evaluate: $-3^2 - (-3)^2$.

Answer −18

EXAMPLE 4   Evaluate: $-2^2 - (-2)^2$.

Solution   This expression contains the operations of raising to a power and subtraction. We are to find the powers first. (Recall that -2^2 means the *opposite* of 2^2.)

$$-2^2 - (-2)^2 = -4 - 4 \qquad \text{Find the powers: } -2^2 = -4 \text{ and } (-2)^2 = 4.$$

$$= -8 \qquad \text{Perform the subtraction.}$$

Self Check 5

Evaluate: $-18 + 4(-7 + 9)$.

Answer −10

EXAMPLE 5   Evaluate: $-15 + 3(-4 + 7)$.

Solution

$$-15 + 3(\mathbf{-4 + 7}) = -15 + 3(\mathbf{3}) \qquad \text{Perform the addition within the parentheses: } -4 + 7 = 3.$$

$$= -15 + 9 \qquad \text{Perform the multiplication: } 3(3) = 9.$$

$$= -6 \qquad \text{Perform the addition.}$$

EXAMPLE 6   Evaluate: $\dfrac{-20 + 3(-5)}{(-4)^2 - 21}$.

Solution   We first evaluate the expressions in the numerator and the denominator, separately.

$$\dfrac{-20 + 3(-5)}{(-4)^2 - 21} = \dfrac{-20 + (\mathbf{-15})}{\mathbf{16} - 21}$$
In the numerator, perform the multiplication: $3(-5) = -15$.
In the denominator, evaluate the power: $(-4)^2 = 16$.

$$= \dfrac{-35}{-5}$$
In the numerator, add: $-20 + (-15) = -35$.
In the denominator, subtract: $16 - 21 = -5$.

$$= 7$$
Perform the division.

Answer 11

EXAMPLE 7   Evaluate: $-5[-1 + (2 - 8)^2]$.

Solution   We begin by working within the innermost pair of grouping symbols and work to the outermost pair.

$$-5[-1 + (\mathbf{2 - 8})^2] = -5[-1 + (\mathbf{-6})^2]$$
Perform the subtraction within the parentheses.

$$= -5(-1 + 36)$$
Evaluate the power within the brackets.

$$= -5(35)$$
Perform the addition within the parentheses.

$$= -175$$
Perform the multiplication.

Answer -56

▪ Absolute value

You will recall that the absolute value of a number is the distance between the number and 0 on the number line. Earlier in this chapter, we evaluated simple absolute value expressions such as $|-3|$ and $|10|$. Absolute value symbols are also used in combination with more complicated expressions, such as $|-4(3)|$ and $|-6 + 1|$. When we apply the rules for the order of operations to evaluate these expressions, *the absolute value symbols are considered to be grouping symbols*, and any operations within them are to be completed first.

EXAMPLE 8   Find each absolute value: **a.** $|-4(3)|$ and   **b.** $|-6 + 1|$.

Solution   We perform the operations within the absolute value symbols first.

a. $|-4(3)| = |-12|$ Perform the multiplication within the absolute value symbol: $-4(3) = -12$.

$$= 12$$ Find the absolute value of -12.

b. $|-6 + 1| = |-5|$ Perform the addition within the absolute value symbol: $-6 + 1 = -5$.

$$= 5$$ Find the absolute value of -5.

Answers **a.** 30,   **b.** 29

> **! COMMENT** Just as $-5(8)$ means $-5 \cdot 8$, the expression $-5|8|$ (read as "negative 5 times the absolute value of 8") means $-5 \cdot |8|$. To evaluate such an expression, we find the absolute value first and then multiply.
>
> $$-5|8| = -5 \cdot 8 \qquad \text{Find the absolute value: } |8| = 8.$$
> $$= -40 \qquad \text{Perform the multiplication.}$$

Self Check 9
Evaluate: $7 - 5|-1 - 6|$.

EXAMPLE 9   Evaluate: $8 - 4|-6 - 2|$.

Solution   We perform the operation within the absolute value symbol first.

$$8 - 4|-6 - 2| = 8 - 4|-8| \qquad \text{Perform the subtraction within the absolute value symbol: } -6 - 2 = -8.$$
$$= 8 - 4(8) \qquad \text{Find the absolute value: } |-8| = 8.$$
$$= 8 - 32 \qquad \text{Perform the multiplication: } 4(8) = 32.$$
$$= -24 \qquad \text{Perform the subtraction.}$$

Answer -28

■ Estimation

Recall that the idea behind estimation is to simplify calculations by using rounded numbers that are close to the actual values in the problem. When an exact answer is not necessary and a quick approximation will do, we can use estimation.

EXAMPLE 10 The stock market.   The Dow Jones Industrial Average is announced at the end of each trading day to give investors an indication of how the New York Stock Exchange performed. A positive number indicates good performance, while a negative number indicates poor performance. Estimate the net gain or loss of points in the Dow for the week shown in Figure 2-21.

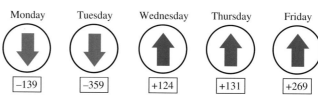

Monday	Tuesday	Wednesday	Thursday	Friday
−139	−359	+124	+131	+269

FIGURE 2-21

Solution   We will approximate each of these numbers. For example, -139 is close to -140, and $+124$ is close to $+120$. To estimate the net gain or loss, we add the approximations.

$$-140 + (-360) + 120 + 130 + 270 = -500 + 520 \qquad \text{Add positive and negative numbers separately to get subtotals.}$$
$$= 20 \qquad \text{Perform the addition.}$$

This estimate tells us that there was a gain of approximately 20 points in the Dow.

Section 2.6   STUDY SET

VOCABULARY   *Fill in the blanks.*

1. When asked to evaluate expressions that contain more than one operation, we should apply the rules for the _____ of operations.

2. In situations where an exact answer is not needed, an approximation or _____ is a quick way of obtaining a rough idea of the size of the actual answer.

3. Absolute value symbols, parentheses, and brackets are types of _____ symbols.

4. If an expression involves two sets of grouping symbols, always begin working within the _____ symbols and then work to the _____.

CONCEPTS

5. Consider $5(-2)^2 - 1$. How many operations need to be performed to evaluate this expression? List them in the order in which they should be performed.

6. Consider $15 - 3 + (-5 \cdot 2)^3$. How many operations need to be performed to evaluate this expression? List them in the order in which they should be performed.

7. Consider $\dfrac{5 + 5(7)}{2 + (4 - 8)}$. In the numerator, what operation should be performed first? In the denominator, what operation should be performed first?

8. In the expression $4 + 2(-7 - 1)$, how many operations need to be performed? List them in the order in which they should be performed.

9. Explain the difference between -3^2 and $(-3)^2$.

10. In the expression $-2 \cdot 3^2$, what operation should be performed first?

NOTATION   *Complete each solution to evaluate the expression.*

11.
$$-8 - 5(-2)^2 = -8 - 5(\ \ )$$
$$= -8 - \ \ $$
$$= -8 + (\ \ )$$
$$= \ \ $$

12.
$$2 + (5 - 6 \cdot 2) = 2 + (5 - \ \ )$$
$$= 2 + [5 + (\ \ )]$$
$$= 2 + (\ \ )$$
$$= \ \ $$

13.
$$[-4(2 + 7)] - 6 = [-4(\ \ )] - 6$$
$$= \ \ - 6$$
$$= \ \ $$

14.
$$\dfrac{|-9 + (-3)|}{9 - 6} = \dfrac{|\ \ |}{3}$$
$$= \dfrac{\ \ }{3}$$
$$= \ \ $$

PRACTICE   *Evaluate each expression.*

15. $(-3)^2 - 4^2$

16. $-7 + 4 \cdot 5$

17. $3^2 - 4(-2)(-1)$

18. $2^3 - 3^3$

19. $(2 - 5)(5 + 2)$

20. $-3(2)^2 4$

21. $-10 - 2^2$

22. $-50 - 3^3$

23. $\dfrac{-6 - 8}{2}$

24. $\dfrac{-6 - 6}{-2 - 2}$

25. $\dfrac{-5 - 5}{2}$

26. $\dfrac{-7 - (-3)}{2 - 4}$

27. $-12 \div (-2)2$

28. $-60(-2) \div 3$

29. $-16 - 4 \div (-2)$

30. $-24 + 4 \div (-2)$

31. $|-5(-6)|$

32. $|-7 - 9|$

33. $|-4 - (-6)|$

34. $|-2 + 6 - 5|$

35. $5|3|$

36. $5|4|$

37. $-6|-7|$

38. $-6|-4|$

39. $(7 - 5)^2 - (1 - 4)^2$

40. $5^2 - (-9 - 3)$

41. $-1(2^2 - 2 + 1^2)$

42. $(-7 - 4)^2 - (-1)$

43. $-50 - 2(-3)^3$

44. $(-2)^3 - (-3)(-2)$

45. $-6^2 + 6^2$

46. $-9^2 + 9^2$

47. $3\left(\dfrac{-18}{3}\right) - 2(-2)$

48. $2\left(\dfrac{-12}{3}\right) + 3(-5)$

49. $6 + \dfrac{25}{-5} + 6 \cdot 3$

50. $-5 - \dfrac{24}{6} + 8(-2)$

51. $\dfrac{1 - 3^2}{-2}$

52. $\dfrac{-3 - (-7)}{2^2 - 3}$

53. $\dfrac{-4(-5) - 2}{-6}$

54. $\dfrac{(-6)^2 - 1}{-4 - 3}$

55. $-3\left(\dfrac{32}{-4}\right) - (-1)^5$

56. $-5\left(\dfrac{16}{-4}\right) - (-1)^4$

57. $6(2^3)(-1)$

58. $2(3^3)(-2)$

59. $2 + 3[5 - (1 - 10)]$

60. $12 - 2[1 - (-8 + 2)]$

61. $-7(2 - 3 \cdot 5)$

62. $-4(1 + 3 \cdot 5)$

63. $-[6 - (1 - 4)^2]$

64. $-[9 - (9 - 12)^2]$

65. $15 + (-3 \cdot 4 - 8)$

66. $11 + (-2 \cdot 2 + 3)$

67. $|-3 \cdot 4 + (-5)|$

68. $|-8 \cdot 5 - 2 \cdot 5|$

69. $|(-5)^2 - 2 \cdot 7|$

70. $|8 \div (-2) - 5|$

71. $-2 + |6 - 4^2|$

72. $-3 - 4|6 - 7|$

73. $2|1 - 8| \cdot |-8|$

74. $2(5) - 6(|-3|)^2$

75. $-2(-34)^2 - (-605)$

75. $11 - (-15)(24)^2$

77. $-60 - \dfrac{1,620}{-36}$

78. $\dfrac{2^5 - 4^6}{-42 + 58}$

Make an estimate.

79. $-379 + (-103) + 287$

80. $\dfrac{-67 - 9}{-18}$

81. $-39 \cdot 8$

82. $-568 - (-227)$

83. $-3,887 + (-5,106)$

84. $-333(-4)$

85. $\dfrac{6,267}{-5}$

86. $-36 + (-78) + 59 + (-4)$

APPLICATIONS

87. TESTING In an effort to discourage her students from guessing on multiple-choice tests, a professor uses the grading scale shown in the table in the next column. If unsure of an answer, a student does best to skip the question, because incorrect responses are penalized very heavily. Find the test score of a student who gets 12 correct and 3 wrong and leaves 5 questions blank.

Response	Value
Correct	+3
Incorrect	−4
Left blank	−1

88. THE FEDERAL BUDGET See below. Suppose you were hired to write a speech for a politician who wanted to highlight the improvement in the federal government's finances during 1990s. Would it be better for the politician to refer to the average budget deficit/surplus for the last half, or for the last four years of that decade? Explain your reasoning.

U.S. Budget Deficit/Surplus
($ billions)

Deficit	Year	Surplus
−164	1995	
−107	1996	
−22	1997	
	1998	+70
	1999	+123

89. SCOUTING REPORTS The illustration shows a football coach how successful his opponent was running a "28 pitch" the last time the two teams met. What was the opponent's average gain with this play?

Play: 28 pitch

Gain 16 yd	Gain 10 yd	Loss 2 yd	No gain
Gain 4 yd	Loss 4 yd	TD Gain 66 yd	Loss 2 yd

90. SPREADSHEETS The table shows the data from a chemistry experiment in spreadsheet form. To obtain a result, the chemist needs to add the values in row 1, double that sum, and then divide that number by the smallest value in column C. What is the final result of these calculations?

	A	B	C	D
1	12	−5	6	−2
2	15	4	5	−4
3	6	4	−2	8

Estimate the answer to each question.

91. OIL PRICES The price per barrel of crude oil fluctuates with supply and demand. It can rise and fall quickly. The line graph shows how many cents the price per barrel rose or fell each day for a week. For example, on Monday the price rose 68 cents, and on Tuesday it rose an additional 91 cents. Estimate the net gain or loss in the value of a barrel of crude oil for the week.

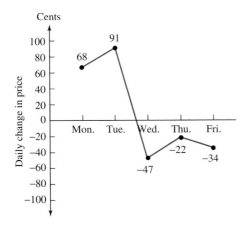

92. ESTIMATION Quickly determine a reasonable estimate of the exact answer in each of the following situations.

a. A diver, swimming at a depth of 34 feet below sea level, spots a sunken ship beneath him. He dives down another 57 feet to reach it. What is the depth of the sunken ship?

b. A dental hygiene company offers a money-back guarantee on its tooth whitener kit. When the kit is returned by a dissatisfied customer, the company loses the $11 it cost to produce it, because it cannot be resold. How much money has the company lost because of this return policy if 56 kits have been mailed back by customers?

c. A tram line makes a 7,561-foot descent from a mountaintop in 18 equal stages. How much does it descend in each stage?

WRITING

93. When evaluating expressions, why are rules for the order of operations necessary?

94. In the rules for the order of operations, what does the phrase *as they occur from left to right* mean?

95. Name a situation in daily life where you use estimation.

96. List some advantages and some disadvantages of the process of estimation.

REVIEW

97. Round 5,456 to the nearest thousand.

98. Evaluate: $6^2 - (10 - 8)$.

99. How do we find the perimeter of a rectangle?

100. Is this statement true or false? "When measuring perimeter, we use square units."

101. An elevator has a weight capacity of 1,000 pounds. Seven people, with an average weight of 140 pounds, are in it. Is it overloaded?

102. List the factors of 36.

Signed Numbers

In mathematics, we work with both positive and negative numbers. We study negative numbers because they are necessary to describe many situations in daily life.

Represent each of these situations using a signed number.

1. Stocks fell 5 points.

2. The river was 12 feet over flood stage.

3. 30 seconds before going on the air

4. A business $6 million in the red

5. 10 degrees above normal

6. The year 2000 B.C.

7. $205 overdrawn

8. 14 units under their quota

The number line can be used to illustrate positive and negative numbers.

9. On the number line, label the location of the positive integers and the negative integers.

10. On the number line, graph 2 and its opposite.

11. Two numbers, −6 and −4, are graphed on the number line. What can you say about their relative sizes?

12. The absolute value of −3, written $|-3|$, is the distance between −3 and 0 on the number line. Show this distance on the number line.

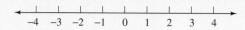

In the space provided, summarize how addition, multiplication, and division are performed with two integers having like signs and with two integers having unlike signs. Then explain the method that is used to subtract integers.

13. Addition

Like signs:

Unlike signs:

14. Multiplication

Like signs:

Unlike signs:

15. Division

Like signs:

Unlike signs:

16. Subtraction with integers

ACCENT ON TEAMWORK

SECTION 2.1

OPPOSITES Have everyone in the class get in pairs. The object of the game is for one member of a team, using one-word clues, to get the other member to say each of the words in Column I. This is done by giving your partner a word clue that has the opposite meaning. For example, if you want your partner to say the word *up*, give the clue *down*. Go through all of the words in Column I. Keep track of how many words are guessed correctly.

Then switch assignments. The person who first gave the clues now receives the clues. Go through all of the words in Column II.

Column I		Column II	
below	surplus	win	positive
over	overdrawn	gain	increase
ahead	profit	forward	debt
before	deduct	retreat	withdrawal
less	liabilities	rise	accelerate

SECTION 2.2

ADDING INTEGERS To illustrate how to add $-5 + 3$, think of a hole 5 feet deep (-5) which then has 3 feet of dirt (3) added to it. The figure on the right shows that the resulting hole would be 2 feet deep (-2). So $-5 + 3 = -2$.

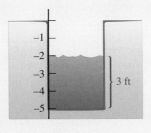

Draw a similar picture to help find each sum.

a. $-4 + 1$
b. $-5 + 4$
c. $-6 + 6$
d. $-3 + (-1)$
e. $-3 + (-3)$
f. $-3 + 5$

SECTION 2.3

SUBTRACTING INTEGERS Write a subtraction problem in which the difference of two negative numbers is
a. a positive number.   **b.** a negative number.

SECTION 2.4

MULTIPLYING INTEGERS

a. Complete the multiplication table that follows.
b. Construct another table with a top row of -1 through -10 and a first column of 1 through 10.

Multiplication Table										
·	1	2	3	4	5	6	7	8	9	10
-1										
-2										
-3										
-4										
-5										
-6										
-7										
-8										
-9										
-10										

c. Construct a table with a top row of -1 through -10 and a first column of -1 through -10.

SECTION 2.5

OPERATIONS WITH TWO INTEGERS For each operation listed in the table, tell whether the answer is always positive, always negative, or may be positive or negative.

Signs of the two integers	Add	Subtract	Multiply	Divide
Both positive				
Both negative				
One positive, one negative				

SECTION 2.6

ESTIMATION Estimate the answer to each problem.

a. $405 - 567$
b. $-2,564 - 2,456$
c. $989 - 898$
d. $-23,250 + 22,750$
e. $56(-87)$
f. $-40 - 30 - 45$
g. $608 \div (-2)$
h. $-94 + 90 - 45$

CHAPTER REVIEW

An Introduction to the Integers

CONCEPTS

The *number line* is a horizontal or vertical line used to represent numbers graphically.

A *negative number* is less than 0. A *positive number* is greater than 0.

Integers:
$\{\ldots, -2, -1, 0, 1, 2, \ldots\}$

Inequality symbols:

> is greater than

< is less than

≥ is greater than or equal to

≤ is less than or equal to

The *absolute value* of a number is the distance between it and 0 on the number line.

On the number line, two numbers the same distance away from 0, but on different sides of it, are called *opposites*.

REVIEW EXERCISES

Graph each set of numbers.

1. $\{-3, -1, 0, 4\}$

2. The integers greater than −3 but less than 4

Insert one of the symbols > or < in the blank to make a true statement.

3. −7 ▢ 0

4. −20 ▢ −19

Insert one of the symbols ≥ or ≤ in the blank to make a true statement.

5. $|-16|$ ▢ −16

6. 56 ▢ 56

7. WATER PRESSURE Salt water exerts a pressure of 14.7 pounds per square inch at a depth of 33 feet. Express the depth using a signed number.

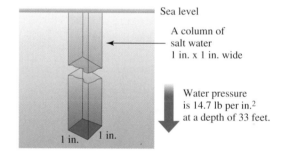

Sea level

A column of salt water 1 in. x 1 in. wide

Water pressure is 14.7 lb per in.² at a depth of 33 feet.

1 in. 1 in.

Represent each of these situations using a signed number.

8. A deficit of $1,200

9. 10 seconds before going on the air

Evaluate each expression.

10. $|-4|$

11. $|0|$

12. $|-43|$

13. $-|12|$

Explain the meaning of each red − sign.

14. −5

15. −(−5)

16. −(−5)

17. 5 − (−5)

The opposite of the opposite of a number is that number.

Find each of the following.

18. $-(-12)$

19. The opposite of 8

20. The opposite of -8

21. -0

SECTION 2.2

Adding Integers

To add two integers with *like signs*, add their absolute values and attach their common sign to that sum.

To add two integers with *unlike signs*, subtract their absolute values, the smaller from the larger. Attach the sign of the number with the larger absolute value to that result.

Use a number line to find each sum.

22. $4 + (-2)$

23. $-1 + (-3)$

Add.

24. $-6 + (-4)$

25. $-23 + (-60)$

26. $-1 + (-4) + (-3)$

27. $-4 + 3$

28. $-28 + 140$

29. $9 + (-20)$

30. $3 + (-2) + (-4)$

31. $(-2 + 1) + [(-5) + 4]$

Addition property of 0: The sum of any number and 0 is that number.

Add.

32. $-4 + 0$

33. $0 + (-20)$

34. $-8 + 8$

35. $73 + (-73)$

Numbers are said to be *additive inverses* if their sum is 0.

Give the additive inverse of each number.

36. -11

37. 4

38. DROUGHT During a drought, the water level in a reservoir fell to a point 100 feet below normal. After two rainy months, it rose a total of 35 feet. How far below normal was the water level after the rain?

SECTION 2.3

Subtracting Integers

Rule for subtraction: To subtract two integers, add the opposite of the second integer to the first integer.

Subtract.

39. $5 - 8$

40. $-9 - 12$

41. $-4 - (-8)$

42. $-6 - 106$

43. $-8 - (-2)$

44. $7 - 1$

45. $0 - 37$

46. $0 - (-30)$

47. Fill in the blanks: Subtracting a number is the same as _____ the _____ of that number.

Evaluate each expression.

48. $-9 - 7 + 12$

49. $7 - [(-6) - 2]$

50. $1 - (2 - 7)$

51. $-12 - (6 - 10)$

52. Subtract 27 from -50.

53. Evaluate: $2 - [-(-3)]$.

54. GOLD MINING Some miners discovered a small vein of gold at a depth of 150 feet. This prompted them to continue their exploration. After descending another 75 feet, they came upon a much larger find. Use a signed number to represent the depth of the second discovery.

To find the change in a quantity, subtract the earlier value from the later value.

55. RECORD TEMPERATURES The lowest and highest recorded temperatures for Alaska and Virginia are shown here. For each state, find the difference in temperature between the record high and low.

Alaska: Low $-80°$ Jan. 23, 1971   Virginia: Low $-30°$ Jan. 22, 1985
High $100°$ June 27, 1915   High $110°$ July 15, 1954

SECTION 2.4

Multiplying Integers

The product of two integers with like signs is positive. The product of two integers with unlike signs is negative.

Multiply.

56. $-9 \cdot 5$

57. $-3(-6)$

58. $7(-2)$

59. $(-8)(-47)$

60. $-20 \cdot 5$

61. $-1(-1)$

62. $-1(25)$

63. $(5)(-30)$

64. $(-6)(-2)(-3)$

65. $4(-3)3$

66. $0(-7)$

67. $(-1)(-1)(-1)(-1)$

68. TAX DEFICITS A state agency's prediction of a tax shortfall proved to be two times worse than the actual deficit of $3 million. The federal prediction of the same shortfall was even more inaccurate — three times the amount of the actual deficit. Complete the illustration, which summarizes these incorrect forecasts.

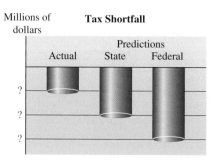

An exponent is used to represent repeated multiplication.

Find each power.

69. $(-5)^2$

70. $(-2)^5$

71. $(-8)^2$

72. $(-4)^3$

73. When $(-5)^9$ is evaluated, will the result be positive or negative?

74. Explain the difference between -2^2 and $(-2)^2$ and then evaluate each.

When a negative integer is raised to an even power, the result is positive. When it is raised to an odd power, the result is negative.

Dividing Integers

75. Fill in the blanks: We know that $\dfrac{-15}{5} = -3$ because ▢ (▢) = ▢ .

The quotient of two numbers with *like* signs is positive.

The quotient of two integers with *unlike* signs is negative.

Divide.

76. $\dfrac{-14}{7}$

77. $\dfrac{25}{-5}$

78. $-64 \div 8$

79. $\dfrac{-202}{-2}$

If 0 is divided by any nonzero number, the quotient is 0.

Find each quotient, if possible.

80. $\dfrac{0}{-5}$

81. $\dfrac{-4}{0}$

Division by 0 is undefined.

82. $\dfrac{-673}{-673}$

83. $\dfrac{-10}{-1}$

84. PRODUCTION TIME Because of improved production procedures, the time needed to produce an electronic component dropped by 12 minutes over the past six months. If the drop in production time was uniform, how much did it change each month over this period?

Order of Operations and Estimation

The rules for the order of operations:

1. Perform all calculations within parentheses and other grouping symbols.
2. Evaluate all exponential expressions.
3. Perform all the multiplications and divisions, working from left to right.
4. Perform all the additions and subtractions, working from left to right.

Always work from the *innermost* set of grouping symbols to the *outermost* set.

A fraction bar is a grouping symbol.

An *estimation* is an approximation that gives a quick idea of what the actual answer would be.

Evaluate each expression.

85. $2 + 4(-6)$

86. $7 - (-2)^2 + 1$

87. $2 - 5(4) + (-25)$

88. $-3(-2)^3 - 16$

89. $-2(5)(-4) + \dfrac{|-9|}{3^2}$

90. $-4^2 + (-4)^2$

91. $-12 - (8 - 9)^2$

92. $7|-8| - 2(3)(4)$

93. $-4\left(\dfrac{15}{-3}\right) - 2^3$

94. $-20 + 2(12 - 5 \cdot 2)$

95. $-20 + 2[12 - (-7 + 5)^2]$

96. $8 - |-3 \cdot 4 + 5|$

97. $\dfrac{10 + (-6)}{-3 - 1}$

98. $\dfrac{3(-6) - 11 + 1}{4^2 - 3^2}$

Estimate each answer.

99. $-89 + 57 + (-42)$

100. $\dfrac{-507}{-26}$

101. $(-681)(9)$

102. $317 - (-775)$

1. Insert one of the symbols $>$ or $<$ in the blank to make the statement true.
   a. -8 ___ -9
   b. -8 ___ $|-8|$
   c. The opposite of 5 ___ 0.

2. List the integers.

3. SCHOOL ENROLLMENT According to the projections in the table, which high school will face the greatest shortage in the year 2010?

High schools with shortage of classroom seats by 2010	
Sylmar	−669
San Fernando	−1,630
Monroe	−2,488
Cleveland	−350
Canoga Park	−586
Polytechnic	−2,379
Van Nuys	−1,690
Reseda	−462
North Hollywood	−1,004
Hollywood	−774

4. Use a number line to find the sum: $-3 + (-2)$.

   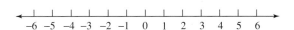

5. Add.
   a. $-65 + 31$
   b. $-17 + (-17)$
   c. $[6 + (-4)] + [-6 + (-4)]$

6. Subtract.
   a. $-7 - 6$
   b. $-7 - (-6)$
   c. $0 - 15$
   d. $-60 - 50 - 40$

7. Find each product.
   a. $-10 \cdot 7$
   b. $-4(-2)(-6)$
   c. $(-2)(-2)(-2)(-2)$
   d. $-55(0)$

8. Write the related multiplication statement for $\dfrac{-20}{-4} = 5$.

9. Find each quotient, if possible.
   a. $\dfrac{-32}{4}$
   b. $\dfrac{8}{6-6}$
   c. $\dfrac{-5}{1}$
   d. $\dfrac{0}{-6}$

10. BUSINESS TAKEOVERS Six investors are contemplating taking over a company that has potential, but they must repay the debt taken on by the company over the past three quarters. If they plan equal ownership, how much will each have to contribute to retire the debt?

    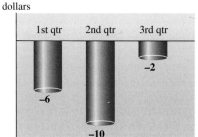

11. GEOGRAPHY The lowest point on the African continent is the Qattarah Depression in the Sahara Desert, 436 feet below sea level. The lowest point on the North American continent is Death Valley, California, 282 feet below sea level. Find the difference in these elevations.

12. Evaluate each expression.

    a. $-(-6)$ **b.** $|-7|$

    c. $-|-9+3|$ **d.** $2|-66|$

13. Find each power.

    a. $(-4)^2$ **b.** -4^2

    c. $(-4+3)^5$

Evaluate each expression.

14. $-18 \div 2 \cdot 3$

15. $4 - (-3)^2 + 6$

16. $-3 + \left(\dfrac{-16}{4} \right) - 3^3$

17. $-10 + 2[6 - (-2)^2(-5)]$

18. $\dfrac{4(-6) - 2^2}{-3 - 4}$

19. TEMPERATURE CHANGE In a lab, the temperature of a fluid was reduced 6° per hour for 12 hours. What signed number represents the change in temperature?

20. GAUGES Many automobiles have an ammeter like that shown. If the headlights, which draw a current of 8 amps, and the radio, which draws a current of 4 amps, are both turned on, which way will the arrow move? What will be the new reading?

21. CARD GAMES After the first round of a card game, Tommy had a score of 8. When he lost the second round, he had to deduct the value of the cards left in his hand from his first-round score. (See the illustration below.) What was his score after two rounds of the game? For scoring, face cards are counted as 10 points and aces as 1 point.

22. Multiplication means repeated addition. Use this fact to show that the product of a positive and a negative number, such as $5(-4)$, is negative.

23. Explain why the absolute value of a number can never be negative.

24. Is the inequality $12 \geq 12$ true or false? Explain why.

Consider the numbers in the set $\{-2, -1, 0, 1, 2\frac{3}{2}, 5, 9\}$.

1. List each natural number.

2. List each whole number.

3. List each negative number.

4. List each integer.

Consider the number 7,326,549.

5. Which digit is in the thousands column?

6. Which digit is in the hundred thousands column?

7. Round to the nearest hundred.

8. Round to the nearest ten thousand.

9. BIDS A school district received the bids shown in the table. Which company should be awarded the contract?

Citrus Unified School District Bid 02-9899 CABLING AND CONDUIT INSTALLATION	
Datatel	$2,189,413
Walton Electric	$2,201,999
Advanced Telecorp	$2,175,081
CRF Cable	$2,174,999
Clark & Sons	$2,175,801

10. NUCLEAR POWER The table gives the number of nuclear power plants in the United States for the years 1978–2003, in five-year increments. Construct a bar graph using the data.

Year	1978	1983	1988	1993	1998	2003
Plants	70	80	108	109	104	104

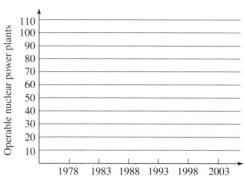

Source: *The World Almanac, 2005*

Perform each operation.

11. $237 + 549$

12. $6,375 - 2,569$

13. $\begin{array}{r} 5,369 \\ -\ \ 685 \end{array}$

14. $\begin{array}{r} 7,899 \\ +5,237 \end{array}$

15. Find the perimeter and the area of the rectangular garden.

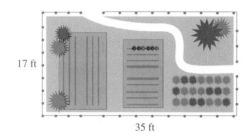

17 ft

35 ft

16. In a shipment of 147 pieces of furniture, 27 pieces were sofas, 55 were leather chairs, and the rest were wooden chairs. Find the number of wooden chairs.

Perform each operation.

17. $435 \cdot 27$

18. $1,261 \div 97$

19. $\begin{array}{r} 4,587 \\ \times\ \ \ \ 67 \end{array}$

20. $38\overline{)17,746}$

21. SHIPPING There are 12 tennis balls in one dozen, and 12 dozen in one gross. How many tennis balls are there in a shipment of 12 gross?

22. Find all of the factors of 18.

Identify each number as a prime, a composite,
an even, or an odd.

23. 17

24. 18

25. 0

26. 1

27. Find the prime factorization of 504.

28. Write the expression $11 \cdot 11 \cdot 11 \cdot 11$ using an exponent.

Evaluate each expression.

29. $5^2 \cdot 7$

30. $16 + 2[14 - 3(5 - 4)^2]$

31. $25 + 5 \cdot 5$

32. $\dfrac{16 - 2 \cdot 3}{2 + (9 - 6)}$

33. SPEED CHECKS A traffic officer monitored several cars on a city street. She found that the speeds of the cars were as follows:

   38, 42, 36, 38, 48, 44

On average, were the drivers obeying the 40-mph speed limit?

34. LIE DETECTOR TESTS A burglar scored -18 on a polygraph test, a score that indicates deception. However, on a second test, he scored $+3$, a score that is inconclusive. Find the difference in the scores.

Graph each set on a number line.

35. $\{-2, -1, 0, 2\}$

36. The integers greater than -4 but less than 2

Evaluate each expression.

37. $-2 + (-3)$

38. $-15 + 10$

39. $|-3 - 1| - 5$

40. $15 - |-3 - 4|$

41. $(-8)(-3)^3$

42. $5(-7)$

43. $\dfrac{-14}{-7}$

44. $\dfrac{45}{-9}$

45. $5 + (-3)(-7)$

46. $-2[-6(5) - 5]$

47. $\dfrac{10 - (-5)}{1 - 2 \cdot 3}$

48. $\dfrac{3(-6) - 10}{3^2 - 4^2}$

49. -9^2

50. $(-9)^2$

51. BUYING A BUSINESS When 12 investors decided to buy a bankrupt company, they agreed to assume equal shares of the company's debt of $1,512,444. How much was each person's share?

52. BANKING A student has $48 in his checking account. He then writes a check for $105 to purchase books. The bank honors the check, but charges the student an additional $22 service fee for being overdrawn. What is the student's new checking account balance?

For Exercises 53 and 54, quickly determine a reasonable estimate of the exact answer.

53. CAMPING Hikers make a 1,150-foot descent into a canyon in 12 stages. How many feet do they descend in each stage?

54. RECALLS An automobile maker has to recall 18,250 cars because they have a faulty engine mount. If it costs $195 to repair each car, how much of a loss will the company suffer because of the recall?

Fractions and Mixed Numbers

Digital Vision/Getty Images

It has been said that "A penny saved is a penny earned." Smart shoppers certainly know this is true. A lot of money can be saved by waiting for big-ticket items, such as appliances, furniture, and jewelry, to go on sale before purchasing them. Many stores advertise from 1/3 to 1/2 off on these items at various times during the year. In such cases, an informed shopper needs a working knowledge of fractions to calculate discounts and sale prices.

To learn more about fractions and discount shopping, visit The Learning Equation on the Internet at http://tle.brookscole.com. (The log-in instructions are in the Preface.) For Chapter 3, the online lessons are:

- *TLE* Lesson 7: Adding and Subtracting Fractions
- *TLE* Lesson 8: Multiplying Fractions and Mixed Numbers

Check Your Knowledge

1. For the fraction $\dfrac{3}{4}$, 3 is the _____ and 4 is the _____.

2. In the formula for the area of a triangle, $A = \dfrac{1}{2}bh$, b stands for the length of the _____ and h stands for the _____.

3. Two numbers are called _____ if their product is 1.

4. The _____ common denominator for a set of fractions is the smallest number each denominator will divide exactly.

5. Fractions that represent the same amount, such as $\dfrac{2}{3}$ and $\dfrac{4}{6}$, are called _____ fractions.

6. What fractional part of a day is 8 hours? What is the remaining fractional part of the day?

7. Simplify each fraction.

   a. $\dfrac{24}{30}$ b. $\dfrac{144}{192}$

8. Multiply:

   a. $\dfrac{1}{2}\left(\dfrac{3}{4}\right)$ b. $-\dfrac{1}{2}\left(\dfrac{2}{3}\right)$

9. How many minutes are there in $\dfrac{3}{8}$ of an hour?

10. Divide:

    a. $\dfrac{2}{3} \div \dfrac{1}{3}$ b. $\dfrac{1}{2} \div 3$

11. Add:

    a. $\dfrac{2}{5} + \dfrac{7}{8}$ b. $-\dfrac{3}{5} + 1$

12. Subtract:

    a. $\dfrac{1}{3} - \dfrac{1}{4}$ b. $1 - \dfrac{3}{8}$

13. Express $\dfrac{5}{6}$ as an equivalent fraction with denominator 36.

14. Graph: $-\dfrac{9}{8}, \dfrac{1}{9}$, and $1\dfrac{1}{2}$.

    $\xleftarrow{\qquad}$ $-5\ -4\ -3\ -2\ -1\ \ 0\ \ 1\ \ 2\ \ 3\ \ 4\ \ 5$ $\xrightarrow{\qquad}$

15. Find the area of a triangle with base is $\dfrac{7}{4}$ cm and height $\dfrac{5}{7}$ cm.

16. Add: $123\dfrac{4}{5} + 189\dfrac{2}{3}$

17. Subtract: $13\dfrac{1}{3} - 8\dfrac{7}{8}$

18. Find the perimeter of a triangle with sides $3\dfrac{1}{4}$ in., $2\dfrac{1}{5}$ in., $4\dfrac{3}{8}$ in.

19. Evaluate: $\dfrac{1}{2} - \dfrac{2}{3} \div \dfrac{1}{2} + \dfrac{1}{6}$

20. Simplify the complex fraction: $\dfrac{\frac{3}{5}}{\frac{1}{3}}$

21. Simplify the complex fraction: $\dfrac{\frac{1}{5} - \frac{1}{6}}{\frac{1}{4} - \frac{1}{3}}$

22. Four-fifths of the people questioned at a beach were using sunscreen. Twelve of the persons surveyed were not using sunscreen. How many people took part in the survey?

23. Josie and five friends purchased ten lottery tickets together. If one of their tickets is chosen, they will win $10\dfrac{1}{2}$ million dollars. What would Josie's share of the prize be if they won?

Study Skills Workshop

LISTENING IN CLASS AND TAKING NOTES

Attending all of your class meetings is crucial to your success in a math course. Your instructor will be giving explanations and examples that may not be found in your textbook, as well as information about test dates and homework assignments. Because this information is not found anywhere else, and because it is *normal* to forget material shortly after it is presented, it is important that you keep a written record of what occurred in class. (*Note:* Auditory learners may also want to keep a recorded version of class notes. Ask your instructor for permission to record lectures).

Listening in Class. Listening in class may be a little different from listening to your favorite CDs or MP3 files, because it requires that you be an *active* listener. It is usually impossible to write down everything that your instructor says in class, but you want to get what's important. In order to catch the significant material, be waiting for cues from your instructor: pauses in lecture and statements such as "this is really important" or "this is a question that shows up frequently on tests" are indications that you should be paying special attention. Listen with a pencil in hand, ready to record these examples, definitions, and concepts.

Taking Notes. Usually when giving a lecture, your instructor will tell you the important chapter(s) and section(s) in your textbook. Bring your textbook to every class and open it to the section that your instructor is covering. If possible, find out what section you will be covering before your next lecture and read this section *before* class. As you read, try to identify which terms and definitions your instructor will think are important. When your instructor refers to definitions found in the text, it is not necessary for you to write them down in entirety, but just jot the term being defined and the page number on which it appears. If topics are being discussed that are not in the text, record them using abbreviations and phrases rather than complete sentences.

Don't worry about making your class notes really neat — you should be "reworking" your notes after class (this will be discussed in more detail in your next study skills workshop). However, do organize your notes as much as possible as you write them. Write down the examples your instructor gives in step-by-step detail. Label examples and definitions that your instructor covers in detail as "key concepts." If you miss a step, or if you don't understand a step in the example, ask your instructor to explain. The examples should look a good deal like the problems that you will see in your homework assignment, so it is important that you write them down as completely as possible. After class, sit down with a classmate and compare notes, to see whether either of you missed anything.

ASSIGNMENT

1. Find out from your instructor which section(s) will be covered in your next class. Read the section(s) and make a list of terms and definitions that you predict your instructor will think are most important.
2. In your next class, bring your textbook and keep it open to the sections being covered. If your instructor mentions a term or definition that is found in your text, write the term and the page number on which it appears in your notes. Write every example that your instructor gives, making sure that you have included all of the steps.
3. Find at least one classmate with whom you can review notes. Make an appointment to compare your class notes as soon as possible after the class. Did you find differences in your notes?

Whole numbers are used to count objects. When we need to represent parts of a whole, fractions can be used.

3.1 Fractions

- Basic facts about fractions   • Equivalent fractions   • Simplifying a fraction
- Expressing a fraction in higher terms

There is no better place to start a study of fractions than with *the fundamental property of fractions.* This property is the foundation for two procedures that we use when working with fractions. But first, we review some basic facts about fractions.

▌ Basic facts about fractions

1. A Fraction Can Be Used to Indicate Division. As we have seen, the fraction $\frac{8}{4}$ represents the division $8 \div 4$. Thus,

$$\text{Dividend} \longrightarrow \frac{8}{4} = 2 \longleftarrow \text{Quotient}$$
$$\text{Divisor} \longrightarrow$$

2. A Fraction Can Be Used to Indicate Equal Parts of a Whole. In our everyday lives, we often deal with parts of a whole. For example, we talk about parts of an hour, parts of an inch, and parts of a pound.

half of an hour   three-eighths of an inch of rain   a quarter-pound hamburger

3. A Fraction Is Composed of a Numerator, a Denominator, and a Fraction Bar.

$$\text{Fraction bar} \longrightarrow \frac{3}{4} \begin{array}{l} \longleftarrow \text{Numerator} \\ \longleftarrow \text{Denominator} \end{array}$$

The denominator (in this case, 4) tells us that a whole was divided into four equal parts. The numerator tells us that we are considering three of those equal parts.

4. Fractions Can Be Proper or Improper. If the numerator of a fraction is less than its denominator, the fraction is called a **proper fraction.** A proper fraction is less than 1. Fractions whose numerators are greater than or equal to their denominators are called **improper fractions.** An improper fraction is greater than or equal to 1.

Proper fractions	**Improper fractions**
$\frac{1}{4}, \frac{2}{3},$ and $\frac{98}{99}$	$\frac{7}{2}, \frac{98}{97}, \frac{16}{16},$ and $\frac{5}{1}$

5. The Denominator of a Fraction Cannot Be 0. $\frac{7}{0}, \frac{23}{0},$ and $\frac{0}{0}$ are meaningless expressions. (Recall that $\frac{7}{0}, \frac{23}{0},$ and $\frac{0}{0}$ represent *division* by 0, and a number cannot be divided by 0.) However, $\frac{0}{7} = 0$ and $\frac{0}{23} = 0$.

6. Fractions Can Be Negative. There are times when a negative fraction is needed to describe a quantity. For example, if an earthquake causes a road to sink one-half inch, the amount of movement can be represented by $-\frac{1}{2}$ inch.

Negative fractions can be written in three ways. The negative sign can appear in the numerator, in the denominator, or in front of the fraction.

$$\frac{-1}{2} = \frac{1}{-2} = -\frac{1}{2} \qquad \frac{-15}{8} = \frac{15}{-8} = -\frac{15}{8}$$

EXAMPLE 1   **a.** In Figure 3-1, what fractional part of the barrel is full? **b.** What fractional part is empty?

FIGURE 3-1

Solution   The barrel has been divided into three equal parts.

a. Two of the three parts are full. Therefore, the barrel is $\frac{2}{3}$ full.

b. One of the three equal parts is not filled. The barrel is $\frac{1}{3}$ empty.

The fractions $\frac{2}{3}$ and $\frac{1}{3}$ are both proper fractions.

Self Check 1

a. According to the calendar below, what fractional part of the month has passed?   **b.** What fractional part remains?

DECEMBER

X	X	X	X	X	X	X
X	X	10	11	12	13	14
15	16	17	18	19	20	21
22	23	24	25	26	27	28
29	30	31				

Answers **a.** $\frac{11}{31}$, **b.** $\frac{20}{31}$

Fractions are often referred to as **rational numbers.** All integers are rational numbers, because every integer can be written as a fraction with a denominator of 1. For example,

$$2 = \frac{2}{1}, \qquad -5 = \frac{-5}{1}, \qquad \text{and} \qquad 0 = \frac{0}{1}$$

Since every integer is also a rational number, the integers are a subset of the rational numbers.

! **COMMENT** Note that not all rational numbers are integers. For example, the rational number $\frac{7}{8}$ is not an integer.

Equivalent fractions

Fractions can look different but still represent the same number. For example, let's divide the rectangle in Figure 3-2(a) in two ways. In Figure 3-2(b), we divide it into halves (2 equal-sized parts). In Figure 3-2(c), we divide it into fourths (4 equal-sized parts). Notice that one-half of the figure is the same size as two-fourths of the figure.

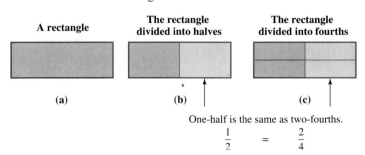

One-half is the same as two-fourths.
$$\frac{1}{2} = \frac{2}{4}$$

FIGURE 3-2

The fractions $\frac{1}{2}$ and $\frac{2}{4}$ look different, but Figure 3-2 shows that they represent the same amount. We say that they are **equivalent fractions.**

> **Equivalent fractions**
> Two fractions are **equivalent** if they represent the same number.

■ Simplifying a fraction

If we replace a fraction with an equivalent one that contains smaller numbers, we are **simplifying** or **reducing the fraction.** To simplify a fraction, we use the following fact.

> **The fundamental property of fractions**
> Multiplying or dividing the numerator and the denominator of a fraction by the same nonzero number does not change the value of the fraction.

As an example, we consider $\frac{24}{28}$. It is apparent that 24 and 28 have a common factor of 4. By the fundamental property of fractions, we can divide the numerator and denominator of this fraction by 4.

Divide the numerator by 4.

$$\frac{24}{28} = \frac{24 \div 4}{28 \div 4}$$

Divide the denominator by 4.

$$= \frac{6}{7} \qquad \text{Perform each division: } 24 \div 4 = 6 \text{ and } 28 \div 4 = 7.$$

Thus, $\frac{24}{28} = \frac{6}{7}$. We say that $\frac{24}{28}$ and $\frac{6}{7}$ are equivalent fractions, because they represent the same number.

In practice, we show the previous simplification in a slightly different way.

$$\frac{24}{28} = \frac{4 \cdot 6}{4 \cdot 7} \qquad \text{Once you see a common factor of the numerator and the denominator, factor each of them so that it shows. In this case, 24 and 28 share a common factor of 4.}$$

$$= \frac{\overset{1}{4} \cdot 6}{\underset{1}{4} \cdot 7} \qquad \text{Divide the numerator and the denominator by 4 by drawing slashes through the common factors. Use small 1's to represent the result of each division of 4 by 4. Note that } \frac{4}{4} = 1.$$

$$= \frac{6}{7} \qquad \text{Multiply in the numerator and in the denominator: } 1 \cdot 6 = 6 \text{ and } 1 \cdot 7 = 7.$$

In the second step of the previous simplification, we say that we *divided out the common factor of 4.*

> **Simplifying a fraction**
> We can **simplify** a fraction by factoring its numerator and denominator and then dividing out all common factors in the numerator and denominator.

When a fraction can be simplified no further, we say that it is written in **lowest terms.**

> **Lowest terms**
> A fraction is in **lowest terms** if the only factor common to the numerator and denominator is 1.

EXAMPLE 2   Simplify to lowest terms: $\dfrac{25}{75}$.

Solution   The numerator and the denominator have a common factor of 25.

$$\dfrac{25}{75} = \dfrac{25 \cdot 1}{3 \cdot 25} \qquad \text{Factor 25 as } 25 \cdot 1 \text{ and 75 as } 3 \cdot 25.$$

$$= \dfrac{\overset{1}{\cancel{25}} \cdot 1}{3 \cdot \underset{1}{\cancel{25}}} \qquad \text{Divide out the common factor of 25.}$$

$$= \dfrac{1}{3} \qquad \text{Multiply in the numerator and in the denominator: } 1 \cdot 1 = 1 \text{ and } 1 \cdot 3 = 3.$$

Self Check 2

Simplify to lowest terms: $\dfrac{60}{80}$.

Answer $\dfrac{3}{4}$

EXAMPLE 3   Simplify: $\dfrac{90}{126}$.

Solution   To find the common factors that will divide out, we prime factor 90 and 126.

$$\dfrac{90}{126} = \dfrac{2 \cdot 3 \cdot 3 \cdot 5}{2 \cdot 3 \cdot 3 \cdot 7} \qquad \begin{array}{l}\text{Use the tree method or the division method to prime factor 90 and} \\ \text{126: } 90 = 2 \cdot 3 \cdot 3 \cdot 5 \text{ and } 126 = 2 \cdot 3 \cdot 3 \cdot 7.\end{array}$$

$$= \dfrac{\overset{1}{\cancel{2}} \cdot \overset{1}{\cancel{3}} \cdot \overset{1}{\cancel{3}} \cdot 5}{\underset{1}{\cancel{2}} \cdot \underset{1}{\cancel{3}} \cdot \underset{1}{\cancel{3}} \cdot 7} \qquad \text{Divide out the common factors of 2, 3, and 3.}$$

$$= \dfrac{5}{7} \qquad \text{Multiply in the numerator and in the denominator.}$$

Self Check 3

Simplify: $\dfrac{42}{150}$.

Answer $\dfrac{7}{25}$

! COMMENT   Negative fractions are simplified in the same way as positive fractions. Just remember to write a negative sign − in front of each step of solution.

$$-\dfrac{45}{72} = -\dfrac{5 \cdot \overset{1}{\cancel{9}}}{8 \cdot \underset{1}{\cancel{9}}} = -\dfrac{5}{8}$$

EXAMPLE 4   Simplify: $-\dfrac{225}{150}$.

Solution   To find the common factors that will divide out, we find the prime factorizations of 225 and 150.

$$-\dfrac{225}{150} = -\dfrac{3 \cdot 3 \cdot 5 \cdot 5}{2 \cdot 3 \cdot 5 \cdot 5} \qquad \text{Factor 225 as } 3 \cdot 3 \cdot 5 \cdot 5 \text{ and 150 as } 2 \cdot 3 \cdot 5 \cdot 5.$$

$$= -\dfrac{3 \cdot \overset{1}{\cancel{3}} \cdot \overset{1}{\cancel{5}} \cdot \overset{1}{\cancel{5}}}{2 \cdot \underset{1}{\cancel{3}} \cdot \underset{1}{\cancel{5}} \cdot \underset{1}{\cancel{5}}} \qquad \text{Divide out the common factors of 3, 5, and 5.}$$

$$= -\dfrac{3}{2} \qquad \text{Multiply in the numerator and in the denominator.}$$

Self Check 4

Simplify: $-\dfrac{88}{12}$.

Answer $-\dfrac{22}{3}$

■ Expressing a fraction in higher terms

It is sometimes necessary to replace a fraction with an equivalent fraction that involves larger numbers. This is called **expressing the fraction in higher terms** or **building** the fraction.

For example, to write $\frac{3}{8}$ as an equivalent fraction with a denominator of 40, we can use the fundamental property of fractions and multiply the numerator and denominator by 5.

$$\frac{3}{8} = \frac{3 \cdot 5}{8 \cdot 5}$$

⌐ Multiply the numerator by 5.

⌐ Multiply the denominator by 5.

$$= \frac{15}{40}$$ Perform the multiplications in the numerator and in the denominator.

Therefore, $\frac{3}{8} = \frac{15}{40}$.

Self Check 5
Write $\frac{2}{3}$ as an equivalent fraction with a denominator of 24.

Answer $\frac{16}{24}$

EXAMPLE 5   Write $\frac{5}{7}$ as an equivalent fraction with a denominator of 28.

Solution   We need to multiply the denominator by 4 to obtain 28. By the fundamental property of fractions, we must multiply the numerator by 4 as well.

$$\frac{5}{7} = \frac{5 \cdot 4}{7 \cdot 4}$$ Multiply the numerator and denominator by 4.

$$= \frac{20}{28}$$ Perform the multiplication in the numerator and in the denominator.

Self Check 6
Write 5 as a fraction with a denominator of 3.

Answer $\frac{15}{3}$

EXAMPLE 6   Write -4 as a fraction with a denominator of 6.

Solution   First, express -4 as a fraction: $-4 = -\frac{4}{1}$. To obtain a denominator of 6, we need to multiply the numerator and denominator by 6.

$$-\frac{4}{1} = -\frac{4 \cdot 6}{1 \cdot 6}$$

$$= -\frac{24}{6}$$ Perform the multiplication: $4 \cdot 6 = 24$ and $1 \cdot 6 = 6$.

Section 3.1   STUDY SET

■ **VOCABULARY**   *Fill in the blanks.*

1. For the fraction $\frac{7}{8}$, 7 is the _____ and 8 is the _____.

2. When we express 15 as $5 \cdot 3$, we say that we have _____ 15.

3. A _____ fraction is less than 1. An _____ fraction is greater than or equal to 1.

4. A fraction is said to be in _____ terms if the only factor common to the numerator and denominator is 1.

5. Two fractions are _____ if they have the same value.

6. A _____ can be used to indicate the number of equal parts of a whole.

7. Multiplying the numerator and denominator of a fraction by a number to obtain an equivalent fraction that involves larger numbers is called expressing the fraction in _____ terms or _____ the fraction.

8. We can _____ a fraction that is not in lowest terms by applying the fundamental property of fractions. We _____ out common factors of the numerator and denominator.

CONCEPTS

9. What common factor (other than 1) do the numerator and the denominator have?

   a. $\dfrac{2}{16}$ **b.** $\dfrac{6}{9}$ **c.** $\dfrac{10}{15}$ **d.** $\dfrac{14}{35}$

10. Given: $\dfrac{15}{35} = \dfrac{3 \cdot \overset{1}{\cancel{5}}}{\cancel{5} \cdot 7}$.

In this work, what do the slashes and 1's mean?

11. What concept studied in this section is shown below?

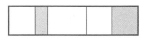

12. Why can't we say that $\frac{2}{5}$ of the figure in the illustration is shaded?

13. a. Explain the difference in the two approaches used to simplify $\frac{20}{28}$.

$$\dfrac{\overset{1}{\cancel{4}} \cdot 5}{\cancel{4} \cdot 7} \quad \text{and} \quad \dfrac{\overset{1}{\cancel{2}} \cdot \overset{1}{\cancel{2}} \cdot 5}{\underset{1}{\cancel{2}} \cdot \underset{1}{\cancel{2}} \cdot 7}$$

   b. Are the results the same?

14. What concept studied in this section does this statement illustrate?

$$\dfrac{5}{10} = \dfrac{4}{8} = \dfrac{3}{6} = \dfrac{2}{4} = \dfrac{1}{2}$$

15. Why isn't this a valid application of the fundamental property of fractions?

$$\dfrac{10}{11} = \dfrac{2 + 8}{2 + 9} = \dfrac{\overset{1}{\cancel{2}} + 8}{\underset{1}{\cancel{2}} + 9} = \dfrac{9}{10}$$

16. Write the fraction $\dfrac{7}{-8}$ in two other ways.

17. Write each number as a fraction.

   a. 8   **b.** -25

18. Fill in the missing numbers in the following statement.

$$\dfrac{\blacksquare \cdot 5}{\blacksquare \cdot 9} = \dfrac{15}{27}$$

NOTATION   *Complete each solution.*

19. Simplify: $\dfrac{18}{24}$.

$$\dfrac{18}{24} = \dfrac{2 \cdot \blacksquare \cdot 3}{2 \cdot 2 \cdot \blacksquare \cdot 3}$$

$$= \dfrac{\overset{1}{\cancel{}} \cdot 3 \cdot \overset{1}{\cancel{}}}{\underset{1}{\cancel{}} \cdot 2 \cdot 2 \cdot \underset{1}{\cancel{}}}$$

$$= \dfrac{3}{4}$$

20. Simplify: $\dfrac{60}{90}$.

$$\dfrac{60}{90} = \dfrac{2 \cdot \blacksquare}{3 \cdot \blacksquare}$$

$$= \dfrac{\blacksquare \cdot \overset{1}{\cancel{30}}}{\blacksquare \cdot \underset{1}{\cancel{30}}}$$

$$= \dfrac{2}{3}$$

PRACTICE   *Simplify each fraction to lowest terms, if possible.*

21. $\dfrac{3}{9}$ **22.** $\dfrac{5}{20}$

23. $\dfrac{7}{21}$ **24.** $\dfrac{6}{30}$

25. $\dfrac{20}{30}$ **26.** $\dfrac{12}{30}$

27. $\dfrac{15}{6}$ **28.** $\dfrac{24}{16}$

29. $-\dfrac{28}{56}$ **30.** $-\dfrac{45}{54}$

31. $-\dfrac{90}{105}$ **32.** $-\dfrac{26}{78}$

33. $\dfrac{60}{108}$ **34.** $\dfrac{75}{125}$

35. $\dfrac{180}{210}$ **36.** $\dfrac{76}{28}$

37. $\dfrac{55}{67}$ **38.** $\dfrac{41}{51}$

39. $\dfrac{36}{96}$ **40.** $\dfrac{\cdot 48}{120}$

41. $\dfrac{25}{35}$ **42.** $\dfrac{16}{20}$

43. $\dfrac{12}{15}$ **44.** $\dfrac{10}{15}$

45. $\dfrac{16}{17}$ **46.** $\dfrac{14}{25}$

47. $\dfrac{77}{88}$ **48.** $\dfrac{10}{210}$

49. $-\dfrac{10}{30}$ **50.** $-\dfrac{14}{28}$

51. $\dfrac{150}{250}$ **52.** $\dfrac{160}{240}$

53. $\dfrac{3,500}{2,800}$ **54.** $\dfrac{3,500}{2,500}$

55. $\dfrac{56}{28}$ **56.** $\dfrac{32}{8}$

Write each fraction as an equivalent fraction with the indicated denominator.

57. $\dfrac{7}{8}$, denominator 40   **58.** $\dfrac{3}{4}$, denominator 24

59. $\dfrac{4}{5}$, denominator 35   **60.** $\dfrac{5}{7}$, denominator 49

61. $\dfrac{5}{6}$, denominator 54   **62.** $\dfrac{11}{16}$, denominator 32

63. $\dfrac{1}{2}$, denominator 30   **64.** $\dfrac{1}{3}$, denominator 60

65. $\dfrac{2}{7}$, denominator 14   **66.** $\dfrac{3}{10}$, denominator 50

67. $\dfrac{9}{10}$, denominator 60   **68.** $\dfrac{2}{3}$, denominator 27

69. $-\dfrac{5}{4}$, denominator 20   **70.** $-\dfrac{9}{4}$, denominator 44

71. $-\dfrac{2}{15}$, denominator 45   **72.** $-\dfrac{5}{12}$, denominator 36

Write each number as a fraction with the indicated denominator.

73. 3 as fifths   **74.** 4 as thirds

75. 6 as eighths   **76.** 3 as sixths

77. 4 as ninths   **78.** 7 as fourths

79. -2 as halves   **80.** -10 as ninths

■ **APPLICATIONS**   *Use the concept of fraction in answering each question.*

81. COMMUTING How much of the commute from home to work has the motorist made?

Home   Work

82. TIME CLOCKS For each clock, how much of the hour has passed?

a.    b.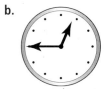

c. 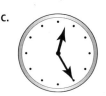   d.

83. SINKHOLES The illustration shows a side view of a depression in the sidewalk near a sinkhole. Describe the movement of the sidewalk using a signed number. (On the tape measure, one inch is divided into 16 equal parts.)

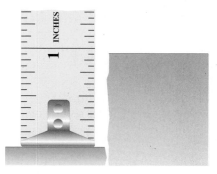

84. POLITICAL PARTIES The illustration on the next page shows the political party affiliation of the governors of the 50 states, as of January 1, 2000.

  a. What fraction were Democrats?

  b. What fraction were Republicans?

  c. What fraction were neither Democrat nor Republican?

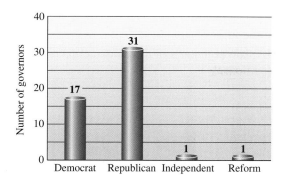

85. PERSONNEL RECORDS Complete the table by finding the amount of the job that will be completed by each person working alone for the given number of hours.

Name	Total time to complete the job alone	Time worked alone	Amount of job completed
Bob	10 hours	7 hours	
Ali	8 hours	1 hour	

86. GAS TANKS How full does the gauge indicate the gas tank is? How much of the tank has been used?

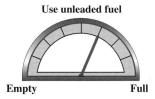

87. MUSIC The illustration shows a side view of the finger position needed to produce a length of string (from the bridge to the fingertip) that gives low C on a violin. To play other notes, fractions of that length are used. Locate these finger positions on the illustration.

a. $\frac{1}{2}$ of the length gives middle C.

b. $\frac{3}{4}$ of the length gives F above low C.

c. $\frac{2}{3}$ of the length gives G.

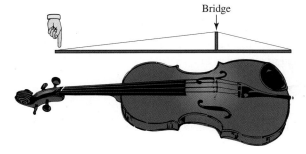

88. RULERS The illustration shows a ruler. First, tell how many spaces there are between the numbers 0 and 1. Then tell to what number the arrow is pointing.

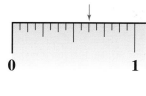

89. MACHINERY The operator of of a machine is to turn the dial from setting A to setting B. Express this in two different ways, using fractions of one complete revolution.

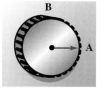

90. EARTH'S ROTATION The Earth rotates about its vertical axis once every 24 hours.

a. What is the significance of $\frac{1}{24}$ of a rotation to us on Earth?

b. What significance does $\frac{24}{24}$ of a revolution have?

91. SUPERMARKET DISPLAYS The amount of space to be given each type of snack food in a supermarket display case is expressed as a fraction. Complete the model of the display, showing where the adjustable shelves should be located, and label where each snack food should be stocked.

$\frac{3}{8}$: potato chips

$\frac{2}{8}$: peanuts

$\frac{1}{8}$: pretzels

$\frac{2}{8}$: tortilla chips

SNACKS

92. MEDICAL CENTERS Hospital designers have located a nurse's station at the center of a circular building. Show how to divide the surrounding office space so that each medical department has the proper fractional amount allocated to it. (Use the circle graph provided.) Label each department.

$\frac{2}{12}$: Radiology

$\frac{5}{12}$: Pediatrics

$\frac{1}{12}$: Laboratory

$\frac{3}{12}$: Orthopedics

$\frac{1}{12}$: Pharmacy

Nurse's station

93. CAMERAS When the shutter of a camera stays open longer than $\frac{1}{125}$ second, any movement of the camera will probably blur the picture. With this in mind, if a photographer is taking a picture of a fast-moving object, should she select a shutter speed of $\frac{1}{60}$ or $\frac{1}{250}$?

94. GDP The Gross Domestic Product is the official measure of the size of the U.S. economy. It represents the market value of all goods and services that have been bought during a given period of time. The GDP for the second quarter of 2004 is listed below. What is meant by the phrase *second quarter of 2004*?

Second quarter of 2004	$11,657,500,000,000

WRITING

95. Explain the concept of equivalent fractions.

96. What does it mean for a fraction to be in lowest terms?

97. Explain the difference between three-fourths and three-fifths of a pizza.

98. Explain both parts of the fundamental property of fractions.

REVIEW

99. Evaluate: $3 + 4 \cdot 2$.

100. Multiply: $(-3)(-4)$.

101. Round 564,112 to the nearest thousand.

102. Give the definition of a prime number.

103. Add: $2 + (-3) + (-5)$.

104. Evaluate the power: $(-10)^2$.

3.2 Multiplying Fractions

- Multiplying fractions   • Simplifying when multiplying fractions
- Powers of a fraction   • Applications

In the next three sections, we discuss how to add, subtract, multiply, and divide fractions. We begin with the operation of multiplication.

Multiplying fractions

Suppose that a television network is going to take out a full-page ad to publicize its fall lineup of shows. The prime-time shows are to get $\frac{3}{5}$ of the ad space and daytime programming the remainder. Of the space devoted to prime time, $\frac{1}{2}$ is to be used to promote weekend programs. How much of the newspaper page will be used to advertise weekend prime-time programs?

The ad for the weekend prime-time shows will occupy $\frac{1}{2}$ of $\frac{3}{5}$ of the page. This can be expressed as $\frac{1}{2} \cdot \frac{3}{5}$. We can calculate $\frac{1}{2} \cdot \frac{3}{5}$ using a three-step process.

Step 1: We divide the page into fifths and shade three of them. This represents the fraction $\frac{3}{5}$, the amount of the page used to advertise prime-time shows.

Step 2: Next, we find $\frac{1}{2}$ of the shaded part of the page by dividing the page into halves, using a vertical line.

Step 3: Finally, we highlight (in purple) $\frac{1}{2}$ of the shaded parts determined in step 2. The highlighted parts are 3 out of 10 or $\frac{3}{10}$ of the page. They represent the amount of the page used to advertise the weekend prime-time shows. This leads us to the conclusion that $\frac{1}{2} \cdot \frac{3}{5} = \frac{3}{10}$.

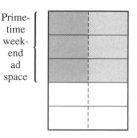

Prime-time weekend ad space

Two observations can be made from this result.

- The numerator of the answer is the product of the numerators of the original fractions.

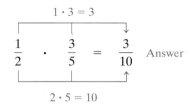

$$1 \cdot 3 = 3$$
$$\frac{1}{2} \quad \cdot \quad \frac{3}{5} \quad = \quad \frac{3}{10} \quad \text{Answer}$$
$$2 \cdot 5 = 10$$

- The denominator of the answer is the product of the denominators of the original fractions.

These observations suggest the following rule for multiplying two fractions.

> **Multiplying fractions**
>
> To multiply two fractions, multiply the numerators and multiply the denominators. Then simplify the result, if possible.

EXAMPLE 1   Multiply: $\dfrac{7}{8} \cdot \dfrac{3}{5}$.

Solution

$$\frac{7}{8} \cdot \frac{3}{5} = \frac{7 \cdot 3}{8 \cdot 5} \qquad \text{Multiply the numerators and multiply the denominators.}$$

$$= \frac{21}{40} \qquad \text{Perform the multiplications: } 7 \cdot 3 = 21 \text{ and } 8 \cdot 5 = 40.$$

Self Check 1

Multiply: $\dfrac{5}{9} \cdot \dfrac{2}{3}$.

Answer $\dfrac{10}{27}$

The rules for multiplying integers also hold for multiplying fractions. When we multiply two fractions with *like* signs, the product is positive. When we multiply two fractions with *unlike* signs, the product is negative.

EXAMPLE 2   Multiply: $-\dfrac{3}{4}\left(\dfrac{1}{8}\right)$.

Solution

$$-\frac{3}{4}\left(\frac{1}{8}\right) = -\frac{3 \cdot 1}{4 \cdot 8} \qquad \begin{array}{l}\text{Multiply the numerators and multiply the denominators. Since} \\ \text{the fractions have unlike signs, the product is negative.}\end{array}$$

$$= -\frac{3}{32} \qquad \text{Perform the multiplications: } 3 \cdot 1 = 3 \text{ and } 4 \cdot 8 = 32.$$

Self Check 2

Multiply: $\dfrac{5}{6}\left(-\dfrac{1}{3}\right)$.

Answer $-\dfrac{5}{18}$

Simplifying when multiplying fractions

After multiplying two fractions, we should simplify the result, if possible.

EXAMPLE 3   Multiply: $\dfrac{5}{8} \cdot \dfrac{4}{5}$.

Solution

$$\dfrac{5}{8} \cdot \dfrac{4}{5} = \dfrac{5 \cdot 4}{8 \cdot 5} \qquad \text{Multiply the numerators and multiply the denominators.}$$

$$= \dfrac{5 \cdot 4}{4 \cdot 2 \cdot 5} \qquad \text{In the denominator, factor 8 as } 4 \cdot 2 \text{ so that we can simplify the fraction.}$$

$$= \dfrac{\overset{1}{\cancel{5}} \cdot \overset{1}{\cancel{4}}}{\underset{1}{\cancel{4}} \cdot 2 \cdot \underset{1}{\cancel{5}}} \qquad \text{Divide out the common factors of 4 and 5.}$$

$$= \dfrac{1}{2} \qquad \text{Perform the multiplications: } 1 \cdot 1 = 1 \text{ and } 1 \cdot 2 \cdot 1 = 2.$$

EXAMPLE 4   Multiply: $45\left(-\dfrac{1}{14}\right)\left(-\dfrac{7}{10}\right)$.

Solution   This expression involves three factors, and one of them is an integer. When multiplying fractions and integers, we express each integer as a fraction with a denominator of 1.

Recall from Section 2.4 that a product is positive when there are an even number of negative factors. Since $45(-\frac{1}{14})(-\frac{7}{10})$ has *two* negative factors, the product is positive.

$$45\left(-\dfrac{1}{14}\right)\left(-\dfrac{7}{10}\right) = \dfrac{45}{1}\left(\dfrac{1}{14}\right)\left(\dfrac{7}{10}\right) \qquad \begin{array}{l}\text{Write 45 as a fraction: } 45 = \frac{45}{1}. \text{ Since the} \\ \text{product has an even number of negative} \\ \text{factors, the answer is positive. We can drop} \\ \text{both } - \text{ signs and continue.}\end{array}$$

$$= \dfrac{45 \cdot 1 \cdot 7}{1 \cdot 14 \cdot 10} \qquad \begin{array}{l}\text{Multiply the numerators and multiply the} \\ \text{denominators.}\end{array}$$

$$= \dfrac{5 \cdot 3 \cdot 3 \cdot 1 \cdot 7}{1 \cdot 7 \cdot 2 \cdot 5 \cdot 2} \qquad \begin{array}{l}\text{Factor 45 as } 5 \cdot 3 \cdot 3, \text{ factor 14 as } 7 \cdot 2, \text{ and} \\ \text{factor 10 as } 5 \cdot 2.\end{array}$$

$$= \dfrac{\overset{1}{\cancel{5}} \cdot 3 \cdot 3 \cdot 1 \cdot \overset{1}{\cancel{7}}}{1 \cdot \underset{1}{\cancel{7}} \cdot 2 \cdot \underset{1}{\cancel{5}} \cdot 2} \qquad \text{Divide out the common factors of 5 and 7.}$$

$$= \dfrac{9}{4} \qquad \begin{array}{l}\text{Multiply in the numerator and in the} \\ \text{denominator.}\end{array}$$

Powers of a fraction

If the base of an exponential expression is a fraction, the exponent tells us how many times to write that fraction as a factor. For example,

$$\left(\dfrac{2}{3}\right)^2 = \dfrac{2}{3} \cdot \dfrac{2}{3} = \dfrac{2 \cdot 2}{3 \cdot 3} = \dfrac{4}{9} \qquad \dfrac{2}{3} \text{ is used as a factor 2 times.}$$

EXAMPLE 5 Find each power: **a.** $\left(-\dfrac{2}{3}\right)^2$ and **b.** $-\left(\dfrac{2}{3}\right)^2$.

Self Check 5

Find each power: **a.** $\left(-\dfrac{3}{4}\right)^3$ and **b.** $-\left(\dfrac{3}{4}\right)^2$.

Solution Exponents are used to indicate repeated multiplication.

a. $\left(-\dfrac{2}{3}\right)^2 = \left(-\dfrac{2}{3}\right)\left(-\dfrac{2}{3}\right)$ Write $-\dfrac{2}{3}$ as a factor 2 times.

$= \dfrac{2 \cdot 2}{3 \cdot 3}$ The product of two fractions with like signs is positive. Multiply the numerators and the denominators.

$= \dfrac{4}{9}$ Multiply in the numerator and in the denominator.

b. $-\left(\dfrac{2}{3}\right)^2 = -\dfrac{2}{3} \cdot \dfrac{2}{3}$ Write $\dfrac{2}{3}$ as a factor 2 times.

$= -\dfrac{2 \cdot 2}{3 \cdot 3}$ Multiply the numerators and the denominators.

$= -\dfrac{4}{9}$

Answers **a.** $-\dfrac{27}{64}$, **b.** $-\dfrac{9}{16}$

▌ Applications

EXAMPLE 6 **House of Representatives.** In the United States House of Representatives, a bill was introduced that would require a $\frac{3}{5}$ vote of the 435 members to authorize any tax increase. Under this requirement, how many representatives would have to vote for a tax increase before it could become law?

Solution   To solve this problem, we must find $\frac{3}{5}$ of 435.

$\dfrac{3}{5}$ of $435 = \dfrac{3}{5} \cdot \dfrac{435}{1}$ Here, the word *of* means to multiply. Write 435 as a fraction: $435 = \frac{435}{1}$.

$= \dfrac{3 \cdot 435}{5 \cdot 1}$ Multiply the numerators and multiply the denominators.

$= \dfrac{3 \cdot 3 \cdot 5 \cdot 29}{5 \cdot 1}$ Prime factor 435 as $3 \cdot 5 \cdot 29$.

$= \dfrac{3 \cdot 3 \cdot \overset{1}{\cancel{5}} \cdot 29}{\underset{1}{\cancel{5}} \cdot 1}$ Divide out the common factor of 5.

$= \dfrac{261}{1}$ Multiply in the numerator: $3 \cdot 3 \cdot 1 \cdot 29 = 261$. Multiply in the denominator: $1 \cdot 1 = 1$.

$= 261$ Simplify: $\frac{261}{1} = 261$.

It would take 261 representatives voting in favor to pass a tax increase.

As Figure 3-3 shows, a triangle has three sides. The length of the base of the triangle can be represented by the letter b and the height by the letter h. The height of a triangle is always perpendicular (makes a square corner) to the base. This is denoted by the symbol ⌐.

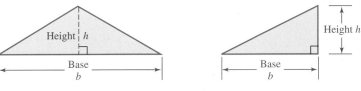

FIGURE 3-3

Recall that the area of a figure is the amount of surface that it encloses. The area of a triangle can be found by using the following formula.

Area of a triangle

The area A of a triangle is one-half the product of its base b and its height h.

$$\text{Area} = \frac{1}{2}(\text{base})(\text{height}) \qquad \text{or} \qquad A = \frac{1}{2} \cdot b \cdot h$$

The formula $A = \frac{1}{2} \cdot b \cdot h$ can be written more simply as $A = \frac{1}{2}bh$. The formula for the area of a triangle can also be written as $A = \dfrac{bh}{2}$.

EXAMPLE 7   Geography.   Approximate the area of the state of Virginia using the triangle in Figure 3-4.

Solution   We can approximate the area of the state by finding the area of the triangle.

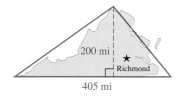

FIGURE 3-4

$$A = \frac{1}{2}bh \qquad\qquad \text{This is the formula for the area of a triangle.}$$

$$= \frac{1}{2} \cdot 405 \cdot 200 \qquad \frac{1}{2}bh \text{ means } \frac{1}{2} \cdot b \cdot h. \text{ Substitute 405 for } b \text{ and 200 for } h.$$

$$= \frac{1}{2} \cdot \frac{405}{1} \cdot \frac{200}{1} \qquad \text{Write 405 and 200 as fractions.}$$

$$= \frac{1 \cdot 405 \cdot 200}{2} \qquad \text{Multiply the numerators. Multiply the denominators.}$$

$$= \frac{1 \cdot 405 \cdot 100 \cdot \overset{1}{\cancel{2}}}{\underset{1}{\cancel{2}}} \qquad \text{Factor 200 as } 100 \cdot 2. \text{ Then divide out the common factor of 2.}$$

$$= 40{,}500 \qquad\qquad 405 \cdot 100 = 40{,}500.$$

The area of the state of Virginia is approximately 40,500 square miles.

Section 3.2   STUDY SET

■ VOCABULARY   *Fill in the blanks.*

1. The word *of* in mathematics usually means _____.
2. The _____ of a triangle is the amount of surface that it encloses.
3. The result of a multiplication problem is called the _____.
4. To _____ a fraction means to divide out common factors of the numerator and denominator.
5. When working with triangles, the letter *b* stands for the length of the _____ and *h* stands for the _____.
6. The _____ for the area of a triangle is $A = \frac{1}{2}bh$.

■ CONCEPTS

7. Find the result when multiplying $\frac{1}{3} \cdot \frac{1}{2}$.
8. Write each of the following as a fraction.
   a. 4   **b.** -3
9. Use the following rectangle to find $\frac{1}{3} \cdot \frac{1}{4}$.

   a. Using vertical lines, divide the given rectangle into four equal parts and lightly shade one of them. What fractional part of the rectangle did you shade?
   b. To find $\frac{1}{3}$ of the shaded portion, use two horizontal lines to divide the given rectangle into three equal parts and lightly shade one of them. Into how many equal parts is the rectangle now divided? How many parts have been shaded twice? What is $\frac{1}{3} \cdot \frac{1}{4}$?

10. In the following solution, what initial mistake did the student make that caused him to have to work with such large numbers?

$$\frac{44}{63} \cdot \frac{27}{55} = \frac{44 \cdot 27}{63 \cdot 55}$$
$$= \frac{1,188}{3,465}$$

11. **a.** Is the product of two numbers with unlike signs positive or negative?
    b. Is the product of two numbers with like signs positive or negative?
12. **a.** Multiply $\frac{9}{10}$ and 20.
    b. When we multiply two numbers, is the product always larger than both those numbers?

13. Determine whether each statement is true or false.
    a. $\left(\frac{1}{2}\right)^2 = \frac{1}{2} \cdot \frac{1}{2}$
    b. $\left(\frac{1}{2}\right)^3 = \frac{1}{2} + \frac{1}{2} + \frac{1}{2}$

14. What is the numerator of the result for the multiplication problem shown here?

$$\frac{4}{15} \cdot \frac{3}{4} = \frac{\overset{1}{4} \cdot \overset{1}{3}}{5 \cdot \underset{1}{3} \cdot \underset{1}{4}}$$

■ NOTATION   *Complete each solution.*

15. Multiply: $\frac{5}{8} \cdot \frac{7}{15}$.

$$\frac{5}{8} \cdot \frac{7}{15} = \frac{5 \cdot \blacksquare}{8 \cdot \blacksquare}$$
$$= \frac{5 \cdot 7}{8 \cdot 5 \cdot \blacksquare}$$
$$= \frac{\overset{1}{5} \cdot 7}{8 \cdot \underset{1}{5} \cdot \blacksquare}$$
$$= \frac{7}{24}$$

16. Multiply: $\frac{7}{12} \cdot \frac{4}{21}$.

$$\frac{7}{12} \cdot \frac{4}{21} = \frac{7 \cdot 4}{\blacksquare \cdot \blacksquare}$$
$$= \frac{7 \cdot 4}{4 \cdot \blacksquare \cdot \blacksquare \cdot 3}$$
$$= \frac{\overset{1}{7} \cdot \overset{1}{4}}{\underset{\blacksquare}{4} \cdot 3 \cdot \underset{\blacksquare}{7} \cdot 3}$$
$$= \frac{1}{9}$$

■ PRACTICE   *Multiply and write answers in lowest terms.*

17. $\frac{1}{4} \cdot \frac{1}{2}$ 18. $\frac{1}{3} \cdot \frac{1}{5}$

19. $\frac{3}{8} \cdot \frac{7}{16}$ 20. $\frac{5}{9} \cdot \frac{2}{7}$

21. $\frac{2}{3} \cdot \frac{6}{7}$ 22. $\frac{5}{12} \cdot \frac{3}{4}$

23. $\frac{14}{15} \cdot \frac{11}{8}$ 24. $\frac{5}{16} \cdot \frac{8}{3}$

25. $-\dfrac{15}{24}\cdot\dfrac{8}{25}$

26. $-\dfrac{20}{21}\cdot\dfrac{7}{16}$

27. $\left(-\dfrac{11}{21}\right)\left(-\dfrac{14}{33}\right)$

28. $\left(-\dfrac{16}{35}\right)\left(-\dfrac{25}{48}\right)$

29. $\dfrac{7}{10}\left(\dfrac{20}{21}\right)$

30. $\left(\dfrac{7}{6}\right)\dfrac{9}{49}$

31. $\dfrac{3}{4}\cdot\dfrac{4}{3}$

32. $\dfrac{4}{5}\cdot\dfrac{5}{4}$

33. $\dfrac{1}{3}\cdot\dfrac{15}{16}\cdot\dfrac{4}{25}$

34. $\dfrac{3}{15}\cdot\dfrac{15}{7}\cdot\dfrac{14}{27}$

35. $\left(\dfrac{2}{3}\right)\left(-\dfrac{1}{16}\right)\left(-\dfrac{4}{5}\right)$

36. $\left(\dfrac{3}{8}\right)\left(-\dfrac{2}{3}\right)\left(-\dfrac{12}{27}\right)$

37. $\dfrac{5}{6}\cdot 18$

38. $6\left(-\dfrac{2}{3}\right)$

39. $15\left(-\dfrac{4}{5}\right)$

40. $-2\left(-\dfrac{7}{8}\right)$

41. $\dfrac{3}{16}\cdot 4\cdot\dfrac{2}{3}$

42. $5\cdot\dfrac{7}{5}\cdot\dfrac{3}{14}$

43. $\dfrac{5}{3}\left(-\dfrac{6}{15}\right)(-4)$

44. $\dfrac{5}{6}\left(-\dfrac{2}{3}\right)(-12)$

45. $-\dfrac{2}{3}\left(-\dfrac{6}{5}\right)\left(-\dfrac{20}{3}\right)$

46. $-\dfrac{3}{2}\left(-\dfrac{5}{6}\right)\left(\dfrac{12}{5}\right)$

47. $-\dfrac{11}{12}\cdot\dfrac{18}{55}\cdot 5$

48. $-\dfrac{24}{5}\cdot\dfrac{7}{12}\cdot\dfrac{1}{14}$

49. $\dfrac{3}{4}\left(\dfrac{5}{7}\right)\left(\dfrac{2}{3}\right)\left(\dfrac{7}{3}\right)$

50. $-\dfrac{5}{4}\left(\dfrac{8}{15}\right)\left(\dfrac{2}{3}\right)\left(\dfrac{7}{2}\right)$

Find each power.

51. $\left(\dfrac{2}{3}\right)^2$

52. $\left(\dfrac{3}{5}\right)^2$

53. $\left(-\dfrac{5}{9}\right)^2$

54. $\left(-\dfrac{5}{6}\right)^2$

55. $\left(\dfrac{4}{3}\right)^2$

56. $\left(\dfrac{3}{2}\right)^2$

57. $\left(-\dfrac{3}{4}\right)^3$

58. $\left(-\dfrac{2}{5}\right)^3$

59. Complete the multiplication table of fractions.

·	$\dfrac{1}{2}$	$\dfrac{1}{3}$	$\dfrac{1}{4}$	$\dfrac{1}{5}$	$\dfrac{1}{6}$
$\dfrac{1}{2}$					
$\dfrac{1}{3}$					
$\dfrac{1}{4}$					
$\dfrac{1}{5}$					
$\dfrac{1}{6}$					

60. Complete the table by finding the original fraction, given its square.

Original fraction squared	Original fraction
$\dfrac{1}{9}$	
$\dfrac{1}{100}$	
$\dfrac{4}{25}$	
$\dfrac{16}{49}$	
$\dfrac{81}{36}$	
$\dfrac{9}{121}$	

Find the area of each triangle.

61.

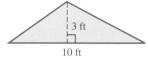

3 ft, 10 ft

62.

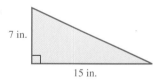

7 in., 15 in.

63.   **64.**

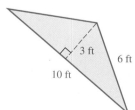   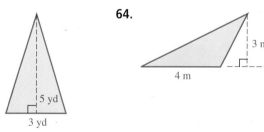

5 yd, 3 yd   3 m, 4 m

65.

3 ft, 6 ft, 10 ft

66.

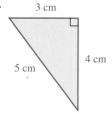

3 cm, 4 cm, 5 cm

APPLICATIONS

67. THE CONSTITUTION Article V of the United States Constitution requires a two-thirds vote of the House of Representatives to propose a constitutional amendment. The House has 435 members. Find the number of votes needed to meet this requirement.

68. GENETICS Gregor Mendel (1822–1884), an Augustinian monk, is credited with developing a heredity model that became the foundation of modern genetics. In his experiments, he crossed purple-flowered plants with white-flowered plants and found that $\frac{3}{4}$ of the offspring plants had purple flowers and $\frac{1}{4}$ had white flowers. According to this concept, when the group of offspring plants shown flower, how many will have purple flowers?

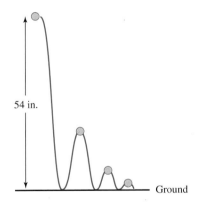

69. TENNIS BALLS A tennis ball is dropped from a height of 54 inches. Each time it hits the ground, it rebounds one-third of the previous height it fell. Find the three missing rebound heights.

54 in.

Ground

70. ELECTIONS The illustration shows the final election returns for a city bond measure.

  a. How many votes were cast?

  b. Find two-thirds of the number of votes cast.

  c. Did the bond measure pass?

> **Measure 1**
> 100% of the precincts reporting
>
> Fire–Police–paramedics General Obligation Bonds
> (Requires two-thirds vote)
>
> Yes 125,599   No 62,801

71. COOKING Use the recipe below, along with the concept of multiplication with fractions, to find how much sugar and molasses are needed to make one dozen cookies.

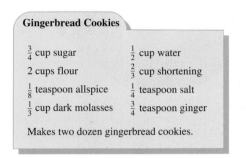

> **Gingerbread Cookies**
>
> $\frac{3}{4}$ cup sugar $\frac{1}{2}$ cup water
>
> 2 cups flour $\frac{2}{3}$ cup shortening
>
> $\frac{1}{8}$ teaspoon allspice $\frac{1}{4}$ teaspoon salt
>
> $\frac{1}{3}$ cup dark molasses $\frac{3}{4}$ teaspoon ginger
>
> Makes two dozen gingerbread cookies.

72. THE EARTH'S SURFACE The surface of the Earth covers an area of approximately 196,800,000 square miles. About $\frac{3}{4}$ of that area is covered by water. Find the number of square miles of the surface covered by water.

73. BOTANY In an experiment, monthly growth rates of three types of plants doubled when nitrogen was added to the soil. Complete the illustration by charting the improved growth rate next to each normal growth rate.

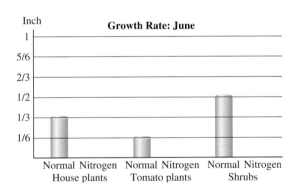

74. STAMPS The best designs in a contest to create a wildlife stamp are shown. To save on paper costs, the postal service has decided to choose the stamp that has the smaller area. Which one did the postal service choose?

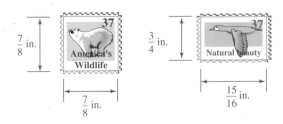

75. STARS AND STRIPES The illustration shows a folded U.S. flag. When it is placed on a table as part of an exhibit, how much area will it occupy?

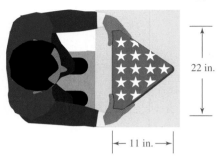

22 in.

11 in.

76. WINDSURFING Estimate the area of the sail on the windsurfing board.

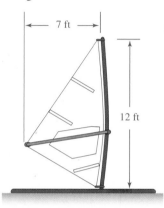

7 ft

12 ft

77. TILE DESIGN A design for bathroom tile is shown. Find the amount of area on a tile that is blue.

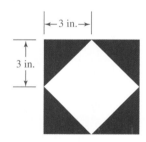

3 in.

3 in.

78. GEOGRAPHY Estimate the area of the state of New Hampshire, using the triangle in the illustration.

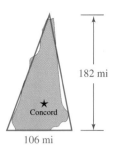

182 mi

★ Concord

106 mi

WRITING

79. In mathematics, the word *of* usually means multiply. Give three real-life examples of this usage.

80. Explain how you could multiply the number 5 and another number and obtain an answer that is less than 5.

81. A MAJORITY The definition of the word *majority* is as follows: "a number greater than one-half of the total." Explain what it means when a teacher says, "A majority of the class voted to postpone the test until Monday." Give an example.

82. What does area measure?

REVIEW

83. Round 6,794 to the nearest hundred.

84. Find the distance covered by a motorist traveling at 60 miles per hour for 3 hours.

85. Evaluate: $5 + 7 \cdot 3$.

86. Multiply: 324
                   $\times$ 45

87. Find the prime factorization of 125.

88. Evaluate: $-2(-3)(-4)$.

3.3 Dividing Fractions

- Division with fractions
- Reciprocals
- A rule for dividing fractions

In this section, we discuss how to divide fractions. We examine problems involving positive and negative fractions. The skills you learned in Section 3.2 will be useful in this section.

Division with fractions

Suppose that the manager of a candy store buys large bars of chocolate and divides each one into four equal parts to sell. How many fourths can be obtained from 5 bars?

We are asking, "How many $\frac{1}{4}$'s are there in 5?" To answer the question, we need to use the operation of division. We can represent this division as $5 \div \frac{1}{4}$. See Figure 3-5.

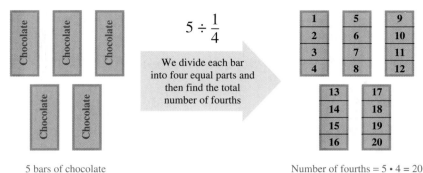

FIGURE 3-5

There are 20 fourths in the 5 bars of chocolate. This result leads to the following observations.

- This division problem involves a fraction: $5 \div \frac{1}{4}$.
- Although we were asked to find $5 \div \frac{1}{4}$, we solved the problem using *multiplication* instead of *division*: $5 \cdot 4 = 20$.

Later in this section, we will see that these observations suggest a rule for dividing fractions. But before we can discuss that rule, we need to introduce a new term.

Reciprocals

Division with fractions involves working with **reciprocals.** To present the concept of reciprocal, we consider the problem $\frac{7}{8} \cdot \frac{8}{7}$.

$$\frac{7}{8} \cdot \frac{8}{7} = \frac{7 \cdot 8}{8 \cdot 7} \qquad \text{Multiply the numerators and multiply the denominators.}$$

$$= \frac{\overset{1}{7} \cdot \overset{1}{8}}{\underset{1}{8} \cdot \underset{1}{7}} \qquad \text{Divide out the common factors of 7 and 8.}$$

$$= \frac{1}{1} \qquad \text{Multiply in the numerator and multiply in the denominator.}$$

$$= 1 \qquad \tfrac{1}{1} = 1.$$

The product of $\frac{7}{8}$ and $\frac{8}{7}$ is 1.

Whenever the product of two numbers is 1, we say that those numbers are *reciprocals.* Therefore, $\frac{7}{8}$ and $\frac{8}{7}$ are reciprocals. To find the reciprocal of a fraction, *we invert the numerator and the denominator.*

Reciprocals

Two numbers are called **reciprocals** if their product is 1.

❗ COMMENT Zero does not have a reciprocal, because the product of 0 and a number can never be 1.

For each number, find its reciprocal and show that their product is 1.

a. $\dfrac{3}{5}$

b. $-\dfrac{5}{6}$

c. 8

Answers **a.** $\dfrac{5}{3}$, **b.** $-\dfrac{6}{5}$, **c.** $\dfrac{1}{8}$

EXAMPLE 1   For each number, find its reciprocal and show that their product is 1:

a. $\dfrac{2}{3}$, **b.** $-\dfrac{3}{4}$, and **c.** 5.

Solution

a. The reciprocal of $\dfrac{2}{3}$ is $\dfrac{3}{2}$.

$$\frac{2}{3} \cdot \frac{3}{2} = \frac{\overset{1}{\cancel{2}} \cdot \overset{1}{\cancel{3}}}{\underset{1}{\cancel{3}} \cdot \underset{1}{\cancel{2}}} = 1$$

b. The reciprocal of $-\dfrac{3}{4}$ is $-\dfrac{4}{3}$.

$$-\frac{3}{4}\left(-\frac{4}{3}\right) = \frac{\overset{1}{\cancel{3}} \cdot \overset{1}{\cancel{4}}}{\underset{1}{\cancel{4}} \cdot \underset{1}{\cancel{3}}} = 1 \qquad \text{The product of two fractions with like signs is positive.}$$

c. $5 = \dfrac{5}{1}$, so the reciprocal of 5 is $\dfrac{1}{5}$.

$$5 \cdot \frac{1}{5} = \frac{5}{1} \cdot \frac{1}{5} = \frac{\overset{1}{\cancel{5}} \cdot 1}{1 \cdot \underset{1}{\cancel{5}}} = 1$$

■ A rule for dividing fractions

In the candy store example, we saw that we can find $5 \div \frac{1}{4}$ by computing $5 \cdot 4$. That is, division by $\frac{1}{4}$ (a fraction) is the same as multiplication by 4 (its reciprocal).

$$5 \div \frac{1}{4} = 5 \cdot 4$$

This observation suggests a general rule for dividing fractions.

> **Dividing fractions**
>
> To divide fractions, multiply the first fraction by the reciprocal of the second fraction.

For example, to find $\frac{5}{7} \div \frac{3}{4}$, we multiply $\frac{5}{7}$ by the reciprocal of $\frac{3}{4}$.

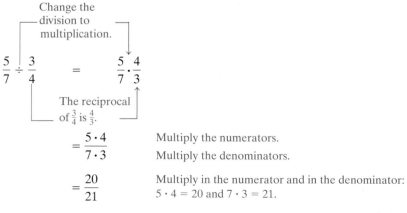

$$= \frac{5 \cdot 4}{7 \cdot 3} \qquad \text{Multiply the numerators.}$$
$$\phantom{= \frac{5 \cdot 4}{7 \cdot 3}} \qquad \text{Multiply the denominators.}$$

$$= \frac{20}{21} \qquad \begin{array}{l}\text{Multiply in the numerator and in the denominator:}\\ 5 \cdot 4 = 20 \text{ and } 7 \cdot 3 = 21.\end{array}$$

Therefore, $\dfrac{5}{7} \div \dfrac{3}{4} = \dfrac{20}{21}$.

EXAMPLE 2 Divide: $\frac{1}{3} \div \frac{4}{5}$.

Solution

$$\frac{1}{3} \div \frac{4}{5} = \frac{1}{3} \cdot \frac{5}{4} \quad \text{Multiply } \frac{1}{3} \text{ by the reciprocal of } \frac{4}{5}, \text{ which is } \frac{5}{4}.$$

$$= \frac{1 \cdot 5}{3 \cdot 4} \quad \text{Multiply the numerators and multiply the denominators.}$$

$$= \frac{5}{12} \quad \text{Multiply in the numerator and in the denominator.}$$

Self Check 2

Divide: $\frac{2}{3} \div \frac{7}{8}$.

Answer $\frac{16}{21}$

EXAMPLE 3 Divide: $\frac{9}{16} \div \frac{3}{20}$.

Solution

$$\frac{9}{16} \div \frac{3}{20} = \frac{9}{16} \cdot \frac{20}{3} \quad \text{Multiply } \frac{9}{16} \text{ by the reciprocal of } \frac{3}{20}, \text{ which is } \frac{20}{3}.$$

$$= \frac{9 \cdot 20}{16 \cdot 3} \quad \text{Multiply the numerators and multiply the denominators.}$$

$$= \frac{\overset{1}{\cancel{3}} \cdot 3 \cdot 5 \cdot \overset{1}{\cancel{4}}}{\underset{1}{\cancel{4}} \cdot 4 \cdot \underset{1}{\cancel{3}}} \quad \text{Factor 9 as } 3 \cdot 3, \text{ factor 20 as } 5 \cdot 4, \text{ and factor 16 as } 4 \cdot 4. \text{ Then divide out the common factors of 3 and 4.}$$

$$= \frac{15}{4} \quad \text{Multiply in the numerator and in the denominator.}$$

Self Check 3

Divide: $\frac{4}{5} \div \frac{8}{25}$.

Answer $\frac{5}{2}$

EXAMPLE 4 **Surfboard designs.** Most surfboards are made of foam plastic covered with several layers of fiberglass to keep them water-tight. How many layers are needed to build up a finish three-eighths of an inch thick if each layer of fiberglass has a thickness of one-sixteenth of an inch?

Solution We need to know how many one-sixteenths there are in three-eighths. To answer this question, we use division and find $\frac{3}{8} \div \frac{1}{16}$.

$$\frac{3}{8} \div \frac{1}{16} = \frac{3}{8} \cdot \frac{16}{1} \quad \text{Multiply } \frac{3}{8} \text{ by the reciprocal of } \frac{1}{16}, \text{ which is } \frac{16}{1}.$$

$$= \frac{3 \cdot 16}{8 \cdot 1} \quad \text{Multiply the numerators and multiply the denominators.}$$

$$= \frac{3 \cdot \overset{1}{\cancel{8}} \cdot 2}{\underset{1}{\cancel{8}} \cdot 1} \quad \text{Factor 16 as } 8 \cdot 2. \text{ Then divide out the common factor of 8.}$$

$$= \frac{6}{1} \quad \text{Multiply in the numerator and in the denominator.}$$

$$= 6 \quad \text{Simplify: } \frac{6}{1} = 6.$$

The number of layers of fiberglass to be applied is 6.

Self Check 5

Divide: $\dfrac{2}{3} \div \left(-\dfrac{7}{6}\right)$.

EXAMPLE 5   Divide: $\dfrac{1}{6} \div \left(-\dfrac{1}{18}\right)$.

Solution   When working with divisions involving negative fractions, we use the same rules as for multiplying numbers with like or unlike signs.

$$\frac{1}{6} \div \left(-\frac{1}{18}\right) = \frac{1}{6}\left(-\frac{18}{1}\right) \qquad \text{Multiply } \frac{1}{6} \text{ by the reciprocal of } -\frac{1}{18}, \text{ which is } -\frac{18}{1}.$$

$$= -\frac{1 \cdot 18}{6 \cdot 1} \qquad \begin{array}{l}\text{The product of two fractions with unlike signs is negative.}\\ \text{Multiply the numerators and multiply the denominators.}\end{array}$$

$$= -\frac{1 \cdot \overset{1}{\cancel{6}} \cdot 3}{\underset{1}{\cancel{6}} \cdot 1} \qquad \text{Factor 18 as } 6 \cdot 3. \text{ Then divide out the common factor of 6.}$$

$$= -\frac{3}{1} \qquad \text{Multiply in the numerator and in the denominator.}$$

$$= -3$$

Answer $-\dfrac{4}{7}$

Self Check 6

Divide: $-\dfrac{35}{16} \div (-7)$.

EXAMPLE 6   Divide: $-\dfrac{21}{36} \div (-3)$.

Solution

$$-\frac{21}{36} \div (-3) = -\frac{21}{36}\left(-\frac{1}{3}\right) \qquad \text{Multiply } -\frac{21}{36} \text{ by the reciprocal of } -3, \text{ which is } -\frac{1}{3}.$$

$$= \frac{21 \cdot 1}{36 \cdot 3} \qquad \begin{array}{l}\text{The product of two fractions with like signs is}\\ \text{positive. Multiply the numerators and multiply the}\\ \text{denominators.}\end{array}$$

$$= \frac{7 \cdot 3 \cdot 1}{36 \cdot 3} \qquad \text{Factor 21 as } 7 \cdot 3.$$

$$= \frac{7 \cdot \overset{1}{\cancel{3}} \cdot 1}{36 \cdot \underset{1}{\cancel{3}}} \qquad \text{Divide out the common factor of 3.}$$

$$= \frac{7}{36} \qquad \text{Multiply in the numerator and in the denominator.}$$

Answer $\dfrac{5}{16}$

Section 3.3   STUDY SET

VOCABULARY   *Fill in the blanks.*

1. Two numbers are called _____ if their product is 1.

2. The result of a division problem is called the _____.

CONCEPTS

3. Complete this statement.

$$\frac{1}{2} \div \frac{2}{3} = \boxed{} \cdot \boxed{}$$

4. Find the reciprocal of each number.

   a. $\dfrac{2}{5}$ **b.** -3

5. Using horizontal lines, divide each rectangle on the right into thirds. What division problem does this illustrate? What is the quotient of that problem?

6. Using horizontal lines, divide each rectangle below into fifths. What division problem does this illustrate? What is the quotient of that problem?

7. Multiply $\frac{4}{5}$ and its reciprocal. What is the result?

8. Multiply $-\frac{3}{5}$ and its reciprocal. What is the result?

9. a. Find $15 \div 3$.

   b. Rewrite $15 \div 3$ as multiplication by the reciprocal of 3, and find the result.

   c. Complete this statement: Division by 3 is the same as multiplication by ▊.

10. a. Find $10 \div \frac{1}{5}$.

   b. Complete this statement: Division by $\frac{1}{5}$ is the same as multiplication by ▊.

NOTATION *Complete each solution.*

11. $\dfrac{25}{36} \div \dfrac{10}{9} = \dfrac{25}{36} \cdot \dfrac{▊}{▊}$

$$= \dfrac{25 \cdot ▊}{36 \cdot ▊}$$

$$= \dfrac{5 \cdot ▊ \cdot 9}{4 \cdot 9 \cdot 2 \cdot 5}$$

$$= \dfrac{\overset{1}{\cancel{5}} \cdot 5 \cdot \overset{1}{\cancel{9}}}{4 \cdot \underset{1}{\cancel{9}} \cdot 2 \cdot \underset{1}{\cancel{5}}}$$

$$= \dfrac{5}{8}$$

12. $\dfrac{4}{9} \div \dfrac{8}{27} = \dfrac{4}{9} \cdot \dfrac{▊}{▊}$

$$= \dfrac{4 \cdot ▊}{9 \cdot ▊}$$

$$= \dfrac{4 \cdot 3 \cdot 9}{9 \cdot 4 \cdot 2}$$

$$= \dfrac{\overset{1}{\cancel{4}} \cdot 3 \cdot \overset{1}{\cancel{9}}}{\underset{1}{\cancel{9}} \cdot \underset{1}{\cancel{4}} \cdot 2}$$

$$= \dfrac{3}{2}$$

PRACTICE *Find each quotient.*

13. $\dfrac{1}{2} \div \dfrac{3}{5}$ **14.** $\dfrac{5}{7} \div \dfrac{5}{6}$

15. $\dfrac{3}{16} \div \dfrac{1}{9}$ **16.** $\dfrac{5}{8} \div \dfrac{2}{9}$

17. $\dfrac{4}{5} \div \dfrac{4}{5}$ **18.** $\dfrac{2}{3} \div \dfrac{2}{3}$

19. $\left(-\dfrac{7}{4}\right) \div \left(-\dfrac{21}{8}\right)$ **20.** $\left(-\dfrac{15}{16}\right) \div \left(-\dfrac{5}{8}\right)$

21. $3 \div \dfrac{1}{12}$ **22.** $9 \div \dfrac{3}{4}$

23. $120 \div \dfrac{12}{5}$ **24.** $360 \div \dfrac{36}{5}$

25. $-\dfrac{4}{5} \div (-6)$ **26.** $-\dfrac{7}{8} \div (-14)$

27. $\dfrac{15}{16} \div 180$ **28.** $\dfrac{7}{8} \div 210$

29. $-\dfrac{9}{10} \div \dfrac{4}{15}$ **30.** $-\dfrac{3}{4} \div \dfrac{3}{2}$

31. $\dfrac{9}{10} \div \left(-\dfrac{3}{25}\right)$ **32.** $\dfrac{11}{16} \div \left(-\dfrac{9}{16}\right)$

33. $-\dfrac{1}{8} \div 8$ **34.** $-\dfrac{1}{15} \div 15$

35. $\dfrac{15}{32} \div \dfrac{15}{32}$ **36.** $-\dfrac{1}{64} \div \left(-\dfrac{1}{64}\right)$

37. $\dfrac{4}{5} \div \dfrac{3}{2}$ **38.** $\dfrac{2}{3} \div \dfrac{3}{2}$

39. $\dfrac{1}{8} \div \dfrac{3}{4}$ **40.** $\dfrac{1}{9} \div \dfrac{3}{5}$

41. $\dfrac{13}{16} \div \dfrac{1}{2}$ **42.** $\dfrac{7}{8} \div \dfrac{6}{7}$

43. $-\dfrac{15}{32} \div \dfrac{3}{4}$ **44.** $-\dfrac{7}{10} \div \dfrac{4}{5}$

45. $-\dfrac{15}{32} \div \dfrac{5}{64}$ **46.** $-\dfrac{28}{15} \div \dfrac{21}{10}$

47. $\dfrac{25}{7} \div \left(-\dfrac{30}{21}\right)$ **48.** $\dfrac{39}{25} \div \left(-\dfrac{13}{10}\right)$

APPLICATIONS

49. MARATHONS Each lap around a stadium track is $\frac{1}{4}$ mile. How many laps would a runner have to complete to get a 26-mile workout?

50. COOKING A recipe calls for $\frac{3}{4}$ cup of flour, and the only measuring container you have holds $\frac{1}{8}$ cup. How many $\frac{1}{8}$ cups of flour would you need to add to follow the recipe?

51. LASER TECHNOLOGY Using a laser, a technician slices thin pieces of aluminum off the end of a rod that is $\frac{7}{8}$ inch long. How many $\frac{1}{64}$-inch-wide slices can be cut from this rod?

52. FURNITURE A production process applies several layers of a clear acrylic coat to outdoor furniture to help protect it from the weather. If each protective coat is $\frac{3}{32}$ inch thick, how many applications will be needed to build up $\frac{3}{8}$ inch of clear finish?

53. UNDERGROUND CABLE In the illustration, which construction proposal will require the fewest days to install underground TV cable from the broadcasting station to the subdivision?

Proposal	Amount of cable installed per day	Comments
Route 1	$\frac{3}{5}$ of a mile	Longer than Route 2
Route 2	$\frac{2}{5}$ of a mile	Terrain very rocky

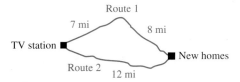

54. PRODUCTION PLANNING The materials used to make a pillow are shown. Examine the inventory list to decide how many pillows can be manufactured in one production run with the materials in stock.

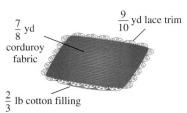

Factory Inventory List

Materials	Amount in stock
Lace trim	135 yd
Corduroy fabric	154 yd
Cotton filling	98 lb

55. 3 × 5 CARDS Ninety 3 × 5 cards are stacked next to a ruler as shown.

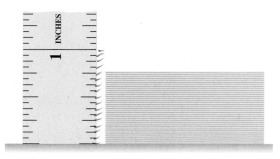

a. Into how many parts is 1 inch divided on the ruler?

b. How thick is the stack of cards?

c. How thick is one 3 × 5 card?

56. COMPUTER PRINTERS The illustration shows how the letter E is formed by a dot matrix printer. What is the height of a dot?

57. FORESTRY A set of forestry maps divides the 6,284 acres of an old-growth forest into $\frac{4}{5}$-acre sections. How many sections do the maps contain?

58. HARDWARE A hardware chain purchases large amounts of nails and packages them in $\frac{9}{16}$-pound bags for sale. How many of these bags of nails can be obtained from 2,871 pounds of nails?

WRITING

59. Explain how to divide two fractions.

60. Explain why 0 does not have a reciprocal.

61. Write an application problem that could be solved by finding $10 \div \frac{1}{5}$.

62. Explain why dividing a fraction by 2 is the same as finding $\frac{1}{2}$ of it.

REVIEW   Fill in the blanks.

63. The symbol $<$ means ____ _____ _____.

64. The statement $9 \cdot 8 = 8 \cdot 9$ illustrates the _____ property of multiplication.

65. Zero is neither _____ nor negative.

66. The sum of two negative numbers is _____.

67. True or false: If equal amounts are subtracted from the numerator and the denominator of a fraction, the result will be an equivalent fraction.

68. Graph each of these numbers on a number line: $-2, 0, -4$, and the opposite of 1.

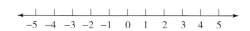

69. Evaluate: $-5 + (-3)(8)$.

70. Find each power.

a. 3^5 **b.** $(-2)^5$

3.4 Adding and Subtracting Fractions

- Fractions with the same denominator • Fractions with different denominators
- Finding the LCD • Comparing fractions

In arithmetic, we can add or subtract only objects that are similar. For example, we can add dollars to dollars, but we cannot add dollars to oranges. This concept is important when adding or subtracting fractions.

Fractions with the same denominator

Consider the problem $\frac{3}{5} + \frac{1}{5}$. When we write it in words, it is apparent that we are adding similar objects.

three-**fifths**   +   one-**fifth**
└── Similar objects ──┘

Because the denominators of $\frac{3}{5}$ and $\frac{1}{5}$ are the same, we say that they have a **common denominator.** Since the fractions have a common denominator, we can add them. Figure 3-6 illustrates the addition process.

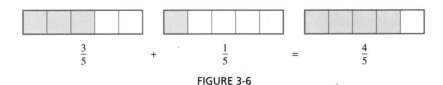

$$\frac{3}{5} \qquad + \qquad \frac{1}{5} \qquad = \qquad \frac{4}{5}$$

FIGURE 3-6

We can make some observations about the addition shown in the figure.

The *sum* of the numerators is the numerator of the answer.

$$\frac{3}{5} + \frac{1}{5} = \frac{4}{5}$$

The answer is a fraction that has the *same* denominator as the two fractions that were added.

These observations suggest the following rule.

Adding or subtracting fractions with the same denominators

To add (or subtract) fractions with the same denominators, add (or subtract) their numerators, write that result over the common denominator, and simplify the result, if possible.

Self Check 1

Add: $\dfrac{5}{12} + \dfrac{1}{12}$.

EXAMPLE 1 Add: $\dfrac{1}{8} + \dfrac{5}{8}$.

Solution

$$\dfrac{1}{8} + \dfrac{5}{8} = \dfrac{1+5}{8} \qquad \text{Add the numerators. Write the sum over the common denominator 8.}$$

$$= \dfrac{6}{8} \qquad \text{Perform the addition: } 1 + 5 = 6. \text{ The fraction can be simplified.}$$

$$= \dfrac{\overset{1}{\cancel{2}} \cdot 3}{\underset{1}{\cancel{2}} \cdot 4} \qquad \text{Factor 6 as } 2 \cdot 3 \text{ and 8 as } 2 \cdot 4. \text{ Divide out the common factor of 2.}$$

$$= \dfrac{3}{4} \qquad \text{Multiply in the numerator and in the denominator.}$$

Answer $\dfrac{1}{2}$

Self Check 2

Subtract: $-\dfrac{9}{11} - \left(-\dfrac{3}{11}\right)$.

EXAMPLE 2 Subtract: $-\dfrac{7}{3} - \left(-\dfrac{2}{3}\right)$.

Solution

$$-\dfrac{7}{3} - \left(-\dfrac{2}{3}\right) = -\dfrac{7}{3} + \dfrac{2}{3} \qquad \text{Add the opposite of } -\dfrac{2}{3}, \text{ which is } \dfrac{2}{3}.$$

$$= \dfrac{-7}{3} + \dfrac{2}{3} \qquad \text{Write } -\dfrac{7}{3} \text{ as } \dfrac{-7}{3}.$$

$$= \dfrac{-7 + 2}{3} \qquad \text{Add the numerators. Write the sum over the common denominator 3.}$$

$$= \dfrac{-5}{3} \qquad \text{Perform the addition: } -7 + 2 = -5.$$

$$= -\dfrac{5}{3} \qquad \text{Rewrite the fraction: } \dfrac{-5}{3} = -\dfrac{5}{3}.$$

Answer $-\dfrac{6}{11}$

Self Check 3

Perform the operations:

$$\dfrac{2}{9} + \dfrac{2}{9} + \dfrac{2}{9}$$

EXAMPLE 3 Perform the operations: $\dfrac{18}{25} - \dfrac{2}{25} - \dfrac{1}{25}$.

Solution

$$\dfrac{18}{25} - \dfrac{2}{25} - \dfrac{1}{25} = \dfrac{16}{25} - \dfrac{1}{25} \qquad \text{Work left to right. Perform the subtraction: } \dfrac{18}{25} - \dfrac{2}{25} = \dfrac{16}{25}.$$

$$= \dfrac{15}{25} \qquad \text{Perform the subtraction.}$$

$$= \dfrac{\overset{1}{\cancel{5}} \cdot 3}{\underset{1}{\cancel{5}} \cdot 5} \qquad \text{Factor 15 as } 5 \cdot 3 \text{ and 25 as } 5 \cdot 5. \text{ Divide out the common factor of 5.}$$

$$= \dfrac{3}{5} \qquad \text{Simplify.}$$

Answer $\dfrac{2}{3}$

▌ Fractions with different denominators

Now we consider the problem $\frac{3}{5} + \frac{1}{3}$. Since the denominators are different, we cannot add these fractions in their present form.

three-**fifths**   +   one-**third**

      ↳ Not similar objects ↴

To add these fractions, we need to find a common denominator. The smallest common denominator (called the **least** or **lowest common denominator**) usually is the easiest common denominator to work with.

> **Least common denominator**
>
> The **least common denominator (LCD)** for a set of fractions is the smallest number each denominator will divide exactly.

In the problem $\frac{3}{5} + \frac{1}{3}$, the denominators are 5 and 3. The numbers 5 and 3 divide many numbers exactly: 30, 45, and 60, to name a few. But the smallest number that 5 and 3 divide exactly is 15. This is the LCD. We now build each fraction into a fraction with a denominator of 15 by applying the fundamental property of fractions.

$$\frac{3}{5} + \frac{1}{3} = \frac{3 \cdot 3}{5 \cdot 3} + \frac{1 \cdot 5}{3 \cdot 5}$$

↑ We need to multiply this denominator by 3 to obtain 15.   ↑ We need to multiply this denominator by 5 to obtain 15.

$$= \frac{9}{15} + \frac{5}{15} \qquad \text{Perform the multiplications in the numerators and denominators.}$$

$$= \frac{9 + 5}{15} \qquad \text{Add the numerators and write the sum over the common denominator 15.}$$

$$= \frac{14}{15} \qquad \text{Perform the addition: } 9 + 5 = 14.$$

Figure 3-7 shows $\frac{3}{5}$ and $\frac{1}{3}$ expressed as equivalent fractions with a denominator of 15. Once the denominators are the same, the fractions can be added easily.

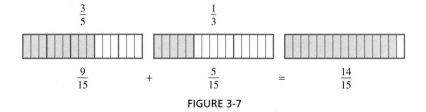

FIGURE 3-7

To add or subtract fractions with different denominators, we follow these steps.

> **Adding or subtracting fractions with different denominators**
>
> **1.** Find the LCD.
> **2.** Express each fraction as an equivalent fraction with a denominator that is the LCD.
> **3.** Add or subtract the resulting fractions. Simplify the result if possible.

Self Check 4

Add: $\dfrac{1}{2} + \dfrac{2}{5}$.

EXAMPLE 4 Add: $\dfrac{1}{8} + \dfrac{2}{3}$.

Solution Since the smallest number the denominators 8 and 3 divide exactly is 24, the LCD is 24.

$$\dfrac{1}{8} + \dfrac{2}{3} = \dfrac{1 \cdot 3}{8 \cdot 3} + \dfrac{2 \cdot 8}{3 \cdot 8} \qquad \text{Express each fraction in terms of 24ths.}$$

$$= \dfrac{3}{24} + \dfrac{16}{24} \qquad \text{Perform the multiplications in the numerators and denominators.}$$

$$= \dfrac{3 + 16}{24} \qquad \text{Add the numerators and write the sum over the common denominator 24.}$$

$$= \dfrac{19}{24} \qquad \text{Perform the addition in the numerator: } 3 + 16 = 19.$$

Answer $\dfrac{9}{10}$

Self Check 5

Subtract: $\dfrac{6}{7} - \dfrac{3}{5}$.

EXAMPLE 5 Subtract: $\dfrac{7}{16} - \dfrac{1}{3}$.

Solution Since the smallest number the denominators 16 and 3 divide exactly is 48, the LCD is 48.

$$\dfrac{7}{16} - \dfrac{1}{3} = \dfrac{7 \cdot 3}{16 \cdot 3} - \dfrac{1 \cdot 16}{3 \cdot 16} \qquad \text{Express each fraction in terms of 48ths.}$$

$$= \dfrac{21}{48} - \dfrac{16}{48} \qquad \text{Perform the multiplications in the numerators and denominators.}$$

$$= \dfrac{21 - 16}{48} \qquad \text{Subtract the numerators and keep the common denominator.}$$

$$= \dfrac{5}{48} \qquad \text{Perform the subtraction.}$$

Answer $\dfrac{9}{35}$

Self Check 6

Add: $-6 + \dfrac{2}{9}$.

EXAMPLE 6 Add: $-5 + \dfrac{1}{4}$.

Solution

$$-5 + \dfrac{1}{4} = \dfrac{-5}{1} + \dfrac{1}{4} \qquad \text{Write } -5 \text{ as } \tfrac{-5}{1}. \text{ The smallest number that 1 and 4 divide exactly is 4, so the LCD is 4.}$$

$$= \dfrac{-5 \cdot 4}{1 \cdot 4} + \dfrac{1}{4} \qquad \text{Express } \tfrac{-5}{1} \text{ in terms of 4ths.}$$

$$= \dfrac{-20}{4} + \dfrac{1}{4} \qquad \text{Multiply in the numerator and in the denominator.}$$

$$= \dfrac{-20 + 1}{4} \qquad \text{Write the sum of the numerators over the common denominator 4.}$$

$$= \dfrac{-19}{4} \qquad \text{Perform the addition: } -20 + 1 = -19.$$

$$= -\dfrac{19}{4} \qquad \text{Rewrite: } \dfrac{-19}{4} = -\dfrac{19}{4}.$$

Answer $-\dfrac{52}{9}$

Finding the LCD

When we add or subtract fractions with different denominators, the least common denominator is not always obvious. We now develop a method that can be used to find the LCD.

Let's find the least common denominator of $\frac{3}{8}$ and $\frac{1}{10}$. We have learned that both 8 and 10 must divide the LCD exactly. Therefore, the LCD must include the prime factorization of 8, which is $2 \cdot 2 \cdot 2$, and the prime factorization of 10, which is $2 \cdot 5$. In Figure 3-8, we see that $2 \cdot 2 \cdot 2 \cdot 5$ is the smallest number that meets both of these requirements. Notice that the common factor of 2 is "shared" by both prime factorizations.

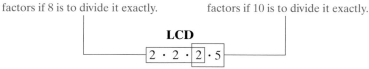

$$8 = 2 \cdot 2 \cdot 2, \qquad\qquad 2 \cdot 5 = 10$$

The LCD must have this set of factors if 8 is to divide it exactly.   The LCD must have this set of factors if 10 is to divide it exactly.

LCD

$$2 \cdot 2 \cdot \boxed{2} \cdot 5$$

FIGURE 3-8

Finding the LCD

To find the LCD for a set of fractions,

1. Prime factor each of the denominators.

2. Select one of the prime factorizations and compare the others to it, one at a time. It must contain each of the other factorizations. If it does not, multiply it by any of the prime factors it lacks.

3. The LCD is the product of the prime factors that result from this comparison.

You will find that the product that gives the LCD contains each of the prime factors the greatest number of times they appear in any single prime factorization.

EXAMPLE 7   Add: $\dfrac{11}{15} + \dfrac{3}{10}$.

Solution   To find the LCD, we factor each denominator: $15 = 3 \cdot 5$ and $10 = 2 \cdot 5$. Then we compare them. The factorization $3 \cdot 5$ does not contain the factorization $2 \cdot 5$; it lacks the factor of 2. Therefore, we multiply $3 \cdot 5$ by 2 to obtain the LCD.

$$15 = 3 \cdot 5 \text{———} \boxed{3 \cdot \boxed{5} \cdot 2} \text{———} 2 \cdot 5 = 10$$

LCD = 30

$$\frac{11}{15} + \frac{3}{10} = \frac{11 \cdot 2}{15 \cdot 2} + \frac{3 \cdot 3}{10 \cdot 3} \qquad \text{Express each fraction in terms of 30ths.}$$

$$= \frac{22}{30} + \frac{9}{30} \qquad \text{Perform the multiplications in the numerators and denominators.}$$

$$= \frac{22 + 9}{30} \qquad \text{Write the sum of the numerators over the common denominator 30.}$$

$$= \frac{31}{30} \qquad \text{Perform the addition: } 22 + 9 = 31.$$

Self Check 7

Add: $\dfrac{3}{8} + \dfrac{5}{6}$.

Answer $\dfrac{29}{24}$

Self Check 8

Subtract: $\dfrac{33}{35} - \dfrac{11}{14}$.

EXAMPLE 8 Subtract: $\dfrac{20}{21} - \dfrac{5}{6}$.

Solution To find the LCD, we factor each denominator: $21 = 7 \cdot 3$ and $6 = 3 \cdot 2$. Then we compare them. The factorization $7 \cdot 3$ does not contain the factorization $3 \cdot 2$; it lacks the factor of 2. Therefore, we multiply $7 \cdot 3$ by 2 to obtain the LCD.

$$21 = 7 \cdot 3 \quad\rule{0.5cm}{0.4pt}\quad \boxed{7 \cdot \boxed{3} \cdot 2} \quad\rule{0.5cm}{0.4pt}\quad 3 \cdot 2 = 6$$

$$\textbf{LCD = 42}$$

$$\dfrac{20}{21} - \dfrac{5}{6} = \dfrac{20 \cdot 2}{21 \cdot 2} - \dfrac{5 \cdot 7}{6 \cdot 7} \qquad \text{Express each fraction in terms of 42nds.}$$

$$= \dfrac{40}{42} - \dfrac{35}{42} \qquad \text{Perform the multiplications in the numerators and denominators.}$$

$$= \dfrac{40 - 35}{42} \qquad \text{Write the difference of the numerators over the common denominator 42.}$$

$$= \dfrac{5}{42} \qquad \text{Perform the subtraction: } 40 - 35 = 5.$$

Answer $\dfrac{11}{70}$

EXAMPLE 9 Television viewing habits. Students on a college campus were asked to estimate to the nearest hour how much television they watched each day. The results are given in the pie chart in Figure 3-9. For example, the chart tells us that $\frac{1}{4}$ of those responding watched 1 hour per day. Find the fraction of the student body that watches from 0 to 2 hours daily.

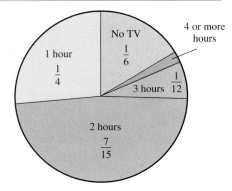

FIGURE 3-9

Solution To answer this question, we need to add $\frac{1}{6}, \frac{1}{4}$, and $\frac{7}{15}$. To find the LCD, we prime factor each of the denominators.

$$4 = 2 \cdot 2 \quad\rule{0.5cm}{0.4pt}\quad \boxed{2 \cdot \boxed{2} \cdot \boxed{3} \cdot 5} \quad\rule{0.5cm}{0.4pt}\quad 3 \cdot 5 = 15$$

$$6 = 2 \cdot 3$$

$$\textbf{LCD} = 2 \cdot 2 \cdot 3 \cdot 5 = \textbf{60}$$

$$\dfrac{1}{6} + \dfrac{1}{4} + \dfrac{7}{15} = \dfrac{1 \cdot 10}{6 \cdot 10} + \dfrac{1 \cdot 15}{4 \cdot 15} + \dfrac{7 \cdot 4}{15 \cdot 4} \qquad \text{Express each fraction in terms of 60ths.}$$

$$= \dfrac{10}{60} + \dfrac{15}{60} + \dfrac{28}{60} \qquad \text{Perform the multiplication in the numerators and denominators.}$$

$$= \dfrac{10 + 15 + 28}{60} \qquad \text{Add the numerators. Write the sum over the common denominator 60.}$$

$$= \dfrac{53}{60} \qquad \text{Perform the addition: } 10 + 15 + 28 = 53.$$

The fraction of the student body that watches 0 to 2 hours of television daily is $\frac{53}{60}$.

Comparing fractions

If fractions have the same denominator, the fraction with the larger numerator is the larger fraction. If their denominators are different, we need to write the fractions with a common denominator before we can make a comparison.

> **Comparing fractions**
>
> To compare unlike fractions, write the fractions as equivalent fractions with the same denominator — preferably the LCD. Then compare their numerators. The fraction with the larger numerator is the larger fraction.

EXAMPLE 10   Which fraction is larger: $\dfrac{5}{6}$ or $\dfrac{7}{8}$?

Solution   To compare these fractions, we express each with the LCD of 24.

$$\frac{5}{6} = \frac{5 \cdot 4}{6 \cdot 4} \qquad \frac{7}{8} = \frac{7 \cdot 3}{8 \cdot 3} \qquad \text{Express each fraction in terms of 24ths.}$$

$$= \frac{20}{24} \qquad\qquad = \frac{21}{24} \qquad \begin{array}{l}\text{Perform the multiplications in the numerators and}\\\text{denominators.}\end{array}$$

Next, we compare the numerators. Since $21 > 20$, we conclude that $\frac{21}{24}$ is greater than $\frac{20}{24}$. Thus, $\frac{7}{8} > \frac{5}{6}$.

Self Check 10

Which fraction is larger: $\dfrac{7}{12}$ or $\dfrac{3}{5}$?

Answer $\dfrac{3}{5}$

Budgets

THINK IT THROUGH

"Putting together a budget is crucial if you don't want to spend your way into serious problems. You're also developing a habit that can serve you well throughout your life." Liz Pulliam Weston, MSN Money

The circle graph below shows a suggested budget for new college graduates as recommended by Springboard, a nonprofit consumer credit counseling service. What fraction of net take-home pay should be spent on housing?

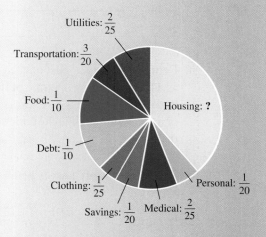

Section 3.4 STUDY SET

VOCABULARY *Fill in the blanks.*

1. The _____ common denominator for a set of fractions is the smallest number each denominator will divide exactly.

2. _____ fractions, such as $\frac{1}{2}$ and $\frac{2}{4}$, are fractions that represent the same amount.

3. To express a fraction in _____ terms, we multiply the numerator and denominator by the same number.

4. _____ a fraction is the process of multiplying the numerator and the denominator of the fraction by the same number.

CONCEPTS

5. Fill in the blanks. To add two fractions with a common denominator, we add the _____ and then write that result over the _____ denominator.

6. **a.** Add the indicated fractions.

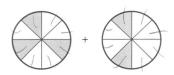

   b. Subtract the indicated fractions.

7. Why must we do some preliminary work before performing the following addition?

$$\frac{2}{9} + \frac{2}{5}$$

8. Why must we do some preliminary work before performing the following subtraction?

$$\frac{5}{6} - \frac{5}{18}$$

9. By what are the numerator and the denominator of the following fractions being multiplied?

   a. $\dfrac{5 \cdot 4}{6 \cdot 4}$ **b.** $\dfrac{1 \cdot 5}{3 \cdot 5}$

10. Consider $\frac{3}{4}$. By what should we multiply the numerator and denominator of $\frac{3}{4}$ to express it as an equivalent fraction with the following denominator?

    a. 12   **b.** 36

11. Consider the following prime factorizations:

$$24 = 3 \cdot 2 \cdot 2 \cdot 2$$
$$90 = 5 \cdot 3 \cdot 3 \cdot 2$$

    For any one factorization, what is the greatest number of times

    a. a 5 appears?

    b. a 3 appears?

    c. a 2 appears?

12. **a.** List the first ten multiples of 9 and the first ten multiples of 12.

    b. Find the smallest number that 9 and 12 divide exactly.

13. The denominators of two fractions involved in a subtraction problem have the prime-factored forms $2 \cdot 2 \cdot 5$ and $2 \cdot 3 \cdot 5$. What is the LCD for the fractions?

14. The denominators of three fractions involved in a subtraction problem have the prime-factored forms $2 \cdot 2 \cdot 5$, $2 \cdot 3 \cdot 5$, and $2 \cdot 3 \cdot 3 \cdot 5$. What is the LCD for the fractions?

15. **a.** Divide the figure on the left into fourths and shade one part. Divide the figure on the right into thirds and shade one part. Which shaded part is larger?

    b. Express the shaded part of each figure in part a as a fraction. Show that one of those fractions is larger than the other by expressing both in terms of a common denominator and then comparing them.

16. Place a $>$ or $<$ symbol in the blank to make a true statement.

    a. $\dfrac{32}{35}$ ___ $\dfrac{31}{35}$ **b.** $\dfrac{7}{8}$ ___ $\dfrac{31}{32}$

NOTATION *Complete each solution.*

17. $\dfrac{2}{5} + \dfrac{1}{3} = \dfrac{2 \cdot \blacksquare}{5 \cdot \blacksquare} + \dfrac{1 \cdot 5}{3 \cdot 5}$

$= \dfrac{\blacksquare}{15} + \dfrac{\blacksquare}{15}$

$= \dfrac{\blacksquare + \blacksquare}{15}$

$= \dfrac{11}{15}$

18. $\dfrac{7}{8} - \dfrac{2}{3} = \dfrac{7 \cdot 3}{\blacksquare \cdot 3} - \dfrac{2 \cdot 8}{\blacksquare \cdot 8}$

$= \dfrac{21}{\blacksquare} - \dfrac{16}{\blacksquare}$

$= \dfrac{21 - 16}{\blacksquare}$

$= \dfrac{5}{24}$

PRACTICE *The denominators of two fractions are given. Find the lowest common denominator.*

19. $18, 6$

20. $15, 3$

21. $8, 6$

22. $10, 4$

23. $8, 20$

24. $14, 21$

25. $15, 12$

26. $25, 30$

Perform each operation. Simplify when necessary.

27. $\dfrac{3}{7} + \dfrac{1}{7}$

28. $\dfrac{16}{25} - \dfrac{9}{25}$

29. $\dfrac{37}{103} - \dfrac{17}{103}$

30. $\dfrac{54}{53} - \dfrac{52}{53}$

31. $\dfrac{12}{25} - \dfrac{1}{25} - \dfrac{1}{25}$

32. $\dfrac{7}{9} + \dfrac{1}{9} + \dfrac{1}{9}$

33. $\dfrac{1}{4} + \dfrac{3}{8}$

34. $\dfrac{2}{3} + \dfrac{1}{6}$

35. $\dfrac{13}{20} - \dfrac{1}{5}$

36. $\dfrac{71}{100} - \dfrac{1}{10}$

37. $\dfrac{4}{5} + \dfrac{2}{3}$

38. $\dfrac{1}{4} + \dfrac{2}{3}$

39. $\dfrac{1}{8} + \dfrac{2}{7}$

40. $\dfrac{1}{6} + \dfrac{5}{9}$

41. $\dfrac{3}{4} - \dfrac{2}{3}$

42. $\dfrac{4}{5} - \dfrac{1}{6}$

43. $\dfrac{5}{6} - \dfrac{3}{4}$

44. $\dfrac{7}{8} - \dfrac{5}{6}$

45. $\dfrac{16}{25} - \left(-\dfrac{3}{10}\right)$

46. $\dfrac{3}{8} - \left(-\dfrac{1}{6}\right)$

47. $-\dfrac{7}{16} + \dfrac{1}{4}$

48. $-\dfrac{17}{20} + \dfrac{4}{5}$

49. $\dfrac{1}{12} - \dfrac{3}{4}$

50. $\dfrac{11}{60} - \dfrac{13}{20}$

51. $-\dfrac{5}{8} - \dfrac{1}{3}$

52. $-\dfrac{7}{20} - \dfrac{1}{5}$

53. $-3 + \dfrac{2}{5}$

54. $-6 + \dfrac{5}{8}$

55. $-\dfrac{3}{4} - 5$

56. $-2 - \dfrac{7}{8}$

57. $\dfrac{5}{24} + \dfrac{3}{16}$

58. $\dfrac{17}{20} - \dfrac{4}{15}$

59. $-\dfrac{11}{15} - \dfrac{2}{9}$

60. $-\dfrac{19}{18} - \dfrac{5}{12}$

61. $\dfrac{7}{25} + \dfrac{1}{15}$

62. $\dfrac{11}{20} - \dfrac{1}{8}$

63. $\dfrac{4}{27} + \dfrac{1}{6}$

64. $\dfrac{8}{9} - \dfrac{7}{12}$

65. $\dfrac{27}{50} + \dfrac{5}{16}$

66. $\dfrac{49}{50} - \dfrac{15}{16}$

67. $\dfrac{7}{30} - \dfrac{19}{75}$

68. $\dfrac{73}{75} - \dfrac{31}{30}$

69. $\dfrac{1}{3} + \dfrac{1}{4} + \dfrac{1}{5}$

70. $\dfrac{1}{10} + \dfrac{1}{8} + \dfrac{1}{5}$

71. $\dfrac{2}{3} + \dfrac{4}{5} + \dfrac{5}{6}$

72. $\dfrac{3}{4} + \dfrac{2}{5} + \dfrac{3}{10}$

73. $-\dfrac{2}{3} + \dfrac{5}{4} + \dfrac{1}{6}$

74. $-\dfrac{3}{4} + \dfrac{3}{8} + \dfrac{7}{6}$

75. Find the difference of $\dfrac{11}{60}$ and $\dfrac{2}{45}$.

76. Find the sum of $\dfrac{9}{48}$ and $\dfrac{7}{40}$.

77. Subtract $\dfrac{5}{12}$ from $\dfrac{2}{15}$.

78. What is the sum of $\dfrac{11}{24}$ and $\dfrac{7}{36}$ increased by $\dfrac{5}{48}$?

APPLICATIONS

79. BOTANY To assess the effects of smog on tree development, botanists cut down a pine tree and measured the width of the growth rings for the last two years.

   a. What was the growth over this two-year period?

   b. What is the difference in the widths of the two rings?

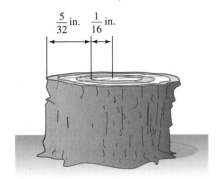

80. MAGAZINE LAYOUTS The page design for a magazine cover includes a blank strip at the top, called a header, and a blank strip at the bottom of the page, called a footer. In the illustration, how much page length is lost because of the header and footer?

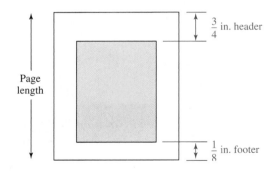

81. DINNERS A family bought two large pizzas for dinner. Some pieces of each pizza were not eaten, as shown. How much pizza was left? Could the family have been fed with just one pizza?

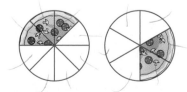

82. GASOLINE BARRELS The contents of two identical-sized barrels are shown in the next column. If they are poured into an empty third barrel that is the same size, how much of the third barrel will they fill?

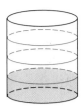

83. WEIGHTS AND MEASURES A consumer protection agency verifies the accuracy of butcher shop scales by placing a known three-quarter-pound weight on the scale and then comparing that to the scale's readout. According to the illustration, by how much is this scale off? Does it result in undercharging or overcharging customers on their meat purchases?

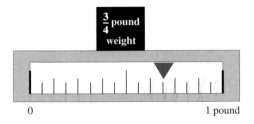

84. WRENCHES A mechanic likes to hang his wrenches above his tool bench in order of narrowest to widest. What is the proper order of the wrenches in the illustration?

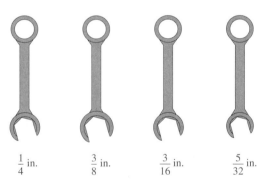

85. HIKING The illustration shows the length of each part of a three-part hike. Rank the lengths from longest to shortest.

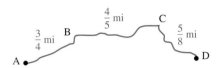

86. FIGURE DRAWING As an aid in drawing the human body, artists divide the body into three parts. Each part is then expressed as a fraction of the total body height. For example, the torso is $\frac{4}{15}$ of the body height. What fraction of body height is the head?

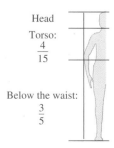

87. STUDY HABITS College students taking a full load were asked to give the average number of hours they studied each day. The results are shown in the pie chart. What fraction of the students study 2 hours or more daily?

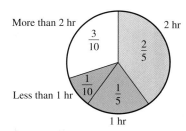

88. MUSICAL NOTES The notes used in music have fractional values. Their names and the symbols used to represent them are shown in Illustration (a). In common time, the values of the notes in each measure must add up to 1. Is the measure in Illustration (b) complete?

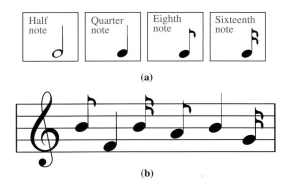

89. GARAGE DOOR OPENERS What is the difference in strength between a $\frac{1}{3}$-hp and a $\frac{1}{2}$-hp garage door opener?

90. DELIVERY TRUCKS A truck can safely carry a one-ton load. Should it be used to deliver one-half ton of sand, one-third ton of gravel, and one-fifth ton of cement in one trip to a job site?

WRITING

91. How are the procedures for expressing a fraction in higher terms and simplifying a fraction to lowest terms similar, and how are they different?

92. Given two fractions, how do we find their lowest common denominator?

93. How do we compare the relative sizes of two fractions with different denominators?

94. What is the difference between a common denominator and the lowest common denominator?

REVIEW

95. Evaluate: $|6 - 10|$.

96. Round 674 to the nearest ten.

97. Evaluate: $3^2 \cdot 4^3$.

98. Find the prime factors of 100.

99. What is the perimeter of a rectangle that has a length of 7 in. and a width of 8 in.?

100. What is the formula for finding the area of a rectangle?

The LCM and the GCF

As we have seen, the **multiples** of a number can be found by multiplying it successively by 1, 2, 3, 4, 5, and so on. The multiples of 4 and the multiples of 6 are shown below.

$1 \cdot 4 = 4$	$1 \cdot 6 = 6$
$2 \cdot 4 = 8$	$2 \cdot 6 = \mathbf{12}$
$3 \cdot 4 = \mathbf{12}$	$3 \cdot 6 = 18$
$4 \cdot 4 = 16$	$4 \cdot 6 = \mathbf{24}$
$5 \cdot 4 = 20$	$5 \cdot 6 = 30$
$6 \cdot 4 = \mathbf{24}$	$6 \cdot 6 = \mathbf{36}$
$7 \cdot 4 = 28$	$7 \cdot 6 = 42$
$8 \cdot 4 = 32$	$8 \cdot 6 = 48$
$9 \cdot 4 = \mathbf{36}$	$9 \cdot 6 = 54$

Common multiples of 4 and 6 are highlighted in red.

Because 12 is the smallest number that is a multiple of both 4 and 6, it is called the **least common multiple (LCM)** of 4 and 6.

Making lists like those shown above can be tedious. A more efficient method to find the least common multiple of several numbers is as follows.

> **Finding the least common multiple**
>
> 1. Write each of the numbers in prime-factored form.
> 2. The least common multiple is a product of prime factors, where each prime factor is used the greatest number of times it appears in any one factorization found in step 1.

Self Check 1
Find the LCM of 18 and 84.

EXAMPLE 1   Find the LCM of 24 and 36.

Solution

Step 1: First, we find the prime factorizations of 24 and 36.

$$24 = 3 \cdot 2 \cdot 2 \cdot 2$$
$$36 = 3 \cdot 3 \cdot 2 \cdot 2$$

Step 2: The prime factorizations of 24 and 36 contain the prime factors 3 and 2. We use each of these factors the greatest number of times it appears in any one factorization.

The greatest number of times 3 appears in any one factorization is two times.   The greatest number of times 2 appears in any one factorization is three times.

$$\text{LCM} = 3 \ \cdot \ 3 \ \cdot \ 2 \ \cdot \ 2 \ \cdot \ 2 = 72$$

Answer 252

The least common multiple of 24 and 36 is 72.

Because 2 divides 36 exactly and because 2 divides 120 exactly, 2 is called a **common factor** of 36 and 120.

$$\frac{36}{2} = 18 \qquad \frac{120}{2} = 60$$

The numbers 36 and 120 have other common factors, such as 3 and 6. The **greatest common factor (GCF)** of 36 and 120 is the largest number that is a factor of both. We follow these steps to find the greatest common factor of several numbers.

> **Finding the greatest common factor**
>
> 1. Write each of the numbers in prime-factored form.
> 2. The greatest common factor is the product of the prime factors that are common to the factorizations found in step 1. If the numbers have no factors in common, the GCF is 1.

Self Check 2
Find the GCF of 60 and 150.

EXAMPLE 2   Find the GCF of 36 and 120.

Solution

Step 1: We find the prime factorizations of 36 and 120.

$$36 = 3 \cdot 3 \cdot 2 \cdot 2$$
$$120 = 5 \cdot 3 \cdot 2 \cdot 2 \cdot 2$$

Step 2: One factor of 3 (highlighted in red) and two factors of 2 (highlighted in blue) are common to the factorizations of 36 and 120. To find the GCF, we form their product.

GCF = **3 · 2 · 2** = 12

The greatest common factor of 36 and 120 is 12.

Answer 30

STUDY SET   THE LCM AND THE GCF

Find the least common multiple of the given numbers.

1. 3, 5
2. 7, 11
3. 8, 14
4. 8, 12
5. 14, 21
6. 16, 20
7. 6, 18
8. 3, 9
9. 44, 60
10. 36, 60
11. 100, 120
12. 120, 180
13. 6, 24, 36
14. 6, 10, 18
15. 18, 54, 63
16. 16, 30, 84

Find the greatest common factor of the given numbers.

17. 6, 9
18. 8, 12
19. 22, 33
20. 15, 20
21. 16, 20
22. 18, 24
23. 25, 100
24. 16, 80
25. 100, 120
26. 120, 180
27. 48, 108
28. 60, 96
29. 18, 24, 36
30. 30, 50, 90
31. 18, 54, 63
32. 28, 42, 84

33. NURSING A nurse, working in an intensive care unit, has to check a patient's vital signs every 45 minutes. Another nurse has to give the same patient his medication every 2 hours. If both nurses are in the patient's room together now, how long will it be until they are once again in the room together?

34. BARBECUES A certain brand of hot dogs comes in packages of 10. A certain brand of hot dog buns comes in packages of 12. For a family reunion barbecue, how many packages of hot dogs and how many packages of hot dog buns should be purchased so that no hot dogs and no buns are wasted?

3.5 Multiplying and Dividing Mixed Numbers

- Mixed numbers • Writing mixed numbers as improper fractions
- Writing improper fractions as mixed numbers • Graphing fractions and mixed numbers
- Multiplying and dividing mixed numbers

In the next two sections, we show how to add, subtract, multiply, and divide *mixed numbers.* These numbers are widely used in daily life.

The recipe calls for $2\frac{1}{3}$ cups of flour.

It took $3\frac{3}{4}$ hours to paint the living room.

The entrance to the park is $1\frac{1}{2}$ miles away.

Mixed numbers

A **mixed number** is the *sum* of a whole number and a proper fraction. For example, $3\frac{3}{4}$ is a mixed number.

$$3\frac{3}{4} \quad = \quad 3 \quad + \quad \frac{3}{4}$$

$\uparrow$ Mixed number $\uparrow$ Whole number $\uparrow$ Proper fraction

! COMMENT Note that $3\frac{3}{4}$ means $3 + \frac{3}{4}$, even though the $+$ symbol is not written. Do not confuse $3\frac{3}{4}$ with $3 \cdot \frac{3}{4}$ or $3(\frac{3}{4})$, which indicate the multiplication of 3 and $\frac{3}{4}$.

In this section, we work with negative as well as positive mixed numbers. For example, the negative mixed number $-4\frac{3}{4}$ could be used to represent $4\frac{3}{4}$ feet below sea level. We think of $-4\frac{3}{4}$ as $-4 - \frac{3}{4}$.

Writing mixed numbers as improper fractions

To see that mixed numbers are related to improper fractions, consider $3\frac{3}{4}$. To write $3\frac{3}{4}$ as an improper fraction, we need to find out how many *fourths* it represents. One way is to use the fundamental property of fractions.

$3\frac{3}{4} = 3 + \frac{3}{4}$ Write the mixed number $3\frac{3}{4}$ as a sum.

$= \frac{3}{1} + \frac{3}{4}$ Write 3 as a fraction: $3 = \frac{3}{1}$.

$= \frac{3 \cdot 4}{1 \cdot 4} + \frac{3}{4}$ Use the fundamental property of fractions to express $\frac{3}{1}$ as a fraction with denominator 4.

$= \frac{12}{4} + \frac{3}{4}$ Perform the multiplications in the numerator and denominator.

$= \frac{15}{4}$ Add the numerators: $12 + 3 = 15$. Write the sum over the common denominator 4.

Thus, $3\frac{3}{4} = \frac{15}{4}$.

We can obtain the same result with far less work. To change $3\frac{3}{4}$ to an improper fraction, we simply multiply 3 by 4 and add 3 to get the numerator, and keep the denominator of 4.

$$3\frac{3}{4} = \frac{3(4) + 3}{4} = \frac{12 + 3}{4} = \frac{15}{4}$$

This example illustrates the following general rule.

> **Writing a mixed number as an improper fraction**
>
> To write a mixed number as an improper fraction, multiply the whole-number part by the denominator of the fraction and add the result to the numerator. Write this sum over the denominator.

EXAMPLE 1   Write the mixed number $5\frac{1}{6}$ as an improper fraction.

Solution

$$5\frac{1}{6} = \frac{5(6) + 1}{6}$$ Multiply 5 by the denominator 6. Add the numerator 1. Write this sum over the denominator 6.

$$= \frac{30 + 1}{6}$$ Perform the multiplication: $5(6) = 30$.

$$= \frac{31}{6}$$ Perform the addition: $30 + 1 = 31$.

Self Check 1
Write the mixed number $3\frac{3}{8}$ as an improper fraction.

Answer $\dfrac{27}{8}$

To write a negative mixed number in fractional form, ignore the − sign and use the method shown in Example 1 on the positive mixed number. Once that procedure is completed, write a − sign in front of the result. For example, $-3\frac{1}{4} = -\frac{13}{4}$.

Writing improper fractions as mixed numbers

To write an improper fraction as a mixed number, we must find two things: the *whole-number part* and the *fractional part* of the mixed number. To develop a procedure to do this, let's consider the improper fraction $\frac{7}{3}$. To find the number of groups of 3 in 7, we can divide 7 by 3. This will find the whole-number part of the mixed number. The remainder is the numerator of the fractional part of the mixed number.

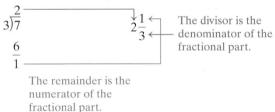

This example suggests the following rule.

Writing an improper fraction as a mixed number

To write an improper fraction as a mixed number, divide the numerator by the denominator to obtain the whole-number part. The remainder over the divisor is the fractional part.

Sleep   THINK IT THROUGH

"Lack of sleep among college students is an old problem, but one that appears to be getting worse, according to some national surveys." CBSnews.com

A 2001 study of over 1,500 college students found that they averaged six hours and forty minutes of sleep per night (Holy Cross Health Survey). James Maas, a professor at Cornell University and author of *Power Sleep*, recommends eight hours of sleep per night for college students. Find the difference between Professor Maas's recommendation and the survey average for college students. Express the result as a mixed number.

Self Check 2

Write $\dfrac{43}{5}$ as a mixed number.

EXAMPLE 2   Write $\dfrac{29}{6}$ as a mixed number.

Solution

$$\begin{array}{r} 4 \\ 6\overline{)29} \\ \underline{24} \\ 5 \end{array}$$

Divide the numerator by the denominator.

The remainder is 5.

Thus, $\dfrac{29}{6} = 4\dfrac{5}{6}$.

Answer $8\dfrac{3}{5}$

■ Graphing fractions and mixed numbers

Earlier, we graphed whole numbers and integers on a number line. Fractions and mixed numbers can also be graphed on a number line.

Self Check 3

Graph: $-1\dfrac{7}{8}$, $-\dfrac{2}{3}$, and $\dfrac{9}{4}$.

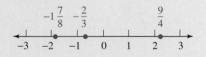

Answer

EXAMPLE 3   Graph: $-2\dfrac{3}{4}$, $-1\dfrac{1}{2}$, $-\dfrac{1}{8}$, and $\dfrac{13}{5}$.

Solution

- Since $-2\dfrac{3}{4} < -2$, the graph of $-2\dfrac{3}{4}$ is to the left of -2 on the number line.
- The number $-1\dfrac{1}{2}$ is between -1 and -2.
- The number $-\dfrac{1}{8}$ is less than 0.
- Expressed as a mixed number, $\dfrac{13}{5} = 2\dfrac{3}{5}$.

■ Multiplying and dividing mixed numbers

> **Multiplying and dividing mixed numbers**
>
> To multiply or divide mixed numbers, first change the mixed numbers to improper fractions. Then perform the multiplication or division of the fractions.

Self Check 4

Multiply: $9\dfrac{3}{5} \cdot 3\dfrac{3}{4}$.

EXAMPLE 4   Multiply: $5\dfrac{1}{5} \cdot 1\dfrac{2}{13}$.

Solution

$$5\dfrac{1}{5} \cdot 1\dfrac{2}{13} = \dfrac{26}{5} \cdot \dfrac{15}{13}$$

Write each mixed number as an improper fraction.

$$= \dfrac{26 \cdot 15}{5 \cdot 13}$$

Multiply the numerators and multiply the denominators.

$$= \frac{13 \cdot 2 \cdot 5 \cdot 3}{5 \cdot 13} \qquad \text{Factor 26 as } 13 \cdot 2 \text{ and 15 as } 5 \cdot 3.$$

$$= \frac{\overset{1}{\cancel{13}} \cdot 2 \cdot \overset{1}{\cancel{5}} \cdot 3}{\underset{1}{\cancel{5}} \cdot \underset{1}{\cancel{13}}} \qquad \text{Divide out the common factors of 13 and 5.}$$

$$= \frac{6}{1} \qquad \text{Multiply in the numerator and denominator.}$$

$$= 6 \qquad \text{Simplify: } \frac{6}{1} = 6.$$

Answer 36

EXAMPLE 5   Divide: $-3\frac{3}{8} \div 2\frac{1}{4}$.

Self Check 5

Divide: $3\frac{4}{15} \div \left(-2\frac{1}{10}\right)$.

Solution

$$-3\frac{3}{8} \div 2\frac{1}{4} = -\frac{27}{8} \div \frac{9}{4} \qquad \text{Write each mixed number as an improper fraction.}$$

$$= -\frac{27}{8} \cdot \frac{4}{9} \qquad \text{Multiply by the reciprocal of } \frac{9}{4}.$$

$$= -\frac{27 \cdot 4}{8 \cdot 9} \qquad \begin{array}{l}\text{The product of two fractions with unlike signs is negative.} \\ \text{Multiply the numerators and multiply the denominators.}\end{array}$$

$$= -\frac{\overset{1}{\cancel{9}} \cdot 3 \cdot \overset{1}{\cancel{4}}}{\underset{1}{\cancel{4}} \cdot 2 \cdot \underset{1}{\cancel{9}}} \qquad \begin{array}{l}\text{Factor 27 as } 9 \cdot 3 \text{ and 8 as } 4 \cdot 2. \text{ Divide out the common} \\ \text{factors of 9 and 4.}\end{array}$$

$$= -\frac{3}{2} \qquad \text{Multiply in the numerator and denominator.}$$

$$= -1\frac{1}{2} \qquad \text{Write } -\frac{3}{2} \text{ as a mixed number.}$$

Answer $-1\frac{5}{9}$

EXAMPLE 6   **Government grants.**   If \$12$\frac{1}{2}$ million is to be divided equally among five cities to fund recreation programs, how much will each city receive?

Solution   To find the amount received by each city, we divide the grant money by 5.

$$12\frac{1}{2} \div 5 = \frac{25}{2} \div \frac{5}{1} \qquad \text{Write } 12\frac{1}{2} \text{ as an improper fraction, and write 5 as a fraction.}$$

$$= \frac{25}{2} \cdot \frac{1}{5} \qquad \text{Multiply by the reciprocal of } \frac{5}{1}.$$

$$= \frac{25 \cdot 1}{2 \cdot 5} \qquad \text{Multiply the numerators and multiply the denominators.}$$

$$= \frac{\overset{1}{\cancel{5}} \cdot 5 \cdot 1}{2 \cdot \underset{1}{\cancel{5}}} \qquad \text{Factor 25 as } 5 \cdot 5. \text{ Divide out the common factor of 5.}$$

$$= \frac{5}{2} \qquad \text{Multiply in the numerator and denominator.}$$

$$= 2\frac{1}{2} \qquad \text{Write } \frac{5}{2} \text{ as a mixed number.}$$

Each city will receive \$2$\frac{1}{2}$ million.

Section 3.5 STUDY SET

VOCABULARY   *Fill in the blanks.*

1. A _____ number is the sum of a whole number and a proper fraction.

2. An _____ fraction is a fraction with a numerator that is greater than or equal to its denominator.

3. To _____ a number means to locate its position on the number line and highlight it using a dot.

4. Multiplying or dividing the _____ and _____ of a fraction by the same nonzero number does not change the value of the fraction.

CONCEPTS

5. What signed number could be used to describe each situation?

   a. A temperature of five and one-half degrees above zero

   b. A sprinkler pipe that is $6\frac{7}{8}$ in. under the sidewalk

6. What signed number could be used to describe each situation?

   a. A rain total two and three-tenths of an inch lower than the average

   b. Three and one-half minutes after liftoff

7. a. In the illustration, the divisions on the face of the meter represent fractions. What value is the arrow registering?

   b. If the arrow moves four marks to the left, what value will it register?

8. a. In the illustration, the divisions on the face of the meter represent fractions. What value is the arrow registering?

   b. If the arrow moves up one mark, what value will it register?

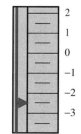

9. What fractions have been graphed on the number line?

10. What mixed numbers have been graphed on the number line?

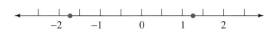

11. DIVING Complete the description of the dive by filling in the blank with a mixed number.

   Forward ▭ somersaults

12. PRODUCT LABELING The following label uses mixed numbers. Write each one as an improper fraction.

Laundry Basket
1³/4 Bushel
•Easy-grip rim is reinforced to handle the biggest loads
23¹/4" L X 18⁷/8" W X 10¹/2" H

13. Draw $\dfrac{17}{8}$ pizzas.

14. a. What mixed number is represented in the illustration below?

   b. What improper fraction is shown in the illustration?

NOTATION *Complete each solution.*

15. Multiply: $-5\frac{1}{4} \cdot 1\frac{1}{7}$

$$-5\frac{1}{4} \cdot 1\frac{1}{7} = -\frac{21}{4} \cdot \frac{\boxed{}}{7}$$

$$= -\frac{21 \cdot \boxed{}}{4 \cdot 7}$$

$$= -\frac{\overset{1}{\cancel{7}} \cdot 3 \cdot \overset{1}{\cancel{/}} \cdot 2}{\underset{1}{\cancel{/}} \cdot \underset{1}{\cancel{7}}}$$

$$= -\frac{\boxed{}}{1}$$

$$= -6$$

16. Divide: $-5\frac{5}{6} \div 2\frac{1}{12}$.

$$-5\frac{5}{6} \div 2\frac{1}{12} = -\frac{\boxed{}}{6} \div \frac{25}{12}$$

$$= -\frac{35}{6} \cdot \frac{12}{\boxed{}}$$

$$= -\frac{35 \cdot 12}{6 \cdot \boxed{}}$$

$$= -\frac{\overset{1}{\cancel{5}} \cdot \boxed{} \cdot \overset{1}{\cancel{6}} \cdot 2}{\underset{1}{\cancel{6}} \cdot \underset{1}{\cancel{5}} \cdot \boxed{}}$$

$$= -\frac{\boxed{}}{5}$$

$$= -2\frac{4}{5}$$

PRACTICE *Write each improper fraction as a mixed number. Simplify the result, if possible.*

17. $\frac{15}{4}$

18. $\frac{41}{6}$

19. $\frac{29}{5}$

20. $\frac{29}{3}$

21. $-\frac{20}{6} = -\frac{10}{3}$

22. $-\frac{28}{8}$

23. $\frac{127}{12}$

24. $\frac{197}{16}$

Write each mixed number as an improper fraction.

25. $6\frac{1}{2}$

26. $8\frac{2}{3}$

27. $20\frac{4}{5}$

28. $15\frac{3}{8}$

29. $-6\frac{2}{9}$

30. $-7\frac{1}{12}$

31. $200\frac{2}{3}$

32. $90\frac{5}{6}$

Graph each set of numbers on the number line.

33. $\left\{ -2\frac{8}{9}, 1\frac{2}{3}, \frac{16}{5} \right\}$

34. $\left\{ -\frac{3}{4}, -3\frac{1}{4}, \frac{5}{2} \right\}$

35. $\left\{ 3\frac{1}{7}, -\frac{98}{99}, -\frac{10}{3} \right\}$

36. $\left\{ -2\frac{1}{5}, \frac{4}{5}, -\frac{11}{3} \right\}$

Multiply.

37. $1\frac{2}{3} \cdot 2\frac{1}{7}$

38. $2\frac{3}{5} \cdot 1\frac{2}{3}$

39. $-7\frac{1}{2}\left(-1\frac{2}{5}\right)$

40. $-4\frac{1}{8}\left(-1\frac{7}{9}\right)$

41. $3\frac{1}{16} \cdot 4\frac{4}{7}$

42. $5\frac{3}{5} \cdot 1\frac{11}{14}$

43. $-6 \cdot 2\frac{7}{24}$

44. $-7 \cdot 1\frac{3}{28}$

45. $2\frac{1}{2}\left(-3\frac{1}{3}\right)$

46. $\left(-3\frac{1}{4}\right)\left(1\frac{1}{5}\right)$

47. $2\frac{5}{8} \cdot \frac{5}{27}$

48. $3\frac{1}{9} \cdot \frac{3}{32}$

49. Find the product of $1\frac{2}{3}$, 6, and $-\frac{1}{8}$.

50. Find the product of $-\frac{5}{6}$, -8, and $-2\frac{1}{10}$.

Find each power.

51. $\left(1\frac{2}{3}\right)^2$

52. $\left(3\frac{1}{2}\right)^2$

53. $\left(-1\frac{1}{3}\right)^3$

54. $\left(-1\frac{1}{5}\right)^3$

Divide.

55. $3\frac{1}{3} \div 1\frac{5}{6}$

56. $3\frac{3}{4} \div 5\frac{1}{3}$

57. $-6\frac{3}{5} \div 7\frac{1}{3}$

58. $-4\frac{1}{4} \div 4\frac{1}{2}$

59. $-20\frac{1}{4} \div \left(-1\frac{11}{16}\right)$

60. $-2\frac{7}{10} \div \left(-1\frac{1}{14}\right)$

61. $6\frac{1}{4} \div 20$

62. $4\frac{2}{5} \div 11$

63. $1\frac{2}{3} \div \left(-2\frac{1}{2}\right)$

64. $2\frac{1}{2} \div \left(-1\frac{5}{8}\right)$

65. $8 \div 3\frac{1}{5}$

66. $15 \div 3\frac{1}{3}$

67. Find the quotient of $-4\frac{1}{2}$ and $2\frac{1}{4}$.

68. Find the quotient of 25 and $-10\frac{5}{7}$.

▮ APPLICATIONS

69. CALORIES A company advertises that its mints contain only $3\frac{1}{5}$ calories apiece. What is the calorie intake if you eat an entire package of 20 mints?

70. CEMENT MIXERS A cement mixer can carry $9\frac{1}{2}$ cubic yards of concrete. If it makes 8 trips to a job site, how much concrete will be delivered to the site?

71. SHOPPING In the illustration, what is the cost of buying the fruit in the scale?

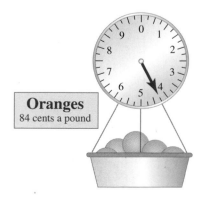

72. PICTURE FRAMES How much molding is needed to make the square picture frame below?

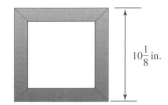

73. SUBDIVISIONS A developer donated to the county 100 of the 1,000 acres of land she owned. She divided the remaining acreage into $1\frac{1}{3}$-acre lots. How many lots were created?

74. CATERING How many people can be served $\frac{1}{3}$-pound hamburgers if a caterer purchases 200 pounds of ground beef?

75. GRAPH PAPER Mathematicians use specially marked paper, called *graph paper,* when drawing figures. It is made up of $\frac{1}{4}$-inch squares. Find the length and width of the piece of graph paper shown.

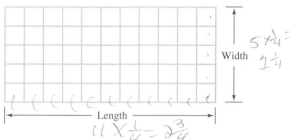

76. LUMBER As shown, 2-by-4's from the lumber yard do not really have dimensions of 2 inches by 4 inches. How wide and how high is the stack of 2-by-4's?

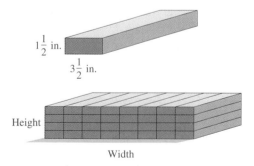

77. EMERGENCY EXITS The following sign marks the emergency exit on a school bus. Find the area of the sign.

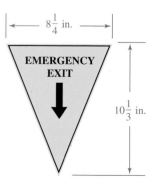

78. HORSE RACING The race tracks on which thoroughbred horses run are marked off in $\frac{1}{8}$-mile-long segments called furlongs. How many furlongs are there in a $1\frac{1}{16}$-mile race?

79. FIRE ESCAPES The fire escape stairway in an office building is shown. Each riser is $7\frac{1}{2}$ inches high. If each story is 105 inches high and the building is 43 stories tall, how many steps are there in the stairway?

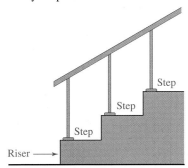

80. LICENSE PLATES Find the area of the license plate below.

81. SHOPPING ON THE INTERNET A mother is ordering a pair of jeans for her daughter from the screen shown. If the daughter's height is $60\frac{3}{4}$ in. and her waist is $24\frac{1}{2}$ in., on what size and what cut should the mother point and click?

Girl's jeans- regular cut

Size	7	8	10	12	14	16
Height	50-52	52-54	54-56	56¼-58½	59-61	61-62
Waist	22¼-22¾	22¾-23¼	23¾-24¼	24¾-25¼	25¾-26¼	26¼-28

Girl's jeans- slim cut

Size	7	8	10	12	14	16
Height	50-52	52-54	54-56	56½-58½	59-61	61-62
Waist	20¾-21¼	21¼-21¾	22¼-22¾	23¼-23¾	24¼-24¾	25-26½

To order:
Point arrow to proper size/cut and click

82. SEWING Use the following table to determine the number of yards of fabric needed

a. to make a size 16 top if the fabric to be used is 60 inches wide.

b. to make size 18 pants if the fabric to be used is 45 inches wide.

8767 Pattern
stitch'n save
by McCall's

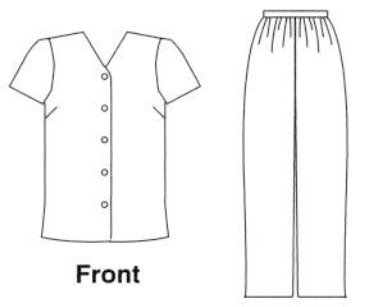

Front

SIZES	8	10	12	14	16	18	20	
Top								
45"	2¼	2⅜	2⅜	2⅜	2½	2⅝	2¾	**Yds**
60"	2	2	2⅛	2⅛	2⅛	2⅛	2⅛	
Pants								
45"	2⅝	2⅝	2⅝	2⅝	2⅝	2⅝	2⅝	**Yds**
60"	1¾	2	2¼	2¼	2¼	2¼	2½	

WRITING

83. Explain the difference between $2\frac{3}{4}$ and $2\left(\frac{3}{4}\right)$.

84. Give three examples of how you use mixed numbers in daily life.

85. Explain the procedure used to write an improper fraction as a mixed number.

86. Explain the procedure used to multiply two mixed numbers.

REVIEW

87. Evaluate: $3^2 \cdot 2^3$.

88. List the first eight natural numbers.

89. Write $8 + 8 + 8 + 8$ as a multiplication.

90. If a square measures 1 inch on each side, what is its area?

91. Simplify: $\frac{115}{25}$.

92. Multiply: $\left(-\frac{6}{5}\right)\left(\frac{35}{14}\right)$.

3.6 Adding and Subtracting Mixed Numbers

- Adding mixed numbers
- Adding mixed numbers in vertical form
- Subtracting mixed numbers

In this section, we discuss methods for adding and subtracting mixed numbers. The first method works well when the whole-number parts of the mixed numbers are small. The second method works well when the whole-number parts of the mixed numbers are large. The third method uses columns as a way to organize the work.

▓ Adding mixed numbers

We can add mixed numbers by writing them as improper fractions. To do so, we follow these steps.

> **Adding mixed numbers: method 1**
> 1. Write each mixed number as an improper fraction.
> 2. Write each improper fraction as an equivalent fraction with a denominator that is the LCD.
> 3. Add the fractions.
> 4. Change the result to a mixed number if desired.

Self Check 1

Add: $3\dfrac{2}{3} + 1\dfrac{1}{5}$.

EXAMPLE 1   Add: $4\dfrac{1}{6} + 2\dfrac{3}{4}$.

Solution

$$4\dfrac{1}{6} + 2\dfrac{3}{4} = \dfrac{25}{6} + \dfrac{11}{4}$$ Write each mixed number as an improper fraction: $4\dfrac{1}{6} = \dfrac{25}{6}$ and $2\dfrac{3}{4} = \dfrac{11}{4}$.

By inspection, we see that the lowest common denominator is 12.

$$= \dfrac{25 \cdot 2}{6 \cdot 2} + \dfrac{11 \cdot 3}{4 \cdot 3}$$ Write each fraction as a fraction with a denominator of 12.

$$= \dfrac{50}{12} + \dfrac{33}{12}$$ Perform the multiplications in the numerators and denominators.

$$= \dfrac{83}{12}$$ Add the numerators: $50 + 33 = 83$. Write the sum over the common denominator 12.

Answer $4\dfrac{13}{15}$

$$= 6\dfrac{11}{12}$$ Write the improper fraction as a mixed number: $\dfrac{83}{12} = 6\dfrac{11}{12}$.

We can also add mixed numbers by adding their whole-number parts and their fractional parts. To do so, we follow these steps.

> **Adding mixed numbers: method 2**
> 1. Write each mixed number as the sum of a whole number and a fraction.
> 2. Use the commutative property of addition to write the whole numbers together and the fractions together.
> 3. Add the whole numbers and the fractions separately.
> 4. Write the result as a mixed number if necessary.

Self Check 2

Find the sum: $275\dfrac{1}{6} + 81\dfrac{3}{5}$.

EXAMPLE 2   Find the sum: $168\dfrac{3}{4} + 85\dfrac{1}{5}$.

Solution

$$168\frac{3}{4} + 85\frac{1}{5} = 168 + \frac{3}{4} + 85 + \frac{1}{5}$$ Write each mixed number as the sum of a whole number and a fraction.

$$= 168 + 85 + \frac{3}{4} + \frac{1}{5}$$ Use the commutative property of addition to change the order of the addition.

$$= 253 + \frac{3}{4} + \frac{1}{5}$$ Add the whole numbers: $168 + 85 = 253$.

$$= 253 + \frac{3 \cdot 5}{4 \cdot 5} + \frac{1 \cdot 4}{5 \cdot 4}$$ Write each fraction as a fraction with denominator 20.

$$= 253 + \frac{15}{20} + \frac{4}{20}$$ Multiply in the numerators and denominators.

$$= 253 + \frac{19}{20}$$ Add the numerators and write the sum over the common denominator 20.

$$= 253\frac{19}{20}$$ Write the sum as a mixed number.

Answer $356\frac{23}{30}$

! COMMENT If we use method 1 to add the mixed numbers in Example 2, the numbers we encounter are cumbersome. As expected, the result is the same: $253\frac{19}{20}$.

$$168\frac{3}{4} + 85\frac{1}{5} = \frac{675}{4} + \frac{426}{5}$$ Write $168\frac{3}{4}$ and $85\frac{1}{5}$ as improper fractions.

$$= \frac{675 \cdot 5}{4 \cdot 5} + \frac{426 \cdot 4}{5 \cdot 4}$$ The LCD is 20.

$$= \frac{3,375}{20} + \frac{1,704}{20}$$

$$= \frac{5,079}{20}$$

$$= 253\frac{19}{20}$$

Generally speaking, the larger the whole-number parts of the mixed numbers get, the more difficult it becomes to add those mixed numbers using method 1.

◾ Adding mixed numbers in vertical form

By working in columns, we can add mixed numbers quickly. The strategy is the same as in Example 2: Add whole numbers to whole numbers and fractions to fractions.

┌ Line up the mixed numbers vertically.

        ┌ Apply the fundamental property of fractions to get an LCD.

           ┌ Add the whole numbers and add the fractions separately.

$$25\frac{3}{4} \quad = \quad 25\frac{3 \cdot 5}{4 \cdot 5} \quad = \quad 25\frac{15}{20}$$

$$+31\frac{1}{5} \quad = \quad +31\frac{1 \cdot 4}{5 \cdot 4} \quad = \quad +31\frac{4}{20}$$

$$\overline{56\frac{19}{20}}$$

EXAMPLE 3   Suspension bridges.   Find the total length of cable that must be ordered if cables a, d, and e of the suspension bridge in Figure 3-10 are to be replaced. (See the table below.)

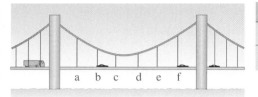

BRIDGE SPECIFICATIONS			
Cable	a	b	c
Length (feet)	$75\frac{1}{12}$	$54\frac{1}{6}$	$43\frac{1}{4}$

FIGURE 3-10

Solution   To find the total length of cable to be ordered, we add the lengths of cables a, d, and e. Because of the symmetric design, cables d and c and cables e and b are the same length.

Length of cable a	plus	length of cable d (or cable c)	plus	length of cable e (or cable b)	equals	the total length needed.
$75\frac{1}{12}$	+	$43\frac{1}{4}$	+	$54\frac{1}{6}$	=	total length

We add the mixed numbers using a vertical format.

$$75\frac{1}{12} \;=\; 75\frac{1}{12} \;=\; 75\frac{1}{12}$$

$$43\frac{1}{4} \;=\; 43\frac{1\cdot 3}{4\cdot 3} \;=\; 43\frac{3}{12}$$

$$+54\frac{1}{6} \;=\; +54\frac{1\cdot 2}{6\cdot 2} \;=\; +54\frac{2}{12}$$

$$172\frac{6}{12} \;=\; 172\frac{1}{2} \quad \text{Simplify: } \frac{6}{12}=\frac{1}{2}.$$

The total length of cable needed for the replacement is $172\frac{1}{2}$ feet.

When we add mixed numbers, the sum of the fractions sometimes yields an improper fraction, as in the next example.

Self Check 4

Add: $76\frac{11}{12} + 49\frac{5}{8}$.

EXAMPLE 4   Add: $45\frac{2}{3} + 96\frac{4}{5}$.

Solution

$$45\frac{2}{3} \;=\; 45\frac{2\cdot 5}{3\cdot 5} \;=\; 45\frac{10}{15}$$

$$+96\frac{4}{5} \;=\; +96\frac{4\cdot 3}{5\cdot 3} \;=\; +96\frac{12}{15}$$

$$141\frac{22}{15}$$

The whole-number part of the answer ↑   ↑ The fractional part of the answer is an improper fraction.

Now write the improper fraction as a mixed number.

$$141\frac{22}{15} = 141 + \frac{22}{15} = 141 + 1\frac{7}{15} = 142\frac{7}{15}$$

Answer $126\frac{13}{24}$

▌Subtracting mixed numbers

Subtracting mixed numbers is similar to adding mixed numbers.

EXAMPLE 5   Cooking.   How much butter is left in a 10-pound tub if $2\frac{2}{3}$ pounds are used for a wedding cake?

Solution   The phrase "How much is left?" suggests subtraction.

$$10 - 2\frac{2}{3} = \frac{10}{1} - \frac{8}{3} \qquad \text{Write 10 as a fraction: } 10 = \frac{10}{1}. \text{ Write } 2\frac{2}{3} \text{ as } \frac{8}{3}.$$

By inspection, we see that the LCD is 3.

$$\frac{10}{1} - 2\frac{2}{3} = \frac{10 \cdot 3}{1 \cdot 3} - \frac{8}{3} \qquad \text{Write the first fraction with a denominator of 3.}$$

$$= \frac{30}{3} - \frac{8}{3} \qquad \text{Perform the multiplications in the first fraction.}$$

$$= \frac{30 - 8}{3} \qquad \text{Subtract the numerators and write the difference over the common denominator.}$$

$$= \frac{22}{3} \qquad \text{Perform the subtraction: } 30 - 8 = 22.$$

$$= 7\frac{1}{3} \qquad \text{Write } \frac{22}{3} \text{ as a mixed number.}$$

There are $7\frac{1}{3}$ pounds of butter left in the tub.

In the next example, the fraction $\frac{1}{5}$ is smaller than $\frac{2}{3}$. Because of this, we have to borrow.

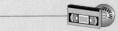

EXAMPLE 6   Subtract: $34\frac{1}{5} - 11\frac{2}{3}$.

Self Check 6

Subtract: $101\frac{3}{4} - 79\frac{15}{16}$.

Solution   We will use the vertical form to subtract. The LCD is 15, so we write each fraction as a fraction with a denominator of 15.

$$\begin{array}{ccccc} 34\frac{1}{5} & = & 34\frac{1 \cdot 3}{5 \cdot 3} & = & 34\frac{3}{15} \\[2mm] -11\frac{2}{3} & = & -11\frac{2 \cdot 5}{3 \cdot 5} & = & -11\frac{10}{15} \end{array}$$

Since $\frac{10}{15}$ is larger than $\frac{3}{15}$, borrow 1 (in the form of $\frac{15}{15}$) from 34 and add it to $\frac{3}{15}$ to obtain $33\frac{3}{15} + \frac{15}{15} = 33\frac{18}{15}$. Then we subtract the fractions and the whole numbers separately.

$$33\frac{3}{15} + \frac{15}{15} = 33\frac{18}{15}$$
$$-11\frac{10}{15} = -11\frac{10}{15}$$
$$\underline{\phantom{-11\frac{10}{15}}}$$
$$22\frac{8}{15}$$

Answer $21\frac{13}{16}$

Self Check 7

Subtract: $2{,}300 - 129\frac{31}{32}.$

Answer $2{,}170\frac{1}{32}$

EXAMPLE 7   Subtract: $419 - 53\frac{11}{16}.$

Solution   We align the numbers vertically and borrow 1 (in the form of $\frac{16}{16}$) from 419. Then we subtract the fractions and subtract the whole numbers separately.

$$419 \quad = \quad 418\frac{16}{16}$$
$$-53\frac{11}{16} = -53\frac{11}{16}$$
$$\underline{\phantom{-53\frac{11}{16}}}$$
$$365\frac{5}{16}$$

Section 3.6 STUDY SET

VOCABULARY   *Fill in the blanks.*

1. By the _____ property of addition, we can add numbers in any order.

2. A _____ number such as $1\frac{7}{8}$ contains a whole-number part and a fractional part.

3. Consider
$$80\frac{1}{3} = \quad 79\frac{1}{3} + \frac{3}{3}$$
$$-24\frac{2}{3} = -24\frac{2}{3}$$
To do the subtraction, we _____ 1 in the form of $\frac{3}{3}$.

4. Fractions that are greater than 1, such as $\frac{11}{8}$, are called _____ fractions.

CONCEPTS

5. **a.** For $76\frac{3}{4}$, list the whole-number part and the fractional part.

   b. Write $76\frac{3}{4}$ as a sum.

6. Use the commutative property of addition to get the whole numbers together.
$$14 + \frac{5}{6} + 53 + \frac{1}{6}$$

7. What property is being highlighted here?
$$25\frac{3 \cdot 5}{4 \cdot 5}$$
$$+31\frac{1 \cdot 4}{5 \cdot 4}$$
$$\underline{\phantom{+31\frac{1 \cdot 4}{5 \cdot 4}}}$$

8. **a.** The denominators of two fractions, expressed in prime-factored form, are $5 \cdot 2$ and $5 \cdot 3$. Find the LCD for the fractions.

   b. The denominators for three fractions, in prime-factored form, are $3 \cdot 5$, $2 \cdot 3$, and $3 \cdot 3$. Find the LCD for the fractions.

9. Simplify.

   a. $9\frac{17}{16}$

   b. $1{,}288\frac{7}{3}$

   c. $16\frac{12}{8}$

   d. $45\frac{24}{20}$

10. Consider

$$108\tfrac{1}{3}$$
$$-\ 99\tfrac{2}{3}$$

    a. Explain why we will have to borrow if we subtract the mixed numbers in this way.

    b. In what form will we borrow a 1 from 108?

NOTATION   *Complete each solution.*

11. $70\tfrac{3}{5} + 39\tfrac{2}{7} = \boxed{} + \tfrac{3}{5} + \boxed{} + \tfrac{2}{7}$

$$= \boxed{} + \boxed{} + \tfrac{3}{5} + \tfrac{2}{7}$$

$$= 109 + \tfrac{3}{5} + \tfrac{2}{7}$$

$$= 109 + \frac{3 \cdot \boxed{}}{5 \cdot \boxed{}} + \frac{2 \cdot \boxed{}}{7 \cdot \boxed{}}$$

$$= 109 + \frac{21}{\boxed{}} + \frac{10}{\boxed{}}$$

$$= 109 + \frac{\boxed{}}{35}$$

$$= 109\tfrac{31}{35}$$

12. $67\tfrac{3}{8} = 67\dfrac{3 \cdot \boxed{}}{8 \cdot \boxed{}} \quad = $

$$-23\tfrac{2}{3} = -23\dfrac{2 \cdot \boxed{}}{3 \cdot \boxed{}} \quad = $$

$$67\tfrac{9}{24} = \boxed{}\tfrac{9}{24} + \dfrac{\boxed{}}{\boxed{}} = 66\dfrac{\boxed{}}{24}$$

$$-23\tfrac{16}{24} = -23\tfrac{16}{24} \qquad = -23\tfrac{16}{24}$$

$$43\dfrac{\boxed{}}{24}$$

PRACTICE   *Find each sum or difference.*

13. $2\tfrac{1}{5} + 2\tfrac{1}{5}$ **14.** $3\tfrac{1}{3} + 2\tfrac{1}{3}$

15. $8\tfrac{2}{7} - 3\tfrac{1}{7}$ **16.** $9\tfrac{5}{11} - 6\tfrac{2}{11}$

17. $3\tfrac{1}{4} + 4\tfrac{1}{4}$ **18.** $2\tfrac{1}{8} + 3\tfrac{3}{8}$

19. $4\tfrac{1}{6} + 1\tfrac{1}{5}$ **20.** $2\tfrac{2}{5} + 3\tfrac{1}{4}$

21. $2\tfrac{1}{2} - 1\tfrac{1}{4}$ **22.** $13\tfrac{5}{6} - 4\tfrac{2}{3}$

23. $2\tfrac{5}{6} - 1\tfrac{3}{8}$ **24.** $4\tfrac{5}{9} - 2\tfrac{1}{6}$

25. $5\tfrac{1}{2} + 3\tfrac{4}{5}$ **26.** $6\tfrac{1}{2} + 2\tfrac{2}{3}$

27. $7\tfrac{1}{2} - 4\tfrac{1}{7}$ **28.** $5\tfrac{3}{4} - 1\tfrac{3}{7}$

29. $56\tfrac{2}{5} + 73\tfrac{1}{3}$ **30.** $44\tfrac{3}{8} + 66\tfrac{1}{5}$

31. $380\tfrac{1}{6} + 17\tfrac{1}{4}$ **32.** $103\tfrac{1}{2} + 210\tfrac{2}{5}$

33. $228\tfrac{5}{9} + 44\tfrac{2}{3}$ **34.** $161\tfrac{7}{8} + 19\tfrac{1}{3}$

35. $778\tfrac{5}{7} - 155\tfrac{1}{3}$ **36.** $339\tfrac{1}{2} - 218\tfrac{3}{16}$

37. $140\tfrac{5}{6} - 129\tfrac{4}{5}$ **38.** $291\tfrac{1}{4} - 289\tfrac{1}{12}$

39. $422\tfrac{13}{16} - 321\tfrac{3}{8}$ **40.** $378\tfrac{3}{4} - 277\tfrac{5}{8}$

Find each difference.

41. $16\tfrac{1}{4} - 13\tfrac{3}{4}$ **42.** $40\tfrac{1}{7} - 19\tfrac{6}{7}$

43. $76\tfrac{1}{6} - 49\tfrac{7}{8}$ **44.** $101\tfrac{1}{4} - 70\tfrac{1}{2}$

45. $140\tfrac{3}{16} - 129\tfrac{3}{4}$ **46.** $211\tfrac{1}{3} - 8\tfrac{3}{4}$

47. $334\tfrac{1}{9} - 13\tfrac{5}{6}$ **48.** $442\tfrac{1}{8} - 429\tfrac{2}{3}$

Find the sum or difference.

49. $7 - \tfrac{2}{3}$ **50.** $6 - \tfrac{1}{8}$

51. $9 - 8\tfrac{3}{4}$ **52.** $11 - 10\tfrac{4}{5}$

53. $4\tfrac{1}{7} - \tfrac{4}{5}$ **54.** $5\tfrac{1}{10} - \tfrac{4}{5}$

55. $6\tfrac{5}{8} - 3$ **56.** $10\tfrac{1}{2} - 6$

57. $\tfrac{7}{3} + 2$ **58.** $\tfrac{9}{7} + 3$

59. $2 + 1\tfrac{7}{8}$ **60.** $3\tfrac{3}{4} + 5$

Find each sum.

61. $12\frac{1}{2} + 5\frac{3}{4} + 35\frac{1}{6}$

62. $31\frac{1}{3} + 20\frac{2}{5} + 10\frac{1}{15}$

63. $58\frac{7}{8} + 340 + 61\frac{1}{4}$

64. $191 + 233\frac{1}{16} + 16\frac{5}{8}$

65. Subtract $9\frac{1}{10}$ from the sum of $7\frac{3}{7}$ and $3\frac{1}{5}$.

66. Subtract $3\frac{2}{3}$ from the sum of $2\frac{5}{12}$ and $1\frac{5}{8}$.

67. Add $12\frac{11}{12}$ to the difference of $5\frac{1}{6}$ and $3\frac{7}{8}$.

68. Add $18\frac{1}{3}$ to the difference of $11\frac{3}{5}$ and $9\frac{11}{15}$.

Find each sum or difference.

69. $-3\frac{3}{4} + \left(-1\frac{1}{2}\right)$ **70.** $-3\frac{2}{3} + \left(-1\frac{4}{5}\right)$

71. $-4\frac{5}{8} - 1\frac{1}{4}$ **72.** $-2\frac{1}{16} - 3\frac{7}{8}$

APPLICATIONS

73. FREEWAY TRAVEL A freeway exit sign is shown. How far apart are the Citrus Ave. and Grand Ave. exits?

> Citrus Ave. $\frac{3}{4}$ mi
>
> Grand Ave. $3\frac{1}{2}$ mi

74. BASKETBALL See the illustration. What is the difference in height between the tallest and the shortest of the starting players?

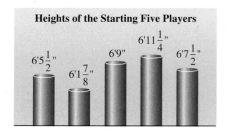

Heights of the Starting Five Players

$6'5\frac{1}{2}''$ $6'1\frac{7}{8}''$ $6'9''$ $6'11\frac{1}{4}''$ $6'7\frac{1}{2}''$

75. TRAIL MIX See the recipe. A camper doubles the amount of sunflower seeds called for in the recipe. How much trail mix will the adjusted recipe yield?

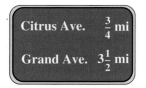

Trail Mix

A healthy snack–great for camping trips

$2\frac{3}{4}$ cups peanuts $\frac{1}{3}$ cup coconut

$\frac{1}{2}$ cup sunflower seeds $2\frac{2}{3}$ cups oat flakes

$\frac{2}{3}$ cup raisins $\frac{1}{4}$ cup pretzels

76. AIR TRAVEL A businesswoman's flight leaves Los Angeles at 8:00 A.M. and arrives in Seattle at 9:45 A.M.

a. Express the duration of the flight as a mixed number.

b. Upon arrival, she boards a commuter plane at 11:15 A.M., arriving at her final destination at 11:45 A.M. Express the length of this flight as a fraction.

c. Find the total time of these two flights.

77. HOSE REPAIRS To repair a bad connector, Ming Lin removes $1\frac{1}{2}$ feet from the end of a 50-foot garden hose. How long is the hose after the repair?

78. SEWING To make some draperies, Liz needs $12\frac{1}{4}$ yards of material for the den and $8\frac{1}{2}$ yards for the living room. If the material comes only in 21-yard bolts, how much will be left over after completing both sets of draperies?

79. SHIPPING A passenger ship and a cargo ship leave San Diego harbor at midnight. During the first hour, the passenger ship travels south at $16\frac{1}{2}$ miles per hour, while the cargo ship is traveling north at a rate of $5\frac{1}{5}$ miles per hour.

a. Complete the following table.

b. How far apart are they at 1:00 A.M.?

	Rate · (mph)	Time traveling (hr)	= Distance traveled (mi)
Passenger ship		1	
Cargo ship		1	

80. HARDWARE Refer to the illustration below. How long should the threaded part of the bolt be?

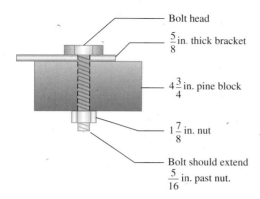

- Bolt head
- $\frac{5}{8}$ in. thick bracket
- $4\frac{3}{4}$ in. pine block
- $1\frac{7}{8}$ in. nut
- Bolt should extend $\frac{5}{16}$ in. past nut.

81. SERVICE STATIONS Use the service station sign to answer the questions.

   a. What is the difference in price between the least and most expensive types of gasoline at the self-service pump?

   b. How much more is the cost per gallon for full service?

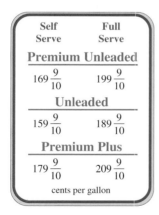

	Self Serve	Full Serve
Premium Unleaded	$169\frac{9}{10}$	$199\frac{9}{10}$
Unleaded	$159\frac{9}{10}$	$189\frac{9}{10}$
Premium Plus	$179\frac{9}{10}$	$209\frac{9}{10}$

cents per gallon

82. SEPTUPLETS On November 19, 1997, at Iowa Methodist Medical Center, Bobbie McCaughey gave birth to seven babies. From the following information find the combined birthweights of the babies.

Kenneth Robert $3\frac{1}{4}$ lb

Nathanial Roy $2\frac{7}{8}$ lb

Kelsey Ann $2\frac{5}{16}$ lb

Brandon James $3\frac{3}{16}$ lb

Natalie Sue $2\frac{5}{8}$ lb

Joel Steven $2\frac{15}{16}$ lb

Alexis May $2\frac{11}{16}$ lb

83. WATER SLIDES An amusement park added a new section to a water slide to create a slide $311\frac{5}{12}$ feet long. How long was the slide before the addition?

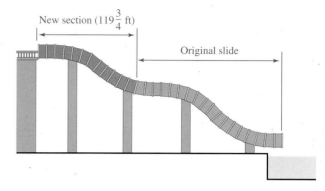

New section ($119\frac{3}{4}$ ft)

Original slide

84. JEWELRY A jeweler is to cut a 7-inch-long gold braid into three pieces. He aligns a 6-inch-long ruler directly below the braid and makes the proper cuts. Find the length of piece 2 of the braid.

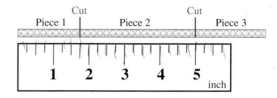

Cut   Cut

Piece 1   Piece 2   Piece 3

1   2   3   4   5   inch

WRITING

85. Of the methods studied to add mixed numbers, which do you like better, and why?

86. When subtracting mixed numbers, when is borrowing necessary? How is it done?

87. Explain how to add $1\frac{3}{8}$ and $2\frac{1}{4}$ if we write each one as an improper fraction.

88. Explain the process of simplifying $12\frac{16}{5}$.

REVIEW

89. Find the mean (average) of 36, 48, and 72.

90. Multiply: $-3(-4)(-5)$.

91. Add: $\frac{2}{5} + \frac{1}{4}$.

92. Which fraction is larger: $\frac{11}{13}$ or $\frac{6}{7}$?

93. Subtract: $-2 - (-8)$.

94. Find the area of a triangle that has a base 6 inches long and a height of 8 inches.

95. What does area measure?

96. Evaluate: $|-12|$.

3.7 Order of Operations and Complex Fractions

- Order of operations • Evaluating formulas • Complex fractions
- Simplifying complex fractions

In this section, we evaluate expressions involving fractions and mixed numbers. We also discuss complex fractions and how to simplify them.

Order of operations

The rules for the order of operations (first discussed in Section 1.7) are used to evaluate numerical expressions that involve more than one operation.

Self Check 1

Evaluate: $\dfrac{7}{8} + \dfrac{3}{2}\left(-\dfrac{1}{4}\right)^2$.

Answer $\dfrac{31}{32}$

EXAMPLE 1 Evaluate: $\dfrac{3}{4} + \dfrac{5}{3}\left(-\dfrac{1}{2}\right)^3$.

Solution The expression involves the operations of raising to a power, multiplication, and addition. By the rules for the order of operations, we must evaluate the power first, the multiplication second, and the addition last.

$$\frac{3}{4} + \frac{5}{3}\left(-\frac{1}{2}\right)^3 = \frac{3}{4} + \frac{5}{3}\left(-\frac{1}{8}\right) \quad \text{Find the power: } \left(-\frac{1}{2}\right)^3 = -\frac{1}{8}.$$

$$= \frac{3}{4} + \left(-\frac{5}{24}\right) \quad \text{Perform the multiplication: } \frac{5}{3}\left(-\frac{1}{8}\right) = -\frac{5}{24}.$$

$$= \frac{3\cdot 6}{4\cdot 6} + \left(-\frac{5}{24}\right) \quad \text{The LCD is 24. Write the first fraction as a fraction with denominator 24.}$$

$$= \frac{18}{24} + \left(-\frac{5}{24}\right) \quad \text{Multiply in the numerator: } 3\cdot 6 = 18. \text{ Multiply in the denominator: } 4\cdot 6 = 24.$$

$$= \frac{13}{24} \quad \text{Add the numerators: } 18 + (-5) = 13. \text{ Write the sum over the common denominator.}$$

If an expression contains grouping symbols, we perform the operations within the grouping symbols first.

EXAMPLE 2 Evaluate: $\left(\dfrac{7}{8} - \dfrac{1}{4}\right) \div \left(-2\dfrac{3}{16}\right)$.

Solution $\left(\dfrac{7}{8} - \dfrac{1}{4}\right) \div \left(-2\dfrac{3}{16}\right) = \left(\dfrac{7}{8} - \dfrac{1\cdot 2}{4\cdot 2}\right) \div \left(-2\dfrac{3}{16}\right)$ Within the first set of parentheses, write $\frac{1}{4}$ as a fraction with denominator 8.

$$= \left(\dfrac{7}{8} - \dfrac{2}{8}\right) \div \left(-2\dfrac{3}{16}\right)$$ Multiply in the numerator: $1\cdot 2 = 2$. Multiply in the denominator: $4\cdot 2 = 8$.

$$= \frac{5}{8} \div \left(-2\frac{3}{16} \right)$$ Subtract the numerators and write the difference over the common denominator: $7 - 2 = 5$.

$$= \frac{5}{8} \div \left(-\frac{35}{16} \right)$$ Write the mixed number as an improper fraction.

$$= \frac{5}{8} \left(-\frac{16}{35} \right)$$ Multiply by the reciprocal of $-\frac{35}{16}$.

$$= -\frac{5 \cdot 16}{8 \cdot 35}$$ The product of two fractions with unlike signs is negative. Multiply the numerators and multiply the denominators.

$$= -\frac{\overset{1}{\cancel{5}} \cdot 2 \cdot \overset{1}{\cancel{8}}}{\underset{1}{\cancel{8}} \cdot \underset{1}{\cancel{5}} \cdot 7}$$ Factor 16 as $2 \cdot 8$ and factor 35 as $5 \cdot 7$. Divide out the common factors of 8 and 5.

$$= -\frac{2}{7}$$ Simplify.

▇ Evaluating formulas

To evaluate a formula, we replace its letters with specific numbers and simplify by using the rules for the order of operations.

EXAMPLE 3   The formula for the area of a trapezoid is $A = \frac{1}{2}h(a + b)$, where A is the area, h is the height, and a and b are the lengths of its bases. Find A when $h = 1\frac{2}{3}$, $a = 2\frac{1}{2}$, and $b = 5\frac{1}{2}$.

Solution

$$A = \frac{1}{2}h(a + b)$$

$$= \frac{1}{2}\left(1\frac{2}{3} \right)\left(2\frac{1}{2} + 5\frac{1}{2} \right)$$ Replace h, a, and b with the given values.

$$= \frac{1}{2}\left(1\frac{2}{3} \right)(8)$$ Perform the addition within the parentheses: $2\frac{1}{2} + 5\frac{1}{2} = 8$.

$$= \frac{1}{2}\left(\frac{5}{3} \right)\left(\frac{8}{1} \right)$$ Write $1\frac{2}{3}$ as an improper fraction and 8 as $\frac{8}{1}$.

$$= \frac{1 \cdot 5 \cdot 8}{2 \cdot 3 \cdot 1}$$ Multiply the numerators and multiply the denominators.

$$= \frac{1 \cdot 5 \cdot \overset{1}{\cancel{2}} \cdot 4}{\underset{1}{\cancel{2}} \cdot 3 \cdot 1}$$ Factor 8 as $2 \cdot 4$ and divide out the common factor of 2.

$$= \frac{20}{3}$$ Multiply in the numerator and the denominator.

$$= 6\frac{2}{3}$$ Write $\frac{20}{3}$ as a mixed number.

The area of the trapezoid is $6\frac{2}{3}$ square units.

Self Check 3

The formula for the area of a triangle is $A = \frac{1}{2}bh$. Find the area of a triangle whose base is $12\frac{1}{2}$ meters long and whose height is $15\frac{1}{3}$ meters.

Answer $95\frac{5}{6}$ m^2

EXAMPLE 4   Masonry.   To build a wall, a mason will use blocks that are $5\frac{3}{4}$ inches high, held together with $\frac{3}{8}$-inch-thick layers of mortar. (See Figure 3-11). If the plans call for 8 layers of blocks, what will be the height of the wall when completed?

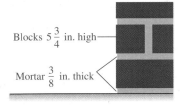

Blocks $5\frac{3}{4}$ in. high

Mortar $\frac{3}{8}$ in. thick

FIGURE 3-11

Solution   To find the height, we must consider 8 layers of blocks and 8 layers of mortar. We will compute the height contributed by one block and one layer of mortar and then multiply that result by 8.

8 times	(	height of 1 block	plus	height of 1 layer of mortar	) equals	height of block wall
8	(	$5\frac{3}{4}$	+	$\frac{3}{8}$	) =	height of wall

$$8\left(5\frac{3}{4}+\frac{3}{8}\right)=8\left(\frac{23}{4}+\frac{3}{8}\right) \quad \text{Write } 5\frac{3}{4} \text{ as the improper fraction } \frac{23}{4}.$$

$$=8\left(\frac{23\cdot 2}{4\cdot 2}+\frac{3}{8}\right) \quad \text{Express } \frac{23}{4} \text{ in terms of 8ths.}$$

$$=8\left(\frac{46}{8}+\frac{3}{8}\right) \quad \text{Perform the multiplication in the numerator and denominator.}$$

$$=\frac{8}{1}\left(\frac{49}{8}\right) \quad \text{Write 8 as } \frac{8}{1}. \text{ Within the parentheses, write the sum of the numerators over the common denominator 8.}$$

$$=\frac{\overset{1}{\cancel{8}}\cdot 49}{1\cdot \underset{1}{\cancel{8}}} \quad \text{Multiply the numerators and the denominators. Divide out the common factor of 8.}$$

$$=49 \quad \text{Simplify: } \frac{49}{1}=49.$$

The wall will be 49 inches high.

Complex fractions

Fractions whose numerators and/or denominators contain fractions are called *complex fractions*. Here is an example.

A fraction in the numerator ⟶ $\dfrac{\dfrac{3}{4}}{\dfrac{7}{8}}$ ⟵ The main fraction bar

A fraction in the denominator ⟶

Complex fraction

A **complex fraction** is a fraction whose numerator or denominator, or both, contain one or more fractions or mixed numbers.

Here are more examples of complex fractions.

$$\frac{-\dfrac{1}{4}-\dfrac{4}{5}}{2\dfrac{4}{5}} \qquad \frac{\dfrac{1}{3}+\dfrac{1}{4}}{\dfrac{1}{3}-\dfrac{1}{4}}$$

⟵ Numerator ⟶

⟵ Main fraction bar ⟶

⟵ Denominator ⟶

■ Simplifying complex fractions

To *simplify* complex fractions means to express them as fractions in simplified form.

> **Simplifying a complex fraction**
>
> Write the numerator and the denominator of the complex fraction as single fractions. Then perform the indicated division of the two fractions and simplify.

This method is based on the fact that the main fraction bar of the complex fraction indicates division.

$$\dfrac{\dfrac{1}{4}}{\dfrac{2}{5}} \longleftarrow \quad \text{The main fraction bar means “divide the fraction in the numerator by the fraction in the denominator.”} \quad \longrightarrow \quad \dfrac{1}{4} \div \dfrac{2}{5}$$

EXAMPLE 5   Simplify: $\dfrac{\dfrac{1}{4}}{\dfrac{2}{5}}$.

Self Check 5

Simplify: $\dfrac{\dfrac{1}{6}}{\dfrac{3}{8}}$

Solution   Since the numerator and the denominator of this complex fraction are single fractions, we can perform the indicated division.

$$\dfrac{\dfrac{1}{4}}{\dfrac{2}{5}} = \dfrac{1}{4} \div \dfrac{2}{5} \quad \text{Express the complex fraction as an equivalent division problem.}$$

$$= \dfrac{1}{4} \cdot \dfrac{5}{2} \quad \text{Multiply by the reciprocal of } \dfrac{2}{5}.$$

$$= \dfrac{1 \cdot 5}{4 \cdot 2} \quad \text{Multiply the numerators and multiply the denominators.}$$

$$= \dfrac{5}{8}$$

Answer $\dfrac{4}{9}$

EXAMPLE 6   Simplify: $\dfrac{-\dfrac{1}{4} + \dfrac{2}{5}}{\dfrac{1}{2} - \dfrac{4}{5}}$.

Solution   We need to write the numerator and the denominator of the complex fraction as single fractions. To do this, we find $-\frac{1}{4} + \frac{2}{5}$ and $\frac{1}{2} - \frac{4}{5}$.

$$\dfrac{-\dfrac{1}{4} + \dfrac{2}{5}}{\dfrac{1}{2} - \dfrac{4}{5}} = \dfrac{-\dfrac{1 \cdot 5}{4 \cdot 5} + \dfrac{2 \cdot 4}{5 \cdot 4}}{\dfrac{1 \cdot 5}{2 \cdot 5} - \dfrac{4 \cdot 2}{5 \cdot 2}} \quad \begin{array}{l}\text{In the numerator of the complex fraction, express the two fractions in terms of their LCD, which is 20. In the denominator, express the two fractions in terms of their LCD, which is 10.}\end{array}$$

$$= \dfrac{-\dfrac{5}{20} + \dfrac{8}{20}}{\dfrac{5}{10} - \dfrac{8}{10}} \quad \text{Multiply in the numerators and the denominators.}$$

$$= \frac{\dfrac{3}{20}}{-\dfrac{3}{10}}$$

In the numerator of the complex fraction, add the fractions. In the denominator, subtract the fractions.

$$= \frac{3}{20} \div \left(-\frac{3}{10}\right)$$

Express the complex fraction as an equivalent division problem.

$$= \frac{3}{20}\left(-\frac{10}{3}\right)$$

Multiply by the reciprocal of $-\dfrac{3}{10}$.

$$= -\frac{3 \cdot 10}{20 \cdot 3}$$

The product of two fractions with unlike signs is negative. Multiply the numerators and multiply the denominators.

$$= -\frac{\overset{1}{\cancel{3}} \cdot \overset{1}{\cancel{10}}}{2 \cdot \underset{1}{\cancel{10}} \cdot \underset{1}{\cancel{3}}}$$

Factor 20 as 2 · 10. Divide out the common factors, 3 and 10.

$$= -\frac{1}{2}$$

Simplify.

Self Check 7

Simplify: $\dfrac{5 - \dfrac{3}{4}}{1\dfrac{7}{8}}$.

EXAMPLE 7 Simplify: $\dfrac{7 - \dfrac{2}{3}}{4\dfrac{5}{6}}$.

Solution To write the numerator of the complex fraction as a single fraction, we need to find $7 - \dfrac{2}{3}$. To write the denominator of the complex fraction as a single fraction, we need to write $4\dfrac{5}{6}$ as an improper fraction.

$$\frac{7 - \dfrac{2}{3}}{4\dfrac{5}{6}} = \frac{\dfrac{7 \cdot 3}{1 \cdot 3} - \dfrac{2}{3}}{\dfrac{29}{6}}$$

In the numerator of the complex fraction, write 7 as $\frac{7}{1}$. Then express $\frac{7}{1}$ in terms of 3rds. In the denominator, write $4\frac{5}{6}$ as the improper fraction $\frac{29}{6}$.

$$= \frac{\dfrac{21}{3} - \dfrac{2}{3}}{\dfrac{29}{6}}$$

Multiply in the numerator and the denominator.

$$= \frac{\dfrac{19}{3}}{\dfrac{29}{6}}$$

In the numerator of the complex fraction, subtract the fractions.

$$= \frac{19}{3} \div \frac{29}{6}$$

Express the complex fraction as an equivalent division problem.

$$= \frac{19}{3} \cdot \frac{6}{29}$$

Multiply by the reciprocal of $\dfrac{29}{6}$.

$$= \frac{19 \cdot 6}{3 \cdot 29}$$

Multiply the numerators. Multiply the denominators.

$$= \frac{19 \cdot 2 \cdot \overset{1}{\cancel{3}}}{\underset{1}{\cancel{3}} \cdot 29}$$

Factor 6 as 2 · 3. Divide out the common factor of 3.

$$= \frac{38}{29}$$

Simplify.

$$= 1\frac{9}{29}$$

Write $\dfrac{38}{29}$ as a mixed number.

Answer $2\dfrac{4}{15}$

Section 3.7 STUDY SET

VOCABULARY *Fill in the blanks.*

1. $\dfrac{\frac{1}{2}}{\frac{3}{4}}$ is a _____ fraction.

2. To evaluate a formula such as $A = \frac{1}{2}h(a+b)$, we substitute specific _____ for the letters in the formula and simplify.

CONCEPTS

3. What division is represented by this complex fraction?

$$-\dfrac{\frac{2}{3}}{\frac{1}{5}}$$

4. Write this division as a complex fraction.

$$-\frac{7}{8} \div \frac{3}{4}$$

5. What is the LCD for the fractions in the numerator of this complex fraction?

$$\dfrac{\frac{2}{3} - \frac{1}{5}}{\frac{1}{2} + \frac{4}{5}}$$

6. Write the denominator of this complex fraction as an improper fraction.

$$\dfrac{\frac{1}{8} - \frac{3}{16}}{5\frac{3}{4}}$$

7. When this complex fraction is simplified, will the result be positive or negative?

$$\dfrac{-\frac{2}{3}}{\frac{3}{4}}$$

8. To evaluate $\frac{7}{8} + \left(\frac{1}{3}\right)\left(\frac{1}{4}\right)$, what operation should be performed first?

9. To evaluate $\frac{7}{8} + \left(\frac{1}{3} - \frac{1}{4}\right)^2$, what operation should be performed first?

10. What operations are involved in this numerical expression?

$$5\left(6\frac{1}{3}\right) + \left(-\frac{1}{4}\right)^2$$

NOTATION *Complete each solution.*

11. $\dfrac{\frac{1}{8}}{\frac{3}{4}} = \dfrac{1}{8} \div \boxed{}$

$ = \dfrac{1}{8} \cdot \boxed{}$

$ = \dfrac{1 \cdot \boxed{}}{8 \cdot 3}$

$ = \dfrac{1 \cdot \overset{1}{\cancel{4}}}{2 \cdot \underset{1}{\cancel{4}} \cdot 3}$

$ = \dfrac{1}{6}$

12. $\dfrac{1}{12} - \left(\dfrac{1}{2}\right)\left(\dfrac{1}{3}\right) = \dfrac{1}{12} - \dfrac{1 \cdot 1}{2 \cdot \boxed{}}$

$ = \dfrac{1}{12} - \dfrac{1}{\boxed{}}$

$ = \dfrac{1}{12} - \dfrac{1 \cdot \boxed{}}{6 \cdot \boxed{}}$

$ = \dfrac{1}{12} - \dfrac{\boxed{}}{12}$

$ = -\dfrac{1}{12}$

PRACTICE *Evaluate each expression.*

13. $\dfrac{2}{3}\left(-\dfrac{1}{4}\right) + \dfrac{1}{2}$

14. $-\dfrac{7}{8} - \left(\dfrac{1}{8}\right)\left(\dfrac{2}{3}\right)$

15. $\dfrac{4}{5} - \left(-\dfrac{1}{3}\right)^2$

16. $-\dfrac{3}{16} - \left(-\dfrac{1}{2}\right)^3$

17. $-4\left(-\dfrac{1}{5}\right) - \left(\dfrac{1}{4}\right)\left(-\dfrac{1}{2}\right)$

18. $(-3)\left(-\dfrac{2}{3}\right) - (-4)\left(-\dfrac{3}{4}\right)$

19. $1\dfrac{3}{5}\left(\dfrac{1}{2}\right)^2\left(\dfrac{3}{4}\right)$

20. $2\dfrac{3}{5}\left(-\dfrac{1}{3}\right)^2\left(\dfrac{1}{2}\right)$

21. $\dfrac{7}{8} - \left(\dfrac{4}{5} + 1\dfrac{3}{4}\right)$

22. $\left(\dfrac{5}{4}\right)^2 + \left(\dfrac{2}{3} - 2\dfrac{1}{6}\right)$

23. $\left(\dfrac{9}{20} \div 2\dfrac{2}{5}\right) + \left(\dfrac{3}{4}\right)^2$

24. $\left(1\dfrac{2}{3}\cdot 15\right) + \left(\dfrac{7}{9} \div \dfrac{7}{81}\right)$

25. $\left(-\dfrac{3}{4}\cdot\dfrac{9}{16}\right) + \left(\dfrac{1}{2} - \dfrac{1}{8}\right)$

26. $\left(\dfrac{8}{5} - 1\dfrac{1}{3}\right) - \left(-\dfrac{4}{5}\cdot 10\right)$

27. $\left|\dfrac{2}{3} - \dfrac{9}{10}\right| \div \left(-\dfrac{1}{5}\right)$

28. $\left|-\dfrac{3}{16} \div 2\dfrac{1}{4}\right| + \left(-2\dfrac{1}{8}\right)$

29. $\left(2 - \dfrac{1}{2}\right)^2 + \left(2 + \dfrac{1}{2}\right)^2$

30. $\left(1 - \dfrac{3}{4}\right)\left(1 + \dfrac{3}{4}\right)$

Find $\frac{1}{2}$ of the given number and then square that result. Express your answer as an improper fraction.

31. -7 **32.** -5

33. $\dfrac{11}{2}$ **34.** $\dfrac{7}{3}$

Evaluate each formula for $l = 12$, $w = 8\frac{1}{2}$, $b = 10$, and $h = 7\frac{1}{5}$.

35. $A = lw$ **36.** $P = 2l + 2w$

37. $A = \frac{1}{2}bh$ **38.** $V = lwh$

Find the perimeter of each figure.

39.

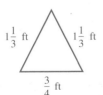

$2\frac{7}{8}$ in.

$1\frac{1}{4}$ in.

40.

$1\frac{1}{3}$ ft $1\frac{1}{3}$ ft

$\frac{3}{4}$ ft

Simplify each complex fraction.

41. $\dfrac{\dfrac{2}{3}}{\dfrac{4}{5}}$ **42.** $\dfrac{\dfrac{3}{5}}{\dfrac{9}{25}}$

43. $\dfrac{\dfrac{14}{15}}{\dfrac{7}{10}}$ **44.** $\dfrac{\dfrac{5}{27}}{\dfrac{5}{9}}$

45. $\dfrac{\dfrac{5}{10}}{\dfrac{10}{21}}$ **46.** $\dfrac{\dfrac{6}{3}}{\dfrac{3}{8}}$

47. $\dfrac{-\dfrac{5}{6}}{-1\dfrac{7}{8}}$ **48.** $\dfrac{-\dfrac{4}{3}}{-2\dfrac{5}{6}}$

49. $\dfrac{\dfrac{1}{2} + \dfrac{1}{4}}{\dfrac{1}{2} - \dfrac{1}{4}}$ **50.** $\dfrac{\dfrac{1}{3} + \dfrac{1}{4}}{\dfrac{1}{3} - \dfrac{1}{4}}$

51. $\dfrac{\dfrac{3}{8} + \dfrac{1}{4}}{\dfrac{3}{8} - \dfrac{1}{4}}$ **52.** $\dfrac{\dfrac{2}{5} + \dfrac{1}{4}}{\dfrac{2}{5} - \dfrac{1}{4}}$

53. $\dfrac{\dfrac{1}{5} + 3}{-\dfrac{4}{25}}$ **54.** $\dfrac{-5 - \dfrac{1}{3}}{\dfrac{1}{6} + \dfrac{2}{3}}$

55. $\dfrac{5\dfrac{1}{2}}{-\dfrac{1}{4} + \dfrac{3}{4}}$ **56.** $\dfrac{4\dfrac{1}{4}}{\dfrac{2}{3} + \left(-\dfrac{1}{6}\right)}$

57. $\dfrac{\dfrac{1}{5} - \left(-\dfrac{1}{4}\right)}{\dfrac{1}{4} + \dfrac{4}{5}}$ **58.** $\dfrac{\dfrac{1}{8} - \left(-\dfrac{1}{2}\right)}{\dfrac{1}{4} + \dfrac{3}{8}}$

59. $\dfrac{\dfrac{1}{3} + \left(-\dfrac{5}{6}\right)}{1\dfrac{1}{3}}$ **60.** $\dfrac{\dfrac{3}{7} + \left(-\dfrac{1}{2}\right)}{1\dfrac{3}{4}}$

APPLICATIONS

61. DELI SHOPS A sandwich shop sells a $\frac{1}{2}$-pound club sandwich made of turkey and ham. The owner buys the turkey in $1\frac{3}{4}$-pound packages and the ham in $2\frac{1}{2}$-pound packages. If he mixes a package of each of the meats together, how many sandwiches can he make from the mixture?

62. SKIN CREAMS Using a formula of $\frac{1}{2}$ ounce of sun block, $\frac{2}{3}$ ounce of moisturizing cream, and $\frac{3}{4}$ ounce of lanolin, a beautician mixes her own brand of skin cream. She packages it in $\frac{1}{4}$-ounce tubes. How many tubes can be produced using this formula?

63. PHYSICAL FITNESS Two people begin their workouts from the same point on a bike path and travel in opposite directions, as shown. How far apart are they in $1\frac{1}{2}$ hours? Use the chart to help organize your work.

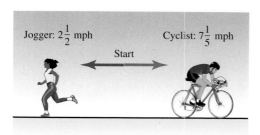

Jogger: $2\frac{1}{2}$ mph   Cyclist: $7\frac{1}{5}$ mph

Start

	Rate (mph) ·	Time (hr) =	Distance (mi)
Jogger			
Cyclist			

64. SLEEP The illustration compares the amount of sleep a 1-month-old baby got to the $15\frac{1}{2}$-hour daily requirement recommended by Children's Hospital of Orange County, California. For the week, how far below the baseline was the baby's daily average?

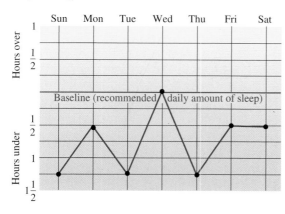

65. POSTAGE RATES Can the following advertising package be mailed for the 1-ounce rate?

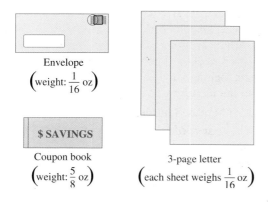

Envelope $\left(\text{weight: } \frac{1}{16} \text{ oz}\right)$

$ SAVINGS

Coupon book $\left(\text{weight: } \frac{5}{8} \text{ oz}\right)$

3-page letter $\left(\text{each sheet weighs } \frac{1}{16} \text{ oz}\right)$

66. PLYWOOD To manufacture a sheet of plywood, several layers of thin laminate are glued together, as shown. Then an exterior finish is affixed to top and bottom. How thick is the finished product?

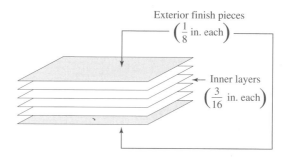

Exterior finish pieces $\left(\frac{1}{8} \text{ in. each}\right)$

Inner layers $\left(\frac{3}{16} \text{ in. each}\right)$

67. PHYSICAL THERAPY After back surgery, a patient followed a walking program to rehabilitate her back muscles, as specified in the table. What was the total distance she walked over this three-week period?

Week	Distance per day
#1	$\frac{1}{4}$ mile
#2	$\frac{1}{2}$ mile
#3	$\frac{3}{4}$ mile

68. READING PROGRAMS To improve reading skills, elementary school children read silently at the end of the school day for $\frac{1}{4}$ hour on Mondays and for $\frac{1}{2}$ hour on Fridays. For the month of January, how many total hours did the children read silently in class?

S	M	T	W	T	F	S
	1	2	3	4	5	6
7	8	9	10	11	12	13
14	15	16	17	18	19	20
21	22	23	24	25	26	27
28	29	30	31			

69. AMUSEMENT PARKS At the end of a ride at an amusement park, a boat splashes into a pool of water. The time (in seconds) that it takes two pipes to refill the pool is given by

$$\frac{1}{\frac{1}{10} + \frac{1}{15}}$$

Find this time.

70. HIKING A scout troop plans to hike from the campground to Glenn Peak. Since the terrain is steep, they plan to stop and rest after every $\frac{2}{3}$ mile. With this plan, how many parts will there be to this hike?

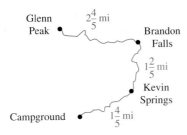

WRITING

71. What is a complex fraction?

72. Explain the method for simplifying complex fractions.

73. Write an application problem involving a complex fraction, and then solve it.

74. What are the rules for the order of operations?

REVIEW

75. Subtract 879 from 1,023.

76. Multiply 879 by 23.

77. Divide 1,665 by 45.

78. List the factors of 24.

79. Evaluate: $2 + 3[-3 - (-4 - 1)]$.

80. What is the sign of the quotient of two numbers with unlike signs?

81. Find the prime factorization of 288.

82. Subtract: $\dfrac{7}{8} - \dfrac{2}{3}$.

The Fundamental Property of Fractions

The **fundamental property of fractions** states that multiplying or dividing the numerator and the denominator of a fraction by the same nonzero number does not change the value of the fraction. This property is used to simplify fractions and to express fractions in higher terms. The following problems review both procedures. Complete each solution.

1. Simplify: $\dfrac{15}{25}$.

   Step 1: The numerator and the denominator share a common factor of ▢.

   Step 2: Apply the fundamental property of fractions. Divide the numerator and the denominator by the common factor ▢.

   $$\frac{15}{25} = \frac{15 \div \ \square}{25 \div \ \square}$$

   Step 3: Perform the divisions to simplify the fraction.

   $$= \frac{3}{\square}$$

2. In practice, we often show the simplifying process described in problem 1 in a different form.

   Step 1: Factor 15 as $3 \cdot \square$ and 25 as $\square \cdot 5$.

   $$\frac{15}{25} = \frac{3 \cdot \square}{\square \cdot 5}$$

   Step 2: The slashes and small 1's indicate that the numerator and the denominator have been divided by ▢.

   $$= \frac{3 \cdot \overset{1}{\cancel{5}}}{\underset{1}{\cancel{5}} \cdot 5}$$

   Step 3: Multiply in the numerator and the denominator.

   $$= \frac{\square}{5}$$

3. When adding or subtracting fractions and mixed numbers, we often need to express a fraction in higher terms. This is called building the fraction. Express $\frac{1}{5}$ as a fraction with denominator 35.

   Step 1: We must multiply the denominator 5 by ▢ to obtain 35.

   Step 2: Use the fundamental property of fractions. Multiply the numerator and the denominator by ▢.

   $$\frac{1}{5} = \frac{1 \cdot \square}{5 \cdot \square}$$

   Step 3: Multiply in the numerator and the denominator.

   $$= \frac{\square}{35}$$

ACCENT ON TEAMWORK

SECTION 3.1

EQUIVALENT FRACTIONS Complete the labeling of each number line using fractions with the same denominator.

a. 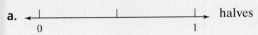 halves

b. fourths

c. eighths

d. sixteenths

FRACTIONS Give everyone in your group a strip of paper that is the same length. Determine ways to fold the strip of paper into

a. fourths   **b.** eighths

c. thirds   **d.** sixths

SECTION 3.2

MULTIPLICATION When we multiply 2 and 4, the answer is greater than 2 and greater than 4. Is this always the case? Is the product of two numbers always greater than either of the two numbers? Explain your answer.

POWERS When we square the number 4, the answer is greater than 4. Is the square of a number always greater than the number? Explain your answer.

SECTION 3.3

DIVIDING SNACKS Devise a way to divide seven brownies equally among six people.

SECTION 3.4

ADDING FRACTIONS Without actually doing the addition, explain why $\frac{3}{7} + \frac{1}{4}$ must be less than 1 and why $\frac{4}{7} + \frac{3}{4}$ must be greater than 1.

COMPARING FRACTIONS

a. When 1 is added to the numerator of a fraction, is the result greater than or less than the original fraction? Explain your reasoning.

b. When 1 is added to the denominator of a fraction, is the result greater than or less than the original fraction? Explain your reasoning.

COMPARING FRACTIONS Think of a fraction. Add 1 to its numerator and add 1 to its denominator. Is the resulting fraction greater than, less than, or equal to the original fraction? Explain your reasoning.

SECTION 3.5

DIVISION WITH MIXED NUMBERS Division can be thought of as repeated subtraction. Use this concept to solve the following problem.

$5\frac{1}{4}$ yards of ribbon needs to be cut into pieces that are $\frac{3}{4}$ of a yard long to form bows. How many bows can be made?

SECTION 3.6

MIXED NUMBERS Two mixed numbers, A and B, are graphed below. Estimate where on the number line the graph of $A + B$ would lie.

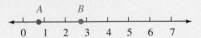

SECTION 3.7

COMPLEX FRACTIONS Write a problem that could be solved by simplifying the following complex fraction.

$$\frac{\frac{7}{8}}{\frac{3}{4}}$$

CHAPTER REVIEW

SECTION 3.1 *Fractions*

CONCEPTS

Fractions can be used to indicate equal parts of a whole.

A fraction is composed of a *numerator*, a *denominator*, and a *fraction bar*.

Equivalent fractions represent the same number.

The *fundamental property of fractions:* Dividing the numerator and denominator of a fraction by the same nonzero number does not change the value of the fraction.

To *simplify* a fraction that is not in lowest terms, divide the numerator and denominator by the same number.

A fraction is in *lowest terms* if the only factor common to the numerator and denominator is 1.

The *fundamental property of fractions:* Multiplying the numerator and denominator of a fraction by the same nonzero number does not change its value.

Expressing a fraction in higher terms results in an equivalent fraction that involves larger numbers or more complex terms.

REVIEW EXERCISES

1. If a woman gets seven hours of sleep each night, what part of a whole day does she spend sleeping?

2. In the illustration, why can't we say that $\frac{3}{4}$ of the figure is shaded?

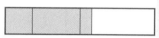

3. Write the fraction $\dfrac{2}{-3}$ in two other ways.

4. What concept about fractions does the illustration demonstrate?

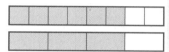

5. Explain the procedure shown here.

$$\frac{4}{6} = \frac{4 \div 2}{6 \div 2} = \frac{2}{3}$$

6. Explain what the slashes and the 1's mean.

$$\frac{4}{6} = \frac{\overset{1}{\cancel{2}} \cdot 2}{\underset{1}{\cancel{2}} \cdot 3} = \frac{2}{3}$$

Simplify each fraction.

7. $\dfrac{15}{45}$

8. $\dfrac{20}{48}$

9. $-\dfrac{63}{84}$

10. $\dfrac{66}{108}$

11. Explain what is being done and why it is valid.

$$\frac{5}{8} = \frac{5 \cdot 2}{8 \cdot 2} = \frac{10}{16}$$

Write each fraction or whole number with the indicated denominator.

12. $\dfrac{2}{3}$, 18

13. $-\dfrac{3}{8}$, 16

14. $\dfrac{7}{15}$, 45

15. 4, 9

Multiplying Fractions

To *multiply two fractions*, multiply their numerators and multiply their denominators.

Multiply.

16. $\dfrac{1}{2}\cdot\dfrac{1}{3}$

17. $\dfrac{2}{5}\left(-\dfrac{7}{9}\right)$

18. $\dfrac{9}{16}\cdot\dfrac{20}{27}$

19. $\dfrac{5}{6}\cdot\dfrac{1}{3}\cdot\dfrac{18}{25}$

20. $\dfrac{3}{5}\cdot 7$

21. $-4\left(-\dfrac{9}{16}\right)$

22. $3\left(\dfrac{1}{3}\right)$

23. $-\dfrac{6}{7}\left(-\dfrac{7}{6}\right)$

Determine whether each statement is true or false.

24. $\dfrac{3}{4}(2) = \dfrac{3(2)}{4}$

25. $-\dfrac{5}{9}(3) = -\dfrac{5}{9(3)}$

Multiply.

26. $\dfrac{3}{5}\cdot\dfrac{10}{27}$

27. $-\dfrac{2}{3}\left(\dfrac{4}{7}\right)$

28. $\dfrac{4}{9}\cdot\dfrac{3}{28}$

29. $9\left(-\dfrac{5}{81}\right)$

An *exponent* indicates repeated multiplication.

Find each power.

30. $\left(\dfrac{3}{4}\right)^2$

31. $\left(-\dfrac{5}{2}\right)^3$

32. $\left(\dfrac{2}{3}\right)^2$

33. $\left(-\dfrac{2}{5}\right)^3$

In mathematics, the word *of* usually means multiply.

34. GRAVITY Objects on the moon weigh only one-sixth as much as on Earth. How much will an astronaut weigh on the moon if he weighs 180 pounds on Earth?

35. Find the area of the triangular sign.

The *area of a triangle:*

$$A = \dfrac{1}{2}bh$$

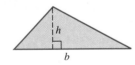

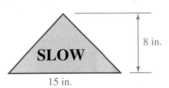

SLOW

8 in.

15 in.

Dividing Fractions

Two numbers are called *reciprocals* if their product is 1.

Find the reciprocal of each number.

36. $\dfrac{1}{8}$

37. $-\dfrac{11}{12}$

38. 5

39. $\dfrac{8}{7}$

To *divide two fractions*, multiply the first by the reciprocal of the second.

Divide.

40. $\dfrac{1}{6} \div \dfrac{11}{25}$

41. $-\dfrac{7}{8} \div \dfrac{1}{4}$

42. $-\dfrac{15}{16} \div (-10)$

43. $8 \div \dfrac{16}{5}$

44. $-\dfrac{3}{8} \div \dfrac{1}{4}$

45. $\dfrac{4}{5} \div \dfrac{1}{2}$

46. $\dfrac{2}{3} \div \left(-\dfrac{3}{2}\right)$

47. $\dfrac{2}{3} \div \left(-\dfrac{1}{9}\right)$

48. GOLD COINS How many $\frac{1}{16}$-ounce coins can be cast from a $\frac{3}{4}$-ounce bar of gold?

SECTION 3.4 — Adding and Subtracting Fractions

To add (or subtract) fractions with like denominators, add (or subtract) their numerators and write the result over the common denominator.

The *LCD* must include the set of prime factors of each of the denominators.

Add or subtract.

49. $\dfrac{2}{7} + \dfrac{3}{7}$ **50.** $-\dfrac{3}{5} - \dfrac{3}{5}$ **51.** $\dfrac{3}{4} - \dfrac{1}{4}$ **52.** $\dfrac{7}{8} + \dfrac{3}{8}$

53. Explain why we cannot immediately add $\dfrac{1}{2} + \dfrac{2}{3}$ without doing some preliminary work.

54. Use prime factorization to find the least common denominator for fractions with denominators of 45 and 30.

To add or subtract fractions with unlike denominators, first express them as equivalent fractions with the same denominator, preferably the LCD.

Add or subtract.

55. $\dfrac{1}{6} + \dfrac{2}{3}$ **56.** $\dfrac{2}{5} + \left(-\dfrac{3}{8} \right)$

57. $-\dfrac{3}{8} - \dfrac{5}{6}$ **58.** $3 - \dfrac{1}{7}$

59. $\dfrac{2}{25} - \dfrac{3}{10}$ **60.** $\dfrac{1}{3} + \dfrac{7}{4}$

61. $\dfrac{13}{6} - 6$ **62.** $\dfrac{1}{3} + \dfrac{1}{4} + \dfrac{1}{5}$

63. MACHINE SHOPS How much must be milled off the $\frac{3}{4}$-inch-thick steel rod below so that the collar will slip over the end of it?

Steel rod

To *compare fractions*, write them as equivalent fractions with the same denominator. Then the fraction with the larger numerator will be the larger fraction.

64. TELEMARKETING In the first hour of work, a telemarketer made 2 sales out of 9 telephone calls. In the second hour, she made 3 sales out of 11 calls. During which hour was the rate of sales to calls better?

SECTION 3.5 — Multiplying and Dividing Mixed Numbers

A *mixed number* is the sum of its whole-number part and its fractional part.

65. What mixed number is represented in the illustration?

66. What improper fraction is represented in the illustration?

To change an *improper fraction* to a mixed number, divide the numerator by the denominator to obtain the whole-number part. Write the remainder over the denominator for the fractional part.

Express each improper fraction as a mixed number or a whole number.

67. $\dfrac{16}{5}$ **68.** $-\dfrac{47}{12}$ **69.** $\dfrac{6}{6}$ **70.** $\dfrac{14}{6}$

To change a mixed number to an improper fraction, multiply the whole number by the denominator and add the result to the numerator. Write this sum over the denominator.

Write each mixed number as an improper fraction.

71. $9\dfrac{3}{8}$ **72.** $-2\dfrac{1}{5}$ **73.** $100\dfrac{1}{2}$ **74.** $1\dfrac{99}{100}$

75. Graph: $-2\frac{2}{3}, \frac{8}{9}$, and $\frac{59}{24}$.

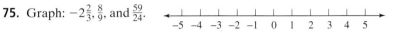

To *multiply* or *divide mixed numbers*, change the mixed numbers to improper fractions and then perform the operations as usual.

Multiply or divide. Write answers as mixed numbers when appropriate.

76. $-5\dfrac{1}{4} \cdot \dfrac{2}{35}$ **77.** $\left(-3\dfrac{1}{2}\right) \div \left(-3\dfrac{2}{3}\right)$

78. $\left(-6\dfrac{2}{3}\right)(-6)$ **79.** $-8 \div 3\dfrac{1}{5}$

80. CAMERA TRIPODS The three legs of a tripod can be extended to become $5\frac{1}{2}$ times their original length. If each leg is $8\frac{3}{4}$ inches long when collapsed, how long will a leg become when it is completely extended?

SECTION 3.6   *Adding and Subtracting Mixed Numbers*

To add (or subtract) mixed numbers, we can change each to an improper fraction and use the method of Section 3.4.

Add or subtract.

81. $1\dfrac{3}{8} + 2\dfrac{1}{5}$ **82.** $3\dfrac{1}{2} + 2\dfrac{2}{3}$

83. $2\dfrac{5}{6} - 1\dfrac{3}{4}$ **84.** $3\dfrac{7}{16} - 2\dfrac{1}{8}$

To add mixed numbers, we can add the whole numbers and the fractions separately.

85. PAINTING SUPPLIES In a project to restore a house, painters used $10\frac{3}{4}$ gallons of primer, $21\frac{1}{2}$ gallons of latex paint, and $7\frac{2}{3}$ gallons of enamel. Find the total number of gallons of paint used.

Vertical form can be used to add or subtract mixed numbers.

Add or subtract.

86. $\begin{array}{r} 133\frac{1}{9} \\ +\ 49\frac{1}{6} \\ \hline \end{array}$ **87.** $\begin{array}{r} 98\frac{11}{20} \\ +14\frac{3}{5} \\ \hline \end{array}$

88. $\begin{array}{r} 50\frac{5}{8} \\ -19\frac{1}{6} \\ \hline \end{array}$ **89.** $\begin{array}{r} 375\frac{3}{4} \\ -\ 59 \\ \hline \end{array}$

If the fraction being subtracted is larger than the first fraction, we need to *borrow* from the whole number.

Subtract.

90. $23\dfrac{1}{3} - 2\dfrac{5}{6}$ **91.** $39 - 4\dfrac{5}{8}$

Order of Operations and Complex Fractions

A *complex fraction* is a fraction whose numerator or denominator, or both, contain one or more fractions or mixed numbers.

To simplify a complex fraction, write the numerator and the denominator of the complex fraction as single fractions. Then perform the indicated division of the two fractions and simplify.

Evaluate each numerical expression.

92. $\dfrac{3}{4} + \left(-\dfrac{1}{3}\right)^2\left(\dfrac{5}{4}\right)$

93. $\left(\dfrac{2}{3} \div \dfrac{16}{9}\right) - \left(1\dfrac{2}{3} \cdot \dfrac{1}{15}\right)$

Simplify each complex fraction.

94. $\dfrac{\dfrac{3}{5}}{-\dfrac{17}{20}}$

95. $\dfrac{\dfrac{2}{3} - \dfrac{1}{6}}{-\dfrac{3}{4} - \dfrac{1}{2}}$

Evaluate the formula P = 2ℓ + 2w for the following values of ℓ and w.

96. $\ell = \dfrac{3}{4}$ and $w = \dfrac{2}{5}$

97. $\ell = 2\dfrac{1}{3}$ and $w = 3\dfrac{1}{4}$

1. See the illustration.

   a. What fractional part of the plant is above ground?

   b. What fractional part of the plant is below ground?

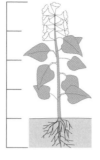

2. Simplify each fraction.

   a. $\dfrac{27}{36}$ **b.** $\dfrac{72}{180}$

3. Multiply: $-\dfrac{3}{4}\left(\dfrac{1}{5}\right)$.

4. COFFEE DRINKERS Of 100 adults surveyed, $\dfrac{2}{5}$ said they started their morning with a cup of coffee. Of the 100, how many would this be?

5. Divide: $\dfrac{4}{3} \div \dfrac{2}{9}$.

6. Subtract: $\dfrac{5}{6} - \dfrac{4}{5}$.

7. Express $\dfrac{7}{8}$ as an equivalent fraction with denominator 24.

8. Graph: $2\dfrac{4}{5}$, $-1\dfrac{1}{7}$, and $\dfrac{7}{6}$.

9. SPORTS CONTRACTS A basketball player signed a nine-year contract for $\$13\dfrac{1}{2}$ million. How much is this per year?

10. Evaluate the formula $A = \dfrac{1}{2}bh$ when $b = \dfrac{7}{3}$ and $h = \dfrac{15}{4}$.

11. Add: $157\dfrac{5}{9} + 103\dfrac{3}{4}$.

12. Subtract: $67\dfrac{1}{4} - 29\dfrac{5}{6}$.

13. BOXING When Oscar De La Hoya fought Pernell Whitaker, the "Tale of the Tape" shown in the illustration appeared in the sports section of many newspapers. What was the difference in the fighters'

   a. weights?

   b. chests (expanded)?

   c. waists?

Tale of the Tape		
De La Hoya		**Whitaker**
24 yr	Age	33 yr
146½ lb	Weight	146½ lb
5-11	Height	5-6
72 in.	Reach	69 in.
39 in.	Chest (Normal)	37 in.
42¼ in.	Chest (Expanded)	39½ in.
31¾ in.	Waist	28 in.

14. Add: $-\dfrac{3}{7} + 2$.

15. SEWING When cutting material for a $10\dfrac{1}{2}$-inch-wide placemat, a seamstress allows $\dfrac{5}{8}$ inch at each end for a hem. How wide should the material be cut?

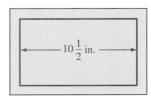

16. Find the perimeter and the area of the triangle below.

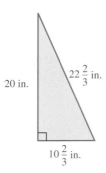

20 in. $22\frac{2}{3}$ in.

$10\frac{2}{3}$ in.

17. NUTRITION A box of Tic Tacs contains 40 of the $1\frac{1}{2}$-calorie flavored breath mints. How many calories are there in a box of Tic Tacs?

18. COOKING How many servings are there in an 8-pound roast, if the suggested serving size is $\frac{2}{3}$ pound?

19. Perform the operations.

$$\left(\frac{2}{3} \cdot \frac{5}{16}\right) - \left(-1\frac{3}{5} \div 4\frac{4}{5}\right)$$

20. Simplify the complex fraction.

$$\frac{-\dfrac{5}{6}}{\dfrac{7}{8}}$$

21. Simplify the complex fraction.

$$\frac{\dfrac{1}{2} + \dfrac{1}{3}}{-\dfrac{1}{6} - \dfrac{1}{3}}$$

22. JOB APPLICANTS Three-fourths of the applicants for a position had previous experience. If 144 people applied, how many had previous experience? How many did not have previous experience?

23. What are the parts of a fraction? What does a fraction represent?

24. Explain what is meant when we say, "The product of any number and its reciprocal is 1."

25. Explain what mathematical concept is being shown.

a. $\dfrac{6}{8} = \dfrac{\overset{1}{\cancel{2}} \cdot 3}{\underset{1}{\cancel{2}} \cdot 4} = \dfrac{3}{4}$

b.

c. $\dfrac{3}{5} = \dfrac{3 \cdot 4}{5 \cdot 4} = \dfrac{12}{20}$

26. Explain why we cannot immediately add $\frac{1}{4}$ and $\frac{3}{8}$ without doing some preliminary work.

Consider the number 5,434,679.

1. Round to the nearest hundred.

2. Round to the nearest ten thousand.

3. THE STOCK MARKET The graph below shows the performance of the Dow Jones Industrial Average on the last trading day of 1999. Estimate the highest mark that the market reached. At what time during the day did that occur?

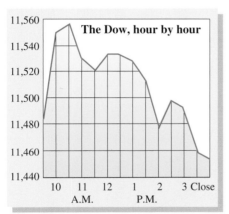

Source: *Los Angeles Times* (December 31, 1999)

4. BANKS One of the world's largest banks, with total assets of $691,920,300,000, is the Bank of Tokyo–Mitsubishi Ltd., Japan. In what place value column is the digit 6 located?

Perform each operation.

5.
$$\begin{array}{r} 4,679 \\ +3,457 \\ \hline \end{array}$$

6.
$$\begin{array}{r} 7,897 \\ -4,378 \\ \hline \end{array}$$

7.
$$\begin{array}{r} 5,345 \\ \times \quad 56 \\ \hline \end{array}$$

8. $35\overline{)34,685}$

Refer to the rectangular swimming pool.

9. Find the perimeter of the pool.

10. Find the area of the pool's surface.

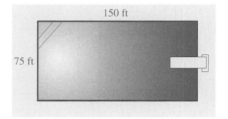

Find the prime factorization of each number.

11. 84

12. 450

13. 360

14. 3,600

Evaluate each expression.

15. $6 + (-2)(-5)$

16. $(-2)^3 - 3^3$

17. $\dfrac{2(-7) + 3(2)}{2(-2)}$

18. $\dfrac{2(3^2 - 4^2)}{-2(3) - 1}$

Simplify each fraction.

19. $\dfrac{21}{28}$

20. $\dfrac{40}{16}$

Perform each operation.

21. $\dfrac{6}{5}\left(-\dfrac{2}{3}\right)$

22. $\dfrac{14}{8} \div \dfrac{7}{2}$

23. $\dfrac{2}{3} + \dfrac{3}{4}$

24. $\dfrac{4}{7} - \dfrac{3}{5}$

Write each mixed number as an improper fraction.

25. $3\dfrac{5}{6}$

26. $-6\dfrac{5}{8}$

Perform each operation.

27. $4\dfrac{2}{3} + 5\dfrac{1}{4}$

28. $14\dfrac{2}{5} - 8\dfrac{2}{3}$

29. FIRE HAZARDS Two terminals in an electrical switch were so close that electricity could jump the gap and start a fire. The illustration shows a newly designed switch that will keep this from happening. By how much was the distance between the ground terminal and the hot terminal increased?

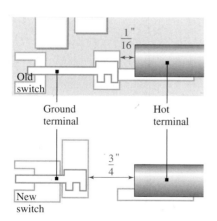

30. SHAVING Advertisements claim that a shaving lotion for men cuts shaving time by a third. If it normally takes a man 90 seconds to shave, how much time will he save if he uses the special lotion? If he uses the special lotion, how long will it take him to shave?

Simplify each expression.

31. $\left(\dfrac{1}{4} - \dfrac{7}{8}\right) \div \left(-2\dfrac{3}{16}\right)$

32. $\dfrac{\dfrac{2}{3} - 7}{4\dfrac{5}{6}}$

33. Describe how to multiply two fractions.

34. Describe how to subtract two fractions that have unlike denominators.

CHAPTER 4

Decimals

CORBIS

Sports fans love to talk facts, figures, and trivia. Many sports records are listed as decimal numbers. For example, did you know that over his 15-year career in the NBA, Michael Jordan averaged 30.1 points per game? Jerry Rice, considered by many to be the best receiver ever to play in the NFL, has averaged 14.8 yards per catch. Swimmer Janet Evans of the United States holds the world record in the 800-meter freestyle. Her time of 8 minutes and 16.22 seconds was set in 1989 and it has stood for over 15 years!

To learn more about decimals and how they are used in sports, visit The Learning Equation on the Internet at http://tle.brookscole.com. (The log-in instructions are in the Preface.) For Chapter 4, the online lessons are:

- *TLE* Lesson 9: Decimal Conversions
- *TLE* Lesson 10: Ordering Numbers

Check Your Knowledge

1. To multiply decimals, multiply them as if they were whole numbers. The number of decimal places in the product is the same as the _____ of the decimal places in the factors.

2. To divide by a decimal, move the decimal point of the divisor so that it becomes a whole number. The decimal point of the dividend is then moved the same number of places to the _____.

3. To write a fraction as a decimal, divide the _____ of the fraction by its denominator.

4. When we find what number is squared to obtain a given number, we are finding the square _____ of the given number.

5. Write 0.084 as a fraction.

6. Brittany purchased a CD player for $39.95, headphones for $17.95, and two CDs for $13.95 each. What was the total cost of her purchases?

7. Perform each operation in your head.
   a. $354{,}278.2 \div 1{,}000$ b. $2.000478 \cdot 10{,}000$

8. A rectangular table is 3.5 ft long and 2.25 ft wide.
   a. What is its area?   b. What is its perimeter?

9. Evaluate: $2.8 + (1.2)(-0.5)^2$.

10. Write each fraction as a decimal. Use an overbar, if necessary.
    a. $\dfrac{3}{20}$ b. $\dfrac{5}{8}$ c. $\dfrac{1}{9}$

11. Divide, and round your answer to the nearest tenth: $\dfrac{25.736}{16.3}$.

12. Graph $-\dfrac{3}{8}$ and $\dfrac{1}{4}$ on the number line. Label each point using its decimal equivalent.

13. Find the exact answer: $\dfrac{1}{6} + 0.25$.

14. A parking lot charges $1.00 for the first half hour and $0.75 for each additional half hour. What is the charge for 3 hours?

15. Graph $-\sqrt{3}$ and $\sqrt{2}$ on the number line. Label each point with its decimal approximation rounded to two decimal places.

16. Evaluate each expression.
    a. $\sqrt{16} + 5\sqrt{9}$ b. $\sqrt{\dfrac{36}{25}} - \sqrt{\dfrac{1}{16}}$

    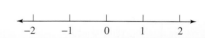

    c. $\sqrt{0.16} - \sqrt{0.49}$

17. Insert the proper symbol $<$ or $>$ to make each statement true.
    a. $-2.7 \ \square \ -2.75$ b. $\sqrt{2} \ \square \ 2$ c. $0.\overline{3} \ \square \ 0.3$

18. Mindy drove 342 miles in 6.5 hours. What was her average speed in miles per hour? Round to the nearest tenth.

Study Skills Workshop

REWORKING NOTES AS A STUDY AID

In the last Study Skills Workshop, you learned the importance of taking notes. In this section, you will learn to rework your notes to use in studying for homework and tests. "Reworking notes" simply means that you look over your class notes, fill in any missing information, and then put the information in a new, easy-to-study format.

Identify Concepts. As soon as possible after class, sit down with your notes and identify the key concepts that were discussed in your lecture. Key concepts are usually identified by definitions, rules, or formulas.

Examples. Once you have identified a key concept, find all the examples that were used to illustrate that concept. Make sure that you understand each example, and fill in any missing steps if that helps you to understand the example. Give the reason for each step in the example.

Format for Reworking. You can rework your notes in a variety of ways, depending on personal preference and learning style. You can use index cards or a study sheet divided into three columns, or if you are an auditory learner (or spend a great deal of time in the car), you can rework your notes in audio form.

- *Index card format:* Make a tab for your cards that states a key concept. On the front of the first index card write the concept. On the back of the card write the definition or formula; you will want to memorize this. Using an index card for each example, write the example on the front of the card. On the back, describe each step of the example, stating the reason for each step. Repeat this process for every concept learned. Store your index cards in a box, like those used for recipe files, and group the cards by concept.
- *Three-column study sheet:* Divide an $8\frac{1}{2}'' \times 11''$ sheet of paper into three columns. Label the left column "Concept," the middle column "Example," and the right column "Steps and Reasons." Rewrite your notes, filling in the columns on your study sheet as labeled.
- *Audiotape or CD:* Using a recorder, state a key concept. Pause for a few seconds, and then state the definition or formula. State an example and then explain each step. Continue to state examples and steps. Repeat for the remaining concepts.

How to Use Reworked Notes for Studying

- *Index cards:* Look at the concept on the first card in a group. Attempt to state the definition or the rule. If you are unable to do so, look at the back of the card and read it. Keep doing this until you can recite from memory. Then look at the example and see if you can work the problem without help of any kind. If you have trouble with this, the steps are written on the back and you need only turn the card over to look at them. Continue to rework the problem until you can do it without looking at the solution.
- *Three-column study sheet:* Go through the concepts and examples, covering the work in the right column while you attempt to work the problem in the middle. If you get stuck, uncover the work to look at the steps. Continue to rework the problem until you can do it without looking at the solution.
- *Audiotape:* In the pause between each phase of the problem, see if you can recite the steps before they are revealed on the tape. Repeat until you can recite all of the steps on your own.

ASSIGNMENT

1. Determine and state which format is the best one for you to rework your notes. If you think of a format that is different from those listed above, describe your format.
2. Rework your notes from the last lecture in your chosen format.

Decimals provide another way to represent fractions and mixed numbers. They are often used in measurement, because it is easy to put them in order and compare them.

4.1 An Introduction to Decimals

- Decimals • The place value system for decimal numbers
- Reading and writing decimals • Comparing decimals • Rounding

This section introduces the **decimal numeration system**—an extension of the place value system that we used with whole numbers.

We can use the decimal numeration system to express the car's mileage. The odometer reads 1,537.6 miles.

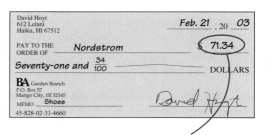

The amount of the check is written using the decimal numeration system.

Decimals

Like fraction notation, decimal notation is used to denote a part of a whole. However, when writing a number in decimal notation, we don't use a fraction bar, nor is a denominator shown. A number written in decimal notation is often called a **decimal.**

In Figure 4-1, a rectangle is divided into ten equal parts. One-tenth of the figure is shaded. We can use either the fraction $\frac{1}{10}$ or the decimal 0.1 to describe the amount of the figure that is shaded. Both are read as "one-tenth."

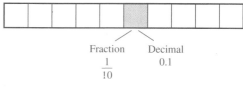

Fraction $\frac{1}{10}$ Decimal 0.1

FIGURE 4-1

$$\frac{1}{10} = 0.1$$

In Figure 4-2, a square is divided into 100 equal parts. One of the 100 parts is shaded; the amount of the figure that is shaded can be represented by the fraction $\frac{1}{100}$ or by the decimal 0.01. Both are read as "one one-hundredth."

Fraction $\frac{1}{100}$

Decimal 0.01

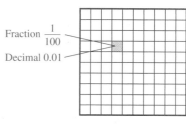

$$\frac{1}{100} = 0.01$$

FIGURE 4-2

■ The place value system for decimal numbers

Decimal numbers are written by placing digits (0, 1, 2, 3, 4, 5, 6, 7, 8, 9) into place value columns that are separated by a **decimal point.** (See Figure 4-3.) The place value names of all the columns to the right of the decimal point end in "th." The "th" tells us that the value of the column is a fraction whose denominator is a power of 10. Columns to the left of the decimal point have a value greater than or equal to 1; columns to the right of the decimal point have a value less than 1. We can show the value represented by each digit of a decimal by using **expanded notation.**

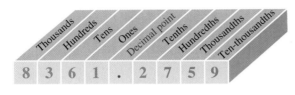

$$8{,}000 + 300 + 60 + 1 + \frac{2}{10} + \frac{7}{100} + \frac{5}{1{,}000} + \frac{9}{10{,}000}$$

Expanded notation

FIGURE 4-3

Decimal points are used to separate the whole-number part of a decimal from its fractional part.

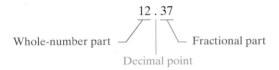

When there is no whole-number part of a decimal, we can show that by entering a zero to the left of the decimal point.

$$.85 \quad = \quad 0.85$$

↑   ↑

No whole-number part   Enter a zero here, if desired.

We can write a whole number in decimal notation by placing a decimal point to its right and then adding a zero, or zeros, to the right of the decimal point.

$$99 \quad = \quad 99.0 \quad = \quad 99.00$$

↑   ↑   ↑

A whole number   Place a decimal point here and enter
a zero, or zeros, to the right of it.

Writing additional zeros to the right of the decimal point *following the last digit* does not change the value of the decimal. Deleting additional zeros to the right of the decimal point following the last digit does not change the value of the decimal.

$$12.37 = 12.370 = 12.3700$$

↑   ↑

These additional zeros do not
change the value of the decimal.

◼ Reading and writing decimals

The decimal 12.37 can be read as "twelve point three seven." Another way of reading a decimal states the whole-number part first and then the fractional part.

Reading a decimal

To read a decimal:

1. Look to the left of the decimal point and say the name of the whole number.

2. The decimal point is then read as "and."

3. Say the fractional part of the decimal as a whole number followed by the name of the place value column of the digit that is farthest to the right.

With this procedure, here is the other way to read 12.37.

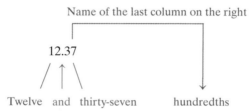

Name of the last column on the right

12.37

Twelve   and   thirty-seven   hundredths

When we read a decimal in this way, it is easy to write it in words and as a mixed number.

Decimal	Words	Mixed number
12.37	Twelve and thirty-seven hundredths	$12\frac{37}{100}$

Self Check 1

Write each decimal in words and then as a mixed number.

a. Sputnik 1, the first artificial satellite, weighed 184.3 pounds.

b. The planet Mercury makes one revolution every 87.9687 days.

Answers a. One hundred eighty-four and three tenths, or $184\frac{3}{10}$
b. Eighty-seven and nine thousand six hundred eighty-seven ten-thousandths, or $87\frac{9,687}{10,000}$

EXAMPLE 1   World records.   Write each decimal in words and then as a fraction or mixed number. **Do not simplify the fraction.**

a. According to the *Guinness Book of World Records*, the fastest qualifying speed at the Indianapolis 500 was 236.986 mph by Aire Luyendyk in 1996.

b. The smallest freshwater fish is the dwarf pygmy goby, found in the Philippines. Adult males weigh 0.00014 ounce.

Solution

a. The whole number part of **236.**986 to the left of the decimal point is 236. The fractional part, stated as a whole number, is 986. The digit the farthest to the right is 6 and it is in the thousandths column. Thus,

   236.986 is two hundred thirty-six and nine hundred eighty-six thousandths,
   or $236\frac{986}{1,000}$.

b. The whole number part of **0.**00014 to the left of the decimal point is 0. The fractional part, stated as a whole number, is 14. The digit the farthest to the right is 4 and it is in the hundred-thousandths column. Thus,

   0.00014 is fourteen hundred-thousandths, or $\frac{14}{100,000}$.

Decimals can be negative. For example, a record low temperature of $-128.6°$ F was recorded in Vostok, Antarctica, on July 21, 1983. This is read as "negative one hundred twenty-eight and six tenths." Written as a mixed number, it is $-128\frac{6}{10}$.

◼ Comparing decimals

The relative sizes of a set of decimals can be determined by scanning their place value columns from left to right, column by column, looking for a difference in the digits. For example,

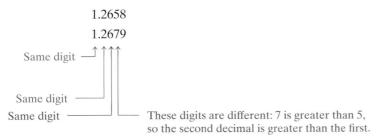

Thus, 1.2679 is greater than 1.2658. We write $1.2679 > 1.2658$.

Comparing positive decimals

To compare two positive decimals:

1. Make sure both numbers have the same number of decimal places to the right of the decimal point. Write any additional zeros necessary to achieve this.
2. Compare the digits of each decimal, column by column, working from left to right.
3. When two digits differ, the decimal with the greater digit is the greater number.

EXAMPLE 2   Which is greater: 54.9 or 54.929?

Solution

54.900   Write two zeros after 9 so that both decimals have the same number of digits to the right of the decimal point.

54.929
  ↑

As we work from left to right, this is the first column in which the digits differ. Since 2 is greater than 0, we can conclude that $54.929 > 54.9$.

Self Check 2
Which is greater: 113.7 or 113.657?

Answer 113.7

Comparing negative decimals

To compare two negative decimals:

1. Make sure both numbers have the same number of decimal places to the right of the decimal point. Write any additional zeros necessary to achieve this.
2. Compare the digits of each decimal, column by column, working from left to right.
3. When two digits differ, the decimal with the smaller digit is the greater number.

Self Check 3
Which is greater: −703.8 or
−703.78?

EXAMPLE 3 Which is greater: −10.45 or −10.419?

Solution

-10.450 Write a zero after 5 to help in the comparison.
-10.419
$\uparrow$

As we work from left to right, this is the first column in which the digits differ. Since 1 is less than 5, we conclude that $-10.419 > -10.45$.

Answer −703.78

Self Check 4
Graph: −1.1, −0.6, 0.8, and 1.9.

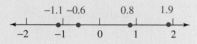

Answer

EXAMPLE 4 Graph: −1.8, −1.23, −0.3, and 1.89.

Solution To graph each decimal, we locate its position on the number line and draw a dot. Since −1.8 is to the left of −1.23, we can write $-1.8 < -1.23$.

▌ Rounding

When working with decimals, we often round answers to a specific number of decimal places.

> **Rounding a decimal**
>
> To round a decimal to a specified decimal place:
>
> **1.** Locate the digit in that place. Call it the *rounding digit.*
> **2.** Look at the *test digit* to the right of the rounding digit.
> **3.** If the test digit is 5 or greater, round up by adding 1 to the rounding digit and dropping all the digits to its right. If the test digit is less than 5, round down by keeping the rounding digit and dropping all the digits to its right.

EXAMPLE 5 Chemistry. In a chemistry class, a student uses a balance to weigh a compound. The digital readout on the scale shows 1.2387 grams. Round this decimal to the nearest thousandth of a gram.

Solution We are asked to round to the nearest thousandth.

┌──── Add 1 to the 8. ────◄
                         1.2387
The rounding digit ──┘└── The test digit is 5 or greater. Therefore,
                          add 1 to the rounding digit and drop
                          all other digits to its right.

The compound weighs approximately 1.239 grams.

EXAMPLE 6   Round each decimal to the indicated place value: **a.** -645.13 to the nearest tenth   and   **b.** 33.097 to the nearest hundredth.

Solution

a. -645.13

   Rounding    Since the test digit is less than 5, drop it and all the digits to its right.
   digit

The result is -645.1.

b. 33.097

   Rounding    Since the test digit is greater than 5, we add 1 to 9 and drop all the
   digit   digits to the right.

       $33.0\overset{10}{9}$ Adding 1 to the 9 requires that we carry a 1 to the tenths column.

When we are asked to round to the nearest hundredth, we must have a digit in the hundredths column, even if it is a zero. Therefore, the result is 33.10.

Self Check 6
Round each decimal to the indicated place value.
a. -708.522 to the nearest tenth
b. 9.1198 to the nearest thousandth

Answers **a.** -708.5, **b.** 9.120

Section 4.1   STUDY SET

VOCABULARY   *Fill in the blanks.*

1. Give the name of each place value column.

   4   7   8   9   •   0   2   6   5

2. We can show the value represented by each digit of the decimal 98.6213 by using _____ notation.

$$98.6213 = 90 + 8 + \frac{6}{10} + \frac{2}{100} + \frac{1}{1,000} + \frac{3}{10,000}$$

3. We can approximate a decimal number using the process called _____.

4. When reading 2.37, we can read the decimal point as "_____" or "_____."

CONCEPTS

5. Consider the decimal 32.415.

   a. Write the decimal in words.

   b. What is its whole-number part?

   c. What is its fractional part?

   d. Write the decimal in expanded notation.

6. Write the following as a decimal.

$$400 + 20 + 8 + \frac{9}{10} + \frac{6}{100}$$

7. Graph: $\dfrac{7}{10}$, -0.7, $-3\dfrac{1}{100}$, and 3.01.

   -5 -4 -3 -2 -1 0   1   2   3   4   5

8. Graph: -1.21, -3.29, and -4.25.

   -5 -4 -3 -2 -1 0   1   2   3   4   5

9. Determine whether each statement is true or false.

   a. $0.9 = 0.90$

   b. $1.260 = 1.206$

   c. $-1.2800 = -1.280$

   d. $0.001 = .0010$

10. Write each fraction as a decimal.

   a. $\dfrac{9}{10}$ **b.** $\dfrac{63}{100}$

   c. $\dfrac{111}{1,000}$ **d.** $\dfrac{27}{10,000}$

11. Represent the shaded part of the square as a fraction and a decimal.

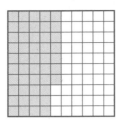

12. Represent the shaded part of the rectangle as a fraction and a decimal.

13. The line segment below is 1 inch long. Show a length of 0.3 inch on it.

_____

14. Read the meter below. What decimal is indicated by the arrow?

NOTATION

15. Construct a decimal number by writing

    0 in the tenths column,

    4 in the thousandths column,

    1 in the tens column,

    9 in the thousands column,

    8 in the hundreds column,

    2 in the hundredths column,

    5 in the ten-thousandths column, and

    6 in the ones column.

16. Represent each situation using a signed number.

    a. A deficit of $15,600.55

    b. A river 6.25 feet above flood stage

    c. A state budget $6.4 million in the red

    d. 3.9 degrees below zero

    e. 17.5 seconds after liftoff

    f. A checking account overdrawn by $33.45

PRACTICE *Write each decimal in words and as a fraction or mixed number.*

17. 50.1   **18.** 0.73

19. −0.0137   **20.** −76.09

21. 304.0003   **22.** 68.91

23. −72.493   **24.** −31.5013

Write each decimal using numbers.

25. Negative thirty-nine hundredths

26. Negative twenty-seven and forty-four hundredths

27. Six and one hundred eighty-seven thousandths

28. Ten and fifty-six ten-thousandths

Round each decimal to the nearest tenth.

29. 506.098   **30.** 0.441

31. 2.718218   **32.** 3,987.8911

Round each decimal to the nearest hundredth.

33. −0.137   **34.** −808.0897

35. 33.0032   **36.** 64.0059

Round each decimal to the nearest thousandth.

37. 3.14159   **38.** 16.0995

39. 1.414213   **40.** 2,300.9998

Round each decimal to the nearest whole number.

41. 38.901   **42.** 405.64

43. 2,988.399   **44.** 10,453.27

Round each amount to the value indicated.

45. $3,090.28

    a. Nearest dollar

    b. Nearest ten cents

46. $289.73

    a. Nearest dollar

    b. Nearest ten cents

Fill in the blanks with the proper symbol <, >, or =.

47. -23.45 ___ -23.1 **48.** -301.98 ___ -302.45

49. $-.065$ ___ $-.066$ **50.** -3.99 ___ -3.9888

Arrange the decimals in order, from least to greatest.

51. 132.64, 132.6499, 132.6401

52. 0.007, 0.00697, 0.00689

APPLICATIONS

53. WRITING CHECKS Complete the check shown by writing in the amount, using a decimal.

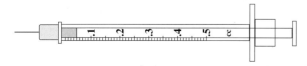

Ellen Russell
455 Santa Clara Ave.
Parker, CO 25413

April 14, 20 03

PAY TO THE ORDER OF   *Citicorp* $ _____

One thousand twenty-five and $\frac{78}{100}$ DOLLARS

BA Downtown Branch
P.O. Box 2456
Colorado Springs, CO 23712
MEMO *Mortgage*   *Ellen Russell*
45-828-02-33-4660

54. MONEY We use a decimal point when working with dollars, but the decimal point is not necessary when working with cents. For each dollar amount in the table, give the equivalent amount expressed as cents.

Dollars	Cents
$0.50	
$0.05	
$0.55	
$5.00	
$0.01	

55. INJECTIONS A syringe is shown. Use an arrow to show to what point the syringe should be filled if a 0.38-cc dose of medication is to be administered. ("cc" stands for "cubic centimeters.")

56. LASERS The laser used in laser vision correction is so precise that each pulse can remove 39 millionths of an inch of tissue in 12 billionths of a second. Write each of these numbers as a decimal.

57. THE METRIC SYSTEM The metric system is widely used in science to measure length (meters), weight (grams), and capacity (liters). Round each decimal to the nearest hundredth.

    a. 1 ft is 0.3048 meter.

    b. 1 mi is 1,609.344 meters.

    c. 1 lb is 453.59237 grams.

    d. 1 gal is 3.785306 liters.

58. WORLD RECORDS As of October 2004, four American women held individual world records in swimming. Their times are given below in the form *minutes: seconds.* Round each to the nearest tenth of a second.

100-meter backstroke	Natalie Coughlin	0:59.58
200-meter breaststroke	Amanda Beard	2:22.44
400-meter freestyle	Janet Evans	4:03.85
800-meter freestyle	Janet Evans	8:16.22
1,500-meter freestyle	Janet Evans	15:52.10

59. GEOLOGY Geologists classify types of soil according to the grain size of the particles that make up the soil. The four major classifications are shown below. Complete the table by classifying each sample.

Clay	0.00008 in. and under
Silt	0.00008 in. to 0.002 in.
Sand	0.002 in. to 0.08 in.
Granule	0.08 in. to 0.15 in.

Sample	Location	Size (in.)	Classification
A	Riverbank	0.009	
B	Pond	0.0007	
C	NE corner	0.095	
D	Dry lake	0.00003	

60. MICROSCOPES A microscope used in a lab is capable of viewing structures that range in size from 0.1 to 0.0001 centimeter. Which of the structures listed in the table would be visible through this microscope?

Structure	Size (cm)
Bacterium	0.00011
Plant cell	0.015
Virus	0.000017
Animal cell	0.00093
Asbestos fiber	0.0002

61. AIR QUALITY The following table shows the cities with the highest one-hour concentrations of ozone (in parts per million) during the summer of 1999. Rank the cities in order, beginning with the city with the highest reading.

Crestline, California	0.170
Galveston, Texas	0.176
Houston, Texas	0.202
Texas City, Texas	0.206
Westport, Connecticut	0.188
White Plains, New York	0.171

Source: *Los Angeles Times* (August 18, 1999)

62. THE DEWEY DECIMAL SYSTEM A system for classifying books in a library is the Dewey Decimal System. Books on the same subject are grouped together by number. For example, books about the arts are assigned numbers between 700 and 799. When stacked on the shelves, the books are to be in numerical order, from left to right. How should the titles in the illustration be rearranged to be in the proper order?

63. THE OLYMPICS The results of the women's all-around gymnastic competition in the 2004 Athens Olympic Games are shown in the following table. Which gymnasts won the gold, silver, and bronze medals?

Name	Country	Score
Nan Zhang	China	38.049
Ana Pavlova	Russia	38.024
Nicoleta Sofronie	Romania	37.948
Carly Patterson	U.S.A.	38.387
Svetlana Khorkina	Russia	38.211
Irina Yarotska	Ukraine	37.687

64. TUNEUPS The six spark plugs from the engine of a Nissan Quest were removed, and the spark plug gap was checked. If vehicle specifications call for the gap to be from 0.031 to 0.035 inch, which of the plugs should be replaced?

Cylinder 1: 0.035 in.
Cylinder 2: 0.029 in.
Cylinder 3: 0.033 in.
Cylinder 4: 0.039 in.
Cylinder 5: 0.031 in.
Cylinder 6: 0.032 in.

65. E-COMMERCE The gain (or loss) in value of one share of Amazon.com stock is shown in the graph below for eleven quarters. (For accounting purposes, a year is divided into four quarters.)

a. In what quarter, of what year, was there the greatest gain? Estimate the gain.

b. In what quarter, of what year, was there the greatest loss? Estimate the loss.

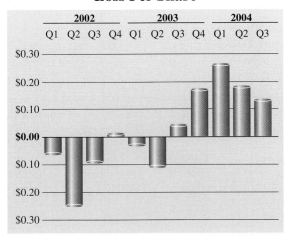

Source: Amazon.com

66. GASOLINE PRICES Refer to the data in the table. Then construct a line graph showing the annual national average retail price per gallon for unleaded regular gasoline for the years 1997 through 2003 (according to *The World Almanac 2005*).

Year	1997	1998	1999	2000	2001	2002	2003
Price (¢)	123.4	105.9	116.5	151.0	146.1	135.8	159.1

Source: *The World Almanac 2005*

WRITING

67. Explain the difference between ten and a tenth.

68. "The more digits a number contains, the larger it is." Is this statement true? Explain.

69. How are fractions and decimals related?

70. Explain the benefits of a monetary system that is based on decimals instead of fractions.

71. The illustration shows the unusual notation many service stations use to express the price of a gallon of gasoline. Explain what this notation means.

REGULAR	UNLEADED	UNLEADED +
$1.79\frac{9}{10}$	$1.89\frac{9}{10}$	$1.99\frac{9}{10}$

72. Write a definition for each of these words.

decade   *decathlon*   *decimal*

REVIEW

73. Add: $75\frac{3}{4} + 88\frac{4}{5}$.

74. Multiply: $\frac{2}{15}\left(-\frac{5}{4}\right)$

75. Evaluate: $\left(\frac{2}{3}\right)^5$.

76. Evaluate: $\dfrac{6 + (5 - 2)^2}{3(2 - 1)}$.

77. Find the area of a triangle with base 16 in. and height 9 in.

78. Express the fraction $\frac{2}{3}$ as an equivalent fraction with a denominator of 12.

79. Add: $-2 + (-3) + 4$.

80. Subtract: $-15 - (-6)$.

4.2 Adding and Subtracting Decimals

- Adding decimals • Subtracting decimals • Adding and subtracting signed decimals

To add or subtract objects, they must be similar. The federal income tax form shown in Figure 4-4 has a vertical line to ensure that dollars are added to dollars and cents added to cents. In this section, we show how decimals are added and subtracted using this type of vertical column format.

Form **1040EZ**	Department of the Treasury—Internal Revenue Service **Income Tax Return for Single and Joint Filers With No Dependents** **2003**			
Income Attach Form(s) W-2 here. Enclose but do not attach any payment.	1	Total wages, salaries, and tips. This should be shown in box 1 of your form(s) W-2.	1	21,056 89
	2	Taxable interest. If the total is over $1,500, you cannot use Form 1040EZ.	2	42 06
	3	Unemployment compensation and Alaska Permanent Fund dividends (see page 14).	3	200 00
	4	Add lines 1, 2, and 3. This is your **adjusted gross income.**	4	21,298 95

FIGURE 4-4

Adding decimals

When adding decimals, we line up the columns so that ones are added to ones, tenths are added to tenths, hundredths are added to hundredths, and so on. As an example, consider the following problem.

Line up the columns and the decimal points vertically. Then add the numbers.

$$
\begin{array}{r}
12.140 \\
3.026 \\
4.000 \\
+\ 0.700 \\
\hline
19.866
\end{array}
$$

Write the decimal point in the result directly under the decimal points in the problem.

> **Adding decimals**
>
> To add decimal numbers:
>
> 1. Line up the decimal points, using the vertical column format.
> 2. Add the numbers as you would add whole numbers.
> 3. Write the decimal point in the result directly below the decimal points in the problem.

Self Check 1

Add: $0.07 + 35 + 0.888 + 4.1$.

EXAMPLE 1   Add: $1.903 + 0.6 + 8 + 0.78$.

Solution

$$
\begin{array}{r}
\overset{2}{1}.903 \\
0.600 \\
8.000 \\
+\ 0.780 \\
\hline
11.283
\end{array}
$$

To make the addition by columns easier, write two zeros after 6, a decimal point and three zeros after 8, and one zero after 0.78.

Carry a 2 (shown in blue) to the ones column.

Answer 40.058

The result is 11.283.

CALCULATOR SNAPSHOT   **Preventing heart attacks**

The bar graph in Figure 4-5 shows the number of grams of fiber in a standard serving of each of several foods. It is believed that men can significantly cut their risk of heart attack by eating at least 25 grams of fiber a day. Does this diet meet or exceed the 25-gram requirement?

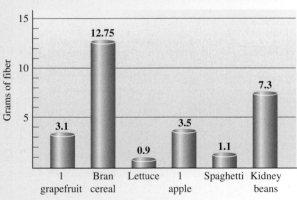

FIGURE 4-5

To find the total fiber intake, we add the fiber content of each of the foods. We can use a scientific calculator to add the decimals.

3.1 $\boxed{+}$ 12.75 $\boxed{+}$ .9 $\boxed{+}$ 3.5 $\boxed{+}$ 1.1 $\boxed{+}$ 7.3 $\boxed{=}$ $\boxed{28.65}$

Since 28.65 > 25, this diet exceeds the daily fiber requirement of 25 grams.

▮ Subtracting decimals

To subtract decimals, we line up the decimal points and corresponding columns so that we subtract like objects — tenths from tenths, hundredths from hundredths, and so on.

Subtracting decimals

To subtract decimal numbers:

1. Line up the decimal points using the vertical column format.
2. Subtract the numbers as you would subtract whole numbers.
3. Write the decimal point in the result directly below the decimal points in the problem.

EXAMPLE 2   Subtract: **a.** 279.6 − 138.7 and **b.** 15.4 − 13.059.

Solution

a.

$$\begin{array}{r} \overset{8\ \ 16}{279.\cancel{6}} \\ -138.7 \\ \hline 140.9 \end{array}$$

To subtract in the tenths column, borrow 1 one in the form of 10 tenths from the ones column. Add 10 to the 6 in the tenths column, which gives 16 (shown in blue).

b.

$$\begin{array}{r} \overset{\ \ \ 9}{3\ \ \cancel{10}\ \ 10} \\ 15.\cancel{4}\ \ \cancel{0}\ \ \cancel{0} \\ -13.0\ \ 5\ \ 9 \\ \hline 2.3\ \ 4\ \ 1 \end{array}$$

Add two zeros to the right of 15.4 to make borrowing easier. First, borrow from the tenths column; then borrow from the hundredths column.

Self Check 2
Subtract:
a. 382.5 − 227.1
b. 30.1 − 27.122

Answers **a.** 155.4, **b.** 2.978

EXAMPLE 3 **Conditioning programs.** A 350-pound football player lost 15.7 pounds during the first week of practice. During the second week, he gained 4.9 pounds. Find his weight after the first two weeks of practice.

Solution   The word *lost* indicates subtraction. The word *gained* indicates addition.

Beginning weight	minus	first-week weight loss	plus	second-week weight gain	equals	weight after two weeks of practice

$$350 - 15.7 + 4.9 = 334.3 + 4.9$$ Working from left to right, perform the subtraction first: 350 − 15.7 = 334.3

$$= 339.2$$ Perform the addition.

The player's weight is 339.2 pounds after two weeks of practice.

Weather balloons   CALCULATOR SNAPSHOT

A giant weather balloon is made of neoprene, a flexible rubberized substance that has an uninflated thickness of 0.011 inch. When the balloon is inflated with helium, the thickness becomes 0.0018 inch. To find the change in thickness, we need to subtract. We can use a scientific calculator to subtract the decimals.

.011 $\boxed{-}$ .0018 $\boxed{=}$ $\boxed{0.0092}$

After the balloon is inflated, the neoprene loses 0.0092 inch in thickness.

◼ Adding and subtracting signed decimals

To add signed decimals, we use the same rules that we used for adding integers.

> **Adding two decimals**
>
> **With like signs:** Add their absolute values and attach their common sign to the sum.
>
> **With unlike signs:** Subtract their absolute values (the smaller from the larger) and attach the sign of the number with the larger absolute value to the sum.

Self Check 4
Add: $-5.04 + (-2.32)$.

EXAMPLE 4   Add: $-6.1 + (-4.7)$.

Solution   Since the decimals are both negative, we add their absolute values and attach a negative sign to the result.

$$-6.1 + (-4.7) = -10.8$$ Add the absolute values, 6.1 and 4.7, to get 10.8. Use their common sign.

Answer -7.36

Self Check 5
Add: $-21.4 + 16.75$.

EXAMPLE 5   Add: $5.35 + (-12.9)$.

Solution   In this example, the signs are unlike. Since -12.9 has the larger absolute value, we subtract 5.35 from 12.9 to get 7.55, and attach a negative sign to the result.

$$5.35 + (-12.9) = -7.55$$

Answer -4.65

Self Check 6
Subtract: $-1.18 - 2.88$.

EXAMPLE 6   Subtract: $-4.3 - 5.2$.

Solution   To subtract signed decimals, we can add the opposite of the decimal that is being subtracted.

$$-4.3 - 5.2 = -4.3 + (-5.2)$$ Add the opposite of 5.2, which is -5.2.
$$= -9.5$$ Add the absolute values, 4.3 and 5.2, to get 9.5. Attach a negative sign to the result.

Answer -4.06

Self Check 7
Subtract: $-2.56 - (-4.4)$.

EXAMPLE 7   Subtract: $-8.37 - (-16.2)$.

Solution

$$-8.37 - (-16.2) = -8.37 + 16.2$$ Add the opposite of -16.2, which is 16.2.
$$= 7.83$$ Subtract the smaller absolute value from the larger, 8.37 from 16.2, to get 7.83. Since 16.2 has the larger absolute value, the result is positive.

Answer 1.84

EXAMPLE 8   Evaluate: $-12.2 - (-14.5 + 3.8)$.

Solution   We perform the addition within the grouping symbols first.

$$-12.2 - (\mathbf{-14.5 + 3.8}) = -12.2 - (\mathbf{-10.7}) \quad \text{Perform the addition:} \\ -14.5 + 3.8 = -10.7.$$

$$= -12.2 + 10.7 \qquad \text{Add the opposite of } -10.7.$$

$$= -1.5 \qquad\qquad \text{Perform the addition.}$$

Self Check 8
Evaluate: $-4.9 - (-1.2 + 5.6)$.

Answer -9.3

Section 4.2   STUDY SET

VOCABULARY   *Fill in the blanks.*

1. The answer to an addition problem is called the
_____.

2. The answer to a subtraction problem is called the
_____.

CONCEPTS

3. To subtract signed decimals, add the _____
of the decimal that is being subtracted.

4. a. Add: $0.3 + 0.17$.

b. Write 0.3 and 0.17 as fractions. Find a common
denominator for the fractions and add them.

c. Express your final answer to part b as a decimal.

d. Compare your answers from part a and part c.

NOTATION

5. Every whole number has an unwritten decimal
_____ to its right.

6. In the subtraction problem below, we must borrow.
How much is borrowed from the 3, and in what form
is it borrowed?

$$\overset{\scriptstyle 2\,11}{29.3\cancel{1}}$$
$$-25.16$$

PRACTICE   *Perform each addition.*

7. $\begin{array}{r} 32.5 \\ +\ 7.4 \\ \hline \end{array}$

8. $\begin{array}{r} 6.3 \\ +13.5 \\ \hline \end{array}$

9. $\begin{array}{r} 21.6 \\ +33.12 \\ \hline \end{array}$

10. $\begin{array}{r} 19.4 \\ +31.95 \\ \hline \end{array}$

11. $12 + 3.9$

12. $0.01 + 3.6$

13. $0.03034 + 0.2003$

14. $19.9 + 19.9$

15. $247.9 + 40 + 0.56$

16. $0.0053 + 1.78 + 6$

17. $45 + 9.9 + 0.12 + 3.02$

18. $505.01 + 23 + 0.989 + 12.07$

Perform each subtraction.

19. $\begin{array}{r} 12.98 \\ -\ 3.45 \\ \hline \end{array}$

20. $\begin{array}{r} 1.6 \\ -0.16 \\ \hline \end{array}$

21. $\begin{array}{r} 78.1 \\ -\ 7.81 \\ \hline \end{array}$

22. $\begin{array}{r} 202.234 \\ -\ 19.34 \\ \hline \end{array}$

23. $5 - 0.023$

24. $30 - 11.98$

25. $24 - 23.81$

26. $7.001 - 5.9$

Perform each addition.

27. $-45.6 + 34.7$

28. $-19.04 + 2.4$

29. $46.09 + (-7.8)$

30. $34.7 + (-30.1)$

31. $-7.8 + (-6.5)$

32. $-5.78 + (-33.1)$

33. $-0.0045 + (-0.031)$

34. $-90.09 + (-0.087)$

Perform each subtraction.

35. $-9.5 - 7.1$

36. $-7.08 - 14.3$

37. $30.03 - (-17.88)$

38. $143.3 - (-64.01)$

39. $-2.002 - (-4.6)$ **40.** $-0.005 - (-8)$

41. $-7 - (-18.01)$ **42.** $-63.04 - (-8.911)$

*Evaluate each expression. Remember to perform any
calculations within grouping symbols first.*

43. $(3.4 - 6.6) + 7.3$

44. $3.4 - (6.6 + 7.3)$

45. $(-9.1 - 6.05) - (-51)$

46. $-9.1 - (-6.05) + 51$

47. $16 - (67.2 + 6.27)$

48. $-43 - (0.032 - 0.045)$

49. $(-7.2 + 6.3) - (-3.1 - 4)$

50. $2.3 + [2.4 - (2.5 - 2.6)]$

51. $|-14.1 + 6.9| + 8$

52. $15 - |-2.3 + (-2.4)|$

53. Find the sum of *two and forty-three hundredths* and
five and six tenths.

54. Find the difference of *nineteen hundredths* and *six
thousandths.*

▮ APPLICATIONS

55. SPORTS PAGES In the sports pages of any
newspaper, decimal numbers are used often.

   a. "German bobsledders set a world record today
with a final run of 53.03, finishing ahead of the
Italian team by only fourteen thousandths of a
second." What was the time for the Italian bobsled
team?

   b. "The women's figure skating title was decided by
only thirty-three hundredths of a point." If the
winner's point total was 102.71, what was the
second-place finisher's total?

56. NURSING The following table shows a patient's
health chart. A nurse failed to fill in certain portions.
(98.6° Fahrenheit is considered normal.) Complete
the chart.

Day of week	Patient's temperature	How much above normal
Monday	99.7°	
Tuesday		2.5°
Wednesday	98.6°	
Thursday	100.0°	
Friday		0.9

57. VEHICLE SPECIFICATIONS Certain dimensions
of a compact car are shown. Find the wheelbase of
the car.

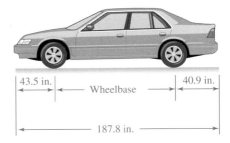

58. pH SCALE The pH scale shown below is used to
measure the strength of acids and bases in chemistry.
Find the difference in pH readings between

   a. bleach and stomach acid.

   b. ammonia and coffee.

   c. blood and coffee.

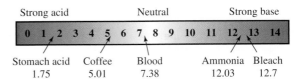

59. BAROMETRIC PRESSURES Barometric pressure
readings are recorded on the weather man. In a
low-pressure area (L on the map), the weather is
often stormy. The weather is usually fair in a high-
pressure area (H). What is the difference in readings
between the areas of highest and lowest pressure?
In what part of the country would you expect the
weather to be fair?

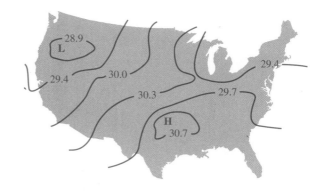

60. QUALITY CONTROL An electronics company
has strict specifications for silicon chips used in a
computer. The company will install only chips that are
within 0.05 centimeter of the specified thickness. The
table on the next page gives that specifications for two
types of chip. Fill in the blanks to complete the chart.

Chip type	Thickness specification	Acceptable range	
		Low	High
A	0.78 cm		
B	0.643 cm		

61. OFFSHORE DRILLING A company needs to construct a pipeline from an offshore oil well to a refinery located on the coast. Company engineers have come up with two plans for consideration, as shown. Use the information in the illustration to complete the table.

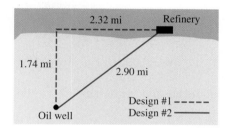

2.32 mi   Refinery
1.74 mi
2.90 mi
Design #1 - - - - -
Design #2 ————
Oil well

	Pipe underwater (mi)	Pipe underground (mi)	Total pipe (mi)
Design 1			
Design 2			

62. TELEVISION The following illustration shows the six most-watched television shows of all time (excluding Super Bowl games).

   a. What was the combined total audience of all six shows?

   b. How many more people watched the last episode of "MASH" than watched the last episode of "Seinfeld"?

   c. How many more people would have had to watch the last "Seinfeld" to move it into a tie for fifth place?

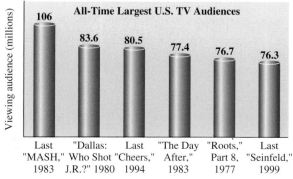

All-Time Largest U.S. TV Audiences

Viewing audience (millions)

106 — Last "MASH," 1983
83.6 — "Dallas: Who Shot J.R.?" 1980
80.5 — Last "Cheers," 1994
77.4 — "The Day After," 1983
76.7 — "Roots," Part 8, 1977
76.3 — Last "Seinfeld," 1999

Source: Nielsen Media Research

63. RECORD HOLDERS The late Florence Griffith-Joyner of the United States still holds the world record in the 100-meter sprint: 10.49 seconds. Jodie Henry of Australia holds the world record in the 100-meter freestyle swim. Jodie's time is 53.52. How much faster did Griffith-Joyner run the 100 meters than Henry swam it?

64. FLIGHT PATHS Find the added distance a plane must travel to avoid flying through the storm.

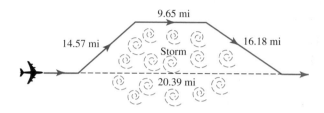

9.65 mi
14.57 mi   16.18 mi
Storm
20.39 mi

65. DEPOSIT SLIPS A deposit slip for a savings account is shown. Find the subtotal and then the total deposit.

Deposit

Cash	242	50
Checks (properly endorsed)	116	10
	47	93
Total from reverse side	359	16
Subtotal		
Less cash	25	00
Total deposit		

66. MOTION Forces such as water current or wind can increase or decrease the speed of an object in motion. Find the speed of each object.

   a. An airplane's speed in still air is 450 mph, and it has a tail wind of 35.5 mph helping it along.

   b. A man can paddle a canoe at 5 mph in still water, but he is going upstream. The speed of the current against him is 1.5 mph.

67. THE HOME SHOPPING NETWORK The illustration on the next page shows a description of a cookware set that was sold on television.

   a. Find the difference between the manufacturer's suggested retail price (MSRP) and the sale price.

   b. Including shipping and handling (S & H), how much will the cookware set cost?

Item 229-442

**Continental 9-piece
Cookware Set**

Stainless steel

MSRP	$149.79
HSN Price	$59.85
On Sale	**$47.85**
S & H	$7.95

68. RETAILING Complete the table by filling in the retail price of each appliance, given its cost to the dealer and the store markup.

Item	Cost	Markup	Retail price
Refrigerator	$510.80	$105.00	
Washing machine	$289.50	$55.50	
Dryer	$263.99	$67.50	

Evaluate each expression.

69. 2,367.909 + 5,789.0253

70. 0.00786 + 0.3423

71. 9,000.09 − 7,067.445

72. 1 − 0.004999

73. 3,434.768 − (908 − 2.3 + .0098)

74. 12 − (0.723 + 3.05611)

WRITING

75. Explain why we line up the decimal points and corresponding columns when adding decimals.

76. Explain why we can write additional zeros to the right of a decimal such as 7.89 without affecting its value.

77. Explain what is wrong with the work shown below.

$$\begin{array}{r} 203.56 \\ 37 \\ +\ \underline{0.43} \\ 204.36 \end{array}$$

78. Consider the addition

$$\begin{array}{r} \overset{2}{2}3.7 \\ 41.9 \\ +\underline{12.8} \\ 78.4 \end{array}$$

Explain the meaning of the small 2 written above the ones column.

REVIEW

79. Add: $44\dfrac{3}{8} + 66\dfrac{1}{5}$.

80. Simplify: $\dfrac{-\dfrac{3}{4}}{\dfrac{5}{16}}$.

81. Multiply: $\dfrac{-15}{26} \cdot 1\dfrac{4}{9}$.

82. Evaluate: $2 + 5[-2 - (6 + 1)]$.

4.3 Multiplying Decimals

- Multiplying decimals
- Multiplying decimals by powers of 10
- Multiplying signed decimals
- Order of operations

We now focus on the operation of multiplication. First, we develop a method used to multiply decimals. Then we use that method to evaluate expressions and to solve problems involving decimals.

Multiplying decimals

To show how to multiply decimals, we examine the multiplication 0.3 · 0.17, finding the product in a roundabout way. First, we write 0.3 and 0.17 as fractions and multiply them. Then we express the resulting fraction as a decimal.

$$0.3 \cdot 0.17 = \frac{3}{10} \cdot \frac{17}{100} \qquad \text{Express 0.3 and 0.17 as fractions.}$$

$$= \frac{3 \cdot 17}{10 \cdot 100} \qquad \text{Multiply the numerators and multiply the denominators.}$$

$$= \frac{51}{1,000} \qquad \text{Multiply in the numerator and denominator.}$$

$$= 0.051 \qquad \text{Write } \frac{51}{1,000} \text{ as a decimal.}$$

From this example, we can make observations about multiplying decimals.

- The digits in the answer are found by multiplying 3 and 17.

$$0.3 \quad \cdot \quad 0.17 \quad = \quad 0.051$$

$$3 \cdot 17 = 51$$

- The answer has 3 decimal places. The *sum* of the number of decimal places in the factors 0.3 and 0.17 is also 3.

$$0.3 \quad \cdot \quad 0.17 \quad = \quad 0.051$$

1 decimal place   2 decimal places   3 decimal places

These observations suggest the following rule for multiplying decimals.

Multiplying decimals

To multiply two decimals:

1. Multiply the decimals as if they were whole numbers.

2. Find the total number of decimal places in both factors.

3. Place the decimal point in the result so that the answer has the same number of decimal places as the total found in step 2.

EXAMPLE 1   Multiply: $5.9 \cdot 3.4$.

Solution   We ignore the decimal points and multiply the decimals as if they were whole numbers. Initially, we think of this problem as 59 times 34.

$$\begin{array}{r} 59 \\ \times 34 \\ \hline 236 \\ 177 \\ \hline 2006 \end{array}$$

To place the decimal point in the product, we find the total number of digits to the right of the decimal points in the factors.

$$\begin{array}{r} 5.9 \quad \leftarrow 1 \text{ decimal place} \\ \times 3.4 \quad \leftarrow 1 \text{ decimal place} \\ \hline 236 \\ 177 \\ \hline 20.06 \end{array}$$

The answer will have $1 + 1 = 2$ decimal places.

Locate the decimal point so that the answer has 2 decimal places.

Self Check 1
Multiply: $2.74 \cdot 4.3$.

Answer 11.782

When multiplying decimals, we do not need to line up the decimal points, as the next example illustrates.

EXAMPLE 2   Multiply: 1.3(0.005).

Solution   We multiply 13 by 5 and find the total number of decimal places in 1.3 and 0.005.

$$\begin{array}{r} 1.3 \leftarrow 1 \text{ decimal place} \\ \times 0.005 \leftarrow 3 \text{ decimal place} \\ \hline 65 \end{array}$$ The answer will have $1 + 3 = 4$ decimal places.

We then place the decimal point in the result.

$$\begin{array}{r} 1.3 \\ \times\ 0.005 \\ \hline 0.0065 \end{array}$$ Add 2 placeholder zeros and position the decimal point so that the product has 4 decimal places.

CALCULATOR SNAPSHOT   **Heating costs**

When billing a household, a gas company converts the amount of natural gas used to units of heat energy called *therms*. The number of therms used by a household in one month and the cost per therm are shown below.

Customer charge . . . . 39 therms @ $0.72264

To find the total charges for the month, we multiply the number of therms by the cost per therm: $39 \cdot 0.72264$.

$$39 \boxed{\times} .72264 \boxed{=}$$ $$\boxed{28.18296}$$

Rounding to the nearest cent, we see that the total charge is $28.18.

EXAMPLE 3   Multiply: 234(3.1).

Solution

$$\begin{array}{r} 234 \leftarrow \text{No decimal places} \\ \times\ 3.1 \leftarrow 1 \text{ decimal place} \\ \hline 23\ 4 \\ 702 \\ \hline 725.4 \end{array}$$ The answer will have $0 + 1 = 1$ decimal place.

Locate the decimal point so that the answer has 1 decimal place.

■ Multiplying decimals by powers of 10

The numbers 10, 100, and 1,000 are called *powers of 10,* because they are the results when we evaluate 10^1, 10^2, and 10^3, respectively. To develop a rule to determine the product when multiplying a decimal and a power of 10, we multiply 8.675 by three different powers of 10.

Overtime

"Employees covered by the Fair Labor Standards Act must receive overtime pay for hours worked in excess of 40 in a workweek of at least 1.5 times their regular rates of pay." United States Department of Labor

The map of the United States shown below is divided into nine regions. The average hourly wage for private industry workers in each region is also listed in the legends below the map. Find the average hourly wage for the region where you live. Then calculate the corresponding average hourly overtime wage for that region.

● West North Central: $16.30	● East South Central: $13.97	● New England: $20.10
○ Mountain: $15.65	○ East North Central: $17.16	○ Middle Atlantic: $19.08
● Pacific: $19.11	● West South Central: $15.22	● South Atlantic: $15.88

Multiply: 8.675 · **10**

```
      8.675
   ×     10
       0000
       8675
      86.750
```

The answer is 86.75.

Multiply: 8.675 · **100**

```
      8.675
   ×    100
       0000
       0000
       8675
     867.500
```

The answer is 867.5

Multiply: 8.675 · **1,000**

```
      8.675
   ×   1000
       0000
       0000
       0000
       8675
     8675.000
```

The answer is 8,675.

We can make observations about the results.

- In each case, the answer contains the same digits as the factor 8.675.

- When we inspect the answers, the decimal point in the first factor 8.675 appears to be moved to the right by the multiplication process. The number of decimal places it moves depends on the power of 10 by which 8.675 is multiplied.

One zero in 10

$8.675 \cdot 10 = 8\ 6.75$

It moves 1 place to the right.

Two zeros in 100

$8.675 \cdot 100 = 8\ 67.5$

It moves 2 places to the right.

Three zeros in 1,000

$8.675 \cdot 1,000 = 8\ 675$

It moves 3 places to the right.

These observations suggest the following rule.

> **Multiplying a decimal by a power of 10**
>
> To multiply a decimal by a power of 10, move the decimal point to the right the same number of places as there are zeros in the power of 10.

Self Check 4
Find each product:
a. 0.721 · 100
b. 6.08 (1,000)

Answers **a.** 72.1,   **b.** 6,080

EXAMPLE 4   Find each product: **a.** 2.81 · 10   and   **b.** 0.076 · 10,000.

Solution

a. 2.81 · 10 = 2 8.1   Since 10 has 1 zero, move the decimal point 1 place to the right.

b. 0.076 · 10,000 = 0760.   Since 10,000 has 4 zeros, move the decimal point 4 places to the right. Write a placeholder zero (shown in blue).

          = 760

EXAMPLE 5   Tachometers.   A tachometer indicates the engine speed of a vehicle, in revolutions per minute (rpm). What engine speed is indicated by the tachometer in Figure 4-6?

FIGURE 4-6

Solution   The needle is pointing to 4.5. The notation "RPM × 1000" on the tachometer instructs us to multiply 4.5 by 1,000 to find the engine speed.

     4.5 · 1,000 = 4 500   Since 1,000 has 3 zeros, move the decimal point 3 places to the right. Write two placeholder zeros.

          = 4,500

The engine speed is 4,500 rpm.

◼ Multiplying signed decimals

Recall that the product of two numbers with like signs is positive, and the product of two numbers with unlike signs is negative.

Self Check 6
Multiply:
a. 6.6(−5.5)
b. (−44.968)(−100)

EXAMPLE 6   Multiply: **a.** −1.8(4.5)   and   **b.** (−1,000)(−59.08).

Solution

a. Since the decimals have unlike signs, their product is negative.

          −1.8(4.5) = −8.1   Multiply the absolute values, 1.8 and 4.5, to get 8.1. Make the result negative.

b. Since the decimals have like signs, their product is positive.

$$(-1,000)(-59.08) = 59,080$$ Multiply the absolute values, 1,000 and 59.08. Since 1,000 has 3 zeros, move the decimal point 3 places to the right. Write a placeholder zero.

Answers **a.** -36.3,   **b.** $4,496.8$

EXAMPLE 7   Evaluate: **a.** $(2.4)^2$ and   **b.** $(-0.05)^2$.

Solution

a. $(2.4)^2 = 2.4 \cdot 2.4$ Write 2.4 as a factor 2 times.

$ = 5.76$ Perform the multiplication.

b. $(-0.05)^2 = (-0.05)(-0.05)$ Write -0.05 as a factor 2 times.

$ = 0.0025$ Perform the multiplication. The product of two decimals with like signs is positive.

Self Check 7
Evaluate:
a. $(-1.3)^2$
b. $(0.09)^2$

Answers **a.** 1.69,   **b.** 0.0081

▌Order of operations

In the remaining examples, we use the rules for the order of operations to evaluate expressions involving decimals.

EXAMPLE 8   Evaluate: $-(0.6)^2 + 5|-3.6 + 1.9|$.

Solution

$-(0.6)^2 + 5|-3.6 + 1.9| = -(0.6)^2 + 5|-1.7|$ Perform the addition within the absolute value symbols.

$ = -(0.6)^2 + 5(1.7)$ Simplify: $|-1.7| = 1.7$.

$ = -0.36 + 5(1.7)$ Find the power: $(0.6)^2 = 0.36$.

$ = -0.36 + 8.5$ Perform the multiplication: $5(1.7) = 8.5$.

$ = 8.14$ Perform the addition.

Self Check 8
Evaluate:
$-2|-4.4 + 5.6| + (-0.8)^2$.

Answer -1.76

Recall that to evaluate a formula, we begin by replacing the letters with specific numbers.

EXAMPLE 9   Evaluate the formula $S = 6.28r(h + r)$ for $h = 3.1$ and $r = 6$.

Solution

$S = 6.28r(h + r)$ $6.28r(h + r)$ means $6.28 \cdot r \cdot (h + r)$.

$ = 6.28(6)(3.1 + 6)$ Replace r with 6 and h with 3.1.

$ = 6.28(6)(9.1)$ Perform the addition within the parentheses: $3.1 + 6 = 9.1$.

$ = 37.68(9.1)$ Perform the multiplication: $6.28(6) = 37.68$.

$ = 342.888$ Perform the multiplication.

Self Check 9
Evaluate $V = 1.3pr^3$ for $p = 3.14$ and $r = 3$.

Answer 110.214

EXAMPLE 10   Weekly earnings.   A cashier's workweek is 40 hours. After his daily shift is over, he can work overtime at a rate 1.5 times his regular rate of $7.50 per hour. How much money will he earn in a week if he works 6 hours of overtime?

Solution   First, we need to find his overtime rate, which is 1.5 times his regular rate of $7.50 per hour.

$$1.5(7.50) = 11.25$$

His overtime rate is $11.25 per hour. To find his total weekly earnings, we use the following fact.

The regular rate	times	40 hours	plus	the overtime rate	times	overtime hours worked	equals	his total earnings.

$$7.50(40) + 11.25(6) = 300 + 67.50 \quad \text{Perform the multiplications.}$$
$$= 367.50 \quad \text{Perform the addition.}$$

The cashier's earnings for the week are $367.50.

Section 4.3   STUDY SET

VOCABULARY   *Fill in the blanks.*

1. In the multiplication problem $2.89 \cdot 15.7$, the numbers 2.89 and 15.7 are called _____. The answer, 45.373, is called the _____.

2. Numbers such as 10, 100, and 1,000 are called _____ of 10.

CONCEPTS   *Fill in the blanks.*

3. To multiply decimals, multiply them as if they were _____ numbers. The number of decimal places in the product is the same as the _____ of the decimal places in the factors.

4. To multiply a decimal by a power of 10, move the decimal point to the _____ the same number of decimal places as the number of _____ in the power of 10.

5. **a.** Multiply $\dfrac{3}{10}$ and $\dfrac{7}{100}$.

   b. Now write both fractions from part a as decimals. Multiply them in that form. Compare your results from parts a and b.

6. **a.** Multiply 0.11 and 0.3.

   b. Now write both decimals in part a as fractions. Multiply them in that form. Compare your results from parts a and b.

NOTATION

7. Suppose that the result of multiplying two decimals is 2.300. Write this result in simpler form.

8. When we move the decimal point to the right, does the decimal number get larger or smaller?

PRACTICE   *Perform each multiplication.*

9. $(0.4)(0.2)$ 10. $(0.2)(0.3)$
11. $(-0.5)(0.3)$ 12. $(0.6)(-0.7)$
13. $(1.4)(0.7)$ 14. $(2.1)(0.4)$
15. $(0.08)(0.9)$ 16. $(0.003)(0.9)$
17. $(-5.6)(-2.2)$ 18. $(-7.1)(-4.1)$
19. $(-4.9)(0.001)$ 20. $(0.001)(-7.09)$
21. $(-0.35)(0.24)$ 22. $(-0.85)(0.42)$
23. $(-2.13)(4.05)$ 24. $(3.06)(-1.82)$
25. $16 \cdot 0.6$ 26. $24 \cdot 0.8$
27. $-7(8.1)$ 28. $-5(4.7)$
29. $0.04(306)$ 30. $0.02(417)$
31. $60.61(-0.3)$ 32. $-70.07 \cdot 0.6$
33. $-0.2(0.3)(-0.4)$ 34. $-0.1(-2.2)(0.5)$
35. $5.5(10)(-0.3)$ 36. $6.2(100)(-0.8)$
37. $4.2 \cdot 10$ 38. $10 \cdot 7.1$
39. $67.164 \cdot 100$ 40. $708.199 \cdot 100$
41. $-0.056(10)$ 42. $-100(0.0897)$

43. 1,000(8.05)

44. 23.7(1,000)

45. 0.098(10,000)

46. 3.63(10,000)

47. $-0.2 \cdot 1,000$

48. $-1,000 \cdot 1.9$

Complete each table.

49.

Decimal	Its square
0.1	
0.2	
0.3	
0.4	
0.5	
0.6	
0.7	
0.8	
0.9	

50.

Decimal	Its cube
0.1	
0.2	
0.3	
0.4	
0.5	
0.6	
0.7	
0.8	
0.9	

Find each power.

51. $(1.2)^2$

52. $(2.3)^2$

53. $(-1.3)^2$

54. $(-2.5)^2$

Evaluate each expression.

55. $-4.6(23.4 - 19.6)$

56. $6.9(9.8 - 8.9)$

57. $(-0.2)^2 + 2(7.1)$

58. $(-6.3)(3) - (1.2)^2$

59. $(-0.7 - 0.5)(2.4 - 3.1)$

60. $(-8.1 - 7.8)(0.3 + 0.7)$

61. $(0.5 + 0.6)^2(-3.2)$

62. $(-5.1)(4.9 - 3.4)^2$

63. $|-2.6| \cdot |-7.2|$

64. $4|-3.1| + 5|-5.5|$

65. $(|-2.6 - 6.7|)^2$

66. $-3|-8.16 + 9.9|$

Evaluate each formula.

67. $A = lw$ for $l = 5.3$ and $w = 7.2$

68. $A = \dfrac{1}{2}bh$ for $b = 7.5$ and $h = 6.8$

69. $P = 2l + 2w$ for $l = 3.7$ and $w = 3.6$

70. $C = 2\pi r$ for $\pi = 3.14$ and $r = 2.5$

■ APPLICATIONS

71. CONCERT SEATING Two types of tickets were sold for a concert. Floor seating costs $12.50 a ticket, and balcony seats cost $15.75.

   a. Complete the following table and find the receipts from each type of ticket.

   b. Find the total receipts from the sale of both types of tickets.

Ticket type	Price	Number sold	Receipts
Floor		1,000	
Balcony		100	

72. CITY PLANNING In the city map, the streets form a grid. The streets are 0.35 mile apart. Find the distance of each trip.

   a. The airport to the Convention Center

   b. City Hall to the Convention Center

   c. The airport to City Hall

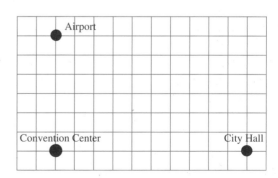

73. STORM DAMAGE After a rainstorm, the saturated ground under a hilltop house began to give way. A survey team noted that the house dropped 0.57 inch initially. In the next two weeks, the house fell 0.09 inch per week. How far did the house fall during this three-week period?

74. WATER USAGE In May, the water level of a reservoir reached its high mark for the year. During the summer months, as water usage increased, the level dropped. In the months of May and June, it fell 4.3 feet each month. In August, because of high temperatures, it fell another 8.7 feet. By September, how far below the year's high mark had the water level fallen?

75. WEIGHTLIFTING The barbell is evenly loaded with iron plates. How much plate weight is loaded on the barbell?

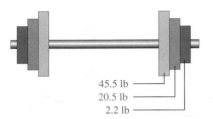

45.5 lb
20.5 lb
2.2 lb

76. PLUMBING BILLS The following invoice for plumbing work is torn. What is the charge for the 4 hours of work? What is the total charge?

Carter Plumbing 100 W. Dalton Ave.	Invoice #210
Standard service charge 4 hr @ $40.55/hr	$25.75
Total	

77. BAKERY SUPPLIES A bakery buys various types of nuts as ingredients for cookies. Complete the table by filling in the cost of each purchase.

Type of nut	Price per pound	Pounds	Cost
Almonds	$5.95	16	
Walnuts	$4.95	25	

78. RETROFITS The illustration shows the width of the three columns of an existing freeway overpass. A computer analysis indicates that each column needs to be increased in width by a factor of 1.4 to ensure stability during an earthquake. According to the analysis, how wide should each of the columns be?

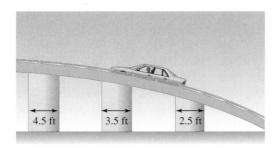

4.5 ft   3.5 ft   2.5 ft

79. SWIMMING POOL CONSTRUCTION Long bricks, called *coping*, can be used to outline the edge of a swimming pool. How many meters of coping will be needed in the construction of the swimming pool shown?

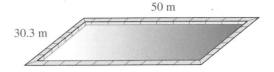

50 m

30.3 m

80. SOCCER A soccer goal measures 24 feet wide by 8 feet high. Major league soccer officials are proposing to increase its width by 1.5 feet and increase its height by 0.75 foot.

  a. What is the area of the goal opening now?

  b. What would the area be if the proposal is adopted?

  c. How much area would be added?

81. BIOLOGY DNA is found in cells. It is referred to as the genetic blueprint. In humans, it determines such traits as eye color, hair color, and height. A model of DNA appears in the illustration. If 1 Å (angstrom) = 0.000000004 inch, determine the three dimensions shown in the illustration.

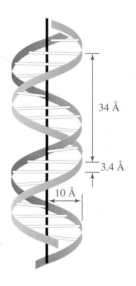

34 Å

3.4 Å

10 Å

82. TACHOMETERS

  a. To what number is the tachometer needle pointing? Give your estimate in decimal form.

  b. What engine speed (in rpm) does the tachometer indicate?

RPM x 1000

Use a calculator to evaluate each problem.

83. $(-9.0089 + 10.0087)(15.3)$

84. $(-4.32)^3 - 78.969$

85. $(18.18 + 6.61)^2 + (5 - 9.09)^2$

86. $304 - 3.780876(100)$

87. ELECTRIC BILLS When billing a household, a utility company charges for the number of kilowatt-hours used. A kilowatt-hour (kwh) is a standard measure of electricity. If the cost of 1 kwh is $0.14277, what is the electric bill for a household that uses 719 kwh in a month? Round the answer to the nearest cent.

88. UTILITY TAXES Some gas companies are required to tax the number of therms used each month by the customer. What are the taxes collected on a monthly usage of 31 therms if the tax rate is $0.00566 per therm? Round the answer to the nearest cent.

WRITING

89. Explain how to determine where to place the decimal point in the answer when multiplying two decimals.

90. List the similarities and differences between whole-number multiplication and decimal multiplication.

91. What is a decimal place?

92. Explain how to multiply a decimal by a power of 10.

REVIEW

93. Round 7,346 to the nearest hundred.

94. Multiply: $3\frac{1}{3}\left(-1\frac{4}{5}\right)$.

95. Write this notation in words: $|-3|$.

96. What is the LCD of fractions with denominators of 4, 5, and 6?

97. Simplify: $-\frac{8}{8}$.

98. Find one-half of 7 and square the result.

4.4 Dividing Decimals

- Dividing a decimal by a whole number • Divisors that are decimals
- Rounding when dividing • Dividing decimals by powers of 10 • Order of operations

In Chapter 1, we used a process called long division to divide whole numbers.

Long division form

$$\begin{array}{r} 2 \leftarrow \text{Quotient} \\ \text{Divisor} \rightarrow 5\overline{)10} \leftarrow \text{Dividend} \\ \underline{10} \\ 0 \leftarrow \text{Remainder} \end{array}$$

In this section, we consider division problems in which the divisor, the dividend, or both are decimals.

Dividing a decimal by a whole number

To use long division to divide 47 by 10, we proceed as follows.

$$\begin{array}{r} 4\frac{7}{10} \\ 10\overline{)47} \\ \underline{40} \\ 7 \end{array}$$ Here the result is written in quotient + $\dfrac{\text{remainder}}{\text{divisor}}$ form.

To perform this same division using decimals, we write 47 as 47.0 and divide as we would divide whole numbers.

$$\begin{array}{r} 4.7 \\ 10\overline{)47.0} \\ 40\downarrow \\ \hline 7\,0 \\ 7\,0 \\ \hline 0 \end{array}$$

Note that the decimal point in the result is placed directly above the decimal point in the dividend.

After subtracting 40 from 47, bring down the 0.

Since $4\frac{7}{10} = 4.7$, either method gives the same result. The second part of this discussion suggests the following method for dividing a decimal by a whole number.

Dividing a decimal by a whole number

To divide a decimal by a whole number:

1. Write the problem in long division form.

2. Divide as if working with whole numbers.

3. Write the decimal point in the result directly above the decimal point in the dividend. If necessary, additional zeros can be written to the right of the dividend to allow the division to proceed.

Self Check 1

Divide: $101.44 \div 32$.

EXAMPLE 1   Divide: $71.68 \div 28$.

Solution

$$\begin{array}{r} \overset{\textstyle\cdot}{} \\ 28\overline{)71.68} \end{array}$$

Write the decimal point in the answer directly above the decimal point in the dividend.

$$\begin{array}{r} 2.56 \\ 28\overline{)71.68} \\ 56\downarrow \\ \hline 15\,6 \\ 14\,0\downarrow \\ \hline 1\,68 \\ 1\,68 \\ \hline 0 \end{array}$$

Divide as if working with whole numbers.

The remainder is 0.

The answer is 2.56.

   We can check this result by multiplying the divisor by the quotient; their product should equal the dividend. Since $2.56 \cdot 28 = 71.68$, the result is correct.

Answer 3.17

Self Check 2

Divide: $3.4 \div 4$.

EXAMPLE 2   Divide: $19.2 \div 5$.

Solution

$$\begin{array}{r} 3.8 \\ 5\overline{)19.2} \\ 15\downarrow \\ \hline 4\,2 \\ 4\,0 \\ \hline 2 \end{array}$$

All the digits in the dividend have been used, but the remainder is not 0.

We can write a zero to the right of 2 in the dividend and continue the division process. Recall that writing additional zeros to the right of the decimal point does not change the value of the decimal.

$$
\begin{array}{r}
3.84 \\
5\overline{)19.20} \\
\underline{15} \\
4\,2 \\
\underline{4\,0} \\
20 \\
\underline{20} \\
0
\end{array}
$$

Write a zero to the right of the 2 and bring it down.

Continue to divide.

The remainder is 0.

The answer is 3.84.

Check: $3.84 \cdot 5 = 19.2$.

Answer 0.85

◼ Divisors that are decimals

When the divisor is a decimal, we change it to a whole number and proceed as in division of whole numbers. To illustrate, we consider the problem $0.36\overline{)0.2592}$, where the divisor is a decimal. First, we express the division in another form.

$$0.36\overline{)0.2592} \qquad \text{can be represented by} \qquad \frac{0.2592}{0.36}$$

To write the divisor, 0.36, as a whole number, its decimal point needs to be moved 2 places to the right. This can be accomplished by multiplying it by 100. However, if the denominator of the fraction is multiplied by 100, the numerator must also be multiplied by 100 so that the fraction maintains the same value.

$$
\frac{0.2592}{0.36} = \frac{0.2592 \cdot \mathbf{100}}{0.36 \cdot \mathbf{100}}
$$

Multiply numerator and denominator by 100.

$$
= \frac{25.92}{36}
$$

Multiplying by 100 moves both decimal points 2 places to the right.

This fraction represents the division problem $36\overline{)25.92}$. From this result, we can make the following observations.

- The division problem $0.36\overline{)0.2592}$ is equivalent to $36\overline{)25.92}$; that is, they have the same answer.

- The decimal points in *both* the divisor and the dividend of the first division problem have been moved two decimal places to the right to create the second division problem.

$$0.36\overline{)0.2592} \qquad \text{becomes} \qquad 36\overline{)25.92}$$

These observations suggest the following rule for division with decimals.

Division with a decimal divisor

To divide with a decimal divisor:

1. Move the decimal point of the divisor so that it becomes a whole number.

2. Move the decimal point of the dividend the same number of places to the right.

3. Divide as if working with whole numbers. Write the decimal point in the answer directly above the decimal point in the dividend.

Divide: $\dfrac{0.6045}{0.65}$.

EXAMPLE 3   Divide: $\dfrac{0.2592}{0.36}$.

Solution

$0\,36\overline{)0\,25.92}$ Move the decimal point 2 places to the right in the divisor and dividend.

$$
\begin{array}{r}
0.72 \\
36\overline{)25.92} \\
25\,2\downarrow \\
\hline
72 \\
72 \\
\hline
0
\end{array}
$$

Now divide as with whole numbers. Write the decimal point in the answer directly above the decimal point in the dividend.

The result is 0.72.

Answer 0.93

Check: $0.72 \cdot 36 = 25.92$.

▌ Rounding when dividing

In Example 3, the division process terminated after we obtained a zero from the second subtraction. Sometimes when we divide, the subtractions never give a zero remainder, and the division process continues forever. In such cases, we can round the result.

EXAMPLE 4   Divide: $\dfrac{2.35}{0.7}$. Round to the nearest hundredth.

Solution   Using long division form, we have $0.7\overline{)2.35}$.

$0\,7\overline{)2\,3.5}$ To write the divisor as a whole number, move the decimal point 1 place to the right. Do the same for the dividend. Place the decimal point in the answer directly above the decimal point in the dividend.

$7\overline{)23.500}$ To round to the hundredths column, we must divide to the thousandths column. We write two zeros on the right of the dividend.

$$
\begin{array}{r}
3.357 \\
7\overline{)23.500} \\
21 \\
\hline
2\,5 \\
2\,1 \\
\hline
40 \\
35 \\
\hline
50 \\
49 \\
\hline
1
\end{array}
$$

After dividing to the thousandths column, round to the hundredths column. The rounding digit is 5. The test digit is 7.

To the nearest hundredth, the answer is 3.36.

The nucleus of a cell CALCULATOR SNAPSHOT

The nucleus of a cell contains vital information about the cell in the form of DNA. The nucleus is very small: A typical animal cell has a nucleus that is only 0.00023622 inch across. How many nuclei would have to be laid end to end to extend to a length of 1 inch?

To find how many 0.00023622-inch lengths there are in 1 inch, we must use division: $1 \div 0.00023622$.

$1 \boxed{\div} .00023622 \boxed{=}$ $\boxed{4233.3418}$

It would take approximately 4,233 nuclei laid end to end to extend to a length of 1 inch.

Dividing decimals by powers of 10

To develop a set of rules for division by a power of 10, we consider the problem $8.13 \div 10$.

$$
\begin{array}{r}
0.813 \\
10)\overline{8.130} \\
\underline{0} \\
8\,1 \\
\underline{8\,0} \\
13 \\
\underline{10} \\
30 \\
\underline{30} \\
0
\end{array}
$$

Write a zero to the right of the 3.

We note that the quotient, 0.813, and the dividend, 8.13, are the same except for the location of the decimal points. The quotient can be easily obtained by moving the decimal point of the dividend 1 place to the left. This observation suggests the following rule for dividing a decimal by a power of 10.

> **Dividing a decimal by a power of 10**
>
> To divide a decimal by a positive power of 10, move the decimal point to the left the same number of places as there are zeros in the power of 10.

EXAMPLE 5 Find the quotient: **a.** $16.74 \div 10$ and **b.** $8.6 \div 10,000$.

Solution

a. $16.74 \div 10 = 1.6\,74$ Since 10 has 1 zero, move the decimal point 1 place to the left.

b. $8.6 \div 10,000 = .0008\,6$ Since 10,000 has 4 zeros, move the decimal point 4 places to the left. Write 3 placeholder zeros.

$= 0.00086$

Self Check 5
Find the quotient:

a. $721.3 \div 100$

b. $\dfrac{1.07}{1,000}$

Answers **a.** 7.213, **b.** 0.00107

Order of operations

In the next example, we use the rules for the order of operations to evaluate an expression that involves division by a decimal.

Self Check 6

Evaluate: $\dfrac{2.7756 + 3(-0.63)}{-0.8}$.

Answer −1.107

EXAMPLE 6   Evaluate: $\dfrac{2(0.351) + 0.5592}{-0.4}$.

Solution

$$\frac{2(0.351) + 0.5592}{-0.4} = \frac{0.702 + 0.5592}{-0.4}$$ Perform the multiplication first:
$2(0.351) = 0.702$.

$$= \frac{1.2612}{-0.4}$$ Perform the addition:
$0.702 + 0.5592 = 1.2612$.

$$= -3.153$$ Perform the division. The quotient of two
numbers with unlike signs is negative.

THINK IT THROUGH   GPA

"In considering all of the factors that are important to employers as they recruit students in colleges and universities nationwide, college major, grade point average, and work-related experience usually rise to the top of the list."
Mary D. Feduccia, Ph.D., Career Services Director, Louisiana State University

A grade point average (GPA) is a weighted average based on the grades received and the number of units (credit hours) taken. A GPA for one semester (or term) is defined as

> *the quotient of the sum of the grade points earned for each class and the sum of the number of units taken. The number of grade points earned for a class is the product of the number of units assigned to the class and the value of the grade received in the class.*

1. Use the table of grade values below to compute the GPA for the student whose semester grade report is shown. Round to the nearest hundredth.

Grade	Value
A	4
B	3
C	2
D	1
F	0

Class	Units	Grade
Geology	4	C
Algebra	5	A
Psychology	3	C
Spanish	2	B

2. If you were enrolled in school last semester (or term), list the classes taken, units assigned, and grades received as shown in the grade report below. Then calculate your GPA.

Section 4.4 STUDY SET

VOCABULARY   *Fill in the blanks.*

1. In the division $2.5\overline{)4.075} = 1.63$, the decimal 4.075 is called the _____, the decimal 2.5 is the _____, and 1.63 is the _____.

2. In $\dfrac{33.6}{0.3}$, the fraction _____ indicates division.

CONCEPTS   *Fill in the blanks.*

3. To divide by a decimal, move the decimal point of the divisor so that it becomes a _____ number. The decimal point of the dividend is then moved the same number of places to the _____. The decimal point in the quotient is written directly _____ the decimal point in the dividend.

4. To divide a decimal by a power of 10, move the decimal point to the _____ the same number of decimal places as the number of zeros in the power of 10.

5. Is this statement true or false?
$$45 = 45.0 = 45.000$$

6. When a positive decimal is divided by 10, is the answer smaller or larger than the original number?

7. To complete the division $7.8\overline{)14.562}$, the decimal points in the divisor and dividend are moved 1 place to the right. This is equivalent to multiplying the numerator and the denominator of $\frac{14.562}{7.8}$ by what number?

8. a. When dividing decimals with like signs, what is the sign of the quotient?

   b. When dividing decimals with unlike signs, what is the sign of the quotient?

9. How can we check the result of this division?
$$\dfrac{1.917}{0.9} = 2.13$$

10. When rounding a decimal to the hundredths column, to what other column must we look at first?

11. A student performed the division
$$4.6\overline{)9,522}$$
and obtained the answer 2.07. Without doing the division, check this result. Is it correct?

12. In the division problem below, explain why we can write the additional zeros (shown in red) after 5.
$$16\overline{)5.50000} \quad \genfrac{}{}{0pt}{}{0.3}{}$$

NOTATION

13. Explain what the arrows are illustrating.
$$4\,67\overline{)32\,08.7}$$

14. What is this arrow illustrating?
$$4\overline{)3.100} \quad \genfrac{}{}{0pt}{}{0.7}{}$$
$$\underline{2\,8\downarrow}$$
$$30$$

PRACTICE   *Perform each division.*

15. $8\overline{)36}$

16. $4\overline{)10}$

17. $-39 \div 4$

18. $-26 \div 8$

19. $49.6 \div 8$

20. $23.5 \div 5$

21. $9\overline{)288.9}$

22. $6\overline{)337.8}$

23. $(-14.76) \div (-6)$

24. $(-13.41) \div (-9)$

25. $\dfrac{-55.02}{7}$

26. $\dfrac{-24.24}{8}$

27. $45\overline{)119.7}$

28. $41\overline{)146.37}$

29. $250.95 \div 35$

30. $241.86 \div 29$

31. $41.6 \div 0.32$

32. $31.8 \div 0.15$

33. $(-199.5) \div (-0.19)$

34. $(-2,381.6) \div (-0.26)$

35. $\dfrac{0.0102}{0.017}$

36. $\dfrac{0.0092}{0.023}$

37. $\dfrac{0.0186}{0.031}$

38. $\dfrac{0.416}{0.52}$

Divide and round each result to the nearest tenth.

39. $3\overline{)16}$

40. $7\overline{)20}$

41. $-5.714 \div 2.4$

42. $-21.21 \div 3.8$

Divide and round each result to the nearest hundredth.

43. $12.243 \div 0.9$

44. $13.441 \div 0.6$

45. $0.04\overline{)0.03164}$

46. $0.08\overline{)0.02201}$

Perform each division mentally.

47. $7.895 \div 100$

48. $23.05 \div 10$

49. $0.064 \div (-100)$

50. $0.0043 \div (-10)$

51. $1000\overline{)34.8}$

52. $100\overline{)678.9}$

53. $\dfrac{45.04}{10}$

54. $\dfrac{22.32}{100}$

Evaluate each expression. Round each result to the nearest hundredth.

55. $\dfrac{-1.2 - 3.4}{3(1.6)}$

56. $\dfrac{(-1.3)^2 + 6.7}{-0.9}$

57. $\dfrac{40.7(-5.3)}{0.4 - 0.61}$

58. $\dfrac{(0.5)^2 - (0.3)^2}{0.005 + 0.1}$

Evaluate each formula.

59. $t = \dfrac{d}{r}$ for $d = 211.75$ and $r = 60.5$

60. $h = \dfrac{2A}{b}$ for $A = 9.62$ and $b = 3.7$

61. $r = \dfrac{d}{t}$ for $d = 219.375$ and $t = 3.75$

62. $\pi = \dfrac{C}{d}$ for $C = 14.4513$ and $d = 4.6$ (Round to the nearest hundredth.)

APPLICATIONS

0.05714

63. BUTCHER SHOPS A meat slicer is designed to trim 0.05-inch-thick pieces from a sausage. If the sausage is 14 inches long, how many slices will result?

64. COMPUTERS A computer can do an arithmetic computation in 0.00003 second. How many of these computations could it do in 60 seconds?

65. HIKING Use the information in the illustration to find the time of arrival for the hiker.

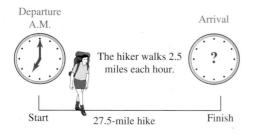

The hiker walks 2.5 miles each hour.

Departure A.M.   Arrival

Start   27.5-mile hike   Finish

66. ELECTRONICS A volume control is shown. If the distance between the Low and High settings is 21 cm, how far apart are the equally spaced volume settings?

Low   Volume Control   High

67. SPRAY BOTTLES Production planners have found that each squeeze of the trigger of a spray bottle emits 0.015 ounce of liquid. How many squeezes would there be in an 8.5-ounce bottle?

68. CAR LOANS See the loan statement. How many more monthly payments must be made to pay off the loan?

American Finance Company		June
Monthly payment:	Paid to date: $547.30	
$42.10	Loan balance: $631.50	

69. HOURLY PAY The illustration shows the average hours worked and the average weekly earnings of U.S. production workers in 1998 and 2003. What did the average production worker earn per hour in 1998 and in 2003? Round to the nearest cent, if necessary.

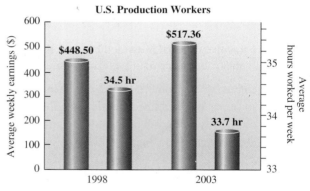

U.S. Production Workers

Source: *The World Almanac 2005*

70. TRAVEL The illustration shows the annual number of person-trips of 50 miles or more (one way) for the years 1999–2003, as estimated by the Travel Industry Association of America. Find the average number of trips per year for this period of time.

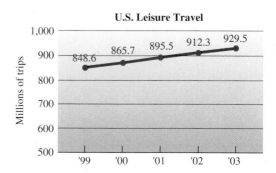

U.S. Leisure Travel

71. OIL WELLS Geologists have mapped out the substances through which engineers must drill to reach an oil deposit. What is the average depth that must be drilled each week if this is to be a four-week project?

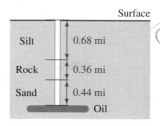

72. INDY 500 The illustration shows the first row of the starting grid for the 2004 Indianapolis 500 automobile race. The drivers' speeds on a qualifying run were used to rank them in this order. What was the average qualifying speed for the drivers in the first row?

B. Rice	D. Wheldon	D. Franchitti
222.024 mph	221.524 mph	221.471 mph

Round to the nearest hundredth.

73. $\dfrac{8.6 + 7.99 + (4.05)^2}{4.56}$

74. $\dfrac{0.33 + (-0.67)(1.3)^3}{0.0019}$

75. $\left(\dfrac{45.9098}{-234.12}\right)^2 - 4$

76. $\left(\dfrac{6.0007}{3.002}\right) - \left(\dfrac{78.8}{12.45}\right)$

WRITING

77. Explain the process used to divide two numbers when both the divisor and the dividend are decimals.

78. Explain why we must sometimes use rounding when we write the answer to a division problem.

79. The division $0.5\overline{)2.005}$ is equivalent to $5\overline{)20.05}$. Explain what *equivalent* means in this case.

80. In $3\overline{)0.7}$, why can additional zeros be placed to the right of 0.7 without affecting the result?

REVIEW

81. Simplify the complex fraction: $\dfrac{\frac{7}{8}}{\frac{3}{4}}$.

82. Express the fraction $\frac{3}{4}$ as an equivalent fraction with a denominator of 36.

83. List the set of integers.

84. Add: $-\dfrac{3}{4} + \dfrac{2}{3}$.

85. Multiply: $7\dfrac{1}{3} \cdot 5\dfrac{1}{4}$.

86. Evaluate: $\left(\dfrac{1}{2}\right)^3 - \left(\dfrac{1}{2}\right)^2$.

87. Multiply: $(-5.8)(-7.25)$.

88. What is the opposite of 5?

Estimation

In this section, we use estimation procedures to approximate the answers to addition, subtraction, multiplication, and division problems involving decimals. Recall that we use rounding when estimating to help simplify the computations so that they can be performed quickly and easily.

EXAMPLE 1 Estimating sums and differences.
a. Estimate to the nearest ten: $261.76 + 432.94$.
b. Estimate using front-end rounding: $381.77 - 57.01$.

Solution
a. We round each number to the nearest ten.

$$261.76 + 432.94$$
$$\downarrow \qquad \downarrow$$
$$260 \;+\; 430 \;=\; 690 \qquad \text{261.76 rounds to 260, and 432.94 rounds to 430.}$$

Self Check 1
a. Estimate to the nearest ten the sum of 526.93 and 284.03.

b. Estimate using front-end rounding: $512.33 - 36.47$.

The estimate is 690. If we compute 261.76 + 432.94, the sum is 694.7. We can see that our estimate is close; it's just 4.7 less than 694.7. This example illustrates the tradeoff when using estimation. The calculations are easier to perform and they take less time, but the answers are not exact.

b. We use front-end rounding.

$$381.77 - 57.01 \qquad \text{Each number is rounded to its largest place value: 381.77 to the}$$
$$\downarrow \qquad \downarrow \qquad \text{nearest hundred and 57.01 to the nearest ten.}$$
$$400 - 60 = 340$$

Answers **a.** 810, **b.** 460

The estimate is 340.

Self Check 2

a. Estimate the product:

$42 \cdot 17.65$

b. Estimate the product:

$182 \cdot 24.04$

c. Estimate the product:

$979.3 \cdot 2.3485$

EXAMPLE 2   Estimating products.

Estimate each product: **a.** $6.41 \cdot 27$,   **b.** $5.2 \cdot 13.91$,   and   **c.** $0.124 \cdot 98.6$.

Solution

a. We use front-end rounding.

$$6.41 \cdot 27 \approx 6 \cdot 30 \qquad \text{The symbol} \approx \text{means "is approximately equal to."}$$

The estimate is 180.

b. We use front-end rounding.

$$5.2 \cdot 13.91 \approx 5 \cdot 10$$

The estimate is 50.

c. Notice that $98.6 \approx 100$.

$$0.124 \cdot 98.6 \approx 0.124 \cdot 100 \qquad \text{To multiply a decimal by 100, move the}$$
$$\text{decimal point 2 places to the right.}$$

Answers **a.** 800, **b.** 4,000,
c. 2,348.5

The estimate is 12.4.

When estimating a quotient, we round the divisor and the dividend so that they will divide evenly. Try to round both numbers up or both numbers down.

Self Check 3

Estimate: $6,429.6 \div 7.19$.

EXAMPLE 3   Estimating quotients.   Estimate: $246.03 \div 4.31$.

Solution   4.31 is close to 4. A multiple of 4 close to 246.03 is 240. (Note that both the divisor and the dividend were rounded down.)

$$246.03 \div 4.31 \approx 240 \div 4 \qquad \text{Do the division in your head.}$$

Answer 900

The estimate is 60.

■ **STUDY SET** *Use the following information about refrigerators to estimate the answer to each question. Remember that answers may vary, depending on the rounding method used.*

Deluxe model	**Standard model**	**Economy model**
Price: $978.88	Price: $739.99	Price: $599.95
Capacity: 25.2 cubic feet	Capacity: 20.6 cubic feet	Capacity: 18.8 cubic feet
Energy cost: $6.79 a month	Energy cost: $5.61 a month	Energy cost: $4.39 a month

1. How much more expensive is the deluxe model than the standard model?

2. A couple wants to buy two standard models, one for themselves and one for their newly married son and daughter-in-law. What is the total cost?

3. How much less storage capacity does the economy model have than the standard model?

4. The owner of a duplex apartment wants to purchase a standard model for one unit and an economy model for the other. What will be the total cost?

5. A stadium manager has a budget of $20,000 to furnish the luxury boxes at a football stadium with refrigerators. How many standard models can she purchase for this amount?

6. How many more cubic feet of storage do you get with the deluxe model as compared to the economy model?

7. Three roommates are planning on purchasing the deluxe model and splitting the cost evenly. How much will each have to pay?

8. What is the energy cost per year to run the deluxe model?

9. If you make a $220 down payment on the standard model, how much of the cost is left to finance?

10. The economy model can be expected to last for 10 years. What would be the total energy cost over that period?

Estimate the answer to each problem. Does the result on the calculator display seem reasonable?

11. $25.9 + 345.1 + 0.09$ $\boxed{347.78}$

12. $8,345.889 - 345.6$ $\boxed{8000.289}$

13. $42,090.8 + 3,303.09$ $\boxed{45393.89}$

14. $10.007 - 0.626$ $\boxed{3.747}$

15. $9.8(8.8)$ $\boxed{86.24}$

16. $\dfrac{24.56}{2.2}$ $\boxed{1.116363636}$

17. $53 \cdot 5.61$ $\boxed{241.23}$

18. $89.11 \div 22.707$ $\boxed{39.24340}$

4.5 Fractions and Decimals

- Writing fractions as equivalent decimals • Repeating decimals
- Rounding repeating decimals • Graphing fractions and decimals
- Problems involving fractions and decimals

In this section, we continue to investigate the relationship between fractions and decimals.

■ Writing fractions as equivalent decimals

To write $\frac{5}{8}$ as a decimal, we use the fact that $\frac{5}{8}$ indicates the division $5 \div 8$. We can convert $\frac{5}{8}$ to decimal form by doing the division.

$$\begin{array}{r} .625 \\ 8\overline{)5.000} \end{array}$$ Write a decimal point and additional zeros to the right of 5.

$$\begin{array}{r} 4\,8 \\ \hline 20 \\ 16 \\ \hline 40 \\ 40 \\ \hline 0 \end{array}$$ ← The remainder is 0.

Thus, $\frac{5}{8} = 0.625$.

Writing a fraction as a decimal

To write a fraction as a decimal, divide the numerator of the fraction by its denominator.

Self Check 1

Write $\dfrac{3}{16}$ as a decimal.

EXAMPLE 1 Write $\dfrac{3}{4}$ as a decimal.

Solution   We divide the numerator by the denominator.

$$\begin{array}{r} .75 \\ 4\overline{)3.00} \end{array}$$ Write a decimal point and two zeros to the right of 3.

$$\begin{array}{r} 2\,8 \\ \hline 20 \\ 20 \\ \hline 0 \end{array}$$ ← The remainder is 0.

Answer 0.1875

Thus, $\frac{3}{4} = 0.75$.

In Example 1, the division process ended because a remainder of 0 was obtained. In this case, we call the quotient, 0.75, a **terminating decimal.**

Repeating decimals

Sometimes, when we are finding a decimal equivalent of a fraction, the division process never gives a remainder of 0. In this case, the result is a **repeating decimal.** Examples of repeating decimals are 0.4444. . . and 1.373737. . . . The three dots tell us that a block of digits repeats in the pattern shown. Repeating decimals can also be written using a bar over the repeating block of digits. For example, 0.4444. . . can be written as $0.\overline{4}$, and 1.373737. . . can be written as $1.\overline{37}$.

❗ COMMENT When using an overbar to write a repeating decimal, use the least number of digits necessary to show the repeating block of digits.

$$0.333. . . = 0.\overline{333} \qquad\qquad 6.7454545. . . = 6.7\overline{454}$$

$$0.333. . . = 0.\overline{3} \qquad\qquad 6.7454545. . . = 6.7\overline{45}$$

EXAMPLE 2   Write $\dfrac{5}{12}$ as a decimal.

Self Check 2

Write $\dfrac{4}{11}$ as a decimal.

Solution   We use division to find the decimal equivalent.

$$
\begin{array}{r}
.4166 \\
12\overline{)5.0000} \\
4\ 8 \\
\hline
20 \\
12 \\
\hline
80 \\
72 \\
\hline
80 \\
72 \\
\hline
8
\end{array}
$$

Write a decimal point and four zeros to the right of 5.

It is apparent that 8 will continue to reappear as the remainder. Therefore, 6 will continue to reappear in the quotient. Since the repeating pattern is now clear, we can stop the division.

Thus, $\dfrac{5}{12} = 0.41\overline{6}$.

Answer $0.\overline{36}$

Every fraction can be written as either a terminating decimal or a repeating decimal. For this reason, the set of fractions (**rational numbers**) form a subset of the set of decimals called the set of **real numbers.** The set of real numbers corresponds to all points on a number line.

Not all decimals are terminating or repeating decimals. For example,

$$0.2020020002\ldots$$

does not terminate, and it has no repeating block of digits. This decimal cannot be written as a fraction with an integer numerator and a nonzero integer denominator. Thus, it is not a rational number. It is an example from the set of **irrational numbers.**

▉ Rounding repeating decimals

When a fraction is written in decimal form, the result is either a terminating or a repeating decimal. Repeating decimals are often rounded to a specified place value.

EXAMPLE 3   Write $\frac{1}{3}$ as a decimal and round to the nearest hundredth.

Solution   First, we divide the numerator by the denominator to find the decimal equivalent of $\frac{1}{3}$.

$$
\begin{array}{r}
0.333 \\
3\overline{)1.000} \\
9 \\
\hline
10 \\
9 \\
\hline
10 \\
9 \\
\hline
1
\end{array}
$$

Write a decimal point and additional zeros to the right of 1.

We see that the division process never gives a remainder of 0. When we write $\frac{1}{3}$ in decimal form, the result is the repeating decimal $0.333\ldots = 0.\overline{3}$.

To find the decimal equivalent of $\frac{1}{3}$ to the nearest hundredth, we proceed as follows.

$$0.333\ldots$$

Round 0.333 to the nearest hundredth by examining the test digit in the thousandths column.

Since 3 is less than 5, we round down, and we have

$$\frac{1}{3} \approx 0.33$$

Read $\approx$ as "is approximately equal to."

Self Check 4

Write $\frac{7}{24}$ as a decimal and round to the nearest thousandth.

EXAMPLE 4   Write $\frac{2}{7}$ as a decimal and round to the nearest thousandth.

Solution

$$
\begin{array}{r}
.2857 \\
7\overline{)2.0000} \\
\underline{1\,4} \\
60 \\
\underline{56} \\
40 \\
\underline{35} \\
50 \\
\underline{49} \\
1
\end{array}
$$

Write a decimal point and additional zeros to the right of 2.

To round to the thousandths column, we must divide to the ten-thousandths column.

$$0.2857$$

Round 0.2857 to the nearest thousandth by examining the test digit in the ten-thousandths column.

Answer 0.292

Since 7 is greater than 5, we round up, and $\frac{2}{7} \approx 0.286$.

CALCULATOR SNAPSHOT   **The fixed-point key**

After performing a calculation, a scientific calculator can round the result to a given decimal place. This is done using the *fixed-point key*. As we did in Example 4, let's find the decimal equivalent of $\frac{2}{7}$ and round to the nearest thousandth. This time, we will use a calculator.

First, we set the calculator to round to the third decimal place (thousandths) by pressing $\boxed{\text{FIX}}$ 3. Then we press 2 $\boxed{\div}$ 7 $\boxed{=}$ $\boxed{ 0.286}$

Thus, $\frac{2}{7} \approx 0.286$. To round to the nearest tenth, we would fix 1; to round to the nearest hundredth, we would fix 2; and so on.

If your calculator does not have a fixed-point key, see the owner's manual.

Self Check 5

Write $8\frac{19}{20}$ in decimal form.

EXAMPLE 5   Write $5\frac{3}{8}$ in decimal form.

Solution   To write a mixed number in decimal form, recall that a mixed number is made up of a whole-number part and a fractional part. Since we can write $5\frac{3}{8}$ as $5 + \frac{3}{8}$, we need only consider how to write $\frac{3}{8}$ as a decimal.

$$\begin{array}{r} .375 \\ 8\overline{)3.000} \\ \underline{2\ 4} \\ 60 \\ \underline{56} \\ 40 \\ \underline{40} \\ 0 \end{array}$$ Write a decimal point and three zeros to the right of 3.

Thus, $5\frac{3}{8} = 5 + \frac{3}{8} = 5 + 0.375 = 5.375$. We would obtain the same result if we changed $5\frac{3}{8}$ to the improper fraction $\frac{43}{8}$ and divided 43 by 8.

Answer 8.95

Graphing fractions and decimals

The number line can be used to show the relationship between fractions and their respective decimal equivalents. Figure 4-7 shows some commonly used fractions that have terminating decimal equivalents. For example, we see from the graph that $\frac{13}{16} = 0.8125$.

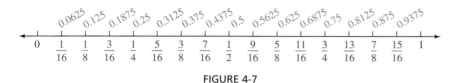

FIGURE 4-7

The number line in Figure 4-8 shows some commonly used fractions that have repeating decimal equivalents.

FIGURE 4-8

Problems involving fractions and decimals

Numerical expressions can contain both fractions and decimals. In the following examples, we show how different methods can be used to solve problems of this type.

EXAMPLE 6   Evaluate $\frac{1}{3} + 0.27$ by working in terms of fractions.

Solution   We write 0.27 as a fraction and add it to $\frac{1}{3}$.

$$\frac{1}{3} + 0.27 = \frac{1}{3} + \frac{27}{100}$$ Replace 0.27 with $\frac{27}{100}$.

$$= \frac{1 \cdot \mathbf{100}}{3 \cdot \mathbf{100}} + \frac{27 \cdot \mathbf{3}}{100 \cdot \mathbf{3}}$$ The LCD for $\frac{1}{3}$ and $\frac{27}{100}$ is 300. Express each fraction in terms of 300ths.

$$= \frac{100}{300} + \frac{81}{300}$$ Multiply in the numerators and in the denominators.

$$= \frac{181}{300}$$ Add the numerators and write the sum over the common denominator 300.

Self Check 6
Evaluate by working in terms of fractions: $0.53 - \frac{1}{6}$.

Answer $\dfrac{109}{300}$

Self Check 7

Estimate the result by working in terms of decimals: $0.53 - \frac{1}{6}$.

Answer approximately 0.36

EXAMPLE 7   Estimate $\frac{1}{3} + 0.27$ by working in terms of decimals.

Solution   We have seen that the decimal equivalent of $\frac{1}{3}$ is the repeating decimal $0.333\ldots$. To add $\frac{1}{3}$ to 0.27, we round $0.333\ldots$ to the nearest hundredth: $\frac{1}{3} \approx 0.33$.

$$\frac{1}{3} + 0.27 \approx 0.33 + 0.27 \qquad \text{Approximate } \frac{1}{3} \text{ with the decimal } 0.33.$$
$$\approx 0.60 \qquad\qquad\quad \text{Perform the addition.}$$

In Examples 6 and 7, we evaluated $\frac{1}{3} + 0.27$ in different ways. In Example 6, we obtained the exact answer, $\frac{181}{300}$. In Example 7, we obtained an approximation, 0.6. It is apparent that the results are in agreement when we write $\frac{181}{300}$ in decimal form: $\frac{181}{300} = 0.60333\ldots$.

Self Check 8

Perform the operations:

$(-0.6)^2 + (2.3)\left(\dfrac{1}{8}\right)$.

Answer 0.6475

EXAMPLE 8   Perform the operations: $\left(\dfrac{4}{5}\right)(1.35) + (0.5)^2$.

Solution   It appears simplest to work in terms of decimals. We use division to find the decimal equivalent of $\frac{4}{5}$.

$$\begin{array}{r} .8 \\ 5\overline{)4.0} \\ \underline{4\,0} \\ 0 \end{array} \qquad \text{Write a decimal point and one zero to the right of the 4.}$$

Now we use the rules for the order of operations to evaluate the given expression.

$$\left(\frac{4}{5}\right)(1.35) + (0.5)^2 = (\mathbf{0.8})(1.35) + (0.5)^2 \quad \begin{array}{l}\text{Replace } \frac{4}{5} \text{ with its decimal equivalent,}\\ 0.8.\end{array}$$
$$= (0.8)(1.35) + 0.25 \quad \text{Find the power: } (0.5)^2 = 0.25.$$
$$= 1.08 + 0.25 \quad\quad \begin{array}{l}\text{Perform the multiplication:}\\ (0.8)(1.35) = 1.08.\end{array}$$
$$= 1.33 \quad\qquad\quad \text{Perform the addition.}$$

EXAMPLE 9   **Shopping.**   During a trip to the grocery store, a shopper purchased $\frac{3}{4}$ pound of fruit, priced at $0.88 a pound, and $\frac{1}{3}$ pound of fresh-ground coffee, selling for $6.60 a pound. Find the total cost of these items.

Solution   To find the cost of each item, we multiply the amount purchased by its unit price. Then we add the two individual costs to obtain the total cost.

Cost of fruit	plus	cost of coffee	equals	total cost.
$\left(\dfrac{3}{4}\right)(0.88)$	$+$	$\left(\dfrac{1}{3}\right)(6.60)$	$=$	total cost

Because 0.88 is divisible by 4 and 6.60 is divisible by 3, we can work with the decimals and fractions in this form; no conversion is necessary.

exact = Fractions

approximation = decimal

$$\left(\frac{3}{4}\right)(0.88) + \left(\frac{1}{3}\right)(6.60) = \left(\frac{3}{4}\right)\left(\frac{0.88}{1}\right) + \left(\frac{1}{3}\right)\left(\frac{6.60}{1}\right)$$

Express 0.88 as $\frac{0.88}{1}$ and 6.60 as $\frac{6.60}{1}$.

$$= \frac{2.64}{4} + \frac{6.60}{3}$$

Multiply the numerators and the denominators.

$$= 0.66 + 2.20$$

Perform each division.

$$= 2.86$$

Perform the addition.

The total cost of the items is $2.86.

Section 4.5   STUDY SET

VOCABULARY   *Fill in the blanks.*

1. The decimal form of the fraction $\frac{1}{3}$ is a _____ decimal, which is written $0.\overline{3}$ or $0.3333.\ldots$

2. The decimal form of the fraction $\frac{2}{5}$ is a _____ decimal, which is written 0.4.

3. The _____ equivalent of $\frac{1}{16}$ is 0.0625.

4. We read ≈ as "is _____ equal to."

CONCEPTS

5. **a.** What division is indicated by the fraction $\frac{7}{8}$?

   b. Fill in: To write a fraction as a decimal, divide the _____ of the fraction by its denominator.

6. Insert the proper symbol < or > in the blank to make the statement true.

   a. $0.\overline{6}$ ___ 0.7   **b.** $0.\overline{6}$ ___ 0.6

7. When rounding 0.272727 . . . to the nearest hundredth, is the result larger or smaller than the original number?

8. Write each decimal in fraction form.

   a. 0.7   **b.** 0.77

9. Graph: $1\frac{3}{4}$, -0.75, $0.\overline{6}$, and $-3.8\overline{3}$.

   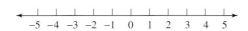

10. Graph: $2\frac{7}{8}$, -2.375, $0.\overline{3}$, and $4.1\overline{6}$.

    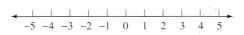

11. Determine whether each statement is true or false.

    a. $\frac{1}{3} = 0.3$ **b.** $\frac{3}{4} = 0.75$

    c. $20\frac{1}{2} = 20.5$ **d.** $\frac{1}{16} = 0.1\overline{6}$

12. When evaluating the expression $0.25 + \left(2.3 + \frac{2}{5}\right)^2$, would it be easier to work in terms of fractions or decimals?

NOTATION

13. Examine the color portion of the long division in the next column.

    a. Will the remainder ever be 0?

    b. What can be deduced about the decimal equivalent of $\frac{5}{6}$?

    $$\begin{array}{r} .833 \\ 6\overline{)5.000} \\ 4\,8 \\ \hline 20 \\ 18 \\ \hline 20 \end{array}$$

14. Write each repeating decimal using an overbar.

    a. 0.888. . .   **b.** 0.323232. . .

    c. 0.56333 . . .   **d.** 0.8898989. . .

PRACTICE   *Write each fraction in decimal form.*

15. $\dfrac{1}{2}$ 16. $\dfrac{1}{4}$

17. $-\dfrac{5}{8}$ 18. $-\dfrac{3}{5}$

19. $\dfrac{9}{16}$ 20. $\dfrac{3}{32}$

21. $-\dfrac{17}{32}$ 22. $-\dfrac{15}{16}$

23. $\dfrac{11}{20}$ 24. $\dfrac{19}{25}$

25. $\dfrac{31}{40}$ 26. $\dfrac{17}{20}$

27. $-\dfrac{3}{200}$ **28.** $-\dfrac{21}{50}$

29. $\dfrac{1}{500}$ **30.** $\dfrac{1}{250}$

Write each fraction in decimal form. Use an overbar.

31. $\dfrac{2}{3}$ **32.** $\dfrac{7}{9}$

33. $\dfrac{5}{11}$ **34.** $\dfrac{4}{15}$

35. $-\dfrac{7}{12}$ **36.** $-\dfrac{17}{22}$

37. $\dfrac{1}{30}$ **38.** $\dfrac{1}{60}$

Write each fraction in decimal form. Round to the nearest hundredth.

39. $\dfrac{7}{30}$ **40.** $\dfrac{14}{15}$

41. $\dfrac{17}{45}$ **42.** $\dfrac{8}{9}$

Write each fraction in decimal form. Round to the nearest thousandth.

43. $\dfrac{5}{33}$ **44.** $\dfrac{5}{12}$

45. $\dfrac{10}{27}$ **46.** $\dfrac{17}{21}$

Write each fraction in decimal form. Round to the nearest hundredth.

47. $\dfrac{4}{3}$ **48.** $\dfrac{10}{9}$

49. $-\dfrac{34}{11}$ **50.** $-\dfrac{25}{12}$

Write each mixed number in decimal form. Round to the nearest hundredth when the result is a repeating decimal.

51. $3\dfrac{3}{4}$ **52.** $5\dfrac{4}{5}$

53. $-8\dfrac{2}{3}$ **54.** $-1\dfrac{7}{9}$

55. $12\dfrac{11}{16}$ **56.** $32\dfrac{1}{8}$

57. $203\dfrac{11}{15}$ **58.** $568\dfrac{23}{30}$

Fill in the correct symbol ($<$ or $>$) to make a true statement. (Hint: Express each number as a decimal.)

59. $\dfrac{7}{8}$ ▢ 0.895 **60.** 4.56 ▢ $4\dfrac{2}{5}$

61. $-\dfrac{11}{20}$ ▢ $-0.\overline{4}$ **62.** $-9.0\overline{9}$ ▢ $-9\dfrac{1}{11}$

Evaluate each expression. Work in terms of fractions.

63. $\dfrac{1}{9} + 0.3$ **64.** $\dfrac{2}{3} + 0.1$

65. $0.9 - \dfrac{7}{12}$ **66.** $0.99 - \dfrac{5}{6}$

67. $\dfrac{5}{11}(0.3)$ **68.** $(0.9)\left(\dfrac{1}{27}\right)$

69. $\dfrac{1}{3}\left(-\dfrac{1}{15}\right)(0.5)$ **70.** $(-0.4)\left(\dfrac{5}{18}\right)\left(-\dfrac{1}{3}\right)$

71. $\dfrac{1}{4}(0.25) + \dfrac{15}{16}$ **72.** $\dfrac{2}{5}(0.02) - (0.04)$

Evaluate each expression to the nearest hundredth.

73. $0.24 + \dfrac{1}{3}$ **74.** $0.02 + \dfrac{5}{6}$

75. $5.69 - \dfrac{5}{12}$ **76.** $3.19 - \dfrac{2}{3}$

77. $\dfrac{3}{4}(0.43) - \dfrac{1}{12}$ **78.** $-\dfrac{2}{5}(0.33) + 0.45$

Evaluate each expression. Work in terms of decimals.

79. $(3.5 + 6.7)\left(-\dfrac{1}{4}\right)$ **80.** $\left(-\dfrac{5}{8}\right)(5.3 - 3.9)$

81. $\left(\dfrac{1}{5}\right)^{2}(1.7)$ **82.** $(2.35)\left(\dfrac{2}{5}\right)^{2}$

83. $7.5 - (0.78)\left(\dfrac{1}{2}\right)$ **84.** $8.1 - \left(\dfrac{3}{4}\right)(0.12)$

85. $\dfrac{3}{8}(-3.2) + (4.5)\left(-\dfrac{1}{9}\right)$

86. $(-0.8)\left(\dfrac{1}{4}\right) + \left(\dfrac{1}{3}\right)(0.39)$

▦ *Evaluate each expression. Round to the nearest hundredth.*

87. $\dfrac{3\dfrac{1}{5} + 2\dfrac{1}{2}}{5.69} + 3\dfrac{1}{4}$ **88.** $4\dfrac{2}{3} - \dfrac{2.7 - \dfrac{7}{8}}{0.12}$

Evaluate each formula. Round to the nearest tenth.

89. $C = \dfrac{5}{9}(F - 32)$ for $F = 64.5$

90. $F = \dfrac{9}{5}C + 32$ for $C = 0.58$

Write each fraction in decimal form.

91. $\dfrac{23}{101}$ **92.** $\dfrac{1}{99}$

93. $\dfrac{1.736}{50}$ **94.** $-\dfrac{11}{128}$

▮ APPLICATIONS

95. DRAFTING The architect's scale has several measuring edges. The edge marked 16 divides each inch into 16 equal parts. Find the decimal form for each fractional part of 1 inch that is highlighted on the scale.

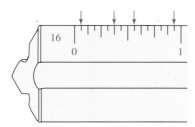

96. FREEWAY SIGNS The freeway sign gives the number of miles to the next three exits. Convert the mileages to decimal notation.

BARRANCA AVE.	$\dfrac{3}{4}$ mi
210 FREEWAY	$2\dfrac{1}{4}$ mi
ADA ST.	$3\dfrac{1}{2}$ mi

97. GARDENING Two brands of replacement line for a lawn trimmer are labeled in different ways. On one package, the line's thickness is expressed as a decimal; on the other, as a fraction. Which line is thicker?

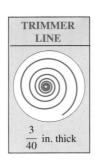

NYLON LINE

Thickness: 0.065 in.

TRIMMER LINE

$\dfrac{3}{40}$ in. thick

98. AUTO MECHANICS While doing a tuneup, a mechanic checks the gap on one of the spark plugs of a car to be sure it is firing correctly. The owner's manual states that the gap should be $\dfrac{2}{125}$ inch. The gauge the mechanic uses to check the gap is in decimal notation; it registers 0.025 inch. Is the spark plug gap too large or too small?

99. HORSE RACING In thoroughbred racing, the time a horse takes to run a given distance is measured using fifths of a second. For example, 55^2 (read "fifty-five and two") means $55\dfrac{2}{5}$ seconds. The illustration lists four split times for a horse. Express the times in decimal form.

| Speedy Flight   Turfway Park, Ky   3-year-old |
| 17 May 97 $1\dfrac{1}{16}$ mile $:23^2$ $:23^4$ $:24^1$ $:32^3$ |

100. GEOLOGY A geologist weighed a rock sample at the site where it was discovered and found it to weigh $17\dfrac{7}{8}$ lb. Later, a more accurate digital scale in the laboratory gave the weight as 17.671 lb. What is the difference in the two measurements?

101. WINDOW REPLACEMENTS The amount of sunlight that comes into a room depends on the area of the windows in the room. What is the area of the window shown?

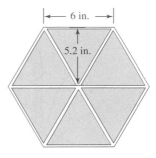

102. FOREST FIRE CONTAINMENT A command post asked each of three fire crews to estimate the length of the fire line they were fighting. Their reports came back in different forms, as shown. Find the perimeter of the fire.

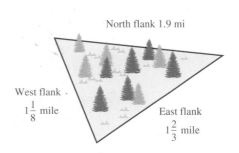

North flank 1.9 mi

West flank $1\dfrac{1}{8}$ mile

East flank $1\dfrac{2}{3}$ mile

WRITING

103. Explain the procedure used to write a fraction in decimal form.

104. Compare and contrast the two numbers 0.5 and $0.\overline{5}$.

105. A student represented the repeating decimal 0.1333. . . as $0.1\overline{333}$. Is this correct? Explain why or why not.

106. Is 0.10100100010000. . . a repeating decimal? Explain why or why not.

REVIEW

107. Add: $-2 + (-3) + 10 + (-6)$.

108. Evaluate: $-3 + 2[-3 + (2 - 7)]$.

109. List the set of the first eight whole numbers.

110. Simplify: $\dfrac{20}{55}$.

111. Multiply: $3\dfrac{1}{3} \cdot 4\dfrac{1}{2}$.

112. Divide: $3\dfrac{1}{3} \div 4\dfrac{1}{2}$.

4.6 Square Roots

- Square roots   • Evaluating numerical expressions containing radicals
- Square roots of fractions and decimals   • Using a calculator to find square roots
- Approximating square roots

We have discussed the relationships between addition and subtraction and between multiplication and division. In this section, we explore the relationship between raising a number to a power and finding a root. Decimals play an important role in this discussion.

▮ Square roots

When we raise a number to the second power, we are squaring it, or finding its **square.**

    The square of 6 is 36, because $6^2 = 36$.

    The square of -6 is 36, because $(-6)^2 = 36$.

The **square root** of a given number is a number whose square is the given number. For example, the square roots of 36 are 6 and -6, because either number when squared yields 36. We can express this concept using symbols.

> **Square root**
>
> A number is a **square root** of a second number if the square of the first number equals the second number.

Self Check 1
Find the square roots of 64.

Answers 8 and -8

EXAMPLE 1   Find the square roots of 49.

Solution   Ask "What number was squared to obtain 49?" The two answers are 7 and -7.

    $7^2 = 49$ and $(-7)^2 = 49$

Thus, 7 and -7 are the square roots of 49.

In Example 1, we saw that 49 has two square roots — one positive and one negative. The symbol $\sqrt{}$ is called a **radical symbol** and is used to indicate a positive square root.

When a number, called the **radicand,** is written under a radical symbol, we have a **radical expression.** Some examples of radical expressions are

$$\sqrt{36} \qquad \sqrt{100} \qquad \sqrt{144} \qquad \sqrt{81}$$

To evaluate (or simplify) a radical expression, we need to find the positive square root of the radicand. For example, if we evaluate $\sqrt{36}$ (read as "the square root of 36"), the result is

$$\sqrt{36} = 6$$

because $6^2 = 36$. The negative square root of 36 is denoted $-\sqrt{36}$, and we have

$$-\sqrt{36} = -6 \quad \text{Read as "the negative square root of 36 is } -6\text{" or "the opposite of the}$$
square root of 36 is -6."

EXAMPLE 2   Evaluate each expression: **a.** $\sqrt{81}$ and   **b.** $-\sqrt{100}$.

Solution

a. $\sqrt{81}$ means the positive square root of 81.

$\sqrt{81} = 9$, because $9^2 = 81$.

b. $-\sqrt{100}$ means the opposite (or negative) of the square root of 100. Since $\sqrt{100} = 10$, we have

$$-\sqrt{100} = -10$$

Self Check 2
Evaluate each expression:
a. $\sqrt{144}$ and   **b.** $-\sqrt{81}$.

Answers **a.** 12,   **b.** −9

! **COMMENT**   Radical expressions such as

$$\sqrt{-36} \qquad \sqrt{-100} \qquad \sqrt{-144} \qquad \sqrt{-81}$$

do not represent real numbers, because there are no real numbers that when squared give a negative number.

Be careful to note the difference between expressions such as $-\sqrt{36}$ and $\sqrt{-36}$. We have seen that $-\sqrt{36}$ is a real number: $-\sqrt{36} = -6$. In contrast, $\sqrt{-36}$ is not a real number.

◾ Evaluating numerical expressions containing radicals

Numerical expressions can contain radical expressions. When applying the rules for the order of operations, we treat a radical expression as we would a power.

EXAMPLE 3   Evaluate: **a.** $\sqrt{64} + \sqrt{9}$ and   **b.** $-\sqrt{25} - \sqrt{4}$.

Solution

a. $\sqrt{64} + \sqrt{9} = 8 + 3$ Evaluate each radical expression first.

$\qquad\qquad\qquad = 11$ Perform the addition.

b. $-\sqrt{25} - \sqrt{4} = -5 - 2$ Evaluate each radical expression first.

$\qquad\qquad\qquad\ = -7$ Perform the subtraction.

Self Check 3
Evaluate:
a. $\sqrt{121} + \sqrt{1}$
b. $-\sqrt{9} - \sqrt{16}$

Answers **a.** 12,   **b.** −7

EXAMPLE 4   Evaluate: **a.** $6\sqrt{100}$ and   **b.** $-5\sqrt{16} + 3\sqrt{9}$.

Solution
a. We note that $6\sqrt{100}$ means $6 \cdot \sqrt{100}$.

$$6\sqrt{100} = 6(10) \qquad \text{Simplify the radical first.}$$
$$= 60 \qquad \text{Perform the multiplication.}$$

b. $-5\sqrt{16} + 3\sqrt{9} = -5(4) + 3(3) \qquad \text{Simplify each radical first.}$
$$= -20 + 9 \qquad \text{Perform the multiplications.}$$
$$= -11 \qquad \text{Perform the addition.}$$

Square roots of fractions and decimals

So far, we have found square roots of whole numbers. We can also find square roots of fractions and decimals.

EXAMPLE 5   Evaluate: **a.** $\sqrt{\dfrac{25}{64}}$ and   **b.** $\sqrt{0.81}$.

Solution
a. $\sqrt{\dfrac{25}{64}} = \dfrac{5}{8}$ because $\left(\dfrac{5}{8}\right)^2 = \dfrac{25}{64}$.

b. $\sqrt{0.81} = 0.9$ because $(0.9)^2 = 0.81$.

Using a calculator to find square roots

We can use a calculator to find square roots.

CALCULATOR SNAPSHOT   **Finding a square root**

We use the $\boxed{\sqrt{}}$ key (square root key) on a scientific calculator to find square roots.
For example, to find $\sqrt{729}$, we enter these numbers and press these keys.

729 $\boxed{\sqrt{}}$ $\boxed{ 27}$

We have found that $\sqrt{729} = 27$. To check this result, we need to square 27. This can
be done by entering 27 and pressing the $\boxed{x^2}$ key. We obtain 729. Thus, 27 is the square
root of 729.

Approximating square roots

Numbers whose square roots are whole numbers are called **perfect squares.** The perfect squares that are less than or equal to 100 are

0, 1, 4, 9, 16, 25, 36, 49, 64, 81, 100

To find the square root of a number that is not a perfect square, we can use a calculator. For example, to find $\sqrt{17}$, we enter 17 and press the square root key.

17 $\boxed{\sqrt{}}$

The display reads 4.123105626. This result is not exact, because $\sqrt{17}$ is a **nonterminating decimal** that never repeats. $\sqrt{17}$ is an irrational number. Together, the rational and the irrational numbers form the set of real numbers. If we round to the nearest thousandth, we have

$$\sqrt{17} \approx 4.123 \quad \text{Read} \approx \text{as "is approximately equal to."}$$

EXAMPLE 6   Use a scientific calculator to find each square root. Round to the nearest hundredth.

a. $\sqrt{373}$ **b.** $\sqrt{56.2}$ **c.** $\sqrt{0.0045}$

Solution

a. From the calculator, we get $\sqrt{373} \approx 19.31320792$. Rounded to the nearest hundredth, $\sqrt{373}$ is 19.31.

b. From the calculator, we get $\sqrt{56.2} \approx 7.496665926$. Rounded to the nearest hundredth, $\sqrt{56.2}$ is 7.50.

c. From the calculator, we get $\sqrt{0.0045} \approx 0.067082039$. Rounded to the nearest hundredth, $\sqrt{0.0045}$ is 0.07.

Self Check 6
Use a scientific calculator to find each square root. Round to the nearest hundredth.

a. $\sqrt{607.8}$

b. $\sqrt{0.076}$

Answers **a.** 24.65, **b.** 0.28

Section 4.6   STUDY SET   All 1-16, 23-40, 71, 78, 80

VOCABULARY   *Fill in the blanks.*

1. When we find what number is squared to obtain a given number, we are finding the square _____ of the given number.

2. Whole numbers such as 25, 36, and 49 are called _____ squares, because their square roots are whole numbers.

3. The symbol $\sqrt{}$ is called a _____ symbol. It indicates that we are to find a _____ square root.

4. The decimal number that represents $\sqrt{17}$ is a _____ decimal — it never ends.

5. In $\sqrt{26}$, the number 26 is called the _____.

6. The symbol $\approx$ means _____.

CONCEPTS   *Fill in the blanks.*

7. The square of 5 is ☐, because $(5)^2 = $ ☐.

8. The square of $\frac{1}{4}$ is ☐, because $\left(\frac{1}{4}\right)^2 = $ ☐.

9. $\sqrt{49} = 7$, because ☐ $= 49$.

10. $\sqrt{4} = 2$, because ☐ $= 4$.

11. $\sqrt{\dfrac{9}{16}} = $ ☐, because $\left(\dfrac{3}{4}\right)^2 = \dfrac{9}{16}$.

12. $\sqrt{0.16} = $ ☐, because $(0.4)^2 = 0.16$.

13. Without evaluating the following square roots, write them in order, from smallest to largest: $\sqrt{23}$, $\sqrt{11}$, $\sqrt{27}$, $\sqrt{6}$.

14. Without evaluating the following square roots, write them in order, from smallest to largest: $-\sqrt{13}$, $-\sqrt{5}$, $-\sqrt{17}$, $-\sqrt{37}$.

15. Evaluate.

   a. $\sqrt{1}$ **b.** $\sqrt{0}$

16. Multiplication can be thought of as the opposite of division. What is the opposite of finding the square root of a number?

Use a calculator.

17. a. Use a calculator to approximate $\sqrt{6}$ to the nearest tenth.

   b. Square the result from part a.

   c. Find the difference between 6 and the answer to part b.

18. a. Use a calculator to approximate $\sqrt{6}$ to the nearest hundredth.

   b. Square the result from part a.

   c. Find the difference between the answer to part b and 6.

19. Graph: $\sqrt{9}$ and $-\sqrt{5}$.

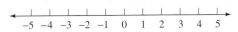

20. Graph: $-\sqrt{3}$ and $\sqrt{7}$.

21. Between what two whole numbers would each square root be located when graphed on a number line?

a. $\sqrt{19}$ **b.** $\sqrt{87}$

22. Between what two whole numbers would each square root be located when graphed on a number line?

a. $\sqrt{50}$ **b.** $\sqrt{33}$

NOTATION *Complete each solution to evaluate the expression.*

23. $-\sqrt{49} + \sqrt{64} = \boxed{} + \boxed{}$
$= 1$

24. $2\sqrt{100} - 5\sqrt{25} = 2(\boxed{}) - 5(\boxed{})$
$= \boxed{} - 25$
$= -5$

PRACTICE *Evaluate each expression without using a calculator.*

25. $\sqrt{16}$ **26.** $\sqrt{64}$

27. $-\sqrt{121}$ **28.** $-\sqrt{144}$

29. $-\sqrt{0.49}$ **30.** $-\sqrt{0.64}$

31. $\sqrt{0.25}$ **32.** $\sqrt{0.36}$

33. $\sqrt{0.09}$ **34.** $\sqrt{0.01}$

35. $-\sqrt{\dfrac{1}{81}}$ **36.** $-\sqrt{\dfrac{1}{4}}$

37. $-\sqrt{\dfrac{16}{9}}$ **38.** $-\sqrt{\dfrac{64}{25}}$

39. $\sqrt{\dfrac{4}{25}}$ **40.** $\sqrt{\dfrac{36}{121}}$

41. $5\sqrt{36} + 1$ **42.** $2 + 6\sqrt{16}$

43. $-4\sqrt{36} + 2\sqrt{4}$ **44.** $-6\sqrt{81} + 5\sqrt{1}$

45. $\sqrt{\dfrac{1}{16}} - \sqrt{\dfrac{9}{25}}$ **46.** $\sqrt{\dfrac{25}{9}} - \sqrt{\dfrac{64}{81}}$

47. $5(\sqrt{49})(-2)$ **48.** $(-\sqrt{64})(-2)(3)$

49. $\sqrt{0.04} + 2.36$ **50.** $\sqrt{0.25} + 4.7$

51. $-3\sqrt{1.44}$ **52.** $-2\sqrt{1.21}$

Use a calculator to complete each square root table. Round to the nearest thousandth when an answer is not exact.

53.

Number	Square root
1	
2	
3	
4	
5	
6	
7	
8	
9	
10	

54.

Number	Square root
10	
20	
30	
40	
50	
60	
70	
80	
90	
100	

Use a calculator to evaluate the expression.

55. $\sqrt{1,369}$ **56.** $\sqrt{841}$

57. $\sqrt{3,721}$ **58.** $\sqrt{5,625}$

Use a calculator to approximate each of the following to the nearest hundredth.

59. $\sqrt{15}$ **60.** $\sqrt{51}$

61. $\sqrt{66}$ **62.** $\sqrt{204}$

Use a calculator to approximate each of the following to the nearest thousandth.

63. $\sqrt{24.05}$ **64.** $\sqrt{70.69}$

65. $-\sqrt{11.1}$ **66.** $\sqrt{0.145}$

Use a calculator to evaluate each radical expression. If an answer is not exact, round to the nearest ten thousandth.

67. $\sqrt{24,000,201}$ **68.** $-\sqrt{4.012009}$

69. $-\sqrt{0.00111}$ **70.** $\sqrt{\dfrac{27}{44}}$

APPLICATIONS *Square roots have been used to express lengths. Solve each problem by evaluating any square roots. You may need to use a calculator. If so, round to the nearest tenth.*

71. CARPENTRY Find the length of the slanted side of each roof truss shown on the next page.

a.

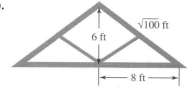

3 ft
√25 ft
← 4 ft →

b.

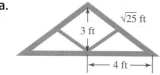

6 ft
√100 ft
← 8 ft →

72. RADIO ANTENNAS How far from the base of the antenna is each guy wire anchored to the ground? (The measurements are in feet.)

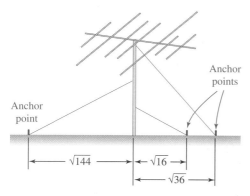

Anchor points

Anchor point

← √144 → ← √16 →
← √36 →

73. BASEBALL The illustration shows some dimensions of a major league baseball field. How far is it from home plate to second base?

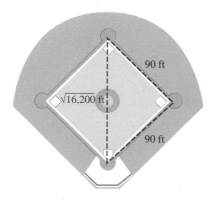

90 ft

√16,200 ft

90 ft

74. SURVEYING Use the imaginary triangles set up by a surveyor to find the length of each lake. (The measurements are in meters.)

a.

Length: √318,096

b.

Length: √93,025

75. BIG-SCREEN TELEVISION The picture screen on a television set is measured diagonally. What size screen is shown?

√1,764 in.

76. LADDERS A painter's ladder is shown. How long are the legs of the ladder?

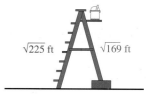

√225 ft   √169 ft

■ **WRITING**

77. When asked to find √16, a student answered 8. Explain his misunderstanding of square root.

78. Explain the difference between the square and the square root of a number.

79. What is a nonterminating decimal? Use an example in your explanation.

80. How would you check whether 17 is a square root of 289?

81. Explain why $\sqrt{-4}$ does not represent a real number.

82. Is there a difference between $-\sqrt{25}$ and $\sqrt{-25}$? Explain.

■ **REVIEW**

83. Multiply: $6.75 \cdot 12.2$.

84. Simplify: $\dfrac{-\frac{2}{3}}{8}$.

85. Evaluate: $5(-2)^2 - \dfrac{16}{4}$.

86. Divide: $5.7\overline{)18.525}$.

87. List the set of whole numbers.

88. Evaluate: $(3.4)^3$.

89. Simplify: $\left(\dfrac{2}{3}\right)^2 - \left(-\dfrac{3}{4}\right)^2$.

90. Insert the proper symbol, < or >, in the blank to make a true statement: -15 ___ -14.

The Real Numbers

A **real number** is any number that can be expressed as a decimal. The set of real numbers corresponds to all points on a number line. All of the types of numbers that we have discussed in this book are real numbers. As we have seen, the set of real numbers is made up of several subsets of numbers.

If possible, list the numbers that belong to each set. If it is not possible to list them, define the set in words.

1. Natural numbers

2. Whole numbers

3. Integers

4. Rational numbers

5. Irrational numbers

This diagram shows how the set of real numbers is made up of two distinct sets: the rational and the irrational numbers. Since every natural number is a whole number, we show the set of natural numbers included in the whole numbers. Because every whole number is an integer, the whole numbers are shown contained in the integers. Since every integer is a rational number, we show the integers included in the rational numbers.

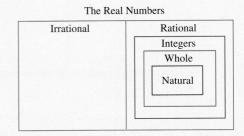

Determine whether each statement is true or false.

6. Every integer is a real number.

7. Every fraction can be written as a terminating decimal.

8. Every real number is a whole number.

9. Some irrational numbers are integers.

10. Some rational numbers are natural numbers.

11. No numbers are both rational and irrational numbers.

12. All real numbers can be graphed on a number line.

13. The set of whole numbers is a subset of the irrational numbers.

14. All decimals either terminate or repeat.

15. Every natural number is an integer.

16. List the numbers in the set $\{-2, -1.2, -\frac{7}{8}, 0, 1\frac{2}{3}, 2.75, \sqrt{23}, 10, 1.161661666\ldots\}$ that are
   a. natural numbers
   b. whole numbers
   c. integers
   d. rational numbers
   e. irrational numbers
   f. real numbers

ACCENT ON TEAMWORK

SECTION 4.1
ROUNDING

a. Find all the three-digit numbers that round to 4.7.
b. Find all the four-digit numbers that round to 8.09.
c. Find all the two-digit numbers that round to 0.0.

SECTION 4.2
VISUAL MODELS Shade a grid like the one in the illustration to compute each addition or subtraction.

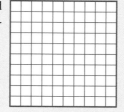

a. 0.62 + 0.24   **b.** 0.45 − 0.41
c. 0.21 + 0.29   **d.** 0.98 − 0.18
e. 0.2 + 0.17   **f.** 0.57 − 0.3

SECTION 4.3
SEQUENCES Multiplication by 2 is used to form the terms of the sequence 3, 6, 12, 24, 48, 96, ... That is, to form the second term (which is 6), we multiply the first term (which is 3) by 2. To get the third term (which is 12), we multiply the second term by 2. To get the fourth term, we multiply the third term by 2, and so on.

What multiplication is used to form the terms of each of the following sequences?

a. 0.2134, 2.134, 21.34, 213.4, 2,134, 21,340, . . .
b. 0.00005, 0.005, 0.5, 50, 5,000, . . .
c. 3, 0.9, 0.27, 0.081, 0.0243, 0.00729, . . .
d. 0.7, 0.07, 0.007, 0.0007, 0.00007, 0.000007, . . .

SECTION 4.4
Read the Think-It-Through feature on page 246 to learn how to compute a semester grade point average (GPA).

A student, taking the classes shown on the grade report below, had a semester GPA of 2.6. Determine what letter grade the student received in Spanish II.

Course no.	Course title	Units	Grade
101	Intro. Accounting	5.0	C
201	Intro. Psychology	3.0	A
102	Spanish II	4.0	?
142	Swimming	1.0	D
080	Keyboarding	2.0	C

SECTION 4.5
EQUIVALENT DECIMALS A student was asked to write several fractions as decimals. His answers, which are all incorrect, are shown below. What was he doing wrong?

$$\frac{2}{5} = 2.5 \qquad \frac{4}{15} = 3.75 \qquad \frac{3}{4} = 1.\overline{3}$$

SECTION 4.6
A SPIRAL OF ROOTS To do this project, you need a blank piece of paper, a ruler, a 3×5 index card, and a pencil. Begin by drawing a triangle with two sides 1 inch long, as shown in the illustration. Use the corner of the 3×5 card to help draw the "sharp corner" (90-degree angle) of

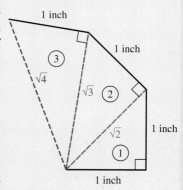

the triangle. Then draw the dashed blue line to complete the first triangle. It is $\sqrt{2}$ inches long.

Next, create a second triangle, using one side of the first triangle and drawing another side 1 inch long as shown. Complete the second triangle by drawing the dashed green line. It is $\sqrt{3}$ inches long. Draw a third triangle in a similar fashion. The dashed purple line is $\sqrt{4} = 2$ inches long. Draw a fourth triangle, a fifth triangle, and so on. If the pattern continues, what is the length of the dashed side of each new triangle?

EVALUATING SQUARE ROOTS A student was asked to evaluate the three square root expressions shown below. Examine his answers and then explain what he is doing wrong.

$$\sqrt{16} = 8 \qquad \sqrt{64} = 32 \qquad \sqrt{100} = 50$$

CHAPTER REVIEW

SECTION 4.1	*An Introduction to Decimals*

CONCEPTS

Decimal notation is used to denote part of a whole.

Expanded notation is used to show the value represented by each digit in the *decimal numeration system.*

$$5.6791 =$$
$$5 + \frac{6}{10} + \frac{7}{100} + \frac{9}{1,000} + \frac{1}{10,000}$$

To express a decimal in words, say:

1. the whole number to the left of the decimal point;

2. "and" for the decimal point;

3. the whole number to the right of the decimal point, followed by the name of the last place value column on the right.

To compare the size of two decimals, compare the digits of each decimal, column by column, working from left to right.

A decimal point and additional zeros may be written to the right of a whole number.

REVIEW EXERCISES

1. Represent the amount of the square that is shaded, using a decimal and a fraction.

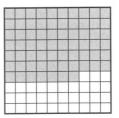

2. In the illustration, shade 0.8 of the rectangle.

3. Write 16.4523 in expanded notation.

Write each decimal in words and then as a fraction or mixed number.

4. 2.3

5. −15.59

6. 0.0601

7. 0.00001

8. Graph: 1.55, −0.8, and −2.7.

9. VALEDICTORIANS At the end of the school year, the five students listed were in the running to be class valedictorian. Rank the students in order by GPA, beginning with the valedictorian.

Name	GPA
Diaz, Cielo	3.9809
Chou, Wendy	3.9808
Washington, Shelly	3.9865
Gerbac, Lance	3.899
Singh, Amani	3.9713

10. True or false: 78 = 78.0.

Place the proper symbol <, >, or = in the blank to make a true statement.

11. 4.5 ▓ 4.6

12. −2.35 ▓ −2.53

13. 10.90 ▓ 10.9

14. 0.027894 ▓ 0.034

To round a decimal, locate the rounding digit and the test digit.

1. If the test digit is less than 5, drop it and all digits to the right of the rounding digit.

2. If it is 5 or greater, add 1 to the rounding digit and drop all digits to its right.

Round each decimal to the specified place value column.

15. 4.578: hundredths

16. 3,706.0895: thousandths

17. −0.0614: tenths

18. 88.12: tenths

Adding and Subtracting Decimals

To add (or subtract) decimals:

1. Line up their decimal points.

2. Add (or subtract) as you would with whole numbers.

3. Write the decimal point in the result directly below the decimal points in the problem.

Perform each addition or subtraction.

19. $19.5 + 34.4 + 12.8$

20. $3.4 + 6.78 + 35 + 0.008$

21. $68.47 − 53.3$

22. $45.08 − 17.37$

Evaluate each expression.

23. $−16.1 + 8.4$

24. $−4.8 − (−7.9)$

25. $−3.55 + (−1.25)$

26. $−15.1 − 13.99$

27. $−8.8 + (−7.3 − 9.5)$

28. $(5 − 0.096) − (−0.035)$

29. SALE PRICE A calculator normally sells for $52.20. If it is being discounted $3.99, what is the sale price?

30. MICROWAVE OVEN A microwave oven is shown. How tall is the window?

Multiplying Decimals

To multiply decimals:

1. Multiply as if working with whole numbers.

2. Place the decimal point in the result so that the answer has the same number of decimal places as the total number of decimal places in the factors.

To multiply a decimal by a power of 10, move the decimal point to the right the same number of places as there are zeros in the power of 10.

Perform each multiplication.

31. $(−0.6)(0.4)$

32. $2.3 \cdot 0.9$

33. $5.5(−3.1)$

34. $32.45(6.1)$

35. $(−0.003)(−0.02)$

36. $7 \cdot 0.6$

Perform each multiplication in your head.

37. $1,000(90.1452)$

38. $(−10)(−2.897)(100)$

Exponents are used to represent repeated multiplication.

Find each power.

39. $(0.2)^2$ **40.** $(-0.15)^2$ **41.** $(3.3)^2$ **42.** $(0.1)^3$

Evaluate each expression.

43. $(0.6 + 0.7)^2 - 12.3$ **44.** $3(7.8) + 2(1.1)^2$

45. Evaluate the formula $A = lw$ for $l = 32.5$ and $w = 21.3$.

To evaluate an algebraic expression, substitute specific numbers for the variables in the expression and apply the rules for the order of operations.

46. WORD PROCESSORS The Page Setup screen for a word processor is shown. Find the area that can be filled with text on an 8.5 in. × 11 in. piece of paper if the margins are set as shown.

—	PAGE SETUP
Margins	
	Preview
Top: 1.0 in.	OK
Bottom: 0.6 in.	Cancel
Left: 0.5 in.	Help
Right: 0.7 in.	

47. AUTO PAINTING A manufacturer uses a three-part process to finish the exterior of the cars it produces.

Step 1: A 0.03-inch-thick rust-prevention undercoat is applied.

Step 2: Three layers of color coat, each 0.015 inch thick, are sprayed on.

Step 3: The finish is then buffed down, losing 0.005 inch of its thickness.

What is the resulting thickness of the automobile's finish?

SECTION 4.4 — *Dividing Decimals*

To divide a decimal by a whole number:

1. Divide as if working with whole numbers.

2. Write the decimal point in the result directly above the decimal point in the dividend.

To divide by a decimal:

1. Move the decimal point in the divisor so that it becomes a whole number.

2. Move the decimal point in the dividend the same number of places to the right.

3. Use the process for dividing a decimal by a whole number.

Perform each division.

48. $12\overline{)15}$ **49.** $-41.8 \div 4$ **50.** $\dfrac{-29.67}{-23}$ **51.** $24.618 \div 6$

Perform each division.

52. $12.47 \div (-4.3)$ **53.** $\dfrac{0.0742}{1.4}$

54. $\dfrac{15.75}{0.25}$ **55.** $\dfrac{-0.03726}{-0.046}$

Divide and round each result to the nearest tenth.

56. $78.98 \div 6.1$ **57.** $\dfrac{-5.338}{0.008}$

58. Evaluate the formula $C = \dfrac{5}{9}(F - 32)$ for $F = 68.4$ and round to the nearest hundredth.

59. THANKSGIVING DINNER The cost of purchasing the ingredients for a Thanksgiving turkey dinner for a family of 5 was $41.70. What was the cost of the dinner per person?

To divide a decimal by a power of 10, move the decimal point to the left the same number of places as there are zeros in the power of 10.

Perform each division in your head.

60. $89.76 \div 100$

61. $\dfrac{0.0112}{-10}$

62. Evaluate the numerical expression: $\dfrac{(1.4)^2 + 2(4.6)}{0.5 + 0.3}$.

63. SERVING SIZE The illustration shows the package labeling on a box of children's cereal. Use the information given to find the number of servings.

Nutrition Facts	
Serving size	1.1 ounce
Servings per container	?
Package weight	15.5 ounces

64. TELESCOPES To change the position of a focusing mirror on a telescope, an adjustment knob is used. The mirror moves 0.025 inch with each revolution of the knob. The mirror needs to be moved 0.2375 inch to improve the sharpness of the image. How many revolutions of the adjustment knob does this require?

SECTION 4.5	*Fractions and Decimals*

To write a fraction as a decimal, divide the numerator by the denominator.

Write each fraction in decimal form.

65. $\dfrac{7}{8}$ **66.** $-\dfrac{2}{5}$ **67.** $\dfrac{9}{16}$ **68.** $\dfrac{3}{50}$

We obtain either a *terminating* or a *repeating* decimal when using division to write a fraction as a decimal.

Write each fraction in decimal form. Use an overbar.

69. $\dfrac{6}{11}$ **70.** $-\dfrac{2}{3}$

An overbar can be used instead of the three dots . . . to represent the repeating pattern in a repeating decimal.

Write each fraction in decimal form. Round to the nearest hundredth.

71. $\dfrac{19}{33}$ **72.** $\dfrac{31}{30}$

Place the proper symbol $<$ or $>$ in the blank to make a true statement.

73. $\dfrac{13}{25}$ ▢ 0.499 **74.** $-0.\overline{26}$ ▢ $-\dfrac{4}{15}$

75. Graph: $1\frac{1}{8}, -\frac{1}{3}, 2\frac{3}{4}$, and $-\frac{9}{10}$.
Label each point using the decimal equivalent of the fraction or mixed number.

Evaluate each numerical expression. Find the exact answer.

76. $\dfrac{1}{3} + 0.4$ **77.** $\dfrac{4}{5}(-7.8)$

78. $\dfrac{1}{2}(9.7 + 8.9)(10)$ **79.** $\dfrac{1}{3}(3.14)(3)^2(4.2)$

80. Evaluate $\dfrac{4}{3}(3.14)(2)^3$. Round the result to the nearest hundredth.

81. ROADSIDE EMERGENCY In case of trouble, truckers carry reflectors to be placed on the highway shoulder to warn approaching cars of a stalled vehicle. What is the area of one of these triangular reflectors?

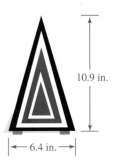

10.9 in.

← 6.4 in. →

A number is a *square root* of a second number if the square of the first is equal to the second.

82. Fill in the blanks.

  a. The symbol $\sqrt{}$ is called a _____ symbol.

  b. $\sqrt{64} = 8$ because $\boxed{} = 64$.

A *radical symbol* $\sqrt{}$ is used to indicate a positive square root. The square root of a *perfect square* is a whole number.

Evaluate each expression without using a calculator.

83. $\sqrt{49}$ **84.** $-\sqrt{16}$ **85.** $\sqrt{100}$ **86.** $\sqrt{0.09}$

87. $\sqrt{\dfrac{64}{25}}$ **88.** $\sqrt{0.81}$ **89.** $-\sqrt{\dfrac{1}{36}}$ **90.** $\sqrt{0}$

91. Between what two whole numbers would $\sqrt{83}$ be located when graphed on a number line?

A square root can be approximated using a calculator.

92. Use a calculator to find $\sqrt{11}$ and round to the nearest tenth. Now square the approximation. How close is it to 11?

93. Graph each square root: $\sqrt{3}, -\sqrt{2},$ and $\sqrt{0}$.

$-5 \quad -4 \quad -3 \quad -2 \quad -1 \quad 0 \quad 1 \quad 2 \quad 3 \quad 4 \quad 5$

When evaluating an expression containing square roots, treat a radical as you would a power when applying the rules for the order of operations.

Evaluate each expression without using a calculator.

94. $-3\sqrt{100}$ **95.** $5\sqrt{0.25}$

96. $-3\sqrt{49} - \sqrt{36}$ **97.** $\sqrt{\dfrac{9}{100}} + \sqrt{1.44}$

 Use a calculator to find each square root to the nearest hundredth.

98. $\sqrt{19}$ **99.** $\sqrt{59}$

1. Express the amount of the square that is shaded, using a fraction and a decimal.

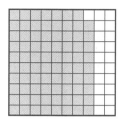

2. WATER PURITY A county health department sampled the pollution content of tap water in five cities, with the results shown. Rank the cities in order, from dirtiest tap water to cleanest.

City	Pollution, parts per million
Monroe	0.0909
Covington	0.0899
Paston	0.0901
Cadia	0.0890
Selway	0.1001

3. Write each decimal in words and then as a fraction or mixed number. **Do not simplify the fraction.**

a. SKATEBOARDING Gary Hardwick of Carlsbad, California, set the skateboard speed record of 62.55 mph in 1998.

b. MONEY A dime weighs 0.08013 ounce.

4. Round to the nearest thousandth: 33.0495.

5. SKATING RECEIPTS At an ice-skating complex, receipts on Friday were $30.25 for indoor skating and $62.25 for outdoor skating. On Saturday, the corresponding amounts were $40.50 and $75.75. Find the total receipts for the two days.

6. Perform each operation in your head.

a. $567.909 \div 1{,}000$ **b.** $0.00458 \cdot 100$

7. FAULT LINES After an earthquake, geologists found that the ground on the west side of the fault line had dropped 0.83 inch. The next week, a strong aftershock caused the same area to sink 0.19 inch deeper. How far did the ground on the west side of the fault drop because of the seismic activity?

8. Perform each operation.

a. $2 + 4.56 + 0.89 + 3.3$

b. $45.2 - 39.079$

c. $(0.32)^2$

d. $-6.7(-2.1)$

9. CENTRAL PARK Central Park, which lies in the middle of Manhattan, is the city's best-known park. If it is 2.5 miles long and 0.5 mile wide, what is its area?

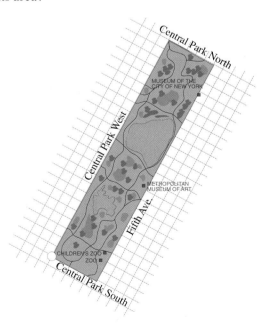

10. **TELEPHONE BOOKS** To print a telephone book, 565 sheets of paper were used. If the book is 2.3 inches thick, what is the thickness of each sheet of paper? (Round to the nearest thousandth of an inch.)

11. Evaluate: $4.1 - (3.2)(0.4)^2$.

12. Write each fraction as a decimal.

    a. $\dfrac{17}{50}$ b. $\dfrac{5}{12}$

13. Perform the division and round to the nearest hundredth: $\dfrac{12.146}{-5.3}$.

14. Divide: $11\overline{)13}$.

15. Graph: $\dfrac{3}{8}$ and $-\dfrac{4}{5}$. Label each point using the decimal equivalent of the given fractions.

16. Find the exact answer: $\dfrac{2}{3} + 0.7$.

17. **CHEMISTRY** In a lab experiment, a chemist mixed three compounds together to form a mixture weighing 4.37 g. Later, she discovered that she had forgotten to record the weight of compound C in her notes. Find the weight of compound C used in the experiment.

	Weight
Compound A	1.86 g
Compound B	2.09 g
Compound C	?
Mixture total	4.37 g

18. **WEDDING COSTS** A printer charges a setup fee of $24 and then 95 cents for each wedding announcement printed (tax included). If a couple has budgeted $100 for printing costs, how many announcements can they have made?

19. Graph: $\sqrt{2}$ and $-\sqrt{5}$.

    ```
    ←―+―+―+―+―+―+―+―+―+―+―→
     -5 -4 -3 -2 -1  0  1  2  3  4  5
    ```

20. Fill in the blank: The $\sqrt{}$ symbol is called a _____ symbol.

21. Fill in the blank: $\sqrt{144} = 12$, because ▉ $= 144$.

22. Evaluate each expression.

    a. $-2\sqrt{25} + 3\sqrt{49}$ b. $\sqrt{\dfrac{1}{36}} - \sqrt{\dfrac{1}{25}}$

23. Insert the proper symbol $<$ or $>$ to make a true statement.

    a. $-6.78 \ \_\_ \ -6.79$ b. $\dfrac{3}{8} \ \_\_ \ 0.3$

    c. $\sqrt{\dfrac{16}{81}} \ \_\_ \ \dfrac{16}{81}$ d. $0.4\overline{5} \ \_\_ \ 0.45$

24. Evaluate each square root.
    a. $-\sqrt{0.04}$ b. $\sqrt{1.69}$

25. Although the decimal 3.2999 contains more digits than 3.3, it is smaller than 3.3. Explain why this is so.

26. What is a repeating decimal? Give an example.

CHAPTERS 1–4   CUMULATIVE REVIEW EXERCISES

1. THE EXECUTIVE BRANCH The annual salaries for the President and the Vice President of the United States are $400,000 and $203,000, respectively. How much more money does the President make than the Vice President during a four-year term?

2. Use 3, 4, and 5 to express the associative property of addition.

3. Divide: $43\overline{)1,203}$.

4. How many thousands are there in one million?

5. Find the prime factorization of 220.

6. List the factors of 20, from smallest to largest.

7. List the set of whole numbers.

8. Add: $-8 + (-5)$.

9. Fill in the blank: Subtraction is the same as _____ the opposite.

10. Complete the solution to evaluate the expression.

$$
\begin{aligned}
(-6)^2 - 2(5 - 4 \cdot 2) &= (-6)^2 - 2(5 - \boxed{}) \\
&= (-6)^2 - 2(\boxed{}) \\
&= \boxed{} - 2(-3) \\
&= 36 - (\boxed{}) \\
&= 36 + \boxed{} \\
&= 42
\end{aligned}
$$

11. Consider the division statement $\dfrac{-15}{-5} = 3$. What is its related multiplication statement?

12. Find the power: $(-1)^5$.

13. Evaluate: $|-7(5)|$.

14. What is the opposite of -102?

15. CHECKING ACCOUNTS After a deposit of $995, a student's checking account was still $105 overdrawn. What was the balance in the account before the deposit?

16. What fraction of the stripes in the flag are white?

17. Although the fractions listed below look different, they all represent the same value. What concept does this illustrate?

$$\frac{1}{2} = \frac{2}{4} = \frac{3}{6} = \frac{4}{8} = \frac{5}{10} = \frac{6}{12}$$

18. Simplify: $\dfrac{90}{126}$.

Perform the operations.

19. $\dfrac{3}{8} \cdot \dfrac{7}{16}$

20. $-\dfrac{15}{8} \div 10$

21. $\dfrac{4}{3} + \dfrac{2}{7}$

22. $-4\dfrac{1}{4}\left(-4\dfrac{1}{2}\right)$

23. $76\dfrac{1}{6} - 49\dfrac{7}{8}$

24. $\dfrac{\dfrac{5}{27}}{-\dfrac{5}{9}}$

25. KITES Find the area of the kite.

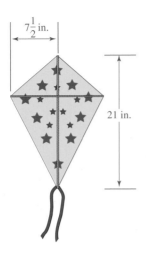

$7\frac{1}{2}$ in.

21 in.

26. Graph each number in the set:

$\{-3\frac{1}{4}, 0.75, -1.5, -\frac{9}{8}, 3.8, \sqrt{4}\}$.

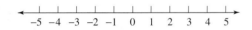

27. GLASS Some electronic and medical equipment uses glass that is only 0.00098 inch thick. Round this number to the nearest thousandth.

28. Place the proper symbol $>$ or $<$ in the box to make the statement true.

356.1978 ▨ 356.22

Perform the operations.

29. $-1.8(4.52)$

30. $\dfrac{-21.28}{-3.8}$

31. $56.012(100)$

32. $\dfrac{0.897}{10,000}$

33. Evaluate: $-9.1 - (-6.05 - 51)$.

34. WEEKLY SCHEDULES Determine the number of hours during a week that an adult spends, on average, watching television.

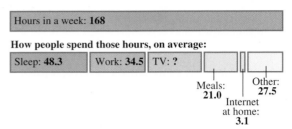

Hours in a week: **168**

How people spend those hours, on average:

Sleep: **48.3** | Work: **34.5** | TV: **?**

Meals: **21.0**

Internet at home: **3.1**

Other: **27.5**

Source: National Sleep Foundation and the U.S. Bureau of Statistics

35. LITERATURE The novel *Fahrenheit 451*, by Ray Bradbury, is a story about censorship and book burning. Use the formula $C = \dfrac{5}{9}(F - 32)$ to convert $451°$ F to degrees Celsius by replacing F with 451. Round to the nearest tenth of a degree.

36. Write $\dfrac{5}{12}$ as a decimal. Use an overbar.

37. CONCESSIONAIRES At a ballpark, a vendor is paid $22 a game plus 35¢ for each bag of peanuts she sells. If she sells 80 bags of peanuts, how much money will she earn working at the ballpark?

Evaluate each expression.

38. $\sqrt{49}$

39. $\sqrt{\dfrac{625}{16}}$

40. $-4\sqrt{36} + 2\sqrt{81}$

CHAPTER 5

Percent

Getty Images

Percents are commonly used to present numerical information. The word percent comes from the Latin phrase *per centum*, which means parts per one hundred. Many instructors use a percent grading scale when evaluating their students' work. For example, if a student correctly answers 85 out of 100 true/false questions on a history exam, the student's grade on the exam can be expressed as 85%.

   To learn more about percent and its many applications, visit The Learning Equation on the Internet at http://tle.brookscole.com. (The log-in instructions are in the Preface.) For Chapter 5, the online lesson is:

• *TLE* Lesson 11: Percent

Check Your Knowledge

1. Percent means parts per _____ _____.

2. In the statement "40 is 50% of 80," 40 is the _____, 50% is the _____, and 80 is the _____.

3. The difference between the original price and the sale price of an item is called the _____.

4. In banking, the original amount of money borrowed or deposited is the _____.

5. Change each fraction to a decimal and to a percent.

   a. $\frac{3}{4}$ **b.** $\frac{5}{8}$ **c.** $\frac{29}{20}$

6. Change each decimal to a percent and to a fraction.

   a. 0.35   **b.** 3.98   **c.** 0.105

7. Change each percent to a decimal and to a fraction.

   a. 25%   **b.** 200%   **c.** 0.5%

8. If a glass is 65% full, what percent of the glass is empty?

9. Change $\frac{2}{3}$ to a percent. Round to the nearest tenth.

10. Find the exact percent equivalent for each fraction.

    a. $\frac{3}{8}$ **b.** $\frac{1}{3}$

11. What is 65% of 500?

12. What percent of 200 is 34?

13. 13 is 25% of what number?

14. What percent of 50 is 125?

15. A pen normally costs $14.95. The sale price is 20% off the normal price. What are the amount of the discount and the sale price?

16. If the sales tax rate is 8.25%, what is the price of an item when the sales tax amount is $2.47? Round to the nearest cent.

17. Find the simple interest on a $2,000.00 savings account for 1 year if the interest rate is 2.3%.

18. ▦ Find the account balance after 2 years on a $1,500.00 investment with earnings of 5.6% compounded quarterly.

19. Only 3 of 46 parking spaces in a parking lot were not taken. What percent of the parking lot spaces were filled? Round to the nearest one percent.

20. If Sheila borrows $1,000 with simple annual interest at 18%, what will be the payoff amount if she pays off the loan after 3 months?

21. If Leslie wants to tip 15% on a meal that cost $35.40, what will the total cost be, including the tip?

22. A quiz has 15 questions. Assuming that the questions are weighted equally, how many questions must George answer correctly to score at least 85%?

Study Skills Workshop

HOMEWORK

Doing a thorough job with your homework is one of the most important steps you can take to be successful in your class. Sitting in class and listening to lectures will help you to place concepts in short-term memory, but in order to do well on tests and in subsequent classes, you want to put these concepts in long-term memory. When done correctly, homework assignments will help with this.

Are You Giving Yourself Enough Time? Recall that in your first Study Skills Workshop assignment you made a study calendar that scheduled 2 hours for study and homework for every hour that you spent in lecture. If you are not keeping this schedule, make changes to ensure that you can spend enough time outside of class to learn new material. Make sure that you spread this time over a period of at least five days per week, rather than in one or two long sessions.

Before You Start Your Homework Problems. In the previous Study Skills Workshop, your assignment was to rework your notes. Always rework the notes that relate to your homework before starting your assignment. After reworking your notes, read the sections in your textbook that pertain to the homework problems, looking especially at the examples. With a pencil in hand, rework the examples, trying to understand each step. Keep a list of anything that you don't understand, both in your notes and in the textbook examples.

Doing Your Homework Problems. Once you have reviewed your notes and the textbook examples, you should be able to successfully manage the bulk of your homework assignment easily. When working on your homework, keep your textbook and notes handy so that you can refer to them if necessary. If you have trouble with a homework question, look through your textbook and notes to see if you can identify an example that is similar to the homework question. See if you can apply the same steps to your homework problem. If there are places where you get stuck, add these to your questions list.

Before Your Homework Is Due. At least one day before your assignment is due, seek help with your question list. You can contact a classmate for assistance, make an appointment with a tutor, or visit your instructor during office hours. Make sure to bring your list and try to pinpoint exactly where in the process you got stuck.

ASSIGNMENT

1. Review your study calendar. Are you following it? If not, what changes can you make to adhere to the rule: two hours of homework and study for every hour of lecture?
2. Find five homework problems in your assignment that are similar to the examples in your textbook. List each problem along with its matching example. Were there any homework problems in your assignment that didn't have an example that was similar?
3. Make a list of homework problems that you had questions about or didn't know how to do. See your tutor or your instructor during office hours with your list of problems and ask one of them to work through these problems with you.

Percents are based on the number 100. They offer us a way to measure and describe many situations in our daily lives.

5.1 Percents, Decimals, and Fractions

- The meaning of percent
- Changing a percent to a decimal
- Changing a fraction to a percent
- Changing a percent to a fraction
- Changing a decimal to a percent

Percents are a popular way to present numeric information. Stores use them to advertise discounts, manufacturers use them to describe the contents of their products, and banks use them to list interest rates for loans and savings accounts. Newspapers are full of statistics presented in percent form. In this section, we introduce percents and show how fractions, decimals, and percents are interrelated.

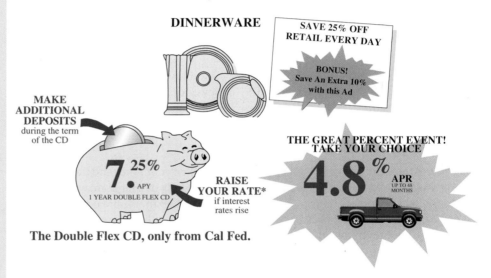

The meaning of percent

A percent tells us the number of parts per one hundred. You can think of a percent as the *numerator* of a fraction that has a denominator of 100.

> **Percent**
>
> **Percent** means parts per one hundred.

In Figure 5-1, there are 100 equal-sized squares, and 93 are shaded. Thus, $\frac{93}{100}$ or 93 percent of the figure is shaded. The word *percent* can be written using the symbol %, so 93% of Figure 5-1 is shaded.

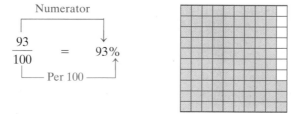

FIGURE 5-1

If the entire grid in Figure 5-1 had been shaded, we would say that 100 out of the 100 squares, or 100%, was shaded. Using this fact, we can determine what percent of the figure is *not* shaded by subtracting the percent of the figure that is shaded from 100%.

$$100\% - 93\% = 7\%$$

So 7% of Figure 5-1 is not shaded.

■ Changing a percent to a fraction

To change a percent to an equivalent fraction, we use the definition of percent.

Changing a percent to a fraction

To change a percent to a fraction, drop the % symbol and write the given number over 100. Then simplify the fraction, if possible.

EXAMPLE 1   Earth. The chemical makeup of Earth's atmosphere is 78% nitrogen, 21% oxygen, and 1% other gases. Write each percent as a fraction.

Solution   We begin with nitrogen.

$$78\% = \frac{78}{100} \qquad \text{Use the definition of percent: 78\% means 78 parts per one hundred.}$$
$$\text{This fraction can be simplified.}$$

$$= \frac{\overset{1}{\cancel{2}} \cdot 39}{\underset{1}{\cancel{2}} \cdot 50} \qquad \text{Factor 78 as } 2 \cdot 39 \text{ and 100 as } 2 \cdot 50.$$
$$\text{Divide out the common factor of 2.}$$

$$= \frac{39}{50}$$

Nitrogen makes up $\frac{78}{100}$, or $\frac{39}{50}$, of Earth's atmosphere.

Oxygen makes up 21%, or $\frac{21}{100}$, of Earth's atmosphere. Other gases make up 1%, or $\frac{1}{100}$, of the atmosphere.

EXAMPLE 2   Unions. In 2003, 12.9% of the U.S. labor force belonged to a union. Write this percent as a fraction.

Solution

$$12.9\% = \frac{12.9}{100} \qquad \text{Drop the \% symbol and write 12.9 over 100.}$$

$$= \frac{12.9 \cdot 10}{100 \cdot 10} \qquad \begin{array}{l}\text{To obtain a whole number in the numerator, multiply by 10.}\\ \text{This will move the decimal point 1 place to the right. Multiply}\\ \text{the denominator by 10 as well.}\end{array}$$

$$= \frac{129}{1,000} \qquad \begin{array}{l}\text{Perform the multiplication in the numerator and in the}\\ \text{denominator.}\end{array}$$

Thus, $12.9\% = \frac{129}{1,000}$. This means that 129 out of every 1,000 workers in the U.S. labor force belonged to a union in 2003.

Self Check 3

Write $83\frac{1}{3}\%$ as a fraction.

EXAMPLE 3   Write $66\frac{2}{3}\%$ as a fraction.

Solution

$$66\frac{2}{3}\% = \frac{66\frac{2}{3}}{100} \qquad \text{Drop the \% symbol and write } 66\frac{2}{3} \text{ over 100.}$$

$$= 66\frac{2}{3} \div 100 \qquad \text{The fraction bar indicates division.}$$

$$= \frac{200}{3} \cdot \frac{1}{100} \qquad \text{Change } 66\frac{2}{3} \text{ to a mixed number and then multiply by the reciprocal of 100.}$$

$$= \frac{2 \cdot 100 \cdot 1}{3 \cdot 100} \qquad \text{Multiply the numerators and the denominators. Factor 200 as } 2 \cdot 100.$$

$$= \frac{2 \cdot \overset{1}{\cancel{100}} \cdot 1}{3 \cdot \underset{1}{\cancel{100}}} \qquad \text{Divide out the common factor of 100.}$$

Answer $\dfrac{5}{6}$

$$= \frac{2}{3} \qquad \text{Multiply in the numerator and in the denominator.}$$

Changing a percent to a decimal

To write a percent as a decimal, recall that a percent can be written as a fraction with denominator 100 and that a denominator of 100 indicates division by 100.

Consider 14.25%, which means 14.25 parts per 100.

$$14.25\% = \frac{14.25}{100} \qquad \text{Use the definition of percent: write 14.25 over 100.}$$

$$= 14.25 \div 100 \qquad \text{The fraction bar indicates division.}$$

$$= 0.14\,25 \qquad \text{To divide a decimal by 100, move the decimal point 2 places to the left.}$$

$$14.25\% = 0.1425$$

This example suggests the following procedure.

> **Changing a percent to a decimal**
>
> To change a percent to a decimal, drop the % symbol and divide by 100 by moving the decimal point 2 places to the left.

Self Check 4

What percent of all music sold is produced on LPs (long-playing vinyl record albums)? Write the percent as a decimal.

EXAMPLE 4
The music industry.
Figure 5-2 shows that the compact disc has become the format of choice among most consumers. What percent of all music sold is produced on CDs? Write the percent as a decimal.

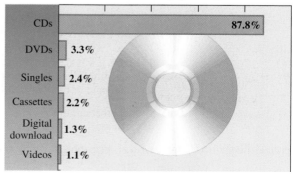

2003 Music Sales, by Format

CDs	87.8%
DVDs	3.3%
Singles	2.4%
Cassettes	2.2%
Digital download	1.3%
Videos	1.1%

Source: *The Recording Industry Association of America*

FIGURE 5-2

Solution From the graph, we see that 87.8% of all music sold is produced on CDs. To write 87.8% as a decimal, we proceed as follows.

$$87.8\% = .87\,8 \qquad \text{Drop the percent symbol and divide by 100 by moving the decimal point 2 places to the left.}$$

$$= 0.878 \qquad \text{Write a zero to the left of the decimal point.}$$

Answer 0.5%, or 0.005

EXAMPLE 5   Write 310% as a decimal.

Self Check 5
Write 600% as a decimal.

Solution The whole number 310 has an understood decimal point to the right of 0.

$$310\% = 310.0\% \qquad \text{Write a decimal point and a zero to the right of 310.}$$

$$= 3.10\,0 \qquad \text{Drop the \% symbol and divide by 100 by moving the decimal point 2 places to the left.}$$

$$= 3.100$$

$$= 3.1 \qquad \text{Drop the unnecessary zeros to the right of the 1.}$$

Answer 6

EXAMPLE 6   States.   The population of the state of Oregon is approximately $1\frac{1}{4}\%$ of the population of the United States. Write this percent as a decimal.

Self Check 6
Write $15\frac{3}{4}\%$ as a decimal.

Solution To change a percent to a decimal, we drop the percent symbol and divide by 100 by moving the decimal point 2 places to the left. In this case, however, there is no decimal point in $1\frac{1}{4}\%$ to move. Since $1\frac{1}{4} = 1 + \frac{1}{4}$, and since the decimal equivalent of $\frac{1}{4}$ is 0.25, we can write $1\frac{1}{4}\%$ in an equivalent form as 1.25%.

$$1\frac{1}{4}\% = 1.25\% \qquad \text{Write } 1\frac{1}{4} \text{ as } 1.25.$$

$$= 0.01\,25 \qquad \text{Drop the \% symbol and divide by 100 by moving the decimal point 2 places to the left.}$$

$$= 0.0125$$

Answer 0.1575

■ Changing a decimal to a percent

To change a percent to a decimal, we drop the % symbol and move the decimal point 2 places to the left. To write a decimal as a percent, we do the opposite: we move the decimal point 2 places to the right and insert a % symbol.

> **Changing a decimal to a percent**
>
> To change a decimal to a percent, multiply the decimal by 100 by moving the decimal point 2 places to the right, and then insert a % symbol.

EXAMPLE 7   Geography.   Land areas make up 0.291 of Earth's surface. Write this decimal as a percent.

Self Check 7
Write 0.5343 as a percent.

Solution

$$0.291 = 0\,29.1\% \qquad \text{Multiply the decimal by 100 by moving the decimal point 2 places to the right, and then insert a \% symbol.}$$

$$= 29.1\%$$

Answer 53.43%

Changing a fraction to a percent

We use a two-step process to change a fraction to a percent. First, we write the fraction as a decimal. Then we change that decimal to a percent.

$$\text{fraction} \quad \longrightarrow \quad \text{decimal} \quad \longrightarrow \quad \text{percent}$$

> **Changing a fraction to a percent**
>
> To change a fraction to a percent:
>
> **1.** Write the fraction as a decimal by dividing its numerator by its denominator.
> **2.** Multiply the decimal by 100 by moving the decimal point 2 places to the right.
> **3.** Insert a % symbol.

Self Check 8

Write $\frac{7}{8}$ as a percent.

EXAMPLE 8 **Television.** The highest-rated television show of all time was a special episode of "M*A*S*H" that aired February 28, 1983. Surveys found that three out of every five American households watched this show. Express the rating as a percent.

Solution Three out of five can be expressed as $\frac{3}{5}$. We need to change this fraction to a decimal.

$$
\begin{array}{r}
0.6 \\
5\overline{)3.0} \\
\underline{3\,0} \\
0
\end{array}
$$

Write 3 as 3.0 and then divide the numerator by the denominator.

$$\frac{3}{5} = 0.6$$ The result is a terminating decimal.

$$0.6 = 0\underset{\curvearrowright}{\,60}.\%$$ Write a placeholder 0 to the right of the 6. Multiply the decimal by 100 by moving the decimal point 2 places to the right, and then insert a % symbol.

$$= 60\%$$

Answer 87.5%

60% of American households watched the special episode of "M*A*S*H."

In Example 8, the result of the division was a terminating decimal. Sometimes when we change a fraction to a decimal, the result of the division is a repeating decimal.

EXAMPLE 9 Write $\frac{5}{6}$ as a percent.

Solution The first step is to change $\frac{5}{6}$ to a decimal.

$$
\begin{array}{r}
0.8333 \\
6\overline{)5.0000} \\
\underline{4\,8} \\
20 \\
\underline{18} \\
20 \\
\underline{18} \\
20
\end{array}
$$

Write 5 as 5.0000. Divide the numerator by the denominator.

$$\frac{5}{6} = 0.8333\ldots$$ The result is a repeating decimal.

$= 0\,83.33\ldots\%$ Change 0.8333... to a percent. Multiply the decimal by 100 by moving the decimal point 2 places to the right, and then insert a % symbol.

$= 83.33\ldots\%$ 83.333... is a repeating decimal.

We must now decide whether we want an approximation or an exact answer. For an approximation, we can round 83.333...% to a specific place value. For an exact answer, we can represent the repeating part of the decimal using an equivalent fraction.

Approximation

$\dfrac{5}{6} = 83.33\ldots\%$

$\approx 83.3\%$ Round to the nearest tenth.

$\dfrac{5}{6} \approx 83.3\%$

Exact answer

$\dfrac{5}{6} = 83.3333\ldots\%$

$= 83\dfrac{1}{3}\%$ Use the fraction $\frac{1}{3}$ to represent .333....

$\dfrac{5}{6} = 83\dfrac{1}{3}\%$

Some percents occur so frequently that it is useful to memorize their fractional and decimal equivalents.

Percent	Decimal	Fraction
1%	0.01	$\dfrac{1}{100}$
10%	0.1	$\dfrac{1}{10}$
20%	0.2	$\dfrac{1}{5}$
25%	0.25	$\dfrac{1}{4}$

Percent	Decimal	Fraction
$33\frac{1}{3}\%$	0.3333...	$\dfrac{1}{3}$
50%	0.5	$\dfrac{1}{2}$
$66\frac{2}{3}\%$	0.6666...	$\dfrac{2}{3}$
75%	0.75	$\dfrac{3}{4}$

Section 5.1   STUDY SET

A111–70, 74, 77–79, 83, 87–91

VOCABULARY   *Fill in the blanks.*

1. _____ means parts per one hundred.

2. When we change a fraction to a decimal, the result is either a _____ or a repeating decimal.

CONCEPTS   *Fill in the blanks.*

3. To write a percent as a fraction, drop the % symbol and write the given number over _____. Then _____ the fraction, if possible.

4. To change a percent to a decimal, drop the % symbol and divide by 100 by moving the decimal point 2 places to the _____.

5. To change a decimal to a percent, multiply the decimal by 100 by moving the decimal point 2 places to the _____, and then insert a % symbol.

6. To write a fraction as a percent, first write the fraction as a _____. Then multiply the decimal by 100 by moving the decimal point 2 places to the _____, and insert a % symbol.

NOTATION

7. **a.** See the illustration on the next page. Express the amount of the figure that is shaded as a decimal, a percent, and a fraction.

   b. What percent of the figure is not shaded?

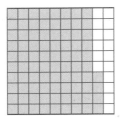

8. In the illustration, each set of 100 squares represents 100%. What percent is shaded?

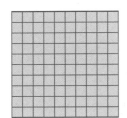 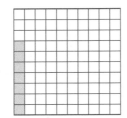

PRACTICE   *Change each percent to a fraction. Simplify when necessary.*

9. 17%

10. 31%

11. 5%

12. 4%

13. 60%

14. 40%

15. 125%

16. 210%

17. $\frac{2}{3}$%

18. $\frac{1}{5}$%

19. $5\frac{1}{4}$%

20. $6\frac{3}{4}$%

21. 0.6%

22. 0.5%

23. 1.9%

24. 2.3%

Change each percent to a decimal.

25. 19%

26. 83%

27. 6%

28. 2%

29. 40.8%

30. 34.2%

31. 250%

32. 600%

33. 0.79%

34. 0.01%

35. $\frac{1}{4}$%

36. $8\frac{1}{5}$%

Change each decimal to a percent.

37. 0.93

38. 0.44

39. 0.612

40. 0.727

41. 0.0314

42. 0.0021

43. 8.43

44. 7.03

45. 50

46. 3

47. 9.1

48. 8.7

Change each fraction to a percent.

49. $\frac{17}{100}$

50. $\frac{29}{100}$ = 29%

51. $\frac{4}{25}$

52. $\frac{47}{50}$ = 94%

53. $\frac{2}{5}$

54. $\frac{21}{50}$ = 42%

55. $\frac{21}{20}$

56. $\frac{33}{20}$ = 165%

57. $\frac{5}{8}$

58. $\frac{3}{8}$ = 37.5%

59. $\frac{3}{16}$

60. $\frac{1}{32}$ = 3.125%

Find the exact equivalent percent for each fraction.

61. $\frac{2}{3}$

62. $\frac{1}{6}$ = 16⅔%

63. $\frac{1}{12}$

64. $\frac{4}{3}$ = 133⅓%

Express each of the given fractions as a percent. Round to the nearest hundredth of a percent.

65. $\frac{1}{9}$

66. $\frac{2}{3}$ = 66.67%

67. $\frac{5}{9}$

68. $\frac{7}{3}$ = 233.33%

APPLICATIONS

69. THE U.N. SECURITY COUNCIL The United Nations has 191 members. The United States, Russia, the United Kingdom, France, and China, along with ten other nations, make up the Security Council.

   a. What fraction of the members of the United Nations belong to the Security Council? $\frac{15}{191}$

   b. Write your answer to part a in percent form. (Round to the nearest one percent.)   8%

70. ECONOMIC FORECASTS One economic indicator of the national economy is the number of orders placed by manufacturers. One month, the number of orders rose one-fourth of 1 percent.

   a. Write this using a % symbol.   ¼%

   b. Express it as a fraction. $\frac{1}{400}$

   c. Express it as a decimal.   .0025

71. PIANO KEYS Of the 88 keys on a piano, 36 are black.

   a. What fraction of the keys are black?

   b. What percent of the keys are black? (Round to the nearest 1 percent.)

72. INTEREST RATES Write as a decimal the interest rate associated with each of these accounts.

    a. Home loan: 7.75% $= 0.0775$

    b. Savings account: 5% $= 0.05$

    c. Credit card: 14.25% $= 0.1425$

73. THE HUMAN SPINE The human spine consists of a group of bones (vertebrae) as shown.

    a. What fraction of the vertebrae are lumbar?

    b. What percent of the vertebrae are lumbar? (Round to the nearest 1 percent.)

    c. What percent of the vertebrae are cervical? (Round to the nearest 1 percent.)

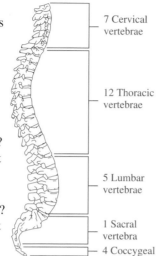

7 Cervical vertebrae

12 Thoracic vertebrae

5 Lumbar vertebrae

1 Sacral vertebra

4 Coccygeal vertebrae

74. REGIONS OF THE COUNTRY The continental United States is divided into seven regions as shown.

    a. What percent of the 50 states are in the Rocky Mountain region? $= 12\%$

    b. What percent of the 50 states are in the Midwestern region? $= 24\%$

    c. What percent of the 50 states are not located in any of the seven regions shown here? 4%

Midwestern States

New England States

Rocky Mountain States

Middle Atlantic States

Southwestern States

Southern States

Pacific Coast States

75. STEEP GRADES Sometimes, signs are used to warn truckers when they are approaching a steep grade on the highway. For a 5% grade, how many feet does the road rise over a 100-foot run?

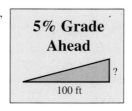

5% Grade Ahead

100 ft

76. COMPANY LOGOS In the illustration, what part of the company's logo is shaded red? Express your answer as a percent, a fraction, and a decimal. Do not round.

Recycling Industries Inc.

77. SOAP Ivory soap claims to be $99\frac{44}{100}\%$ pure. Write this percent as a decimal.

78. DRUNK DRIVING In most states, it is illegal to drive with a blood alcohol concentration of 0.08% or higher. Change this percent to a fraction. Do not simplify. Explain what the numerator and the denominator of the fraction represent. $\frac{8}{10000}$

79. BASKETBALL In the standings, we see that Chicago has won 60 of 67, or $\frac{60}{67}$, of its games. In what form is the team's winning percentage presented in the newspaper? Express it as a percent.

Eastern Conference			
Team	**W**	**L**	**Pct.**
Chicago	60	7	.896

80. WON-LOST RECORDS In sports, when a team wins as many games as it loses, it is said to be playing "500 ball." Examine the standings and explain the significance of the number 500, using concepts studied in this section.

Eastern Conference			
Team	**W**	**L**	**Pct.**
Orlando	33	33	.500

81. HUMAN SKIN The illustration shows roughly what percent each section of the body represents of the total skin area. Determine the missing percent, and then complete the bar graph on the next page.

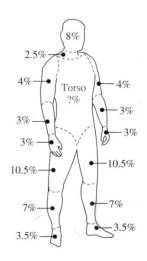

8%

2.5%

4%   Torso ?%   4%

3%

3%

3%

10.5%   10.5%

7%   7%

3.5%

3.5%

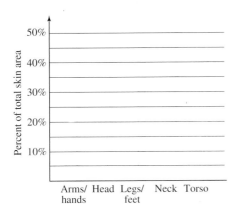

82. RAP MUSIC The table shows what percent rap/
hip-hop music sales were of total U.S. dollar sales
of recorded music for the years 1997–2003. In the
illustration construct a line graph using the given data.

1997	1998	1999	2000	2001	2002	2003
10.1%	9.7%	10.8%	12.9%	11.4%	13.8%	13.3%

Source: *The Recording Industry Association of America*

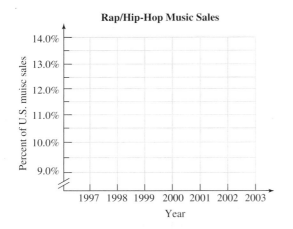

83. THE RED CROSS A fact sheet released by the
American Red Cross stated, "For the past three
fiscal years, an average of 92 cents of every dollar
spent by the Red Cross went to programs and
services to help those in need." What percent of the
money spent by the Red Cross went to programs
and services?

84. TAXES Santa Anita Thoroughbred Racetrack in
Arcadia, California, has to pay a one-third of 1% tax
on all the money wagered at the track. Write the
percent as a fraction.

A calculator may be helpful to solve these problems.

85. BIRTHDAYS If the day of your birthday represents
$\frac{1}{365}$ of a year, what percent of the year is it? Round to
the nearest hundredth of a percent.

86. POPULATION As a fraction, each resident of the
United States represents approximately $\frac{1}{295,000,000}$ of
the U.S. population. Express this as a percent. Round
to one nonzero digit.

WRITING

87. If you were writing advertising, which form do you
think would attract more customers: "25% off" or
"$\frac{1}{4}$ off"? Explain your reasoning.

88. Many coaches ask their players to give a 110% effort
during practices and games. What do you think this
means? Is it possible?

89. Explain how to change a fraction to a percent.

90. Explain how an amusement park could have an
attendance that is 103% of capacity.

91. BOXING Muhammad Ali won 92% of his
professional boxing matches. Does that mean he had
exactly 100 fights and won 92 of them? Explain your
answer.

92. CALCULATORS To change the fraction $\frac{15}{16}$ to a
percent, a student used a calculator to divide 15 by 16.
The display is shown below.

$$\boxed{0.9375}$$

Now what keys should the student press to change
this decimal to a percent?

REVIEW

93. Subtract: $\frac{2}{3} - \frac{3}{4}$.

94. Add: $\frac{1}{3} + \frac{1}{4} + \frac{1}{2}$.

95. Evaluate: $3 - |4 - 8|$.

96. Find the perimeter of a rectangle whose width is
6.5 cm and whose length is 10.5 cm.

97. Find the area of the rectangle in Exercise 96.

98. Subtract: $41 - 10.287$.

5.2 Solving Percent Problems

- Equations • Percent problems • Finding the amount • Finding the percent
- Finding the base • Circle graphs

Percent problems occur in three forms. In this section, we study a method that can be used to solve all three types. This method involves solving simple equations.

Equations

An **equation** is a statement that two quantities are equal. Here are some examples of equations:

$$4 + 4 = 8, \qquad \frac{1}{2} \cdot 12 = 6, \qquad \text{and} \qquad 38 = p \cdot 40$$

Note that every equation contains an = symbol. The following statements are not equations, because they do not contain an = symbol.

$$3 \cdot 5, \qquad \frac{1}{3} + \frac{3}{4}, \qquad \text{and} \qquad 2p + 2l$$

In this section, we *solve* simple equations by dividing both sides by the same number. In Chapter 8, we discuss how to solve equations in greater detail.

Percent problems

The articles on the front page of the newspaper in Figure 5-3 suggest three types of percent problems.

Type 1: In the labor article, if we want to know how many union members voted to accept the new offer, we would ask:

What number is 84% of 500?

Type 2: In the article on drinking water, if we want to know what percent of the wells are safe, we would ask:

38 is what percent of 40?

Type 3: In the article on new appointees, if we want to know how many examiners are on the State Board, we would ask:

6 is 75% of what number?

These percent problems have things in common.

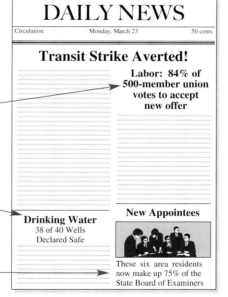

FIGURE 5-3

- Each problem contains the word *is*. Here, *is* can be translated as an = symbol.
- Each of the problems contains a phrase such as *what number* or *what percent*. In other words, there is an unknown quantity.
- Each problem contains the word *of*. In this context, *of* means multiply.

These observations suggest that each of the percent problems can be translated into an equation. The equation, called a **percent equation,** contains an unknown quantity, and the operation of multiplication is involved.

Finding the amount

To solve the first type of problem *(What number is 84% of 500?)*, we multiply.

$$84\% \text{ of } 500 = 84\% \cdot 500 \qquad \text{The word } of \text{ means multiply.}$$
$$= 0.84 \cdot 500 \qquad \text{To perform the multiplication, change 84\% to the decimal 0.84.}$$
$$= 420$$

We have found that 420 is 84% of 500.

In the statement "420 is 84% of 500," the number 420 is called the **amount,** 84% is the percent, and 500 is the **base.** In words, the relationship between the amount, the percent, and the base is *Amount is percent of base.* This relationship is shown in the **percent formula.**

> **The percent formula**
>
> Amount = percent · base

Self Check 1
What number is 240% of 80?

EXAMPLE 1 What number is 160% of 15.8?

Solution In this example, the percent is 160% and the base is 15.8, the number following the word *of.* We can let A stand for the amount and use the percent formula.

Amount	=	percent	·	base	
A	=	160%	·	15.8	Substitute 160% for the percent and 15.8 for the base.

The statement $A = 160\% \cdot 15.8$ is an equation, with the amount being the unknown. For this equation, we can find the unknown amount by multiplication. We call the process of finding the unknown amount *solving the equation.*

$$A = 1.6 \cdot 15.8 \qquad \text{Change 160\% to a decimal.}$$
$$= 25.28 \qquad \text{Perform the multiplication.}$$

Answer 192

Thus, 25.28 is 160% of 15.8.

Finding the percent

In the second type of problem *(38 is what percent of 40?)*, we must find the percent. In this problem, 38 is the amount and 40 is the base. Once again, we form an equation by using the percent formula.

Amount	=	percent	·	base	
38	=	p	·	40	38 is the amount, p is the percent, and 40 is the base.

$$38 = p \cdot 40 \qquad \text{This is the equation to solve.}$$
$$\frac{38}{40} = \frac{p \cdot 40}{40} \qquad \text{To undo the multiplication of } p \text{ by 40, divide both sides by 40.}$$
$$0.95 = p \qquad \tfrac{38}{40} = 0.95 \text{ and } \tfrac{40}{40} = 1.$$
$$p = 95\% \qquad \text{To change a decimal to a percent, move the decimal point 2 places to the right and insert a \% symbol.}$$

Thus, 38 is 95% of 40.

EXAMPLE 2   14 is what percent of 32?

Solution   In this example, 14 is the amount and 32 is the base. Once again, we use the percent formula and let p stand for the percent.

Amount	=	percent	·	base	
14	=	p	·	32	Substitute 14 for the amount and 32 for the base.

The statement $14 = p \cdot 32$ is an equation, with the percent being the unknown. For this equation, we can find the unknown percent by division.

$14 = p \cdot 32$ This is the equation to solve.

$\dfrac{14}{32} = \dfrac{p \cdot 32}{32}$ To undo the multiplication of p by 32, divide both sides by 32.

$0.4375 = p$ Perform the divisions: $\frac{14}{32} = 0.4375$ and $\frac{32}{32} = 1$.

$p = 43.75\%$ To change the decimal to a percent, move the decimal point 2 places to the right and insert a % symbol.

Thus, 14 is 43.75% of 32.

Self Check 2
9 is what percent of 16?

Answer 56.25%

Air bags

CALCULATOR SNAPSHOT

An air bag is estimated to add an additional $500 to the cost of a car. The cost of an air bag is what percent of a car's $16,295 sticker price?

In this problem, 500 is the amount and 16,295 is the base. We can form an equation by using the percent formula.

Amount	=	percent	·	base	
500	=	p	·	16,295	Substitute 500 for the amount and 16,295 for the base. Let p represent the percent.

Then we solve the equation.

$500 = p \cdot 16,295$ This is the equation to solve.

$\dfrac{500}{16,295} = \dfrac{p \cdot 16,295}{16,295}$ To undo the multiplication of p by 16,295, divide both sides by 16,295.

$\dfrac{500}{16,295} = p$ $\dfrac{16,295}{16,295} = 1$.

To perform the division using a calculator, we enter these numbers and press these keys.

$500 \;\boxed{÷}\; 16295 \;\boxed{=}$ $\boxed{\text{0.030684259}}$

This display gives the answer in decimal form. To change it to a percent, we multiply the result by 100 and insert a % symbol. This moves the decimal point 2 places to the right. If we round to the nearest tenth of a percent, the cost of an air bag is about 3.1% of the sticker price.

Finding the base

In the third type of problem *(6 is 75% of what number?)*, we must find the base. Again, we substitute into the percent formula.

Amount	=	percent	·	base
6	=	75%	·	b

6 is the amount, and 75% is the percent. Let b stand for the base.

The statement $6 = 75\% \cdot b$ is an equation with the base being unknown. For this equation, we can find the unknown base by division.

$6 = 0.75 \cdot b$ Change 75% to 0.75. This is the equation to solve.

$\dfrac{6}{0.75} = \dfrac{0.75 \cdot b}{0.75}$ To undo the multiplication of b by 0.75, divide both sides by 0.75.

$8 = b$ Perform the divisions: $\frac{6}{0.75} = 8$ and $\frac{0.75}{0.75} = 1$.

Thus, 6 is 75% of 8.

EXAMPLE 3   **Rentals.**   In an apartment complex, 110 of the units are currently rented. If this represents an 88% occupancy rate, how many units are in the complex?

Solution   An occupancy rate of 88% means that 88% of the units are occupied. In this problem, the percent is 88% and the amount is 110. We are to find the base, which is the total number of units. We form an equation by using the percent formula.

Amount	=	percent	·	base
110	=	88%	·	b

Let b represent the base.

We can then solve the equation to find the base.

$110 = 0.88 \cdot b$ Change 88% to 0.88. This is the equation to solve.

$\dfrac{110}{0.88} = \dfrac{0.88 \cdot b}{0.88}$ To undo the multiplication of b by 0.88, divide both sides by 0.88.

$125 = b$ Perform the divisions: $\frac{110}{0.88} = 125$ and $\frac{0.88}{0.88} = 1$.

The complex has 125 units.

In the next example, the computations are easier if we change the percent to a fraction instead of a decimal.

Self Check 4
150 is $66\frac{2}{3}$% of what number?

EXAMPLE 4   31.5 is $33\frac{1}{3}$% of what number?

Solution   In this example, 31.5 is the amount and $33\frac{1}{3}$ is the percent. To find the base, we form an equation using the percent formula.

Amount	=	percent	·	base
31.5	=	$33\frac{1}{3}$%	·	b

Substitute 31.5 for the amount and $33\frac{1}{3}$% for the percent.

The statement $31.5 = 33\frac{1}{3}\% \cdot b$ is an equation, with the base being the unknown. In this equation, we can find the unknown amount by division.

$$31.5 = 33\frac{1}{3}\% \cdot b \qquad \text{This is the equation to solve.}$$

$$31.5 = \frac{1}{3} \cdot b \qquad \text{Replace } 33\frac{1}{3}\% \text{ with its fractional equivalent: } 33\frac{1}{3}\% = \frac{33\frac{1}{3}}{100} = \frac{1}{3}.$$

$$31.5 \div \frac{\mathbf{1}}{\mathbf{3}} = \frac{1}{3} \cdot b \div \frac{\mathbf{1}}{\mathbf{3}} \qquad \text{To undo the multiplication of } b \text{ by } \frac{1}{3}, \text{ divide both sides by } \frac{1}{3}.$$

$$31.5 \cdot \frac{3}{1} = \frac{1}{3} \cdot b \cdot \frac{3}{1} \qquad \text{To divide by } \frac{1}{3}, \text{ multiply by } \frac{3}{1}.$$

$$94.5 = b \qquad \text{Perform the multiplications: } 31.5 \cdot 3 = 94.5 \text{ and } \frac{1}{3} \cdot \frac{3}{1} = 1.$$

Thus, 31.5 is $33\frac{1}{3}\%$ of 94.5.

Answer 225

■ Circle graphs

Percents are used with **circle graphs,** or **pie charts,** as a way of presenting data for comparison. In Figure 5-4, the entire circle represents the total amount of electricity generated in the United States in 2003. The pie-shaped pieces of the graph show the relative sizes of the energy sources used to generate the electricity. For example, we see that the greatest amount of electricity (51%) was generated from coal. Note that if we add the percents from all categories (51% + 3% + 7% + 16% + 20% + 3%), the sum is 100%.

The 100 tick marks equally spaced around the circle serve as a visual aid when constructing a circle graph. For example, to represent hydropower as 7%, a line was drawn from the center of the circle to a tick mark. Then we counted off 7 ticks and drew a second line from the center to that tick to complete the pie-shaped wedge.

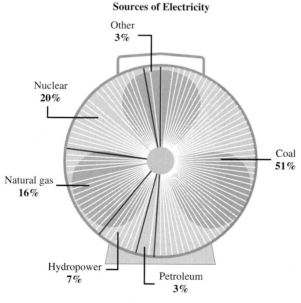

Sources of Electricity

Other 3%

Nuclear 20%

Coal 51%

Natural gas 16%

Hydropower 7%

Petroleum 3%

Source: Energy Information Administration

FIGURE 5-4

EXAMPLE 5   Presidential elections.   Results from the 2004 presidential election are shown in Figure 5-5 on the next page. Use the information to find the number of states won by George W. Bush.

Solution   The circle graph shows that George W. Bush was victorious in 62% of the 50 states. Here, the percent is 62% and the base is 50. We use the percent formula and solve for the amount.

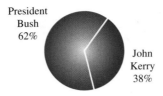

2004 Presidential Election
States won by each candidate

FIGURE 5-5

Amount	=	percent	·	base

A = 62% · 50   Substitute 62% for the percent and 50 for the base.

$A = 0.62 \cdot 50$ Change 62% to a decimal: 62% = 0.62. This is the equation to solve.

$A = 31$ Perform the multiplication.

George W. Bush won 31 states in the 2004 presidential election.

THINK IT THROUGH   **Community College Students**

"Community Colleges are centers of educational opportunity. More than 100 years ago, this unique, American invention put publicly funded higher education at close-to-home facilities and initiated a practice of welcoming all who desire to learn, regardless of wealth, heritage or previous academic experience. Today, the community college continues the process of making higher education available to a maximum number of people at 1,166 public and independent community colleges." The American Association of Community Colleges (AACC)

More than 33,500 students responded to the 2002 Community College Survey of Student Engagement. Some results are shown below. Study each circle graph and then complete its legend.

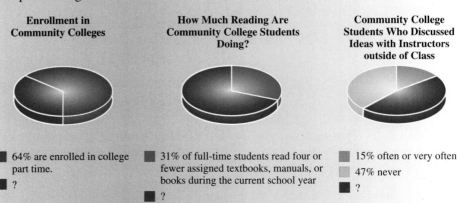

Enrollment in Community Colleges

64% are enrolled in college part time.
?

How Much Reading Are Community College Students Doing?

31% of full-time students read four or fewer assigned textbooks, manuals, or books during the current school year
?

Community College Students Who Discussed Ideas with Instructors outside of Class

15% often or very often
47% never
?

Section 5.2 STUDY SET

VOCABULARY *Fill in the blanks.*

1. In a circle _____, pie-shaped wedges are used to show the division of a whole quantity into its component parts.

2. In the statement "45 is 90% of 50," 45 is the _____, 90% is the _____, and 50 is the _____.

NOTATION *Use the percent formula to write each sentence as an equation.* **Do not solve the equation.**

3. What number is 10% of 50?

4. 16 is 55% of what number?

5. 48 is what percent of 47?

6. 12 is what percent of 20?

CONCEPTS

7. When computing with percents, we must change the percent to a decimal or a fraction. Change each percent to a decimal.
   a. 12%   b. 5.6%
   c. 125%   d. $\frac{1}{4}$%

8. When computing with percents, we must change the percent to a decimal or a fraction. Change each percent to a fraction.
   a. $33\frac{1}{3}$%   b. $66\frac{2}{3}$%
   c. $16\frac{2}{3}$%   d. $83\frac{1}{3}$%

9. Without doing the calculation, tell whether 120% of 55 is more than 55 or less than 55.

10. Without doing the calculation, tell whether 12% of 55 is more than 55 or less than 55.

11. E-MAIL The circle graph shows the types of e-mail messages a typical Internet user receives. *Spam* is the name given "junk" e-mail that is sent to a large number of people to promote products or services. What percent of e-mail messages is spam?

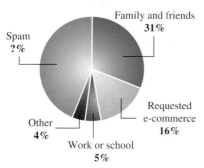

Spam ?%
Family and friends 31%
Requested e-commerce 16%
Work or school 5%
Other 4%

Source: *USA Today*, October 31, 2003

12. HOUSING In the last quarter of 2000, approximately 105.5 million housing units in the United States were occupied. Use the data in the circle graph to determine what percent were owner-occupied.

2000 Housing Inventory

Owner occupied

Renter occupied 33.8%

Source: *The U.S. Census Bureau*

13. What is 100% of 25? 25

14. What percent of 32 is 32?

15. What is 200% of 25? 50

16. What is 300% of 25? 75

PRACTICE *Solve each problem by solving a percent equation.*

17. What number is 36% of 250?

18. What number is 82% of 300?

19. 16 is what percent of 20?

20. 13 is what percent of 25?

21. 7.8 is 12% of what number?

22. 39.6 is 44% of what number?

23. What number is 0.8% of 12?

24. What number is 5.6% of 40?

25. 0.5 is what percent of 40?

26. 0.3 is what percent of 15?

27. 3.3 is 7.5% of what number?

28. 8.4 is 20% of what number?

Write each item as a Type 1, Type 2, or Type 3 problem, and then solve it. For example, Exercise 29 can be written as "What number is $7\frac{1}{4}$% of 60?"

29. Find $7\frac{1}{4}$% of 600.

30. Find $1\frac{3}{4}$% of 800.

31. 102% of 105 is what number?

32. 210% of 66 is what number?

33. $33\frac{1}{3}$% of what number is 33?

34. $66\frac{2}{3}$% of what number is 28?

35. $9\frac{1}{2}$% of what number is 5.7?

36. $\frac{1}{2}$% of what number is 5,000?

37. What percent of 8,000 is 2,500?

38. What percent of 3,200 is 1,400?

Use a circle graph to illustrate the given data. A circle divided into 100 sections is provided to aid in the graphing process.

39. ENERGY Complete the illustration to show what percent of the total U.S. energy produced was provided by each source in 2003.

Renewable	12%
Nuclear	11%
Coal	31%
Natural gas	29%
Petroleum	17%

Source: *Energy Information Administration*

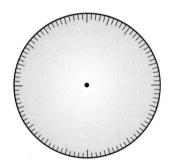

Source: Energy Information Administration

40. GREENHOUSE EFFECT Complete the illustration to show what percent of the total U.S. emissions from human activities in 2002 came from each greenhouse gas.

Carbon dioxide	83%
Nitrous oxide	6%
Methane	9%
PFCs, HFCs	2%

Source: *The World Almanac 2005*

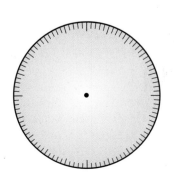

APPLICATIONS

41. CHILD CARE After the first day of registration, 84 children had been enrolled in a new day care center. That represented 70% of the available slots. What was the maximum number of children the center could enroll?

42. RACING PROGRAMS One month before a stock car race, the sale of ads for the official race program was slow. Only 12 pages, or 60% of the available pages, had been sold. What was the total number of pages devoted to advertising in the program?

43. GOVERNMENT SPENDING The illustration shows the breakdown of federal spending for fiscal year 2003. If the total spending was approximately $1,800 billion, how many dollars were spent on Social Security, Medicare, and other retirement programs?

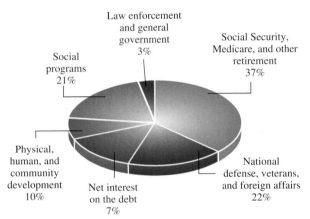

Source: 2004 Federal Income Tax Form 1040

44. GOVERNMENT INCOME Complete the table by finding what percent of total federal government income each source provided in 2003. Round to the nearest percent. Then complete the circle graph on the next page.

Total income, fiscal year 2003: $2,200 billion		
Source of income	**Amount**	**Percent of total**
Social Security, Medicare, unemployment taxes	$726 billion	
Personal income taxes	$814 billion	
Corporate income taxes	$132 billion	
Excise, estate, customs taxes	$154 billion	
Borrowing to cover deficit	$374 billion	

Source: 1999 Federal Income Tax Form

2003 Federal Income Sources

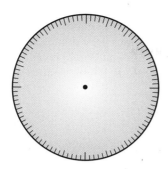

45. THE INTERNET See the illustration. The message at the bottom of the screen indicates that 24% of the 50K bytes of information that the user has decided to view have been downloaded to her computer. How many more bytes of information must be downloaded? (50K stands for 50,000.)

46. REBATES A long-distance telephone company offered its customers a rebate of 20% of the cost of all long-distance calls made in the month of July. One customer's calls are listed in the table. What amount will this customer receive in the form of a rebate?

Date	Time	Place called	Min.	Amount
Jul 4	3:48 P.M.	Denver	47	$3.80
Jul 9	12:00 P.M.	Detroit	68	$7.50
Jul 20	8:59 A.M.	San Diego	70	$9.45

47. PRODUCT PROMOTION To promote sales, a free 6-ounce bottle of shampoo is packaged with every large bottle. Use the information on the package to find how many ounces of shampoo the large bottle contains.

48. NUTRITION FACTS The nutrition label on a package of corn chips is shown.
  a. How many milligrams of sodium are in one serving of chips?
  b. According to the label, what percent of the daily value of sodium is this?
  c. What daily value of sodium intake is deemed healthy?

Nutrition Facts
Serving Size: 1 oz. (28g/About 29 chips)
Servings Per Container: About 11

Amount Per Serving
Calories 160   Calories from Fat 90

	% Daily Value
Total fat 10g	**15%**
Saturated fat 1.5 g	**7%**
Cholesterol 0mg	**0%**
Sodium 240mg	**12%**
Total carbohydrate 15g	**5%**
Dietary fiber 1g	**4%**
Sugars less than 1g	

Protein 2g

49. DRIVER'S LICENSES On the written part of his driving test, a man answered 28 out of 40 questions correctly. If 70% correct is passing, did he pass the test?

50. ALPHABET What percent of the English alphabet do the vowels a, e, i, o, and u make up? (Round to the nearest 1 percent.)

51. MIXTURES Complete the table to find the number of gallons of sulfuric acid in each of two storage tanks.

	Gallons of solution in tank	% sulfuric acid	Gallons of sulfuric acid in tank
Tank 1	60	50%	30%
Tank 2	40	30%	12%

52. GUARANTEES To assure its customers of low prices, the Home Club offers a "10% Plus" guarantee. If the customer finds the same item selling for less somewhere else, he or she receives the difference in price, plus 10% of the difference. A woman bought miniblinds at the Home Club for $120 but later saw the same blinds on sale for $98 at another store. How much can she expect to be reimbursed?

53. MAKING COPIES The zoom key on the control panel of a copier programs it to print a magnified or reduced copy of the original document. If the zoom is set at 180% and the original document contains type that is 1.5 inches tall, what will be the height of the type on the copy?

54. MAKING COPIES The zoom setting for a copier is entered as a decimal: 0.98. Express it as a percent and find the resulting type size on the copy if the original has type 2 inches tall.

55. INSURANCE The cost to repair a car after a collision was $4,000. The automobile insurance policy paid the entire bill except for a $200 deductible, which the driver paid. What percent of the cost did he pay?

56. FLOOR SPACE A house has 1,200 square feet on the first floor and 800 square feet on the second floor. What percent of the square footage of the house is on the first floor?

57. MAJORITIES In Los Angeles City Council races, if no candidate receives more than 50% of the vote, a runoff election is held between the first- and second-place finishers. From the election results in the table, determine whether there must be a runoff election for District 10.

City Council	District 10
Nate Holden	8,501
Madison T. Shockley	3,614
Scott Suh	2,630
Marsha Brown	2,432

58. PORTS In 2002, the busiest port in the United States was the Port of South Louisiana, which handled 216,396,497 tons of goods. Of that amount, 124,908,067 tons were domestic goods and 91,488,430 tons were foreign. What percent of the total was domestic? Round to the nearest tenth of a percent.

WRITING

59. Explain the relationship in a percent problem between the amount, the percent, and the base.

60. Write a real-life situation that can be described by "9 is what percent of 20?"

61. Explain why 150% of a number is more than the number.

62. Explain why "Find 9% of 100" is an easy problem to solve.

REVIEW

63. Add: $2.78 + 6 + 9.09 + 0.3$.

64. Evaluate: $\sqrt{64} + 3\sqrt{9}$.

65. On the number line, which is closer to 5: the number 4.9 or the number 5.001?

66. Multiply: $34.5464 \cdot 1,000$.

67. Find: $(0.2)^3$.

68. Evaluate the formula $d = 4t$ for $t = 25$.

5.3 Applications of Percent

• Taxes • Commissions • Percent of increase or decrease • Discounts

In this section, we discuss applications of percent. Three of them (taxes, commissions, and discounts) are directly related to purchasing. A solid understanding of these concepts will make you a better consumer. The fourth uses percent to describe increases or decreases of such things as unemployment and grocery store sales.

Taxes

The sales receipt in Figure 5-6 on the next page gives a detailed account of what items were purchased, how many of each were purchased, and the price of each item.

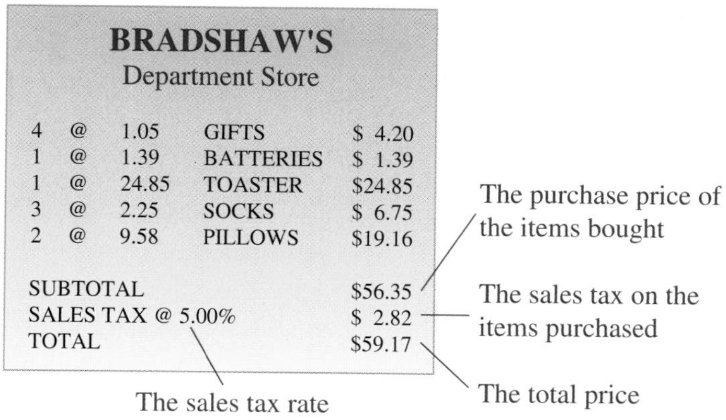

BRADSHAW'S
Department Store

4	@	1.05	GIFTS	$ 4.20
1	@	1.39	BATTERIES	$ 1.39
1	@	24.85	TOASTER	$24.85
3	@	2.25	SOCKS	$ 6.75
2	@	9.58	PILLOWS	$19.16

SUBTOTAL $56.35
SALES TAX @ 5.00% $ 2.82
TOTAL $59.17

The purchase price of the items bought

The sales tax on the items purchased

The sales tax rate

The total price

FIGURE 5-6

The receipt shows that the $56.35 purchase price (labeled *subtotal*) was taxed at a rate of 5%. The sales tax of $2.82 was then added to the subtotal to get the total price of $59.17.

Finding the total price

Total price = purchase price + sales tax

In Example 1, we verify that the amount of sales tax shown on the receipt in Figure 5-6 is correct.

EXAMPLE 1   Sales tax. Find the sales tax on a purchase of $56.35 if the sales tax rate is 5%.

Solution First we use the percent formula to write an equation. The percent is 5%, and the base is 56.35. We are to find the amount of the tax.

Amount	=	percent	·	base	
A	=	5%	·	56.35	Substitute 5% for the percent and 56.35 for the base.

$A = 0.05 \cdot 56.35$ Change 5% to a decimal: 5% = 0.05.

$A = 2.8175$ Perform the multiplication.

Rounding to the nearest cent (hundredths), we find that the sales tax is $2.82. The sales receipt in Figure 5-6 is correct.

In addition to sales tax, we pay many other taxes in our daily lives. Income tax, gasoline tax, and Social Security tax are just a few.

EXAMPLE 2   Withholding tax. A waitress found that $11.04 was deducted from her weekly gross earnings of $240 for federal income tax. What withholding tax rate was used?

Solution First, we use the percent formula to write an equation. The amount is 11.04, and the base is 240. We need to find the tax rate.

Self Check 1
What would the sales tax be if the $56.35 purchase were made in a state that has a 6.25% state sales tax?

Answer $3.52

Self Check 2
A tax of $5,250 was paid on an inheritance of $15,000. What was the inheritance tax rate?

$$\text{Amount} = \text{percent} \cdot \text{base}$$
$$11.04 = p \cdot 240$$

$11.04 = p \cdot 240$	This is the equation to solve.
$\dfrac{11.04}{240} = \dfrac{p \cdot 240}{240}$	To undo the multiplication by 240, divide both sides by 240.
$0.046 = p$	Perform the divisions.
$4.6\% = p$	Change 0.046 to a percent.

Answer 35%

The withholding tax rate was 4.6%.

■ Commissions

Instead of working for a salary or getting paid at an hourly rate, many salespeople are paid on **commission.** They earn an amount based on the goods or services they sell.

Self Check 3
An insurance salesperson receives a 4.1% commission on each $120 premium paid by a client. What is the amount of the commission on this premium?

EXAMPLE 3 Appliance sales. The commission rate for a salesperson at an appliance store is 16.5%. Find his commission from the sale of a refrigerator that cost $499.95.

Solution We use the percent formula to write an equation. The percent is 16.5%, and the base is 499.95. We are to find the amount of the commission.

$$\text{Amount} = \text{percent} \cdot \text{base}$$
$$A = 16.5\% \cdot 499.95$$

$A = 0.165 \cdot 499.95$	Change 16.5% to a decimal: 16.5% = 0.165.
$A = 82.49175$	Use a calculator to perform the multiplication.

Answer $4.92

Rounding to the nearest cent (hundredth), we find that the commission is $82.49.

■ Percent of increase or decrease

Percents can be used to describe how a quantity has changed. For example, consider Figure 5-7, which compares the numbers of hours of work it took the average U.S. worker to earn enough to buy a dishwasher in 1950 and 1998.

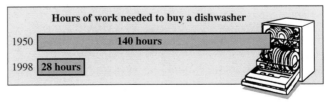

Hours of work needed to buy a dishwasher

| 1950 | 140 hours |
| 1998 | 28 hours |

Source: Federal Reserve Bank of Dallas

FIGURE 5-7

From the figure, we see that the number of hours an average American had to work in order to buy a dishwasher has decreased over the years. To describe this decrease using a percent, we first subtract to find the amount of the decrease.

$$140 - 28 = 112$$ Subtract the hours of work needed in 1998 from the hours of work needed in 1950.

Next, we find what percent of the original number of hours of work needed in 1950 this difference represents.

Amount	=	percent	·	base
112	=	p	·	140

$112 = p \cdot 140$ This is the equation to solve.

$\dfrac{112}{140} = \dfrac{p \cdot 140}{140}$ To undo the multiplication by 140, divide both sides by 140.

$0.8 = p$ Perform the divisions.

$80\% = p$ Change 0.8 to a percent.

From 1950 to 1998, there was an 80% decrease in the number of hours it took the average U.S. worker to earn enough to buy a dishwasher.

Finding the percent of increase or decrease

To find the percent of increase or decrease:

1. Subtract the smaller number from the larger to find the amount of increase or decrease.

2. Find what percent the difference is of the original amount.

EXAMPLE 4   **JFK.**   A 1996 auction included an oak rocking chair used by President John F. Kennedy in the Oval Office. The chair, originally valued at $5,000, sold for $453,500. Find the percent of increase in the value of the rocking chair.

Solution   First we find the amount of increase.

$453,500 - 5,000 = 448,500$ Subtract the original value from the price paid at auction.

The rocking chair increased in value by $448,500. Next, we find what percent of the original value the increase represents. Here, the amount is 448,500, and the base is 5,000.

Amount	=	percent	·	base
448,500	=	p	·	5,000

$448,500 = p \cdot 5,000$ This is the equation to solve.

$\dfrac{448,500}{5,000} = \dfrac{p \cdot 5,000}{5,000}$ To undo the multiplication by 5,000, divide both sides by 5,000.

$89.7 = p$ Perform the divisions.

$8,970\% = p$ Change 89.7 to a percent.

The Kennedy rocking chair increased in value by an amazing 8,970%.

Self Check 4

In one school district, the number of home-schooled children increased from 15 to 150 in 4 years. Find the percent of increase.

Answer 900%

EXAMPLE 5   **Population decline.**   Norfolk, Virginia, experienced a decrease in population over the ten-year period from 1990 to 2000. Use the information in Figure 5-8 on the next page to determine the population of Norfolk in 2000.

Solution   In 1990, the population was 261,000. We are told that the number fell, and we need to find out by how much. To do so, we solve the following percent formula.

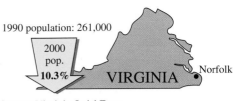

1990 population: 261,000

2000 pop. 10.3%

VIRGINIA   Norfolk

Source: Virginia QuickFacts

FIGURE 5-8

Amount	=	percent	·	base
A	=	10.3%	·	261,000

$A = 0.103 \cdot 261,000$ Change 10.3% to a decimal: 10.3% = 0.103.

$A = 26,883$ Perform the multiplication.

From 1990 to 2000, the population decreased by 26,883. To find the city's population in 2000, we subtract the decrease from the population in 1990.

$261,000 - 26,883 = 234,117$

In 2000, the population of Norfolk, Virginia, was 234,117.

We can solve this problem in another way. If the population of Norfolk decreased by 10.3%, then the population in 2000 was $100\% - 10.3\% = 89.7\%$ of the population in 1990. Using this approach, we can find the 2000 population directly by solving the following percent formula.

Amount	=	percent	·	base
A	=	89.7%	·	261,000

$A = 0.897 \cdot 261,000$ Change 89.7% to a decimal: 89.7% = 0.897.

$A = 234,117$ Perform the multiplication.

As before, we see that the population of Norfolk in 2000 was 234,117.

THINK IT THROUGH   Studying Mathematics

"All students, regardless of their personal characteristics, backgrounds, or physical challenges, must have opportunities to study—and support to learn—mathematics." National Council of Teachers of Mathematics

The table below shows the number of students enrolled in Basic Mathematics classes at two-year colleges.

Year	1970	1975	1980	1985	1990	1995	2000
Enrollment	57,000	100,000	146,000	142,000	147,000	134,000	122,000

Source: 2000 CBMS Survey of Undergraduate Programs

1. Over what five-year span was there the greatest percent increase in enrollment in Basic Mathematics classes? What was the percent increase?

2. Over what five-year span was there the greatest percent decrease in enrollment in Basic Mathematics classes? What was the percent increase?

■ Discounts

The difference between the original price and the sale price of an item is called the **discount.** If the discount is expressed as a percent of the selling price, it is called the **rate of discount.** We will use the information in the advertisement shown in Figure 5-9 to discuss how to find a discount and how to find a discount rate.

FIGURE 5-9

EXAMPLE 6   Shoe sales.   Find the amount of the discount on the pair of men's basketball shoes shown in Figure 5-9. Then find the sale price.

Solution   To find the discount, we find 25% of the regular price, $59.80. Here, the percent is 25%, and the base is 59.80.

Amount	=	percent	·	base
A	=	25%	·	59.80

$A = 0.25 \cdot 59.80$ Change 25% to a decimal: 25% = 0.25.
$A = 14.95$ Perform the multiplication.

The discount is $14.95. To find the sale price, we subtract the amount of the discount from the regular price.

$59.80 - 14.95 = 44.85$

The sale price of the men's basketball shoes is $44.85.

Self Check 6
Sunglasses, regularly selling for $15.40, are discounted 15%. Find the sale price.

Answer $13.09

In Example 6, we used the following formula to find the sale price.

> **Finding the sale price**
>   Sale price = original price − discount

EXAMPLE 7   Discounts.   What is the rate of discount on the ladies' aerobic shoes advertised in Figure 5-9?

Solution   We can think of this as a percent-of-decrease problem. We first compute the amount of the discount. This decrease in price is found using subtraction.

$39.99 - 21.99 = 18$

The shoes are discounted $18. Now we find what percent of the original price the discount is. Here, the amount is 18, and the base is 39.99.

Amount	=	percent	·	base
18	=	p	·	39.99

Self Check 7
An early-bird special at a restaurant offers a $10.99 prime rib dinner for only $7.95 if it is ordered before 6 P.M. Find the rate of discount. Round to the nearest 1 percent.

$$18 = p \cdot 39.99 \qquad \text{This is the equation to solve.}$$

$$\frac{18}{39.99} = \frac{p \cdot 39.99}{39.99} \qquad \begin{array}{l}\text{To undo the multiplication by 39.99, divide both sides} \\ \text{by 39.99.}\end{array}$$

$$0.450113 \approx p \qquad \text{Perform the division.}$$

$$45.0113\% \approx p \qquad \text{Change 0.450113 to a percent.}$$

Answer 28%

Rounded to the nearest 1 percent, the discount rate is 45%.

Section 5.3   STUDY SET

VOCABULARY   *Fill in the blanks.*

1. Some salespeople are paid on _____. It is based on a percent of the total dollar amount of the goods or services they sell.

2. When we use percent to describe how a quantity has increased compared to its original value, we are finding the percent of _____.

3. The difference between the original price and the sale price of an item is called the _____.

4. The _____ of a sales tax is expressed as a percent.

CONCEPTS

5. Fill in: To find the percent decrease, _____ the smaller number from the larger number to find the amount of decrease. Then find what percent that difference is of the _____ amount.

6. NEWSPAPERS The table below shows how the circulations of two daily newspapers changed from 1997 to 2003.

CIRCULATION		
	Miami Herald	**USA Today**
1997	356,803	1,629,665
2003	315,850	2,154,539

Source: *The World Almanac 2005*

   a. What was the *amount of decrease* of the *Miami Herald*'s circulation?

   b. What was the *amount of increase* of *USA Today*'s circulation?

APPLICATIONS   *Solve each problem. If an answer is not exact, round to the nearest 1 percent.*

7. SALES TAX The state sales tax rate in Utah is 4.75%. Find the sales tax on a dining room set that sells for $900.

8. SALES TAX Find the sales tax on a pair of jeans costing $40 if they are purchased in Missouri which has a sales tax rate of 4.225%.

9. ROOM TAX After checking out of a hotel, a man noticed that the hotel bill included an additional charge labeled *room tax*. If the price of the room was $129 plus a room tax of $10.32, find the room tax rate.

10. EXCISE TAX While examining her monthly telephone bill, a woman noticed an additional charge of $1.24 labeled *federal excise tax*. If the basic service charges for that billing period were $42, what is the federal excise tax rate?

11. SALES RECEIPTS Complete the sales receipt below by finding the subtotal, the sales tax, and the total.

NURSERY CENTER		
Your one-stop garden supply		
3 @ 2.99	PLANTING MIX	$ 8.97
1 @ 9.87	GROUND COVER	$ 9.87
2 @ 14.25	SHRUBS	$28.50
SUBTOTAL		$
SALES TAX @ 6.00%		$
TOTAL		$

12. SALES RECEIPTS Complete the sales receipt below by finding the prices, the subtotal, the sales tax, and the total.

McCOY'S FURNITURE		
1 @ 450.00	SOFA	$
2 @ 90.00	END TABLES	$
1 @ 350.00	LOVE SEAT	$
SUBTOTAL		$
SALES TAX @ 4.20%		$
TOTAL		$

13. TAX HIKES In order to raise more revenue, some states raise the sales tax rate. How much additional money will be collected on the sale of a $15,000 car if the sales tax rate is raised 1%?

14. FOREIGN TRAVEL Value-added tax (VAT) is a consumer tax imposed on goods and services. Currently, VAT systems are in place all around the world. (The United States is one of the few industrialized nations not using a value-added tax system.) Complete the table by determining the VAT tax a traveler would pay in each country on a dinner that cost $20.95.

Country	VAT tax rate	Tax on a $20.95 dinner
Canada	7%	
Germany	16%	
England	17.5%	
Sweden	25%	

15. PAYCHECKS Use the information on the paycheck stub to find the tax rate for the federal withholding, worker's compensation, Medicare, and Social Security taxes that were deducted from the gross pay.

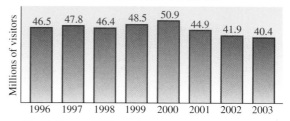

6286244
Issue date: 03-27-05
GROSS PAY $360.00
TAXES
FED. TAX $ 28.80
WORK. COMP. $ 4.32
MEDICARE $ 5.22
SOCIAL SECURITY $ 22.32
NET PAY $299.34

16. GASOLINE TAX In one state, a gallon of unleaded gasoline sells for $1.89. This price includes federal and state taxes that total approximately $0.54. Therefore, the price of a gallon of gasoline, before taxes, is about $1.35. What is the tax rate on gasoline?

17. OVERTIME Factory management wants to reduce the number of overtime hours by 25%. If the total number of overtime hours is 480 this month, what is the target number of overtime hours for next month?

18. COST-OF-LIVING INCREASES If a woman making $32,000 a year receives a cost-of-living increase of 2.4%, how much is her raise? What is her new salary?

19. REDUCED CALORIES A company advertised its new, improved chips as having 36% fewer calories per serving than the original style. How many calories are in a serving of the new chips if a serving of the original style contained 150 calories?

20. POLICE FORCE A police department plans to increase its 80-person force by 5%. How many additional officers will be hired? What will be the new size of the department?

21. TOURISM The graph below shows the number of international travelers to the United States from 1996 to 2003. Between what two years did the greatest percent of decrease in the number of visitors occur? What was that percent of decrease? Round to the nearest percent.

International Travel to the U.S.

Millions of visitors

1996: 46.5   1997: 47.8   1998: 46.4   1999: 48.5   2000: 50.9   2001: 44.9   2002: 41.9   2003: 40.4

Source: *World Almanac 2005*

22. CROP DAMAGE After flooding damaged much of the crop, the cost of a head of lettuce jumped from $0.99 to $2.20. What percent of increase is this?

23. CAR INSURANCE A student paid a car insurance premium of $400 every three months. Then the premium dropped to $360, because she qualified for a good-student discount. What was the percent of decrease in the premium?

24. BUS PASSES To increase the number of riders, a bus company reduced the price of a monthly pass from $112 to $98. What was the percent of decrease?

25. LAKE SHORELINES Because of a heavy spring runoff, the shoreline of a lake increased from 5.8 miles to 7.6 miles. What was the percent of increase in the shoreline?

26. BASEBALL The illustration shows the path of a baseball hit 110 mph, with a launch angle of 35 degrees, at sea level and at Coors Field, home of the Colorado Rockies. What is the percent of increase in the distance the ball travels at Coors Field?

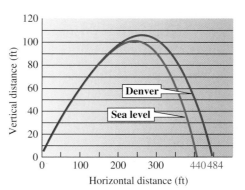

Source: *Los Angeles Times* (September 16, 1996)

27. EARTH MOVING The illustration shows the typical soil volume change during earth moving. (One cubic yard of soil fits in a cube that is 1 yard long, 1 yard wide, and 1 yard high.)

    a. Find the percent of increase in the soil volume as it goes through step 1 of the process.

    b. Find the percent of decrease in the soil volume as it goes through step 2 of the process.

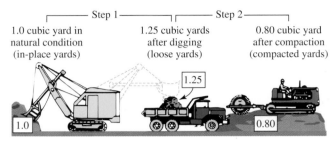

┌──── Step 1 ────┐ ┌──── Step 2 ────┐

1.0 cubic yard in natural condition (in-place yards)   1.25 cubic yards after digging (loose yards)   0.80 cubic yard after compaction (compacted yards)

Source: U.S. Department of the Army

28. PARKING The management of a mall has decided to increase the parking area. The plans are shown below. What will be the percent of increase in the parking area when the project is completed?

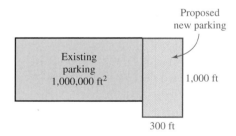

Proposed new parking

Existing parking 1,000,000 ft²

1,000 ft

300 ft

29. REAL ESTATE After selling a house for $98,500, a real estate agent split the 6% commission with another agent. How much did each person receive?

30. MEDICAL SUPPLIES A salesperson for a medical supplies company is paid a commission of 9% for orders less than $8,000. For orders exceeding $8,000, she receives an additional 2% in commission on the total amount. What is her commission on a sale of $14,600?

31. SPORTS AGENTS A sports agent charges her clients a fee to represent them during contract negotiations. The fee is based on a percent of the contract amount. If the agent earned $37,500 when her client signed a $2,500,000 professional football contract, what rate did she charge for her services?

32. ART GALLERIES An art gallery displays paintings for artists and receives a commission from the artist when a painting is sold. What is the commission rate if a gallery received $135.30 when a painting was sold for $820?

33. CONCERT PARKING A concert promoter gets $33\frac{1}{3}$% of the revenue the arena receives from its parking concession the night of the performance. How much can the promoter make if 6,000 cars are anticipated and parking costs $6 a car?

34. PARTIES A homemaker invited her neighbors to a kitchenware party to show off cookware and utensils. As party hostess, she received 12% of the total sales. How much was purchased if she received $41.76 for hosting the party?

35. WATCH SALE Find the regular price and the rate of discount for the watch shown in the ad.

WATCHES SAVE $10! You pay only $29.95

36. STEREO SALE Find the regular price and the rate of discount for the stereo system shown in the ad.

STEREO

Now only $545.88 SAVE $54!

CLOSEOUT! While quantities last

37. RING SALE What does a ring regularly sell for if it has been discounted 20% and is on sale for $149.99? (*Hint:* The ring is selling for 80% of its regular price.)

38. BLINDS SALE What do vinyl blinds regularly sell for if they have been discounted 55% and are on sale for $49.50? (*Hint:* The blinds are selling for 45% of their regular price.)

39. VCR SALE What are the sale price and the discount rate for a VCR with remote that regularly sells for $399.97 and is being discounted $50?

40. CAMCORDER SALE What are the sale price and the discount rate for a camcorder that regularly sells for $559.97 and is being discounted $80?

41. REBATES Find the discount, the discount rate, and the reduced price for a case of motor oil if a shopper receives the manufacturer's rebate mentioned in the ad.

GXT MOTOR OIL MULTI-VIS

Regular price $15.48/case
Mfr's rebate: $3.60

42. DOUBLE COUPONS Find the discount, the discount rate, and the reduced price for a box of cereal that normally sells for $3.29 if a shopper presents the coupon at a store that doubles the value of the coupon.

SAVE 35¢

GREAT HARVEST CEREAL WHOLE GRAIN GOODNESS

Manufacturer's coupon (Limit 1)

43. TV SHOPPING Determine the Home Shopping Network (HSN) price of the ring described in the illustration if it sells it for 55% off of the retail price. Ignore shipping and handling costs.

Item 169-117
2.75 lb ctw
10K
Blue Topaz
Ring
6, 7, 8, 9, 10

Retail value $170

HSN Price
$??.??

S&H $5.95

44. INFOMERCIALS The host of a TV infomercial says that the suggested retail price of a rotisserie grill is $249.95 and that it is now offered "for just 4 easy payments of only $39.95." What is the discount, and what is the discount rate?

WRITING

45. List the pros and cons of working on commission.

46. In Example 6, explain why you get the correct answer for the sale price by finding 75% of the regular price.

47. Explain the difference between a tax and a tax rate.

48. Explain how to find the sale price of an item if you know the regular price and the discount rate.

REVIEW

49. Multiply: $-5(-5)(-2)$.

50. Divide: $\dfrac{3}{5} \div \dfrac{3}{10}$.

51. Evaluate: $\left(\dfrac{1}{2}\right)^2 + \left(\dfrac{1}{3}\right)^2$.

52. Multiply: $0.45 \cdot 675$.

53. Divide: $0.2\overline{)34.68}$

54. Evaluate: $-4 - (-7)$.

55. Evaluate: $|-5 - 8|$.

56. Evaluate: $\sqrt{25} - \sqrt{16}$.

Estimation

We now discuss some estimation methods that you can use when working with percent. To begin, we consider a way to find 10% of a number quickly. Recall that 10% of a number is found by multiplying the number by 10% or 0.1. When multiplying a number by 0.1, we simply move the decimal point 1 place to the left to find the result.

EXAMPLE 1   **10% of a number.**   Find 10% of 234.

Solution   To find 10% of 234, move the decimal point 1 place to the left.

234 = 23.4 0

Thus, 10% of 234 is 23.4, or approximately 23.

To find 15% of a number, first find 10% of the number. Then find half of that to obtain the other 5%. Finally, add the two results.

EXAMPLE 2   Estimating 15% of a number.   Estimate 15% of 78.

Solution

10% of 78 is 7.8, or about 8. ⟶   8
Add half of 8 to get the other 5%. ⟶ + 4
                                    ⎯⎯
                                     12

Thus, 15% of 78 is approximately 12.

To find 20% of a number, first find 10% of it and then double that result. A similar procedure can be used when working with any multiple of 10%.

EXAMPLE 3   Estimating 20% of a number.   Estimate 20% of 3,234.15.

Solution   10% of 3,234.15 is 323.415 or about 323. To find 20%, double that.

Thus, 20% of 3,234.15 is approximately 646.

EXAMPLE 4   1% of a number.   Find 1% of 0.8.

Solution   To find 1% of a number, multiply it by 0.01, because 1% = 0.01. When multiplying a number by 0.01, simply move the decimal point 2 places to the left to find the result.

$$0.8 = .00 \, 8$$

$$= 0.008$$

Thus, 1% of 0.8 is 0.008.

EXAMPLE 5   50% of a number.   Find 50% of 2,800,000,000.

Solution   To find 50% of a number means to find $\frac{1}{2}$ of that number. To find one-half of a number, simply divide it by 2. Thus, 50% of 2,800,000,000 is $2,800,000,000 \div 2 = 1,400,000,000$.

To find 25% of a number, first find 50% of it, and then divide that result by 2.

EXAMPLE 6   Estimating 25% of a number.   Estimate 25% of 16,813.

Solution   16,813 is about 16,800. Half of that is 8,400. Thus, 50% of 16,813 is approximately 8,400.
   To estimate 25% of 16,813, divide 8,400 by 2. Thus, 25% of 16,813 is approximately 4,200.

100% of a number is the number itself. To find 200% of a number, double the number.

EXAMPLE 7   Estimating 200% of a number.   Estimate 200% of 65.198.

Solution   65.198 is about 65. To find 200% of 65, double it. Thus, 200% of 65.198 is approximately 65 · 2 or 130.

STUDY SET   *Estimate each answer.*

1. **COLLEGE COURSES** 20% of the 815 students attending a small college were enrolled in a science course. How many students is this?

2. **SPECIAL OFFERS** In the grocery store, a 65-ounce bottle of window cleaner was marked "25% free." How many ounces are free?

3. **DISCOUNTS** By how much is the price of a VCR discounted if the regular price of $196.88 is reduced by 30%?

4. **TIPPING** A restaurant tip is normally 15% of the cost of the meal. Find the tip on a dinner costing $38.64.

5. **FIRE DAMAGE** An insurance company paid 50% of the $107,809 it cost to rebuild a home that was destroyed by fire. How much did the insurance company pay?

6. **SAFETY INSPECTIONS** Of the 2,580 vehicles inspected at a safety checkpoint, 10% had code violations. How many cars had code violations?

7. **WEIGHTLIFTING** A 158-pound weightlifter can bench press 200% of his body weight. How many pounds can he bench press?

8. **TESTING** On a 120-question true/false test, 5% of a student's answers were wrong. How many questions did she miss?

9. **TRAFFIC STUDIES** According to an electronic traffic monitor, 20% of the 650 motorists who passed it were speeding. How many of these motorists were speeding?

10. **SELLING A HOME** A homeowner has been told she will recoup 70% of her $5,000 investment if she paints her home before selling it. What is the potential payback if she paints her home?

Approximate the percent and then estimate each answer.

11. **NO-SHOWS** The attendance at a seminar was only 31% of what the organizers had anticipated. If 68 people were expected, how many actually attended the seminar?

12. **HONOR ROLL** Of the 900 students in a school, 16% were on the principal's honor roll. How many students were on the honor roll?

13. **INTERNET SURVEYS** The illustration shows an online survey question. How many people voted yes?

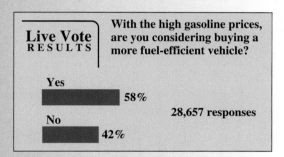

14. **MEDICARE** The Medicare payroll tax rate is 1.45%. How much Medicare tax will be deducted from a paycheck of $596?

15. **VOTING** On election day, 48% of the 6,200 workers at the polls were volunteers. How many volunteers helped with the election?

16. **BUDGETS** Each department at a college was asked to cut its budget by 21%. By how much money should the mathematics department budget be reduced if it is currently $4,515?

5.4 Interest

- Simple interest   - Compound interest

When money is borrowed, the lender expects to be paid back the amount of the loan plus an additional charge for the use of the money. The additional charge is called **interest.** When money is deposited in a bank, the depositor is paid for the use of the money. The money the deposit earns is also called interest. In general, interest is money that is paid for the use of money.

■ Simple interest

Interest is calculated in one of two ways: either as **simple interest** or as **compound interest.** We begin by discussing simple interest. First, we need to introduce some key terms associated with borrowing or lending money.

> **Principal:** the amount of money that is invested, deposited, or borrowed.

> **Interest rate:** a percent that is used to calculate the amount of interest to be paid. It is usually expressed as an annual (yearly) rate.

> **Time:** the length of time (usually in years) that the money is invested, deposited, or borrowed.

The amount of interest to be paid depends on the principal, the rate, and the time. That is why all three are usually mentioned in advertisements for bank accounts, investments, and loans. (See Figure 5-10.)

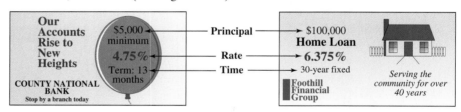

FIGURE 5-10

Simple interest is interest earned on the original principal. It is found by using a formula.

> **Simple interest formula**
>
> Interest = principal · rate · time   or $I = p \cdot r \cdot t$
>
> where the rate r is expressed as an annual rate and the time t is expressed in years. This formula can be written more simply as $I = prt$.

Self Check 1

If $4,200 is invested for 2 years at a rate of 4% annual interest, how much interest is earned?

EXAMPLE 1   If $3,000 is invested for 1 year at a rate of 5%, how much interest is earned?

Solution   We use the formula $I = Prt$ to calculate the interest earned. The principal is $3,000, the interest rate is 5% (or 0.05), and the time is 1 year.

$$P = 3,000 \qquad r = 5\% = 0.05 \qquad t = 1 \text{ year}$$

$I = Prt$	This is the interest formula.
$I = 3,000 \cdot 0.05 \cdot 1$	Substitute the values for P, r, and t.
$I = 150$	Perform the multiplication.

The interest earned in 1 year is $150.

The information given in this problem and the result can be presented in a table.

Principal	Rate	Time	Interest earned
$3,000	5%	1 year	$150

Answer $336

When we use the formula $I = Prt$, the time must be expressed in years. If the time is given in days or months, we rewrite it as a fractional part of a year. For example, a 30-day investment lasts $\frac{30}{365}$ of a year, since there are 365 days in a year. For a 6-month loan, we express the time as $\frac{6}{12}$ or $\frac{1}{2}$ of a year, since there are 12 months in a year.

EXAMPLE 2   To start a business, a couple borrows $5,500 to purchase equipment and supplies. If the loan has a 14% interest rate, how much must they repay at the end of the 90-day period?

Solution   First, we find the amount of interest paid on the loan. We must rewrite the time (90-day period) as a fractional part of a 365-day year.

$$P = 5,500 \qquad r = 14\% = 0.14 \qquad t = \frac{90}{365}$$

$I = Prt$ This is the interest formula.

$I = 5,500 \cdot 0.14 \cdot \dfrac{90}{365}$ Substitute the values for P, r, and t.

$I = \dfrac{5,500}{1} \cdot \dfrac{0.14}{1} \cdot \dfrac{90}{365}$ Write 5,500 and 0.14 as fractions.

$I = \dfrac{69,300}{365}$ Use a calculator to multiply the numerators. Multiply the denominators.

$I \approx 189.86$ Use a calculator to perform the division. Round to the nearest cent.

The interest on the loan is $189.86. To find how much they must pay back, we add the principal and the interest.

$$5,500 + 189.86 = 5,689.86$$

The couple must pay back $5,689.86 at the end of 90 days.

Self Check 2
How much must be repaid if $3,200 is borrowed at a rate of 15% for 120 days?

Answer $3,357.81

Compound interest

Most savings accounts pay **compound interest** rather than simple interest. Compound interest is interest paid on accumulated interest. To illustrate this concept, suppose that $2,000 is deposited in a savings account at a rate of 5% for 1 year. We can use the formula $I = Prt$ to calculate the interest earned at the end of 1 year.

$I = Prt$

$I = 2,000 \cdot 0.05 \cdot 1$ Substitute for P, r, and t.

$I = 100$ Perform the multiplication.

Interest of $100 was earned. At the end of the first year, the account contains the interest ($100) plus the original principal ($2,000), for a balance of $2,100.

Suppose that the money remains in the savings account for another year at the same interest rate. For the second year, interest will be paid on a principal of $2,100.

That is, during the second year, we earn *interest on the interest* as well as on the original $2,000 principal. Using $I = Prt$, we can find the interest earned in the second year.

$$I = Prt$$
$$I = 2{,}100 \cdot 0.05 \cdot 1 \quad \text{Substitute for } P, r, \text{ and } t.$$
$$I = 105 \quad\quad\quad\quad \text{Perform the multiplication.}$$

In the second year, $105 of interest is earned. The account now contains that interest plus the $2,100 principal, for a total of $2,205.

    As Figure 5-11 shows, we calculated the simple interest two times to find the compound interest.

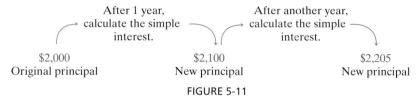

| After 1 year, calculate the simple interest. | | After another year, calculate the simple interest. | |

$2,000	$2,100	$2,205
Original principal	New principal	New principal

FIGURE 5-11

    If we compute the *simple interest* on $2,000, at 5% for 2 years, the interest earned is $I = 2{,}000 \cdot 0.05 \cdot 2 = 200$. Thus, the account balance would be $2,200. Comparing the balances, we find that the account earning compound interest will contain $5 more than the account earning simple interest.

    In the previous example, the interest was calculated at the end of each year, or **annually.** When compounding, we can compute the interest in other time increments, such as **semiannually** (twice a year), **quarterly** (four times a year), or even **daily.**

EXAMPLE 3   Compound interest.

As a gift for her newborn granddaughter, a grandmother opens a $1,000 savings account in the baby's name. The interest rate is 4.2%, compounded quarterly. Find the amount of money the child will have in the bank on her first birthday.

Solution   If the interest is compounded quarterly, the interest will be computed four times in one year. To find the amount of interest $1,000 will earn in the first quarter of the year, we use the simple interest formula, where t is $\frac{1}{4}$ of a year.

Interest earned in the first quarter

$$P = 1{,}000 \quad\quad r = 4.2\% = 0.042 \quad\quad t = \frac{1}{4}$$

$$I = 1{,}000 \cdot 0.042 \cdot \frac{1}{4}$$

$$I = \$10.50$$

The interest earned in the first quarter is $10.50. This now becomes part of the principal for the second quarter.

$$\$1{,}000 + \$10.50 = \$1{,}010.50$$

    To find the amount of interest $1,010.50 will earn in the second quarter of the year, we use the simple interest formula, where t is again $\frac{1}{4}$ of a year.

$$P = 1{,}010.50 \quad\quad r = 0.042 \quad\quad t = \frac{1}{4}$$

$$I = 1{,}010.50 \cdot 0.042 \cdot \frac{1}{4}$$

$$I \approx \$10.61 \quad \text{(Rounded)}$$

The interest earned in the second quarter is $10.61. This becomes part of the principal for the third quarter.

$1,010.50 + $10.61 = $1,021.11

To find the interest $1,021.11 will earn in the third quarter of the year, we proceed as follows.

$$P = \mathbf{1,021.11} \qquad r = \mathbf{0.042} \qquad t = \frac{1}{4}$$

$$I = \mathbf{1,021.11 \cdot 0.042} \cdot \frac{1}{4}$$

$$I \approx \$10.72 \quad \text{(Rounded)}$$

The interest earned in the third quarter is $10.72. This now becomes part of the principal for the fourth quarter.

$1,021.11 + $10.72 = $1,031.83

To find the interest $1,031.83 will earn in the fourth quarter, we again use the simple interest formula.

$$P = \mathbf{1,031.83} \qquad r = \mathbf{0.042} \qquad t = \frac{1}{4}$$

$$I = \mathbf{1,031.83 \cdot 0.042} \cdot \frac{1}{4}$$

$$I \approx \$10.83 \quad \text{(Rounded)}$$

The interest earned in the fourth quarter is $10.83. Adding this to the existing principal, we get

$1,031.83 + $10.83 = $1,042.66

The amount that has accumulated in the account after four quarters, or 1 year, is $1,042.66.

Computing compound interest by hand is tedious. The **compound interest formula** can be used to find the total amount of money that an account will contain at the end of the term.

Compound interest formula

The total amount A in an account can be found using the formula

$$A = P\left(1 + \frac{r}{n}\right)^{nt}$$

where P is the principal, r is the annual interest rate expressed as a decimal, t is the length of time in years, and n is the number of compoundings in one year.

A calculator is often helpful in solving compound interest problems.

Compound interest   CALCULATOR SNAPSHOT

A businessman invests $9,250 at 7.6% interest, to be compounded monthly. To find what the investment will be worth in 3 years, we use the compound interest formula with the following values.

$$P = \$9,250 \quad r = 7.6\% = 0.076 \quad t = 3 \text{ years} \quad n = 12 \text{ times a year (monthly)}$$

(continued)

We apply the compound interest formula.

$$A = P\left(1 + \frac{r}{n}\right)^{nt}$$ This is the compound interest formula.

$$A = 9{,}250\left(1 + \frac{0.076}{12}\right)^{12(3)}$$ Substitute the values of P, r, t, and n. nt means $n \cdot t$.

$$A = 9{,}250\left(1 + \frac{0.076}{12}\right)^{36}$$ Simplify the exponent: $12(3) = 36$.

To evaluate the expression on the right-hand side of the equation, we enter these numbers and press these keys.

9250 ☓ (1 + .076 ÷ 12) y^x 36 = $\boxed{\mathtt{11610.43875}}$

Rounded to the nearest cent, the amount in the account after 3 years will be $11,610.44.

If your calculator does not have parenthesis keys, calculate the sum within the parentheses first. Then find the power. Finally, multiply by 9,250.

Self Check 4

Find the amount of interest $25,000 will earn in 10 years if it is deposited in an account at 5.99% interest, compounded daily.

EXAMPLE 4   A man deposited $50,000 in a long-term account at 6.8% interest, compounded daily. How much money will he be able to withdraw in 7 years if the principal is to remain in the bank?

Solution   "Compounded daily" means that compounding will be done 365 times in a year.

$$P = \$50{,}000 \qquad r = 6.8\% = 0.068 \qquad t = 7 \text{ years} \qquad n = 365 \text{ times a year}$$

$$A = P\left(1 + \frac{r}{n}\right)^{nt}$$ This is the compound interest formula.

$$A = 50{,}000\left(1 + \frac{0.068}{365}\right)^{365(7)}$$ Substitute the values of P, r, t, and n. nt means $n \cdot t$.

$$A = 50{,}000\left(1 + \frac{0.068}{365}\right)^{2{,}555}$$ Multiply in the exponent.

$$A \approx 80{,}477.58$$ Use a calculator. Round to the nearest cent.

The account will contain $80,477.58 at the end of 7 years. To find the amount the man can withdraw, we subtract.

$$80{,}477.58 - 50{,}000 = 30{,}477.58$$

Answer $20,505.20

The man can withdraw $30,477.58 without having to touch the $50,000 principal.

Section 5.4 STUDY SET

VOCABULARY   *Fill in the blanks.*

1. In banking, the original amount of money borrowed or deposited is known as the _____.

2. Borrowers pay _____ to lenders for the use of their money.

3. The percent that is used to calculate the amount of interest to be paid is called the _____ rate.

4. _____ interest is interest paid on accumulated interest.

5. Interest computed on only the original principal is called _____ interest.

6. Percent means parts per _____.

CONCEPTS

7. When we do calculations with percents, they must be changed to decimals or fractions. Change each percent to a decimal.

    a. 7%   **b.** 9.8%   **c.** $6\frac{1}{4}\%$

8. Express each of the following as a fraction of a year. Simplify the fraction.

    a. 6 months   **b.** 90 days

    c. 120 days   **d.** 1 month

9. Complete the table by finding the simple interest earned.

Principal	Rate	Time	Interest earned
$10,000	6%	3 years	

10. Determine how many times a year the interest on a savings account is calculated if the interest is compounded

    a. semiannually   **b.** quarterly

    c. daily   **d.** monthly

11. a. What concept studied in this section is illustrated by the diagram in the illustration?

    b. What was the original principal?

    c. How many times was the interest found?

    d. How much interest was earned on the first compounding?

    e. For how long was the money invested?

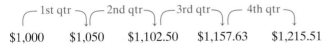

$1,000 $1,050 $1,102.50 $1,157.63 $1,215.51

12. $3,000 is deposited in a savings account that earns 10% interest compounded annually. Complete the series of calculations in the illustration to find how much money will be in the account at the end of 2 years.

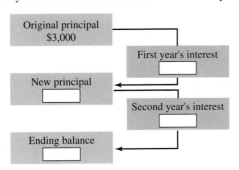

NOTATION

13. In the formula $I = Prt$, what operations are indicated by Prt?

14. In the formula $A = P\left(1 + \dfrac{r}{n}\right)^{nt}$, how many operations must be performed to find A?

APPLICATIONS   *In Problems 15–26, use simple interest.*

15. RETIREMENT INCOME A retiree invests $5,000 in a savings plan that pays 6% per year. What will the account balance be at the end of the first year?

16. INVESTMENTS A developer promised a return of 8% annual interest on an investment of $15,000 in her company. How much could an investor expect to make in the first year?

17. REMODELING A homeowner borrows $8,000 to pay for a kitchen remodeling project. The terms of the loan are 9.2% annual interest and repayment in 2 years. How much interest will be paid on the loan?

18. CREDIT UNIONS A farmer borrowed $7,000 from a credit union. The money was loaned at 8.8% annual interest for 18 months. How much money did the credit union charge him for the use of the money?

19. MEETING PAYROLLS In order to meet end-of-the-month payroll obligations, a small business had to borrow $4,200 for 30 days. How much did the business have to repay if the interest rate was 18%?

20. CAR LOANS To purchase a car, a man takes out a loan for $2,000. If the interest rate is 9% per year, how much interest will he have to pay at the end of the 120-day loan period?

21. SAVINGS ACCOUNTS Find the interest earned on $10,000 at $7\frac{1}{4}\%$ for 2 years. Use the table to organize your work.

P	r	· t	I

22. TUITION A student borrows $300 from an educational fund to pay for books for spring semester. If the loan is for 45 days at $3\frac{1}{2}\%$ annual interest, what will the student owe at the end of the loan period?

23. LOAN APPLICATIONS Complete the following loan application.

Loan Application Worksheet

1. Amount of loan (principal) _$1,200.00_
2. Length of loan (time) _2 YEARS_
3. Annual percentage rate _8%_
4. Interest charged _____
5. Total amount to be repaid _____
6. Check method of repayment:
☐ 1 lump sum   ☑ monthly payments

Borrower agrees to pay _24_ equal payments of _____ to repay loan.

24. LOAN APPLICATIONS Complete the following loan application.

Loan Application Worksheet

1. Amount of loan (principal) _$810.00_
2. Length of loan (time) _9 mos._
3. Annual percentage rate _12%_
4. Interest charged _____
5. Total amount to be repaid _____
6. Check method of repayment:
☐ 1 lump sum   ☑ monthly payments

Borrower agrees to pay _9_ equal payments of _____ to repay loan.

25. LOW-INTEREST LOANS An underdeveloped country receives a low-interest loan from a bank to finance the construction of a water treatment plant. What must the country pay back at the end of 2 years if the loan is for $18 million at 2.3%?

26. REDEVELOPMENT A city is awarded a low-interest loan to help renovate the downtown business district. The $40-million loan, at 1.75%, must be repaid in $2\frac{1}{2}$ years. How much interest will the city have to pay?

A calculator may be helpful in solving these problems.

27. COMPOUNDING ANNUALLY If $600 is invested in an account that earns 8%, compounded annually, what will the account balance be after 3 years?

28. COMPOUNDING SEMIANNUALLY If $600 is invested in an account that earns annual interest of 8%, compounded semiannually, what will the account balance be at the end of 3 years?

29. COLLEGE FUNDS A ninth-grade student opens a savings account that locks her money in for 4 years at an annual rate of 6%, compounded daily. If the initial deposit is $1,000, how much money will be in the account when she begins college in 4 years?

30. CERTIFICATE OF DEPOSITS A 3-year certificate of deposit pays an annual rate of 5%, compounded daily. The maximum allowable deposit is $90,000. What is the most interest a depositor can earn from the CD?

31. TAX REFUNDS A couple deposits an income tax refund check of $545 in an account paying an annual rate of 4.6%, compounded daily. What will the size of the account be at the end of 1 year?

32. INHERITANCES After receiving an inheritance of $11,000, a man deposits the money in an account paying an annual rate of 7.2%, compounded daily. How much money will be in the account at the end of 1 year?

33. LOTTERIES Suppose you won $500,000 in the lottery and deposited the money in a savings account that paid an annual rate of 6% interest, compounded daily. How much interest would you earn each year?

34. CASH GIFTS After receiving a $250,000 cash gift, a university decides to deposit the money in an account paying an annual rate of 5.88%, compounded quarterly. How much money will the account contain in 5 years?

WRITING

35. What is the difference between simple and compound interest?

36. Explain: *Interest is the amount of money paid for the use of money.*

37. On some accounts, banks charge a penalty if the depositor withdraws the money before the end of the term. Why would a bank do this?

38. Explain why it is better for a depositor to open a savings account that pays 5% interest, compounded daily, than one that pays 5% interest, compounded monthly.

REVIEW

39. Simplify: $\sqrt{\frac{1}{4}}$.

40. Find: $\left(\frac{1}{4}\right)^2$.

41. Add: $\frac{3}{7} + \frac{2}{5}$.

42. Subtract: $\frac{3}{7} - \frac{2}{5}$.

43. Multiply: $2\frac{1}{2} \cdot 3\frac{1}{3}$.

44. Divide: $-12\frac{1}{2} \div 5$.

45. Find 6% of 200.

46. Evaluate: $(0.2)^2 - (0.3)^2$.

Percent

Since the word *percent* means *parts per one hundred,* we can think of a percent as the numerator of a fraction that has a denominator of 100.

To change a percent to a fraction, we drop the % symbol and write the given number over 100. Finish each conversion.

$$67\% = \frac{}{100} \qquad 56\% = \frac{56}{} = \frac{14}{25} \qquad 0.05\% = \frac{}{100} = \frac{5}{10,000} = \frac{1}{}$$

To change a percent to a decimal, we drop the % symbol and move the decimal point 2 places to the left. Finish each conversion.

$$67\% = \boxed{} \qquad 56\% = \boxed{} \qquad 0.05\% = \boxed{}$$

To change a fraction to a percent, we write the fraction as a decimal and then move the decimal point 2 places to the right and insert a % symbol. Finish each conversion.

$$\frac{3}{4} = 0.75 = \boxed{} \qquad\qquad \frac{4}{5} = \boxed{} = 80\%$$

$$\frac{5}{8} = \boxed{} = 62.5\% \qquad\qquad \frac{25}{4} = 6.25 = \boxed{}$$

To solve problems involving percent, we use the percent formula.

$$\boxed{\text{Amount}} = \boxed{\text{percent}} \cdot \boxed{\text{base}}$$

Solve each problem.

1. Find 32% of 620.

2. 300 is what percent of 500?

3. 25 is 40% of what number?

4. Find 125% of 850.

5. 106.25 is what percent of 625?

6. 163.84 is 32% of what number?

Percents are used to compute interest. If I is the interest, P is the principal, r is the annual rate (or percent), and t is the length of time in years, the formula for simple interest is

$$\boxed{I = Prt}$$

7. Find the amount of interest that will be earned if $10,000 is invested for 5 years at 6% annual interest.

If A is the amount, P is the principal, r is the annual rate of interest, t is the length of time in years, and n is the number of compoundings in one year, the formula for compound interest is

$$\boxed{A = P\left(1 + \frac{r}{n}\right)^{nt}}$$

8. Find the amount of interest that will be earned if $10,000 is invested for 5 years at 6% annual interest, compounded quarterly.

ACCENT ON TEAMWORK

SECTION 5.1

M & M'S   Give each member of your group a bag of M & M's candies.

a. Determine what percent of the total number of M & M's in your bag are yellow. Do the same for each of the other colors. Enter the results in the table. (Round to the nearest 1 percent.)

M & M's color	Percent
Yellow	
Brown	
Green	
Red	
Blue	

b. Present the data in the table using the circle graph. Compare your graph to the graphs made by the other members of your group. Do the colors occur in the same percentages in each of the bags?

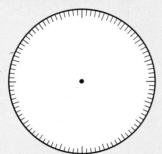

SECTION 5.2

NUTRITION   Have each person in your group bring in a nutrition label like that shown and write the name of the food product on the back. Have the members of the group exchange labels. With the label that you receive, determine what percent of the total calories come from fat.

The USDA recommends that no more than 30% of a person's daily calories should come from fat. Which products exceed the recommendation?

Nutrition Facts

Serving Size 1 meal

Amount Per Serving

Calories 560   Calories from Fat 190

	% Daily Value
Total fat 21g	**32%**
Saturated fat 9 g	**43%**
Cholesterol 60mg	**20%**
Sodium 2110mg	**88%**
Total carbohydrate 67g	**22%**
Dietary fiber 7g	**29%**
Sugars less than 25g	
Protein 27g	

SECTION 5.3

ENROLLMENTS   From your school's admissions office, get the enrollment figures for the last ten years. Calculate the percent of increase (or decrease) in enrollment for each of the following periods.

- Ten years ago to the present
- Five years ago to the present
- One year ago to the present

NEWSPAPER ADS   Have each person in your group find a newspaper advertisement for some item that is on sale. The ad should include only two of the four details listed below.

- The regular price
- The sale price
- The discount
- The discount rate

For example, the following ad gives the regular price and the sale price, but it doesn't give the discount or the discount rate.

Have the members of your group exchange ads. Determine the two missing details on the ad that you receive. In your group, which item had the highest discount rate?

SECTION 5.4

INTEREST RATES   Recall that interest is money that the borrower pays to the lender for the use of the money. The amount of interest that the borrower must pay depends on the interest rate charged by the lender.

Have members of your group call banks, savings and loans, credit unions, and other financial services to get the lending rates for various types of loans.

Find out what rate is charged by credit cards such as VISA, department stores, and gasoline companies. List the interest rates in order, from greatest to least, and present your findings to the class.

CHAPTER REVIEW

| SECTION 5.1 | *Percents, Decimals, and Fractions* |

CONCEPTS

Percent means parts per one hundred.

To change a percent to a fraction, drop the % symbol and put the given number over 100.

To change a percent to a decimal, drop the % symbol and divide by 100 by moving the decimal point 2 places to the left.

To change a decimal to a percent, multiply the decimal by 100 by moving the decimal point 2 places to the right, and then insert a % symbol.

To change a fraction to a percent, write the fraction as a decimal by dividing its numerator by its denominator. Multiply the decimal by 100 by moving the decimal point 2 places to the right, and then insert a % symbol.

REVIEW EXERCISES

Express the amount of each figure that is shaded as a percent, as a decimal, and as a fraction. Each set of squares represents 100%.

1.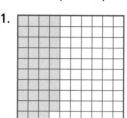

2.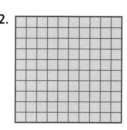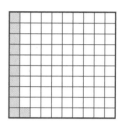

3. In Problem 1, what percent of the figure is not shaded?

Change each percent to a fraction.

4. 15% **5.** 120% **6.** $9\frac{1}{4}\%$ **7.** 0.1%

Change each percent to a decimal.

8. 27% **9.** 8% **10.** 155% **11.** $1\frac{4}{5}\%$

Change each decimal to a percent.

12. 0.83 **13.** 0.625 **14.** 0.051 **15.** 6

Change each fraction to a percent.

16. $\frac{1}{2}$ **17.** $\frac{4}{5}$ **18.** $\frac{7}{8}$ **19.** $\frac{1}{16}$

Find the exact percent equivalent for each fraction.

20. $\frac{1}{3}$ **21.** $\frac{5}{6}$

Change each fraction to a percent. Round to the nearest hundredth of a percent.

22. $\frac{5}{9}$ **23.** $\frac{8}{3}$

24. BILL OF RIGHTS There are 27 amendments to the Constitution of the United States. The first ten are known as the Bill of Rights. What percent of the amendments were adopted after the Bill of Rights? (Round to the nearest 1 percent.)

25. Express one-tenth of a percent as a fraction.

The percent formula:

Amount = percent · base

26. Identify the amount, the base, and the percent in the statement "15 is $33\frac{1}{3}$% of 45."

27. Translate the given sentence into a percent equation:

What number   is   32%   of   96?

Solve each percent problem.

28. What number is 40% of 500?

29. 16% of what number is 20?

30. 1.4 is what percent of 80?

31. $66\frac{2}{3}$% of 3,150 is what number?

32. Find 220% of 55.

33. What is 0.05% of 60,000?

34. RACING The nitro–methane fuel mixture used to power some experimental cars is 96% nitro and 4% methane. How many gallons of each fuel component are needed to fill a 15-gallon fuel tank?

35. HOME SALES After the first day on the market, 51 homes in a new subdivision had already sold. This was 75% of the total number of homes available. How many homes were originally for sale?

36. HURRICANE DAMAGE In a mobile home park, 96 of the 110 trailers were either damaged or destroyed by hurricane winds. What percent is this? (Round to the nearest 1 percent.)

37. TIPPING The cost of dinner for a family of five at a restaurant was $36.20. Find the amount of the tip if it should be 15% of the cost of dinner.

38. AIR POLLUTION Complete the circle graph to show the given data.

A *circle graph* is a way of presenting data for comparison. The sizes of the segments of the circle indicate the percents of the whole represented by each category.

Sources of carbon monoxide air pollution	
Transportation vehicles	63%
Fuel combustion in homes, offices, electrical plants	12%
Industrial processes	8%
Solid waste disposal	3%
Miscellaneous	14%

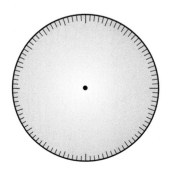

39. EARTH'S SURFACE The surface of the Earth is approximately 196,800,000 square miles. Use the information in the illustration to determine the number of square miles of the Earth's surface that are covered with water.

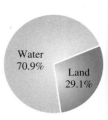

Water 70.9%

Land 29.1%

| **SECTION 5.3** | *Applications of Percent* |

To find the total price of an item:

> Total price = purchase price + sales tax

40. SALES RECEIPTS Complete the sales receipt.

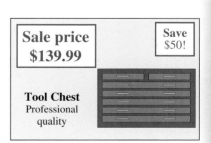

CAMERA CENTER

35mm Canon Camera	$59.99

SUBTOTAL $59.99
SALES TAX @ 5.5%
TOTAL

41. SALES TAX RATES Find the sales tax rate if the sales tax is $492 on the purchase of an automobile priced at $12,300.

Commission is based on a percent of the total dollar amount of the goods or services sold.

42. COMMISSIONS If the commission rate is 6%, find the commission earned by an appliance salesperson who sells a washing machine for $369.97 and a dryer for $299.97.

43. Fill in the blank: Always find the percent of increase or decrease of a quantity with respect to the _____ amount.

To find *percent of increase or decrease:*

1. Subtract the smaller number from the larger to find the amount of increase or decrease.

2. Find what percent the difference is of the original amount.

44. PEACEKEEPING The size of a peacekeeping force was increased from 10,000 to 12,500 troops. Find the percent of increase.

45. GAS MILEAGE Experimenting with a new brand of gasoline in her truck, a woman found that the gas mileage fell from 18.8 to 17.0 miles per gallon. Find the percent of decrease. (Round to the nearest tenth of a percent.)

The difference between the original price and the sale price of an item is called the *discount*.

To find the *sale price:*

> Sale price = original price − discount

46. TOOL CHESTS Use the information in the ad to find the discount, the original price, and the discount rate on the tool chest.

Sale price $139.99

Save $50!

Tool Chest
Professional quality

Simple interest is interest earned on the original principal and is found using the formula

$$I = Prt$$

where P is the principal, r is the annual interest rate, and t is the length of time in years.

Compound interest is interest earned on interest.

The compound interest formula is

$$A = P\left(1 + \frac{r}{n}\right)^{nt}$$

where A is the amount in the account, P is the principal, r is the annual interest rate, n is the number of compoundings in one year, and t is the length of time in years.

47. Find the interest earned on $6,000 invested at 8% per year for 2 years. Use the following table to organize your work.

P	r	t	I

48. CODE VIOLATIONS A business was ordered to correct safety code violations in a production plant. To pay for the needed corrections, the company borrowed $10,000 at 12.5% for 90 days. Find the total amount that had to be paid after 90 days.

49. MONTHLY PAYMENTS A couple borrows $1,500 for 1 year at $7\frac{3}{4}$% and decides to repay the loan by making 12 equal monthly payments. How much will each monthly payment be?

50. Find the amount of money that will be in a savings account at the end of 1 year if $2,000 is the initial deposit and the annual interest rate of 7% is compounded semi-annually. (*Hint:* Find the simple interest twice.)

51. Find the amount that will be in a savings account at the end of 3 years if a deposit of $5,000 earns interest at an annual rate of $6\frac{1}{2}$%, compounded daily.

52. CASH GRANTS Each year a cash grant is given to a deserving college student. The grant consists of the interest earned that year on a $500,000 savings account. What is the cash award for the year if the money is invested at an annual rate of 8.3%, compounded daily?

1. Express the amount of the figure that is shaded as a percent, as a fraction, and as a decimal.

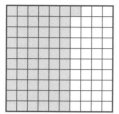

2. In the illustration, each set of 100 squares represents 100%. Express as a percent the amount of the figure that is shaded. Then express that percent as a fraction and as a decimal.

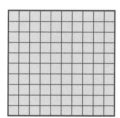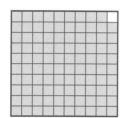

3. Change each percent to a decimal.

   a. 67%   b. 12.3%   c. $9\frac{3}{4}\%$

4. Change each fraction to a percent.

   a. $\frac{1}{4}$ b. $\frac{5}{8}$ c. $\frac{3}{25}$

5. Change each decimal to a percent.
   a. 0.19   b. 3.47   c. 0.005

6. Change each percent to a fraction.
   a. 55%   b. 0.01%   c. 125%

7. Change $\frac{7}{30}$ to a percent. Round to the nearest hundredth of a percent.

8. WEATHER REPORTS A weatherman states that there is a 40% chance of rain. What are the chances that it will not rain?

9. Find the exact percent equivalent for the fraction $\frac{2}{3}$.

10. Find the exact percent equivalent for the fraction $\frac{1}{4}$.

11. SHRINKAGE See the following label on a new pair of jeans.

    a. How much length will be lost due to shrinkage?

    b. What will be the resulting length?

WAIST	INSEAM
33	34

    Expect shrinkage of approximately **3%** in length after the jeans are washed.

12. 65 is what percent of 1,000?

13. TIPPING Find the amount of a 15% tip on a meal costing $25.40.

14. FUGITIVES As of October 2004, 450 of the 479 fugitives who have appeared on the FBI's Ten Most Wanted list have been apprehended. What percent is this? Round to the nearest tenth of a percent.

15. SWIMMING WORKOUTS A swimmer was able to complete 18 laps before a shoulder injury forced him to stop. This was only 20% of a typical workout. How many laps does he normally complete during a workout?

16. COLLEGE EMPLOYEES The 700 employees at a community college fall into three major categories, as shown. How many employees are in administration?

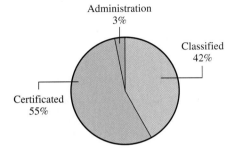

Administration
3%

Classified
42%

Certificated
55%

17. What number is 24% of 600?

18. HAIRCUTS The illustration shows the number of minutes it took the average U.S. worker to earn enough to pay for a man's haircut in 1950 and 1998. Find the percent of decrease, to the nearest one percent.

1950 **63 minutes**

1998 **46 minutes**

19. INSURANCE An insurance sales-person receives a 4% commission on the annual premium of any policy she sells. Find her commission on a homeowner's policy if the premium is $898.

20. COST-OF-LIVING INCREASES A teacher earning $40,000 just received a cost-of-living increase of 3.6%. What is the teacher's new salary?

21. AUTO CARE A car waxing kit, regularly priced at $14.95, is on sale for $3 off. Find the sale price, the discount, and the rate.

22. POPULATION INCREASES After a new freeway was completed, the population of a city it passed through increased from 12,808 to 15,565 in two years. Find the percent of increase. (Round to the nearest one percent.)

23. Find the simple interest on a loan of $3,000 at 5% per year for 1 year.

24. Find the amount of interest earned on an investment of $24,000 paying an annual rate of 6.4% interest, compounded daily for 3 years.

25. POLITICAL ADS Explain what is unclear about the following ad.

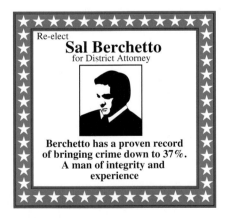

Re-elect
Sal Berchetto
for District Attorney

Berchetto has a proven record of bringing crime down to 37%. A man of integrity and experience

26. In Section 5.4, we discussed *interest*. What is interest?

1. **SHAQUILLE O'NEAL** Use the data in the table to complete the illustration by drawing a line graph to chart the growth of Shaquille O'Neal, the Los Angeles Lakers' center.

Age (yr)	4	6	8	10	12	16	21	28
Weight (lb)	56	82	108	139	192	265	302	315

Based on data from *Los Angeles Times*

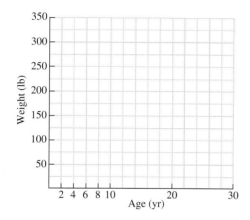

2. Use the numbers 6 and 8 to state the commutative property of multiplication.

3. **a.** Find the factors of 40.
   b. Find the prime factorization of 40.

4. **AUTO INSURANCE** See the premium comparison in the illustration. Find the average (mean) insurance premium for the companies listed.

Allstate	$2,672	Mercury	$1,370
Auto Club	$1,680	State Farm	$2,737
Farmers	$2,485	20th Century	$1,692

Criteria: Six-month premium. Husband, 45, drives a 1995 Explorer, 12,000 annual miles. Wife, 43, drives a 1996 Dodge Caravan, 12,000 annual miles. Son, 17, is an occasional operator. All have clean driving records.

5. **PAINTING** A square tarp has sides 8 feet long. When it is laid out on a floor, how much area will it cover?

6. Evaluate: $-12 - (-5)$.

7. Evaluate: $12 - 2[-8 - 2^4(-1)]$.

8. Find: $|-55|$.

9. Evaluate: $(-3)^2$.

10. Multiply: $(-2)(-3)(-5)$.

11. **FRUIT STORAGE** Use the formula $C = \frac{5}{9}(F - 32)$ to find the missing number in the label shown.

12. Simplify: $2 - 4 \cdot 3$.

13. **SPELLING** What fraction of the letters in the word *Mississippi* are vowels?

14. Simplify: $\frac{10}{15}$.

Perform each operation.

15. $-\frac{16}{35} \cdot \frac{25}{48}$

16. $4\frac{2}{5} \div 11$

17. $\frac{4}{3} + \frac{2}{7}$

18. $34\frac{1}{9} - 13\frac{5}{6}$

19. $\left(-\dfrac{2}{3}\right)^2$ **20.** $\left(-\dfrac{2}{3}\right)^3$

21. $78.1 - 7.81$ **22.** $2.13(-4.05)$

23. $0.752(1,000)$ **24.** $\dfrac{241.86}{2.9}$

25. Round 452.0298 to the nearest hundredth.

26. Round 452.0298 to the nearest thousandth.

27. Write $\dfrac{11}{15}$ as a decimal. Use an overbar.

28. Simplify: $\sqrt{\dfrac{4}{9}}$.

29. Simplify: $3\sqrt{81} - 8\sqrt{49}$.

30. LABOR COSTS A car repair bill is shown. The bill is torn, and one line cannot be read. How many hours of labor did it take to repair the car?

> **_Brian Wood_ Auto Repair**
>
> Parts...$175.00
> Total labor (at $35 an hour)...........................
> Total..$297.50

31. Complete the table.

Percent	Decimal	Fraction
	0.29	
47.3%		
		$\dfrac{7}{8}$

32. 16% of what number is 20?

33. TIPPING Complete the sales draft if a 15% tip, rounded up to the next dollar, is to be left for the waiter.

> **STEAK STAMPEDE**
> **Bloomington, MN**
> **Server #12\ AT**
>
> **VISA**   67463777288
> **NAME**   DALTON/ LIZ
>
> **AMOUNT** $75.18
> **GRATUITY $**_____
> **TOTAL $**_____

34. GENEALOGY Through an extensive computer search, a genealogist determined that worldwide, 180 out of every 10 million people had his last name. What percent is this?

35. SAVINGS ACCOUNTS Find the simple interest earned on $10,000 at $7\frac{1}{4}\%$ for 2 years.

36. Find 100% of 50.

Ratio, Proportion, and Measurement

CORBIS

TLE We are all familiar with the basic American units of measurement: feet, pounds, and gallons. However, most of the countries in the world use a different system of measurement called the *metric system*. In the metric system, the basic unit of length is the meter, the basic unit of weight is the gram, and the basic unit of capacity is the liter.

The metric system was invented by French scientists in the late 18th century. Until then, units of measure differed from country to country and even village to village. The objective was to create worldwide standard units of measurement based on the decimal system rather than fractions.

To learn more about the metric system, visit The Learning Equation on the Internet at http://tle.brookscole.com. (The log-in instructions are in the Preface.) For Chapter 6, the online lesson is:

• *TLE* Lesson 12: Measurement

Check Your Knowledge

1. A _____ is a quotient of two numbers or a quotient of two quantities with the same units.

2. A _____ is a statement that two ratios or rates are equal.

3. Inches, feet, and miles are examples of American units of _____. Meters, grams, and liters are units of measurement in the _____ system.

4. In the American system, temperatures are measured in degrees _____. In the metric system, temperatures are measured in degrees _____.

Reduce ratios to lowest terms.

5. What is the ratio of 14 pounds to 10 pounds?

6. What is the ratio of 8 inches to 3 feet?

7. David drives a hybrid gas–electric automobile. Recently he drove 517 miles using 11 gallons of gasoline. How many miles per gallon (mpg) did he get?

8. Specialty coffee is for sale in 12-ounce bags for $8.95 and in 2-pound bags for $23.95. Which is the better buy?

9. Chelsea threw 20 darts and hit the target 16 times. Find her ratio of hits to misses?

10. Determine whether each statement is a proportion.

    a. $\dfrac{33}{44} = \dfrac{39}{52}$ b. $\dfrac{3.5}{2.7} = \dfrac{10.5}{8.3}$

11. Are the numbers 17, 27 and 125, 199 proportional?

Solve each proportion.

12. $\dfrac{12}{x} = \dfrac{3}{17}$ 13. $-\dfrac{3}{8} = \dfrac{15}{x}$

14. $\dfrac{13.2}{4} = \dfrac{x}{7.8}$ 15. $\dfrac{x}{0.04} = \dfrac{-0.35}{0.07}$

16. Convert 5 meters to centimeters.

17. How many yards are in 29 feet?

18. How many feet are in 5 miles?

19. How many inches are in 7 yards?

20. Convert 100 meters to yards (1 m ≈ 1.0936 yd). Round to one decimal place.

21. Convert 5 feet 9 inches to centimeters (1 in. ≈ 2.54 cm).

22. To estimate the size of the coyote population in a state park, rangers trapped, tagged, and released 50 coyotes. Later, a random sample of 100 coyotes included only 2 tagged animals. Estimate the size of the coyote population.

23. One-half inch on a map corresponds to 25 miles. Two cities are 175 miles apart. How many inches separate the two cities on the map?

24. Convert −40° Fahrenheit to Celsius. $\left(Hint: C = \dfrac{5}{9}(F - 32) \right)$

Study Skills Workshop

STUDYING FOR TESTS

Doing homework regularly is important, but tests require another type of strategy for preparation. Before you take a test, it is important that you have done your best to commit the important concepts to memory.

How Much Time Should I Devote to Preparing for a Test? The time needed to prepare for a test will vary according to how well you have understood and memorized the basic concepts. Plan to prepare at least four days in advance of your test and schedule this on your study calendar.

How Do I Prepare? Four days before the test Know exactly what material the test will cover. Then imagine that you could bring one $8\frac{1}{2}'' \times 11''$ sheet of paper to the test. What would you write on that sheet? Go through each section (and your reworked notes) to identify the key rules and definitions to include on your study sheet. Keep this paper with you all the time until the test, and review it whenever you can.

Three days before the test Go through your reworked notes and find the examples that your instructor gave you in class. Add any examples you might have missed to your paper and continue to look at it whenever you have time.

Two days before the test Use a chapter test in your textbook, or make up your own practice test by choosing a sampling of problems from your text or reworked notes. You can also take a quiz on chapter and section material online using the iLrn Web site rather than create your own test. This site is located at www.iLrn.com, and the quizzes are found in the "Tutorial" section. Try to include the same number of questions that your real test will have. Choose problems that have a solution that can be checked. Then, *with your book closed,* take a *timed* trial. Don't be upset if you can't do everything perfectly; it's only a trial. When you are done, check your answers. Be honest with yourself! Make a list of the topics that were difficult and add these to your study sheet.

One day before the test Get help, if necessary, with yesterday's problems from your instructor during office hours, from a tutor at the tutorial center, or from a classmate. Practice your trial test again, without books or notes. Go back over any problem you didn't get correct. Get plenty of rest the night before your test.

Test day Review your study sheet, if you have time. Focus on how well you have prepared and relax as much as possible. When taking your test, complete the problems that you are sure of first. Skip the problems that you don't understand right away, and return to them later. Be aware of time throughout the test so that you don't spend too long on any one problem.

ASSIGNMENT

1. Four days before your test, prepare your study sheet.
2. Three days before your test, go through all of the examples in your reworked notes.
3. Two days before your test, make a written practice test and time youreslf on it. Then check your answers.
4. The day before the test, fix the problems that you did incorrectly. Use your textbook, notes, classmate, tutor, or instructor for help. Review all of your practice problems until you can do them without any help. Gather the materials that you will need for your test and get enough rest.
5. On test day, arrive in class a few minutes early. Look over your study sheet if you have time. Relax as much as possible, knowing that you have prepared well.

Ratios and proportions are used to compare quantities. They are also used to convert from one unit of measurement to another.

6.1 Ratios

- Ratios • Rates • Unit rates • Unit costs

The concept of *ratio* occurs often in real-life situations.

To prepare fuel for an outboard marine engine, gasoline must be mixed with oil in the ratio of 50 to 1.

To make 14-karat jewelry, gold is combined with other metals in the ratio of 14 to 10.

In this drawing, the eyes-to-nose distance and the nose-to-chin distance are drawn using a ratio of 2 to 3.

In this section, we demonstrate how *ratios* and *rates* are used to express relationships between two quantities.

▌ Ratios

Ratios give us a way to compare numerical quantities.

> **Ratios**
>
> A **ratio** is the quotient of two numbers or the quotient of two quantities that have the same units.

There are three ways to write a ratio: as a fraction, as two numbers separated by the word *to,* or as two numbers separated by a colon. For example, the ratios described in the examples above can be written in three ways.

$$\frac{50}{1}, \qquad 14 \text{ to } 10, \qquad \text{and} \qquad 2{:}3$$

- The fraction $\frac{50}{1}$ is read as "the ratio of 50 to 1."
- 14 to 10 is read as "the ratio of 14 to 10."
- 2:3 is read as "the ratio of 2 to 3."

Self Check 1

Express each ratio as a fraction in lowest terms:

a. the ratio of 9 to 12

b. 3.2:16

EXAMPLE 1 Express each ratio as a fraction in lowest terms: **a.** the ratio of 25 to 10   and   **b.** the ratio of 0.3 to 1.2.

Solution

a. To write the phrase "the ratio of 25 to 10" in fractional form, we write 25 as the numerator and 10 as the denominator. Then we simplify the fraction.

$$\frac{25}{10} = \frac{\overset{1}{\cancel{5}} \cdot 5}{2 \cdot \cancel{5}}$$ Factor 25 as 5 · 5 and 10 as 2 · 5. Then divide out the common factor of 5.

$$= \frac{5}{2}$$

The ratio 25 to 10 can be written as the fraction $\frac{25}{10}$, which simplifies to $\frac{5}{2}$. Because the fractions $\frac{25}{10}$ and $\frac{5}{2}$ represent equal numbers, they are **equal ratios.**

b. The ratio of 0.3 to 1.2 can be written as the fraction $\frac{0.3}{1.2}$. To write this as a ratio of whole numbers, we need to clear it of decimals. We can do so by multiplying the numerator and denominator by 10.

$$\frac{0.3}{1.2} = \frac{0.3 \cdot \mathbf{10}}{1.2 \cdot \mathbf{10}}$$

$$= \frac{3}{12}$$ Perform the multiplications: $0.3 \cdot 10 = 3$ and $1.2 \cdot 10 = 12$.

$$= \frac{1}{4}$$ Simplify the fraction: $\frac{3}{12} = \frac{\overset{1}{\cancel{3}} \cdot 1}{\cancel{3} \cdot 4} = \frac{1}{4}$.

Answers **a.** $\frac{3}{4}$, **b.** $\frac{1}{5}$

EXAMPLE 2 Carry-on luggage.

Airlines allow passengers to carry a piece of luggage onto an airplane only if it will fit in the space shown in Figure 6-1. What is the ratio of the length of the space to its width?

Solution Since the length of the carry-on luggage space is 24 inches and its width is 10 inches, the ratio of the length to the width is $\frac{24 \text{ inches}}{10 \text{ inches}}$. Common factors and common units should be divided out.

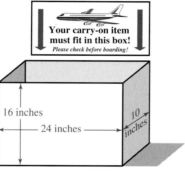

FIGURE 6-1

16 inches · 24 inches · 10 inches

$$\frac{24 \text{ inches}}{10 \text{ inches}} = \frac{\overset{1}{\cancel{2}} \cdot 12 \;\cancel{\text{inches}}}{\underset{1}{\cancel{2}} \cdot 5 \;\cancel{\text{inches}}}$$ Divide out the common factor of 2 and divide out the common units of inches.

$$= \frac{12}{5}$$

The length-to-width ratio of the carry-on space is $\frac{12}{5}$.

Self Check 2
What is the ratio of the height to the length of the carry-on space in Figure 6-1?

Answer $\frac{2}{3}$

EXAMPLE 3 Express the phrase *12 ounces to 1 pound* as a ratio in lowest terms.

Solution Recall that a ratio is a comparison of two quantities with the same units. Since there are 16 ounces in 1 pound, the phrase *12 ounces to 1 pound* can be expressed as the ratio $\frac{12 \text{ ounces}}{16 \text{ ounces}}$, which simplifies to $\frac{3}{4}$.

Self Check 3
Express the phrase *2 feet to 1 yard* as a ratio. (*Hint:* 3 feet = 1 yard.)

Answer $\frac{2}{3}$

Student-to-Instructor Ratio

"A more personal classroom atmosphere can sometimes be an easier adjustment for college freshmen. They are less likely to feel like a number, a feeling that can sometimes impact students' first semester grades."
From *The Importance of Class Size* by Stephen Pemberton

The data below come from a nationwide study of mathematics programs at two-year colleges. Determine which course has the lowest student-to-instructor ratio. (Assume that there is one instructor per section.)

	Basic Mathematics	Elementary Algebra	Intermediate Algebra
Students enrolled	119,350	268,152	243,828
Number of sections	5,425	11,173	9,378

Source: Conference Board of the Mathematical Science, 2000 CBMS Survey of Undergraduate Programs

Rates

When we compare two different kinds of quantities, we call the comparison a **rate,** and we can write it as a fraction. For example, on the label of the can of paint in Figure 6-2, we see that 1 quart of paint is needed for every 200 square feet to be painted. Writing this as a rate, we have

$$\frac{1 \text{ quart}}{200 \text{ square feet}}$$ Read as "1 quart per 200 square feet."

When writing a rate, always include the units.

Keep out of reach of children
Avoid contact with eyes and skin

LATEX SEMI-GLOSS ENAMEL

Dries in one hour
COVERAGE: one quart covers 200 square feet

FIGURE 6-2

> **Rates**
>
> A **rate** is a quotient of two quantities with different units.

Self Check 4
The fastest-growing flowering plant on record grew 12 feet in 14 days. Find the rate of growth over this period.

EXAMPLE 4   Snowfall.   According to the *Guinness Book of World Records,* a total of 78 inches of snow fell at Mile 47 Camp, Cooper River Division, Arkansas, in a 24-hour period in 1963. What was the rate of snowfall?

Solution   We begin by comparing the amount of snow, 78 inches, to the elapsed time, 24 hours. Then we simplify the fraction.

$$\frac{78 \text{ inches}}{24 \text{ hours}} = \frac{\overset{1}{\cancel{6}} \cdot 13 \text{ inches}}{4 \cdot \underset{1}{\cancel{6}} \text{ hours}}$$ Factor 78 and 24. Then divide out the common factor of 6.

The snow fell at a rate of 13 inches per 4 hours: $\frac{13 \text{ inches}}{4 \text{ hours}}$.

Answer $\dfrac{6 \text{ feet}}{7 \text{ days}}$

Unit rates

A **unit rate** is a rate in which the denominator is 1. To illustrate the concept of a unit rate, suppose a driver makes the 354-mile trip from Pittsburgh to Indianapolis in 6 hours. Then the motorist's rate (or more specifically, rate of speed) is given by

$$\frac{354 \text{ miles}}{6 \text{ hours}} = \frac{\overset{1}{\cancel{6}} \cdot 59 \text{ miles}}{\underset{1}{\cancel{6}} \cdot 1 \text{ hours}} = \frac{59 \text{ miles}}{1 \text{ hour}} \qquad \text{Factor 354 as } 6 \cdot 59 \text{ and divide out the common factor of 6.}$$

We can also find the unit rate by dividing 354 by 6.

$$
\begin{array}{r}
59 \\
6\overline{)354} \\
\underline{30} \\
54 \\
\underline{54} \\
0
\end{array}
$$

The unit rate $\frac{59 \text{ miles}}{1 \text{ hour}}$ can be expressed in any of the following forms:

$$59 \, \frac{\text{miles}}{\text{hour}}, \quad 59 \text{ miles per hour}, \quad 59 \text{ miles/hour}, \quad \text{or} \quad 59 \text{ mph}$$

EXAMPLE 5   A student earns $152 for working 16 hours in a bookstore. Find his hourly rate of pay.

Solution   We can write the rate of pay as

$$\text{Rate of pay} = \frac{\$152}{16 \text{ hr}} \qquad \text{Compare the amount of money earned to the number of hours worked.}$$

To find the rate of pay for 1 hour of work, we divide 152 by 16.

$$
\begin{array}{r}
9.5 \\
16\overline{)152.0} \\
\underline{144} \\
8\,0 \\
\underline{8\,0} \\
0
\end{array}
\qquad \text{Write a decimal point and a 0 to the right of 2.}
$$

The unit rate of pay is $\frac{\$9.50}{1 \text{ hour}}$, which can be written as $9.50 per hour.

Self Check 5
Joan earns $436 per 40-hour week managing a dress shop. Find her hourly rate of pay.

Answer $10.90 per hour

EXAMPLE 6   **Energy consumption.**   One household used 795 kilowatt-hours (kwh) of electricity during a 30-day period. Find the rate of energy consumption in kilowatt-hours per day.

Solution   We can write the rate of energy consumption as

$$\text{Rate of energy consumption} = \frac{795 \text{ kwh}}{30 \text{ days}}$$

To find the unit rate, we divide 795 by 30.

$$\text{Unit rate of energy consumption} = \frac{26.5 \text{ kwh}}{1 \text{ day}}$$

The rate of energy consumption was 26.5 kilowatt-hours per day.

Self Check 6
To heat a house for 30 days, a furnace burned 69 therms of natural gas. Find the rate of gas consumption in terms per day.

Answer 2.3 therms per day

CALCULATOR SNAPSHOT   **Computing gas mileage**

A man drove from Houston to St. Louis—a total of 775 miles. Along the way, he stopped for gas three times, pumping 10.5, 11.3, and 8.75 gallons of gas. He started with the tank half full and ended with the tank half full. To find how many miles he got per gallon (mpg), we need to compare the total distance to the total number of gallons of gas consumed.

$$\frac{775 \text{ miles}}{(10.5 + 11.3 + 8.75) \text{ gallons}}$$

We can simplify this rate by entering these numbers and pressing these keys on a scientific calculator.

775 ÷ (10.5 + 11.3 + 8.75) = | 25.368249 |

To the nearest hundredth, he got 25.37 mpg.

Unit costs

If a store sells 5 pounds of coffee for $18.75, a consumer might want to know what the coffee costs per pound. When we find the cost of 1 pound of the coffee, we are finding a *unit cost*. To find the unit cost of an item, we begin by comparing its cost to its quantity.

$$\frac{\$18.75}{5 \text{ pounds}}$$

Then we divide the cost by the number of items.

$$\begin{array}{r} 3.75 \\ 5)\overline{18.75} \end{array}$$

The unit cost of the coffee is $3.75 per pound.

Self Check 7
A fast-food restaurant sells a 12-ounce cola for 72¢ and a 16-ounce cola for 99¢. Which is the better buy?

EXAMPLE 7 Comparison shopping. Olives come packaged in a 10-ounce jar, which sells for $2.49, or in a 6-ounce jar, which sells for $1.53. (See Figure 6-3.) Which is the better buy?

Solution To find the better buy, we must find each unit cost.

10-ounce jar:

$$\frac{\$2.49}{10 \text{ oz}} = \frac{249¢}{10 \text{ oz}}$$ Change $2.49 to 249 cents.

$$= 24.9¢ \text{ per oz}$$ Divide 249 by 10.

6-ounce jar:

$$\frac{\$1.53}{6 \text{ oz}} = \frac{153¢}{6 \text{ oz}}$$ Change $1.53 to 153 cents.

$$= 25.5¢ \text{ per oz}$$ Perform the division.

FIGURE 6-3

One ounce for 24.9¢ is a better buy than one ounce for 25.5¢. The unit cost is less when olives are packaged in 10-ounce jars, so that is the better buy.

Answer the 12-oz cola

Section 6.1 STUDY SET

VOCABULARY *Fill in the blanks.*

1. A _____ is a quotient of two numbers or a quotient of two quantities with the same units.

2. A quotient of two quantities with different units is called a _____.

3. When the price of candy is advertised as $1.75 per pound, we are told its unit _____.

4. A _____ rate is a rate in which the denominator is 1.

CONCEPTS

5. To write the ratio $\frac{15}{24}$ in lowest terms, we divide out any common factors of the numerator and denominator. What common factor do they have?

6. Complete the solution. Write the ratio $\frac{14}{21}$ in lowest terms.

$$\frac{14}{21} = \frac{2 \cdot 7}{\boxed{} \cdot \boxed{}} = \frac{2 \cdot \overset{1}{\cancel{7}}}{\boxed{} \cdot \underset{1}{\cancel{7}}} = \boxed{}$$

7. Consider the ratio $\frac{0.5}{0.6}$. By what number should we multiply numerator and denominator to make this a ratio of whole numbers?

8. What should be done to write $\frac{15 \text{ inches}}{22 \text{ inches}}$ in simplest form?

9. Since a ratio is a comparison of quantities with the same units, how should the ratio $\frac{11 \text{ minutes}}{1 \text{ hour}}$ be rewritten?

10. **a.** Consider the rate $\frac{\$248}{16 \text{ hours}}$. How can we find the unit rate ($ per hour)?

    b. Consider the rate $\frac{\$7.95}{3 \text{ pairs}}$. How can we find the unit cost of a pair of socks?

NOTATION

11. Write the ratio of the flag's length to its width using a fraction, using the word *to*, and using a colon.

9 inches

13 inches

12. The rate $\frac{55 \text{ miles}}{1 \text{ hour}}$ can be expressed as

   • 55 _____ _____ _____ (in three words)

   • 55 _____ / _____ (in two words)

   • 55 __ __ __ (in three letters)

PRACTICE *Write each comparison as a ratio in simplest form, using a fraction.*

13. 5 to 7   14. 3 to 5

15. 17 to 34   16. 19 to 38

17. 22:33   18. 14:21

19. 1.5:2.4   20. 0.9:0.6

21. 7 to 24.5   22. 0.65 to 0.15

23. 4 ounces to 12 ounces   24. 3 inches to 15 inches

25. 12 minutes to 1 hour   26. 8 ounces to 1 pound

27. 3 days to 1 week   28. 4 inches to 2 yards

29. 18 months to 2 years   30. 8 feet to 4 yards

Refer to the monthly budget shown in the illustration. Give each ratio in lowest terms.

31. Find the total amount of the budget.

32. Find the ratio of the amount budgeted for rent to the total budget.

33. Find the ratio of the amount budgeted for food to the total budget.

34. Find the ratio of the amount budgeted for the phone to the total budget.

Item	Amount
Rent	$800
Food	$600
Gas and electric	$180
Phone	$100
Entertainment	$120

*Refer to the list of tax deductions shown in the illustration.
Give each ratio in lowest terms.*

35. Find the total amount of the deductions.

36. Find the ratio of the real estate tax deduction to the total deductions.

37. Find the ratio of the charitable contributions to the total deductions.

38. Find the ratio of the mortgage interest deduction to the union dues deduction.

Item	Amount
Medical expenses	$875
Real estate taxes	$1,250
Charitable contributions	$1,750
Mortgage interest	$4,375
Union dues	$500

Write each rate as a fraction in simplest form.

39. 64 feet in 6 seconds

40. 45 applications for 18 openings

41. 84 made out of 100 attempts

42. 75 days on 20 gallons of water

43. 3,000 students over a 16-year career

44. 16 right compared to 34 wrong

45. 18 beats every 12 measures

46. 1.5 inches as a result of 30 turns

Write each phrase as a unit rate.

47. 60 revolutions in 5 minutes

48. 14 trips every 2 months

49. 12 errors in 8 hours

50. $50,000 paid over 10 years

51. 245 presents for 35 children

52. 108 occurrences in a 12-month period

53. 4,000,000 people living in 12,500 square miles

54. 117.6 pounds of pressure on 8 square inches

Find the unit cost.

55. $3.50 for 50 feet

56. 150 barrels cost $4,950.

57. 65 ounces sell for 78 cents.

58. They charged $48 for 15 minutes.

59. Four of us donated a total of $272.

60. For 7 dozen, you will pay $10.15.

61. $4 billion over a 5-month span

62. 7,020 pesos will buy six stickers.

APPLICATIONS

63. ART HISTORY Leonardo da Vinci drew the human figure shown within a square. (All four sides of a square are the same length.) Find the ratio of the length of the man's outstretched arms to his height.

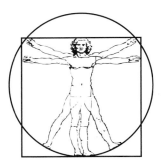

64. FLAGS The checkered flag is composed of squares (All four sides of a square are the same length.) What is the ratio of the width of the flag to its length?

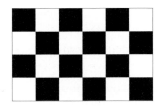

65. GEAR RATIOS Refer to the illustration. Find the ratio of the number of teeth of the larger gear to the number of teeth of the smaller gear.

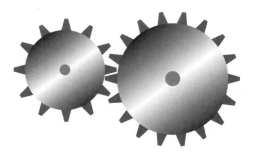

66. BANKRUPTCIES After declaring bankruptcy, a company could reimburse its creditors only 5¢ on the dollar. Write this as a ratio in lowest terms.

67. COOKING A recipe from *Easy Living* magazine is shown. Write the ratio of sugar to milk as a fraction. **Do not simplify the ratio.**

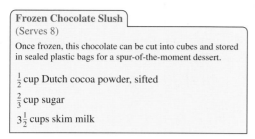

> **Frozen Chocolate Slush**
> (Serves 8)
>
> Once frozen, this chocolate can be cut into cubes and stored in sealed plastic bags for a spur-of-the-moment dessert.
>
> $\frac{1}{2}$ cup Dutch cocoa powder, sifted
>
> $\frac{2}{3}$ cup sugar
>
> $3\frac{1}{2}$ cups skim milk

68. HEARING From the graph, determine the ratio of hearing loss in males as compared to females.

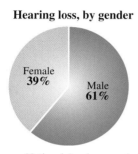

Hearing loss, by gender

Female **39%** Male **61%**

Source: National Academy on Aging

69. SOFTBALL Lisa Fernandez led the U.S. women's softball team in winning a gold medal at the 2004 Olympic games. Her hitting statistics are shown below. What was her rate of hits (H) to at-bats (AB) during the Olympic competition?

	BA	AB	H	R	2B	3B	HR	RBI
Fernandez	.545	22	12	3	3	0	1	8

70. TYPING A secretary typed a document containing 330 words in 5 minutes. How many words per minute did he type?

71. CPR A paramedic performed 125 compressions to 50 breaths on an adult with no pulse. What compressions-to-breaths rate did the paramedic use?

72. INTERNET SALES A web site determined that it had 112,500 hits in one month. Of those visiting the site, 4,500 made purchases. How many did not make a purchase? Find the browser/buyers unit rate for the web site that mouth.

73. AIRLINE COMPLAINTS An airline had 3.29 complaints for every 1,000 passengers. Write this rate as a fraction of whole numbers.

74. FINGERNAILS On average, fingernails grow 0.02 inch per week. Write this rate using whole numbers.

75. FACULTY–STUDENT RATIOS At a college, there are 125 faculty members and 2,000 students. Find the rate of faculty to students. (This is often referred to as the faculty–student *ratio*, even though the units are different.)

76. PARKING METERS A parking meter requires 25¢ for 20 minutes of parking. Find the unit cost.

77. UNIT COSTS A driver pumped 17 gallons of gasoline into his tank at a cost of $32.13. Find the unit cost of the gasoline.

78. UNIT COSTS A 50-pound bag of grass seed costs $222.50. Find the unit cost of grass seed.

79. UNIT COSTS A 12-ounce can of cranberry juice sells for 84¢. Find the unit cost in cents per ounce.

80. UNIT COSTS A 24-ounce package of green beans sells for $1.29. Find the unit cost in cents per ounce.

81. COMPARISON SHOPPING A 6-ounce can of orange juice sells for 89¢, and an 8-ounce can sells for $1.19. Which is the better buy?

82. COMPARISON SHOPPING A 30-pound bag of fertilizer costs $12.25, and an 80-pound bag costs $30.25. Which is the better buy?

83. COMPARISON SHOPPING A certain brand of cold and sinus medication is sold in 20-tablet boxes for $4.29 and in 50-tablet boxes for $9.59. Which is the better buy?

84. COMPARISON SHOPPING Which tire shown is the better buy?

ECONOMY
$30.99
35,000-mile warranty

PREMIUM
$37.50
40,000-mile warranty

85. COMPARING SPEEDS A car travels 345 miles in 6 hours, and a truck travels 376 miles in 6.2 hours. Which vehicle is going faster?

86. READING SPEEDS One seventh-grader read a 54-page book in 40 minutes. Another read an 80-page book in 62 minutes. If the books were equally difficult, which student read faster?

87. EMPTYING TANKS An 11,880-gallon tank can be emptied in 27 minutes. Find the rate of flow in gallons per minute.

88. RATES OF PAY Ricardo worked for 27 hours to help insulate a hockey arena. For his work, he received $337.50. Find his hourly rate of pay.

89. AUTO TRAVEL A car's odometer reads 34,746 at the beginning of a trip. Five hours later, it reads 35,071. How far has the car traveled? What is the average rate of speed?

90. RATES OF SPEED An airplane travels from Chicago to San Francisco, a distance of 1,883 miles, in 3.5 hours. Find the average rate of speed of the plane.

91. GAS MILEAGE One car went 1,235 miles on 51.3 gallons of gasoline, and another went 1,456 miles on 55.78 gallons. Which car got the better gas mileage?

92. ELECTRICITY RATES In one community, a bill for 575 kilowatt-hours of electricity is $38.81. In a second community, a bill for 831 kwh is $58.10. In which community is electricity cheaper?

WRITING

93. Are the ratios 3 to 1 and 1 to 3 the same? Explain why or why not.

94. Give three examples of ratios (or rates) that you have encountered in the past week.

95. How will the topics studied in this section make you a better shopper?

96. What is a unit rate? Give some examples.

REVIEW   *Perform each operation.*

97. $3.05 + 17.17 + 25.317$

98. $3.5\overline{)157.85}$

99. $13.2 + 25.07 \cdot 7.16$

100. $\dfrac{4}{3} - \dfrac{1}{4}$

6.2 Proportions

- Proportions • Means and extremes of a proportion • Solving proportions
- Writing proportions to solve problems

Like any tool, a ladder can be dangerous if used improperly. A safety pamphlet states, "When setting up an extension ladder, use the *4-to-1 rule*—For every 4 feet of ladder height, position the legs of the ladder 1 foot away from the base of the wall." The 4-to-1 rule for ladders can be expressed using a ratio.

$$\frac{4 \text{ feet}}{1 \text{ foot}} = \frac{4 \overset{1}{\cancel{\text{ feet}}}}{1 \underset{1}{\cancel{\text{ foot}}}} = \frac{4}{1}$$

In Figure 6-4, the 4-to-1 rule was used to position the legs of a ladder properly, 3 feet from the base of a 12-foot-high wall. We can write a ratio comparing the ladder's height to its distance from the wall.

$$\frac{12 \text{ feet}}{3 \text{ feet}} = \frac{12 \overset{1}{\cancel{\text{ feet}}}}{3 \underset{1}{\cancel{\text{ feet}}}} = \frac{12}{3}$$

Since this ratio satisfies the 4-to-1 rule, the two ratios $\frac{4}{1}$ and $\frac{12}{3}$ must be equal. Therefore, we have

$$\frac{4}{1} = \frac{12}{3}$$

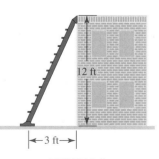

12 ft

|← 3 ft →|

FIGURE 6-4

Such equations, which show that two ratios are equal, are called *proportions*. In this section, we introduce the concept of proportion and use proportions to solve many different types of problems.

Proportions

> **Proportion**
> A **proportion** is a statement that two ratios (or rates) are equal.

Some examples of proportions are

$$\frac{1}{2} = \frac{3}{6} \quad \text{and} \quad \frac{3 \text{ waiters}}{7 \text{ tables}} = \frac{9 \text{ waiters}}{21 \text{ tables}}$$

- The proportion $\frac{1}{2} = \frac{3}{6}$ can be read as "1 is to 2 as 3 is to 6."
- The proportion $\frac{3 \text{ waiters}}{7 \text{ tables}} = \frac{9 \text{ waiters}}{21 \text{ tables}}$ can be read as "3 waiters are to 7 tables as 9 waiters are to 21 tables."

The terms of the proportion $\frac{1}{2} = \frac{3}{6}$ are numbered as follows:

$$\text{First term} \longrightarrow \frac{1}{2} = \frac{3}{6} \longleftarrow \text{Third term}$$
$$\text{Second term} \longrightarrow \quad\quad \longleftarrow \text{Fourth term}$$

Means and extremes of a proportion

In any proportion, the first and fourth terms are called the **extremes.** The second and third terms are called the **means.**

In the proportion $\frac{1}{2} = \frac{3}{6}$, 1 and 6 are the **extremes,** and 2 and 3 are the **means.**

In this proportion, the product of the extremes is equal to the product of the means.

$$1 \cdot 6 = 6 \quad \text{and} \quad 2 \cdot 3 = 6$$

This example illustrates a fundamental property of proportions.

> **Fundamental property of proportions**
> In any proportion, the product of the extremes is equal to the product of the means.

As another example, we consider the proportion $\frac{5}{6} = \frac{10}{12}$.

- The product of the extremes is $5 \cdot 12 = 60$.
- The product of the means is $6 \cdot 10 = 60$.

Again, we see that in a proportion, the product of the extremes (60) is equal to the product of the means (60).

To determine whether an equation is a proportion, we can check whether the product of the extremes is equal to the product of the means.

Self Check 1
Determine whether the equation
is a proportion:

$$\frac{6}{13} = \frac{18}{39}$$

EXAMPLE 1   Determine whether each equation is a proportion: **a.** $\frac{3}{7} = \frac{9}{21}$ and   **b.** $\frac{8}{3} = \frac{13}{5}$.

Solution   In each case, we check to see whether the product of the extremes is equal to the product of the means.

a. The product of the extremes is $3 \cdot 21 = 63$. The product of the means is $7 \cdot 9 = 63$.

$$3 \cdot 21 = 63 \qquad 7 \cdot 9 = 63$$
$$\frac{3}{7} = \frac{9}{21}$$

Since the products are equal, the equation is a proportion: $\frac{3}{7} = \frac{9}{21}$. The product of the extremes and the product of the means are also known as **cross products.**

b. The product of the extremes is $8 \cdot 5 = 40$. The product of the means is $3 \cdot 13 = 39$.

$$8 \cdot 5 = 40 \qquad 3 \cdot 13 = 39$$
$$\frac{8}{3} = \frac{13}{5}$$

Answer yes

Since the cross products are not equal, the equation is not a proportion: $\frac{8}{3} \neq \frac{13}{5}$.

When two pairs of numbers such as 2, 3 and 8, 12 form a proportion, we say that they are **proportional.** To show that 2, 3 and 8, 12 are proportional, we check to see whether the equation

$$\frac{2}{3} = \frac{8}{12}$$

is a proportion. To do so, we find the product of the extremes and the product of the means.

$$2 \cdot 12 = 24 \qquad 3 \cdot 8 = 24$$

Since the cross products are equal, the equation is a proportion, and the numbers are proportional.

Self Check 2
Determine whether 6, 11 and
54, 99 are proportional.

EXAMPLE 2   Determine whether 3, 7 and 36, 91 are proportional.

Solution   We check to see whether $\frac{3}{7} = \frac{36}{91}$ is a proportion by finding two products.

$$3 \cdot 91 = 273 \qquad \text{This is the product of the extremes.}$$
$$7 \cdot 36 = 252 \qquad \text{This is the product of the means.}$$

Answer yes

Since the cross products are not equal, the numbers are not proportional.

Solving proportions

Suppose that we know three terms in the following proportion.

$$\frac{?}{5} = \frac{24}{20}$$

To find the missing term, we represent it by the letter x, multiply the extremes and multiply the means, set them equal, and solve for x.

$$\frac{x}{5} = \frac{24}{20}$$

$x \cdot 20 = 5 \cdot 24$ In a proportion, the product of the extremes is equal to the product of the means.

$x \cdot 20 = 120$ Perform the multiplication: $5 \cdot 24 = 120$.

$\dfrac{x \cdot 20}{20} = \dfrac{120}{20}$ To undo the multiplication by 20, divide both sides by 20.

$x = 6$ Perform the divisions.

The first term is 6. To check this result, we substitute 6 for x in $\frac{x}{5} = \frac{24}{20}$ and find the cross products.

Check:

$$\frac{6}{5} \overset{?}{=} \frac{24}{20} \qquad 6 \cdot 20 = 120$$
$$5 \cdot 24 = 120$$

Since the cross products are equal, x is 6.

EXAMPLE 3   Solve the proportion $\dfrac{12}{18} = \dfrac{3}{x}$ for x.

Solution

$$\frac{12}{18} = \frac{3}{x}$$

$12 \cdot x = 18 \cdot 3$ In a proportion, the product of the extremes equals the product of the means.

$12 \cdot x = 54$ Multiply: $18 \cdot 3 = 54$.

$\dfrac{12 \cdot x}{12} = \dfrac{54}{12}$ To undo the multiplication by 12, divide both sides by 12.

$x = \dfrac{9}{2}$ Simplify: $\dfrac{54}{12} = \dfrac{\overset{1}{\cancel{6}} \cdot 9}{2 \cdot \underset{1}{\cancel{6}}} = \dfrac{9}{2}$.

Thus, x is $\dfrac{9}{2}$. Check this result in the proportion.

Self Check 3

Solve the proportion $\dfrac{15}{x} = \dfrac{20}{32}$ for x. Check the result.

Answer 24

EXAMPLE 4   Find the third term of the proportion $\dfrac{3.5}{7.2} = \dfrac{x}{15.84}$.

Solution

$$\frac{3.5}{7.2} = \frac{x}{15.84}$$

$3.5(15.84) = 7.2 \cdot x$ In a proportion, the product of the extremes equals the product of the means.

$55.44 = 7.2 \cdot x$ Multiply: $3.5(15.84) = 55.44$.

$\dfrac{55.44}{7.2} = \dfrac{7.2 \cdot x}{7.2}$ To undo the multiplication by 7.2, divide both sides by 7.2.

$7.7 = x$ Perform the divisions.

Thus, x is 7.7. Check the result in the proportion.

Self Check 4

Find the second term of the proportion $\dfrac{6.7}{x} = \dfrac{33.5}{38}$.

Answer 7.6

Solving proportions with a calculator

To solve the proportion in Example 4 with a calculator, we can proceed as follows.

$$\frac{3.5}{7.2} = \frac{x}{15.84}$$

$$\frac{3.5(15.84)}{7.2} = x \qquad \text{Multiply both sides by 15.84 to isolate } x.$$

We can find x by entering these numbers and pressing these keys on a scientific calculator.

$$3.5 \boxed{\times} 15.84 \boxed{\div} 7.2 \boxed{=}$$

$$\boxed{7.7}$$

Thus, $x = 7.7$.

◼ Writing proportions to solve problems

We can use proportions to solve many real-world problems. If we are given a ratio (or rate) comparing two quantities, the words of the problem can be translated to a proportion, and we can solve it to find the unknown.

Self Check 5

If 9 tickets to a concert cost $112.50, find the cost of 15 tickets.

EXAMPLE 5   Grocery shopping.   If 5 apples cost $1.15, find the cost of 16 apples.

Solution   Let c represent the cost of 16 apples. If we compare the number of apples to their cost, we know that the two rates are equal.

5 apples is to $1.15 as 16 apples is to $c.

$$\begin{array}{c} \text{5 apples} \rightarrow \\ \text{Cost of 5 apples} \rightarrow \end{array} \frac{5}{1.15} = \frac{16}{c} \begin{array}{c} \leftarrow \text{16 apples} \\ \leftarrow \text{Cost of 16 apples} \end{array}$$

To find the cost of 16 apples, we solve the proportion for c.

$5 \cdot c = 1.15(16)$ In a proportion, the product of the extremes is equal to the product of the means.

$5 \cdot c = 18.4$ Perform the multiplication: $1.15(16) = 18.4$.

$\dfrac{5 \cdot c}{5} = \dfrac{18.4}{5}$ To undo the multiplication by 5, divide both sides by 5.

$c = 3.68$ Perform the divisions.

Sixteen apples will cost $3.68. To check the result, we substitute 3.68 for c in the proportion and find the cross products.

Check:

$$\frac{5}{1.15} \stackrel{?}{=} \frac{16}{3.68} \qquad \begin{array}{l} 5 \cdot 3.68 = \mathbf{18.4} \\ 1.15 \cdot 16 = \mathbf{18.4} \end{array}$$

Answer $187.50

The cross products are equal. The result 3.68 checks.

In Example 5, we could have compared the cost of the apples to the number of apples: \$1.15 is to 5 apples as \$$c$ is to 16 apples. This would have led to the proportion

$$\text{Cost of 5 apples} \rightarrow \frac{1.15}{5} = \frac{c}{16} \leftarrow \text{Cost of 16 apples}$$
$$\text{5 apples} \rightarrow \qquad\qquad \leftarrow \text{16 apples}$$

If we solve this proportion for c, we obtain the same result: $c = 3.68$.

! COMMENT When solving problems using proportions, make sure that the units of the numerators are the same and the units of the denominators are the same. For Example 5, it would be incorrect to write

$$\text{Cost of 5 apples} \rightarrow \frac{1.15}{5} = \frac{16}{c} \leftarrow \text{16 apples}$$
$$\text{5 apples} \rightarrow \qquad\qquad \leftarrow \text{Cost of 16 apples}$$

EXAMPLE 6   Scale drawings.   A **scale** is a ratio (or rate) that compares the size of a model, drawing, or map to the size of an actual object. The airplane in Figure 6-5 is drawn using a scale of 1 inch: 6 feet. This means that 1 inch on the drawing is actually 6 feet on the plane. The distance from wing tip to wing tip (the wingspan) on the drawing is 5 inches. What is the actual wingspan of the plane?

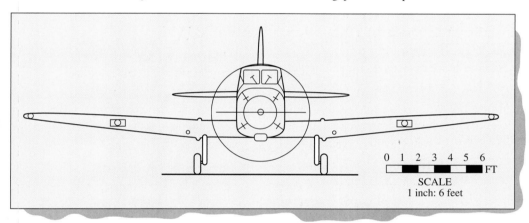

FIGURE 6-5

Solution   Let w represent the actual wingspan of the plane. Since 1 inch corresponds to 6 feet as 5 inches corresponds to w feet, we can write the proportion.

$$\text{Measure on drawing} \rightarrow \frac{1}{6} = \frac{5}{w} \leftarrow \text{Measure on drawing}$$
$$\text{Measure on plane} \rightarrow \qquad\qquad \leftarrow \text{Measure on plane}$$

$$1 \cdot w = 6 \cdot 5 \qquad \text{In a proportion, the product of the extremes is equal to the product of the means.}$$

$$w = 30 \qquad \text{Perform the multiplications.}$$

The actual wingspan of the plane is 30 feet. Check the result by finding the cross products.

EXAMPLE 7   Baking.   A recipe for rhubarb cake calls for $1\frac{1}{4}$ cups of sugar for every $2\frac{1}{2}$ cups of flour. How many cups of flour are needed if the baker intends to use 3 cups of sugar?

Self Check 7
How many cups of sugar will be needed to make several rhubarb cakes that will require a total of 25 cups of flour?

Solution   Let f represent the number of cups of flour to be mixed with the sugar. The ratios of the cups of sugar to the cups of flour are equal. We have $1\frac{1}{4}$ cups sugar is to $2\frac{1}{2}$ cups flour as 3 cups sugar is to f cups flour. We can write the proportion.

$$\begin{array}{l} 1\frac{1}{4} \text{ cups sugar} \rightarrow \\ \\ 2\frac{1}{2} \text{ cups flour} \rightarrow \end{array} \dfrac{1\dfrac{1}{4}}{2\dfrac{1}{2}} = \dfrac{3}{f} \begin{array}{l} \leftarrow 3 \text{ cups sugar} \\ \\ \leftarrow f \text{ cups flour} \end{array}$$

$$\dfrac{1.25}{2.5} = \dfrac{3}{f} \qquad \text{Change the fractions to decimals: } 1\tfrac{1}{4} = 1.25 \text{ and } 2\tfrac{1}{2} = 2.5.$$

$$1.25 \cdot f = 2.5 \cdot 3 \qquad \text{In a proportion, the product of the extremes is equal to the product of the means.}$$

$$1.25 \cdot f = 7.5 \qquad \text{Perform the multiplication: } 2.5 \cdot 3 = 7.5.$$

$$\dfrac{1.25 \cdot f}{\mathbf{1.25}} = \dfrac{7.5}{\mathbf{1.25}} \qquad \text{To undo the multiplication by 1.25, divide both sides by 1.25.}$$

$$f = 6 \qquad \text{Perform the divisions.}$$

The baker should use 6 cups of flour.

EXAMPLE 8   Manufacturing.   In a manufacturing process, 15 parts out of 90 were found to be defective. How many defective parts will be expected in a run of 120 parts?

Solution   Let d represent the expected number of defective parts. In each run, the ratio of the defective parts to the total number of parts should be the same: 15 defective parts is to 90 as d defective parts is to 120.

$$\begin{array}{l} 15 \text{ defective parts} \rightarrow \\ 90 \text{ parts} \rightarrow \end{array} \dfrac{15}{90} = \dfrac{d}{120} \begin{array}{l} \leftarrow \quad d \text{ defective parts} \\ \leftarrow \quad 120 \text{ parts} \end{array}$$

$$15 \cdot 120 = 90 \cdot d \qquad \text{In a proportion, the product of the extremes is equal to the product of the means.}$$

$$1{,}800 = 90 \cdot d \qquad \text{Perform the multiplication: } 15 \cdot 120 = 1{,}800.$$

$$\dfrac{1{,}800}{\mathbf{90}} = \dfrac{90 \cdot d}{\mathbf{90}} \qquad \text{To undo the multiplication by 90, divide both sides by 90.}$$

$$20 = d \qquad \text{Divide: } \tfrac{1{,}800}{90} = 20.$$

The expected number of defective parts is 20.

Section 6.2 STUDY SET

VOCABULARY   *Fill in the blanks.*

1. A _____ is a statement that two ratios or rates are equal.

2. In $\frac{1}{2} = \frac{5}{10}$, the terms 1 and 10 are called the _____ of the proportion and the _____ 2 and 5 are called the _____ of the proportion.

3. The product of the extremes and the product of the means of a proportion are also known as _____ products.

4. When two pairs of numbers form a proportion, we say that the numbers are _____.

CONCEPTS   *Fill in the blanks.*

5. The equation $\frac{2}{5} = \frac{4}{10}$ will be a proportion if the product ▦ · 10 is equal to the product ▦ · 4.

6. ▦ · 10 = ▦   2 · ▦ = ▦

$$\frac{9}{2} = \frac{45}{10}$$

7. Write each statement as a proportion.

   a. 5 is to 8 as 15 is to 24.

   b. 3 teacher's aides are to 25 children as 12 teacher's aides are to 100 children.

8. Consider the proportion $\frac{3}{4} = \frac{15}{20}$. What are the two cross products?

9. For every 15 feet of chain link fencing, 4 support posts are used. How many support posts will be needed for 300 feet of chain link fence? Which of the following proportions could be used to solve this problem?

   i. $\frac{15}{4} = \frac{300}{x}$ **ii.** $\frac{15}{4} = \frac{x}{300}$

   iii. $\frac{4}{15} = \frac{300}{x}$ **iv.** $\frac{4}{15} = \frac{x}{300}$

10. Write a problem that could be solved using the following proportion.

   Ounces of cashews → $\dfrac{4}{639} = \dfrac{10}{x}$ ← Ounces of cashews
   Calories →   ← Calories

NOTATION   *Complete each solution.*

11. Solve for x: $\dfrac{12}{18} = \dfrac{x}{24}$.

$$12 \cdot 24 = \boxed{}$$
$$\boxed{} = 18 \cdot x$$
$$\frac{288}{\boxed{}} = \frac{18 \cdot x}{\boxed{}}$$
$$16 = x$$

12. Solve for x: $\dfrac{14}{x} = \dfrac{49}{17.5}$.

$$14 \cdot \boxed{} = x \cdot 49$$
$$\boxed{} = x \cdot 49$$
$$\frac{245}{\boxed{}} = \frac{x \cdot 49}{\boxed{}}$$
$$5 = x$$

PRACTICE   *Determine whether each statement is a proportion.*

13. $\dfrac{9}{7} = \dfrac{81}{70}$ **14.** $\dfrac{5}{2} = \dfrac{20}{8}$

15. $\dfrac{7}{3} = \dfrac{14}{6}$ **16.** $\dfrac{13}{19} = \dfrac{65}{95}$

17. $\dfrac{9}{19} = \dfrac{38}{80}$ **18.** $\dfrac{40}{29} = \dfrac{29}{22}$

19. $\dfrac{10.4}{3.6} = \dfrac{41.6}{14.4}$ **20.** $\dfrac{13.23}{3.45} = \dfrac{39.96}{11.35}$

21. $\dfrac{\frac{2}{3}}{\frac{5}{8}} = \dfrac{\frac{4}{5}}{\frac{9}{16}}$ **22.** $\dfrac{\frac{3}{2}}{\frac{8}{9}} = \dfrac{\frac{1}{4}}{\frac{4}{27}}$

23. $\dfrac{4\frac{1}{6}}{\frac{12}{7}} = \dfrac{2\frac{3}{16}}{\frac{9}{10}}$ **24.** $\dfrac{2\frac{1}{2}}{\frac{4}{5}} = \dfrac{3\frac{3}{4}}{\frac{9}{10}}$

Solve for the variable in each proportion. **Check each result.**

25. $\dfrac{2}{3} = \dfrac{x}{6}$ **26.** $\dfrac{3}{6} = \dfrac{x}{8}$

27. $\dfrac{5}{10} = \dfrac{3}{c}$ **28.** $\dfrac{7}{14} = \dfrac{2}{x}$

29. $\dfrac{6}{x} = \dfrac{8}{4}$ **30.** $\dfrac{4}{x} = \dfrac{2}{8}$

31. $\dfrac{x}{8} = \dfrac{9}{2}$ **32.** $\dfrac{x}{2} = \dfrac{18}{6}$

33. $\dfrac{x}{2.5} = \dfrac{3.7}{9.25}$ **34.** $\dfrac{8.5}{x} = \dfrac{4.25}{1.7}$

35. $\dfrac{-0.8}{2} = \dfrac{x}{5}$ **36.** $\dfrac{0.9}{0.3} = \dfrac{-6}{x}$

37. $\dfrac{-0.4}{1.2} = \dfrac{-6}{x}$ **38.** $\dfrac{-5}{x} = \dfrac{-2}{4.4}$

39. $\dfrac{x}{5.2} = \dfrac{-4.65}{7.8}$ **40.** $\dfrac{-8.6}{2.4} = \dfrac{x}{6}$

41. $\dfrac{4{,}000}{x} = \dfrac{3.2}{2.8}$ **42.** $\dfrac{0.4}{1.6} = \dfrac{96.7}{x}$

43. $\dfrac{\frac{1}{2}}{\frac{1}{5}} = \dfrac{x}{2\frac{1}{4}}$ **44.** $\dfrac{x}{4\frac{1}{10}} = \dfrac{3\frac{3}{4}}{1\frac{7}{8}}$

APPLICATIONS   *Set up and solve a proportion.*

45. SCHOOL LUNCHES A manager of a school cafeteria orders 750 pudding cups. What will the order cost if she purchases them wholesale, 6 cups for $1.75?

46. CLOTHES SHOPPING As part of a spring clearance, a men's store put dress shirts on sale, 2 for $25.98. How much will a businessman pay if he buys five shirts?

47. GARDENING Three packets of garden seeds sell for 98¢. A Girl Scout troop leader needs to purchase three dozen packets. What will they cost?

48. COOKING A recipe for spaghetti sauce requires four 16-ounce bottles of ketchup to make 2 gallons of sauce. How many bottles of ketchup are needed to make 10 gallons of sauce?

49. BUSINESS PERFORMANCE The following bar graph shows the yearly costs incurred and the revenue received by a business. How do the ratios of costs to revenue for 2003 and 2004 compare?

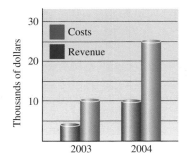

50. RAMPS Write a ratio of the rise to the run for each ramp shown. Set the ratios equal. Is the resulting proportion true? Is one ramp steeper than the other?

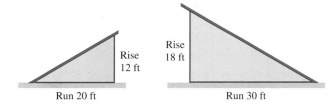

51. MIXING PERFUMES A perfume is to be mixed in the ratio of 3 drops of pure essence to 7 drops of alcohol. How many drops of pure essence should be mixed with 56 drops of alcohol?

52. MAKING COLOGNE A cologne can be made by mixing 2 drops of pure essence with 5 drops of distilled water. How much water should be used with 15 drops of pure essence?

53. LAB WORK In a red blood cell count, a drop of the patient's diluted blood is placed on a grid like that shown in the next column. Instead of counting each and every red blood cell in the 25-square grid, a technician counts only the number of cells in the five highlighted squares. Then he or she uses a proportion to estimate the total red blood cell count. If there are 195 red blood cells in the blue squares, about how many red blood cells are in the entire grid?

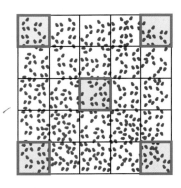

54. DOSAGES The proper dosage of a certain medication for a 30-pound child is shown. At this rate, what would be the dosage for a 45-pound child?

55. MAKING COOKIES A recipe for chocolate chip cookies calls for $1\frac{1}{4}$ cups of flour and 1 cup of sugar. The recipe will make $3\frac{1}{2}$ dozen cookies. How many cups of flour will be needed to make 12 dozen cookies?

56. MAKING BROWNIES A recipe for brownies calls for 4 eggs and $1\frac{1}{2}$ cups of flour. If the recipe makes 15 brownies, how many cups of flour will be needed to make 130 brownies?

57. COMPUTER SPEED Using the *Mathematica 3.0* program, a Dell Dimension XPS R350 (Pentium II) computer can perform a set of 15 calculations in 2.85 seconds. How long will it take the computer to perform 100 such calculations?

58. QUALITY CONTROL Out of a sample of 500 men's shirts, 17 were rejected because of crooked collars. How many crooked collars would you expect to find in a run of 15,000 shirts?

59. FUEL CONSUMPTION A "high mobility multipurpose wheeled vehicle" is better known as a Hummer. Under normal conditions, a Hummer can travel 325 miles on a full tank (25 gallons) of diesel. How far can it travel on its auxiliary tank, which holds 17 gallons of diesel?

60. ANNIVERSARY GIFTS A florist sells a dozen long-stemmed red roses for $57.99. In honor of their 16th wedding anniversary, a man wants to buy 16 roses for his wife. What will the roses cost?

61. PAYCHECKS Billie earns $412 for a 40-hour week. If she missed 10 hours of work last week, how much did she get paid?

62. STAFFING A school board has determined that there should be 3 teachers for every 50 students. Complete the table by filling in the number of teachers needed at each school.

	Glenwood High	Goddard Junior High	Sellers Elementary
Enrollment	2,700	1,900	850
Teachers			

63. BLUEPRINTS The scale for the drawing in the blueprint tells the reader that a $\frac{1}{4}$-inch length ($\frac{1}{4}''$) on the drawing corresponds to an actual size of 1 foot ($1'0''$). Suppose the length of the kitchen is $2\frac{1}{2}$ inches on the blueprint. How long is the actual kitchen?

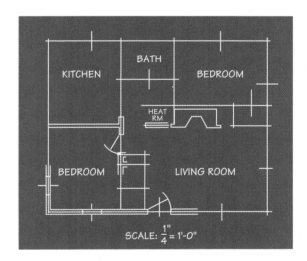

64. DRAFTING In a scale drawing, a 280-foot antenna tower is drawn 7 inches high. The building next to it is drawn 2 inches high. How tall is the actual building?

65. MODEL RAILROADS An HO-scale model railroad engine is 9 inches long. If HO scale is 87 feet to 1 foot, how long is a real engine?

66. MODEL RAILROADS An N-scale model railroad caboose is 4 inches long. If N scale is 169 feet to 1 foot, how long is a real caboose?

67. CAROUSELS The ratio in the illustration in the next column indicates that 1 inch on the model carousel is equivalent to 160 inches on the actual carousel. How wide should the model be if the actual carousel is 35 feet wide?

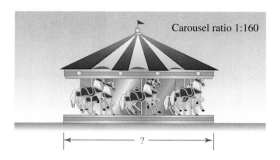

Carousel ratio 1:160

68. MIXING FUELS The instructions on a can of oil intended to be added to lawn mower gasoline read as shown. Are these instructions correct? (*Hint:* There are 128 ounces in 1 gallon.)

Recommended	Gasoline	Oil
50 to 1	6 gal	16 oz

WRITING

69. Explain the difference between a ratio and a proportion.

70. Explain how to determine whether $\frac{3.2}{3.7} = \frac{5.44}{6.29}$ is a true proportion.

71. The following paragraph is from a book about dollhouses. What concept from this section is mentioned?

Today, the internationally recognized scale for dollhouses and miniatures is 1 in. = 1 ft. This is small enough to be defined as a miniature, yet not too small for all details of decoration and furniture to be seen clearly.

72. Write a problem about a situation you encounter in your daily life that could be solved by using a proportion.

REVIEW

73. Change $\frac{9}{10}$ to a percent.

74. Change $\frac{7}{8}$ to a percent.

75. Change $33\frac{1}{3}\%$ to a fraction.

76. Find 30% of 1,600.

77. Find $\frac{1}{2}\%$ of 520.

78. SHOPPING Bill purchased a shirt on sale for $17.50. Find the original cost of the shirt if it was marked down 30%.

6.3 American Units of Measurement

- American units of length • Converting units of length • American units of weight
- American units of capacity • Units of time

Two common systems of measurement are the American (or English) system and the metric system. We discuss American units in this section and metric units in the next section. Some common American units are *inches, feet, miles, ounces, pounds, tons, cups, pints, quarts,* and *gallons.* These units are used when measuring length, weight, and capacity.

- A newborn baby is 20 inches long.
- The distance from St. Louis to Memphis is 285 miles.
- First-class postage for a letter that weighs less than 1 ounce is 37¢.
- The largest pumpkin ever grown weighed 1,092 pounds.
- Milk is sold in quart and gallon containers.

▌ American units of length

A ruler is one of the most common devices used for measuring distances or lengths. Figure 6-6 shows only a portion of a ruler; most rulers are 12 inches (1 foot) long. Since 12 inches = 1 foot, a ruler is divided into 12 equal distances of 1 inch. Each inch is divided into halves of an inch, quarters of an inch, eighths of an inch, and sixteenths of an inch. Several distances are measured using the ruler shown in Figure 6-6.

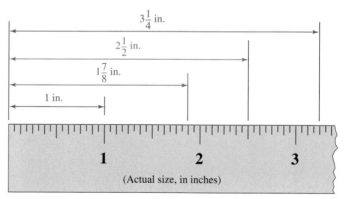

FIGURE 6-6

Each point on a ruler, like each point on a number line, has a number associated with it: the distance between the point and 0.

Self Check 1
To the nearest $\frac{1}{4}$ inch, find the width of the circle.

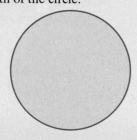

EXAMPLE 1   To the nearest $\frac{1}{4}$ inch, find the length of the nail in Figure 6-7.

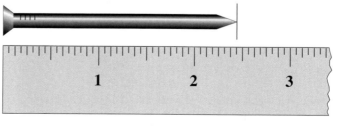

FIGURE 6-7

Solution We place the end of the ruler by one end of the nail and note that the other end of the nail is closer to the $2\frac{1}{2}$-inch mark than to the $2\frac{1}{4}$-inch mark on the ruler. To the nearest quarter-inch, the nail is $2\frac{1}{2}$ inches long.

Answer $1\frac{1}{4}$ in.

EXAMPLE 2 To the nearest $\frac{1}{8}$ inch, find the length of the paper clip in Figure 6-8.

Solution We place the end of the ruler by one end of the paper clip and note that the other end is closer to the $1\frac{3}{8}$-inch mark than to the $1\frac{1}{2}$-inch mark on the ruler. To the nearest eighth of an inch, the paper clip is $1\frac{3}{8}$ inches long.

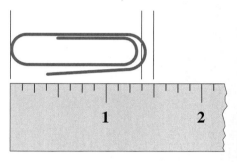

FIGURE 6-8

Self Check 2
To the nearest $\frac{1}{8}$ inch, find the length of the jumbo paper clip.

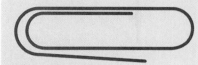

Answer $1\frac{7}{8}$ in.

Converting units of length

American units of length are related in the following ways.

American units of length	
1 foot (ft) = 12 inches (in.)	1 yard (yd) = 36 inches
1 yard = 3 feet	1 mile (mi) = 5,280 feet

To convert from one unit to another, we use *unit conversion factors*. To find the unit conversion factor between yards and feet, we begin with this fact:

3 ft = 1 yd

If we divide both sides of this equation by 1 yard, we get

$$\frac{3 \text{ ft}}{1 \text{ yd}} = \frac{1 \text{ yd}}{1 \text{ yd}}$$

$$\frac{3 \text{ ft}}{1 \text{ yd}} = 1 \qquad \text{A number divided by itself is } 1: \frac{1 \text{ yd}}{1 \text{ yd}} = 1.$$

The fraction $\frac{3 \text{ ft}}{1 \text{ yd}}$ is called a **unit conversion factor,** because its value is 1. It can be read as "3 feet per yard." Since this fraction is equal to 1, multiplying a length by this fraction does not change its measure; it changes only the *units* of measure.

To convert units of length, we use the following unit conversion factors.

To convert from	Use the unit conversion factor	To convert from	Use the unit conversion factor
feet to inches	$\frac{12 \text{ in.}}{1 \text{ ft}}$	inches to feet	$\frac{1 \text{ ft}}{12 \text{ in.}}$
yards to feet	$\frac{3 \text{ ft}}{1 \text{ yd}}$	feet to yards	$\frac{1 \text{ yd}}{3 \text{ ft}}$
yards to inches	$\frac{36 \text{ in.}}{1 \text{ yd}}$	inches to yards	$\frac{1 \text{ yd}}{36 \text{ in.}}$
miles to feet	$\frac{5,280 \text{ ft}}{1 \text{ mi}}$	feet to miles	$\frac{1 \text{ mi}}{5,280 \text{ ft}}$

Self Check 3
Convert 9 yards to feet.

EXAMPLE 3 Convert 7 yards to feet.

Solution To convert from yards to feet, we must use a unit conversion factor that relates feet to yards. Since there are 3 feet per yard, we multiply 7 yards by the unit conversion factor $\frac{3 \text{ ft}}{1 \text{ yd}}$.

$$7 \text{ yd} = \frac{7 \text{ yd}}{1} \cdot \frac{3 \text{ ft}}{1 \text{ yd}} \qquad \text{Write 7 yd as a fraction: } 7 \text{ yd} = \frac{7 \text{ yd}}{1}. \text{ Then multiply by 1: } \frac{3 \text{ ft}}{1 \text{ yd}} = 1.$$

$$= \frac{7 \overset{1}{\cancel{\text{yd}}}}{1} \cdot \frac{3 \text{ ft}}{1 \underset{1}{\cancel{\text{yd}}}} \qquad \text{The units of yards divide out.}$$

$$= 7 \cdot 3 \text{ ft}$$

$$= 21 \text{ ft} \qquad \text{Multiply: } 7 \cdot 3 = 21.$$

Seven yards is equal to 21 feet.

Answer 27 ft

Notice that in Example 3, we eliminated the units of yards and introduced the units of feet by multiplying by the appropriate unit conversion factor. In general, a unit conversion factor is a fraction with the following form:

$$\frac{\text{Unit we want to introduce}}{\text{Unit we want to eliminate}} \quad \begin{array}{l} \leftarrow \text{Numerator} \\ \leftarrow \text{Denominator} \end{array}$$

Self Check 4
Convert 1.5 feet to inches.

EXAMPLE 4 Convert $1\frac{3}{4}$ feet to inches.

Solution To convert from feet to inches, we must use a unit conversion factor that relates inches to feet. Since there are 12 inches per foot, we multiply $1\frac{3}{4}$ feet by the unit conversion factor $\frac{12 \text{ in.}}{1 \text{ ft}}$.

$$1\frac{3}{4} \text{ ft} = \frac{7}{4} \text{ ft} \cdot \frac{12 \text{ in.}}{1 \text{ ft}} \qquad \text{Write } 1\frac{3}{4} \text{ as an improper fraction: } 1\frac{3}{4} = \frac{7}{4}. \text{ Multiply by 1: } \frac{12 \text{ in.}}{1 \text{ ft}} = 1.$$

$$= \frac{7}{4} \overset{1}{\cancel{\text{ft}}} \cdot \frac{12 \text{ in.}}{1 \underset{1}{\cancel{\text{ft}}}} \qquad \text{The units of feet divide out.}$$

$$= \frac{7 \cdot 12}{4 \cdot 1} \text{ in.} \qquad \text{Multiply the fractions.}$$

$$= 21 \text{ in.} \qquad \text{Simplify: } \frac{7 \cdot 12}{4 \cdot 1} = \frac{7 \cdot 3 \cdot \overset{1}{\cancel{4}}}{\underset{1}{\cancel{4}} \cdot 1} = 7 \cdot 3 = 21.$$

$1\frac{3}{4}$ feet is equal to 21 inches.

Answer 18 in.

Sometimes we must use two unit conversion factors to eliminate the given units while introducing the desired units. The following example illustrates this concept.

Finding the length of a football field in miles

A football field (including the end zones) is 120 yards long. To find this distance in miles, we set up the problem so that the units of yards divide out and leave us with units of miles. Since there are 3 feet per yard and 5,280 feet per mile, we multiply 120 yards by $\frac{3 \text{ ft}}{1 \text{ yd}}$ and $\frac{1 \text{ mi}}{5,280 \text{ ft}}$.

$$120 \text{ yd} = 120 \text{ yd} \cdot \frac{3 \text{ ft}}{1 \text{ yd}} \cdot \frac{1 \text{ mi}}{5,280 \text{ ft}} \qquad \text{Use two unit conversion factors:}$$
$$\frac{3 \text{ ft}}{1 \text{ yd}} = 1 \text{ and } \frac{1 \text{ mi}}{5,280 \text{ ft}} = 1.$$

$$= \frac{120 \overset{1}{\cancel{\text{yd}}}}{1} \cdot \frac{3 \overset{1}{\cancel{\text{ft}}}}{1 \underset{1}{\cancel{\text{yd}}}} \cdot \frac{1 \text{ mi}}{5,280 \underset{1}{\cancel{\text{ft}}}} \qquad \text{Divide out the units of yards and feet.}$$

$$= \frac{120 \cdot 3}{5,280} \text{ mi} \qquad \text{Multiply the fractions.}$$

We can do this arithmetic using a scientific calculator by entering these numbers and pressing these keys.

$$120 \boxed{\times} 3 \boxed{\div} 5280 \boxed{=} \qquad \boxed{\text{0.0681818}}$$

To the nearest hundredth, a football field is 0.07 mile long.

American units of weight

American units of weight are related in the following ways.

American units of weight
1 pound (lb) = 16 ounces (oz)   1 ton = 2,000 pounds

To convert units of weight, we use the following unit conversion factors.

To convert from	Use the unit conversion factor	To convert from	Use the unit conversion factor
pounds to ounces	$\frac{16 \text{ oz}}{1 \text{ lb}}$	ounces to pounds	$\frac{1 \text{ lb}}{16 \text{ oz}}$
tons to pounds	$\frac{2,000 \text{ lb}}{1 \text{ ton}}$	pounds to tons	$\frac{1 \text{ ton}}{2,000 \text{ lb}}$

EXAMPLE 5 Convert 40 ounces to pounds.

Solution Since there is 1 pound per 16 ounces, we multiply 40 ounces by the unit conversion factor $\frac{1 \text{ lb}}{16 \text{ oz}}$.

$$40 \text{ oz} = \frac{40 \text{ oz}}{1} \cdot \frac{\textbf{1 lb}}{\textbf{16 oz}} \qquad \text{Write 40 oz as a fraction: } 40 \text{ oz} = \frac{40 \text{ oz}}{1}. \text{ Then multiply by 1:}$$
$$\frac{1 \text{ lb}}{16 \text{ oz}} = 1.$$

$$= \frac{40 \overset{1}{\cancel{\text{oz}}}}{1} \cdot \frac{1 \text{ lb}}{16 \underset{1}{\cancel{\text{oz}}}} \qquad \text{The units of ounces divide out.}$$

$$= \frac{40}{16} \text{ lb} \qquad \text{Multiply the fractions.}$$

Self Check 5
Convert 60 ounces to pounds.

There are two ways to complete the solution. First, we can divide out the common factors of the numerator and denominator and write the result as a mixed number.

$$\frac{40}{16}\text{ lb} = \frac{\overset{1}{\cancel{8}}\cdot 5}{\underset{1}{\cancel{8}}\cdot 2}\text{ lb} = \frac{5}{2}\text{ lb} = 2\frac{1}{2}\text{ lb}$$

A second approach is to divide the numerator by the denominator and express the result as a decimal.

$$\frac{40}{16}\text{ lb} = 2.5\text{ lb}\qquad \text{Perform the division: } 40 \div 16 = 2.5.$$

Forty ounces is equal to $2\frac{1}{2}$ lb (or 2.5 lb).

Answer $3\frac{3}{4}$ lb $= 3.75$ lb

Self Check 6
Convert 60 pounds to ounces.

EXAMPLE 6   Convert 25 pounds to ounces.

Solution   Since there are 16 ounces per pound, we multiply 25 pounds by the unit conversion factor $\frac{16\text{ oz}}{1\text{ lb}}$.

$$25\text{ lb} = \frac{25\text{ lb}}{1}\cdot\frac{\mathbf{16\text{ oz}}}{\mathbf{1\text{ lb}}}\qquad \text{Multiply by 1: } \frac{16\text{ oz}}{1\text{ lb}} = 1.$$

$$= \frac{25\overset{1}{\cancel{\text{lb}}}}{1}\cdot\frac{16\text{ oz}}{1\underset{1}{\cancel{\text{lb}}}}\qquad \text{The units of pounds divide out.}$$

$$= 25\cdot 16\text{ oz}$$

$$= 400\text{ oz}\qquad \text{Multiply: } 25\cdot 16 = 400.$$

Twenty-five pounds is equal to 400 ounces.

Answer 960 oz

CALCULATOR SNAPSHOT   **Finding the weight of a car in pounds**

A BMW 323Ci convertible weighs 1.78 tons. To find its weight in pounds, we set up the problem so that the units of tons divide out and leave us with units of pounds. Since there are 2,000 pounds per ton, we multiply by $\frac{2{,}000\text{ lb}}{1\text{ ton}}$.

$$1.78\text{ tons} = \frac{1.78\text{ tons}}{1}\cdot\frac{2{,}000\text{ lb}}{1\text{ ton}}\qquad \text{Multiply by 1: } \frac{2{,}000\text{ lb}}{1\text{ ton}} = 1.$$

$$= \frac{1.78\overset{1}{\cancel{\text{tons}}}}{1}\cdot\frac{2{,}000\text{ lb}}{1\underset{1}{\cancel{\text{ton}}}}\qquad \text{Divide out the units of tons.}$$

$$= 1.78\cdot 2{,}000\text{ lb}$$

We can do this multiplication using a scientific calculator by entering these numbers and pressing these keys.

1.78 $\boxed{\times}$ 2000 $\boxed{=}$ $\boxed{\text{3560}}$

The convertible weighs 3,560 pounds.

American units of capacity

American units of capacity are related as follows.

American units of capacity	
1 cup (c) = 8 fluid ounces (fl oz)	1 pint (pt) = 2 cups (c)
1 quart (qt) = 2 pints (pt)	1 gallon (gal) = 4 quarts (qt)

To convert units of capacity, we use the following unit conversion factors.

To convert from	Use the unit conversion factor	To convert from	Use the unit conversion factor
cups to ounces	$\frac{8 \text{ fl oz}}{1 \text{ c}}$	ounces to cups	$\frac{1 \text{ c}}{8 \text{ fl oz}}$
pints to cups	$\frac{2 \text{ c}}{1 \text{ pt}}$	cups to pints	$\frac{1 \text{ pt}}{2 \text{ c}}$
quarts to pints	$\frac{2 \text{ pt}}{1 \text{ qt}}$	pints to quarts	$\frac{1 \text{ qt}}{2 \text{ pt}}$
gallons to quarts	$\frac{4 \text{ qt}}{1 \text{ gal}}$	quarts to gallons	$\frac{1 \text{ gal}}{4 \text{ qt}}$

EXAMPLE 7 Cooking. If a recipe calls for 3 pints of milk, how many fluid ounces of milk should be used?

Solution Since there are 2 cups per pint and 8 fluid ounces per cup, we multiply 3 pints by unit conversion factors of $\frac{2 \text{ c}}{1 \text{ pt}}$ and $\frac{8 \text{ fl oz}}{1 \text{ c}}$.

$$3 \text{ pt} = \frac{3 \text{ pt}}{1} \cdot \frac{\textbf{2 c}}{\textbf{1 pt}} \cdot \frac{\textbf{8 fl oz}}{\textbf{1 c}} \qquad \text{Use two unit conversion factors: } \frac{2 \text{ c}}{1 \text{ pt}} = 1 \text{ and } \frac{8 \text{ fl oz}}{1 \text{ c}} = 1.$$

$$= \frac{3 \overset{1}{\cancel{\text{pt}}}}{1} \cdot \frac{2 \overset{1}{\cancel{\text{c}}}}{1 \underset{1}{\cancel{\text{pt}}}} \cdot \frac{8 \text{ fl oz}}{1 \underset{1}{\cancel{\text{c}}}} \qquad \text{Divide out the units of pints and cups.}$$

$$= 3 \cdot 2 \cdot 8 \text{ fl oz}$$

$$= 48 \text{ fl oz}$$

Since 3 pints is equal to 48 fluid ounces, 48 fluid ounces of milk should be used.

Self Check 7
How many pints are in 1 gallon?

Answer 8 pt

Units of time

Units of time are related in the following ways.

Units of time	
1 minute (min) = 60 seconds (sec)	1 hour (hr) = 60 minutes
1 day = 24 hours	

To convert units of time, we use the following unit conversion factors.

To convert from	Use the unit conversion factor	To convert from	Use the unit conversion factor
minutes to seconds	$\frac{60 \text{ sec}}{1 \text{ min}}$	seconds to minutes	$\frac{1 \text{ min}}{60 \text{ sec}}$
hours to minutes	$\frac{60 \text{ min}}{1 \text{ hr}}$	minutes to hours	$\frac{1 \text{ hr}}{60 \text{ min}}$
days to hours	$\frac{24 \text{ hr}}{1 \text{ day}}$	hours to days	$\frac{1 \text{ day}}{24 \text{ hr}}$

Self Check 8
A solar eclipse (eclipse of the sun) can last as long as 450 seconds. Express this time in minutes.

EXAMPLE 8   Astronomy.   A lunar eclipse occurs when the Earth is between the sun and the moon in such a way that Earth's shadow darkens the moon. (See Figure 6-9, which is not to scale.) A total lunar eclipse can last as long as 105 minutes. Express this time in hours.

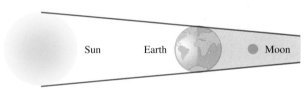

FIGURE 6-9

Solution   Since there is 1 hour for every 60 minutes, we multiply 105 by the unit conversion factor $\frac{1 \text{ hr}}{60 \text{ min}}$.

$$105 \text{ min} = \frac{105 \text{ min}}{1} \cdot \frac{\textbf{1 hr}}{\textbf{60 min}} \qquad \text{Multiply by 1: } \frac{1 \text{ hr}}{60 \text{ min}} = 1.$$

$$= \frac{105 \text{ m\!i\!n}}{1} \cdot \frac{1 \text{ hr}}{60 \text{ m\!i\!n}} \qquad \text{The units of minutes divide out.}$$

$$= \frac{105}{60} \text{ hr} \qquad \text{Multiply the fractions.}$$

$$= \frac{\overset{1}{3} \cdot \overset{1}{5} \cdot 7}{2 \cdot 2 \cdot \underset{1}{3} \cdot \underset{1}{5}} \text{hr} \qquad \begin{array}{l}\text{Prime factor 105 and 60. Then divide out common} \\ \text{factors of the numerator and denominator.}\end{array}$$

$$= \frac{7}{4} \text{ hr}$$

$$= 1\frac{3}{4} \text{ hr} \qquad \text{Write } \frac{7}{4} \text{ as a mixed number.}$$

Answer $7\frac{1}{2}$ min

A total lunar eclipse can last as long as $1\frac{3}{4}$ hours.

Section 6.3 STUDY SET

VOCABULARY *Fill in the blanks.*

1. Inches, feet, and miles are examples of American units of _____.
2. A ruler is used for measuring _____.
3. The value of any unit conversion factor is ▢.
4. Ounces, pounds, and tons are examples of American units of _____.
5. Some examples of American units of _____ are cups, pints, quarts, and gallons.
6. Some units of _____ are seconds, hours, and days.

CONCEPTS *Fill in the blanks.*

7. 12 in. = ▢ ft
8. ▢ ft = 1 yd
9. 1 mi = ▢ ft
10. 1 yd = ▢ in.
11. ▢ ounces = 1 pound
12. ▢ pounds = 1 ton
13. 1 cup = ▢ fluid ounces
14. 1 pint = ▢ cups
15. 2 pints = ▢ quart
16. 4 quarts = ▢ gallon
17. 1 day = ▢ hours
18. 2 hours = ▢ minutes

19. Determine which measurements the arrows point to on the ruler.

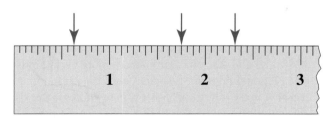

20. Determine which measurements the arrows point to on the ruler, to the nearest $\frac{1}{8}$ inch.

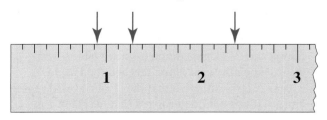

21. Write a unit conversion factor to convert
    a. pounds to tons   b. quarts to pints

22. Write the two unit conversion factors used to convert
    a. inches to yards
    b. days to minutes

23. Match each item with its proper measurement.
    a. Length of the U.S. coastline
    b. Height of a Barbie doll
    c. Span of the Golden Gate Bridge
    d. Width of a football field

    i. $11\frac{1}{2}$ in.
    ii. 4,200 ft
    iii. 53.5 yd
    iv. 12,383 mi

24. Match each item with its proper measurement.
    a. Weight of the men's shot put used in track and field
    b. Weight of an African elephant
    c. Amount of gold that is worth $500

    i. $1\frac{1}{2}$ oz
    ii. 16 lb
    iii. 7.2 tons

25. Match each item with its proper measurement.
    a. Amount of blood in an adult
    b. Size of the Exxon Valdez oil spill in 1989
    c. Amount of nail polish in a bottle
    d. Amount of flour to make 3 dozen cookies

    i. $\frac{1}{2}$ fluid oz
    ii. 2 cups
    iii. 5 qt
    iv. 10,080,000 gal

26. Match each item with its proper measurement.
    a. Length of first U.S. manned space flight
    b. A leap year
    c. Time difference between New York and Fairbanks, Alaska
    d. Length of Wright Brothers' first flight

    i. 12 sec
    ii. 15 min
    iii. 4 hr
    iv. 366 days

NOTATION   *Complete each solution.*

27. Convert 12 yards to inches.

$$12 \text{ yd} = 12 \text{ yd} \cdot \frac{\boxed{} \text{ in.}}{1 \text{ yd}}$$
$$= 12 \cdot \boxed{} \text{ in.}$$
$$= 432 \text{ in.}$$

28. Convert 1 ton to ounces.

$$1 \text{ ton} = 1 \text{ ton} \cdot \frac{\boxed{} \text{ lb}}{1 \text{ ton}} \cdot \frac{\boxed{} \text{ oz}}{1 \text{ lb}}$$
$$= 1 \cdot 2{,}000 \cdot 16 \text{ oz}$$
$$= \boxed{} \text{ oz}$$

29. Convert 12 pints to gallons.

$$12 \text{ pt} = 12 \text{ pt} \cdot \frac{1 \text{ qt}}{\boxed{} \text{ pt}} \cdot \frac{1 \text{ gal}}{\boxed{} \text{ qt}}$$
$$= \boxed{} \cdot \frac{1}{2} \cdot \frac{1}{4} \text{ gal}$$
$$= 1.5 \text{ gal}$$

30. Convert 37,440 minutes to days.

$$37{,}440 \text{ min} = 37{,}440 \text{ min} \cdot \frac{1 \text{ hr}}{\boxed{} \text{ min}} \cdot \frac{1 \text{ day}}{\boxed{} \text{ hr}}$$
$$= \frac{\boxed{}}{60 \cdot 24} \text{ days}$$
$$= 26 \text{ days}$$

PRACTICE   *Use a ruler with a scale in inches to measure each object to the nearest $\frac{1}{8}$ inch.*

31. The width of a dollar bill

32. The length of a dollar bill

33. The length (top to bottom) of this page

34. The length of the word supercalifragilisticexpialidocious

Perform each conversion.

35. 4 feet to inches

36. 7 feet to inches

37. $3\frac{1}{2}$ feet to inches

38. $2\frac{2}{3}$ feet to inches

39. 24 inches to feet

40. 54 inches to feet

41. 8 yards to inches

42. 288 inches to yards

43. 90 inches to yards

44. 12 yards to inches

45. 56 inches to feet

46. 44 inches to feet

47. 5 yards to feet

48. 21 feet to yards

49. 7 feet to yards

50. $4\frac{2}{3}$ yards to feet

51. 15,840 feet to miles

52. 2 miles to feet

53. $\frac{1}{2}$ mile to feet

54. 1,320 feet to miles

55. 80 ounces to pounds

56. 8 pounds to ounces

57. 7,000 pounds to tons

58. 2.5 tons to ounces

59. 12.4 tons to pounds

60. 48,000 ounces to tons

61. 3 quarts to pints

62. 20 quarts to gallons

63. 16 pints to gallons

64. 3 gallons to fluid ounces

65. 32 fluid ounces to pints

66. 2 quarts to fluid ounces

67. 240 minutes to hours

68. 2,400 seconds to hours

69. 7,200 minutes to days

70. 691,200 seconds to days

APPLICATIONS

71. THE GREAT PYRAMID The Great Pyramid in Egypt is about 450 feet high. Express this distance in yards.

72. THE WRIGHT BROTHERS In 1903, Orville Wright made the world's first sustained flight. It lasted 12 seconds, and the plane traveled 120 feet. Express the length of the flight in yards.

73. THE GREAT SPHINX The Great Sphinx of Egypt is 240 feet long. Express this in inches.

74. HOOVER DAM The Hoover Dam in Nevada is 726 feet high. Express this distance in inches.

75. THE SEARS TOWER The Sears Tower in Chicago has 110 stories and is 1,454 feet tall. To the nearest hundredth, express this height in miles.

76. NFL RECORDS Emmit Smith, the former Dallas Cowboys and Arizona Cardinals running back, holds the National Football League record for yards rushing in a career: 18,355. How many miles is this? Round to the nearest tenth of a mile.

77. NFL RECORDS When Dan Marino of the Miami Dolphins retired, it was noted that Marino's career passing total was nearly 35 miles! How many yards is this?

78. LEWIS AND CLARK The trail traveled by the Lewis and Clark expedition is shown on the next page. When the expedition reached the Pacific Ocean, Clark estimated that they had traveled 4,162 miles. (It was later determined that his guess was within 40 miles of the actual distance.) Express Clark's estimate of the distance in feet.

87. CAMPING How many ounces of camping stove fuel will fit in the container shown?

FUEL 2½ gal

79. WEIGHT OF WATER One gallon of water weighs about 8 pounds. Express this weight in ounces.

80. WEIGHT OF A BABY A newborn baby weighed 136 ounces. Express this weight in pounds.

81. HIPPOS An adult hippopotamus can weigh as much as 9,900 pounds. Express this weight in tons.

82. ELEPHANTS An adult elephant can consume as much as 495 pounds of grass and leaves in one day. How many ounces is this?

83. BUYING PAINT A painter estimates that he will need 17 gallons of paint for a job. To take advantage of a closeout sale on quart cans, he decides to buy the paint in quarts. How many cans will he need to buy?

84. CATERING How many cups of apple cider can be dispensed from a 10-gallon container of cider?

85. SCHOOL LUNCHES Each student attending Eagle River Elementary School receives 1 pint of milk for lunch each day. If 575 students attend the school, how many gallons of milk are used each day?

86. RADIATORS The radiator capacity of a piece of earth-moving equipment is 39 quarts. If the radiator is drained and new coolant put in, how many gallons of new coolant will be used?

88. HIKING A college student walks 11 miles in 155 minutes. To the nearest tenth, how many hours does he walk?

89. SPACE TRAVEL The astronauts of the Apollo 8 mission, which was launched on December 21, 1968, were in space for 147 hours. How many days did the mission take?

90. AMELIA EARHART In 1935, Amelia Earhart became the first woman to fly across the Atlantic Ocean alone, establishing a new record for the crossing: 13 hours and 30 minutes. How many minutes is this?

WRITING

91. Explain how to find the unit conversion factor that will convert feet to inches.

92. Explain how to find the unit conversion factor that will convert pints to gallons.

REVIEW   *Round each number as indicated.*

93. 3,673.263; nearest hundred

94. 3,673.263; nearest ten

95. 3,673.263; nearest hundredth

96. 3,673.263; nearest tenth

97. 0.100602; nearest thousandth

98. 0.100602; nearest hundredth

99. 0.09999; nearest tenth

100. 0.09999; nearest one

6.4 Metric Units of Measurement

- Metric units of length • Converting units of length • Metric units of mass
- Metric units of capacity • Cubic centimeters

The metric system is the system of measurement used by most countries in the world. All countries, including the United States, use it for scientific purposes. The metric system, like our decimal numeration system, is based on the number 10. For this reason, converting from one metric unit to another is easier than with the American system.

■ Metric units of length

The basic metric unit of length is the **meter** (m). One meter is approximately 39 inches, slightly more than 1 yard. Figure 6-10 shows the relative sizes of a yardstick and a meterstick.

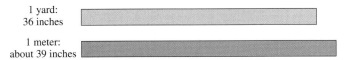

1 yard: 36 inches

1 meter: about 39 inches

FIGURE 6-10

Larger and smaller units are created by adding prefixes to the front of this basic unit, *meter*.

deka means tens	*deci* means tenths
hecto means hundreds	*centi* means hundredths
kilo means thousands	*milli* means thousandths

Metric units of length

1 dekameter (dam) = 10 meters.
1 dam is a little less than 11 yards.

1 hectometer (hm) = 100 meters.
1 hm is about 1 football field long,
plus one end zone.

1 kilometer (km) = 1,000 meters.
1 km is about $\frac{3}{5}$ mile.

1 decimeter (dm) = $\frac{1}{10}$ of 1 meter
1 dm is about the length of your palm.

1 centimeter (cm) = $\frac{1}{100}$ of 1 meter.
1 cm is about as wide as the nail of your
little finger.

1 millimeter (mm) = $\frac{1}{1,000}$ of 1 meter.
1 mm is about the thickness of a dime.

Figure 6-11 shows a portion of a metric ruler, scaled in centimeters, and a ruler scaled in inches. The rulers are used to measure several lengths.

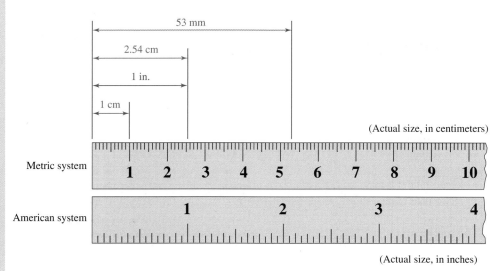

53 mm

2.54 cm

1 in.

1 cm

(Actual size, in centimeters)

Metric system

American system

(Actual size, in inches)

FIGURE 6-11

EXAMPLE 1   To the nearest centimeter, find the length of the nail in Figure 6-12.

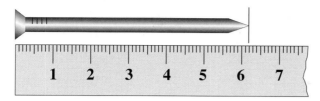

FIGURE 6-12

Solution   We place the end of the ruler by one end of the nail and note that the other end of the nail is closer to the 6-cm mark than to the 7-cm mark on the ruler. To the nearest centimeter, the nail is 6 cm long.

Self Check 1
To the nearest centimeter, find the width of the circle.

Answer 3 cm

EXAMPLE 2   To the nearest millimeter, find the length of the paper clip in Figure 6-13.

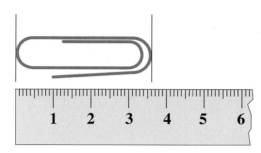

FIGURE 6-13

Solution   On the ruler, each centimeter has been divided into 10 millimeters. We place the end of the ruler by one end of the paper clip and note that the other end is closer to the 36-mm mark than to the 37-mm mark on the ruler. To the nearest millimeter, the paper clip is 36 mm long.

Self Check 2
To the nearest millimeter, find the length of the jumbo paper clip.

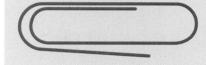

Answer 47 mm

Converting units of length

Metric units of length are related as shown in Table 6-1.

Metric units of length		
1 kilometer (km) = 1,000 meters	or	1 meter = $\frac{1}{1,000}$ kilometer
1 hectometer (hm) = 100 meters	or	1 meter = $\frac{1}{100}$ hectometer
1 dekameter (dam) = 10 meters	or	1 meter = $\frac{1}{10}$ dekameter
1 decimeter (dm) = $\frac{1}{10}$ meter	or	1 meter = 10 decimeters
1 centimeter (cm) = $\frac{1}{100}$ meter	or	1 meter = 100 centimeters
1 millimeter (mm) = $\frac{1}{1,000}$ meter	or	1 meter = 1,000 millimeters

TABLE 6-1

We can use the information in the table to write unit conversion factors that can be used to convert metric units of length. For example, in the table we see that

1 meter = 100 centimeters

From this fact, we can write two unit conversion factors.

$$\frac{1 \text{ m}}{100 \text{ cm}} = 1 \quad \text{and} \quad \frac{100 \text{ cm}}{1 \text{ m}} = 1$$

To obtain the first unit conversion factor, divide both sides of the equation 1 m = 100 cm by 100 cm. To obtain the second unit conversion factor, divide both sides by 1 m.

One advantage of the metric system is that multiplying or dividing by a unit conversion factor involves multiplying or dividing by a power of 10.

Self Check 3
Convert 860 centimeters to meters.

EXAMPLE 3   Convert 350 centimeters to meters.

Solution   Since there is 1 meter per 100 centimeters, we multiply 350 centimeters by the unit conversion factor $\frac{1 \text{ m}}{100 \text{ cm}}$.

$$350 \text{ cm} = \frac{350 \text{ cm}}{1} \cdot \frac{\textbf{1 m}}{\textbf{100 cm}} \qquad \text{Multiply by 1: } \frac{1 \text{ m}}{100 \text{ cm}} = 1.$$

$$= \frac{350 \overset{1}{\cancel{\text{cm}}}}{1} \cdot \frac{1 \text{ m}}{100 \underset{1}{\cancel{\text{cm}}}} \qquad \text{The units of centimeters divide out.}$$

$$= \frac{350}{100} \text{ m}$$

$$= 3.5 \text{ m} \qquad \text{Divide by 100 by moving the decimal point 2 places to the left.}$$

Answer 8.6 m

Thus, 350 centimeters = 3.5 meters.

In Example 3, we converted 350 centimeters to meters using a unit conversion factor. We can also make this conversion by recognizing that all units of length in the metric system are powers of 10 of a meter. Converting from one unit to another is as easy as multiplying by the correct power of 10 or, simply moving a decimal point the correct number of places to the right or left. For example, in the chart below, we see that to convert from centimeters to meters, we move 2 places to the left.

km   hm   dam   **m**   dm   **cm**   mm

To go from centimeters to meters,
we must move 2 places to the left.

If we write 350 centimeters as 350.0 centimeters, we can convert to meters by moving the decimal point 2 places to the left.

350.0 centimeters = 3.50 0 meters = 3.5 meters

With the unit conversion factor method or the chart method, we get 350 cm = 3.5 m.

❗ COMMENT   When using a chart to help make a metric conversion, be sure to list the units from largest to smallest when reading from left to right.

EXAMPLE 4   Convert 2.4 meters to millimeters.

Self Check 4
Convert 5.3 meters to
millimeters.

Solution   Since there are 1,000 millimeters per meter, we multiply 2.4 meters by the unit conversion factor $\frac{1,000 \text{ mm}}{1 \text{ m}}$.

$$2.4 \text{ m} = \frac{2.4 \text{ m}}{1} \cdot \frac{\mathbf{1,000 \text{ mm}}}{\mathbf{1 \text{ m}}} \qquad \text{Multiply by 1: } \frac{1,000 \text{ mm}}{1 \text{ m}} = 1.$$

$$= \frac{2.4 \overset{1}{\cancel{\text{m}}}}{1} \cdot \frac{1,000 \text{ mm}}{1 \underset{1}{\cancel{\text{m}}}} \qquad \text{The units of meters divide out.}$$

$$= 2.4 \cdot 1,000 \text{ mm}$$

$$= 2,400 \text{ mm} \qquad \text{Multiply by 1,000 by moving the decimal point 3 places to the right.}$$

Thus, 2.4 meters = 2,400 millimeters.

We can also make this conversion using a chart.

km   hm   dam   **m**   dm   cm   **mm**

From the chart, we see that we should move the decimal point 3 places to the right to convert from meters to millimeters.

2.4 meters = 2 400. millimeters = 2,400 millimeters

Answer 5,300 mm

EXAMPLE 5   Convert 3.2 kilometers to centimeters.

Self Check 5
Convert 5.15 kilometers to
centimeters.

Solution   To convert to centimeters, we set up the problem so that the units of kilometers divide out and leave us with units of centimeters. Since there are 1,000 meters per kilometer and 100 centimeters per meter, we multiply 3.2 kilometers by $\frac{1,000 \text{ m}}{1 \text{ km}}$ and $\frac{100 \text{ cm}}{1 \text{ m}}$.

$$3.2 \text{ km} = \frac{3.2 \overset{1}{\cancel{\text{km}}}}{1} \cdot \frac{\mathbf{1,000} \overset{1}{\cancel{\text{m}}}}{\mathbf{1} \underset{1}{\cancel{\text{km}}}} \cdot \frac{\mathbf{100 \text{ cm}}}{\mathbf{1} \cancel{\text{m}}} \qquad \text{The units of kilometers and meters divide out.}$$

$$= 3.2 \cdot 1,000 \cdot 100 \text{ cm}$$

$$= 320,000 \text{ cm} \qquad \text{Multiply by 1,000 and 100 by moving the decimal point 5 places to the right.}$$

Thus, 3.2 kilometers = 320,000 centimeters.

Using a chart, we see that the decimal point should be moved 5 places to the right to convert kilometers to centimeters.

km   hm   dam   m   dm   **cm**   mm

3.2 kilometers = 3 20000. centimeters = 320,000 centimeters

Answer 515,000 cm

■ Metric units of mass

The **mass** of an object is a measure of the amount of material in the object. When an object is moved about in space, its mass does not change. One basic unit of mass in the metric system is the **gram** (g). A gram is defined to be the mass of water contained in a cube having sides 1 centimeter long. (See Figure 6-14 on the next page.)

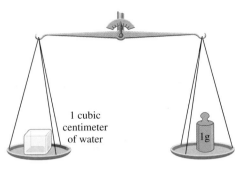

FIGURE 6-14

The **weight** of an object is determined by the Earth's gravitational pull on the object. Since gravitational pull on an object decreases as the object gets farther from Earth, the object weighs less as it gets farther from Earth's surface. This is why astronauts experience weightlessness in space. However, since most of us remain near Earth's surface, we will use the words *mass* and *weight* interchangeably. Thus, a mass of 30 grams is said to weigh 30 grams.

Metric units of mass are related as shown in Table 6-2.

Metric units of mass

1 kilogram (kg) = 1,000 grams	or	1 gram = $\frac{1}{1,000}$ kilogram
1 hectogram (hg) = 100 grams	or	1 gram = $\frac{1}{100}$ hectogram
1 dekagram (dag) = 10 grams	or	1 gram = $\frac{1}{10}$ dekagram
1 decigram (dg) = $\frac{1}{10}$ gram	or	1 gram = 10 decigrams
1 centigram (cg) = $\frac{1}{100}$ gram	or	1 gram = 100 centigrams
1 milligram (mg) = $\frac{1}{1,000}$ gram	or	1 gram = 1,000 milligrams

TABLE 6-2

Here are examples of these units of mass:

- An average bowling ball weighs about 6 kilograms.
- A raisin weighs about 1 gram.
- A certain vitamin tablet contains 450 milligrams of calcium.

We can use the information in Table 6-2 to write unit conversion factors that can be used to convert metric units of mass. For example, in the table we see that

$$1 \text{ kilogram} = 1,000 \text{ grams}$$

From this fact, we can write two unit conversion factors.

$$\frac{1 \text{ kg}}{1,000 \text{ g}} = 1 \quad \text{and} \quad \frac{1,000 \text{ g}}{1 \text{ kg}} = 1$$

To obtain the first unit conversion factor, divide both sides of the equation 1 kg = 1,000 g by 1,000 g. To obtain the second unit conversion factor, divide both sides by 1 kg.

Self Check 6
Convert 5 kilograms to grams.

EXAMPLE 6   Convert 7.2 kilograms to grams.

Solution   To convert to grams, we set up the problem so that the units of kilograms divide out and leave us with the units of grams. Since there are 1,000 grams per 1 kilogram, we multiply 7.2 kilograms by $\frac{1,000 \text{ g}}{1 \text{ kg}}$.

$$7.2 \text{ kg} = \frac{7.2 \overset{1}{\cancel{\text{kg}}}}{1} \cdot \frac{\mathbf{1,000\ g}}{\mathbf{1\ \underset{1}{\cancel{\text{kg}}}}} \qquad \text{Divide out the units of kilograms.}$$

$$= 7.2 \cdot 1,000 \text{ g}$$

$$= 7,200 \text{ g} \qquad \text{Perform the multiplication by moving the decimal point 3 places to the right.}$$

Thus, 7.2 kilograms = 7,200 grams.

To use a chart to make the conversion, we list the metric units of weight from the largest (kilograms) to the smallest (milligrams).

kg   hg   dag   **g**   dg   cg   mg

From the chart, we see that we must move the decimal point 3 places to the right to change kilograms to grams.

7.2 kilograms = 7 200. grams = 7,200 grams

Answer 5,000 g

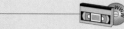

EXAMPLE 7   Medications. A bottle of Verapamil, a drug taken for high blood pressure, contains 30 tablets. If each tablet contains 180 mg of active ingredient, how many centigrams of active ingredient are in the bottle?

Solution   Since there are 30 tablets and each one contains 180 mg of active ingredient, there are

$$30 \cdot 180 \text{ mg} = 5,400 \text{ mg}$$

of active ingredient in the bottle. To convert milligrams to centigrams, we multiply 5,400 milligrams by $\frac{1 \text{ g}}{1,000 \text{ mg}}$ and $\frac{100 \text{ cg}}{1 \text{ g}}$.

$$5,400 \text{ mg} = \frac{5,400 \overset{1}{\cancel{\text{mg}}}}{1} \cdot \frac{\mathbf{1} \overset{1}{\cancel{\text{g}}}}{\mathbf{1,000} \underset{1}{\cancel{\text{mg}}}} \cdot \frac{\mathbf{100\ cg}}{\mathbf{1} \underset{1}{\text{g}}} \qquad \text{Divide out the units of milligrams and grams.}$$

$$= \frac{5,400 \cdot 100}{1,000} \text{ cg} \qquad \text{Multiply the fractions.}$$

$$= 540 \text{ cg} \qquad \text{Simplify}$$

There are 540 centigrams of active ingredient in the bottle.

Using a chart, we see that we must move the decimal point 1 place to the left to convert from milligrams to centigrams.

kg   hg   dag   g   dg   **cg**   **mg**

5,400 milligrams = 540.0 centigrams = 540 centigrams

Self Check 7
One brand name for Verapamil is Isoptin. If a bottle of Isoptin contains 90 tablets, each containing 200 mg of active ingredient, how many centigrams of active ingredient are in the bottle?

Answer 1,800 cg

■ Metric units of capacity

In the metric system, one basic unit of capacity is the **liter** (L), which is defined to be the capacity of a cube with sides 10 centimeters long. (See Figure 6-15.) A liter of liquid is slightly more than 1 quart.

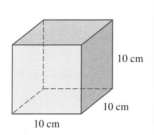

10 cm
10 cm
10 cm

FIGURE 6-15

Metric units of capacity are related as shown in Table 6-3.

Metric units of capacity		
1 kiloliter (kL) = 1,000 liters	or	1 liter = $\frac{1}{1,000}$ kiloliter
1 hectoliter (hL) = 100 liters	or	1 liter = $\frac{1}{100}$ hectoliter
1 dekaliter (daL) = 10 liters	or	1 liter = $\frac{1}{10}$ dekaliter
1 deciliter (dL) = $\frac{1}{10}$ liter	or	1 liter = 10 deciliters
1 centiliter (cL) = $\frac{1}{100}$ liter	or	1 liter = 100 centiliters
1 milliliter (mL) = $\frac{1}{1,000}$ liter	or	1 liter = 1,000 milliliters

TABLE 6-3

Here are examples of these units of capacity:

- Soft drinks are sold in 2-liter plastic bottles.
- The fuel tank of a certain minivan can hold about 75 liters of gasoline.
- Chemists use glass cylinders, scaled in milliliters, to measure liquids.

We can use the information in Table 6-3 to write unit conversion factors that can be used to convert metric units of capacity. For example, in the table we see that

1 liter = 100 centiliters

From this fact, we can write two unit conversion factors.

$$\frac{1 \text{ L}}{100 \text{ cL}} = 1 \quad \text{and} \quad \frac{100 \text{ cL}}{1 \text{ L}} = 1$$

Self Check 8
How many milliliters are in two 2-liter bottles of cola?

EXAMPLE 8   Soft drinks.   How many centiliters are in three 2-liter bottles of cola?

Solution   Three 2-liter bottles of cola contain 6 liters of cola. To convert to centiliters, we set up the problem so that liters divide out and leave us with centiliters. Since there are 100 centiliters per 1 liter, we multiply 6 liters by the unit conversion factor $\frac{100 \text{ cL}}{1 \text{ L}}$.

$$6 \text{ L} = 6 \text{ L} \cdot \frac{\textbf{100 cL}}{\textbf{1 L}} \qquad \text{Multiply by 1: } \frac{100 \text{ cL}}{1 \text{ L}} = 1.$$

$$= \frac{6 \overset{1}{\cancel{L}}}{1} \cdot \frac{100 \text{ cL}}{1 \underset{1}{\cancel{L}}} \qquad \text{The units of liters divide out.}$$

$$= 6 \cdot 100 \text{ cL}$$

$$= 600 \text{ cL}$$

Thus, there are 600 centiliters in three 2-liter bottles of cola.

To make this conversion using a chart, we list the metric units of capacity in order from largest (kiloliter) to smallest (milliliter).

kL   hL   daL   **L**   dL   **cL**   mL

From the chart, we see that we should move the decimal point 2 places to the right to convert from liters to centiliters.

Answer 4,000 mL

6 liters = 6 00. centiliters = 600 centiliters

■ Cubic centimeters

Another metric unit of capacity is the **cubic centimeter,** which is represented by the notation cm³ or, more simply, cc. One milliliter and one cubic centimeter represent the same capacity.

$$1 \text{ mL} = 1 \text{ cm}^3 = 1 \text{ cc}$$

The units of cubic centimeters are used frequently in medicine. For example, when a nurse administers an injection containing 5 cc of medication, the dosage can also be expressed using milliliters.

$$5 \text{ cc} = 5 \text{ mL}$$

When a doctor orders that a patient be put on 1,000 cc of dextrose solution, the request can be expressed in different ways.

$$1,000 \text{ cc} = 1,000 \text{ mL} = 1 \text{ liter}$$

Section 6.4 STUDY SET

■ VOCABULARY   *Fill in the blanks.*

1. *Deka* means _____.
2. *Hecto* means _____.
3. *Kilo* means _____.
4. *Deci* means _____.
5. *Centi* means _____.
6. *Milli* means _____.
7. Meters, grams, and liters are units of measurement in the _____ system.
8. The _____ of an object is determined by the Earth's gravitational pull on the object.

■ CONCEPTS

9. To the nearest centimeter, determine which measurements the arrows point to on the ruler.

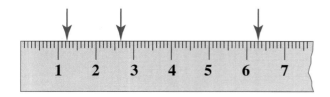

10. To the nearest millimeter, determine which measurements the arrows point to on the ruler.

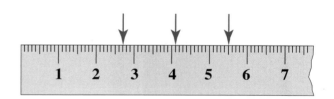

11. Write a unit conversion factor to convert
    a. meters to kilometers
    b. grams to centigrams
    c. liters to milliliters

12. Use the chart to determine how many decimal places and in which direction to move the decimal point when converting the following.
    a. Kilometers to centimeters

        km hm dam m dm cm mm

    b. Milligrams to grams

        kg hg dag g dg cg mg

    c. Hectoliters to centiliters

        kL hL daL L dL cL mL

13. Match each item with its proper measurement.
    a. Thickness of a phone book   i.   6,275 km
    b. Length of the Amazon River   ii. 2 m
    c. Height of a soccer goal   iii. 6 cm

14. Match each item with its proper measurement.

   a. Weight of a giraffe

   b. Weight of a paper clip

   c. Active ingredient in an aspirin tablet

    i. 800 kg

    ii. 1 g

    iii. 325 mg

15. Match each item with its proper measurement.

   a. Amount of blood in an adult

   b. Cola in an aluminum can

   c. Kuwait's daily production of crude oil

    i. 290,000 kL

    ii. 6 L

    iii. 355 mL

16. Of the objects in the illustration, which can be used to measure the following?

   a. Millimeters

   b. Milligrams

   c. Milliliters

Balance

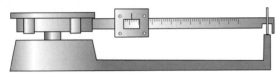

Beaker

Micrometer

Fill in the blanks.

17. 1 dekameter = ▢ meters

18. 1 decimeter = ▢ meter

19. 1 centimeter = ▢ meter

20. 1 kilometer = ▢ meters

21. 1 millimeter = ▢ meter

22. 1 hectometer = ▢ meters

23. 1 gram = ▢ milligrams

24. 100 centigrams = ▢ gram

25. 1 kilogram = ▢ grams

26. 1 milliliter = ▢ cubic centimeter

27. 1 liter = ▢ cubic centimeters

28. 1 kiloliter = ▢ liters

29. 1 centiliter = ▢ liter

30. 1 milliliter = ▢ liter

31. 100 liters = ▢ hectoliter

32. 10 deciliters = ▢ liter

NOTATION *Complete each solution.*

33. Convert 20 centimeters to meters.

$$20 \text{ cm} = 20 \text{ cm} \cdot \frac{\boxed{} \text{ m}}{100 \text{ cm}}$$

$$= \frac{20}{\boxed{}} \text{ m}$$

$$= 0.2 \text{ m}$$

34. Convert 300 centigrams to grams.

$$300 \text{ cg} = 300 \text{ cg} \cdot \frac{\boxed{} \text{ g}}{100 \text{ cg}}$$

$$= \frac{\boxed{}}{100} \text{ g}$$

$$= 3 \text{ g}$$

35. Convert 2 kilometers to decimeters.

$$2 \text{ km} = 2 \text{ km} \cdot \frac{\boxed{} \text{ m}}{1 \text{ km}} \cdot \frac{10 \text{ dm}}{\boxed{} \text{ m}}$$

$$= 2 \cdot \boxed{} \cdot 10 \text{ dm}$$

$$= 20,000 \text{ dm}$$

36. Convert 3 deciliters to milliliters.

$$3 \text{ dL} = 3 \text{ dL} \cdot \frac{1 \text{ L}}{\boxed{} \text{ dL}} \cdot \frac{\boxed{} \text{ mL}}{1 \text{ L}}$$

$$= \frac{\boxed{} \cdot 1,000}{10} \text{ mL}$$

$$= 300 \text{ mL}$$

PRACTICE   *Use a metric ruler to measure each object to the nearest millimeter.*

37. The length of a dollar bill

38. The width of a dollar bill

Use a metric ruler to measure each object to the nearest centimeter.

39. The length (top to bottom) of this page

40. The length of the word antidisestablishmentarianism

Convert each measurement between the given metric units.

41. 3 m = _____ cm

42. 5 m = _____ cm

43. 5.7 m = _____ cm

44. 7.36 km = _____ dam

45. 0.31 dm = _____ cm

46. 73.2 m = _____ dm

47. 76.8 hm = _____ mm

48. 165.7 km = _____ m

49. 4.72 cm = _____ dm

50. 0.593 cm = _____ dam

51. 453.2 cm = _____ m

52. 675.3 cm = _____ m

53. 0.325 dm = _____ m

54. 0.0034 mm = _____ m

55. 3.75 cm = _____ mm

56. 0.074 cm = _____ mm

57. 0.125 m = _____ mm

58. 134 m = _____ hm

59. 675 dam = _____ cm

60. 0.00777 cm = _____ dam

61. 638.3 m = _____ hm

62. 6.77 cm = _____ m

63. 6.3 mm = _____ cm

64. 6.77 mm = _____ cm

65. 695 dm = _____ m

66. 6,789 cm = _____ dm

67. 5,689 m = _____ km

68. 0.0579 km = _____ mm

69. 576.2 mm = _____ dm

70. 65.78 km = _____ dam

71. 6.45 dm = _____ km

72. 6.57 cm = _____ mm

73. 658.23 m = _____ km

74. 0.0068 hm = _____ km

75. 3 g = _____ mg

76. 5 g = _____ cg

77. 2 kg = _____ g

78. 4,000 g = _____ kg

79. 1,000 kg = _____ g

80. 2 kg = _____ cg

81. 500 mg = _____ g

82. 500 mg = _____ cg

83. 3 kL = _____ L

84. 500 mL = _____ L

85. 500 cL = _____ mL

86. 400 L = _____ hL

87. 10 mL = _____ cc

88. 2,000 cc = _____ L

APPLICATIONS

89. SPEED SKATING American Eric Heiden won an unprecedented five gold medals by capturing the men's 500-m, 1,000-m, 1,500-m, 5,000-m, and 10,000-m races at the 1980 Winter Olympic Games in Lake Placid, New York. Convert each race length to kilometers.

90. THE SUEZ CANAL The 163-km-long Suez Canal connects the Mediterranean Sea with the Red Sea. It provides a shortcut for ships operating between European and American ports. Convert the length of the Suez Canal to meters.

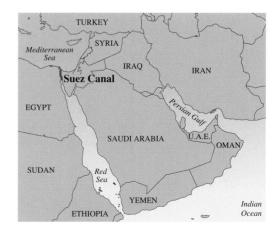

91. THE HANCOCK CENTER The John Hancock Center in Chicago has 100 stories and is 343 meters high. Give this height in hectometers.

92. WEIGHT OF A BABY A baby weighs 4 kilograms. Give this weight in centigrams.

93. HEALTH CARE Blood pressure is measured by a *sphygmomanometer*. The measurement is read at two points and is expressed, for example, as 120/80. This indicates a *systolic* pressure of 120 millimeters of mercury and a *diastolic* pressure of 80 millimeters of mercury. Convert each measurement to centimeters of mercury.

94. JEWELRY A gold chain weighs 1,500 milligrams. Give this weight in grams.

95. CONTAINERS How many deciliters of root beer are in two 2-liter bottles?

96. BOTTLING How many liters of wine are in a 750-mL bottle?

97. BUYING OLIVES The net weight of a bottle of olives is 284 grams. Find the smallest number of bottles that must be purchased to have at least 1 kilogram of olives.

98. BUYING COFFEE A can of Cafe Vienna has a net weight of 133 grams. Find the smallest number of cans that must be packaged to have at least 1 metric ton of coffee. (*Hint:* 1 metric ton = 1,000 kg.)

99. MEDICINE A bottle of hydrochlorothiazine contains 60 tablets. If each tablet contains 50 milligrams of active ingredient, how many grams of active ingredient are in the bottle?

100. INJECTIONS The illustration shows a 3cc syringe. Express its capacity using units of milliliters.

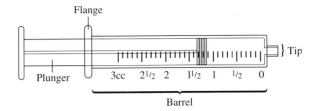

WRITING

101. To change 3.452 kilometers to meters, we can move the decimal point in 3.452 three places to the right to get 3,452 meters. Explain why.

102. To change 7,532 grams to kilograms, we can move the decimal point in 7,532 three places to the left to get 7.532 kilograms. Explain why.

103. A centimeter is one hundredth of a meter. Make a list of other words that begin with the prefix *centi* or *cent* and write a definition for each.

104. List the advantages of the metric system of measurement as compared to the American system. There have been several attempts to bring the metric system into general use in the United States. Why do you think these efforts have been unsuccessful?

REVIEW

105. Find 7% of $342.72

106. $32.16 is 8% of what amount?

107. Divide: $3\frac{1}{7} \div 2\frac{1}{2}$.

108. Simplify: $3\frac{1}{7} + 2\frac{1}{2} \cdot 3\frac{1}{3}$.

6.5 Converting between American and Metric Units

- Converting between American and metric units
- Comparing American and metric units of temperature

It is often necessary to convert between American units and metric units. For example, we must convert units to answer the following questions.

- Which is higher: Pikes Peak (elevation 14,110 feet) or the Matterhorn (elevation 4,478 meters)?
- Does a 2-pound tub of butter weigh more than a 1-kilogram tub?
- Is a quart of soda pop more or less than a liter of soda pop?

In this section, we discuss how to answer such questions.

Converting between American and metric units

We can convert between American and metric units of length using the table below.

Equivalent lengths	
American to metric	**Metric to American**
1 in. ≈ 2.54 cm	1 cm ≈ 0.3937 in.
1 ft ≈ 0.3048 m	1 m ≈ 3.2808 ft
1 yd ≈ 0.9144 m	1 m ≈ 1.0936 yd
1 mi ≈ 1.6093 km	1 km ≈ 0.6214 mi

EXAMPLE 1   Clothing labels. Figure 6-16 shows a label sewn into some pants made in Mexico for sale in the United States. Express the waist size to the nearest inch.

Solution We need to convert from metric to American units. From the table, we see that there is 0.3937 inch in 1 centimeter. To make the conversion, we substitute 0.3937 inch for 1 centimeter.

81 centimeters = 81 (**centimeters**)
  ≈ 81(**0.3937 in.**)   Substitute 0.3937 inch for 1 centimeter.
  ≈ 31.8897 in.   Perform the multiplication.

To the nearest inch, the waist size is 32 inches.

WAIST: 81 cm
INSEAM: 76 cm
RN-80811
SEE REVERSE FOR CARE
MADE IN MEXICO

FIGURE 6-16

Self Check 1
Refer to Figure 6-16. What is the inseam length, to the nearest inch?

Answer 30 in.

EXAMPLE 2   Mountain elevations. Pikes Peak, one of the most famous peaks in the Rocky Mountains, has an elevation of 14,110 feet. The Matterhorn, in the Swiss Alps, rises to an elevation of 4,478 meters. Which mountain is higher?

Solution To make a comparison, the elevations must be expressed in the same units. We will convert the elevation of Pikes Peak, which is given in feet, to meters.

14,110 feet = 14,110 (**feet**)
  ≈ 14,110(**0.3048 m**)   Substitute 0.3048 meter for 1 foot.
  ≈ 4,300.728 m   Perform the multiplication.

Since the elevation of Pikes Peak is about 4,301 meters, we can conclude that the Matterhorn, with an elevation of 4,478 meters, is higher.

Self Check 2
Which is longer: a 500-meter race or a 550-yard race?

Answer the 550-yard race

We can convert between American units of weight and metric units of mass by using the accompanying table.

Equivalent weights and masses	
American to metric	**Metric to American**
1 oz ≈ 28.35 g	1 g ≈ 0.035 oz
1 lb ≈ 0.454 kg	1 kg ≈ 2.2 lb

Self Check 3
Change 20 kilograms to pounds.

EXAMPLE 3   Change 50 pounds to grams.

Solution

$$50 \text{ lb} = 50(\textbf{1 lb})$$
$$= 50(\textbf{16 oz}) \qquad \text{Substitute 16 ounces for 1 pound.}$$
$$= 50(16)(\textbf{1 oz})$$
$$\approx 50(16)(\textbf{28.35 g}) \qquad \text{Substitute 28.35 grams for 1 ounce.}$$
$$\approx 22,680 \text{ g} \qquad \text{Perform the multiplication.}$$

Thus, 50 pounds is equal to 22,680 grams.

Answer 44 lb

Self Check 4
Who weighs more: a person who
weighs 165 pounds or one who
weighs 76 kilograms?

EXAMPLE 4   **Packaging.**   Does a 2-pound tub of butter weigh more than a
1-kilogram tub?

Solution   To decide which contains more butter, we can change 2 pounds to kilograms.

$$2 \text{ lb} = 2(\textbf{1 lb})$$
$$\approx 2(\textbf{0.454 kg}) \qquad \text{Substitute 0.454 kilogram for 1 pound.}$$
$$\approx 0.908 \text{ kg} \qquad \text{Perform the multiplication.}$$

Since a 2-pound tub weighs only 0.908 kilogram, the 1-kilogram tub weighs more.

Answer the person who weighs
76 kg

We can convert between American and metric units of capacity by using the accompanying table.

Equivalent capacities	
American to metric	**Metric to American**
1 fl oz ≈ 0.030 L	1 L ≈ 33.8 fl oz
1 pt ≈ 0.473 L	1 L ≈ 2.1 pt
1 qt ≈ 0.946 L	1 L ≈ 1.06 qt
1 gal ≈ 3.785 L	1 L ≈ 0.264 gal

Self Check 5
A student bought a 355-mL can
of cola. How many ounces of
cola does the can contain?

EXAMPLE 5   **Soft drinks.**   A bottle of 7UP contains 750 milliliters. Convert
this measure to quarts.

Solution   We convert milliliters to liters and then liters to quarts.

$$750 \text{ mL} = 750 \text{ mL} \cdot \frac{1 \text{ L}}{1{,}000 \text{ mL}} \qquad \text{Use a unit conversion factor: } \frac{1 \text{ L}}{1{,}000 \text{ mL}} = 1.$$

$$= \frac{750}{1{,}000} \text{ L} \qquad \text{The units of mL divide out.}$$

$$= \frac{3}{4} \text{ L} \qquad \text{Simplify the fraction: } \frac{750}{1{,}000} = \frac{3 \cdot \overset{1}{\cancel{250}}}{4 \cdot \underset{1}{\cancel{250}}} = \frac{3}{4}.$$

$$\approx \frac{3}{4}(\textbf{1.06 qt}) \qquad \text{Substitute 1.06 quart for 1 liter.}$$

$$\approx 0.795 \text{ qt} \qquad \text{Perform the arithmetic.}$$

The bottle contains 0.795 quart.

Answer 12 oz

From the table of equivalent capacities, we see that 1 liter is equal to 1.06 quarts. Thus, a liter of soda pop is more than a quart of soda pop.

EXAMPLE 6   Comparison shopping.   A 2-quart bottle of soda pop is priced at $1.89, and a 1-liter bottle is priced at 97¢. Which is the better buy?

Solution   We can convert 2 quarts to liters and find the price per liter of the 2-quart bottle.

$$2 \text{ qt} = 2(\mathbf{1 \text{ qt}})$$
$$\approx 2(\mathbf{0.946 \text{ L}}) \quad \text{Substitute 0.946 liter for 1 quart.}$$
$$\approx 1.892 \text{ L} \quad \text{Perform the multiplication.}$$

Thus, the 2-quart bottle contains 1.892 liters. To find the price per liter of the 2-quart bottle, we divide $\frac{\$1.89}{1.892}$.

$$\frac{\$1.89}{1.892} \approx \$0.998942918$$

Since the price per liter of the 2-quart bottle is a little more than 99¢, the 1-liter bottle priced at 97¢ is the better buy.

Studying in Other Countries   THINK IT THROUGH

"Over the past decade, the number of U.S. students studying abroad has more than doubled." From *The Open Doors 2003 Report*

In 2001/2002, a record number of 160,920 college students received credit for study abroad. Since students traveling to other countries are almost certain to come into contact with the metric system of measurement, they need to have a basic understanding of metric units.

Suppose a student studying overseas needs to purchase the following school supplies. For each item in red, choose the appropriate metric units.

1. $8\frac{1}{2}$ in. × 11 in. notebook paper:

    216 meters × 279 meters   216 centimeters × 279 centimeters

        216 millimeters × 279 millimeters

2. A backpack that can hold 20 pounds of books:

    9 kilograms   9 grams   9 milligrams

3. $\frac{3}{4}$ fluid ounce bottle of Liquid Paper correction fluid:

    22.5 hectoliters   2.5 liters   22.5 milliliters

▪ Comparing American and metric units of temperature

In the American system, we measure temperature using **degrees Fahrenheit** (°F). In the metric system, we measure temperature using **degrees Celsius** (°C). These two scales are shown on the thermometers in Figure 6-17. From the figure, we can see that

- 212° F ≈ 100° C   Water boils.
- 32° F ≈ 0° C   Water freezes.
- 5° F ≈ −15° C   A cold winter day
- 95° F ≈ 35° C   A hot summer day

As we have seen, there is a formula that enables us to convert from degrees Fahrenheit to degrees Celsius. There is also a formula to convert from degrees Celsius to degrees Fahrenheit.

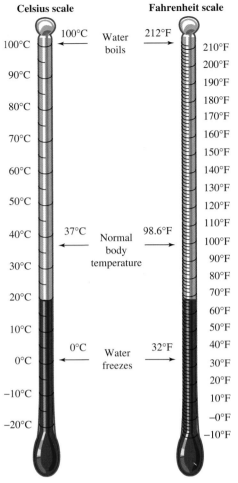

FIGURE 6-17

Conversion formulas for temperature

If *F* is the temperature in degrees Fahrenheit and *C* is the corresponding temperature in degrees Celsius, then

$$C = \frac{5}{9}(F - 32) \quad \text{and} \quad F = \frac{9}{5}C + 32$$

Self Check 7

Hot coffee is 110° F. To the nearest tenth of a degree, express this temperature in degrees Celsius.

EXAMPLE 7   Bathing.   Warm bath water is 90° F. Find the equivalent temperature in degrees Celsius.

Solution   We substitute 90 for *F* in the formula $C = \frac{5}{9}(F - 32)$ and simplify.

$$C = \frac{5}{9}(F - 32)$$

$$= \frac{5}{9}(90 - 32) \quad \text{Substitute 90 for } F.$$

$$= \frac{5}{9}(58) \quad \text{Subtract: } 90 - 32 = 58.$$

$$= 32.222\ldots \quad \text{Perform the arithmetic.}$$

Answer 43.3° C

To the nearest tenth of a degree, the equivalent temperature is 32.2° C.

EXAMPLE 8 A dishwasher manufacturer recommends that dishes be rinsed in hot water with a temperature of 60° C. Express this temperature in degrees Fahrenheit.

Solution We substitute 60 for C in the formula $F = \dfrac{9}{5}C + 32$ and simplify.

$$F = \frac{9}{5}C + 32$$

$$= \frac{9}{5}(60) + 32 \qquad \text{Substitute 60 for } C.$$

$$= \frac{540}{5} + 32 \qquad \text{Multiply: } \frac{9}{5}(60) = \frac{540}{5}.$$

$$= 108 + 32 \qquad \text{Perform the division.}$$

$$= 140 \qquad \text{Perform the addition.}$$

The manufacturer recommends that dishes be rinsed in 140° F water.

Self Check 8
To determine whether a baby has a fever, her mother takes her temperature with a Celsius thermometer. If the reading is 38.8° C, does the baby have a fever? (*Hint:* Normal body temperature is 98.6° F.)

Answer yes

Section 6.5 STUDY SET

VOCABULARY *Fill in the blanks.*

1. In the American system, temperatures are measured in degrees _____. In the metric system, temperatures are measured in degrees _____.

2. Inches and centimeters are units used to measure _____. Gallons and liters are units used to measure _____.

CONCEPTS

3. Which is longer?
   a. A yard or a meter?
   b. A foot or a meter?
   c. An inch or a centimeter?
   d. A mile or a kilometer?

4. Which is heavier?
   a. An ounce or a gram?
   b. A pound or a kilogram?

5. Which is the greater unit of capacity?
   a. A pint or a liter?
   b. A quart or a liter?
   c. A gallon or a liter?

6. a. What formula is used for changing degrees Celsius to degrees Fahrenheit?
   b. What formula is used for changing degrees Fahrenheit to degrees Celsius?

NOTATION *Complete each solution.*

7. Change 4,500 feet to kilometers.

$$4{,}500 \text{ ft} = 4{,}500(\boxed{} \text{ m})$$
$$= \boxed{} \text{ m}$$
$$= 1.3716 \text{ km}$$

8. Change 3 kilograms to ounces.

$$3 \text{ kg} = 3(\boxed{} \text{ lb})$$
$$= 3(2.2)(\boxed{} \text{ oz})$$
$$= 105.6 \text{ oz}$$

9. Change 8 liters to gallons.

$$8 \text{ L} = 8(\boxed{} \text{ gal})$$
$$= 2.112 \text{ gal}$$

10. Change 70°C to degrees Fahrenheit.

$$F = \frac{9}{5}C + 32$$
$$= \frac{9}{5}(\boxed{}) + 32$$
$$= \boxed{} + 32$$
$$= 158° \text{ F}$$

PRACTICE *Make each conversion. Since most conversions are approximate, answers will vary depending on the method used.*

11. 3 ft = $\boxed{}$ cm

12. 7.5 yd = $\boxed{}$ m

13. 3.75 m = $\boxed{}$ in.

14. 2.4 km = $\boxed{}$ mi

15. 12 km = $\boxed{}$ ft

16. 3,212 cm = $\boxed{}$ ft

17. 5,000 in. = [   ] m   **18.** 25 mi = [   ] km

19. 37 oz = [   ] kg   **20.** 10 lb = [   ] kg

21. 25 lb = [   ] g   **22.** 7.5 oz = [   ] g

23. 0.5 kg = [   ] oz   **24.** 35 g = [   ] lb

25. 17 g = [   ] oz   **26.** 100 kg = [   ] lb

27. 3 fl oz = [   ] L   **28.** 2.5 pt = [   ] L

29. 7.2 L = [   ] fl oz   **30.** 5 L = [   ] qt

31. 0.75 qt = [   ] mL   **32.** 3 pt = [   ] mL

33. 500 mL = [   ] qt   **34.** 2,000 mL = [   ] gal

35. 50° F = [   ] C   **36.** 67.7° F = [   ] C

37. 50° C = [   ] F   **38.** 36.2° C = [   ] F

39. −10° C = [   ] F   **40.** −22.5° C = [   ] F

41. −5° F = [   ] C   **42.** −10° F = [   ] C

APPLICATIONS   *Since most conversions are approximate, answers will vary depending on the method used.*

43. THE MIDDLE EAST The distance between Jerusalem and Bethlehem is 8 kilometers. To the nearest mile, give this distance in miles.

44. THE DEAD SEA The Dead Sea is 80 kilometers long. To the nearest mile, give this distance in miles.

45. CHEETAHS A cheetah can run 112 kilometers per hour. Express this speed in mph.

46. LIONS A lion can run 50 mph. Express this speed in kilometers per hour.

47. MOUNT WASHINGTON The highest peak of the White Mountains of New Hampshire is Mount Washington, at 6,288 feet. To the nearest tenth, give this height in kilometers.

48. TRACK AND FIELD Track meets are held on an oval track. One lap around the track is usually 400 meters. However, some older tracks in the United States are 440-yard ovals. Are these two types of tracks the same length? If not, which is longer?

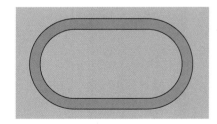

49. HAIR GROWTH When hair is short, its rate of growth averages about $\frac{3}{4}$ inch per month. How many centimeters is this a month?

50. WHALES An adult male killer whale can weigh as much as 12,000 pounds and be as long as 25 feet. Change these measurements to kilograms and meters.

51. WEIGHTLIFTING The table lists the personal best bench press records for two of the world's best powerlifters. Change each metric weight to pounds. Round to the nearest pound.

Name	Hometown	Bench press
Liz Willet	Ferndale, Washington	187 kg
Brian Siders	Charleston, W. Virginia	338 kg

52. WORDS OF WISDOM Refer to the wall hanging. Convert the first metric weight to ounces and the second to pounds. What famous saying results?

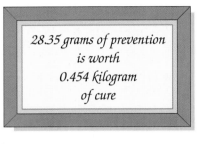

28.35 grams of prevention is worth 0.454 kilogram of cure

53. OUNCES AND FLUID OUNCES

   a. There are 310 calories in 8 ounces of broiled chicken. Convert 8 ounces to grams.

   b. There are 112 calories in a glass of fresh Valencia orange juice that holds 8 fluid ounces. Convert 8 fluid ounces to liters.

54. TRACK AND FIELD A shot-put weighs 7.264 kilograms. Give this weight in pounds.

55. POSTAL REGULATIONS You can mail a package weighing up to 70 pounds via priority mail. Can you mail a package that weighs 32 kilograms by priority mail?

56. NUTRITION Refer to the nutrition label for a packet of oatmeal shown. Change each circled weight to ounces.

Nutrition Facts
Serving Size: 1 Packet (46g)
Servings Per Container: 10

Amount Per Serving
Calories 170   Calories from Fat 20

	% Daily Value
Total fat 2g	3%
Saturated fat (0.5g)	2%
Polyunsaturated Fat 0.5g	
Monounsaturated Fat 1g	
Cholesterol 0mg	0%
Sodium (250mg)	10%
Total carbohydrate 35g	12%
Dietary fiber 3g	12%
Soluble Fiber 1g	
Sugars 16g	
Protein (4g)	

57. COMPARISON SHOPPING Which is the better buy: 3 quarts of root beer for $4.50 or 2 liters of root beer for $3.60?

58. COMPARISON SHOPPING Which is the better buy: 3 gallons of antifreeze for $10.35 or 12 liters of antifreeze for $10.50?

59. HOT SPRINGS The thermal springs in Hot Springs National Park in central Arkansas emit water as warm as 143° F. Change this temperature to degrees Celsius.

60. COOKING MEAT Meats must be cooked at high enough temperatures to kill harmful bacteria. According to the USDA and the FDA, the internal temperature for cooked roasts and steaks should be at least 145° F, and whole poultry should be 180° F. Convert these temperatures to degrees Celsius. Round up to the next degree.

61. TAKING A SHOWER When you take a shower, which water temperature would you choose: 15° C, 28° C, or 50° C?

62. DRINKING WATER To get a cold drink of water, which temperature would you choose: −2° C, 10° C, or 25° C?

63. SNOWY WEATHER At which temperatures might it snow: −5° C, 0° C, or 10° C?

64. AIR CONDITIONING At which outside temperature would you be likely to run the air conditioner: 15° C, 20° C, or 30° C?

WRITING

65. Explain how to change kilometers to miles.

66. Explain how to change 50° C to degrees Fahrenheit.

67. The United States is the only industrialized country in the world that does not officially use the metric system. Some people claim this is costing American businesses money. Do you think so? Why?

68. What is meant by the phrase *a table of equivalent measures*?

REVIEW *Perform each operation.*

69. $\dfrac{3}{5} + \dfrac{4}{3}$

70. $\dfrac{3}{5} - \dfrac{4}{3}$

71. $\dfrac{3}{5} \cdot \dfrac{4}{3}$

72. $\dfrac{3}{5} \div \dfrac{4}{3}$

73. $3.25 + 4.8$

74. $3.25 - 4.8$

75. $3.25 \cdot 4.8$

76. $4.8\overline{)15.6}$

Proportions

A **proportion** is a statement that two ratios or rates are equal.

Fill in the blanks as we set up a proportion to solve a problem.

1. TEACHER'S AIDES For every 15 children on the playground, a child care center is required to have 2 teacher's aides supervising. How many teacher's aides will be needed to supervise 75 children?

   Step 1: Let x = the number of

   _____.

   If we compare the number of children to the number of teacher's aides, we know that the two rates must be equal.

   > 15 children are to ▢ aides as ▢ children are to ▢ aides.

   We can express this as a proportion.

   $$\text{Number of children} \rightarrow \frac{15}{\boxed{}} = \frac{75}{\boxed{}} \leftarrow \text{Number of children}$$
   $$\text{Number of aides} \rightarrow \qquad\qquad \leftarrow \text{Number of aides}$$

   In the proportion $\frac{15}{2} = \frac{75}{x}$, 15 and x are the *extremes* and 2 and 75 are the *means*. After setting up the proportion, we solve it using the fact that the product of the extremes is equal to the product of the means.

   Step 2: Solve for x: $\dfrac{15}{2} = \dfrac{75}{x}$.

   $\boxed{} \cdot x = 2 \cdot \boxed{}$ The product of the extremes equals the product of the means.

   $15 \cdot x = \boxed{}$ Perform the multiplication.

   $\dfrac{15 \cdot x}{\boxed{}} = \dfrac{150}{\boxed{}}$ Divide both sides by 15.

   $x = \boxed{}$ Perform the divisions.

   To supervise 75 children, 10 teacher's aides are needed.

   Step 3: To check the result, we substitute 10 for x in $\frac{15}{2} = \frac{75}{x}$ and find the cross products.

   $15 \cdot 10 = \mathbf{150} \qquad 2 \cdot 75 = \mathbf{150}$

   $$\frac{15}{2} \overset{?}{=} \frac{75}{10}$$

   Since the cross products are equal, 10 is the solution.

Set up and solve each problem using a proportion.

2. PARKING A city code requires that companies provide 10 parking spaces for every 12 employees. How many spaces will be needed if a company employs 450 people?

3. MOTION PICTURES Every 2 seconds, 3 feet of motion picture film pass through the projector. How many feet of film are there in a movie that runs for 120 minutes?

4. BEAUTY SUPPLIES A 0.5-fluid-ounce bottle of nail polish costs $4.50. What would be the cost of one gallon of the nail polish? (*Hint:* 1 gallon = 128 fluid ounces)

ACCENT ON TEAMWORK

SECTION 6.1

ART HISTORY The illustration shows a drawing made by Leonardo Da Vinci (1452–1519) of a human figure within a square. We see that the man's height and the span of his outstretched arms are the same. That is, their ratio is 1 to 1. Use a tape measure to determine this ratio for each member of your group. Work in terms of inches. Are any ratios exactly 1 to 1?

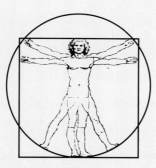

SECTION 6.2

ENLARGEMENTS Duplicating or enlarging a picture can be done by the *grid-transfer* method. To enlarge the picture of a cook shown in illustration (a), begin by copying the markings in the square in the lower right-hand corner onto its corresponding square in the enlargement grid in illustration (b). One by one, copy the contents of each square of the original picture to its counterpart in the enlargement.

SECTION 6.3

DISNEY CLASSICS The movie *20,000 Leagues Under the Sea* is a science fiction thriller about Captain Nemo and the crew of the submarine Nautilus. Use the fact that 1 mile $= \frac{1}{3}$ league to express a depth of 20,000 leagues in feet.

SECTION 6.4

METRIC MISHAP In 1999, NASA lost the $125-million Mars Climate Orbiter because a Lockheed Martin engineering team used American units of measurement, while NASA's team used the metric system. Use the Internet to research this incident and make a report to your class.

SECTION 6.5

TRUTH IN LABELING Have each member of your group bring in two items—one whose product label indicates capacity and another that indicates weight. For example, a bottle of shampoo could contain 15 fluid ounces (444 mL) or a can of soup could weigh 1 pound 3 ounces (539 g). Exchange your items with another person in your group. Check the accuracy of each label by converting from American units to metric units.

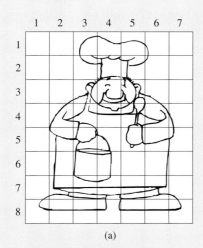

(a)

The original picture is drawn on a grid of $\frac{1}{4}$-inch squares. The enlargement is drawn using a grid of $\frac{1}{2}$-inch squares. By how much was the original picture enlarged?

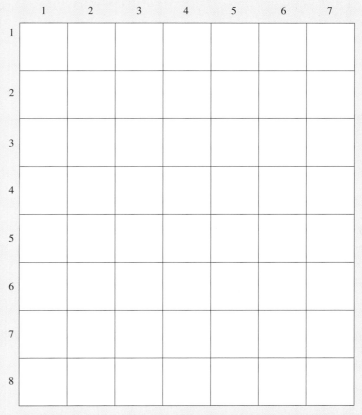

(b)

CHAPTER REVIEW

| SECTION 6.1 | *Ratios* |

CONCEPTS

A *ratio* is a quotient of two numbers or a quotient of two quantities that have the same units.

A *rate* is a comparison of two quantities with different units.

REVIEW EXERCISES

Express each phrase as a ratio in lowest terms.

1. The ratio of 4 inches to 12 inches

2. The ratio of 8 ounces to 2 pounds

3. 21:14

4. 24 to 36

5. AIRCRAFT Specifications for a Boeing B-52 Stratofortress are given in the illustration. What is the ratio of the airplane's wingspan to its length?

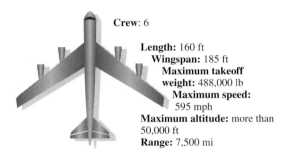

Crew: 6
Length: 160 ft
Wingspan: 185 ft
Maximum takeoff weight: 488,000 lb
Maximum speed: 595 mph
Maximum altitude: more than 50,000 ft
Range: 7,500 mi

6. PAY SCALES Find the hourly rate of pay for a student who earned $333.25 for working 43 hours.

A *unit cost* is a comparison of the cost of an item to its quantity.

7. COMPARISON SHOPPING Mixed nuts come packaged in a 12-ounce can, which sells for $4.95, or an 8-ounce can, which sells for $3.25. Which is the better buy?

| SECTION 6.2 | *Proportions* |

A *proportion* is a statement that two ratios (or rates) are equal.

Consider the proportion $\dfrac{5}{15} = \dfrac{25}{75}$.

8. Which term is the fourth term?

9. Which term is the second term?

In any proportion, the product of the *extremes* is equal to the product of the *means*.

Determine whether each of the following statements is a proportion.

10. $\dfrac{15}{29} = \dfrac{105}{204}$

11. $\dfrac{17}{7} = \dfrac{204}{84}$

When two pairs of numbers form a proportion, we say that the numbers are *proportional*.

Determine whether the numbers are proportional.

12. 5, 9 and 20, 36

13. 7, 13 and 29, 54

Solve each proportion.

14. $\dfrac{12}{18} = \dfrac{3}{x}$

15. $\dfrac{4}{x} = \dfrac{2}{8}$

16. $\dfrac{4.8}{6.6} = \dfrac{x}{9.9}$

17. $\dfrac{-0.08}{x} = \dfrac{0.04}{0.06}$

18. TRUCKS A Dodge Ram pickup truck can go 35 miles on 2 gallons of gas. How far can it go on 11 gallons?

19. QUALITY CONTROL In a manufacturing process, 12 parts out of 66 were found to be defective. How many defective parts will be expected in a run of 1,650 parts?

20. SCALE DRAWINGS The illustration shows an architect's drawing of a kitchen using a scale of $\frac{1}{8}$ inch to 1 foot ($\frac{1}{8}''$: $1'0''$). On the drawing, the length of the kitchen is $1\frac{1}{2}$ inches. How long is the actual kitchen? (The symbol $''$ means inch and $'$ means foot.)

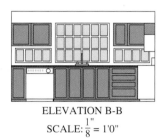

ELEVATION B-B
SCALE: $\frac{1''}{8} = 1'0''$

SECTION 6.3 — *American Units of Measurement*

21. Use a ruler to measure the length of the computer mouse to the nearest quarter of an inch.

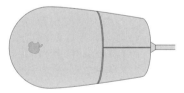

22. Write two unit conversion factors using the fact that 1 mile = 5,280 ft.

Make each conversion.

23. 5 yards to feet

24. 6 yards to inches

25. 66 inches to feet

26. 25.5 feet to inches

27. 9,240 feet to miles

28. 1 mile to yards

Make each conversion.

29. 32 ounces to pounds

30. 17.2 pounds to ounces

31. 3 tons to ounces

32. 4,500 pounds to tons

Common American units of
capacity are *fluid ounces, cups,
pints, quarts,* and *gallons.*

1 c = 8 fl oz
1 pt = 2 c
1 qt = 2 pt
1 gal = 4 qt

Units of time are *seconds,
minutes, hours,* and *days.*

1 min = 60 sec
1 hr = 60 min
1 day = 24 hr

Make each conversion.

33. 5 pints to fluid ounces

34. 8 cups to gallons

35. 17 quarts to cups

36. 176 fluid ounces to quarts

37. 5 gallons to pints

38. 3.5 gallons to cups

Make each conversion.

39. 20 minutes to seconds

40. 900 seconds to minutes

41. 200 hours to days

42. 6 hours to minutes

43. 4.5 days to hours

44. 1 day to seconds

45. SKYSCRAPERS The Sears Tower in Chicago is 1,454 feet high. Express this distance in yards.

46. BOTTLING A magnum is a 2 quart bottle of wine. How many magnums will be needed to hold 50 gallons of wine?

SECTION 6.4   *Metric Units of Measurement*

Common metric units of length
are *millimeter, centimeter,
decimeter, meter, dekameter,
hectometer,* and *kilometer.*

$1 \text{ mm} = \frac{1}{1,000} \text{ m}$

$1 \text{ cm} = \frac{1}{100} \text{ m}$

$1 \text{ dm} = \frac{1}{10} \text{ m}$

1 dam = 10 m

1 hm = 100 m

1 km = 1,000 m

47. Use a metric ruler to measure the length of the computer mouse to the nearest centimeter.

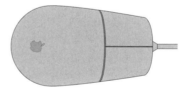

48. Write two unit conversion factors using the fact that 1 km = 1,000 m.

Make each conversion.

49. 475 centimeters to meters

50. 8 meters to millimeters

51. 3 dekameters to kilometers

52. 2 hectometers to decimeters

53. 5 kilometers to hectometers

54. 2,500 meters to hectometers

Common metric units of mass
are *milligrams, centigrams,
grams,* and *kilograms.*

$1 \text{ mg} = \frac{1}{1,000} \text{ g}$

$1 \text{ cg} = \frac{1}{100} \text{ g}$

$1 \text{ g} = \frac{1}{1,000} \text{ kg}$

Make each conversion.

55. 7 centigrams to milligrams

56. 800 centigrams to grams

57. 5,425 grams to kilograms

58. 5,425 grams to milligrams

59. 7,500 milligrams to grams

60. 5,000 centigrams to kilograms

61. TYLENOL A bottle of Extra Strength Tylenol contains 100 caplets of 500 milligrams each. How many grams of Tylenol are in the bottle?

Common metric units of
capacity are *milliliters,
centiliters, deciliters, liters,
hectoliters,* and *kiloliters.*

$$1 \text{ mL} = \tfrac{1}{1,000} \text{ L}$$

$$1 \text{ cL} = \tfrac{1}{100} \text{ L}$$

$$1 \text{ dL} = \tfrac{1}{10} \text{ L}$$

$$1 \text{ L} = 1,000 \text{ cc}$$

$$1 \text{ hL} = 100 \text{ L}$$

$$1 \text{ kL} = 1,000 \text{ L}$$

Make each conversion.

62. 150 centiliters to liters

63. 3,250 liters to kiloliters

64. 1 hectoliter to deciliters

65. 400 milliliters to centiliters

66. 2 kiloliters to hectoliters

67. 4 deciliters to milliliters

68. SURGERY A dextrose solution is being administered
to a patient intravenously using the apparatus shown. How
many milliliters of solution does the IV bag hold?

SECTION 6.5   *Converting between American and Metric Units*

We can convert between
American and metric units
using the following:

1 in. ≈ 2.54 cm
1 ft ≈ 0.3048 m
1 yd ≈ 0.9144 m
1 mi ≈ 1.6093 km
1 cm ≈ 0.3937 in.
1 m ≈ 3.2808 ft
1 m ≈ 1.0936 yd
1 km ≈ 0.6214 mi

1 oz ≈ 28.35 g
1 lb ≈ 0.454 kg
1 g ≈ 0.035 oz
1 kg ≈ 2.2 lb

1 fl oz ≈ 0.030 L
1 pt ≈ 0.473 L
1 qt ≈ 0.946 L
1 gal ≈ 3.785 L
1 L ≈ 33.8 fl oz
1 L ≈ 2.1 pt
1 L ≈ 1.06 qt
1 L ≈ 0.264 gal

Two units used to measure
temperature are degrees
Fahrenheit and degrees Celsius.

$$C = \frac{5}{9}(F - 32)$$

$$F = \frac{9}{5}C + 32$$

69. SWIMMING Olympic-size swimming pools are 50 meters long. Express this
distance in feet.

70. HIGH-RISE BUILDINGS The Sears Tower is 443 meters high, and the Empire
State Building is 1,250 feet high. Which building is taller?

71. WESTERN SETTLERS The Oregon Trail was an overland route pioneers used in
the 1840s through the 1870s to reach the Oregon Territory. It stretched 1,930 miles
from Independence, Missouri, to Oregon City, Oregon. Find this distance to the
nearest kilometer.

72. AIR JORDAN Michael Jordan is 6 feet, 6 inches tall. Express his height in
centimeters.

Make each conversion.

73. 30 ounces to grams

74. 15 kilograms to pounds

75. 25 pounds to grams (Round to
the nearest thousand.)

76. 2,000 pounds to kilograms
(Round to the nearest ten.)

77. POLAR BEARS At birth, polar bear cubs weigh less than human babies — about
910 grams. Convert this to pounds.

78. BOTTLED WATER LaCroix® bottled water can be purchased in bottles
containing 17 fluid ounces. Mountain Valley® water can be purchased in half-liter
bottles. Which bottle contains more water?

79. COMPARISON SHOPPING One gallon of bleach costs $1.39. A 5-liter economy
bottle costs $1.80. Which is the better buy?

80. Change 77° F to degrees Celsius.

81. Which water temperature is appropriate for swimming: 10° C, 30° C, 50° C, or 70° C?

Write each phrase as a ratio in lowest terms.

1. The ratio of 6 feet to 8 feet

2. The ratio of 8 ounces to 3 pounds

3. COMPARISON SHOPPING Two pounds of coffee can be purchased for $3.38, and a 5-pound can can be purchased for $8.50. Which is the better buy?

4. UTILITY COSTS A household used 675 kilowatt-hours of electricity during a 30-day month. Find the rate of electric consumption in kilowatt-hours per day.

5. CHECKERS What is the ratio of the number of red squares to the number of black squares for the checker-board shown? Express your answer in three ways: as a fraction, using a colon, and using the word *to*.

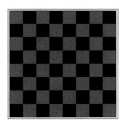

Determine whether each statement is a proportion.

6. $\dfrac{25}{33} = \dfrac{350}{460}$

7. $\dfrac{2.2}{3.5} = \dfrac{1.76}{2.8}$

8. Are the numbers 7, 15 and 245, 525 proportional?

Solve each proportion.

9. $\dfrac{x}{3} = \dfrac{35}{7}$

10. $\dfrac{15.3}{x} = \dfrac{3}{12.4}$

11. $\dfrac{0.07}{-0.5} = \dfrac{x}{1.5}$

12. $\dfrac{25}{0.1} = \dfrac{50}{x}$

13. SHOPPING If 13 ounces of tea costs $2.79, how much would you expect to pay for 16 ounces?

14. COOKING A recipe calls for $\frac{2}{3}$ cup of sugar and 2 cups of flour. How much sugar should be used with 5 cups of flour?

15. Convert 180 inches to feet.

16. TOOLS If a 25-foot tape measure is completely extended, how many yards does it stretch?

17. Convert 10 pounds to ounces.

18. A car weighs 1.6 tons. Find its weight in pounds.

19. How many fluid ounces are in a 1-gallon carton of milk?

20. LITERATURE An excellent work of early science fiction is the book *Around the World in 80 Days* by Jules Verne (1828–1905). Convert 80 days to minutes.

21. A quart and a liter of fruit punch are shown. Which is the 1-liter carton?

22. The figures below show the relative lengths of a yardstick and a meterstick. Which one represents the meterstick?

23. An ounce and a gram are placed on the balance shown. On which side is the gram?

24. SPEED SKATING American Bonnie Blair won gold medals in the women's 500-meter speed skating competitions at the 1988, 1992, and 1994 Winter Olympic Games. Convert the race length to kilometers.

25. How many centimeters are in 5 meters?

26. Convert 8,000 centigrams to kilograms.

27. Convert 70 liters to milliliters.

28. PRESCRIPTIONS A bottle contains 50 tablets, each containing 150 mg of medicine. How many grams of medicine does the bottle contain?

29. Which is the longer distance: a 100-yard race or an 80-meter race?

30. Which person is heavier: Jim, who weighs 160 pounds, or Ricardo, who weighs 72 kilograms?

31. COMPARISON SHOPPING A 2-quart bottle of soda pop costs $1.73, and a 1-liter bottle costs 89¢. Which is the better buy? (*Hint:* 1 quart = 0.946 liter.)

32. COOKING MEAT The USDA recommends that turkey be cooked to a temperature of 83° C. Change this to degrees Fahrenheit. To be safe, round up to the next degree. (*Hint:* $F = \frac{9}{5}C + 32$.)

33. What is a scale drawing? Give an example.

34. Explain the benefits of the metric system of measurement as compared to the American system.

1. Write 64,502 in expanded notation.

2. Divide: $37)\overline{743}$.

3. ENLISTMENTS The table shows how the U.S. Army fared in reaching its recruiting goals in 2004. Complete the table. Use a negative number to denote a shortfall of enlistees.

	Goal	Enlistments	Outcome
Active	77,000	77,587	
Reserve	21,000	21,278	
National Guard	56,000	49,210	

Source: U.S. Army

4. Evaluate each expression, if possible.

   a. $0 + (-8)$ **b.** $\dfrac{-8}{0}$

   c. $0 - |-8|$ **d.** $\dfrac{0}{-8}$

   e. $0 - (-8)$ **f.** $0(-8)$

5. GOLF Tiger Woods won the 100th U.S. Open in June 2000 by the largest margin in the history of that tournament. If he shot 12 under par (-12) and the second-place finisher, Miguel Angel Jimenez, shot 3 over par ($+3$), what was Tiger's margin of victory?

6. Evaluate: -3^2 and $(-3)^2$.

7. Evaluate: $2 + 3[5(-6) - (1 - 10)]$.

Perform each operation.

8. $25 \cdot 100$

9. $25 \div 100$

10. $1.3 \cdot 2.7$

11. $85.34 \div 3.4$

12. $24 - 23.81$

13. PHONE BOOKS A driver left a warehouse in the morning with 500 new telephone books loaded on his truck. His delivery route consisted of office buildings, each of which was to receive 5 of the books. The driver returned at the end of the day with 105 books on the truck. To how many office buildings did he deliver?

14. What is the formula for the area of a triangle?

15. Simplify: $\dfrac{16}{20}$.

16. Express $\dfrac{9}{10}$ as an equivalent fraction with a denominator of 60.

17. Divide: $-\dfrac{7}{8} \div \dfrac{7}{8}$.

18. What is $\dfrac{1}{2}$ of $\dfrac{1}{2}$?

19. MOTORS What is the difference in horsepower (hp) between the two motors shown?

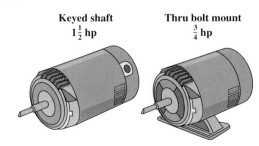

Keyed shaft
$1\frac{1}{2}$ hp

Thru bolt mount
$\frac{3}{4}$ hp

Determine whether each expression is a proportion.

20. $\dfrac{25}{33} = \dfrac{350}{460}$

21. $\dfrac{22}{35} = \dfrac{176}{280}$

22. Write the third term of the proportion $\dfrac{3}{5} = \dfrac{6}{10}$.

23. Are the numbers 7, 15 and 245, 525 proportional?

24. Solve the proportion: $\dfrac{x}{3} = \dfrac{35}{7}$.

25. GLOBAL WARMING The average temperature in the United States during March through May of 2000 was a record-setting 55.5°F. That was 0.4 degree warmer than the previous record, set in 1910. What was the previous spring temperature record?

26. Evaluate: $3\sqrt{25} + 4\sqrt{4}$.

27. Write $\dfrac{1}{12}$ as a decimal.

28. 16 is what percent of 24?

29. What is the formula for simple interest?

30. Complete the table.

Percent	Decimal	Fraction
	0.99	
1.3%		
		$\frac{5}{16}$

31. GUITAR SALE What are the regular price and the rate of discount for the guitar shown?

Save on the Standard Strat

Fender

Now Only
$299⁹⁹
Save $128

32. Express the phrase "3 inches to 15 inches" as a ratio in lowest terms.

33. SURVIVAL GUIDE
   a. A person can go without food for about 40 days. How many hours is this?
   b. A person can go without water for about 3 days. How many minutes is that?
   c. A person can go without breathing oxygen for about 8 minutes. How many seconds is that?

34. Convert 40 ounces to pounds.

35. Convert 2.4 meters to millimeters.

36. Convert 320 grams to kilograms.

37. **a.** Which holds more: a 2-liter bottle or a 1-gallon bottle?
   b. Which is longer: a meterstick or a yardstick?

38. BUILDING MATERIALS Which is the better buy: a 94-pound bag of cement for $4.48 or a 45-kilogram bag of cement for $4.56?

Descriptive Statistics

CORBIS

Every ten years, the federal government conducts a survey, called a *census*, to gather information about the population of the United States. Across the country, members of households, as well as individuals, are asked to complete a questionnaire that asks their age, gender, and race. The Census Bureau collects the data and then uses a branch of mathematics called *statistics* to examine the information. A census report is published that presents the results in tables and graphs.

To learn more about statistics, visit The Learning Equation on the Internet at http://tle.brookscole.com. (The log-in instructions are in the Preface.) For Chapter 7, the online lesson is:

• *TLE* Lesson 13: Statistics

Check Your Knowledge

1. A histogram is a type of _____ graph.

2. A _____ is like a bar graph, but the bars are composed of pictures, where each picture represents a quantity.

3. A _____ polygon is a special line graph formed from a histogram.

4. The sum of several values divided by the number of values is called the _____ of the distribution.

5. The _____ of several values written in increasing order is the middle value.

6. The value that appears most often in a distribution is called the _____ of the distribution.

Refer to the graph, which depicts the blood types of a sample of individuals.

7. What type of graph is shown?

8. How many individuals have blood type B?

9. How many individuals are in the sample group?

10. What percent of the group have blood type A?

11. Which blood type group is the largest? The smallest?

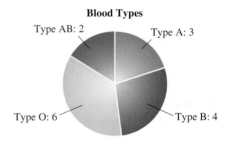

Blood Types

Type AB: 2   Type A: 3

Type O: 6   Type B: 4

Refer to the graph on the right, which depicts a class grade distribution.

12. What type of graph is shown?

13. How many individuals have earned grades in the 60–70 range?

14. How many students are in the class?

15. What percent of the group earned grades in the 70–80 range?

16. How many students earned at least 80?

17. What percent of the class has grades below a 70?

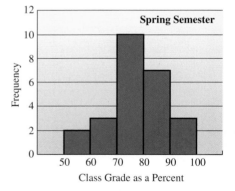

Spring Semester

Frequency / Class Grade as a Percent

A group of college men were surveyed to determine how many pairs of shoes they own. The results of the survey were:

5, 8, 3, 5, 7, 9, 4, 9, 10, 7, 9

18. Find the mean number of pairs owned. Round to the nearest one pair.

19. Find the median number of pairs owned.

20. Find the mode of the list.

21. Which of the above three measures of central tendency would be affected by the removal of 10 from the list?

22. Which of the above three measures of central tendency would be affected by the removal of 3 from the list?

Study Skills Workshop

HELP OUTSIDE OF CLASS

Have you ever had the experience where you understand everything that your instructor is saying in class, only to go home and try a homework problem and be completely stumped? This is a common complaint among math students. The key to being a successful math student is to take care of these problems before you go on to tackle new material; that is why you should know what resources are available outside of class.

Instructor Office Hours. Your instructor may hold office hours at particular days and times during the week. The purpose of these hours is to be available to help students with questions. Usually these hours are listed in your syllabus, and you do not need to make appointments to see your instructor at these times. When you visit your instructor during office hours, have a list of questions that are giving you trouble and try to pinpoint exactly where in the process you are getting stuck. This will help your instructor answer your questions efficiently and effectively.

Tutorial Centers. Many colleges have tutorial centers where students can make appointments to meet with a tutor, either one-on-one or in a group of students taking the same class. Tutorial centers usually offer their services for free and have regular hours of operation. Even if you are not having major difficulty with your homework, if you lack confidence or feel high levels of anxiety, it might be a good idea to schedule regular meetings with a tutor. When you visit your tutor, bring your list of questions that detail where in the process you're having difficulty.

Math Labs. Some colleges have math labs or learning centers where students are able to drop in at their convenience to have their math questions answered or where they can hang out and work on their homework. If something like this is available at your college, try to organize your study calendar to spend some time there doing your homework. It is very helpful to have a knowledgeable person available to answer your questions at the moment they come up.

Study Groups. Study groups are groups of classmates who meet outside of class to discuss homework problems or study for tests. Study groups work best when they are relatively small (no more than four members), when they meet regularly, and when they follow these guidelines:

- Members should have attempted all homework problems before meeting.
- No one person in the group should be responsible for doing all of the work or all of the explaining.
- The group should meet in a place where members can spread out and talk. Do not plan to meet in a quiet area of the library.
- Members should practice verbalizing and explaining processes and concepts to others in the study group. The best way to really learn a topic is by teaching it to someone else.

ASSIGNMENT

1. List your instructor's office hours and location. Next, pay a visit to your instructor at his/her office this week, even if you don't have any homework questions.
2. List the hours that your college's tutorial center is open, the location of the center, and what procedure you need to follow to make an appointment with a tutor.
3. Does your college have a math lab or learning center? If so, list its hours of operation, location, and rules.
4. Find at least two other students who can meet at a time that is mutually convenient for a study group. Plan to meet two days before your next homework assignment is due and follow the guidelines given above. After your group has met, evaluate how well it worked. Is there anything that your group might do to make it better the next time?

Newspapers and magazines often present information in the form of graphs and tables. In this chapter, we show how information can be obtained by reading graphs. We then discuss three measures of central tendency: The mean, the median, and the mode.

7.1 Reading Graphs and Tables

- Reading data from tables   • Reading bar graphs   • Reading pictographs
- Reading circle graphs   • Reading line graphs   • Reading histograms and frequency polygons

It is often said that a picture is worth a thousand words. In this section, we show how to read information from mathematical pictures called *graphs*.

■ Reading data from tables

The **table** in Figure 7-1(a), the **bar graph** in Figure 7-1(b), and the **circle graph** or **pie chart** in Figure 7-1(c) all show the results of a survey of viewers' opinions. In the bar graph, the length of each bar represents the percent of responses in each category. In the circle graph, the size of each region represents the percent of responses. The two graphs tell the story more quickly and more clearly than the table of numbers.

Ratings of Prime-Time News Coverage

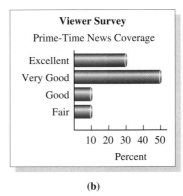

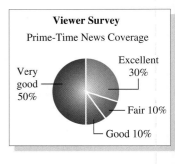

(a)   (b)   (c)

FIGURE 7-1

It is easy to see from either graph that the largest percent of those surveyed rated the programming *very good,* and that the responses *good* and *fair* were tied for last. The same information is available in the table of Figure 7-1(a), but it is not as easy to see at a glance.

Data are often presented in tables, with information organized in rows and columns. To read a table, we must find the intersection of the row and the column that contains the needed information.

Postal rates (in 2004) for priority mail appear in Table 7-1 on the next page. To find the cost of mailing an $8\frac{1}{2}$-pound package by priority mail to postal zone 4, we find the *row* of the postage table for a package that does not exceed 9 pounds. We then find the *column* for zone 4. At the intersection of this row and this column, we read the number 11.70. This means that it would cost $11.70 to mail the package.

Postage Rates for Priority Mail 2004							
Weight Not Over (pounds)	Zones Local, 1, 2, & 3	Zone 4	Zone 5	Zone 6	Zone 7	Zone 8
1	$3.85	$3.85	$3.85	$3.85	$3.85	$3.85
2	3.95	4.55	4.90	5.05	5.40	5.75
3	4.75	6.05	6.85	7.15	7.85	8.55
4	5.30	7.05	8.05	8.50	9.45	10.35
5	5.85	8.00	9.30	9.85	11.00	12.15
6	6.30	8.85	9.90	10.05	11.30	12.30
7	6.80	9.80	10.65	11.00	12.55	14.05
8	7.35	10.75	11.45	11.95	13.80	15.75
9	7.90	**11.70**	12.20	12.90	15.05	17.50
10	8.40	12.60	13.00	14.00	16.30	19.20
11	8.95	13.35	13.75	15.15	17.55	20.90
12	9.50	14.05	14.50	16.30	18.80	22.65

TABLE 7-1

▊ Reading bar graphs

EXAMPLE 1 Income.

The bar graph in Figure 7-2 shows the total income generated by three sectors of the economy in each of three years. The height of each bar, representing income in billions of dollars, is measured on the scale on the vertical *axis*. The years appear on the horizontal axis. Read the graph to answer the following questions.

a. What income was generated by retail sales in 1990?

b. Which sector of the economy consistently generated the most income?

c. By what amount did income from the wholesale sector increase from 1980 through 2000?

National Income by Industry
(in billions of dollars)

Source: *The World Almanac 2004*

FIGURE 7-2

Self Check 1
What income was generated by the services sector in 1990?

Solution

a. The second group of bars indicates income in 1990, and the middle bar of that group shows sales in the retail sector. Since the vertical axis is scaled in units of $125 billion, the height of that bar is approximately 250 + 125 = 375, which represents $375 billion.

b. In each group, the rightmost bar is the tallest. That bar, according to the key, represents income from the services sector of the economy. Therefore, services consistently generated the most income.

c. According to the color key, the leftmost bar in each group shows income from the wholesale sector. That sector generated about $125 billion in 1980 and $500 billion in 2000. The amount of increase in income is the difference of these two quantities.

$$\$500 \text{ billion} - \$125 \text{ billion} = \$375 \text{ billion}$$

Wholesale income increased by $375 billion between 1980 and 2000.

Answer $1,000 billion

Self Check 2
Which model has shown the greatest decrease in sales?

EXAMPLE 2   Automobile sales.   The bar graph in Figure 7-3 shows the number of cars of various models purchased in Dale County for two consecutive years.

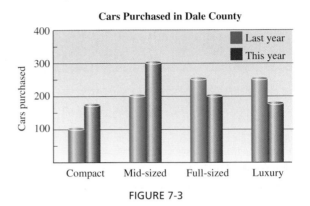

FIGURE 7-3

a. Which models have shown a decrease in sales?

b. Which model showed the greatest increase in sales?

Solution

a. In each pair, the bar on the left gives last year's sales. The bar on the right represents this year's sales. Only for full-sized and luxury cars is the left bar taller than the right bar. This means that full-sized and luxury cars have decreased in sales.

b. Sales of compact and mid-sized cars have increased over last year, because for these models the right bar is taller than the left bar. The difference in the heights of the bars represents the amount of increase. That increase is greater for mid-sized cars. Of all models, mid-sized cars have shown the greatest increase in sales.

Answer luxury cars

▎ Reading pictographs

A **pictograph** is like a bar graph, but the bars are composed of pictures, where each picture represents a quantity. In Figure 7-4, each picture represents 50 pizzas ordered during exam week. The top bar contains three complete pizzas and one partial pizza. This indicates that the men in the men's residence hall ordered $3 \cdot 50$, or 150 pizzas, plus approximately $\frac{1}{4}$ of 50, or about 13 pizzas. This totals 163 pizzas. The women in the women's residence hall ordered $4\frac{1}{2} \cdot 50$, or 225 pizzas.

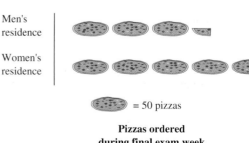

= 50 pizzas

**Pizzas ordered
during final exam week**

FIGURE 7-4

Reading circle graphs

EXAMPLE 3 **Gold production.** The circle graph in Figure 7-5 gives information about world gold production. The entire circle represents the world's total production, and the sizes of the segments of the circle represent the parts of that total contributed by various nations and regions. Use the graph to answer the following questions.

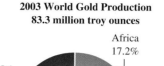
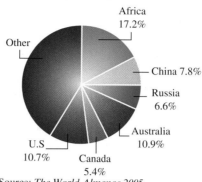

2003 World Gold Production
83.3 million troy ounces

Africa 17.2%
Other
China 7.8%
Russia 6.6%
Australia 10.9%
U.S 10.7%
Canada 5.4%

Source: *The World Almanac 2005*

FIGURE 7-5

a. What percent of the total was the combined production of the United States and Canada?

b. What percent of the total production came from sources other than those listed?

c. If the world's total production of gold was 56.3 million ounces during the year of the survey, how many ounces did Australia produce?

Solution

a. According to the graph, the United States produced 10.7% and Canada produced 5.4% of the total. Together, they produced (10.7 + 5.4)%, or 16.1% of the total.

b. To find the percent of gold produced by countries that are not listed, we add the contributions of all the listed sources and subtract that total from 100%.

$$100\% - (17.2\% + 7.8\% + 6.6\% + 10.9\% + 5.4\% + 10.7\%) = 100\% - 58.6\%$$
$$= 41.4\%$$

The countries that are not listed produced 41.4% of the world's total production of gold.

c. From the graph, we see that Australia produced 10.9% of the world's gold. Since the world total was 83.3 million ounces, Australia's share (in millions of ounces) was

$$10.9\% \text{ of } 83.3 = (0.109)(83.3)$$
$$= 9.0797$$

Rounded to the nearest tenth of a million, Australia produced 9.1 million ounces of gold.

Self Check 3
To the nearest tenth of a million, how many ounces of gold did Russia produce?

Answer 5.5 million ounces

Reading line graphs

Another graph, called a **line graph,** is used to show how quantities change with time. From such a graph, we can determine when a quantity is increasing and when it is decreasing.

EXAMPLE 4 **Automobile production.** The line graph in Figure 7-6 on the next page shows how U.S. automobile production has changed since 1900. Look at the graph and answer the following questions.

a. How many automobiles were manufactured in 1940?

b. How many were manufactured in 1950?

c. Over which 20-year span did automobile production increase most rapidly?

Self Check 4
How many more cars were produced in 1960 than in 1940?

d. When did production decrease?

e. Why is a broken line used for a portion of the graph?

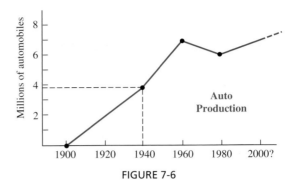

FIGURE 7-6

Solution

a. To find the number of autos produced in 1940, we follow the dashed line from the label 1940 straight up to the graph and then directly over to the scale. There, we read about 3.8. Since the scale indicates millions of automobiles, approximately 3.8 million autos were produced in 1940.

b. To find the number of autos produced in 1950, we find the point halfway between 1940 and 1960. From there, we move up to the graph and then sideways to the scale, where we read about 5. Approximately 5 million autos were manufactured in 1950.

c. Since the upward tilt of the graph is the greatest between 1940 and 1960, auto production increased most rapidly in those years.

d. Between 1960 and 1980, the graph drops, indicating that auto production decreased during that period.

e. Because beyond the year 2000 was still in the future when the graph was made, the production levels were *projections,* and the broken line indicates that the numbers are only estimates.

Answer about 3.2 million

Self Check 5

In Figure 7-7, what is train 1 doing at time D?

EXAMPLE 5   The graph in Figure 7-7 shows the movements of two trains. The horizontal axis represents time, and the vertical axis represents the distance that the trains have traveled.

a. How are the trains moving at time A?

b. At what time (A, B, C, D, or E) are both trains stopped?

c. At what times have both trains gone the same distance?

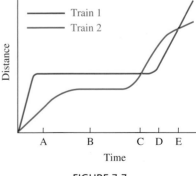

FIGURE 7-7

Solution   The movement of train 1 is represented by the red line, and that of train 2 is represented by the blue line.

a. At time A, the blue line is rising. This shows that the distance traveled by train 2 is increasing: At time A, train 2 is moving.
At time A, the red line is horizontal. This indicates that the distance traveled by train 1 is not changing: At time A, train 1 is stopped.

b. To find the time at which both trains are stopped, we find the time at which both the red and the blue lines are horizontal. At time B, both trains are stopped.

c. At any time, the height of a line gives the distance a train has traveled. Both trains have traveled the same distance whenever the two lines are the same height — that is, at any time when the lines intersect. This occurs at times C and E.

Answer Train 1, which had been stopped, is beginning to move.

◼ Reading histograms and frequency polygons

A pharmaceutical company is sponsoring a series of reruns of old Westerns. The marketing department must choose from three advertisements.

1. Children talking about Chipmunk Vitamins

2. A college student catching a quick breakfast and a TurboPill Vitamin

3. A grandmother talking about Seniors Vitamins

A survey of the viewing audience records the age of each viewer, counting the number in the 6-to-15-year-old age group, the 16-to-25-year-old age group, and so on. The graph of the data is displayed in a special type of bar graph called a **histogram** as shown in Figure 7-8. The vertical axis, labeled *Frequency*, indicates the number

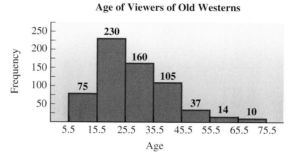

Age of Viewers of Old Westerns

FIGURE 7-8

of viewers in each age group. For example, the histogram shows that 105 viewers are in the 36-to-45-year-old age group.

A histogram is a bar graph with three important features.

1. The bars of a histogram touch.

2. Data values never fall at the edge of a bar.

3. The widths of each bar are equal and represent a range of values.

The width of each bar in Figure 7-8 represents an age span of 10 years. Since most viewers are in the 16-to-25-year-old age group, the marketing department decides to advertise TurboPills in commercials that appeal to active young adults.

EXAMPLE 6   Carry-on luggage.

An airline weighs the carry-on luggage of 2,260 passengers. See the histogram in Figure 7-9.

a. How many passengers carried luggage in the 8-to-11-pound range?

b. How many carried luggage in the 12-to-19-pound range?

Weight of Carry-on Luggage

FIGURE 7-9

Solution

a. The second bar, with edges at 7.5 and 11.5 pounds, corresponds to the 8-to-11-pound range. Use the height of the bar (or the number written there) to determine that 430 passengers carried such luggage.

b. The 12-to-19-pound range is covered by two bars. The total number of passengers with luggage in this range is 970 + 540, or 1,510.

A special line graph, called a **frequency polygon,** can be constructed from the histogram in Figure 7-9 by joining the center points at the top of each bar. (See Figure 7-10.) On the horizontal axis, we write the coordinate of the middle value of each bar. After erasing the bars, we get the frequency polygon shown in Figure 7-11.

Weight of Carry-on Luggage

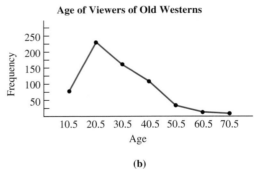

FIGURE 7-10

Weight of Carry-on Luggage

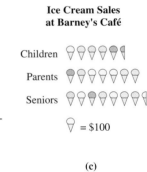

FIGURE 7-11

Section 7.1 STUDY SET

VOCABULARY *Refer to graphs a through f in the illustration. Fill in the blanks with the correct letter.*

1. Graph _____ is a bar graph.
2. Graph _____ is a circle graph.
3. Graph _____ is a pictograph.
4. Graph _____ is a line graph.
5. Graph _____ is a histogram.
6. Graph _____ is a frequency polygon.

Coupons Distributed (in billions)

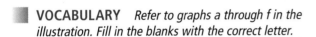
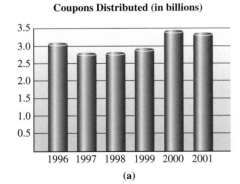

(a)

Age of Viewers of Old Westerns

(b)

Ice Cream Sales at Barney's Café

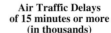

Children

Parents

Seniors

∇ = $100

(c)

Commuting Miles per Week

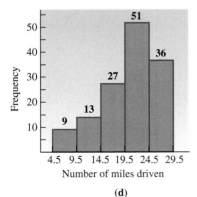

(d)

2003 U.S. Energy Production by Source (in quadrillion BTUs)

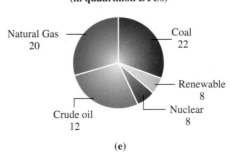

Natural Gas 20

Coal 22

Renewable 8

Nuclear 8

Crude oil 12

(e)

Air Traffic Delays of 15 minutes or more (in thousands)

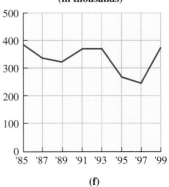

(f)

CONCEPTS *Fill in the blanks.*

7. The _____ of a histogram touch. The widths of the bars of a histogram are _____ and represent a range of values.

8. A _____ polygon can be constructed from a histogram by joining the center points at the top of each bar.

APPLICATIONS *Refer to the postal rate table, Table 7-1, on page 391.*

9. PRIORITY MAIL Find the cost of using priority mail to send a package weighing $7\frac{1}{4}$ pounds to zone 3.

10. PRIORITY MAIL Find the cost of sending (priority mail) a package weighing $2\frac{1}{4}$ pounds to zone 5.

11. COMPARING POSTAGE Juan wants to send a package weighing 6 pounds 1 ounce to a friend living in zone 2. Fourth-class postage would be $2.79. How much could he save by sending the package fourth class instead of priority mail?

12. SENDING TWO PACKAGES Jenny wants to send a birthday gift and an anniversary gift to her brother, who lives in zone 6. One package weighs 2 pounds 9 ounces, and the other weighs 3 pounds 8 ounces. If she uses priority mail, how much will she save by sending both gifts as one package instead of two? (*Hint:* 16 ounces = 1 pound.)

Refer to the federal income tax tables.

13. FILING A JOINT RETURN Raul has an adjusted income of $57,100, is married, and files jointly. Compute his tax.

14. FILING A SINGLE RETURN Herb is single and has an adjusted income of $79,250. Compute his tax.

15. TAX-SAVING STRATEGIES Angelina is single and has an adjusted income of $53,000. If she gets married, she will gain other deductions that will reduce her income by $2,000, and she can file a joint return. How much will she save in tax by getting married?

16. FILING STATUS A man with an adjusted income of $53,000 married a woman with an adjusted income of $75,000. They filed a joint return. Would they have saved on their taxes if they had both stayed single?

Revised 2003 Tax Rate Schedules

| | If TAXABLE INCOME | | The TAX is | | |
| | | | THEN | | |
	Is Over	But Not Over	This Amount	Plus This %	Of the Excess Over
SCHEDULE X —					
Single	$0	$7,000	$0.00	10%	$0.00
	$7,000	$28,400	$700.00	15%	$7,000
	$28,400	$68,800	$3,910.00	25%	$28,400
	$68,800	$143,500	$14,010.00	28%	$68,800
	$143,500	$311,950	$34,926.00	33%	$143,500
	$311,950	—	$90,514.50	35%	$311,950
SCHEDULE Y-1 —					
	$0	$14,000	$0.00	10%	$0.00
	$14,000	$56,800	$1,400.00	15%	$14,000
Married Filing Jointly or Qualifying Widow(er)	$56,800	$114,650	$7,820.00	25%	$56,800
	$114,650	$174,700	$22,282.50	28%	$114,650
	$174,700	$311,950	$39,096.50	33%	$174,700
	$311,950	—	$84,389.00	35%	$311,950

Refer to the following graph.

17. Which source supplied the least amount of energy in 1975?

18. Which energy source remained essentially unchanged between 1975 and 2000?

19. What percent of electrical energy was produced by crude oil in 1975?

20. Which sources provided about 10% of the U.S. energy needs in 2000?

21. Which source supplied the greatest amount of energy in 2000?

22. On which sources of energy does the U.S. rely more heavily since 1975?

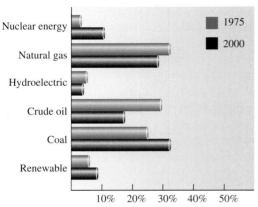

Changing Sources of Electricity

Source: *The World Almanac 2005*

Refer to the following graph.

23. The world production of lead in 1970 was approximately equal to the production of zinc in another year. In what other year was that?

24. The world production of zinc in 1990 was approximately equal to the production of lead in another year. In what other year was that?

25. In what year was the production of zinc less than one-half that of lead?

26. In what year was the production of zinc more than twice that of lead?

27. By how many metric tons did the production of zinc increase between 1970 and 1980?

28. By how many metric tons did the production of lead decrease between 1980 and 1990?

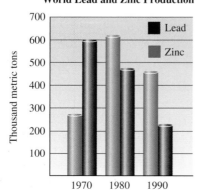

World Lead and Zinc Production

Refer to the following graph.

29. In which categories of moving violations have arrests decreased since last month?

30. Last month, which violation occurred most often?

31. This month, which violation occurred least often?

32. Which violation has shown the greatest decrease in number of arrests since last month?

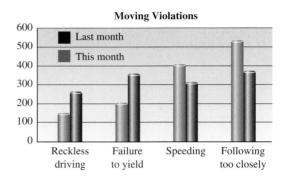

Moving Violations

Refer to the pictograph.

33. Which group (children, parents, or seniors) spent the most money on ice cream at Barney's Café?

34. How much money did parents spend on ice cream?

35. How much more money did seniors spend than parents?

36. How much more money did seniors spend than children?

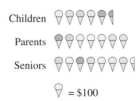

Ice Cream Sales at Barney's Café

Refer to the circle graph on the next page.

37. Two of the seven languages considered are spoken by groups of about the same size. Which languages are they?

38. Of the languages in the graph, which is spoken by the greatest number of people?

39. Do more people speak Russian or English?

40. What percent of the world's population speak Russian or English?

41. What percent of the world's population speak a language other than these seven?

42. What percent of the world's population do not speak either French or German?

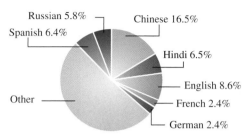

World Languages
and the percents of the population
that speak them

Russian 5.8%   Chinese 16.5%
Spanish 6.4%
Hindi 6.5%
Other
English 8.6%
French 2.4%
German 2.4%

Refer to the following graph.

43. What percent of total energy production is from nuclear energy?

44. What percent of energy production comes from renewable sources?

45. What percent of total energy production comes from coal and crude oil combined?

46. By what *percent* does energy produced from coal exceed that produced from crude oil?

47. By what *percent* does energy produced from coal exceed that produced from nuclear sources?

48. If production of nuclear energy tripled in the next 10 years and other sources remained the same, what percent of total energy sources would nuclear energy be?

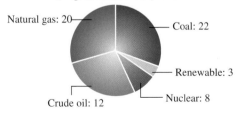

U.S. Energy Production Sources
(in quadrillion BTUs) 2003

Natural gas: 20   Coal: 22
Renewable: 3
Nuclear: 8
Crude oil: 12

Refer to the line graph in the next column.

49. What were the average weekly earnings in mining for the year 1975?

50. What were the average weekly earnings in construction for the year 1980?

51. In the period between 1982 and 1984, which salary was increasing most rapidly?

52. In approximately what year did miners begin to earn more than construction workers?

53. In the period from 1970 to 1995, which workers received the greatest increase in wages?

54. In what five-year interval did wages in mining increase most rapidly?

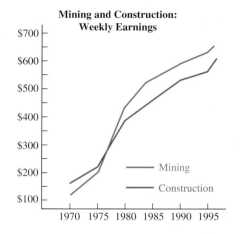

Mining and Construction:
Weekly Earnings

$700
$600
$500
$400
$300
$200   Mining
$100   Construction

1970 1975 1980 1985 1990 1995

Refer to the following line graph.

55. Which runner ran faster at the start of the race?

56. Which runner stopped to rest first?

57. Which runner dropped the baton and had to go back to get it?

58. At what times (A, B, C, or D) was runner 1 stopped and runner 2 running?

59. Describe what was happening at time D.

60. Which runner won the race?

Five-Mile Run

Finish

Distance

Runner 1
Runner 2
Time
Start
A   B   C   D

61. COMMUTING MILES An insurance company has collected data on the number of miles its employees drive to and from work. The data are presented in the histogram on the next page. How many employees commute between 14.5 and 19.5 miles per week?

62. COMMUTING The employees of a marketing firm were surveyed to determine the number of miles that they commute to work each week. How many employees commute 14 miles or less per week?

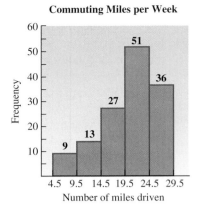

Commuting Miles per Week

63. NIGHT SHIFT STAFFING A hospital administrator surveyed the medical staff to determine the number of room calls during the night. She constructed the frequency polygon below. On how many nights were there about 30 room calls?

64. NIGHT SHIFT STAFFING Refer to Problem 63 and the graph below. On how many nights were there about 60 room calls?

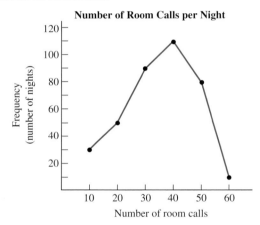

Number of Room Calls per Night

65. MAKING BAR GRAPHS Use the data in the table below to make a bar graph showing the number of U.S. farms in the years 1950 through 2000.

66. MAKING LINE GRAPHS Use the data in the table to make a line graph showing the average acreage of U.S. farms for the years 1950 through 2000.

Year	Number of U.S. farms (in millions)	Average size of U.S. farms (acres)
1950	5.6	213
1960	4.0	297
1970	2.9	374
1980	2.4	426
1990	2.1	460
2000	2.2	434

Source: U.S. Dept. of Agriculture

67. MAKING LINE GRAPHS The coupon in the table provides savings for shoppers. Make a line graph that relates the original price (in dollars, on the horizontal axis) to the sale price (on the vertical axis).

SAVE!	On purchases of
$10	$100–$250
$25	$250–$500
$50	over $500

68. MAKING HISTOGRAMS To study the effect of fluoride in preventing tooth decay, researchers counted the number of fillings in the teeth of 28 patients and recorded these results:

3, 7, 11, 21, 16, 22 18, 8, 12, 3, 7, 2, 8, 19, 12, 19, 12, 10, 13, 10, 14, 15, 14, 14, 9, 10, 12, 13

Tally the results by completing the table. Then make a histogram. The first bar extends from 0.5 to 5.5, the second bar from 5.5 to 10.5, and so on.

Number of fillings	Frequency
1–5	
6–10	
11–15	
16–20	
21–25	

WRITING *Write a paragraph using your own words.*

69. What kind of presentation (table, bar graph, line graph, pie chart, pictograph, or histogram) is most appropriate for displaying each type of information?

- The percent of students, classified by major
- The percent of biology majors each year since 1970
- The number of hours spent studying for finals
- Various ethnic populations of the ten largest cities
- Average annual salary of corporate executives for ten major industries

Explain your choices.

70. A histogram is a special type of bar graph. Explain.

REVIEW *Perform the operations.*

71. $5 - 3 \cdot 4$

72. $3(6 - 9) + 4$

73. $\left(\dfrac{1}{2} + \dfrac{1}{3}\right)^2$

74. $5^2 + |6 - 10|$

75. Write the prime numbers between 10 and 30.

76. Write the first ten composite numbers.

77. Write the even numbers less than 6 that are not prime.

78. Write the prime numbers between 0 and 10.

7.2 Mean, Median, and Mode

- The mean (the arithmetic average) • The median • The mode

Graphs are not the only way of describing *distributions* (lists) of numbers compactly. We can often find *one* number that represents the center of all of the numbers in a collection of data. We have already seen one such typical number: the *mean*, or the *average*. There are two others: the *median* and the *mode*. In this section, we discuss these three measures of **central tendency.**

The mean (the arithmetic average)

A student has taken five tests this semester, scoring 87, 73, 89, 92, and 84. To find out how well she is doing, she calculates the **mean,** or the **arithmetic average,** of these grades, by finding the sum of the grades and then dividing by 5.

$$\text{Mean score} = \frac{87 + 73 + 89 + 92 + 84}{5}$$

$$= \frac{425}{5}$$

$$= 85$$

The mean score is 85. Some exams were better and some were worse, but 85 is a good indication of her performance in the class.

> **The mean (arithmetic average)**
>
> The **mean,** or the **arithmetic average,** of several values is given by the formula
>
> $$\text{Mean (or average)} = \frac{\text{sum of the values}}{\text{number of values}}$$

EXAMPLE 1 **Store sales.** The week's sales in three departments of the Tog Shoppe are given in Table 7-2 on the next page. Find the mean of the daily sales in the women's department for this week.

Self Check 1
Find the mean daily sales in all three departments of the Tog Shoppe on Wednesday.

	Men's department	Women's department	Children's department
Monday	$2,315	$3,135	$1,110
Tuesday	2,020	2,310	890
Wednesday	1,100	3,206	1,020
Thursday	2,000	2,115	880
Friday	955	1,570	1,010
Saturday	850	2,100	1,000

Solution Use a calculator to add the sales in the women's department for the week. Then divide the sum of those six values by 6.

$$\text{Mean sales in the women's department} = \frac{3{,}135 + 2{,}310 + 3{,}206 + 2{,}115 + 1{,}570 + 2{,}100}{6}$$

$$= \frac{14{,}436}{6}$$

$$= 2{,}406$$

Answer $1,775.33

The mean of the week's daily sales in the women's department is $2,406.

CALCULATOR SNAPSHOT   **Finding the mean**

Most scientific calculators do statistical calculations and can easily find the mean of a set of numbers. To use a statistical calculator in statistical mode to find the mean in Example 1, try these keystrokes:

- Set the calculator to statistical mode.
- Reset the calculator to clear the *statistical registers*.
- Enter each number, followed by the $\boxed{\Sigma+}$ key instead of the $\boxed{+}$ key. That is, enter 3,135, press $\boxed{\Sigma+}$, enter 2,310, press $\boxed{\Sigma+}$, and so on.
- When all data are entered, find the mean by pressing the $\boxed{\bar{x}}$ key. You may need to press $\boxed{2^{nd}}$ first. The mean is 2,406.

Because keystrokes vary among calculator brands, you might have to check the owner's manual if these instructions don't work.

Self Check 2
If Bob drove 3,360 miles in February 2005, how many miles did he drive per day, on average?

EXAMPLE 2 Driving. In January, Bob drove a total of 4,805 miles. On the average, how many miles did he drive per day?

Solution To find the average number of miles driven per day, we divide the total number of miles by the number of days. Because there are 31 days in January, we divide 4,805 by 31.

$$\text{Average number of miles per day} = \frac{\text{total miles driven}}{\text{number of days}}$$

$$= \frac{4{,}805}{31}$$

$$= 155$$

On average, Bob drove 155 miles per day.

Answer 120

◾ The median

The mean is not always representative of the values in a list. For example, suppose that the weekly earnings of four workers in a small business are $280, $300, $380, and $240, and the owner of the company pays himself $5,000. The mean salary is

$$\text{Mean salary} = \frac{280 + 300 + 380 + 240 + 5{,}000}{5}$$

$$= \frac{6{,}200}{5}$$

$$= 1{,}240$$

The owner could say, "Our employees earn an average of $1,240 per week." Clearly, the mean does not fairly represent the typical worker's salary.

A better measure of the company's typical salary is the *median:* the salary in the middle when all the numbers are arranged by size.

240   280   300   380   5,000
                 ↑
        The middle salary

The typical worker earns $300 per week, far less than the mean salary.

If there is an even number of values in a list, there is no middle value. In that case, the median is the mean of the two numbers closest to the middle. For example, there is no middle number in the list 2, 5, 6, 8, 13, 17. The two numbers closest to the middle are 6 and 8. The median is the mean of 6 and 8, which is $\frac{6 + 8}{2}$, or 7.

The median

The **median** of several values is the middle value. To find the median:

1. Arrange the values in increasing order.

2. If there is an odd number of values, the median is the value in the middle.

3. If there is an even number of values, the median is the average of the two values that are closest to the middle.

EXAMPLE 3   Grades.   On an exam, there were three scores of 59, four scores of 77, and scores of 43, 47, 53, 60, 68, 82, and 97. Find the median score.

Solution   We arrange the 14 scores in increasing order.

43   47   53   59   59   59   **60   68**   77   77   77   77   82   97

Self Check 3
Find the median of these values:
7.5, 2.1, 9.8, 5.3, 6.2.

Since there is an even number of scores, the median is the mean of the two scores closest to the middle: the 60 and the 68.

$$\text{The median is } \frac{60 + 68}{2}, \text{ or } 64.$$

Answer 6.2

▋ The mode

A hardware store displays 20 outdoor thermometers. Twelve of them read 68°, and the other eight have different readings. To choose an accurate thermometer, should we choose one with a reading that is closest to the *mean* of all 20, or to their *median*? Neither. Instead, we should choose one of the 12 that all read the same, figuring that any of those that agree will likely be correct.

By choosing that temperature that appears most often, we have chosen the *mode* of the 20 numbers.

> **The mode**
>
> The **mode** of several values is the single value that occurs most often. The mode of several values is also called the **modal value.**

Self Check 4
Find the mode of these values:
2, 3, 4, 6, 2, 4, 3, 4, 3, 4, 2, 5

EXAMPLE 4   Find the mode of these values: 3, 6, 5, 7, 3, 7, 2, 4, 3, 5, 3, 7, 8, 7, 3, 7, 6, 3, 4.

Solution   To find the mode of the numbers in the list, we make a chart of the distinct numbers that appear and make tally marks to record the number of times they occur.

2	3	4	5	6	7	8
/	### /	//	//	//	###	/

Answer 4

Because 3 occurs more times than any other number, it is the mode.

EXAMPLE 5   Machinist's tools.   The diameters (distances across) of eight stainless steel bearings were found using the vernier calipers shown in Figure 7-12. Find   **a.** the mean,   **b.** the median,   and   **c.** the mode   of the set of measurements listed below.

3.43 cm, 3.25 cm, 3.48 cm, 3.39 cm, 3.54 cm, 3.48 cm, 3.23 cm, 3.24 cm

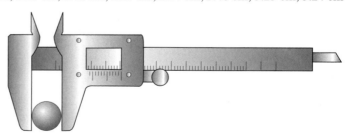

FIGURE 7-12

Solution

a. To find the mean, we add the measurements and divide by the number of values, which is 8.

$$\text{Mean} = \frac{3.43 + 3.25 + 3.48 + 3.39 + 3.54 + 3.48 + 3.23 + 3.24}{8} = 3.38 \text{ cm}$$

b. To find the median, we first arrange the measurements in increasing order.

3.23, 3.24, 3.25, 3.39, 3.43 , 3.48, 3.48, 3.54

Because there is an even number of measurements, the median will be the sum of the middle two values (3.39 and 3.43) divided by 2.

$$\text{Median} = \frac{3.39 + 3.43}{2} = \frac{6.82}{2} = 3.41 \text{ cm}$$

c. Since the measurement 3.48 cm occurs most often, it is the mode.

The Value of an Education   THINK IT THROUGH

"Additional education makes workers more productive and enables them to increase their earnings." Virginia Governor, Mark R. Warner, 2004

As college costs increase, some people wonder if it is worth it to spend years working toward a degree when that same time could be spent earning money. The following median income data makes it clear that, over time, additional education is well worth the investment. Use the given facts to complete the bar graph.

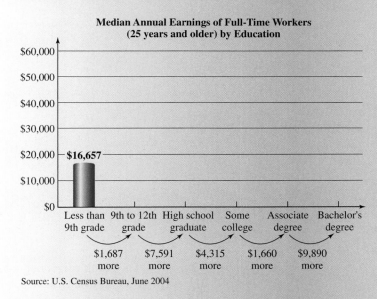

Source: U.S. Census Bureau, June 2004

Section 7.2   STUDY SET

VOCABULARY   *Fill in the blanks.*

1. The sum of the values in a distribution of numbers divided by the number of values in the distribution is called the _____ of the distribution.

2. The value that appears most often in a distribution is called the _____ of the distribution.

3. The _____ of several values arranged in increasing order is the middle value.

4. The mean, median, and mode are three measures of _____ tendency.

■ **CONCEPTS**   *Fill in the blanks.*

5. The mean of several values is given by

$$\text{Mean} = \frac{\text{the sum of the values}}{\rule{3cm}{0.4pt}}$$

6. Consider the following list of values:

4, 5, 5, 6, 8, 9, 9, 15

$$\text{Median} = \frac{\rule{0.7cm}{0.4pt} + \rule{0.7cm}{0.4pt}}{2} = \rule{0.7cm}{0.4pt}$$

■ **PRACTICE**   *Find the mean of each list of data.*

7. 3, 4, 7, 7, 8, 11, 16

8. 13, 15, 17, 17, 15, 13

9. 5, 9, 12, 35, 37, 45, 60, 77

10. 0, 0, 3, 4, 7, 9, 12

11. 15, 7, 12, 19, 27, 17, 19, 35, 20

12. 45, 67, 42, 35, 86, 52, 91, 102

Find the median of each list of data.

13. 2, 5, 9, 9, 9, 17, 29

14. 16, 18, 27, 29, 35, 47

15. 4, 7, 2, 11, 5, 4, 9, 17

16. 0, 0, 3, 4, 0, 0, 3, 4, 5

17. 18, 17, 2, 9, 21, 23, 21, 2

18. 5, 13, 5, 23, 43, 56, 32, 45

Find the mode (if any) of each list of data.

19. 3, 5, 7, 3, 5, 4, 6, 7, 2, 3, 1, 4

20. 12, 12, 17, 17, 12, 13, 17, 12

21. 5, 9, 12, 35, 37, 45, 60

22. 0, 3, 0, 2, 7, 0, 6, 0, 3, 4, 2, 0

23. 23.1, 22.7, 23.5, 22.7, 34.2, 22.7

24. $\frac{1}{2}, \frac{1}{3}, \frac{1}{3}, 2, \frac{1}{2}, 2, \frac{1}{5}, \frac{1}{2}, 5, \frac{1}{3}$

■ **APPLICATIONS**

25. SOFT DRINKS A survey of soft-drink machines indicates the following prices for a can (in cents): 50, 60, 50, 50, 70, 75, 50, 45, 50, 50, 65, 75, 60, 75, 100, 50, 80, 75. Find the mean price of a soft drink.

26. COMPUTER SUPPLIES Several computer stores reported differing prices for toner cartridges for a laser printer (in dollars): 51, 55, 73, 75, 72, 70, 53, 59, 75. Find the mean price of a toner cartridge.

27. SOFT DRINKS Find the median price for a soft drink. (See Exercise 25.)

28. COMPUTER SUPPLIES Find the median price for a toner catridge. (See Exercise 26.)

29. SOFT DRINKS Find the modal price for a soft drink. (See Exercise 25.)

30. COMPUTER SUPPLIES Find the mode of the prices for a toner cartridge. (See Exercise 26.)

31. CHANGING TEMPERATURES Temperatures are recorded at hourly intervals, as listed in the table. Find the average temperature of the period from midnight to 11:00 A.M.

Time	Temperature	Time	Temperature
12:00 A.M.	53	12:00 noon	71
1:00	53	1:00 P.M.	75
2:00	57	2:00	77
3:00	58	3:00	77
4:00	59	4:00	79
5:00	59	5:00	72
6:00	60	6:00	70
7:00	62	7:00	64
8:00	64	8:00	61
9:00	66	9:00	59
10:00	68	10:00	53
11:00	70	11:00	51

32. SEMESTER GRADES Frank's algebra grade is based on the average of four exams, which count equally. His grades are 75, 80, 90, and 85. Find his average.

33. AVERAGE TEMPERATURES Find the average temperature for the 24-hour period. (See Exercise 31.)

34. FINAL EXAMS If Frank's professor decided to count the fourth examination double, what would Frank's average be? (See Exercise 32.)

35. FLEET MILEAGE An insurance company's sales force uses 37 cars. Last June, those cars logged a total of 98,790 miles. On the average, how many miles did each car travel that month?

36. BUDGETING FOR GROCERIES The Hinrichs family spent $519 on groceries last April. On the average, how much did they spend each day?

37. DAILY MILEAGE Find the average number of miles driven daily for each car in Exercise 35.

38. GROCERY COSTS See Exercise 36. The Hinrichs family has five members. What is the average spent for groceries for one family member for one day?

39. EXAM AVERAGES Roberto received the same score on each of five exams, and his mean score is 85. Find his median score and his modal score.

40. EXAM SCORES The scores on the first exam of the students in a history class were 57, 59, 61, 63, 63, 63, 87, 89, 95, 99, and 100. Kia got a score of 70 and claims that "70 is better than average." Which of the three measures of central tendency is she better than: the mean, the median, or the mode?

41. COMPARING GRADES A student received scores of 37, 53, and 78 on three quizzes. His sister received scores of 53, 57, and 58. Who had the better average? Whose grades were more consistent?

42. What is the average of all of the integers from −100 to 100, inclusive?

43. OCTUPLETS In December 1998, Nkem Chukwu gave birth to eight babies in Texas Children's Hospital. Find the mean and the median of their birth weights.

Ebuka (girl) 24 oz	Odera (girl) 11.2 oz
Chidi (girl) 27 oz	Ikem (boy) 17.5 oz
Echerem (girl) 28 oz	Jioke (boy) 28.5 oz
Chima (girl) 26 oz	Gorom (girl) 18 oz

44. ICE SKATING Listed below are Tara Lipinski's artistic impression scores for the long program of the women's figure skating competition at the 1998 Winter Olympics. Find the mean, median, and mode. Round to the nearest tenth.

Australia	5.8	Germany	5.8	Ukraine	5.9
Hungary	5.8	U.S.	5.8	Poland	5.8
Austria	5.9	Russia	5.9	France	5.9

45. COMPARISON SHOPPING A survey of grocery stores found the price of a 15-ounce box of Cheerios cereal ranging from $3.89 to $4.39. (See below.) What are the mean, median, and mode of the prices listed?

$4.29	$3.89	$4.29	$4.09	$4.24	$3.99
$3.98	$4.19	$4.19	$4.39	$3.97	$4.29

46. EARTHQUAKES The magnitudes of 1999's major earthquakes are listed below. Find the mean, median, and mode. Round to the nearest tenth.

1/19/99	New Ireland, Papua New Guinea	7.0
2/6/99	Santa Cruz Islands, S. Pacific Sea	7.3
3/4/99	Celebes Sea, Indonesia	7.1
4/5/99	New Britain, Papua New Guinea	7.4
4/8/99	E. Russia/N.E. China border	7.1
5/10/99	New Britain, Papua New Guinea	7.1
5/16/99	New Britain, Papua New Guinea	7.1
8/17/99	Izmit region, western Turkey	7.4
9/21/99	Taiwan	7.6
9/30/99	Oaxaca, Mexico	7.4
11/12/99	Bolu Province, northwest Turkey	7.2

47. FUEL EFFICIENCY The ten most fuel-efficient cars in 2002, based on manufacturer's estimated city and highway average miles per gallon (mpg), are shown in the table. Find the mean, median, and mode of both sets of data.

Model	mpg city/hwy
Honda Insight	61/68
Toyota Prius	52/45
Honda Civic Hybrid	47/51
VW Jetta Wagon	42/50
VW Golf	42/49
VW Jetta Sedan	42/49
VW Beetle	42/49
Honda Civic Coupe	36/44
Toyota Echo	34/41
Chevy Prizm	32/41

Source: edmonds.com

48. SPORT FISHING The report shown below lists the fishing conditions at Pyramid Lake for a Saturday in January. Find the median and the mode of the weights of the striped bass caught at the lake.

Pyramid Lake— Some striped bass are biting but are on the small side. Striking jigs and plastic worms. Water is cold: 38°. Weights of fish caught (lb): 6, 9, 4, 7, 4, 3, 3, 5, 6, 9, 4, 5, 8, 13, 4, 5, 4, 6, 9

49. Refer to the data in the table.

   a. Find the mean. Round to the nearest tenth of a percent.

   b. Find the median.

   c. Find the mode.

U.S. Unemployment Rate (1990–2002) in percent						
1990	**1991**	**1992**	**1993**	**1994**	**1995**	**1996**
5.6	6.8	7.5	6.9	6.1	5.6	5.4
1997	**1998**	**1999**	**2000**	**2001**	**2002**	
4.9	4.5	4.2	4.0	4.7	5.8	

Source: U.S. Department of Labor

50. SCHOLASTIC APTITUDE TEST The mean SAT verbal test scores of college-bound seniors for the years 1993–2003 are listed below. Find the mean, median, and mode. Round to the nearest one point.

1993	**1994**	**1995**	**1996**	**1997**	**1998**
500	499	504	505	505	505
1999	**2000**	**2001**	**2002**	**2003**	
505	505	506	504	507	

Source: *The World Almanac 2004*

WRITING

51. Explain how to find the mean, the median, and the mode of several numbers.

52. The mean, median, and mode are used to measure the central tendency of a list of numbers. What is meant by central tendency?

REVIEW

53. Find the prime factorization of 81.

54. Find the LCD for two fractions whose denominators are 36 and 81.

Perform each operation.

55. $\dfrac{3}{4} \cdot \dfrac{2}{9}$

56. $\dfrac{2}{15} \div \dfrac{4}{5}$

57. $\dfrac{18}{5} + \dfrac{12}{5}$

58. $\dfrac{7}{12} - \dfrac{5}{12}$

59. $\dfrac{8}{5} + \dfrac{3}{10}$

60. $\dfrac{5}{6} - \dfrac{1}{12}$

Mean, Median, and Mode

To indicate the center of a distribution of numbers, we can use the mean, the median, or the mode.

- The **mean** of a distribution is the sum of the values in the distribution divided by the number of values in the distribution.

$$\text{Mean} = \frac{\text{sum of the values in the distribution}}{\text{number of values in the distribution}}$$

- The **median** of several values written in increasing order is the middle value. Just as many scores are above the median as are below it. If there are an even number of values in the distribution, the median is the mean of the two values that are closest to the middle.

- The **mode** of a distribution is the value that occurs most often.

Consider the following distribution: 3, 7, 4, 12, 15, 23, 17, 21, 15, 20.

1. Calculate the mean.

2. Find the median.

3. Find the mode.

4. Are the mean, the median, and the mode the same number?

Consider the distribution 4, 2, 6, 8, 6, 10.

5. Calculate the mean.

6. Find the median.

7. Find the mode.

8. Are the mean, the median, and the mode the same number?

Construct a distribution with the following characteristics.

9. The mean is greater than the mode.

10. The mean is less than the median.

11. The mode is less than the median.

12. The mode is greater than the median.

ACCENT ON TEAMWORK

SECTION 7.1

DAILY HIGH AND LOW TEMPERATURES Make a bar graph that shows the daily high and low temperatures for your city for a two-week period. You can find this information in a newspaper. From your graph, answer the following questions.

a. What was the highest high temperature?

b. What was the lowest high temperature?

c. What was the highest low temperature?

d. What was the lowest low temperature?

e. What was the difference between the highest high temperature and the lowest low temperature?

f. Were any trends apparent from the graph?

SECTION 7.2

MEAN, MEDIAN, AND MODE

1. Find the mean, median, and mode of the following set of values.

    2.3   2.3   3.6   3.8   4.5

    a. Is the mean of the set of values one of the values in the set?

    b. Is the median of the set of values one of the values in the set?

    c. Is the mode of the set of values one of the values in the set?

2. Construct a set of values (not all the same number) such that

    mean = median = mode

3. Construct a set of values such that

    mean < median < mode

4. Construct a set of values such that

    mean > median > mode

CHAPTER REVIEW

Reading Graphs and Tables

CONCEPTS

Numerical information can be presented in the form of tables, bar graphs, pictographs, circle graphs, and line graphs.

REVIEW EXERCISES

Refer to the table.

1. WIND-CHILL TEMPERATURES Find the wind-chill temperature on a 10° F day when a 15-mph wind is blowing.

2. WIND SPEEDS The wind-chill temperature is −25° F, and the outdoor temperature is 15° F. How fast is the wind blowing?

Determining the Wind-Chill Temperature

Wind speed	Actual temperature													
	35° F	30° F	25° F	20° F	15° F	10° F	5° F	0° F	−5° F	−10° F	−15° F	−20° F	−25° F	−30° F
5 mph	33°	27°	21°	16°	12°	7°	0°	−5°	−10°	−15°	−21°	−26°	−31°	−36°
10 mph	22	16	10	3	−3	−9	−15	−22	−27	−34	−40	−46	−52	−58
15 mph	16	9	−2	−5	−11	−18	−25	−31	−38	−45	−51	−58	−65	−72
20 mph	12	4	−3	−10	−17	−24	−31	−39	−46	−53	−60	−67	−74	−81
25 mph	8	1	−7	−15	−22	−29	−36	−44	−51	−59	−66	−74	−81	−88
30 mph	6	−2	−10	−18	−25	−33	−41	−49	−56	−64	−71	−79	−86	−93
35 mph	4	−4	−12	−20	−27	−35	−43	−52	−58	−67	−74	−82	−89	−97
40 mph	3	−5	−13	−21	−29	−37	−45	−53	−60	−69	−76	−84	−92	−100
45 mph	2	−6	−14	−22	−30	−38	−46	−54	−62	−70	−78	−85	−93	−102

Refer to the graph below.

3. How many coupons were distributed in 2000?

4. Between what years did the number of coupons distributed remain essentially unchanged?

5. Between what two years did the number of coupons distributed increase the most?

6. Between what two years was there the greatest decrease in the number of coupons distributed?

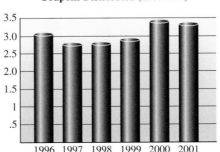

Coupons Distributed (in billions)

411

Refer to the graph below.

7. How many eggs were produced in Wisconsin in 1985?

8. How many eggs were produced in Nebraska in 1987?

9. In what year was the egg production of Wisconsin equal to that of Nebraska?

10. What was the total egg production of Wisconsin and Nebraska in 1988?

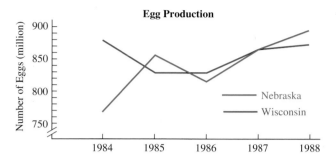

Refer to the graph below.

A *histogram* is a bar graph with these features:

1. The bars of the histogram touch.

2. Data values never fall at the edge of a bar.

3. The widths of the bars are equal and represent a range of values.

A *frequency polygon* is a special line graph formed from a histogram.

11. A survey of the television viewing habits of 320 households produced the following histogram. How many households watch between 6 and 15 hours of TV each week?

12. How many households watch 11 hours or more each week?

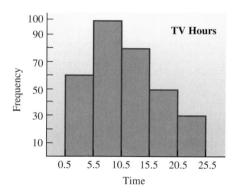

SECTION 7.2   *Mean, Median, and Mode*

The *mean* (or average) is given by the formula

$$\text{Mean} = \frac{\text{sum of the values}}{\text{number of values}}$$

13. EARNING AN A Jose worked hard this semester, earning grades of 87, 92, 97, 100, 100, 98, 90, and 98. If he needs a 95 average to earn an A in the class, did he make it?

To find the *median* of several values:

1. Arrange the values in increasing order.

2. If there is an odd number of values, the median is the value in the middle.

3. If there is an even number of values, the median is the average of the two values that are closest to the middle.

The *mode* of several numbers is the single value that occurs most often.

14. GRADE SUMMARIES The students in a mathematics class had final averages of 43, 83, 40, 100, 40, 36, 75, 39, and 100. When asked how well her students did, their teacher answered, "43 was typical." What measure was the teacher using?

15. PRETZEL PACKAGING Samples of SnacPak pretzels were weighed to find out whether the package claim "Net weight 1.2 ounces" was accurate. The tally appears in the table. Find the modal weight.

Weights of SnacPak Pretzels	
Ounces	Number
0.9	1
1.0	6
1.1	18
1.2	23
1.3	2

16. Find the mean weight of the samples in Exercise 15.

17. BLOOD SAMPLES A medical laboratory technician examined a blood sample under a microscope and measured the sizes (in microns) of the white blood cells. The data are listed below. Find the mean, median, and mode.

7.8   6.9   7.9   6.7   6.8   8.0   7.2   6.9   7.5

18. TOBACCO SETTLEMENTS In November 1998, the country's four largest tobacco companies reached an agreement with 46 states to pay $206.4 billion to cover public health costs related to smoking. The payments to each of the New England states are shown below. Find the median payment.

Connecticut	$3.63 billion	New Hampshire	$1.3 billion
Maine	$1.5 billion	Rhode Island	$1.4 billion
Massachusetts	$8.0 billion	Vermont	$0.81 billion

Refer to the graph below. Keeping one prisoner for one month costs $2,266.

1. How much money is spent monthly, per prisoner, to pay the prison staff?

2. How much money is spent monthly, per prisoner, on office costs?

3. What percent of the monthly allotment is spent on one prisoner's food?

4. What percent of the monthly allotment is spent on one prisoner's recreation and training?

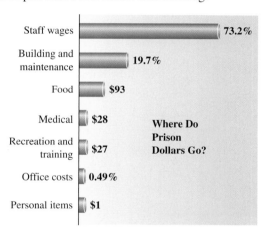

Staff wages	**73.2%**
Building and maintenance	**19.7%**
Food	**$93**
Medical	**$28**
Recreation and training	**$27**
Office costs	**0.49%**
Personal items	**$1**

Where Do Prison Dollars Go?

Refer to the graph below.

5. Approximately what percent of all employees are in the food and clothing industries?

6. Among workers in food and clothing, 2.4 million are in food and the rest are in clothing. What percent of all workers are in clothing?

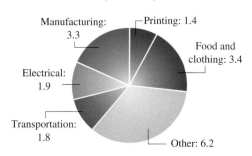

Employees in Industry
(in millions)

Manufacturing: 3.3
Printing: 1.4
Food and clothing: 3.4
Electrical: 1.9
Transportation: 1.8
Other: 6.2

Refer to the line graph below.

7. How many air traffic delays occurred in 1995?

8. Which year was worst for air traffic delays?

9. In 1999, about 2% of the air traffic delays were due to controller equipment glitches. How many flights were delayed for that reason?

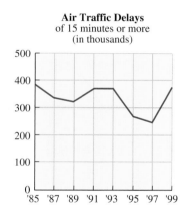

Air Traffic Delays
of 15 minutes or more
(in thousands)

500
400
300
200
100
0
'85 '87 '89 '91 '93 '95 '97 '99

10. Refer to the circle graph below. What is the missing percent for the weather delays?

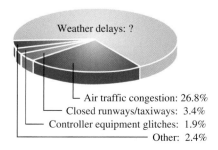

Causes of Air Traffic Delays

Weather delays: ?
Air traffic congestion: 26.8%
Closed runways/taxiways: 3.4%
Controller equipment glitches: 1.9%
Other: 2.4%

Refer to the illustration and choose the best answer from the following statements.

A. Both bicyclists are moving, and bicyclist 1 is faster than 2.

B. Both bicyclists are moving, and bicyclist 2 is faster than 1.

C. Bicyclist 1 is stopped, and bicyclist 2 is not.

D. Bicyclist 2 is stopped, and bicyclist 1 is not.

E. Both bicyclists are stopped.

11. Indicate what is happening at time A.

12. Indicate what is happening at time B.

13. Indicate what is happening at time C.

14. Which bicyclist won the race?

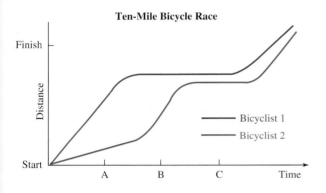

Ten-Mile Bicycle Race

Refer to this information: The hours served last month by the individual volunteers at a homeless shelter were 4, 6, 8, 2, 8, 10, 11, 9, 5, 12, 5, 18, 7, 5, 1, and 9.

15. Find the median of the hours of volunteer service.

16. Find the mean of the hours of volunteer service.

17. Find the mode of the hours of volunteer service.

18. If the one value of 18 were removed from the list, which would be most affected: the mean or the median?

19. RATINGS The seven top-rated cable television programs for the week of February 8–14 are given below. What are the mean, median, and mode of the ratings? Round to the nearest tenth.

Show/day/time/network	Rating
1. "WCW Monday," Mon. 9 p.m., TNT	4.5
2. "WCW Monday," Mon. 10 p.m., TNT	4.4
3. "WCW Monday," Mon. 8 p.m., TNT	3.9
4. "WWF Special," Sat. 9 p.m., USA	3.6
5. "WWF Wrestling," Sun. 7 p.m., USA	3.1
6. "Dog Show," Tues. 8 p.m., USA	3.1
7. "WWF Special," Sat. 8 p.m., USA	2.9

20. STATISTICS The bar graph has an asterisk * that refers readers to a note at the bottom. In your own words, complete the explanation of the term *median*.

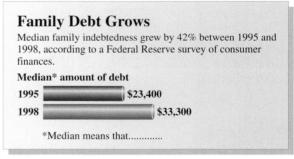

Family Debt Grows

Median family indebtedness grew by 42% between 1995 and 1998, according to a Federal Reserve survey of consumer finances.

Median* amount of debt

1995 $23,400
1998 $33,300

*Median means that.............

Source: *Los Angeles Times* (February 1, 2000)

1. GASOLINE In 1999, gasoline consumption in the United States was three hundred fifty-eight million, six hundred thousand gallons a day. Write this number in standard notation.

2. Round 49,999 to the nearest thousand.

Perform each operation.

3. $\begin{array}{r} 38,908 \\ +15,696 \\ \hline \end{array}$

4. $\begin{array}{r} 9,700 \\ -5,491 \\ \hline \end{array}$

5. $\begin{array}{r} 345 \\ \times\ 67 \\ \hline \end{array}$

6. $23\overline{)2,001}$

7. Explain how to check the following result using addition.

$$\begin{array}{r} 1,142 \\ -\ 459 \\ \hline 683 \end{array}$$

8. THE VIETNAMESE CALENDAR An animal represents each Vietnamese lunar year. Recent Years of the Cat are listed below. If the cycle continues, what year will be the next Year of the Cat?

1915 1927 1939 1951 1963 1975 1987 1999

9. Consider the multiplication statement 4 · 5 = 20. Show that multiplication is repeated addition.

10. ROOM DIVIDERS Four pieces of plywood, each 22 inches wide and 62 inches high, are to be covered with fabric, front and back, to make the room divider shown. How many square inches of fabric will be used?

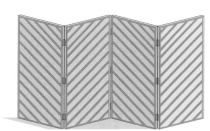

11. a. Find the factors of 18.

    b. Find the prime factorization of 18.

12. List the first ten prime numbers.

13. Why isn't 27 a prime number?

14. Evaluate: $(9 - 2)^2 - 3^3$.

15. Divide: $\dfrac{-315}{-1}$.

16. Simplify: $-(-6)$.

17. Graph the integers greater than -3 but less than 4.

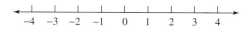

18. Find the absolute value: $|-5|$.

19. Is the statement $-12 > -10$ true or false?

20. ANNUAL NET INCOME Use the following data for the Polaroid Corporation to construct a line graph.

Year	1995	1996	1997	1998	1999
Total net income ($ millions)	−139	15	−127	−51	9

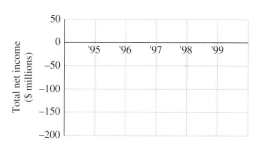

Perform each operation.

21. $-25 + 5$

22. $25 - (-5)$

23. $-25(5)(-1)$

24. $\dfrac{-25}{-5}$

25. Evaluate: $\dfrac{(-6)^2 - 1^5}{-4 - 3}$.

26. Evaluate: $-3 + 3(-4 - 4 \cdot 2)^2$.

27. Evaluate: -3^2 and $(-3)^2$.

28. PLANETS Mercury orbits closer to the sun than does any other planet. Temperatures on Mercury can get as high as $810°$ F and as low as $-290°$ F. What is the temperature range?

29. 360 is 45% of what number?

30. 90 is what percent of 600?

31. Write an expression illustrating division by 0 and an expression illustrating division of 0. Which is undefined?

32. Add: $\dfrac{1}{2} + \dfrac{2}{3}$.

33. Subtract: $\dfrac{1}{2} - \dfrac{2}{3}$.

34. TENNIS Find the length of the handle on the tennis racquet shown.

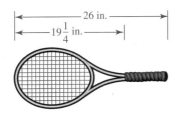

35. Multiply: $\dfrac{4}{5} \cdot \dfrac{2}{7}$.

36. Divide: $2\dfrac{4}{5} \div 2\dfrac{2}{3}$.

37. Complete the table.

	Rate (mph)	Time (hr)	Distance traveled (mi)
Truck	55	4	

38. Multiply: $3.45 \cdot 100$.

39. Multiply: $(0.31)(2.4)$.

40. Divide: $0.72\overline{)536.4}$.

41. Change $\dfrac{8}{11}$ to a decimal.

42. CLASS TIME In a chemistry course, students spend a total of 300 minutes in lab and lecture each week. If $\dfrac{7}{15}$ of the time is spent in lab each week, how many minutes are spent in lecture each week?

43. WEEKLY SCHEDULES Refer to the illustration. Determine the number of hours during a week that an adult spends, on average, on the Internet at home.

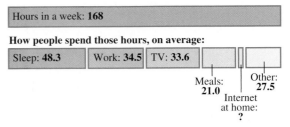

Based on data from the National Sleep Foundation and the United States Bureau of Statistics

44. TEAM GPA The grade point averages of the players on a badminton team are listed below. Find the mean, median, and mode of the team's GPAs.

3.04	4.00	2.75	3.23	3.87	2.20
3.02	2.25	2.99	2.56	3.58	2.75

An Introduction to Algebra

Getty Images

Most manufacturing companies have a production process that uses different machines, materials, and workers. For example, printing presses, binders, and packaging machines are needed to produce textbooks. Paper, ink, glue, and covers must be ordered in the correct quantities. And teams of workers must be scheduled to arrive at the proper times.

Planners often use algebra to organize the entire production process. They use mathematics to make sure production runs are efficient and, most important, profitable.

To learn more about the use of algebra by production planners, visit The Learning Equation on the Internet at http://tle.brookscole.com. (The log-in instructions are in the Preface.) For Chapter 8, the online lessons are:

• *TLE* Lesson 14: Writing and Evaluating Expressions
• *TLE* Lesson 15: Solving Equations

Check Your Knowledge

1. An equation is a statement that two expressions are _____. An equation contains an ____ symbol.

2. A _____ is a letter that represents a number. A _____ is a number that is fixed and does not change in value.

3. According to the _____ property of equality: If equal quantities are multiplied by the same nonzero quantity, the results will be equal quantities.

4. A _____ is a general rule that mathematically describes a relationship between two or more variables.

5. An algebraic _____ is a combination of variables, numbers, and the operation symbols for addition, subtraction, multiplication, and division. Algebraic _____ are simplified, and _____ are solved.

6. In the term $3x$, 3 is called the _____ and x is called the _____ part.

7. Terms with exactly the same variables and exponents are called _____ terms.

Solve each equation.

8. $x - 2 = 1$

9. $y + 3 = 5$

10. $4n = 64$

11. $\dfrac{z}{4} = 12$

12. $y - 70 = 50$

13. $10x = 1400$

14. There are twice as many students in Max's astronomy class as in his math class. If there are 34 students in the astronomy class, how many are in the math class?

15. Jo is 3 years older than Jill. Jo is 21. How old is Jill?

16. Evaluate $\dfrac{20 - x}{x + 1}$ for $x = 2$.

17. Michelle began her morning bicycle ride at 6:00 A.M. and returned at 9:00 A.M. She averaged 15 miles per hour for the ride. Find the distance she rode that morning.

Simplify by removing parentheses.

18. $3(2x + 7)$

19. $-3(2x - 3)$

20. Combine like terms.
   a. $2x - 2y + x + 2y$
   b. $x + x - x - 2x$

21. Simplify.
   a. $-x(-3)$
   b. $2a \cdot 2b \cdot 3$
   c. $3(x - 3) - 2(x - 5)$

22. Solve each equation. **Check each solution.**
   a. $3x - 3 = 2x + 3$
   b. $4y + 7 - y = 10$
   c. $3(2x - 5) = 5(x - 1)$

23. Write an algebraic expression for
   a. the value of q quarters
   b. the value of f five-dollar bills

24. Simplify.
   a. $x^2 \cdot x \cdot x \cdot x^4$
   b. $2x(3x^5)$
   c. $(3xy^3)^2$
   d. $(x^3)^4 (x^2)^3$

Study Skills Workshop

GETTING READY FOR THE FINAL EXAM

Final exams are stressful for many students. Most math teachers give comprehensive finals (covering material from the beginning to the end of the course), and the number of topics to study might be overwhelming. Relax! If you have been following the guidelines presented in these workshops, much of your work is already done. Preparing for your final exam is much like preparing for your other tests, but because you will have more topics, it may require a little more time. Planning for this at least a week in advance will give you plenty of time and will reduce the amount of anxiety you may feel.

Remember Those Study Sheets? In the Chapter 6 workshop you learned to prepare study sheets (or index cards, or audiotapes) for each test. Use these to begin your final exam preparation. Review each study sheet and make a note of only those things that you did not remember. Try to reduce all of your study materials to a list that will fit on one sheet of paper. This will be your final exam study sheet. Carry it with you all week and review it whenever you have the opportunity.

Returned Tests. Go over all of your returned tests and correct any mistakes that you made. Make a list of any test problems that you still are unsure about.

Make a Practice Final. Just like the practice tests you made, find problems for which you can find a solution (examples in your text, odd homework problems, and examples from your notes are all good sources for these problems). Make sure you get a good representation of problems from all of the chapters that you studied. Make a practice final that has about the same number of questions as your real final exam will have.

Get Help From Others. It's a good idea to compare notes with other students in your class; you might even swap practice finals. If you have problems from old tests that you can't answer, see either a tutor or your instructor during office hours.

Final Exam Day. Make sure you have all of the materials you will need for your final exam. Scratch paper, pencils with good erasers, and possibly Scantron forms are things that you might need. Be sure you are rested, and plan to arrive early. Make sure you know the correct time for your final exam, because finals are often given at times that are different from your class meeting time. Always work on problems that you are sure of first and save the most difficult ones for last.

ASSIGNMENT

1. Make a calendar a week before your final for studying for this exam and all other exams that you will have to take. Plan to study at least two hours each day.
2. Prepare a study sheet for your final exam.
3. Make a practice final. Find at least one other student in your class with whom you can swap practice tests. At least two days before your final, take a practice final under conditions that are close to those that you will have in your real exam (closed book, timed, etc.).
4. See your tutor or instructor with any unanswered questions at least one day before the final.
5. The night before your final, make sure you have all of the materials that you will need. Get a good night's sleep.
6. On the day of the final, go over problems that you feel confident about. Do not try to learn new material on the day of the final if you feel stressed.

Algebra is the language of mathematics. In this chapter, you will learn more about thinking and writing in this language, using its most important tool — a variable.

8.1 Solving Equations by Addition and Subtraction

- Equations • Checking solutions • Solving equations • Problem solving with equations

The language of mathematics is *algebra*. The word *algebra* comes from the title of a book written by the Arabian mathematician Al-Khowarazmi around A.D. 800. Its title, *Ihm aljabr wa'l muqabalah,* means restoration and reduction, a process then used to solve equations. In this section, we begin discussing equations, one of the most powerful ideas in algebra.

Equations

An **equation** is a statement indicating that two expressions are equal. Three examples of equations are

$$x + 5 = 21, \qquad 16 + 5 = 21, \qquad \text{and} \qquad 10 + 5 = 21$$

> **Equations**
>
> **Equations** are mathematical sentences that contain an = symbol.

In the equation $x + 5 = 21$, the expression $x + 5$ is called the **left-hand side,** and 21 is called the **right-hand side.** The letter x is the **variable** (or the **unknown**).

An equation can be true or false. For example, $16 + 5 = 21$ is a true equation, whereas $10 + 5 = 21$ is a false equation. An equation containing a variable can be true or false, depending upon the value of the variable. If x is 16, the equation $x + 5 = 21$ is true, because

16 + 5 = 21   Substitute 16 for x.

However, this equation is false for all other values of x.

Any number that makes an equation true when substituted for its variable is said to *satisfy* the equation. Such numbers are called **solutions.** Because 16 is the only number that satisfies $x + 5 = 21$, it is the only solution of the equation.

Checking solutions

Self Check 1
Is 8 a solution of $x + 17 = 25$?

EXAMPLE 1   Verify that 18 is a solution of the equation $x - 3 = 15$.

Solution   We substitute 18 for x in the equation and verify that both sides of the equation are equal.

$x - 3 = 15$ This is the given equation.

18 $- 3 \overset{?}{=} 15$ Substitute 18 for x. Read $\overset{?}{=}$ as "is possibly equal to."

$15 = 15$ Perform the subtraction.

Answer yes

Since $15 = 15$ is a true equation, 18 is a solution of $x - 3 = 15$.

EXAMPLE 2 Is 23 a solution of $32 = y + 10$?

Solution We substitute 23 for y and simplify.

$32 = y + 10$ This is the given equation.

$32 \stackrel{?}{=} 23 + 10$ Substitute 23 for y.

$32 \neq 33$ Perform the addition.

Since the left-hand and right-hand sides are not equal, 23 is not a solution of $32 = y + 10$.

Self Check 2
Is 5 a solution of $20 = y + 17$?

Answer no

▉ Solving equations

Since the solution of an equation is usually not given, we must develop a process to find it. This process is called *solving the equation*. To develop an understanding of the properties and procedures used to solve an equation, we will examine $x + 2 = 5$ and make some observations as we solve it in a practical way.

We can think of the scales shown in Figure 8-1(a) as representing the equation $x + 2 = 5$. The weight (in grams) on the left-hand side of the scales is $x + 2$, and the weight (in grams) on the right-hand side is 5. Because these weights are equal, the scales are in balance. To find x, we need to isolate it. That can be accomplished by removing 2 grams from the left-hand side of the scales. Common sense tells us that we must also remove 2 grams from the right-hand side if the scales are to remain in balance. In Figure 8-1(b), we can see that x grams are balanced by 3 grams. We say that we have *solved* the equation and that the *solution* is 3.

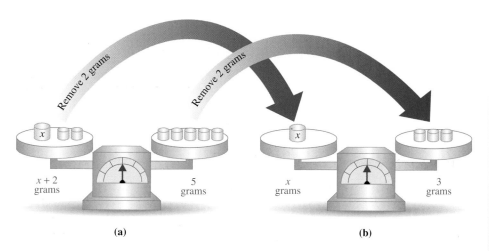

(a)   (b)

FIGURE 8-1

From this example, we can make observations about solving an equation.

- To find the value of x, we needed to isolate it on the left-hand side of the scales.

- To isolate x, we had to undo the addition of 2 grams. This was accomplished by subtracting 2 grams from the left-hand side.

- We wanted the scales to remain in balance. When we subtracted 2 grams from the left-hand side, we subtracted the same amount from the right-hand side.

The observations suggest a property of equality: *If the same quantity is subtracted from equal quantities, the results will be equal quantities.* We can express this property in symbols.

> **Subtraction property of equality**
>
> Let a, b, and c represent numbers.
>
> If $a = b$, then $a - c = b - c$.

When we use this property, the resulting equation is equivalent to the original equation.

> **Equivalent equations**
>
> Two equations are **equivalent equations** when they have the same solutions.

In the previous example, we found that $x + 2 = 5$ is equivalent to $x = 3$. This is true because these equations have the same solution, 3.

We now show how to solve $x + 2 = 5$ using an algebraic approach.

Self Check 3

Solve $x + 7 = 14$ and check the result.

EXAMPLE 3   Solve: $x + 2 = 5$.

Solution   To isolate x on the left-hand side of the equation, we undo the addition of 2 by subtracting 2 from both sides of the equation.

$$x + 2 = 5 \qquad \text{This is the equation to solve.}$$

$$x + 2 - 2 = 5 - 2 \qquad \text{Subtract 2 from both sides.}$$

$$x = 3 \qquad \begin{array}{l}\text{On the left-hand side, subtracting 2 undoes the addition of 2 and}\\ \text{leaves } x. \text{ On the right-hand side, } 5 - 2 = 3.\end{array}$$

We check by substituting 3 for x in the original equation and simplifying. If 3 is the solution, we will obtain a true statement.

Check: $x + 2 = 5$

$$3 + 2 \stackrel{?}{=} 5 \qquad \text{Substitute 3 for } x.$$

$$5 = 5 \qquad \text{Perform the addition.}$$

Answer 7

Since the resulting equation is true, 3 is the solution.

A second property that we use to solve equations involves addition. It is based on the following idea: *If the same quantity is added to equal quantities, the results will be equal quantities.* In symbols, we have the following property.

> **Addition property of equality**
>
> Let a, b, and c represent numbers.
>
> If $a = b$, then $a + c = b + c$.

We can think of the scales shown in Figure 8-2(a) as representing the equation $x - 2 = 3$. To find x, we need to use the addition property of equality and add 2 grams of weight to each side. The scales will remain in balance. From the scales in Figure 8-2(b), we can see that x grams are balanced by 5 grams. The solution of $x - 2 = 3$ is therefore 5.

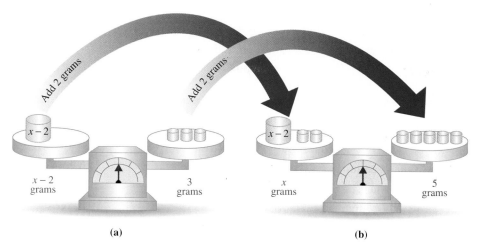

FIGURE 8-2

To solve $x - 2 = 3$ algebraically, we apply the addition property of equality. We can isolate x on the left-hand side of the equation by adding 2 to both sides.

$$x - 2 = 3$$
$$x - 2 + \mathbf{2} = 3 + \mathbf{2} \qquad \text{To undo the subtraction of 2, add 2 to both sides.}$$
$$x = 5 \qquad \begin{array}{l}\text{On the left-hand side, adding 2 undoes the subtraction of 2 and} \\ \text{leaves } x. \text{ On the right-hand side, } 3 + 2 = 5.\end{array}$$

To check this result, we substitute 5 for x in the original equation and simplify.

Check: $x - 2 = 3$

$$\mathbf{5} - 2 \overset{?}{=} 3 \qquad \text{Substitute 5 for } x.$$
$$3 = 3 \qquad \text{Perform the subtraction.}$$

Since this is a true statement, 5 is the solution.

EXAMPLE 4   Solve: $19 = y - 7$.

Solution   To isolate the variable y on the right-hand side, we use the addition property of equality. We can undo the subtraction of 7 by adding 7 to both sides.

$$19 = y - 7$$
$$19 + \mathbf{7} = y - 7 + \mathbf{7} \qquad \text{Add 7 to both sides.}$$
$$26 = y \qquad \begin{array}{l}\text{On the left-hand side, } 19 + 7 = 26. \text{ On the right-hand side,} \\ \text{adding 7 undoes the subtraction of 7 and leaves } y.\end{array}$$
$$y = 26 \qquad \begin{array}{l}\text{When we state a solution, it is common practice to write the} \\ \text{variable first. If } 26 = y, \text{ then } y = 26.\end{array}$$

We check by substituting 26 for y in the original equation and simplifying.

Check: $19 = y - 7 \qquad \text{This is the original equation.}$

$$19 \overset{?}{=} \mathbf{26} - 7 \qquad \text{Substitute 26 for } y.$$
$$19 = 19 \qquad \text{Perform the subtraction.}$$

Since this is a true statement, 26 is the solution.

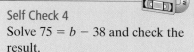

Self Check 4
Solve $75 = b - 38$ and check the result.

Answer 113

▉ Problem solving with equations

The key to problem solving is to understand the problem and then to devise a plan for solving it. The following list of steps provides a good strategy to follow.

> **Strategy for problem solving**
>
> 1. **Analyze the problem** by reading it carefully to understand the given facts. What information is given? What vocabulary is given? What are you asked to find? Often a diagram will help you visualize the facts of a problem.
> 2. **Form an equation** by picking a variable to represent the quantity to be found. Then express all other unknown quantities as expressions involving that variable. Key words or phrases can be helpful. Finally, translate the words of the problem into an equation.
> 3. **Solve the equation.**
> 4. **State the conclusion.**
> 5. **Check the result** in the words of the problem.

We will now use this five-step strategy to solve problems. The purpose of the following examples is to help you learn the strategy, even though you can probably solve these examples without it.

EXAMPLE 5   Financial data.   Figure 8-3 shows the 1999 quarterly net income for Nike, the athletic shoe company. What was the company's total net income for 1999?

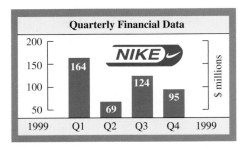

FIGURE 8-3

Analyze the problem

- We are given the net income for each quarter.
- We are asked to find the total net income.

Form an equation

Let n = the total net income for 1999. To form an equation involving n, we look for a key word or phrase in the problem.

Key word: *total*   **Translation:** *addition*

Now we translate the words of the problem into an equation.

Total net income	is	1st qtr. net income	plus	2nd qtr. net income	plus	3rd qtr. net income	plus	4th qtr. net income.
n	$=$	164	$+$	69	$+$	124	$+$	95

Solve the equation

$n = 164 + 69 + 124 + 95$ We are working in millions of dollars.

$ = 452$ Perform the additions.

State the conclusion

Nike's total net income was $452 million.

Check the result

We can check the result by estimation. To estimate, we round the net income from each quarter and add.

$$160 + 70 + 120 + 100 = 450$$

The answer, 452, is reasonable.

EXAMPLE 6   Small business.   Last year a hairdresser lost 17 customers who moved away. If he now has 73 customers, how many did he have originally?

Analyze the problem

- We know that he started with some unknown number of customers, and after 17 moved away, 73 were left.
- We are asked to find the number of customers he had before any moved away.

Form an equation

We can let c = the original number of customers. To form an equation involving c, we look for a key word or phrase in the problem.

    Key phrase: *moved away*   **Translation:** *subtraction*

Now we translate the words of the problem into an equation.

The original number of customers	minus	17	is	the remaining number of customers.
c	$-$	17	$=$	73

Solve the equation

$$c - 17 = 73$$
$$c - 17 + \mathbf{17} = 73 + \mathbf{17} \qquad \text{To undo the subtraction of 17, add 17 to both sides.}$$
$$c = 90 \qquad \text{Simplify each side of the equation.}$$

State the conclusion

He originally had 90 customers.

Check the result

The hairdresser had 90 customers. After losing 17, he has $90 - 17$, or 73 left. The answer, 90, checks.

EXAMPLE 7   Buying a house.   Sue wants to buy a house that costs $87,000. Since she has only $15,000 for a down payment, she will have to borrow some additional money by taking a mortgage. How much will she have to borrow?

Analyze the problem

- The house costs $87,000.
- Sue has $15,000 for a down payment.
- We must find how much money she needs to borrow.

Form an equation

We can let x = the money that she needs to borrow. To form an equation involving x, we look for a key word or phrase in the problem.

Key phrase: *borrow some additional money*   **Translation:** *addition*

Now we translate the words of the problem into an equation.

The amount Sue has	plus	the amount she borrows	is	the total cost of the house.
15,000	+	x	=	87,000

Solve the equation

$$15,000 + x = 87,000$$
$$15,000 + x - \mathbf{15,000} = 87,000 - \mathbf{15,000}$$ To undo the addition of 15,000, subtract 15,000 from both sides.
$$x = 72,000$$ Perform the subtractions.

State the conclusion

She must borrow $72,000.

Check the result

With a $72,000 mortgage, she will have $15,000 + $72,000, which is the $87,000 that is necessary to buy the house. The answer, 72,000, checks.

Section 8.1   STUDY SET

VOCABULARY   *Fill in the blanks.*   Remember these

1. An equation is a statement that two expressions are ~~equal~~. An equation contains an $=$ symbol.
2. A ~~solution~~ of an equation is a number that satisfies the equation.
3. In $30 = t - 12$, the ~~right-hand~~ side of the equation is $t - 12$.
4. A letter that is used to represent a number is called a ~~variable~~
5. ~~equivalent~~ equations have exactly the same solutions.
6. To solve an equation, we ~~isolate~~ the variable on one side of the equals symbol.

CONCEPTS   *Complete the properties of equality.*

7. If $x = y$ and c is any number, then $x + c = \boxed{} + \boxed{}$.
8. If $x = y$ and c is any number, then $x - c = \boxed{} - \boxed{}$.

9. In $x + 6 = 10$, what operation is performed on the variable? How do we undo that operation to isolate the variable?

10. In $9 = y - 5$, what operation is performed on the variable? How do we undo that operation to isolate the variable?

NOTATION   *Complete each solution to solve the given equation.*

11. $x + 8 = 24$
$$x + 8 - \boxed{} = 24 - \boxed{}$$
$$x = 16$$
Check: $x + 8 = 24$
$$\boxed{} + 8 \stackrel{?}{=} 24$$
$$\boxed{} = 24$$
So $\boxed{}$ is the solution.

12.
$$x - 8 = 24$$
$$x - 8 + \boxed{} = 24 + \boxed{}$$
$$x = 32$$

Check: $x - 8 = 24$

$$\boxed{} - 8 \quad 24$$

$$\boxed{} = 24$$

So $\boxed{}$ is the solution.

■ PRACTICE *Determine whether each statement is an equation.*

13. $x = 2$
14. $y - 3$
15. $7x < 8$
16. $7 + x = 2$
17. $x + y = 0$
18. $3 - 3y > 2$
19. $1 + 1 = 3$
20. $5 = a + 2$

For each equation, is the given number a solution?

21. $x + 2 = 3$; 1
22. $x - 2 = 4$; 6
23. $a - 7 = 0$; 7
24. $x + 4 = 4$; 0
25. $8 - y = y$; 5
26. $10 - c = c$; 5
27. $x + 32 = 0$; 16
28. $x - 1 = 0$; 4
29. $z + 7 = z$; 7
30. $n - 9 = n$; 9
31. $x = x$; 0
32. $x = 2$; 0

Use the addition or subtraction property of equality to solve each equation. Check each answer.

33. $x - 7 = 3$
34. $y - 11 = 7$
35. $a - 2 = 5$
36. $z - 3 = 9$
37. $1 = b - 2$
38. $0 = t - 1$
39. $x - 4 = 0$
40. $c - 3 = 0$
41. $y - 7 = 6$
42. $a - 2 = 4$
43. $70 = x - 5$
44. $66 = b - 6$
45. $312 = x - 428$
46. $x - 307 = 113$
47. $x - 117 = 222$
48. $y - 27 = 317$
49. $x + 9 = 12$
50. $x + 3 = 9$
51. $y + 7 = 12$
52. $c + 11 = 22$
53. $t + 19 = 28$
54. $s + 45 = 84$
55. $23 + x = 33$
56. $34 + y = 34$
57. $5 = 4 + c$
58. $41 = 23 + x$
59. $99 = r + 43$
60. $92 = r + 37$
61. $512 = x + 428$
62. $x + 307 = 513$
63. $x + 117 = 222$
64. $y + 38 = 321$

65. $3 + x = 7$
66. $b - 4 = 8$
67. $y - 5 = 7$
68. $z + 9 = 23$
69. $0.4 + a = 1.2$
70. $0.5 + x = 1.3$
71. $x - 1.3 = 3.4$
72. $x - 2.3 = 1.9$
73. $b + \dfrac{1}{3} = \dfrac{1}{2}$
74. $c + \dfrac{3}{5} = \dfrac{3}{4}$
75. $p - \dfrac{1}{8} = \dfrac{1}{3}$
76. $q + \dfrac{2}{7} = \dfrac{3}{5}$

■ APPLICATIONS *Complete each solution.*

77. ARCHAEOLOGY A 1,700-year-old manuscript is 425 years older than the clay jar in which it was found. How old is the jar?

Analyze the problem

- The manuscript is _____ old.
- The manuscript is _____ older than the jar.
- We are asked to find _____.

Form an equation Since we want to find the age of the jar, we can let $x = $ _____. Now we look for a key word or phrase in the problem.

    Key phrase: _____

    Translation: _____

Now we translate the words of the problem into an equation.

The age of the manuscript	is	425	plus	the age of the jar.
$\boxed{}$	=	425	+	$\boxed{}$

Solve the equation

$$\boxed{} = 425 + x$$
$$1{,}700 - \boxed{} = 425 + x - \boxed{}$$
$$\boxed{} = x$$

State the conclusion _____.

Check the result If the jar is $\boxed{}$ years old, then the manuscript is $1{,}275 + 425 = \boxed{}$ years old. The answer checks.

78. BANKING After a student wrote a $1,500 check to pay for a car, he had a balance of $750 in his account. How much did he have in the account before he wrote the check?

Analyze the problem

- A _____ check was written.
- The balance became _____.
- We are asked to find
  _____.

Form an equation Since we want to find his balance before he wrote the check, we let $x =$ _____. Now we look for a key word or phrase in the problem.

Key phrase: _____

Translation: _____

Now we translate the words of the problem into an equation.

The original balance in the account	minus	$1,500	is	$750.
x	$-$	$1,500$	$=$	750

Solve the equation

$$\boxed{} - 1,500 = 750$$
$$x - 1,500 + \boxed{} = 750 + \boxed{}$$
$$x = \boxed{}$$

State the conclusion _____.

Check the result The original balance was ⬚. After writing a check, he had a balance of $2,250 − $1,500, or ⬚. The answer checks.

Let a variable represent the unknown quantity. Then write and solve an equation to answer the question.

79. ELECTIONS The illustration shows the votes received by the three major candidates running for President of the United States in 1996. Find the total number of votes cast for them.

Bill Clinton (D)	47,401,185
Bob Dole (R)	39,197,469
H. Ross Perot (RF)	8,085,294

80. HIT RECORDS The oldest artist to have a number 1 single was Louis Armstrong at age 67, with *Hello Dolly*. The youngest artist to have the number 1 single was 12-year-old Jimmy Boyd, with *I Saw Mommy Kissing Santa Claus*. What is the difference in their ages?

81. PARTY INVITATIONS Three of Mia's party invitations were lost in the mail, but 59 were delivered. How many invitations did she send?

82. HEARING PROTECTION The sound intensity of a jet engine is 110 decibels. What noise level will an airplane mechanic experience if the ear plugs she is wearing reduce the sound intensity by 29 decibels?

83. FAST FOOD The franchise fee and startup costs for a Taco Bell restaurant are $287,000. If an entrepreneur has $68,500 to invest, how much money will she need to borrow to open her own Taco Bell restaurant?

84. BUYING GOLF CLUBS A man needs $345 for a new set of golf clubs. How much more money does he need if he now has $317?

85. CELEBRITY EARNINGS *Forbes* magazine estimates that in 2003, Celine Dion earned $28 million. If this was $152 million less than Oprah Winfrey's earnings, how much did Oprah earn in 2003?

86. HELP WANTED From the following ad from the classified section of a newspaper, determine the value of the benefit package. ($45K means $45,000.)

> **★ACCOUNTS PAYABLE★**
> 2-3 yrs exp as supervisor. Degree a +. High vol company. Good pay, $45K & xlnt benefits; total compensation worth $52K. Fax resume.

87. POWER OUTAGES The electrical system in a building automatically shuts down when the meter shown reads 85. By how much must the current reading increase to cause the system to shut down?

88. VIDEO GAMES After a week of playing Sega's *Sonic Adventure,* a boy scored 11,053 points in one game — an improvement of 9,485 points over the very first time he played. What was his score for his first game?

89. AUTO REPAIR A woman paid $29 less to have her car fixed at a muffler shop than she would have paid at a gas station. At the gas station, she would have paid $219. How much did she pay to have her car fixed?

90. BUS RIDERS A man had to wait 20 minutes for a bus today. Three days ago, he had to wait 15 minutes longer than he did today, because four buses passed by without stopping. How long did he wait three days ago?

■ WRITING

91. Explain what it means for a number to satisfy an equation.

92. Explain how to tell whether a number is a solution of an equation.

93. Explain what Figure 8-1 (page 423) is trying to show.

94. Explain what Figure 8-2 (page 425) is trying to show.

95. When solving equations, we *isolate* the variable. Write a sentence in which the word *isolate* is used in a different context.

96. Think of a number. Add 8 to it. Now subtract 8 from that result. Explain why we will always obtain the original number.

▮ REVIEW

97. Round 325,784 to the nearest ten

98. Find the power: 1^5.

99. Evaluate: $2 \cdot 3^2 \cdot 5$.

100. Represent $4 + 4 + 4$ as a multiplication.

101. Evaluate: $8 - 2(3) + 1^3$.

102. Write 1,055 in words.

8.2 Solving Equations by Division and Multiplication

- The division property of equality
- The multiplication property of equality
- Problem solving with equations

In the previous section, we solved equations of the forms

$$x - 4 = 10 \qquad \text{and} \qquad x + 5 = 16$$

by using the addition and subtraction properties of equality. In this section, we describe how to solve equations of the forms

$$2x = 8 \qquad \text{and} \qquad \frac{x}{3} = 25$$

by using the division and multiplication properties of equality.

▮ The division property of equality

To solve many equations, we must divide both sides of the equation by the same nonzero number. The resulting equation will be equivalent to the original one. This idea is summed up in the division property of equality: *If equal quantities are divided by the same nonzero quantity, the results will be equal quantities.*

Division property of equality

Let a, b, and c represent numbers, and c is not 0.

$$\text{If } a = b, \text{ then } \frac{a}{c} = \frac{b}{c}.$$

We will now consider how to solve the equation $2x = 8$. Recall that $2x$ means $2 \cdot x$. Therefore, the given equation can be rewritten as $2 \cdot x = 8$. We can think of the scales in Figure 8-4(a) on the next page as representing the equation $2 \cdot x = 8$. The weight (in grams) on the left-hand side of the scales is $2 \cdot x$, and the weight (in grams) on the right-hand side is 8. Because these weights are equal, the scales are in balance. To find x, we need to isolate it. That can be accomplished by using the division property of equality to remove half of the weight from each side. The scales will remain in balance. From the scales shown in Figure 8-4(b), we see that x grams are balanced by 4 grams.

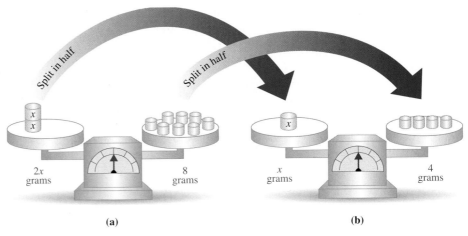

FIGURE 8-4

We now show how to solve $2x = 8$ using an algebraic approach.

Self Check 1
Solve $17x = 153$ and check the result.

EXAMPLE 1   Solve $2x = 8$ and check the result.

Solution   Recall that $2x = 8$ means $2 \cdot x = 8$. To isolate x on the left-hand side of the equation, we undo the multiplication by 2 by dividing both sides of the equation by 2.

$$2x = 8 \qquad \text{This is the equation to solve.}$$

$$\frac{2x}{2} = \frac{8}{2} \qquad \text{To undo the multiplication by 2, divide both sides by 2.}$$

$$x = 4 \qquad \begin{array}{l}\text{When } x \text{ is multiplied by 2 and that product is then divided by 2, the result is } x.\\ \text{Perform the division: } 8 \div 2 = 4.\end{array}$$

To check this result, we substitute 4 for x in $2x = 8$.

Check: $2x = 8$

$2 \cdot 4 \overset{?}{=} 8 \qquad$ Substitute 4 for x.

$8 = 8 \qquad$ Perform the multiplication.

Answer 9

Since $8 = 8$ is a true statement, 4 is the solution.

▌ The multiplication property of equality

We can also multiply both sides of an equation by the same nonzero number to get an equivalent equation. This idea is summed up in the multiplication property of equality: *If equal quantities are multiplied by the same nonzero quantity, the results will be equal quantities.*

Multiplication property of equality

Let a, b, and c represent numbers, and c is not 0.

If $a = b$, then $c \cdot a = c \cdot b$ \qquad or, more simply, \qquad $ca = cb$.

We can think of the scales shown in Figure 8-5(a) on the next page as representing the equation $\frac{x}{3} = 25$. The weight on the left-hand side of the scales is $\frac{x}{3}$ grams, and the weight on the right-hand side is 25 grams. Because these weights are equal, the scales are in balance. To find x, we can use the multiplication property of equality to triple (or multiply by 3) the weight on each side. The scales will remain in balance. From the scales shown in Figure 8-5(b), we can see that x grams are balanced by 75 grams.

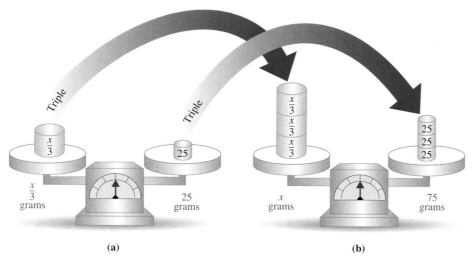

FIGURE 8-5

We now show how to solve $\frac{x}{3} = 25$ using an algebraic approach.

EXAMPLE 2   Solve $\frac{x}{3} = 25$ and check the result.

Solution   To isolate x on the left-hand side of the equation, we undo the division of the variable by 3 by multiplying both sides by 3.

$\quad \dfrac{x}{3} = 25$ 　　This is the equation to solve.

$\mathbf{3} \cdot \dfrac{x}{3} = \mathbf{3} \cdot 25$ 　　To undo the division by 3, multiply both sides by 3.

$\quad x = 75$ 　　When x is divided by 3 and that quotient is then multiplied by 3, the result is x. Perform the multiplication: $3 \cdot 25 = 75$.

To check the result, we substitute 75 for x in $\frac{x}{3} = 25$.

Check: $\dfrac{x}{3} = 25$

$\dfrac{75}{3} \overset{?}{=} 25$ 　　Substitute 75 for x.

$25 = 25$ 　　Perform the division: $75 \div 3 = 25$. The answer checks.

Since $25 = 25$ is a true statement, 75 is the solution.

Self Check 2

Solve $\dfrac{x}{12} = 24$ and check the result.

Answer 288

▌ Problem solving with equations

As before, we can use equations to solve problems. Remember that the purpose of these early examples is to help you learn the strategy, even though you can probably solve the examples without it.

EXAMPLE 3   Buying electronics.

The owner of an apartment complex bought six television sets that were on sale for $499 each. What was the total cost?

Analyze the problem

- Six television sets were bought.
- They cost $499 each.
- We are asked to find the total cost.

Form an equation

We let c = the total cost of the TVs. To form an equation, we look for a key word or phrase in the problem. We can add 499 six times or multiply 499 by 6. Since it is easier, we will multiply.

Key phrase: *six TVs, each costing $499*   **Translation:** *multiplication*

Now we translate the words of the problem into an equation.

The total number of TVs	multiplied by	the cost of each TV	is	the total cost.
6	$\cdot$	499	$=$	c

Solve the equation

$$6 \cdot 499 = c$$
$$2{,}994 = c \quad \text{Perform the multiplication.}$$

State the conclusion

The total cost will be $2,994.

Check the result

We can check by estimation. Since each TV costs a little less than $500, we would expect the total cost to be a little less than 6 · $500, or $3,000. An answer of $2,994 is reasonable.

EXAMPLE 4   Splitting an inheritance.

If seven brothers inherit $343,000 and split the money evenly, how much will each brother get?

Analyze the problem

- There are 7 brothers.
- They split $343,000 evenly.
- We are asked to find how much each brother will get.

Form an equation

We can let g = the number of dollars each brother will get. To form an equation, we look for a key word or phrase in the problem.

Key phrase: *split the money evenly*   **Translation:** *division*

Now we translate the words of the problem into an equation.

The total amount of the inheritance	divided by	the number of brothers	is	the share each brother will get.
343,000	÷	7	=	g

Solve the equation

$$\frac{343,000}{7} = g \qquad \text{343,000 ÷ 7 can be written as } \tfrac{343,000}{7}.$$

$$49,000 = g \qquad \text{Perform the division.}$$

State the conclusion

Each brother will get $49,000.

Check the result

If we multiply $49,000 by 7, we get $343,000.

EXAMPLE 5 Traffic violations.

For speeding, a motorist had to pay a fine of $592. The violation occurred on a highway posted with signs like the one shown in Figure 8-6. What would the fine have been if such signs were not posted?

TRAFFIC FINES DOUBLED IN CONSTRUCTION ZONE

FIGURE 8-6

Analyze the problem

- The motorist was fined $592.
- The fine was double what it would normally have been.
- We are asked to find what the fine would have been, had the area not been posted.

Form an equation

We can let f = the amount that the fine would normally have been. To form an equation, we look for a key word or phrase in the problem or analysis.

Key word: *double*   **Translation:** *multiply by 2*

Now we translate the words of the problem into an equation.

Two	times	the normal speeding fine	is	the new fine.
2	·	f	=	592

Solve the equation

$$2f = 592 \qquad \text{Write } 2 \cdot f \text{ as } 2f.$$

$$\frac{2f}{2} = \frac{592}{2} \qquad \text{To undo the multiplication by 2, divide both sides by 2.}$$

$$f = 296 \qquad \text{Perform the division: 562 ÷ 2 = 296.}$$

State the conclusion

The fine would normally have been $296.

Check the result

If we double $296, we get 2 ($296) = $592. The answer checks.

EXAMPLE 6   Entertainment costs.   A five-piece band worked on New Year's Eve. If each player earned $120, what fee did the band charge?

Analyze the problem

- There were 5 players in the band.
- Each player made $120.
- We are asked to find the band's fee. We know that the fee divided by the number of players will give each person's share.

Form an equation

We can let f = the band's fee. To form an equation, we look for a key word or phrase. In this case, we find it in the analysis of the problem.

Key phrase: *divided by*   **Translation:** *division*

Now we translate the words of the problem into an equation.

The band's fee	divided by	the number in the band	is	each person's share.
f	$\div$	5	=	120

Solve the equation

$$\frac{f}{5} = 120 \qquad \text{Write } f \div 5 \text{ as } \frac{f}{5}.$$

$$5 \cdot \frac{f}{5} = 5 \cdot 120 \qquad \text{To undo the division by 5, multiply both sides by 5.}$$

$$f = 600 \qquad \text{Perform the multiplication: } 5 \cdot 120 = 600.$$

State the conclusion

The band's fee was $600.

Check the result

If we divide $600 by 5, we get each person's share: $120.

Section 8.2   STUDY SET

VOCABULARY   *Fill in the blanks.*   remember these

1. According to the ___division___ property of equality: If equal quantities are divided by the same nonzero quantity, the results will be equal quantities.

2. According to the ___multiplication___ property of equality: If equal quantities are multiplied by the same nonzero quantity, the results will be equal quantities.

CONCEPTS   *Fill in the blanks.*

3. If we multiply x by 6 and then divide that product by 6, the result is ▨ .

4. If we divide x by 8 and then multiply that quotient by 8, the result is ▨ .

5. If $x = y$, then $\dfrac{x}{z} = $ ▨   where $z \neq 0$.

6. If $x = y$, then $zx = $ ▨   where $z \neq 0$.

7. In the equation $4t = 40$, what operation is being performed on the variable? How do we undo it?

8. In the equation $\frac{t}{15} = 1$, what operation is being performed on the variable? How do we undo it?

9. Name the first step in solving each of the following equations.

   a. $x + 5 = 10$ **b.** $x - 5 = 10$

   c. $5x = 10$ **d.** $\frac{x}{5} = 10$

10. For each of the following equations, check the given possible solution.

   a. $16 = t - 8;\ 33$

   b. $16 = t + 8;\ 8$

   c. $16 = 8t;\ 128$

   d. $16 = \frac{t}{8};\ 2$

NOTATION *Complete each solution to solve the equation.*

11. $3x = 12$

   $\dfrac{3x}{} = \dfrac{12}{}$

   $x = 4$

   Check: $3x = 12$

   $3 \cdot \boxed{} \overset{?}{=} 12$

   $\boxed{} = 12$

   So $\boxed{}$ is the solution.

12. $\dfrac{x}{5} = 9$

   $\boxed{} \cdot \dfrac{x}{5} = \boxed{} \cdot 9$

   $x = 45$

   Check: $\dfrac{x}{5} = 9$

   $\dfrac{\boxed{}}{5} \overset{?}{=} 9$

   $\boxed{} = 9$

   So $\boxed{}$ is the solution.

PRACTICE *Use the division or the multiplication property of equality to solve each equation.* **Check each answer.**

13. $3x = 3$ 14. $5x = 5$

15. $2x = 192$ 16. $4x = 120$

17. $17y = 51$ 18. $19y = 76$

19. $34y = 204$ 20. $18y = 90$

21. $100 = 100x$ 22. $35 = 35y$

23. $16 = 8r$ 24. $44 = 11m$

25. $0.4p = 6.4$ 26. $0.36 = 0.12g$

27. $5x = \dfrac{2}{3}$ 28. $6m = \dfrac{3}{4}$

29. $\dfrac{x}{7} = 2$ 30. $\dfrac{x}{12} = 4$

31. $\dfrac{y}{14} = 3$ 32. $\dfrac{y}{13} = 5$

33. $\dfrac{a}{15} = 5$ 34. $\dfrac{b}{25} = 5$

35. $\dfrac{c}{13} = 3$ 36. $\dfrac{d}{100} = 11$

37. $1 = \dfrac{x}{50}$ 38. $1 = \dfrac{x}{25}$

39. $7 = \dfrac{t}{7}$ 40. $4 = \dfrac{m}{4}$

41. $9z = 90$ 42. $3z = 6$

43. $7x = 21$ 44. $13x = 52$

45. $86 = 43t$ 46. $288 = 96t$

47. $21s = 21$ 48. $31x = 155$

49. $\dfrac{d}{20} = 2$ 50. $\dfrac{x}{16} = 4$

51. $400 = \dfrac{t}{3}$ 52. $250 = \dfrac{y}{2}$

53. $\dfrac{y}{0.3} = 1.2$ 54. $\dfrac{t}{3.5} = 0.7$

55. $\dfrac{q}{0.5} = \dfrac{1}{2}$ 56. $\dfrac{n}{0.8} = \dfrac{3}{4}$

APPLICATIONS *Complete each solution.*

57. **NOBEL PRIZE** In 1998, three Americans, Louis Ignarro, Robert Furchgott, and Dr. Fred Murad, were awarded the Nobel Prize for Medicine. They shared the prize money. If each person received $318,500, what was the Nobel Prize cash award?

Analyze the problem

- ▨ people shared the cash award.
- Each person received ▨▨▨▨.
- We are asked to find the

  _____.

Form an equation

Since we want to find what the Nobel Prize cash award was, we let $c =$ _____. To form an equation, we look for a key word or phrase in the problem.

Key Phrase: _____

Translation: _____

Now we translate the words of the problem into an equation.

The Nobel Prize cash award	divided by	the number of recipients	was	$318,500.
▨	÷	3	=	318,500

Solve the equation

$$\frac{x}{3} = 318{,}500$$

$$▨ \cdot \frac{x}{3} = ▨ \cdot 318{,}500$$

$$x = ▨▨▨▨$$

State the conclusion

_____

Check the result

If we divide the Nobel Prize cash award by 3, we have

$$\frac{▨▨▨}{3} = ▨▨▨▨.$$ This was the amount each person received. The answer checks.

58. INVESTING An investor has watched the value of his portfolio double in the last 12 months. If the current value of his portfolio is $274,552, what was its value one year ago?

Analyze the problem

- The value of the portfolio ▨▨▨ in 12 months.
- The current value is ▨▨▨▨.
- We must find _____.

Form an equation

We can let $x =$ _____. We now look for a key word or phrase in the problem.

Key phrase: _____

Translation: _____

Now we translate the words of the problem into an equation.

2	times	the value of the portfolio one year ago	is	the current value of the portfolio.
2	·	▨	=	$274,552

Solve the equation

$$2x = ▨▨▨$$

$$\frac{2x}{▨} = \frac{274{,}552}{▨}$$

$$x = ▨▨▨$$

State the conclusion

_____

Check the result

If the value of the portfolio one year ago was ▨▨▨▨ and it doubled, its current value would be ▨▨▨. The answer checks.

Let a variable represent the unknown quantity. Then write and solve an equation to answer the question.

59. SPEED READING An advertisement for a speed reading program claimed that successful completion of the course could triple a person's reading rate. If Alicia can currently read 130 words a minute, at what rate can she expect to read after taking the classes?

60. COST OVERRUNS Lengthy delays and skyrocketing costs caused a rapid-transit construction project to go over budget by a factor of 10. The final audit showed the project costing $540 million. What was the initial cost estimate?

61. STAMPS Large sheets of commemorative stamps honoring Marilyn Monroe are to be printed. On each sheet, there are 112 stamps, with 8 stamps per row. How many rows of stamps are on a sheet?

62. SPREADSHEETS The grid shown on the next page is a Microsoft Excel spreadsheet. The rows are labeled with numbers, and the columns are labeled with letters. Each empty box of the grid is called a *cell*. Suppose a certain project calls for a spreadsheet with 294 cells, using columns A through F. How many rows will need to be used?

Microsoft Excel-Book 1					
File **Edit** **View** **Insert** **Format** **Tools**					
	A	B	C	D	E	F
1						
2						
3						
4						
5						
6						
7						
8						

Sheet 1 / Sheet 2 / Sheet 3 / Sheet 4 / Sheet 5 /

On Earth

63. PHYSICAL EDUCATION A high school PE teacher had the students in her class form three-person teams for a basketball tournament. Thirty-two teams participated in the tournament. How many students were in the PE class?

64. LOTTO WINNERS The grocery store employees listed below pooled their money to buy $120 worth of lottery tickets each week, with the understanding they would split the prize equally if they happened to win. One week they did have the winning ticket and won $480,000. What was each employee's share of the winnings?

Sam M. Adler	Ronda Pellman	Manny Fernando
Lorrie Jenkins	Tom Sato	Sam Lin
Kiem Nguyen	H. R. Kinsella	Tejal Neeraj
Virginia Ortiz	Libby Sellez	Alicia Wen

65. ANIMAL SHELTERS The number of phone calls to an animal shelter quadrupled after the evening news aired a segment explaining the services the shelter offered. Before the publicity, the shelter received 8 calls a day. How many calls did the shelter receive each day after being featured on the news?

66. OPEN HOUSES The attendance at an elementary school open house was only half of what the principal had expected. If 120 people visited the school that evening, how many had she expected to attend?

67. GRAVITY The weight of an object on Earth is 6 times greater than what it is on the moon. The situation shown in the next column took place on Earth. If it took place on the moon, what weight would the scale register?

68. INFOMERCIALS The number of orders received each week by a company selling skin care products increased fivefold after a Hollywood celebrity was added to the company's infomercial. After adding the celebrity, the company received about 175 orders each week. How many orders were received each week before the celebrity took part?

WRITING

69. Explain what Figure 8-4 (page 432) is trying to show.

70. Explain what Figure 8-5 (page 433) is trying to show.

71. What does it mean to solve an equation?

72. Think of a number. Double it. Now divide it by 2. Explain why you always obtain the original number.

REVIEW

73. Find the perimeter of a rectangle with sides measuring 8 cm and 16 cm.

74. Find the area of a rectangle with sides measuring 23 inches and 37 inches.

75. Find the prime factorization of 120.

76. Find the prime factorization of 150.

77. Evaluate: $3^2 \cdot 2^3$.

78. Evaluate: $5 + 6 \cdot 3$.

79. FUEL ECONOMY Five basic models of automobiles made by Saturn have city mileage ratings of 24, 22, 28, 29, and 27 miles per gallon. What is the average (mean) city mileage for the five models?

80. Solve: $x - 4 = 20$.

8.3   Algebraic Expressions and Formulas

- Algebraic expressions   • Evaluating algebraic expressions   • Formulas
- Formulas from business   • Formulas from science

An algebraic expression is a combination of variables and numbers with the operations of addition, subtraction, multiplication, and division. In this section, we replace the variables in these expressions with numbers and then evaluate them. We also study formulas.

Algebraic expressions

In the equation $1{,}700 = 425 + x$, the expression $425 + x$ is called an **algebraic expression.**

> ### Algebraic expressions
> Variables and numbers can be combined with the operations of addition, subtraction, multiplication, and division to create **algebraic expressions.**

Here are some examples of algebraic expressions.

$5(2a)$	This algebraic expression is a combination of the numbers 5 and 2, the variable a, and the operation of multiplication.
$x + 2x + 3x$	This algebraic expression involves the variable x, the numbers 2 and 3, and the operations of addition and multiplication.
$\dfrac{10 - y}{-3}$	This algebraic expression is a combination of the numbers 10 and -3, the variable y, and the operations of subtraction and division.
$5(r - 6)$	This algebraic expression is a combination of the numbers 5 and 6, the variable r, and the operations of subtraction and multiplication.

Evaluating algebraic expressions

The manufacturer's instructions for installing a kitchen garbage disposal include the diagram in Figure 8-7. Word phrases are used to represent the lengths of the pieces of pipe needed to connect the disposal to the drain line.

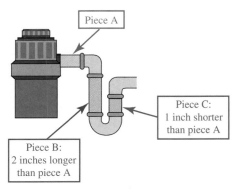

FIGURE 8-7

The instructions tell us that the lengths of pieces B and C are related to the length of piece A. If we let x = the length of piece A, then the lengths of the two other pieces of pipe can be represented using algebraic expressions, as shown in Figure 8-8.

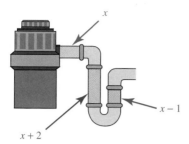

FIGURE 8-8

Since piece B is 2 inches longer than piece A,

$x + 2$ = the length of piece B

Since piece C is 1 inch shorter than piece A,

$x - 1$ = the length of piece C

See Table 8-1, which shows part of the manufacturer's instruction sheet. Suppose that model #201 is being installed. The table tells us that piece A should be 3 inches long. We then find the lengths of the other two pieces of pipe by replacing x with 3 in each of the algebraic expressions.

Model	Length of piece A
#101	2 inches
#201	3 inches
#301	4 inches

TABLE 8-1

To find the length of piece B:

Replace x with 3.

$x + 2 = \mathbf{3} + 2$
$\quad\quad = 5$

Piece B should be 5 inches long.

To find the length of piece C:

Replace x with 3.

$x - 1 = \mathbf{3} - 1$
$\quad\quad = 2$

Piece C should be 2 inches long.

When we replace the variable in an algebraic expression with a number and simplify, we are **evaluating the algebraic expression.** In the previous example, we say that we *substituted* 3 for x to find the lengths of the other two pieces of pipe.

In the next three examples, we replace the variables with positive or negative numbers. It is a good idea to write parentheses around the number when it is inserted into the algebraic expression in place of the variable.

EXAMPLE 1 Evaluate each expression for $x = 3$: **a.** $2x - 1$ and
b. $\dfrac{-x - 15}{6}$.

Solution

a. $2x - 1 = 2(\mathbf{3}) - 1$ Substitute 3 for x. Use parentheses to show the multiplication.

$\quad\quad\quad = 6 - 1$ Perform the multiplication first: $2(3) = 6$.

$\quad\quad\quad = 5$ Perform the subtraction.

Self Check 1
Evaluate each expression for $y = 5$:
a. $5y - 4$
b. $\dfrac{-y - 15}{5}$

b. $\dfrac{-x - 15}{6} = \dfrac{-(\mathbf{3}) - 15}{6}$ Substitute 3 for x. Use parentheses.

$= \dfrac{-3 - 15}{6}$ The opposite of 3 is -3.

$= \dfrac{-3 + (-15)}{6}$ Add the opposite of 15.

$= \dfrac{-18}{6}$ Perform the addition: $-3 + (-15) = -18$.

Answers a. 21, **b.** -4 $= -3$ Perform the division.

Self Check 2

Evaluate each expression for $t = -3$:

a. $-2t + 4t^2$

b. $-t + 2(t + 1)$

EXAMPLE 2   Evaluate each expression for $a = -2$: **a.** $-3a + 4a^2$ and **b.** $-a + 3(1 + a)$.

Solution

a. $-3a + 4a^2 = -3(-\mathbf{2}) + 4(-\mathbf{2})^2$ Replace a with -2. Use parentheses.

$= -3(-2) + 4(4)$ Find the power: $(-2)^2 = 4$.

$= 6 + 16$ Perform the multiplications: $-3(-2) = 6$ and $4(4) = 16$.

$= 22$ Perform the addition.

b. $-a + 3(1 + a) = -(-\mathbf{2}) + 3[1 + (-\mathbf{2})]$ Substitute -2 for a.

$= -(-2) + 3(-1)$ Perform the addition within the brackets.

$= 2 + (-3)$ Perform the multiplication: $3(-1) = -3$.

Answers a. 42, **b.** -1 $= -1$ Perform the addition.

To evaluate algebraic expressions containing two or more variables, we need to know the value of each variable.

Self Check 3

Evaluate $(5rs + 4s)^2$ for $r = -1$ and $s = 5$.

EXAMPLE 3   Evaluate $(8hg + 6g)^2$ for $h = -1$ and $g = 5$.

Solution

$(8hg + 6g)^2 = [8(-\mathbf{1})(\mathbf{5}) + 6(\mathbf{5})]^2$ Replace h with -1 and g with 5.

$= (-40 + 30)^2$ Perform the multiplications within the brackets: $8(-1)(5) = -40$ and $6(5) = 30$.

$= (-10)^2$ Perform the addition within the parentheses.

Answer 25 $= 100$ Find the power.

▌ Formulas

A **formula** is a mathematical expression that is used to state a relationship between two or more variables. Formulas are used in many fields: economics, physical education, anthropology, biology, automotive repair, and nursing, to name just a few. In this section, we consider seven formulas from business, science, and mathematics.

Study Time   THINK IT THROUGH

"Your success in school is dependent on your ability to study effectively and efficiently. The results of poor study skills are wasted time, frustration, and low or failing grades." Effective Study Skills, Dr. Bob Kizlik, 2004

For a course that meets for h hours each week, the formula $H = 2h$ gives the suggested number of hours H that a student should study the course outside of class each week. If a student expects difficulty in a course, the formula can be adjusted upward to $H = 3h$. Use the formulas to complete the table below.

If a course meets for:	Suggested study time outside of class (hours per week)	Expanded study time outside of class (hours per week)
2 hours per week		
3 hours per week		
4 hours per week		
5 hours per week		

▮ Formulas from business

A Formula to Find the Sale Price. If a car that normally sells for $12,000 is discounted $1,500, you can find the sale price using the formula

Sale price   =   original price   –   discount

Using the variables s to represent the sale price, p the original price, and d the discount, we can write this formula as

$$s = p - d$$

To find the sale price of the car, we substitute 12,000 for p, substitute 1,500 for d, and evaluate the expression on the right-hand side of the equation.

$$s = p - d$$
$$= 12{,}000 - 1{,}500 \quad \text{Substitute 12,000 for } p \text{ and 1,500 for } d.$$
$$= 10{,}500 \quad \text{Perform the subtraction.}$$

The sale price of the car is $10,500.

A Formula to Find the Retail Price. To make a profit, a merchant must sell a product for more than he paid for it. The price at which he sells the product, called the *retail price,* is the *sum* of what the item cost him and the markup.

Retail price   =   cost   +   markup

Using the variables r to represent the retail price, c the cost, and m the markup, we can write this formula as

$$r = c + m$$

As an example, suppose that a store owner buys a lamp for $35 and then marks up the cost $20 before selling it. We can find the retail price of the lamp using this formula.

$$r = c + m$$
$$= \mathbf{35} + \mathbf{20} \qquad \text{Substitute 35 for } c \text{ and 20 for } m.$$
$$= 55 \qquad\qquad \text{Perform the addition.}$$

The retail price of the lamp is $55.

A Formula to Find Profit. The profit a business makes is the *difference* of the revenue (the money it takes in) and the costs.

Profit	=	revenue	−	costs

Using the variables p to represent the profit, r the revenue, and c the costs, we have the formula

$$\boxed{p = r - c}$$

As an example, suppose that a charity telethon took in $14 million in donations but had expenses totaling $2 million. We can find the profit made by the charity by subtracting the expenses (costs) from the donations (revenue).

$$p = \mathbf{r} - \mathbf{c}$$
$$= \mathbf{14} - \mathbf{2} \qquad \text{Substitute 14 for } r \text{ and 2 for } c.$$
$$= 12 \qquad\qquad \text{Perform the subtraction.}$$

The charity made a profit of $12 million from the telethon.

▌ Formulas from science

A Formula to Find the Distance Traveled. If we know the rate (speed) at which we are traveling and the time we will be moving at that rate (say, 3 hours), we can find the distance traveled using the formula

Distance	=	rate	·	time

Writing the formula using variables, we have

$$\boxed{d = rt}$$

If we replace the rate r with 55 and the time t with 3, we can use this formula to find the distance traveled.

$$d = \mathbf{rt}$$
$$= \mathbf{55}(\mathbf{3}) \qquad \text{Substitute 55 for } r \text{ and 3 for } t.$$
$$= 165 \qquad\quad \text{Perform the multiplication.}$$

The distance traveled in 3 hours at a rate of 55 miles per hour is 165 miles.

Self Check 4

Nevada's speed limit for trucks is 75 mph. How far would a truck travel in 3 hours at top speed?

EXAMPLE 4 Speed limits. Three speed limits for trucks are shown. At each of these speeds, how far would a truck travel in 3 hours?

Solution To find the distance traveled by a truck in Ohio, we write

$$d = \mathbf{rt}$$
$$= \mathbf{55}(\mathbf{3}) \qquad \text{55 mph is the rate } r, \text{ and 3 hours is the time } t.$$
$$= 165 \qquad\quad \text{Perform the multiplication. The units of the answer are miles.}$$

At 55 mph, a truck would travel 165 miles in 3 hours. We can use a table as shown on the next page to display the calculations for each state.

	r	t	= d
Ohio	55	3	165
Indiana	60	3	180
Kentucky	65	3	195

This column gives the
distance traveled, in miles.

Answer 225 mi

! COMMENT When using $d = rt$ to find distance, make sure that the units are similar. For example, if the rate is given in miles per hour, the time must be expressed in hours.

A Formula for Converting Degrees Fahrenheit to Degrees Celsius.

Electronic message boards in front of some banks flash two temperature readings. This is because temperature can be measured using the Fahrenheit or the Celsius scale. The Fahrenheit scale is used in the American system of measurement and the Celsius scale in the metric system. The two scales are shown on the thermometers in Figure 8-9. This should help you to see how the two scales are related. There is a formula to convert a Fahrenheit reading F to a Celsius reading C:

$$C = \frac{5}{9}(F - 32)$$

or

$$C = \frac{5}{9}F - \frac{160}{9}$$

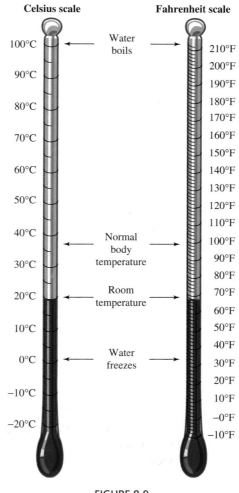

FIGURE 8-9

EXAMPLE 5 The thermostat in an office building was set at 77° F. Use the first formula to convert this setting to degrees Celsius.

Self Check 5
The record high temperature for New Mexico is 122° F, on June 27, 1994. Convert this temperature to degrees Celsius.

Solution

$$C = \frac{5}{9}(F - 32)$$

$$= \frac{5}{9}(77 - 32) \quad \text{Replace } F \text{ with 77.}$$

$$= \frac{5}{9}(45) \quad \text{Perform the subtraction: } 77 - 32 = 45.$$

$$= \frac{225}{9} \quad \text{Perform the multiplication: } 5(45) = 225.$$

$$= 25 \quad \text{Perform the division.}$$

The thermostat is set at 25° C.

Answer 50° C.

A Formula to Find the Distance an Object Falls. The distance an object falls (in feet) when it is dropped from a height is related to the time (in seconds) that it has been falling by the formula

Distance fallen	=	16	·	(time)^2

Using the variables d to represent the distance and t the time, we have

$$d = 16t^2$$

Self Check 6

Find the distance a rock fell in 3 seconds if it was dropped over the edge of the Grand Canyon.

EXAMPLE 6   Find the distance a camera fell in 6 seconds if it was dropped by a vacationer taking a hot-air balloon ride.

Solution   We can use the formula $d = 16t^2$ to find the distance the camera fell.

$$d = 16t^2$$

$$= 16(6)^2 \quad \text{Replace } t \text{ with 6.}$$

$$= 16(36) \quad \text{Find the power: } 6^2 = 36.$$

$$= 576 \quad \text{Perform the multiplication.}$$

Answer 144 ft

The camera fell 576 feet.

Section 8.3   STUDY SET

Remember These

VOCABULARY   *Fill in the blanks.*

1. A _formula_ is an equation that states a relationship between two or more variables.

2. An algebraic _expression_ is a combination of variables, numbers, and the operation symbols for addition, subtraction, multiplication, and division.

3. To evaluate an algebraic expression, we _substitute_ specific numbers for the variables in the expression and apply the rules for the order of operations.

4. A _variable_ is a letter that stands for a number.

CONCEPTS

5. Show the misunderstanding that occurs if we don't write parentheses around -8 when evaluating the expression $2x + 10$ for $x = -8$.

6. **a.** Which of the formulas studied in this section involve a *difference* of two quantities?

   b. Which of the formulas studied in this section involve a *product*?

7. PLAYGROUND EQUIPMENT The plans for building a children's swing set are shown. The builder can choose a size that is appropriate for the children who will use it.

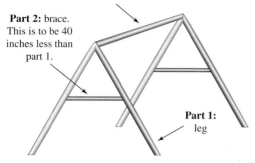

Part 3: crossbar.
This is to be 16 inches longer than part 1.

Part 2: brace.
This is to be 40 inches less than part 1.

Part 1: leg

a. Choose a variable to represent the length of one of the parts of the swing set. Then write algebraic expressions that represent the lengths of the other two parts.

b. If the builder chooses to have part 1 be 60 inches long, how long should parts 2 and 3 be?

8. SET DESIGN A television studio art department plans to construct a series of set decorations out of plywood, using the plan shown.

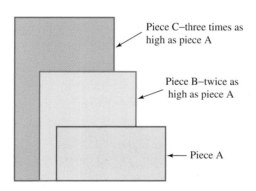

Piece C–three times as high as piece A

Piece B–twice as high as piece A

Piece A

a. Choose a variable to represent the height of one piece of plywood. Then write algebraic expressions that represent the heights of the other two pieces.

b. For the foreground, designers will make piece A 15 inches high. How high should pieces B and C be?

c. For the background, piece A will be 30 inches high. How high should pieces B and C be?

9. A ticket outlet adds a service charge of $2 to the price of every ticket it sells. Complete the table.

Ticket price	Service charge	Total cost
20	2	
25	2	
p	2	
$10p$	2	

10. A tire shop charges $5 to balance each new tire it sells. Complete the table.

Price of new tire	Balancing charge	Total cost
$45	$5	
$65	$5	
$t	$5	
$5t	$5	

11. Complete the table by finding the distance traveled in each instance. Be sure to give the units for each of your answers.

	Rate (mph)	Time (hr)	Distance traveled
Bus	25	2	
Bike	12	4	
Walking	3	t	
Running	5	1	
Car	x	3	

12. Each semester, a college charges its students a campus service fee of $10, a parking fee of $12, and an enrollment fee of $13 per unit. Complete the table.

Units taken	Enrollment fee ($)	Campus service fee ($)	Parking fee ($)	Total charges ($)
4	52	10	12	
10	130	10	12	
12	156	10	12	
u	$13u$	10	12	

13. DASHBOARDS The illustration shows part of a dashboard. Explain what each instrument measures. How are these measurements mathematically related?

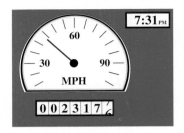

14. What occupation might use a formula that finds
    a. target heart rate after a workout
    b. gas mileage of a car
    c. age of a fossil
    d. equity in a home
    e. dose to administer
    f. cost-of-living index

15. SPREADSHEETS A car travels at a rate of 65 mph for 15 minutes. What is wrong with the following thinking?

$$d = rt$$
$$= 65(15)$$
$$= 975$$

The car travels 975 miles in 15 minutes.

16. SPREADSHEETS The data from a chemistry experiment are shown. To obtain a result, the chemist must use the formula

SUM(A1:D1) − **MINIMUM**(D1:D3)

If A1:D1 means from cell A1 up to and including cell D1, find the result.

	A	B	C	D
1	12	−5	6	−2
2	15	4	5	−4
3	6	4	−2	8

NOTATION

17. Use variables to write the formula that relates each of the quantities listed.
    a. Rate, distance, time
    b. Centigrade temperature, Fahrenheit temperature
    c. Time, the distance an object falls when dropped

18. Use variables to write the formula that relates each of the quantities listed.
    a. Original price, sale price, discount
    b. Number of values, average, sum of values
    c. Cost, profit, revenue
    d. Markup, retail price, cost

PRACTICE *Evaluate each expression for the given value of the variable.*

19. $3x + 5$ for $x = 4$ 20. $1 + 7a$ for $a = 2$

21. $-p$ for $p = -4$ 22. $-j$ for $j = -9$

23. $-4t$ for $t = -10$ 24. $-12m$ for $m = -6$

25. $\dfrac{x - 8}{2}$ for $x = -4$ 26. $\dfrac{-10 + y}{-4}$ for $y = -6$

27. $2(p + 9)$ for $p = -12$ 28. $3(r - 20)$ for $r = 15$

29. $x^2 - x - 7$ for $x = -5$ 30. $a^2 + 3a - 9$ for $a = -3$

31. $8s - s^3$ for $s = -2$ 32. $5r + r^3$ for $r = 1$

33. $4x^2$ for $x = 5$ 34. $3f^2$ for $f = 3$

35. $3b - b^2$ for $b = -4$ 36. $5a - a^2$ for $a = -3$

37. $\dfrac{24 + k}{3k}$ for $k = 3$ 38. $\dfrac{4 - h}{h - 4}$ for $h = -1$

Evaluate each expression for the given values of the variables.

39. $\dfrac{x}{y}$ for $x = 30$ and $y = -10$

40. $\dfrac{e}{3f}$ for $e = 24$ and $f = -8$

41. $-x - y$ for $x = -1$ and $y = 8$

42. $-a - 5b$ for $a = -9$ and $b = 6$

43. $x(5h - 1)$ for $x = -2$ and $h = 2$

44. $c(2k - 7)$ for $c = -3$ and $k = 4$

45. $b^2 - 4ac$ for $b = -3$, $a = 4$, and $c = -1$

46. $3r^2h$ for $r = 4$ and $h = 2$

47. $x^2 - y^2$ for $x = 5$ and $y = -2$

48. $x^3 - y^3$ for $x = -1$ and $y = 2$

49. $\dfrac{50 - 6s}{-t}$ for $s = 5$ and $t = 4$

50. $\dfrac{7v - 5r}{-r}$ for $v = 8$ and $r = 4$

51. $-5abc + 1$ for $a = -2$, $b = -1$, and $c = 3$

52. $-rst + 2t$ for $r = -3$, $s = -1$, and $t = -2$

53. $5s^2t$ for $s = -3$ and $t = -1$

54. $-3k^2t$ for $k = -2$ and $t = -3$

Use the appropriate formula to help answer each question.

55. It costs a snack bar owner 20 cents to make a snow cone. If the markup is 50 cents, what is the price of a snow cone?

56. Find the distance covered by a jet if it travels for 3 hours at 550 mph.

57. A school carnival brought in revenues of $13,500 and had costs of $5,300. What was the profit?

58. For the month of June, a florist's cost of doing business was $3,795. If June revenues totaled $5,115, what was her profit for the month?

59. A jewelry store buys bracelets for $18 and marks them up $5. What is the retail price of a bracelet?

60. A shopkeeper marks up the cost of every item she carries by the amount she paid for the item. If a fan costs her $27, what does she charge for the fan?

61. Find the distance covered by a car traveling 60 miles per hour for 5 hours.

62. Find the sale price of a pair of skis that normally sells for $200 but is discounted $35.

63. Find the Celsius temperature reading if the Fahrenheit reading is 14°.

64. Find the Celsius temperature reading if the Fahrenheit reading is 113°.

65. The area of a trapezoid is given by the formula $A = \frac{1}{2}h(b_1 + b_2)$, where h is the height and b_1 and b_2 are the bases. Find the area of a trapezoid whose bases measure 32 meters and 16 meters and whose height is 10 meters.

66. On its first night of business, a pizza parlor brought in $445. The owner estimated his costs that night to be $295. What was the profit?

67. Find the distance a ball has fallen 2 seconds after being dropped from a tall building.

68. A store owner buys a pair of pants for $25 and marks them up $15 for sale. What is the retail price of the pants?

APPLICATIONS

69. FINANCIAL STATEMENTS Use the data in the illustration to complete the financial statement for Avon Products, Inc.

Avon Products, Inc.

■ Revenue ■ Cost of goods sold

$ millions	1998	1997	1996
Revenue	5,213	5,079	4,814
Cost of goods sold	2,053	2,051	1,921

Based on data from *Hoover's online.*

Annual Financials: Income Statement All dollar amounts in millions			
	Dec. '98	Dec. '97	Dec. '96
Revenue			
Cost of goods sold			
Gross profit			

70. THERMOMETER SCALES A thermometer manufacturer wishes to scale a thermometer in both degrees Celsius and degrees Fahrenheit. Find the missing Celsius degree measures in the illustration.

Fahrenheit   Celsius

86° —   — ?
59° —   — ?
23° —   — ?

71. SPREADSHEETS A store manager wants to use a spreadsheet to post the prices of sale items for the checkers at the cash registers. If column B lists the regular price and column C lists the discount, write a formula using column names to have the computer find the sale price. Then fill in column D.

	A	B	C	D
1	Bath towel set	$25	$5	
2	Pillows	$15	$3	
3	Comforter	$53	$11	

72. DEALER MARKUPS A car dealer marks up the cars he sells $500 above factory invoice (that is, $500 over what it costs him to purchase the car from the factory). Complete the table.

Model	Factory invoice ($)	Markup ($)	Price ($)
Minivan	15,600		
Pickup	13,200		
Convertible	x		

73. FOOTBALL The results of each carry that a running back had during a game are recorded in the table. Find his average yards gained per carry for the game.

Play	Result
Sweep	Gain of 16 yd
Pitch	Gain of 10 yd
Dive	Gain of 4 yd
Dive	No gain
Sweep	Loss of 4 yd

74. QUARTERLY REPORTS The financial performance of a computer company in each of the four quarters of the year is shown. Find the company's quarterly financial average. The figures are in millions of dollars.

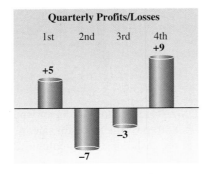

Quarterly Profits/Losses

CALCULATOR *Use a calculator to answer each question.*

75. SHIPPING The illustration shows the number of ships that use the docking facilities of a port one week in July. Find the average number of ships that docked there per day for that week.

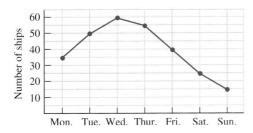

76. FALLING OBJECTS In the table in the next column, find the distance in feet traveled by a falling object in 1, 2, 3, and 4 seconds. Enter the results in the middle column. Then find the distance the object traveled over each time interval and enter it in the right-hand column.

Time falling	Distance traveled (ft)	Time intervals
1 sec		Distance traveled from 0 sec to 1 sec
2 sec		Distance traveled from 1 sec to 2 sec
3 sec		Distance traveled from 2 sec to 3 sec
4 sec		Distance traveled from 3 sec to 4 sec

77. CUSTOMER SURVEYS As customers were leaving a restaurant, they were asked to rate the service they had received. Good service was rated with a 5, fair service with a 3, and poor service with a 1. The tally sheet compiled by the questioner is shown. What was the restaurant's average score on this survey?

Type of service	Point value	Number
Good	5	~~卌~~ ~~卌~~ ~~卌~~ ~~卌~~ ~~卌~~ ~~卌~~ ~~卌~~ ~~卌~~ ~~卌~~ ~~卌~~ ‖‖
Fair	3	~~卌~~ ~~卌~~ ~~卌~~ ~~卌~~ ~~卌~~ ‖
Poor	1	~~卌~~ ‖‖‖‖

78. DISTANCE TRAVELED

a. When in orbit, the space shuttle travels at a rate of approximately 17,250 mph. How far does it travel in one day?

b. The speed of light is approximately 186,000 miles per second. How far will light travel in 1 minute?

c. The speed of a sound wave in air is about 1,100 feet per second at normal temperatures. How far does it travel in half a minute?

WRITING

79. Explain the process of evaluating an algebraic expression.

80. Explain how we can use a stopwatch to find the distance traveled by a falling object.

81. Write a definition for each of these business words: *revenue, markup,* and *profit.*

82. What is a formula?

REVIEW

83. Which of these are prime numbers: 9, 15, 17, 33, 37, 41?

84. How can this repeated multiplication be rewritten in simpler form: $2 \cdot 2 \cdot 2 \cdot 2 \cdot 2$?

85. Evaluate: $|-2 + (-5)|$.

86. Multiply: $-3(-2)(4)$.

87. In the equation $\frac{x}{3} = -4$, what operation is performed on the variable?

88. True or false: $-25.001 < -24.999$

89. Subtract: $-3 - (-6)$.

90. Which is undefined: division of zero or division by zero?

8.4 Simplifying Algebraic Expressions and the Distributive Property

- Simplifying algebraic expressions involving multiplication   • The distributive property
- Distributing a factor of -1 • Extending the distributive property

In mathematics, it is often useful to replace something complicated with something simpler in form. In this section, we simplify algebraic expressions. The results are equivalent but simpler expressions.

Simplifying algebraic expressions involving multiplication

To **simplify an algebraic expression,** we use properties of algebra to write the given expression in a simpler form. Two of the properties used to simplify algebraic expressions are the associative and the commutative properties of multiplication. Recall that the associative property enables us to change the grouping of the factors involved in a multiplication. The commutative property allows us to change the order of the factors.

To write $6(5x)$ in a simpler form, we can begin by rewriting it as $6 \cdot (5 \cdot x)$.

$6(5x) = 6 \cdot (5 \cdot x)$ $5x = 5 \cdot x$.

$\quad\quad = (6 \cdot 5)x$ Use the associative property of multiplication. Instead of grouping 5 with x, group it with 6.

$\quad\quad = 30x$ Perform the operation within the parentheses first: $6 \cdot 5 = 30$.

We say that $6(5x)$ simplifies to $30x$; that is, $6(5x) = 30x$.

To verify that $6(5x)$ and $30x$ are **equivalent expressions** (they represent the same number), we can evaluate each expression for several choices of x. For each value of x, the results should be the same.

If $x = 10$		If $x = -3$	
$6(5x) = 6[5(\mathbf{10})]$	$30x = 30(\mathbf{10})$	$6(5x) = 6[5(\mathbf{-3})]$	$30x = 30(\mathbf{-3})$
$= 6[50]$	$= 300$	$= 6[-15]$	$= -90$
$= 300$		$= -90$	

EXAMPLE 1   Simplify: **a.** $-2(7x)$ and **b.** $-12t(-6)$.

Solution   We use the associative property to regroup the factors of the expression so that the numbers are separated from the variable. After multiplying the numbers, the result is multiplied by the variable.

a. $-2(7x) = (-2 \cdot 7)x$ Apply the associative property of multiplication to regroup the factors.

$= -14x$ Perform the operation within the parentheses first: $-2 \cdot 7 = -14$.

b. $-12t(-6) = -12(-6)t$ Apply the commutative property of multiplication. Change the order of the factors.

$= [-12(-6)]t$ Apply the associative property of multiplication to group the numbers together. Use brackets.

$= 72t$ Perform the operation within the brackets first: $-12(-6) = 72$.

EXAMPLE 2   Simplify: **a.** $-4m(-5n)$ and **b.** $2(6y)(-4z)$.

Solution

a. $-4m(-5n) = [-4(-5)](m \cdot n)$ Group the numbers and variables separately, using the commutative and associative properties of multiplication.

$= 20mn$ Perform the multiplication within the brackets: $-4(-5) = 20$. Write $m \cdot n$ as mn.

b. $2(6y)(-4z) = [2(6)(-4)](y \cdot z)$ Use the commutative and associative properties to change the order and regroup the factors.

$= -48yz$ Perform the multiplication within the brackets: $2(6)(-4) = -48$. Write $y \cdot z$ as yz.

! COMMENT Be careful when using the terms *simplify* and *solve*. In mathematics, we *simplify expressions* and we *solve equations*. We do not simplify equations, nor do we solve expressions.

The distributive property

Another property of algebra that is used to simplify algebraic expressions is the **distributive property.** To introduce this property, we will examine the expression $2(5 + 3)$, which can be evaluated in two ways.

Method 1: Rules for the Order of Operations. Because of the grouping symbols in $2(5 + 3)$, the rules for the order of operations require that we compute the *sum* within the parentheses first.

$2(\mathbf{5 + 3}) = 2(\mathbf{8})$ Perform the addition within the parentheses first: $5 + 3 = 8$.

$= 16$ Perform the multiplication.

Method 2: The Distributive Property. The distributive property allows us to evaluate the expression $2(5 + 3)$ in another way. We can distribute the factor of 2 across to the 5 and across to the 3. We then find each of those products separately and add the results.

Distribute the multiplication by 2.

$$2(5 + 3) = 2(5 + 3)$$

Each number within the parentheses is multiplied by the factor outside the parentheses.

First product Second product
↓   ↓

$$= 2(5) + 2(3)$$

$$= 10 + 6$$ Perform the multiplications first: $2(5) = 10$ and $2(3) = 6$.

$$= 16$$ Perform the addition.

Notice that the result using each method is 16.

We now state the distributive property in symbols.

The distributive property

If a, b, and c represent numbers,

$$a(b + c) = ab + ac$$

Since subtraction is the same as adding the opposite, the distributive property also holds for subtraction.

The distributive property

If a, b, and c represent numbers,

$$a(b - c) = ab - ac$$

EXAMPLE 3 Use the distributive property to remove the parentheses:
a. $3(s + 7)$ and   **b.** $6(x - 1)$.

Solution

a. $3(s + 7) = 3s + 3(7)$ Distribute the multiplication by 3.

$$= 3s + 21$$ Perform the multiplication.

After applying the distributive property to $3(s + 7)$ to obtain $3s + 21$, we say that we have *removed the parentheses*.

b. $6(x - 1) = 6x - 6(1)$ Distribute the multiplication by 6.

$$= 6x - 6$$ Perform the multiplication.

Self Check 3
Remove the parentheses:
a. $5(h + 4)$
b. $9(a - 3)$

Answers **a.** $5h + 20$, **b.** $9a - 27$

! COMMENT The fact that an expression contains parentheses does not automatically mean that the distributive property should be used to simplify it. For example, the distributive property does not apply to expressions such as $5(4x)$ or $5(4 \cdot x)$, where a product is multiplied by 5. The distributive property applies to expressions such as $5(4 + x)$ or $5(4 - x)$, where a sum or difference is multiplied by a number.

Self Check 4
Remove the parentheses:
a. $-4(6y + 8)$
b. $-7(2 - 8m)$

EXAMPLE 4   Remove the parentheses: **a.** $-3(4x + 2)$ and   **b.** $-9(3 - 2t)$.

Solution

a. $-3(4x + 2) = -3(4x) + (-3)(2)$ Distribute the multiplication by -3.

$\qquad\qquad\quad = -12x + (-6)$ Perform the multiplications.

$\qquad\qquad\quad = -12x - 6$ Write the answer in simpler form. Adding -6
                      is the same as subtracting 6.

b. $-9(3 - 2t) = -9(3) - (-9)(2t)$ Apply the distributive property.

$\qquad\qquad\quad = -27 - (-18t)$ Perform the multiplications.

$\qquad\qquad\quad = -27 + 18t$ Write the answer in simpler form. Add the
                      opposite of $-18t$.

Answers a. $-24y - 32$,
b. $-14 + 56m$

! COMMENT It is common practice to write answers in simplified form. For example, the answer to Example 4, part a, is expressed as $-12x - 6$ because it involves fewer symbols than $-12x + (-6)$. The answer to Example 4, part b, is given as $-27 + 18t$ instead of $-27 - (-18t)$.

Since multiplication is commutative, we can write the distributive property in either of the following forms.

$$(b + c)a = ba + ca$$
$$(b - c)a = ba - ca$$

Self Check 5
Multiply:
a. $(8 + 7x)5$
b. $(5 - c)3$

EXAMPLE 5   Multiply: **a.** $(5 + 3r)7$ and   **b.** $(4 - x)2$.

Solution

a. $(5 + 3r)7 = (5)7 + (3r)7$ Distribute the multiplication by 7.

$\qquad\qquad = 35 + 21r$ Perform the multiplications.

b. $(4 - x)2 = (4)2 - (x)2$ Distribute the multiplication by 2.

$\qquad\qquad = 8 - 2x$ Perform the multiplications.

Answers a. $40 + 35x$,
b. $15 - 3c$

▌ Distributing a factor of -1

At first glance, $-(x + 8)$ doesn't appear to be in the proper form to apply the distributive property; the number in front of the parentheses appears to be missing. But the negative sign in front of the parentheses actually represents the number -1.

The negative sign represents -1.

$-(x + 8) = -1(x + 8)$

$\qquad\quad = -1(x) + (-1)(8)$ Use the distributive property. Distribute -1.

$\qquad\quad = -x + (-8)$ Perform the multiplications.

$\qquad\quad = -x - 8$ Adding -8 is the same as subtracting 8.

EXAMPLE 6   Simplify: $-(-6 - 2e)$.

Solution

$$-(-6 - 2e) = -1(-6 - 2e)$$ Rewrite the negative sign in front of the parentheses as -1.

$$= -1(-6) - (-1)(2e)$$ Distribute the multiplication by -1.

$$= 6 - (-2e)$$ Perform the multiplications.

$$= 6 + 2e$$ Add the opposite of $-2e$.

Self Check 6
Simplify: $-(-2t + 4)$.

Answer $2t - 4$

After working several problems like Example 6, you will see that it is not necessary to show each of the steps. The result can be obtained quickly by *changing the sign
of each quantity within the parentheses and dropping the parentheses.*

▌ Extending the distributive property

The distributive property can be extended to situations where there are more than
two terms within parentheses.

> **The extended distributive property**
>
> If a, b, c, and d represent numbers,
>
> $$a(b + c + d) = ab + ac + ad$$

EXAMPLE 7   Remove the parentheses: $-6(-3x - 6y + 8)$.

Solution   We distribute the multiplication by -6.

$$-6(-3x - 6y + 8) = -6(-3x) - (-6)(6y) + (-6)(8)$$

$$= 18x - (-36y) + (-48)$$ Perform the multiplications.

$$= 18x + 36y + (-48)$$ Write the subtraction as addition of the opposite of $-36y$, which is $36y$.

$$= 18x + 36y - 48$$ Adding -48 is the same as subtracting 48.

Section 8.4   STUDY SET

remember these

▌ **VOCABULARY**   *Fill in the blanks.*

1. The <u>distributive</u> property tells us how to multiply
   $5(x + 7)$. After performing the multiplication to
   obtain $5x + 35$, we say that the parentheses have
   been <u>removed</u>.

2. To <u>simplify</u> an algebraic expression means to use
   algebraic properties to write it in simpler form.

3. When an algebraic expression is simplified, the result
   is an <u>equivalent</u> expression.

4. We <u>simplify</u> expressions and we <u>solve</u> equations.

CONCEPTS

5. State the distributive property using the variables x, y, and z.

6. Use the variables r, s, and t to state the distributive property.

7. The following expressions are examples of the *right* and *left* distributive properties:

$$5(w + 7) \quad \text{and} \quad (w + 7)5$$

Which of the two do you think would be termed the right distributive property?

8. For each of the following expressions, tell whether the distributive property applies.

a. $2(5t)$ **b.** $2(t + 5)$

c. $5(2 \cdot t)$ **d.** $(2t)5$

e. $(2)(-t)5$ **f.** $(5 - t)(-2)$

9. The distributive property can be demonstrated using the illustration. Fill in the blanks: Two groups of 6 plus three groups of 6 is ▊ groups of 6. Therefore,

$$\boxed{} \cdot 2 + \boxed{} \cdot 3 \quad = \quad 6(\boxed{} + \boxed{})$$

10. a. Simplify: $2(5x)$.

    b. Remove the parentheses: $2(5 + x)$.

11. Write an equivalent expression for $-(y + 9)$ without using parentheses.

12. Explain what the arrows illustrate.

$$-9(y - 7)$$

NOTATION   *Complete each solution.*

13. $-5(7n) = (\boxed{} \cdot 7)n$

$$= -35n$$

14. $4(2a + b - 1) = 4(\boxed{}) + 4(\boxed{}) - \boxed{}(1)$

$$= 8a + 4b - 4$$

Fill in the blanks.

15. a. $2(x + 4) = 2x \boxed{} 8$

    b. $2(x - 4) = 2x \boxed{} 8$

    c. $-2(x + 4) = -2x \boxed{} 8$

    d. $-2(-x - 4) = 2x \boxed{} 8$

16. $-(x + 10) = \boxed{}(x + 10)$

17. Write each expression in simpler form, using fewer mathematical symbols.

a. $-(-x)$

b. $x - (-5)$

c. $5x - 10y + (-15)$

d. $5 \cdot x$

18. Determine what number is to be distributed.

a. $-6(x - 2)$ **b.** $(t + 1)(-5)$

c. $(a + 24)8$ **d.** $-(z - 16)$

PRACTICE   *Simplify each expression.*

19. $2(6x)$ **20.** $4(7b)$

21. $-5(6y)$ **22.** $-12(6t)$

23. $-10(-10t)$ **24.** $-8(-6k)$

25. $(4s)3$ **26.** $(9j)7$

27. $2c \cdot 7$ **28.** $11f \cdot 9$

29. $-5 \cdot 8h$ **30.** $-8 \cdot 4d$

31. $-7x(6y)$ **32.** $13a(-2b)$

33. $4r \cdot 4s$ **34.** $7x \cdot 7y$

35. $2x(5y)(3)$ **36.** $4(3z)(4)$

37. $5r(2)(-3b)$ **38.** $4d(5)(-3e)$

39. $5 \cdot 8c \cdot 2$ **40.** $3 \cdot 6j \cdot 2$

41. $(-1)(-2e)(-4)$ **42.** $(-1)(-5t)(-1)$

Use the distributive property to remove parentheses.

43. $4(x + 1)$ **44.** $5(y + 3)$

45. $4(4 - x)$ **46.** $5(7 + k)$

47. $-2(3e + 3)$ **48.** $-5(7t + 2)$

49. $-8(2q - 6)$ **50.** $-5(3p - 8)$

51. $-4(-3 - 5s)$ **52.** $-6(-1 - 3d)$

53. $(7 + 4d)6$ **54.** $(8r + 2)7$

55. $(5r - 6)(-5)$ **56.** $(3z - 7)(-8)$

57. $(-4 - 3d)6$ **58.** $(-4 - 2j)5$

59. $3(3x - 7y + 2)$ **60.** $5(4 - 5r + 8s)$

61. $-3(-3z - 3x - 5y)$ **62.** $-10(5e + 4a + 6t)$

Write each expression without using parentheses.

63. $-(x + 3)$ **64.** $-(5 + y)$

65. $-(4t + 5)$ **66.** $-(8x + 4)$

67. $-(-3w - 4)$ **68.** $-(-6 - 4y)$

69. $-(5x - 4y + 1)$ **70.** $-(6r - 5f + 1)$

Each expression is the result of an application of the distributive property. What was the original algebraic expression?

71. $2(4x) + 2(5)$ **72.** $3(3y) + 3(7)$

73. $-4(5) - 3x(5)$ **74.** $-8(7) - (4s)(7)$

75. $-3(4y) - (-3)(2)$ **76.** $-5(11s) - (-5)(11t)$

77. $3(4) - 3(7t) - 3(5s)$ **78.** $2(7y) + 2(8x) - 2(4)$

▌ WRITING

79. Explain what it means to simplify an algebraic expression. Give an example.

80. Explain the commutative and associative properties of multiplication.

81. Explain how to apply the distributive property.

82. Explain why the distributive property applies to $2(3 + x)$ but does not apply to $2(3x)$.

▌ REVIEW

83. Simplify: $|-6 + 1|$. **84.** Simplify: $-1 - (-4)$.

85. Identify the operation associated with each word: *product*, *quotient*, *difference*, and *sum*.

86. What are the steps used to find the mean (average) of a set of scores?

87. Insert the proper inequality symbol, $>$ or $<$:

$$-6 \quad \blacksquare \quad -7$$

88. Fill in the blank: To factor a number means to express it as the _____ of other whole numbers.

89. Which of the following involve area: carpeting a room, fencing a yard, walking around a lake, painting a wall?

90. Write seven squared and seven cubed.

8.5 Combining Like Terms

- Terms of an algebraic expression • Coefficients of a term • Terms and factors
- Like terms • Combining like terms • Perimeter formulas

In this section, we show how the distributive property can be used to simplify algebraic expressions that involve addition and subtraction. We also review the concept of perimeter and write the formulas for the perimeters of a rectangle and a square using variables.

▌ Terms of an algebraic expression

Addition signs break algebraic expressions into smaller parts called **terms.** The expression $3x + 8$ contains two terms: $3x$ and 8.

The addition sign breaks the expression into two terms.

> **Terms**
>
> A **term** is a number or a product of a number and one or more variables.

Some examples of terms are

$$6b^2, \quad -15x, \quad -4ac, \quad x, \quad -y, \quad 12$$

If an algebraic expression involves subtraction, the subtraction can be expressed as addition of the opposite. For example, $5x - 6$ can be written in the equivalent form $5x + (-6)$. We then see that $5x - 6$ contains two terms, $5x$ and -6.

Self Check 1

List the terms of each expression:

a. $-12y^2 + y + 10$

b. $-4ab$

c. $9 - m - 6m + 12$

Answers a. $-12y^2$, y, 10,
b. $-4ab$,   c. 9, $-m$, $-6m$, 12

EXAMPLE 1   List the terms of each expression: **a.** $-3x^2 + 5x + 8$, **b.** $-24rs$,   and   **c.** $x - 5 - 3x + 10$.

Solution

a. $-3x^2 + 5x + 8$ contains three terms: $-3x^2$, $5x$, and 8.

b. $-24rs$ is one term.

c. $x - 5 - 3x + 10$ can be written as $x + (-5) + (-3x) + 10$. It contains four terms: x, -5, $-3x$, and 10.

▌ Coefficients of a term

A term of an algebraic expression can consist of a single number (called a **constant**), a single variable, or a product of numbers and variables.

> **Numerical coefficients**
>
> In a term that is the product of a number and one or more variables, the numerical factor is called the **numerical coefficient** (or simply the **coefficient**) of the term. The coefficient of a constant term is the constant itself.

In the term $3x$, 3 is the *coefficient* and x is the *variable* part. Some more examples are shown in Table 8-2.

Term	Coefficient	Variable part
$6b^2$	6	b^2
$-15x$	-15	x
$-4ac$	-4	ac
x	1	x
$-y$	-1	y
25 (constant)	25	none

TABLE 8-2

Notice that when there is no number in front of a variable, the coefficient is understood to be 1. For example, the coefficient of the term x is 1. If there is only a negative (or opposite) sign in front of the variable, the coefficient is understood to be -1. Therefore, $-y$ can be thought of as $-1y$.

Self Check 2

Identify the coefficient of each term in the expression $6y^3 - y + 7$.

EXAMPLE 2   Identify the coefficient of each term in the expression $-5x^2 + x - 15$.

Solution   First, we write the subtraction as addition of the opposite.

$$-5x^2 + x - 15 = -5x^2 + x + (-15)$$

We see that the expression has three terms.

Term	Coefficient	
$-5x^2$	-5	
x	1	
-15	-15	We say that -15 is a constant term.

Answers $6, -1, 7$

■ Terms and factors

It is important to be able to distinguish between the *terms* of an algebraic expression and the *factors* of a term. Terms are separated by an addition sign $+$, and factors are numbers or variables that are multiplied together.

Consider the expression $-2x + 3xy$, which contains two terms.

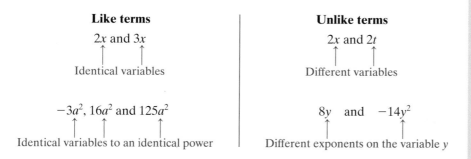

The first term   The second term
$$-2x + 3xy$$

The first term contains two factors: -2 and x.

The second term contains three factors: $3, x,$ and y.

EXAMPLE 3 Determine whether y is a factor or a term of each expression:
a. $8 + y$ and **b.** $8y$.

Solution

a. Since y is added to 8, y is a term of $8 + y$.

b. Since y is multiplied by 8, y is a factor of $8y$.

Self Check 3
Determine whether x is a factor or a term of each expression:
a. $-5x$ and **b.** $x + 5$.

Answers **a.** factor, **b.** term

■ Like terms

The expression $5t - 6t + 10$ contains three terms. The variable parts of two of the terms, $5t$ and $-6t$, are identical. We say those two terms are **like** or **similar terms.**

Like terms (similar terms)

Like terms, or **similar terms,** are terms with exactly the same variables and exponents. Any constants in an expression are considered like terms.

Like terms	**Unlike terms**
$2x$ and $3x$	$2x$ and $2t$
Identical variables	Different variables
$-3a^2, 16a^2$ and $125a^2$	$8y$ and $-14y^2$
Identical variables to an identical power	Different exponents on the variable y

! COMMENT When looking for like terms, don't look at the coefficients of the terms. Consider only their variable parts.

Self Check 4

List the like terms in each expression:

a. $10c + 8 + 7$

b. $8 + 8x + x^2 + x^3$

c. $4r + 8t - 3r + t$

EXAMPLE 4   List the like terms in each of the following expressions: **a.** $5a + 6 + 3a$,   **b.** $3x^2 - 3x^5 + 2$,   and   **c.** $-5x^2 + 3 - 1 + x^2$.

Solution

a. In $5a + 6 + 3a$, the terms $5a$ and $3a$ have the same variable and the same exponent. They are like terms.

b. $3x^2 - 3x^5 + 2$ does not contain any like terms.

c. $-5x^2 + 3 - 1 + x^2$ contains two pairs of like terms. $-5x^2$ and x^2 are like terms, because the variable and the exponent are the same. The numbers 3 and -1 are like terms.

Answers **a.** 8 and 7,   **b.** none, **c.** $4r$ and $-3r$, $8t$ and t

▪ Combining like terms

If we are to add or subtract objects, they must be similar. For example, fractions that are to be added must have a common denominator. When adding decimals, we align columns to be sure that we add tenths to tenths, hundredths to hundredths, and so on. The same is true when we work with terms of an algebraic expression. They can be added or subtracted only if they are like terms.

The following expression cannot be simplified, because its terms are unlike.

$$3x + 4y$$
↑   ↑

Unlike terms
The variable parts are not identical.

The following expression can be simplified, because it contains like terms.

$$3x + 4x$$
↑   ↑

Like terms
The variable parts are identical.

To simplify an expression containing like terms, we use the distributive property. For example, we can simplify $3x + 4x$ as follows.

$3x + 4x = (3 + 4)x$ Distribute the factor x to 3 and 4.

        $= 7x$ Perform the addition within the parentheses: $3 + 4 = 7$.

We say that we have *simplified* the expression $3x + 4x$. The result is the equivalent expression $7x$. Simplifying the sum (or difference) of like terms is called **combining like terms.**

Self Check 5

Simplify each expression by combining like terms:

a. $-5b + 10b$

b. $12c - 9c$

EXAMPLE 5   Simplify each expression by combining like terms: **a.** $-3x + 7x$ and   **b.** $6y - 4y$.

Solution

a. $-3x + 7x = (-3 + 7)x$ Use the distributive property. The factor x was distributed to -3 and 7.

          $= 4x$ Perform the addition within the parentheses: $-3 + 7 = 4$.

b. $6y - 4y = (6 - 4)y$ Use the distributive property. The factor y was distributed.

Answers **a.** $5b$,   **b.** $3c$

          $= 2y$ Perform the subtraction within the parentheses: $6 - 4 = 2$.

The results of Example 5 suggest the following rule.

> **Combining like terms**
>
> To add or subtract like terms, combine their coefficients and keep the same variables with the same exponents.

EXAMPLE 6   Simplify by combining like terms: **a.** $-7x + (-9x)$ and **b.** $6m + 3m$.

Solution

a. $-7x + (-9x) = -16x$ Add the coefficients of the like terms: $-7 + (-9) = -16$.

b. $6m + 3m = 9m$ Add the coefficients: $6 + 3 = 9$. Keep the variable m.

Self Check 6
Simplify by combining like terms:
a. $-5n + (-2n)$
b. $15r + 4r$

Answers **a.** $-7n$, **b.** $19r$

EXAMPLE 7   Simplify by combining like terms: **a.** $5n - 3n$, **b.** $16h - 24h$, and **c.** $-4d - (-5d)$.

Solution

a. $5n - 3n = 2n$ Subtract: $5 - 3 = 2$. Keep the variable n.

b. $16h - 24h = -8h$ Subtract: $16 - 24 = 16 + (-24) = -8$. Keep the variable h.

c. $-4d - (-5d) = -4d + 5d$ Add the opposite of $-5d$.

$\qquad\qquad\quad = 1d$ Add: $-4 + 5 = 1$. Keep the variable d.

$\qquad\qquad\quad = d$ $1d = d$.

Self Check 7
Simplify by combining like terms:
a. $25c - 5c$
b. $9w - 15w$
c. $-7p - (-6p)$

Answers **a.** $20c$, **b.** $-6w$, **c.** $-p$

❗ COMMENT Expressions that involve subtraction from 0 are often incorrectly simplified. For example, $0 - 6x \neq 6x$. To simplify $0 - 6x$, we can use the fact that subtraction is the same as addition of the opposite.

$0 - 6x = 0 + (-6x)$ Add the opposite of $6x$, which is $-6x$.

$\qquad\quad = -6x$ When we add 0 to any number, the number remains the same.

EXAMPLE 8   Simplify: $8s - 8S - 5s + S$.

Solution   The lowercase s and the capital S are different variables. We rearrange the terms so that like terms are next to each other.

$8s - 8S - 5s + S = 8s - 5s - 8S + S$ Use the commutative property of addition to get the like terms together.

$\qquad\qquad\qquad\quad = 3s - 7S$ Combine like terms: $8 - 5 = 3$ and keep s, $-8 + 1 = -7$ and keep S.

EXAMPLE 9   Simplify: $4(x + 3) - 2(x - 1)$.

Solution

$4(x + 3) - 2(x - 1) = 4x + 12 - 2x + 2$ Use the distributive property twice.

$\qquad\qquad\qquad\quad = 4x - 2x + 12 + 2$ Use the commutative property of addition to get like terms together.

$\qquad\qquad\qquad\quad = 2x + 14$ Combine like terms: $4x - 2x = 2x$ and $12 + 2 = 14$.

Self Check 9
Simplify: $6(y + 4) - 5(y - 6)$.

Answer $y + 54$

The expressions in Examples 8 and 9 contained two sets of like terms. In each solution, we rearranged terms so that like terms were next to each other. However, with practice you will be able to combine like terms without having to write them next to each other.

Self Check 10
Simplify: $-6(3y - 2) + 4y - 1$.

EXAMPLE 10   Simplify: $-5(2x - 2) + 8x - 6$.

Solution

$$-5(2x - 2) + 8x - 6 = -10x + 10 + 8x - 6 \qquad \text{Distribute } -10.$$
$$= -2x + 4 \qquad\qquad \text{Combine like terms:}$$
$$\text{} \qquad\qquad\qquad\qquad\qquad -10x + 8x = -2x \text{ and } 10 - 6 = 4.$$

Answer $-14y + 11$

■ Perimeter formulas

To develop the formula for the perimeter of a rectangle, we let l = the length of the rectangle and w = the width of the rectangle. (See Figure 8-10.) Then

$$P = l + w + l + w \qquad \text{The perimeter is the distance around the rectangle.}$$
$$= 2l + 2w \qquad\qquad \text{Combine like terms: } l + l = 2l \text{ and } w + w = 2w.$$

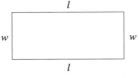

FIGURE 8-10

The perimeter of a rectangle

The perimeter P of a rectangle with length l and width w is given by

$$P = 2l + 2w$$

To develop the formula for the perimeter of a square, we let s = the length of a side of the square. (See Figure 8-11.) Then

$$P = s + s + s + s \qquad \text{Add the lengths of the four sides.}$$
$$= 4s \qquad\qquad\qquad \text{Combine like terms. Recall that } s = 1s.$$

FIGURE 8-11

The perimeter of a square

The perimeter P of a square with sides of length s is given by

$$P = 4s$$

EXAMPLE 11 Energy conservation.

See Figure 8-12. Find the cost to weatherstrip the front door and window of the house if the material costs 20¢ a foot.

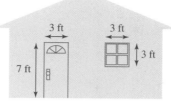

FIGURE 8-12

Analyze the problem

To find the cost of the weatherstripping, we must find the perimeters of the door and the window. The door is in the shape of a rectangle, and the window is in the shape of a square.

Form an equation

Let P = the total perimeter, and translate the words of the problem into an equation.

The total perimeter	is	the perimeter of the door	plus	the perimeter of the window.
P	$=$	$2l + 2w$	$+$	$4s$

Write the formulas for the perimeter of a rectangle and a square.

Solve the equation

$$P = 2l + 2w + 4s$$
$$= 2(7) + 2(3) + 4(3)$$ Substitute 7 for l, 3 for w, and 3 for s.
$$= 14 + 6 + 12$$ Perform the multiplications.
$$= 32$$ Perform the additions.

State the conclusion

The total perimeter is 32 feet. At 20¢ a foot, the total cost will be $(32 \cdot 20)$¢. This is 640¢, or $6.40.

Check the result

We can check the results by estimation. The perimeter is approximately 30 feet, and $30 \cdot 20 = 600$¢, which is $6. The answer, $6.40, seems reasonable.

Section 8.5 STUDY SET

VOCABULARY   *Fill in the blanks.*

1. A ~~term~~ is a number or a product of a number and one or more variables.

2. In the term $5t$, 5 is called the ~~coefficient~~ and t is called the ~~variable~~ part.

3. The ~~perimeter~~ of a geometric figure is the distance around it.

4. When we write $9x + x$ as $10x$, we say we have ~~combined~~ like terms.

5. $2(x + 3) = 2x + 2(3)$ is an example of the use of the ~~distributive~~ property.

6. Terms with exactly the same variables and exponents are called ~~like~~ terms.

7. Simplifying the _sum_ (or difference) of like terms is called combining like terms.

8. The numbers multiplied together to form a product are called _factors._

CONCEPTS

9. Determine whether x is used as a factor or as a term.

  a. $12 + x$

  b. $7x$

  c. $12y + 12x - 6$

  d. $-36xy$

10. Determine whether $6y$ is used as a factor or as a term.

  a. $6yz$ **b.** $10 + 6y$

  c. $9xy + 6y$ **d.** $6y - 18$

11. What is the coefficient of each term?

  a. $11x$ **b.** $8t$

  c. $-4x^2$ **d.** a

  e. $-x$ **f.** $102xy$

12. What is the coefficient of the second term of each expression?

  a. $5x^2 + 6x + 7$

  b. $xy - x + y + 10$

  c. $9y^2 + y + 8$

  d. $5x^3 - 4x^2 + 3x + 1$

13. Complete the table.

Term	Coefficient	Variable part
$6m$		
$-75t$		
w		
$4bh$		

14. Simplify each pair of expressions, if possible.

  a. $5(2x)$ and $5 + 2x$

  b. $6(-7x)$ and $6 - 7x$

  c. $2(3x)(3)$ and $2 + 3x + 3$

  d. $x \cdot x$ and $x + x$

15. When simplifying an algebraic expression, some students use underlining.

$$\underline{3y} + \underline{4} + \underline{5y} + \underline{8}$$

What purpose does the underlining serve?

16. Determine whether each statement is true or false.

  a. $x = 1x$ **b.** $2x + 0 = 2x$

  c. $-y = -1y$ **d.** $0 - 4c = 4c$

17. The illustration shows the distance (in miles) that two men live from the office. Find the total distance the men travel from home to office.

18. The heights of two trees are shown. Find the sum of their heights.

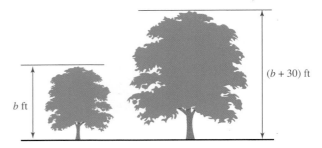

19. **a.** What does the diagram illustrate?

$$9x + 5x = 14x$$

  b. What does the diagram illustrate?

$$12k - 4k = 8k$$

20. For each expression, identify any like terms.

  a. $3a + 8 + 2a$ **b.** $10 - 13h + 12$

  c. $3x^2 + 3x + 3$ **d.** $9y^2 - 9m - 8y^2$

NOTATION   _Complete each solution._

21. $5x + 7x = (5 + \boxed{})x$

           $= 12x$

22. $12w - 16w = (\boxed{} - 16)w$

              $= -4w$

23. $2(x - 1) + 3x = 2x - \boxed{} + 3x$

               $= \boxed{} - 2$

24. $-3(1 - b) - b = \boxed{} + \boxed{} - b$

              $= -3 + \boxed{}$

25. In the formula $P = 2l + 2w$,

  a. what does P represent?

  b. what does $2l$ mean?

  c. what does $2w$ mean?

26. In the formula $P = 4s$,

  a. what does P represent?

  b. what does $4s$ mean?

PRACTICE *Identify the terms of each expression.*

27. $3x^2 + 5x + 4$

28. $y^2 + 12y + 6$

29. $5 + 5t - 8t + 4$

30. $3x - y - 5x + y$

What exponent must appear in each box to make the terms like terms?

31. $3x^{\square}, -6x^2$

32. $7a^3, 21a^{\square}$

33. $-8h^5, -5h^{\square}$

34. $25n^4, -15n^{\square}$

Simplify by combining like terms, if possible.

35. $6t + 9t$

36. $7r + 5r$

37. $5s - s$

38. $8y - y$

39. $-5x + 6x$

40. $-8m + 6m$

41. $-5d + 9d$

42. $-4a + 12a$

43. $3e - 7e$

44. $2s - 4s$

45. $h - 7$

46. $j - 8$

47. $4z - 10z$

48. $3w - 18w$

49. $-3x - 4x$

50. $-7y - 9y$

51. $2t - 2t$

52. $7r - 7r$

53. $-6s + 6s$

54. $19c + (-19c)$

55. $x + x + x + x$

56. $s - s - s$

57. $2x + 2y$

58. $5a - 5b$

59. $0 - 2y$

60. $0 - 7x$

61. $3a - 0$

62. $10t + 0$

63. $6t + 9 + 5t + 3$

64. $5x + 3 + 5x + 4$

65. $3w - 4 - w - 1$

66. $6y + 6 - y - 1$

67. $-4r + 8R + 2R - 3r + R$

68. $12a - A - a - 8A - a$

69. $-45d - 12a - 5d + 12a$

70. $-m - n - 8m + n$

71. $4x - 3y - 7 + 4x - 2 - y$

72. $2a + 8 - b - 5 + 5a - 9b$

Simplify each expression.

73. $4(x + 1) + 5(6 + x)$

74. $7(1 + y) + 8(2y + 3)$

75. $5(3 - 2s) + 4(2 - 3s)$

76. $6(t - 3) + 9(2 - t)$

77. $-4(6 - 4e) + 3(e + 1)$

78. $-5(7 - 4t) + 3(2 + 5t)$

79. $3t - (t - 8)$

80. $6n - (4n + 1)$

81. $-2(2 - 3x) - 3(x - 4)$

82. $-3(1 - y) - 5(2y - 6)$

83. $-4(-4y + 5) - 6(y + 2)$

84. $-3(-6y - 8) - 4(5 - y)$

APPLICATIONS

85. MOBILE HOME DESIGNS The design of a mobile home calls for a 6-inch-wide strip of stained pine around the outside of each exterior wall, as shown in brown. If the pine strip costs 80¢ a linear foot, how much will be spent on the pine used for the trim?

Pine strip

10 ft

60 ft

10 ft

86. LANDSCAPE DESIGNS A landscape architect has designed a planter surrounding two birch trees, as shown. The planter is to be outlined with redwood edging in the shape of a rectangle and two squares. If the material costs 17¢ a running foot, how much will the redwood cost for this project?

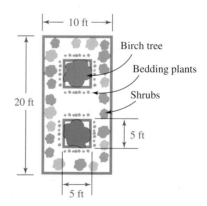

10 ft

Birch tree

Bedding plants

Shrubs

20 ft

5 ft

5 ft

87. PARTY PREPARATIONS The appropriate size of a dance floor for a given number of dancers can be determined from the table on the next page. Find the perimeter of each of the dance floors listed.

Slow dancers	Fast dancers	Size of floor (in feet)
8	5	9×9
14	9	12×12
22	15	15×15
32	20	18×18
50	30	21×21

88. COASTAL DRILLING The following map shows an area of the California coast where oil drilling is planned. Use the scale to estimate the lengths of the sides of the area highlighted on the map. Then find its perimeter.

▌ WRITING

89. Explain what it means for two terms to be like terms.

90. Explain what it means to say that the coefficient of x is an understood 1.

91. Explain the difference between a term and a factor. Give some examples.

92. The formula for the perimeter of a rectangle is $P = 2l + 2w$. Explain why we can write this formula in the form $P = 2(l + w)$.

▌ REVIEW

93. Solve: $-4t = -8$.

94. Evaluate: $(-1)(-1)(-1)$.

95. Find the prime factorization of 100.

96. Write $3 \cdot 3 \cdot 3 \cdot 3 \cdot 3$ as an exponential expression.

97. Fill in the blanks to make a true statement. The _____ _____ of a number is the distance between it and 0 on the number line.

98. State the division property of equality in words.

8.6 Simplifying Expressions to Solve Equations

- Checking solutions • Solving equations • Variables on both sides of an equation
- Removing parentheses • A strategy for solving equations

We must often simplify algebraic expressions to solve equations. Sometimes it is necessary to combine like terms in order to isolate the variable on one side of the equation. At other times, it is necessary to apply the distributive property to write an equation in a form that can be solved. In this section, we discuss both situations.

▌ Checking solutions

To *solve an equation* means to find all values of the variable that make the equation a true statement.

Self Check 1

Is -4 a solution of $-4w - 6 = -2 - 3w$?

EXAMPLE 1   Determine whether -3 is a solution of $5x - 1 = 6x + 8$.

Solution   If -3 is a solution, we will obtain a true statement when -3 is substituted for x.

$$5x - 1 = 6x + 8 \qquad 5x \text{ means } 5 \cdot x, 6x \text{ means } 6 \cdot x.$$

$$5(-3) - 1 \stackrel{?}{=} 6(-3) + 8 \qquad \text{Substitute } -3 \text{ for } x.$$

$$-15 - 1 \stackrel{?}{=} -18 + 8 \qquad \text{Perform the multiplications.}$$

$$-16 \neq -10 \qquad \text{Simplify each side.}$$

Since $-16 \neq -10$, we conclude that -3 is not a solution of $5x - 1 = 6x + 8$.

Answer yes

■ Solving equations

Sometimes several properties of equality must be applied one after another to solve an equation.

EXAMPLE 2 Solve: $-12x + 5 = 17$. Check the result.

Self Check 2
Solve: $-3a - 15 = 30$.

Solution On the left-hand side of the equation, x is multiplied by -12 and then 5 is added to that product. To isolate x, we use the rules for the order of operations *in reverse.*

- To undo the addition of 5, we subtract 5 from both sides.
- To undo the multiplication by -12, we divide both sides by -12.

$$-12x + 5 = 17$$

$$-12x + 5 - 5 = 17 - 5 \qquad \text{Subtract 5 from both sides.}$$

$$-12x = 12 \qquad \text{Perform the subtractions: } 5 - 5 = 0 \text{ and } 17 - 5 = 12.$$

$$\frac{-12x}{-12} = \frac{12}{-12} \qquad \text{Divide both sides by } -12.$$

$$x = -1 \qquad \text{Perform the division } \frac{12}{-12} = -1.$$

Verify that -1 is the solution by performing a check.

Answer -15

EXAMPLE 3 Solve: $100 = t - 20 + t$.

Self Check 3
Solve: $155 = d - 1 + d$.

Solution

$$100 = t - 20 + t$$

$$100 = 2t - 20 \qquad \text{Combine like terms: } t + t = 2t.$$

$$100 + 20 = 2t - 20 + 20 \qquad \text{To undo the subtraction of 20, add 20 to both sides.}$$

$$120 = 2t \qquad \text{Perform the additions.}$$

$$\frac{120}{2} = \frac{2t}{2} \qquad \text{To undo the multiplication by 2, divide both sides by 2.}$$

$$60 = t \qquad \text{Perform the divisions.}$$

$$t = 60 \qquad \text{Interchange the sides of the equation.}$$

Verify that 60 satisfies the original equation.

Answer 78

Marriage

"Adults are marrying later today than they did in the past. The postponement of marriage has led to a substantial increase in the number of young, never-married adults." Jason Fields, U.S. Census Bureau, 2000

1. Consider the following problem:

> In 1950, the U.S. median age at first marriage for men was 2.5 years more than that for women. The sum of the median ages of the bride and groom was 43.1 years. What was the median age at first marriage for men and for women in 1950?

If we let x = the median age at first marriage for women in 1950, then $x + 2.5$ = the median age at first marriage for men in 1950. Write an equation showing that the sum of the median ages was 43.1 and then solve it to answer the problem.

2. Use a similar approach to solve the following problem:

> In 2002, the U.S. median age at first marriage for men was 1.6 years more than that for women. The sum of the median age of the bride and groom was 52.2 years. What was the median age at first marriage for men and for women in 2002?

▮ Variables on both sides of an equation

When solving an equation, we want to isolate the variable on one side of the equation. If variables appear on both sides, we can use the addition (or subtraction) property of equality to get all variable terms on one side and all constant terms on the other.

Self Check 4
Solve: $9B = 3B - 18$.

EXAMPLE 4   Solve: $8y = 2y + 12$.

Solution   There are variable terms (highlighted in blue) on both sides of the equation. To isolate y on the left-hand side of the equation, we use the subtraction property of equality to eliminate $2y$ on the right-hand side.

$$8y = 2y + 12$$

$8y - 2y = 2y + 12 - 2y$ To eliminate $2y$ from the right-hand side, subtract $2y$ from both sides.

$6y = 12$ Combine like terms: $8y - 2y = 6y$ and $2y - 2y = 0$.

$\dfrac{6y}{6} = \dfrac{12}{6}$ To undo the multiplication by 6, divide both sides by 6.

$y = 2$ Perform the divisions.

Answer -3

Verify that 2 satisfies the equation.

❗ COMMENT   When we solve equations, it doesn't matter whether the variable is isolated on the right or the left side of the equation. In Example 4, we could have isolated y on the right-hand side, but this approach would have involved more steps.

Self Check 5
Solve: $72 - 13d = 12d - 3$.

EXAMPLE 5   Solve: $-9 - 6t = 5t + 2$.

Solution   There are variable terms (highlighted in blue) on both sides of the equation. Either we can subtract $5t$ from both sides to isolate t on the left side, or we can add $6t$ to both sides to isolate t on the right side. It appears that the computations will be easier if we add $6t$ to both sides.

$$-9 - 6t = 5t + 2$$

$-9 - 6t + 6t = 5t + 2 + 6t$ To eliminate $-6t$ from the left-hand side, add $6t$ to both sides.

$-9 = 11t + 2$ Combine like terms: $-6t + 6t = 0$ and $5t + 6t = 11t$.

$-9 - 2 = 11t + 2 - 2$ To undo the addition of 2, subtract 2 from both sides.

$-11 = 11t$ Perform the subtractions: $-9 - 2 = -11$ and $2 - 2 = 0$.

$\dfrac{-11}{11} = \dfrac{11t}{11}$ To undo the multiplication by 11, divide both sides by 11.

$-1 = t$ Perform the divisions.

$t = -1$ Interchange the sides of the equation.

Verify that -1 satisfies the equation.

Answer 3

◼ Removing parentheses

At times, we must use the distributive property to solve an equation.

EXAMPLE 6 Solve: $3(x + 15) = 45$.

Solution

$$3(x + 15) = 45$$

$3x + 3(15) = 45$ Distribute the multiplication by 3.

$3x + 45 = 45$ Perform the multiplication.

$3x + 45 - 45 = 45 - 45$ To undo the addition of 45, subtract 45 from both sides.

$3x = 0$ Perform the subtractions: $45 - 45 = 0$.

$\dfrac{3x}{3} = \dfrac{0}{3}$ To undo the multiplication by 3, divide both sides by 3.

$x = 0$ Perform the divisions.

Verify that 0 satisfies the equation.

Self Check 6
Solve: $7(t + 5) = -70$.

Answer -15

◼ A strategy for solving equations

To summarize, when solving an equation, we must isolate the variable on one side of the = symbol. At times, this requires that we remove the parentheses and/or combine like terms. The following steps should be applied, in order, when solving an equation.

> **Strategy for solving equations**
> 1. Use the distributive property to remove any parentheses.
> 2. Combine like terms on either side of the equation.
> 3. Apply the addition or subtraction properties of equality to get the variables on one side of the = symbol and the constant terms on the other.
> 4. Continue to combine like terms when possible.
> 5. Undo the operations of multiplication and division to isolate the variable.
> 6. Check the result.

You won't always have to use all six steps to solve a given equation. If a step doesn't apply, skip it and go to the next step.

Self Check 7
Solve: $4x = -13 - (3x + 8)$.

EXAMPLE 7   Solve: $2x = 2 - (4x + 14)$.

Solution

$2x = 2 - (4x + 14)$	
$2x = 2 - \mathbf{1}(4x + 14)$	Rewrite: $2 - (4x + 14) = 2 - 1(4x + 14)$.
$2x = 2 - 4x - 14$	Use the distributive property to remove parentheses: $-1(4x + 14) = -4x - 14$.
$2x = -12 - 4x$	Combine like terms: $2 - 14 = -12$.
$2x + \mathbf{4x} = -12 - 4x + \mathbf{4x}$	To eliminate $-4x$ from the right-hand side, add $4x$ to both sides.
$6x = -12$	Combine like terms: $2x + 4x = 6x$ and $-4x + 4x = 0$.
$\dfrac{6x}{6} = \dfrac{-12}{6}$	To undo the multiplication by 6, divide both sides by 6.
$x = -2$	Perform the divisions.

Answer -3

Verify that -2 satisfies the equation.

Section 8.6 STUDY SET

VOCABULARY   *Fill in the blanks.*

1. To __solve__ an equation means to find all values of the variable that make the equation a true statement.

2. To __check__ a solution means to substitute that value into the original equation to see whether a true statement results.

3. In $2(x + 4)$, to remove parentheses means to apply the __distributive__ property.

4. Algebraic expressions are simplified, and __equations__ are solved.

5. The phrase "__combined__ like terms" refers to the operations of addition and subtraction.

6. A __variable__ is a letter that represents a number. A __constant__ is a number that is fixed and does not change in value.

CONCEPTS

7. Explain why -5 is not a solution of $5x - 3x = -9$.

8. **a.** For each equation, circle the terms involving variables:

   $5x + 3x = 8$ $5t = 3t + 8$ $7 = 5h + 3h - 1$

   b. Which equation has variables on both sides?

9. To solve $6k = 5k - 18$, we need to eliminate $5k$ from the right-hand side. To do this, what should we subtract from both sides?

10. Determine the first step in solving each equation.

    a. $2x + 4x = 36$

    b. $6x = x + 10$

    c. $5(x + 1) = 15$

    d. $50 = x + 4 + x$

11. Consider the equation $2x - 8 = 4x - 14$.

    a. To solve this equation by isolating x on the left side, what should we subtract from both sides?

    b. To solve this equation by isolating x on the right side, what should we subtract from both sides?

12. Fill in the blanks.

    $6 - (d - 4) = 8$
    $6 - \blacksquare(d - 4) = 8$
    $6 - \blacksquare + \blacksquare = 8$

13. **a.** Simplify: $3t - t - 8$.

    b. Solve: $3t - t = -8$.

    c. Evaluate $3t - t - 8$ for $t = -4$.

14. a. Evaluate $2(x + 1)$ for $x = -4$.

 b. Simplify: $2(x + 1) - 4$.

 c. Solve: $2(x + 1) = -4$.

NOTATION *Complete each solution to solve the equation.*

15. $4x - 2x = -20$

$$\boxed{} = -20$$

$$\frac{2x}{\boxed{}} = \frac{-20}{\boxed{}}$$

$$x = \boxed{}$$

16. $8y - 6 = -2 + 10y$

$$8y - 6 - \boxed{} = -2 + 10y - \boxed{}$$

$$-6 = -2 + \boxed{}$$

$$-6 + \boxed{} = -2 + 2y + \boxed{}$$

$$-4 = \boxed{}$$

$$\frac{-4}{\boxed{}} = \frac{2y}{\boxed{}}$$

$$-2 = y$$

$$y = \boxed{}$$

17. $5(x - 9) = 5$

$$5x - 5() = 5$$

$$5x - \boxed{} = 5$$

$$5x - 45 + \boxed{} = 5 + 45$$

$$\boxed{} = 50$$

$$\frac{5x}{\boxed{}} = \frac{50}{\boxed{}}$$

$$x = \boxed{}$$

18. $-2(-1 - x) = 16$

$$2 + \boxed{} = 16$$

$$2 + 2x - \boxed{} = 16 - \boxed{}$$

$$2x = \boxed{}$$

$$\frac{2x}{\boxed{}} = \frac{\boxed{}}{\boxed{}}$$

$$x = 7$$

PRACTICE *For each equation, determine whether the given number is a solution.*

19. $5f + 8 = 4f + 11$; 3

20. $3r + 8 = 5r - 2$; 5

21. $2(x - 1) = 33$; 12

22. $-6(x + 4) = -40$; 8

Solve each equation.

23. $7x - 6 = 8$

24. $4x - 2 = 26$

25. $-2x + 3 = 31$

26. $-6x + 7 = 49$

27. $60 = 3v - 5v$

28. $28 = x - 3x$

29. $-28 = -m + 2m$

30. $-120 = -p + 4p$

31. $x + x + 6 = 90$

32. $c + c + 1 = 51$

33. $T + T - 17 = 57$

34. $r + r - 15 = 95$

35. $600 = m - 12 + m$

36. $403 = x - 3 + x$

37. $1,500 = b + 30 + b$

38. $8,000 = h + 100 + h$

39. $7x = 3x + 8$

40. $4x = 2x + 14$

41. $x - 14 = 2x$

42. $2x - 7 = 3x$

43. $9t - 40 = 14t$

44. $5r - 24 = 8r$

45. $25 + 4j = 9j$

46. $36 + 5j = 9j$

47. $-48 + 12t = 16t$

48. $-28 + 7t = 21t$

49. $-5g - 40 = -15g$

50. $-20s - 20 = -40s$

51. $3s + 1 = 4s - 7$

52. $6v + 2 = 7v - 3$

53. $50a - 1 = 60a - 101$

54. $25y - 2 = 75y - 202$

55. $-7 + 5r = 83 - 10r$

56. $-20 + t = 44 - 7t$

57. $100 - y = 100 + y$

58. $-60 + z = -60 - z$

59. $2(x + 6) = 4$

60. $9(y - 1) = 27$

61. $-16 = 2(t + 2)$

62. $-10 = 5(y - 7)$

63. $-3(2w - 3) = 9$

64. $-4(5t + 2) = -8$

65. $-(c - 4) = 3$

66. $-(6 - 2x) = -8$

67. $4(p - 2) = 0$

68. $10(4s - 4) = 0$

69. $2(4y + 8) = 3(2y - 2)$

70. $3(7 - y) = 3(2y + 1)$

71. $16 - (x + 3) = -13$

72. $10 - (w + 4) = -12$

73. $5 - (7 - y) = -5$

74. $10 - (x - 5) = 40$

75. $2x + 3(x - 4) = 23$

76. $5j + 6(j + 1) = 226$

77. $10q + 3(q - 7) = 18$

78. $2q + 6(q - 4) = 24$

WRITING

79. Explain the error in the work below.

$$2x = 4x - x$$
$$2x = 4$$
$$\frac{2x}{2} = \frac{4}{2}$$
$$x = 2$$

80. Consider $3x = 2x + 9$. Why is it necessary to eliminate one of the variable terms in order to solve for x?

81. What does it mean to *solve an equation*?

82. Explain how to determine whether a number is a solution of an equation.

REVIEW

83. Subtract: $-7 - 9$.

84. Which numbers are not factors of 28: 4, 6, 7, 8?

85. Evaluate: $\dfrac{-8 + 2}{-2 + 4}$.

86. Simplify: $\dfrac{\frac{3}{2} + 5}{3 - \frac{1}{2}}$.

87. Simplify: $-(-5)$.

88. Using x and y, illustrate the commutative property of addition.

89. What is the sign of the product of two negative integers?

90. Evaluate: $3 + 4[-4 - 3(-2)]$.

8.7 Exponents

- The product rule for exponents
- The power rule for exponents
- The power rule for products

In previous chapters, we have applied the commutative, associative, and distributive properties to simplify algebraic expressions. We now discuss how to simplify expressions that involve exponents.

The product rule for exponents

We have seen that exponents are used to represent repeated multiplication. For example, x^5 is an exponential expression with base x and an exponent of 5. It is called a *power of x*. Applying the definition of an exponent, we see that

$$x^5 = \underbrace{x \cdot x \cdot x \cdot x \cdot x}_{5 \text{ factors of } x}$$

Self Check 1

Write the repeated multiplication represented by each expression:

a. t^4

b. $(12f)^3$

c. $(-2x)^2$

EXAMPLE 1 Write the repeated multiplication represented by each expression: **a.** n^3, **b.** $(3h)^2$, and **c.** $(-6b)^4$.

Solution

a. For n^3, the base is n and the exponent is 3. Therefore,

$$n^3 = n \cdot n \cdot n$$

b. For $(3h)^2$, the base is $3h$ and the exponent is 2. Therefore,

$$(3h)^2 = 3h \cdot 3h$$

c. For $(-6b)^4$, the base is $-6b$ and the exponent is 4. Therefore,

$$(-6b)^4 = (-6b)(-6b)(-6b)(-6b)$$

The expression $x^3 \cdot x^5$ is the *product* of two powers of x. To develop a rule for multiplying them, we will use the fact that an exponent indicates repeated multiplication.

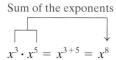

$x^3 \cdot x^5 = \underbrace{(x \cdot x \cdot x)}_{3 \text{ factors of } x}\underbrace{(x \cdot x \cdot x \cdot x \cdot x)}_{5 \text{ factors of } x}$ x^3 means to write x as a factor 3 times.
   x^5 means to write x as a factor 5 times.

$= \underbrace{(x \cdot x \cdot x \cdot x \cdot x \cdot x \cdot x \cdot x)}_{8 \text{ factors of } x}$ Perform the multiplication to get 8 factors of x.

$= x^8$ Since x is used as a factor 8 times, we can write the product as x^8.

Notice that the exponent of the result is the *sum* of the exponents in $x^3 \cdot x^5$.

Sum of the exponents

$x^3 \cdot x^5 = x^{3+5} = x^8$

This observation suggests the following rule.

The product rule for exponents

For any number x and any positive integers m and n,

$$x^m \cdot x^n = x^{m+n}$$

To multiply two exponential expressions with the same base, add the exponents and keep the common base.

EXAMPLE 2   Simplify each product: **a.** $3^4 \cdot 3^7$ **b.** $(y^2)(y^4)$,   and   **c.** $x^2 x^4 x^9$.

Solution

a. $3^4 \cdot 3^7 = 3^{4+7}$ Since the bases are the same, add the exponents and keep the common base, which is 3.

$= 3^{11}$ Perform the addition: $4 + 7 = 11$.

b. $(y^2)(y^4) = y^{2+4}$ Since the bases are the same, add the exponents and keep the common base, which is y.

$= y^6$ Perform the addition: $2 + 4 = 6$.

c. $x^2 x^4 x^9 = x^{2+4+9}$ Since the bases are the same, add the exponents and keep the common base, which is x.

$= x^{15}$ Perform the addition: $2 + 4 + 9 = 15$.

! COMMENT   We cannot use the product rule for exponents to simplify an expression such as $x^4 + x^3$, because it is not a product. Nor can we use it to simplify $x^4 \cdot y^3$, because the bases are not the same.

The product rule for exponents can be used to simplify more complicated algebraic expressions involving multiplication.

Self Check 3
Simplify each product:
a. $4m \cdot 6m^5$
b. $8r^3(-5r^2)$

EXAMPLE 3   Simplify each product: **a.** $3a(5a^2)$ and   **b.** $-2t^2 \cdot 6t^6$.

Solution

a. $3a(5a^2) = (3 \cdot 5)(a \cdot a^2)$ Apply the commutative and associative properties to change the order and regroup the factors.

$= (3 \cdot 5)(a^1 \cdot a^2)$ Recall that $a = a^1$.

$= 15a^{1+2}$ Perform the multiplication: $3 \cdot 5 = 15$. Then add the exponents and keep the common base, which is a.

$= 15a^3$ Perform the addition: $1 + 2 = 3$.

b. $-2t^2 \cdot 6t^6 = (-2 \cdot 6)(t^2 \cdot t^6)$ Change the order of the factors and regroup them.

$= -12t^{2+6}$ Perform the multiplication: $-2 \cdot 6 = -12$. Then add the exponents and keep the common base.

Answers a. $24m^6$, **b.** $-40r^5$

$= -12t^8$ Perform the addition: $2 + 6 = 8$.

Exponential expressions often contain more than one variable.

Self Check 4
Simplify the following:
a. $c^3d^2 \cdot cd^5$
b. $-7a^2b^3(8a^4b^5)$

EXAMPLE 4   Simplify the following: **a.** $n^2m \cdot n^8m^3$ and   **b.** $4xy^2(-3x^2y^3)$.

Solution

a. $n^2m \cdot n^8m^3 = (n^2 \cdot n^8)(m \cdot m^3)$ Change the order and group the factors with like bases.

$= n^{2+8} \cdot m^{1+3}$ Add the exponents of the like bases. Recall that $m = m^1$.

$= n^{10}m^4$ Perform the additions.

b. $4xy^2(-3x^2y^3) = [4(-3)](x \cdot x^2)(y^2 \cdot y^3)$ Group the factors with like bases.

$= -12 \cdot x^{1+2} \cdot y^{2+3}$ Perform the multiplication. Add the exponents of the like bases.

Answers a. c^4d^7, **b.** $-56a^6b^8$

$= -12x^3y^5$ Perform the additions.

The power rule for exponents

To develop the power rule for exponents, we consider the expression $(x^2)^5$. Notice that the base, x^2, is raised to a power. Therefore, we are working with a power of a power. We will again rely on the definition of an exponent to find a rule for simplifying this exponential expression.

$(x^2)^5 = x^2 \cdot x^2 \cdot x^2 \cdot x^2 \cdot x^2$ The exponent 5 tell us to write the base x^2 five times.

$= x^{2+2+2+2+2}$ Since the bases are alike, add the exponents and keep the common base.

$= x^{10}$ Perform the addition: $2 + 2 + 2 + 2 + 2 = 10$.

Notice that the exponent of the result is the *product* of the exponents in $(x^2)^5$.

Product of the exponents

$$(x^2)^5 = x^{2 \cdot 5} = x^{10}$$

This observation suggests the following rule.

> **The power rule for exponents**
>
> For any number x and any positive integers m and n,
>
> $$(x^m)^n = x^{m \cdot n} \qquad \text{or, more simply,} \qquad (x^m)^n = x^{mn}$$
>
> To raise an exponential expression to a power, keep the base and multiply the exponents.

EXAMPLE 5   Simplify each expression: **a.** $(2^3)^7$ and   **b.** $(b^5)^3$.

Solution

a. $(2^3)^7 = 2^{3 \cdot 7}$ Apply the power rule for exponents by keeping the base and multiplying the exponents.

$\qquad\quad = 2^{21}$ Perform the multiplication: $3 \cdot 7 = 21$.

b. $(b^5)^3 = b^{5 \cdot 3}$ Keep the base and multiply the exponents.

$\qquad\quad = b^{15}$ Perform the multiplication: $5 \cdot 3 = 15$.

Self Check 5
Simplify each expression:
a. $(4^2)^6$
b. $(y^6)^4$

Answers **a.** 4^{12},   **b.** y^{24}

In some cases, when we simplify algebraic expressions involving exponents, two rules of exponents must be applied.

EXAMPLE 6   Simplify: **a.** $(n^3)^4(n^2)^5$ and   **b.** $(n^2n^3)^5$.

Solution

a. $(n^3)^4(n^2)^5 = n^{3 \cdot 4} \cdot n^{2 \cdot 5}$ Keep each base and multiply their exponents.

$\qquad\qquad\quad = n^{12} \cdot n^{10}$ Perform the multiplications: $3 \cdot 4 = 12$ and $2 \cdot 5 = 10$.

$\qquad\qquad\quad = n^{12+10}$ Since the bases are alike, keep the base and add the exponents.

$\qquad\qquad\quad = n^{22}$ Perform the addition: $12 + 10 = 22$.

b. $(n^2n^3)^5 = (n^{2+3})^5$ Work within the parentheses first. Since the bases are alike, keep the base and add the exponents.

$\qquad\quad = (n^5)^5$ Perform the addition: $2 + 3 = 5$.

$\qquad\quad = n^{5 \cdot 5}$ Keep the base and multiply the exponents.

$\qquad\quad = n^{25}$ Perform the multiplication: $5 \cdot 5 = 25$.

Self Check 6
Simplify:
a. $(x^4)^2(x^3)^3$
b. $(x^4x^2)^3$

Answers **a.** x^{17},   **b.** x^{18}

▌ The power rule for products

The exponential expression $(2x)^4$ has an exponent of 4 and a base of $2x$. The base $2x$ is a product, since $2x = 2 \cdot x$. Therefore, $(2x)^4$ is a power of a product. To find a rule to simplify it, we will again use the definition of exponent.

$(2x)^4 = 2x \cdot 2x \cdot 2x \cdot 2x$ Write the base, $2x$, as a factor 4 times.

$\qquad = (2 \cdot 2 \cdot 2 \cdot 2)(x \cdot x \cdot x \cdot x)$ Apply the commutative and associative properties of multiplication to change the order and group like factors.

$\qquad = 2^4x^4$ The factor 2 and the factor x are both repeated 4 times. Apply the definition of an exponent.

The result has factors of 2 and x. In the original problem, the factors were within the parentheses. Each is now raised to the fourth power.

Each factor within the parentheses ends up being raised to the 4th power.

$$(2x)^4 = 2^4x^4$$

This observation suggests the following rule.

The power rule for products

For any numbers x and y, and any positive integer m,

$$(xy)^m = x^m y^m$$

To raise a product to a power, raise each factor of the product to that power.

Self Check 7

Simplify each expression:

a. $(10c)^2$

b. $(5rs)^3$

Answers **a.** $100c^2$, **b.** $125r^3s^3$

EXAMPLE 7   Simplify each expression: **a.** $(8a)^2$ and   **b.** $(2bx)^3$.

Solution

a. $(8a)^2 = 8^2a^2$ To raise $8a$ to the 2nd power, raise each factor of the product to the 2nd power.

$= 64a^2$ Find the power: $8^2 = 64$.

b. $(2bx)^3 = 2^3b^3x^3$ To raise $2bx$ to the 3rd power, raise each factor of the product to the 3rd power.

$= 8b^3x^3$ Find the power: $2^3 = 8$.

Self Check 8

Simplify each expression:

a. $(3n^2)^3$

b. $(6h^2s^9)^2$

Answers **a.** $27n^6$, **b.** $36h^4s^{18}$

EXAMPLE 8   Simplify each expression: **a.** $(10a^2)^3$ and   **b.** $(3c^5d^3)^4$.

Solution

a. $(10a^2)^3 = 10^3(a^2)^3$ To raise $10a^2$ to the 3rd power, raise each factor of the product, 10 and a^2, to the 3rd power.

$= 10^3a^{2\cdot3}$ To raise a^2 to a power, keep the base and multiply the exponents.

$= 10^3a^6$ Perform the multiplication: $2 \cdot 3 = 6$.

$= 1,000a^6$ Find the power: $10^3 = 1,000$.

b. $(3c^5d^3)^4 = 3^4(c^5)^4(d^3)^4$ To raise $3c^5d^3$ to the 4th power, raise each factor of the product, 3, c^5, and d^3, to the 4th power.

$= 3^4c^{5\cdot4}d^{3\cdot4}$ To raise c^5 and d^3 to powers, keep the bases and multiply their exponents.

$= 3^4c^{20}d^{12}$ Perform the multiplications: $5 \cdot 4 = 20$ and $3 \cdot 4 = 12$.

$= 81c^{20}d^{12}$ Find the power: $3^4 = 81$.

Self Check 9

Simplify: $(4y^3)^2(3y^4)^3$.

EXAMPLE 9   Simplify: $(2a^2)^2(4a^3)^3$.

Solution

$$(2a^2)^2(4a^3)^3 = 2^2(a^2)^2 \cdot 4^3(a^3)^3$$ To raise $2a^2$ and $4a^3$ to powers, raise the factors of each product to the appropriate power.

$$= 2^2 a^{2 \cdot 2} \cdot 4^3 \cdot a^{3 \cdot 3}$$ To raise a^2 and a^3 to powers, keep the bases and multiply the exponents.

$$= 2^2 a^4 \cdot 4^3 a^9$$ Perform the multiplications: $2 \cdot 2 = 4$ and $3 \cdot 3 = 9$.

$$= (2^2 \cdot 4^3)(a^4 \cdot a^9)$$ Change the order of the factors and group like bases.

$$= (2^2 \cdot 4^3)(a^{4+9})$$ To multiply $a^4 \cdot a^9$, keep the base and add the exponents.

$$= (2^2 \cdot 4^3)a^{13}$$ Perform the addition: $4 + 9 = 13$.

$$= (4 \cdot 64)a^{13}$$ Find the powers: $2^2 = 4$ and $4^3 = 64$.

$$= 256a^{13}$$ Perform the multiplication: $4 \cdot 64 = 256$.

Answer $432y^{18}$

Section 8.7 STUDY SET

VOCABULARY *Fill in the blanks.*

1. In x^n, x is called the _base_ and n is called the _exponent_.

2. x^2 is the second _power_ of x, or we can read it as "x _squared_."

3. $x^m \cdot x^n$ is the product of two exponential expressions with _like_ bases.

4. $(x^m)^n$ is a power of a _power_.

5. $(2x)^n$ is a _product_ raised to a power.

6. In x^{m+n}, $m + n$ is the _sum_ of m and n.

CONCEPTS

7. Represent each multiplication using exponents.
   a. $x \cdot x \cdot x \cdot x \cdot x \cdot x \cdot x$
   b. $x \cdot x \cdot y \cdot y \cdot y$
   c. $3 \cdot 3 \cdot 3 \cdot 3 \cdot a \cdot a \cdot b \cdot b \cdot b$

8. Write each exponential expression as a repeated multiplication.
   a. $a^3 b^5$
   b. $(x^2)^3$
   c. $(2a)^6$

9. Write a product of two exponential expressions with like variable bases. Then simplify it using a rule of exponents.

10. Write a power of a product and then simplify it using a rule of exponents.

11. Write a power of a power and then simplify it using a rule of exponents.

12. What algebraic property enables us to change the order of the factors of a multiplication?

13. Complete each rule for exponents.
    a. $x^m x^n =$
    b. $(x^m)^n =$
    c. $(xy)^n =$

14. In each case, tell how the expression has been improperly simplified.
    a. $2^3 \cdot 2^4 = 2^{12}$
    b. $3^3 \cdot 3^4 = 9^7$
    c. $(2^3)^4 = 2^7$

15. Write each expression without an exponent.
    a. 2^1 b. $(-10)^1$ c. x^1

16. Find each power.
    a. 2^3 b. 4^3 c. 5^3

17. Simplify each expression, if possible.
    a. $x \cdot x$ and $x + x$
    b. $x \cdot x^2$ and $x + x^2$
    c. $x^2 \cdot x^2$ and $x^2 + x^2$

18. Simplify each expression, if possible.
    a. $a \cdot a$ and $a - a$
    b. $2a \cdot a$ and $2a - a$
    c. $2a \cdot 3a$ and $2a - 3a$

19. Simplify each expression, if possible.
    a. $4x \cdot x$ and $4x + x$
    b. $4x \cdot 3x$ and $4x + 3x$
    c. $4x^2 \cdot 3x$ and $4x^2 + 3x$

20. Simplify each expression, if possible.

   a. $a(-2a)$ and $a - 2a$

   b. $-2a(3a)$ and $-2a + 3a$

   c. $-2a(-3a)$ and $-2a - 3a$

21. Evaluate the exponential expression x^{m+n} for $x = 3$, $m = 2$, and $n = 1$.

22. Evaluate the exponential expression $(x^m)^n$ for $x = 2$, $m = 3$, and $n = 2$.

NOTATION   Complete each solution.

23. $x^5 \cdot x^7 = x^{\blacksquare + \blacksquare}$

       $= x^{12}$

24. $(x^5)^4 = x^{\blacksquare \cdot \blacksquare}$

       $= x^{20}$

25. $(2x^4)(8x^3) = (2 \cdot 8)(\boxed{} \cdot \boxed{})$

           $= 16x^{\blacksquare + \blacksquare}$

           $= 16x^7$

26. $(2x^2)^3 = 2^3(x^2)^3$

       $= 2^3 x^{\blacksquare \cdot \blacksquare}$

       $= 2^3 x^{\blacksquare}$

       $= 8x^6$

PRACTICE   Write each expression using one exponent.

27. $x^2 \cdot x^3$ x^5

28. $t^4 \cdot t^3$ 10^7

29. $x^3 x^7$ x^{10}

30. $y^2 y^5$ 9^7

31. $f^5(f^8)$ f^{13}

32. $g^6(g^2)$ 9^8

33. $n^{24} \cdot n^8$ n^{32}

34. $m^9 \cdot m^{61}$ m^{70}

35. $l^4 \cdot l^5 \cdot l$ l^{10}

36. $w^4 \cdot w \cdot w^3$ w^8

37. $x^6(x^3)x^2$

38. $y^5(y^2)(y^3)$ y^{10}

39. $2^4 \cdot 2^8$ 2^{12}

40. $3^4 \cdot 3^2$ $3^6/9^{10}$

41. $5^6(5^2)$ 5^8

42. $(8^3)8^4$ 8^7

Simplify each product.

43. $2x^2 \cdot 4x$

44. $5y \cdot 6y^3$ $30y^4$

45. $5t \cdot t^9$ $5t^{10}$

46. $f^4 \cdot 3f$

47. $-6x^3(4x^2)$

48. $-7y^5(5y^3)$

49. $-x \cdot x^3$

50. $8x^6(-x)$

51. $6y(2y^3)3y^4$

52. $2d(5d^4)(d^2)$

53. $-2t^3(-4t^2)(-5t^5)$

54. $-7k^5(-3k^3)(-2k^9)$

55. $xy^2 \cdot x^2y$

56. $s^2t \cdot st$

57. $b^3 \cdot c^2 \cdot b^5 \cdot c^6$

58. $h^3 \cdot f^3 \cdot f^2 \cdot h^4$

59. $x^4y(xy)$

60. $(ab)(ab^2)$

61. $a^2b \cdot b^3a^2$

62. $w^2y \cdot yw^4$

63. $x^5y \cdot y^6$

64. $a^7 \cdot b^2a^4$

65. $3x^2y^3 \cdot 6xy$

66. $25a^3b \cdot 2ab^5$

67. $xy^2 \cdot 16x^3$

68. $mn^4 \cdot 8n^3$

69. $-6f^2t(4f^4t^3)$

70. $(-5a^2b^2)(5a^3b^6)$

71. $ab \cdot ba \cdot a^2b$

72. $xy \cdot y^2x \cdot x^2y$

73. $-4x^2y(-3x^2y^2)$

74. $-2rt^4(-5r^2t^2)$

Simplify each expression.

75. $(x^2)^4$ x^8

76. $(y^6)^3$ y^{18}

77. $(m^{50})^{10}$ m^{500}

78. $(n^{25})^4$ n^{100}

79. $(2a)^3$ $8a^3$

80. $(3x)^3$ $27x^3$

81. $(xy)^4$ x^4y^4

82. $(ab)^8$ a^8b^8

83. $(3s^2)^3$ $27s^6$

84. $(5f^6)^2$ $25f^{12}$

85. $(2s^2t^3)^2$ $4s^4t^6$

86. $(4h^5y^6)^2$ $16h^{10}y^{12}$

87. $(x^2)^3(x^4)^2$ $x^6x^8 = x^{14}$

88. $(a^5)^2(a^3)^3$ $= a^{10}(a^9) = a^{19}$

89. $(c^5)^3 \cdot (c^3)^5$ c^{30}

90. $(y^2)^8 \cdot (y^8)^2$ y^{32} $\frac{y^8}{3y}$

91. $(2a^4)^2(3a^3)^2$ 36^{14}

92. $(5x^3)^2(2x^4)^3$

93. $(3a^3)^3(2a^2)^3$

94. $(6t^5)^2(2t^2)^2$

95. $(x^2x^3)^{12}$

96. $(a^3a^3)^3$

97. $(2b^4b)^5$

98. $(3y^2y^5)^3$

WRITING

99. Explain the difference between x^2 and $2x$.

100. Explain why the rules of exponents do not apply to $x^2 + x^3$.

101. One of the rules of exponents is that the power of a product is the product of the powers. Use a specific example to explain this rule.

102. To find the result when *multiplying* two exponential expressions with like bases, we must *add* the exponents. Explain why this is so.

REVIEW

103. JEWELRY A lot of what we refer to as gold jewelry is actually made of a combination of gold and another metal. For example, 18-karat gold is $\frac{18}{24}$ gold by weight. Simplify this fraction.

104. After evaluation, what is the sign of $(-13)^5$?

105. Divide: $\dfrac{-25}{-5}$.

106. How much did the temperature change if it went from $-4°$ F to $-17°$ F?

107. Evaluate: $2\left(\dfrac{12}{-3}\right) + 3(5)$.

108. Solve: $-10 = x + 1$.

109. Solve: $-x = -12$.

110. Divide: $\dfrac{0}{10}$.

Variables

One of the major objectives of this course is for you to become comfortable working with **variables.** You will recall that a variable is a letter that stands for a number.

In application problems, we let the variable represent an unknown quantity such as the number of customers a hairdresser used to have, the age of a jar, and the cash award given a Nobel Prize winner. We then write an equation to describe the situation mathematically and solve the equation to find the value represented by the variable.

In Problems 1–6, suppose that you are going to answer the question: What quantity should be represented by a variable? State your response in the form "Let $x = \ldots$."

1. The monthly cost to lease a van is $120 less than the monthly cost to buy it. To buy it, the monthly payments are $290. How much does it cost to lease the van each month?

2. One piece of pipe is 10 feet longer than another. Together, their lengths total 24 feet. How long is the shorter piece of pipe?

3. The length of a rectangular field is 50 feet. What is its width if it has a perimeter of 200 feet?

4. If one hose can fill a vat in 2 hours and another can fill it in 3 hours, how long will it take to fill the vat if both hoses are used?

5. Find the distance traveled by a motorist in three hours if her average speed was 55 miles per hour.

6. In what year was a couple married if their 50th anniversary was in 1988?

Variables can also be used to state properties of mathematics in a concise, "shorthand" notation. In Problems 7–14, state each property using mathematical symbols and the given variable(s).

7. Use the variables a and b to state that two numbers can be added in either order to get the same sum.

8. Use the variable x to state that when 0 is subtracted from a number, the result is the same number.

9. Use the variable b to state that the result when dividing a number by 1 is the same number.

10. Use the variable x to show that the sum of a number and 1 is greater than the number.

11. Using the variable n, state the fact that when 1 is subtracted from any number, the difference is less than the number.

12. State the fact that the product of any number and 0 is 0, using the variable a.

13. Use the variables r, s, and t to state that the way we group three numbers when adding them does not affect the answer.

14. Using the variable n, state the fact that when a number is multiplied by 1, the result is the number.

ACCENT ON TEAMWORK

SECTION 8.1

SOLVING EQUATIONS Borrow a scale and some weights from the chemistry department. Use them as part of a class presentation to explain how the subtraction property of equality is used to solve the equation $x + 2 = 5$. Then use the addition property to solve $x - 3 = 7$.

SECTION 8.2

SOLVING EQUATIONS Use a scale and some weights to explain to the class how to solve $5x = 20$ and $\frac{x}{3} = 4$.

SECTION 8.3

AREA OF A SQUARE Examine the illustrations. What patterns do you see as the square increases in size? Draw the next four squares of this sequence, labeling them in a similar way.

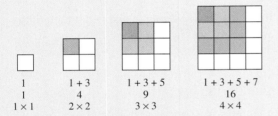

1	1 + 3	1 + 3 + 5	1 + 3 + 5 + 7
1	4	9	16
1×1	2×2	3×3	4×4

SECTION 8.4

THE DISTRIBUTIVE PROPERTY The illustration shows three rectangles that are divided into squares. Since the area of the rectangle on the left-hand side of the equals sign can be found by multiplying its width by its length, its area is $4(5 + 3)$ square units. Evaluating this expression, we see that the area shaded in blue is $4(8) = 32$ square units.

The area on the right-hand side of the equals sign is the sum of the areas of the two rectangles: $4(5) + 4(3)$. Evaluating this expression, we see that the area shaded in red is also 32 square units: $4(5) + 4(3) = 20 + 12 = 32$. Therefore,

$$4(5 + 3) = 4(5) + 5(3)$$

Create a similar demonstration of the distributive property using rectangles with dimensions different from those in the illustration.

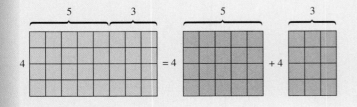

SECTION 8.5

LIKE TERMS If we are to add or subtract objects, they must be similar. Simplify each of the following expressions by combining like terms. You will have to change some of the units so that you are working with like terms. To do so, use the following conversion facts.

- There are 12 inches in one foot.
- There are 36 inches in one yard.
- There are 3 feet in one yard.
- There are 5,280 feet in one mile.

a. 1 foot + 6 inches
b. 1 yard + 11 inches
c. 1 mile − 1 foot
d. 12 feet − 1 yard
e. 1 yard + 1 foot + 5 inches
f. 2 yards + 2 feet − 2 inches
g. 6 inches + 3 feet − 4 inches + 2 feet

SECTION 8.6

SOLVING EQUATIONS We have seen how scales can be used to illustrate the steps used to solve an equation.

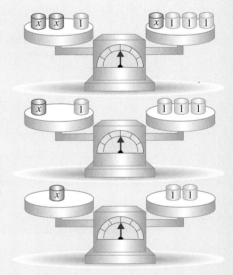

a. What equation is being solved in the illustration? What is the solution?
b. Draw a similar series of pictures showing the solution of each of the following equations.
  1. $3x = 2x + 4$
  2. $3x + 1 = 2x + 4$
  3. $2x + 6 = 4x$
  4. $2x + 6 = 4x + 2$

SECTION 8.7

RULES FOR EXPONENTS Have one student in your group write the three rules for exponents introduced in Section 8.7 on separate 3×5 cards. Have another student write a word description of the rules on separate cards. Finally, have a third student write an example of the use of the rules on separate cards.

When the three sets of cards are completed, put them together, shuffle them, and then work together as a group to match the symbolic description, the word description, and the example for each of the rules.

CHAPTER REVIEW

Solving Equations by Addition and Subtraction

CONCEPTS

An *equation* is a statement that two expressions are equal.

In the equation $x + 5 = 7$, x is the *variable* or the *unknown*. Two equations with exactly the same solutions are called *equivalent equations.*

To solve an equation, isolate the variable on one side of the equation by undoing the operation performed on it. This is accomplished using the opposite operation.

If the same number is added to both sides of an equation, an equivalent equation results:

If $a = b$, then $a + c = b + c$.

If the same number is subtracted from both sides of an equation, an equivalent equation results:

If $a = b$, then $a - c = b - c$.

To solve a problem, follow these steps:

1. Analyze the problem.
2. Form an equation.
3. Solve the equation.
4. State the conclusion.
5. Check the result.

REVIEW EXERCISES

Determine whether the given number is a solution of the equation.

1. $x + 2 = 13$; 5

NO

2. $x - 3 = 1$; 4

Yes

Identify the variable in each equation.

3. $y - 12 = 50$

y

4. $114 = 4 - t$

t

Solve each equation and check the result.

5. $x - 7 = 2$

6. $x - 11 = 20$

7. $225 = y - 115$

8. $101 = p - 32$

Solve each equation and check the result.

9. $x + 9 = 18$

10. $b + 12 = 26$

11. $175 = p + 55$

12. $212 = m + 207$

13. FINANCING A newly married couple made a $25,500 down payment on a $122,750 house. How much did they need to borrow?

14. DOCTOR'S CLIENTELE After moving his office, a doctor lost 13 patients. If he had 172 patients left, how many did he have originally?

Solving Equations by Division and Multiplication

If both sides of an equation are divided by the same nonzero number, an equivalent equation results:

If $a = b$, then $\dfrac{a}{c} = \dfrac{b}{c}$

$(c \neq 0)$.

Solve each equation and check the result.

15. $3x = 12$

16. $15y = 45$

17. $105 = 5r$

18. $224 = 16q$

If both sides of an equation are multiplied by the same nonzero number, an equivalent equation results:

If $a = b$, then $a \cdot c = b \cdot c$ ($c \neq 0$).

To solve a problem, follow these steps:

1. Analyze the problem.
2. Form an equation.
3. Solve the equation.
4. State the conclusion.
5. Check the result.

Solve each equation and check the result.

19. $\dfrac{x}{7} = 3$

20. $\dfrac{a}{3} = 12$

21. $15 = \dfrac{s}{21}$

22. $25 = \dfrac{d}{17}$

23. CARPENTRY If you cut an 18-foot board into three equal pieces, how long will each piece be?

24. JEWELRY Four sisters split the cost of a gold chain evenly. How much did the chain cost if each sister's share was $32?

| **SECTION 8.3** | *Algebraic Expressions and Formulas* |

When we replace the variable, or variables, in an algebraic expression with a specific number and then apply the rules for the order of operations, we are *evaluating the algebraic expression.*

RETAINING WALLS   The illustration shows the design for a retaining wall. The relationships between the lengths of its important parts are given in words.

25. Choose a variable to represent the height of the wall. Write algebraic expressions to represent the lengths of the upper and lower bases.

26. Suppose engineers determine that a 10-foot-high wall is needed. Find the lengths of the upper and lower bases.

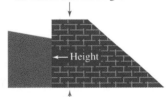

The length of the upper base is 5 ft less than the height.

←— Height

The length of the lower base is 3 ft less than twice the height.

Evaluate each algebraic expression.

27. $-2x + 6$ for $x = -3$
    $6 + 6 = 12$

28. $\dfrac{6 - a}{1 + a}$ for $a = -2$ -8

29. $b^2 - 4ac$ for $a = 4$, $b = 6$, and $c = -4$ 100

30. $\dfrac{-2k^3}{1 - 2 - 3}$ for $k = -2$ -4

A *formula* is a general rule that describes a known relationship between two or more variables.

Distance = rate · time

31. DISTANCE TRAVELED Complete the table by finding the distance traveled for a given time at a given rate.

	Rate (mph)	Time (hr)	Distance traveled (mi)
Monorail	65	2	
Subway	38	3	
Train	x	6	
Bus	55	t	

Formulas from business:

Sale price = original price
− discount

Retail price = cost + markup

Profit = revenue − costs

32. SALE PRICES Find the sale price of a trampoline that normally sells for $315 if a $37 discount is being offered.

33. RETAIL PRICES Find the retail price of a car if the dealer pays $14,505 and the markup is $725.

ANNUAL PROFITS The bar graph shows the revenue and costs for a company for the years 2002 to 2004 in millions of dollars.

34. In which year was there the most revenue?

35. Which year had the largest profit?

36. What can you say about costs over this three-year span?

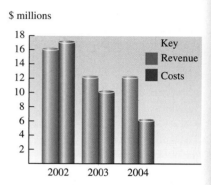

Formulas from science:

$$C = \frac{5}{9}(F - 32)$$

Distance fallen = 16 · (time)²

37. TEMPERATURE CONVERSIONS At a summer resort, visitors can relax by taking a dip in a swimming pool or a lake. The pool water is kept at a constant temperature of 77° F. The water in the lake is 23° C. Which water is warmer, and by how many degrees Celsius?

38. DISTANCE FALLEN A steelworker accidentally dropped his hammer while working atop a new high-rise building. How far will the hammer fall in 3 seconds?

SECTION 8.4

Simplifying Algebraic Expressions and the Distributive Property

To *simplify an algebraic expression,* we use properties of algebra to write the expression in simpler form.

Simplify each expression.

39. $-2(5x)$ −10x

40. $-7x(-6y)$ 42xy

41. $4d \cdot 3e \cdot 5$ 60de

42. $(4s)8$ 32s

43. $-1(-e)(2)$ 2e

44. $7x \cdot 7y$ 49xy

45. $4 \cdot 3k \cdot 7$ 84k

46. $(-10t)(-10)$ 100t

The *distributive property:*
If *a, b,* and *c* are numbers, then

$a(b + c) = ab + ac$
$(b + c)a = ba + ca$
$a(b − c) = ab − ac$
$(b − c)a = ba − ca$
$a(b + c + d)$
  $= ab + ac + ad$

Multiply to remove the parentheses.

47. $4(y + 5)$ 4y+20

48. $-5(6t + 9)$ −30t−45

49. $(-3 - 3x)7$ −21−21x

50. $-3(4e - 8x - 1)$ −12e+24x+3

Write an equivalent expression without parentheses.

51. $-(6t - 4)$ −6t+4

52. $-(5 + x)$ −5−x

53. $-(6t - 3s + 1)$ −6t+3s−1

54. $-(-5a - 3)$ 5a+3

483

Combining Like Terms

A *term* is a number or a product of a number and one or more variables.

In a term that is a product of a number and one or more variables, the factor that is a number is called the *coefficient* of the term.

Like terms, or *similar terms*, are terms with exactly the same variables and exponents.

To combine like terms, combine their coefficients and keep the variables and exponents.

Identify the second term and the coefficient of the third term.

55. $5x^2 - 4x + 8$ **56.** $7y - 3y + x - y$

Determine whether x is used as a factor or a term.

57. $5x - 6y^2$ **58.** $x + 6$ **59.** $6xy$ **60.** $-36 - x + b$

Determine whether the following are like terms.

61. $4x, -5x$ **62.** $4x, 4x^2$ **63.** $3xy, xy$ **64.** $-5b^2c, -5bc^2$

Simplify by combining like terms.

65. $3x + 4x$ **66.** $6r - 9r$

67. $-3t - 6t$ **68.** $2z + (-5z)$

69. $6x - x$ **70.** $-6y - 7y - (-y)$

71. $5w - 8 - 4w + 3$ **72.** $-5x - 5y - x + 7y$

Simplify by combining like terms.

73. $-45d - 2a + 4a - d$ **74.** $5y + 8h - 3 + 7h + 5y + 2$

Simplify each expression.

75. $7(y + 6) + 3(2y + 2)$ **76.** $-4(t - 7) - (t + 6)$

77. $5x - 2(x - 6)$ **78.** $6f + 7(12 - 8f)$

The perimeter of a rectangle is given by

$$P = 2l + 2w$$

The perimeter of a square is given by

$$P = 4s$$

79. HOLIDAY LIGHTS To decorate a house, lights will be hung around the entire home, as shown. They will also be placed around the two 5-foot-by-5-foot windows in the front. How many feet of lights will be needed?

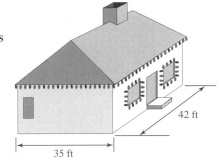

42 ft

35 ft

Simplifying Expressions to Solve Equations

To *solve an equation* means to find all values of the variable that, when substituted into the original equation, make the equation a true statement.

80. Is -3 a solution of $-4x + 6 = 2(x + 12)$? Explain.

Solve each equation. **Check all answers.**

81. $3x + 4 = -8$ **82.** $5a - 1 - 3a = -32 - 5$

83. $7x = 3x - 12$ **84.** $5(y - 15) = 0$

85. $-3(2x + 4) - 4 = -40$ **86.** $y + 170 = 5y - 170$

87. $-6(2x + 3) = -(5x - 3)$ **88.** $2(3 - 2x) - 5(1 - 3x) = -21$

89. List the steps of the strategy for solving equations.

Exponents represent repeated multiplication.

90. a. What repeated multiplication does $(4h)^3$ represent?
   b. Write this expression using exponents: $5 \cdot 5 \cdot d \cdot d \cdot d \cdot m \cdot m \cdot m \cdot m$

The *product rule* for exponents:

$x^m x^n = x^{m+n}$

Simplify each expression.

91. $h^6 h^4$ **92.** $t^3(t^5)$

93. $w^2 \cdot w \cdot w^4$ **94.** $4^7 \cdot 4^5$

Simplify each product.

95. $2b^2 \cdot 4b^5$ **96.** $-6x^3(4x)$

97. $-2f^2(-4f)(3f^4)$ **98.** $-ab \cdot b \cdot a$

99. $xy^4 \cdot xy^2$ **100.** $(mn)(mn)$

101. $3z^3 \cdot 9m^3 z^4$ **102.** $-5cd(4c^2 d^5)$

The *power rule* for exponents:

$(x^m)^n = x^{m \cdot n}$

Simplify each expression.

103. $(v^3)^4$ **104.** $(3y)^3$ **105.** $(5t^4)^2$ **106.** $(2a^4 b^5)^3$

The *power rule for products:*

$(xy)^m = x^m y^m$

Simplify each expression.

107. $(c^4)^5(c^2)^3$ **108.** $(3s^2)^3(2s^3)^2$

109. $(c^4 c^3)^2$ **110.** $(2xx^2)^3$

Solve each equation.

1. $10 = x + 6$

2. $y - 12 = 18$

3. $5t = 55$

4. $\dfrac{q}{3} = 27$

5. $500 + x = 700$

6. $100x = 2,400$

7. PARKING After many student complaints, a college decided to commit funds to double the number of parking spaces on campus. This increase would bring the total number of spaces up to 6,200. How many parking spaces does the college have at this time?

8. LIBRARIES A library building is 6 years shy of its 200th birthday. How old is the building at this time?

9. Evaluate $\dfrac{x - 16}{x}$ for $x = 4$.

10. DISTANCE TRAVELED Find the distance traveled by a motorist who departed from home at 9:00 A.M. and arrived at his destination at noon, traveling at a rate of 55 miles per hour.

Simplify by removing the parentheses.

11. $5(5x + 1)$

12. $-6(7 - x)$

13. $-(6y + 4)$

14. $3(2a + 3b - 7)$

15. Determine whether x is used as a factor or as a term.
   a. $5xy$ **b.** $8y + x + 6$

16. Simplify by combining like terms.
   a. $-20y + 6x - 8y + 4x$
   b. $-t - t - t$

17. Identify each term in $8x^2 - 4x - 6$.

18. Simplify.

   a. $7x + 4x$ **b.** $3c \cdot 4e \cdot 2$

   c. $6x - x$ **d.** $-5y(-6)$

19. Simplify: $4(y + 3) - 5(2y + 3)$.

Solve each equation.

20. $15a - 10 = 20$

21. $3x - 4 - 3 = 2x + 3x + 11$

22. $6r + 3 - 2r = -9$

23. $2(4x - 1) = 3(4 - 2x)$.

24. a. What is the value (in cents) of k dimes?

   b. What is the value of $p + 2$ twenty-dollar bills?

25. Simplify each expression.

   a. h^2h^4 **b.** $-7x^3(4x^2)$

   c. $b^2 \cdot b \cdot b^5$ **d.** $-3g^2k^3(-8g^3k^{10})$

26. Simplify each expression.

   a. $(f^3)^5$ **b.** $(2a^2b)^2$

   c. $(x^2)^3(x^3)^3$ **d.** $(x^2x^3)^3$

27. Explain the difference between an expression and an equation.

28. Explain what it means to *solve an equation*.

29. Show how to check whether -7 is a solution of $2x + 1 = -15$.

30. Explain what is wrong with the following work: $5^4 \cdot 5^3 = 25^7$.

Consider the number 7,535,670.

1. Round to the nearest hundred.

2. Round to the nearest ten thousand.

Perform each operation.

3. 5,679
    +3,458

4. 7,697
    −4,375

5. 5,345
    ×   46

6. $35\overline{)30,625}$

Refer to the rectangular swimming pool below.

7. Find the perimeter of the pool.

8. Find the area of the surface of the pool.

80 ft

50 ft

Find the prime factorization of each number.

9. 168

10. 225

11. 180

12. 720

Evaluate each expression.

13. $8 + (-2)(-5)$

14. $(-2)^4 - 3^3$

15. $\dfrac{2(-7) + 3(2)}{2(-4)}$

16. $\dfrac{2(3^2 + 4^2)}{-2(3) + 1}$

Perform the operations.

17. $\dfrac{10}{21} \cdot \dfrac{3}{10}$

18. $\dfrac{22}{25} \div \dfrac{11}{5}$

19. $\dfrac{11}{12} + \dfrac{2}{3}$

20. $\dfrac{11}{12} - \dfrac{2}{3}$

Evaluate each expression when x = 4.

21. $2x - 1$

22. $\dfrac{9x}{2} - x$

23. $3x - x^3$

24. $x + 2(x - 7)$

Simplify each expression.

25. $-3(5x)$ **26.** $-4x(-7x)$

27. $-2(3x - 4)$ **28.** $-5(3x - 2y + 4)$

Combine like terms.

29. $-3x + 8x$ **30.** $4a^2 - (-3a^2)$

31. $4x - 3y - 5x + 2y$ **32.** $-2(3x - 4) + 2x$

Solve each equation and check the answer.

33. $3x + 2 = -13$ **34.** $-5z - 7 = 18$

35. $\dfrac{y}{4} - 1 = -5$ **36.** $\dfrac{n}{5} + 1 = -11$

37. $6x - 12 = 2x + 4$

38. $3(2y - 8) = -2(y - 4)$

39. OBSERVATION HOURS To get a Masters degree in learning disabilities, a student must have 100 hours of observation time. If the student has already observed for 37 hours, how many more 3-hour shifts must he observe?

40. GEOMETRY A rectangle is 84 feet long. If its perimeter is 210 feet, find its dimensions.

Simplify each product.

41. $p^3 p^5$ **42.** $-3t^2 \cdot 5t^7$

43. $(-x^2y^3)(x^3y^4)$ **44.** $(3a^2)^4$

45. $(-n^2)^3$ **46.** $(3x^3y)^2$

47. $(2p^3)^2(3p^2)^3$ **48.** $(-x^2)^3(2x)^3$

CHAPTER 9

An Introduction to Geometry

Getty Images

Many people enjoy tackling home-improvement projects themselves. These projects run more smoothly and are more cost efficient if careful planning is done in advance. This planning often requires some mathematics, and in particular, geometry. For example, to purchase the correct amount of materials to fence a yard, you need to know how to calculate *perimeter*. If you are painting a bedroom, you need to calculate the total *area* of the wall surfaces to determine the number of gallons of paint to buy.

To learn more about perimeter and area, visit The Learning Equation on the Internet at http://tle.brookscole.com. (The log-in instructions are in the Preface.) For Chapter 9, the online lesson is:

• *TLE* Lesson 16: Perimeter and Area

Check Your Knowledge

1. _____ lines do not intersect. If two lines intersect and form right angles, they are _____.

2. A polygon with four sides is called a _____. A _____ is a polygon with three sides.

3. A triangle that includes a 90° angle is called a _____ triangle. The longest side of such a triangle is the _____.

4. Triangles that are the same size and the same shape are called _____ triangles. If two triangles are the same shape but not necessarily the same size, they are called _____ triangles.

5. The distance around a polygon is called the _____. The measure of the surface enclosed by a polygon is called its _____.

6. A segment drawn from the center of a circle to a point on the circle is called a _____. The distance around a circle is called its _____. A _____ is a chord that passes through the center of a circle.

7. The space contained within a geometric solid is called its _____.

8. Match the descriptions with the angles.

    a. 37° _____   I.   right angle

    b. 90° ____   II. straight angle

    c. 125° _____   III. acute angle

    d. 180° _____   IV. obtuse angle

9. In the notation $\angle ABC$, which point is the vertex of the angle?

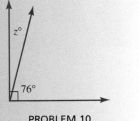

PROBLEM 10

10. Refer to the illustration on the left. Find z.

11. Refer to the illustration on the left. Find y.

12. Find the supplement of an angle measuring 119°.

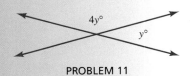

PROBLEM 11

13. Refer to the illustration, in which the horizontal lines are parallel.

    a. Find m($\angle 1$).   **b.** Find m($\angle 2$).

    c. Find m($\angle 3$).   **d.** Find m($\angle 4$).

    e. Find x.

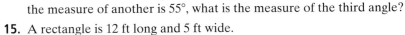

14. If the measure of one angle of a triangle is 45° and the measure of another is 55°, what is the measure of the third angle?

15. A rectangle is 12 ft long and 5 ft wide.

    a. Find length of the diagonal of the rectangle.

    b. Find the area of the rectangle.

16. Find the area of a triangle with base 4.5 in. and height 7.8 in.

17. Find the area of a circle if the diameter is 10 ft.

18. Find the volume of a sphere that is 4 ft in diameter.

19. Find the volume of a rectangular solid with length 5 ft, width 4 ft, and height 10 ft.

20. Find the volume of a cylinder 8 ft in diameter and 10 ft tall.

Study Skills Workshop

PREPARING FOR YOUR NEXT MATH CLASS

According to Dr. Benjamin Bloom, 50% of your overall achievement in your next math class is based on the math knowledge that you have coming into that class [Bloom, *Human Characteristics and School Learning* (New York: McGraw-Hill, 1976)]. So, if you want to be successful in your next math class, it's a good idea to do some preparation in advance. How much you will have to do depends on how much time has passed between your math courses, how well you remember material, and how well you did in your previous math course. It is always a good idea to take your next math course as soon as possible after this class is finished. The exception to this is that you should not take a math course during an accelerated term (like summer semesters at many colleges), unless you are very confident in your math abilities. It is a good idea to keep your textbook from this class until you finish your next class; resist the urge to make a couple of dollars by selling your book at textbook "buy-backs" on your campus. The textbook that you used this term is a good resource between terms and throughout your next course.

Here are some activities you can do to prepare for your next class:

- *Review old tests.* Try to work problems from old tests without peeking at the answers. If you get stuck, you can look — but then repeat the problem until you can do it without looking at the solution.

- *Use packaged software or online programs.* Using tutorials on the CD-ROM that came packaged with your text or on the iLrn website can help you to refresh concepts.

- *Review chapter tests from your old textbook.* Try taking the tests found at the end of each chapter in your textbook from this course. Check your answers and review the book for help in correcting the problems that you did incorrectly.

- *Try to get your new text before class starts.* If you have time, reading ahead in your new text is a good way to get a head start in your new class.

- *Consider getting a tutor.* If you didn't do as well as you would have liked in this class, consider hiring a tutor in the time between terms to learn some of the concepts that you missed in this course.

Geometry comes from the Greek words geo *(meaning earth) and* metron *(meaning measure).*

9.1 Some Basic Definitions

- Points, lines, and planes • Angles • Adjacent and vertical angles
- Complementary and supplementary angles

In this chapter, we study two-dimensional geometric figures such as rectangles and circles. In daily life, it is often necessary to find the perimeter or area of one of these figures. For example, to find the amount of fencing that is needed to enclose a circular garden, we must find the perimeter of a circle (called its *circumference*). To find the amount of paint needed to paint a room, we must find the area of its four rectangular walls.

We also study three-dimensional figures such as cylinders and spheres. To find the amount of space enclosed within these figures, we must find their volumes.

Points, lines, and planes

Geometry is based on three undefined words: **point, line,** and **plane.** Although we will make no attempt to define these words formally, we can think of a point as a geometric figure that has position but no length, width, or depth. Points are always labeled with capital letters. Point *A* is shown in Figure 9-1(a).

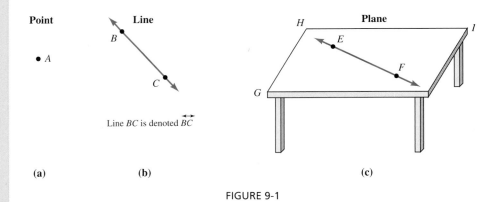

FIGURE 9-1

A line is infinitely long but has no width or depth. Figure 9-1(b) shows line *BC*, passing through points *B* and *C*. A plane is a flat surface, like a table top, that has length and width but no depth. In Figure 9-1(c), line *EF* lies in the plane *GHI*.

As Figure 9-1(b) illustrates, points *B* and *C* determine exactly one line, the line *BC*. In Figure 9-1(c), the points *E* and *F* determine exactly one line, the line *EF*. In general, any two points determine exactly one line.

Other geometric figures can be created by using parts or combinations of points, lines, and planes.

Line segment

The **line segment** *AB*, denoted as $\overline{AB}$, is the part of a line that consists of points *A* and *B* and all points in between. (See Figure 9-2 on the next page.) Points *A* and *B* are the **endpoints** of the segment.

FIGURE 9-2

Every line segment has a **midpoint,** which divides the segment into two parts of equal length. In Figure 9-3, M is the midpoint of segment AB, because the measure of $\overline{AM}$, denoted as m($\overline{AM}$), is equal to the measure of $\overline{MB}$, denoted as m($\overline{MB}$).

$$\text{m}(\overline{AM}) = 4 - 1$$
$$= 3$$

and

$$\text{m}(\overline{MB}) = 7 - 4$$
$$= 3$$

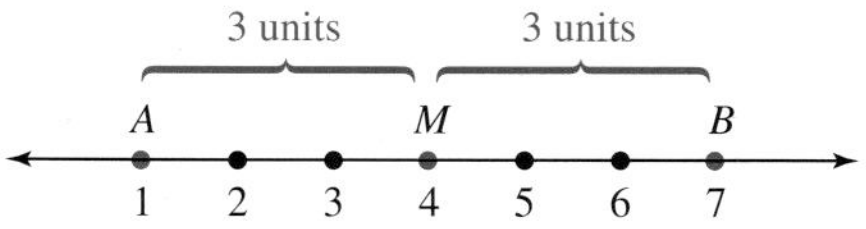

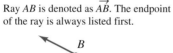

FIGURE 9-3

Since the measure of both segments is 3 units, m($\overline{AM}$) = m($\overline{MB}$).

When two line segments have the same measure, we say that they are **congruent.** Since m($\overline{AM}$) = m($\overline{MB}$), we can write

$$\overline{AM} \cong \overline{MB} \quad \text{Read} \cong \text{as "is congruent to."}$$

Another geometric figure is the *ray,* as shown in Figure 9-4.

Ray

A **ray** is the part of a line that begins at some point (say, A) and continues forever in one direction. Point A is the **endpoint** of the ray.

Ray AB is denoted as $\overrightarrow{AB}$. The endpoint of the ray is always listed first.

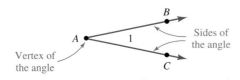

FIGURE 9-4

▌ Angles

Angle

An **angle** is a figure formed by two rays with a common endpoint. The common endpoint is called the **vertex,** and the rays are called **sides.**

The angle in Figure 9-5 can be denoted as

$$\angle BAC, \qquad \angle CAB, \qquad \angle A, \qquad \text{or} \qquad \angle 1 \quad \text{The symbol } \angle \text{ means angle.}$$

FIGURE 9-5

❗ COMMENT When using three letters to name an angle, be sure the letter name of the vertex is the middle letter.

One unit of measurement of an angle is the **degree.** It is $\frac{1}{360}$ of a full revolution. We can use a **protractor** to measure angles in degrees. (See Figure 9-6.)

Angle	Measure in degrees
$\angle ABC$	30°
$\angle ABD$	60°
$\angle ABE$	110°
$\angle ABF$	150°
$\angle ABG$	180°

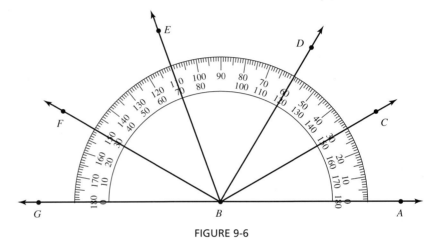

FIGURE 9-6

If we read the protractor from left to right, we can see that the measure of $\angle GBF$ (denoted as m($\angle GBF$)) is 30°.

When two angles have the same measure, we say that they are congruent. Since m($\angle ABC$) = 30° and m($\angle GBF$) = 30°, we can write

$$\angle ABC \cong \angle GBF$$

We classify angles according to their measure, as in Figure 9-7.

Classification of angles

Acute angles: Angles whose measures are greater than 0° but less than 90°.

Right angles: Angles whose measures are 90°.

Obtuse angles: Angles whose measures are greater than 90° but less than 180°.

Straight angles: Angles whose measures are 180°.

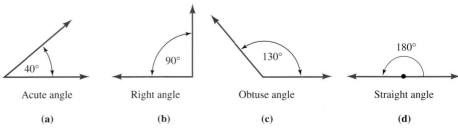

FIGURE 9-7

EXAMPLE 1 Classify each angle in Figure 9-8 as an acute angle, a right angle, an obtuse angle, or a straight angle.

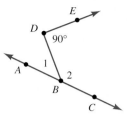

Solution

Since m($\angle 1$) < 90°, it is an acute angle.

Since m($\angle 2$) > 90°, but less than 180°, it is an obtuse angle.

Since m($\angle BDE$) = 90°, it is a right angle.

Since m($\angle ABC$) = 180°, it is a straight angle.

FIGURE 9-8

■ Adjacent and vertical angles

Two angles that have a common vertex and are side-by-side are called **adjacent angles.**

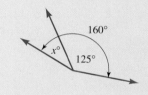

EXAMPLE 2 Two angles with measures of $x°$ and 35° are adjacent angles. Use the information in Figure 9-9 to find x.

Solution We can use algebra to solve this problem. Since the sum of the measures of the angles is 80°, we have

$$x + 35 = 80$$
$$x + 35 - 35 = 80 - 35$$ To undo the addition of 35, subtract 35 from both sides.
$$x = 45$$ Perform the subtractions: $35 - 35 = 0$ and $80 - 35 = 45$.

Thus, x is 45.

FIGURE 9-9

Self Check 2
In the figure below, find x.

Answer 35

When two lines intersect, pairs of nonadjacent angles are called **vertical angles.** In Figure 9-10(a), lines l_1 (read as "line l sub 1") and l_2 (read as "line l sub 2") intersect. $\angle 1$ and $\angle 3$ are vertical angles, as are $\angle 2$ and $\angle 4$.

To illustrate that vertical angles always have the same measure, we refer to Figure 9-10(b) with angles having measures of $x°$, $y°$, and 30°. Since the measure of any straight angle is 180°, we have

$$30 + x = 180 \qquad \text{and} \qquad 30 + y = 180$$
$$x = 150 \qquad\qquad\qquad y = 150$$ To undo the addition of 30, subtract 30 from both sides.

Since x and y are both 150, $x = y$.

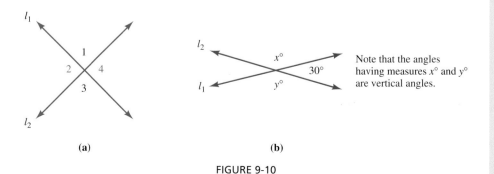

Note that the angles having measures $x°$ and $y°$ are vertical angles.

(a) (b)

FIGURE 9-10

> **Property of vertical angles**
> Vertical angles are congruent (have the same measure).

Self Check 3
In Figure 9-11, find:
a. m(∠2)
b. m(∠4)

EXAMPLE 3   In Figure 9-11, find: **a.** m(∠1) and **b.** m(∠3).

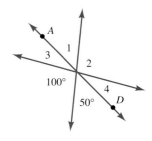

FIGURE 9-11

Solution
a. The 50° angle and ∠1 are vertical angles. Since vertical angles are congruent, m(∠1) = 50°.

b. Since AD is a line, the sum of the measures of ∠3, the 100° angle, and the 50° angle is 180°. If m(∠3) = x, we have

$$x + 100 + 50 = 180$$
$$x + 150 = 180 \quad \text{Perform the addition: } 100 + 50 = 150.$$
$$x = 30 \quad \text{Subtract 150 from both sides.}$$

Answers a. 100°, **b.** 30°

Thus, m(∠3) = 30°.

Self Check 4
In the figure below, find y.

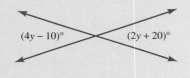

EXAMPLE 4   In Figure 9-12, find x.

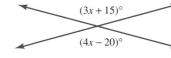

FIGURE 9-12

Solution   Since the angles are vertical angles, they have equal measures.

$$4x - 20 = 3x + 15$$
$$x - 20 = 15 \quad \text{To eliminate } 3x \text{ from the right-hand side, subtract } 3x \text{ from both sides.}$$
$$x = 35 \quad \text{To undo the subtraction of 20, add 20 to both sides.}$$

Answer 15

Thus, x is 35.

■ Complementary and supplementary angles

> **Complementary and supplementary angles**
> Two angles are **complementary angles** when the sum of their measures is 90°.
> Two angles are **supplementary angles** when the sum of their measures is 180°.

Angles with measures of 60° and 30° are examples of complementary angles, because the sum of their measures is 90°. Each angle is the complement of the other.

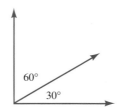

Angles of 130° and 50° are supplementary, because the sum of their measures is 180°. Each angle is the supplement of the other.

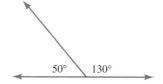

! COMMENT The definition of supplementary angles requires that the sum of two angles be 180°. Three angles of 40°, 60°, and 80° are not supplementary even though their sum is 180°.

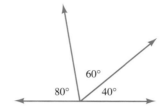

EXAMPLE 5

a. Find the complement of a 35° angle.

b. Find the supplement of a 105° angle.

Solution

a. See Figure 9-13. Let x represent the complement of the 35° angle. Since the angles are complementary, we have

$x + 35 = 90$ The sum of the angles' measures must be 90°.

$x = 55$ To undo the addition of 35, subtract 35 from both sides.

The complement of 35° is 55°.

b. See Figure 9-14. Let y represent the supplement of the 105° angle. Since the angles are supplementary, we have

$y + 105 = 180$ The sum of the angles' measures must be 180°.

$y = 75$ To undo the addition of 105, subtract 105 from both sides.

The supplement of 105° is 75°.

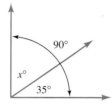

FIGURE 9-13

FIGURE 9-14

Self Check 5

a. Find the complement of a 50° angle.

b. Find the supplement of a 50° angle.

Answers **a.** 40°, **b.** 130°

Section 9.1 STUDY SET

VOCABULARY *Fill in the blanks.*

1. A line _____ has two endpoints.

2. Two points _____ at most one line.

3. A _____ divides a line segment into two parts of equal length.

4. An angle is measured in _____.

5. A _____ is used to measure angles.

6. An _____ angle is less than 90°.

7. A _____ angle measures 90°.

8. An _____ angle is greater than 90° but less than 180°.

9. The measure of a straight angle is _____.

10. Adjacent angles have the same vertex and are _____.

11. The sum of two _____ angles is 180°.

12. The sum of two complementary angles is _____.

CONCEPTS   *Refer to the illustration and determine whether each statement is true. If a statement is false, explain why.*

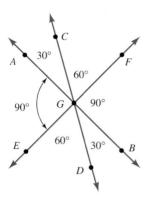

13. $\overrightarrow{GF}$ has point G as its endpoint.

14. $\overline{AG}$ has no endpoints.

15. Line CD has three endpoints.

16. Point D is the vertex of $\angle DGB$.

17. m($\angle AGC$) = m($\angle BGD$)

18. $\angle AGF \cong \angle BGE$

19. $\angle FGB \cong \angle EGA$

20. $\angle AGC$ and $\angle CGF$ are adjacent angles.

Refer to the illustration above and determine whether each angle is an acute angle, a right angle, an obtuse angle, or a straight angle.

21. $\angle AGC$

22. $\angle EGA$

23. $\angle FGD$

24. $\angle BGA$

25. $\angle BGE$

26. $\angle AGD$

27. $\angle DGC$

28. $\angle DGB$

Refer to the illustration below and determine whether each statement is true. If a statement is false, explain why.

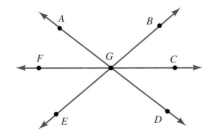

29. $\angle AGF$ and $\angle DGC$ are vertical angles.

30. $\angle FGE$ and $\angle BGA$ are vertical angles.

31. m($\angle AGB$) = m($\angle BGC$)

32. $\angle AGC \cong \angle DGF$

Refer to the illustration below and determine whether each pair of angles are congruent.

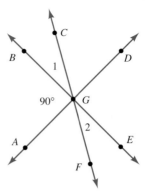

33. $\angle 1$ and $\angle 2$

34. $\angle FGB$ and $\angle CGE$

35. $\angle AGB$ and $\angle DGE$

36. $\angle CGD$ and $\angle CGB$

37. $\angle AGF$ and $\angle FGE$

38. $\angle AGB$ and $\angle BGD$

Refer to the illustration above and determine whether each statement is true.

39. $\angle 1$ and $\angle CGD$ are adjacent angles.

40. $\angle 2$ and $\angle 1$ are adjacent angles.

41. $\angle FGA$ and $\angle AGC$ are supplementary.

42. $\angle AGB$ and $\angle BGC$ are complementary.

43. $\angle AGF$ and $\angle 2$ are complementary.

44. $\angle AGB$ and $\angle EGD$ are supplementary.

45. $\angle EGD$ and $\angle DGB$ are supplementary.

46. $\angle DGC$ and $\angle AGF$ are complementary.

NOTATION   *Fill in the blanks.*

47. The symbol $\angle$ means _____.

48. The symbol $\overline{AB}$ is read as "_____ AB."

49. The symbol $\overrightarrow{AB}$ is read as "_____ AB."

50. The symbol ▮ is read as "is congruent to."

PRACTICE   *Refer to the illustration below and find the length of each segment.*

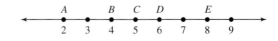

51. $\overline{AC}$

52. $\overline{BE}$

53. $\overline{CE}$

54. $\overline{BD}$

55. $\overline{CD}$

56. $\overline{DE}$

Refer to the illustration above and find each midpoint.

57. Find the midpoint of $\overline{AD}$.

58. Find the midpoint of $\overline{BE}$.

Use a protractor to measure each angle.

59.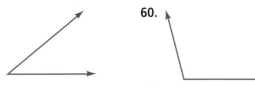

60.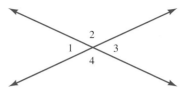

61.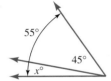

62.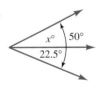

Refer to the illustration below in which m(∠1) = 50°. Find the measure of each angle or sum of angles.

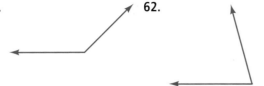

75. ∠4   **76.** ∠3

77. m(∠1) + m(∠2) + m(∠3)

78. m(∠2) + m(∠4)

Find x.

63.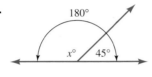

64.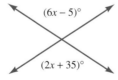

65.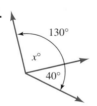

66.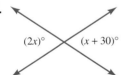

Refer to the illustration below, in which m(∠1) + m(∠3) + m(∠4) = 180°, ∠3 ≅ ∠4, and ∠4 ≅ ∠5. Find the measure of each angle.

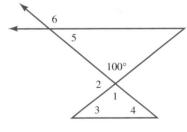

79. ∠1   **80.** ∠2

81. ∠3   **82.** ∠6

67.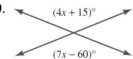

68.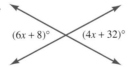

69. $(4x + 15)°$ $(7x - 60)°$

70. $(6x + 8)°$ $(4x + 32)°$

APPLICATIONS

83. BASEBALL Use the following definition to draw the strike zone for the player shown.

> *The strike zone is that area over home plate the upper limit of which is a horizontal line at the midpoint between the top of the shoulders and the top of the uniform pants and the lower level is a line at the hollow beneath the kneecap.*

Let x represent the unknown angle measure. Draw a diagram, write an appropriate equation, and solve it for x.

71. Find the complement of a 30° angle.

72. Find the supplement of a 30° angle.

73. Find the supplement of a 105° angle.

74. Find the complement of a 75° angle.

84. PHYSICS The illustration shows a 15-pound block that is suspended with two ropes, one of which is horizontal. Classify each numbered angle in the illustration as acute, obtuse, or right.

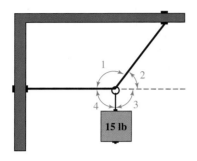

85. SYNTHESIZERS Refer to the illustration. Find *x* and *y*.

86. AVIATION How many degrees from the horizontal position are the wings of the airplane?

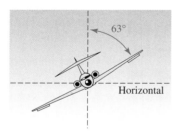

87. GARDENING What angle does the handle of the lawn mower make with the ground?

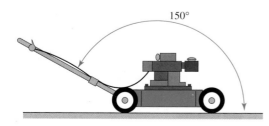

88. MUSICAL INSTRUMENTS Suppose that you are a beginning band teacher describing the correct posture needed to play various instruments. Use the following diagrams to approximate the angle measure at which each instrument should be held in relation to the student's body: **a.** flute **b.** clarinet **c.** trumpet

a.   b.   c.

WRITING

89. PHRASES Explain what you think each of these phrases means. How is geometry involved?

   a. The president did a complete 180-degree flip on the subject of a tax cut.

   b. The rollerblader did a "360" as she jumped off the ramp.

90. In the statements below, the ° symbol is used in two different ways. Explain the difference.

   $85° \text{ F}$ and $m(\angle A) = 85°$

91. What is a protractor?

92. Explain the difference between a ray and a line segment.

93. Explain why an angle measuring 105° cannot have a complement.

94. Explain why an angle measuring 210° cannot have a supplement.

REVIEW

95. Find: 2^4.

96. Add: $\dfrac{1}{2} + \dfrac{2}{3} + \dfrac{3}{4}$.

97. Subtract: $\dfrac{3}{4} - \dfrac{1}{8} - \dfrac{1}{3}$.

98. Multiply: $\dfrac{5}{8} \cdot \dfrac{2}{15} \cdot \dfrac{6}{5}$.

99. Divide: $\dfrac{12}{17} \div \dfrac{4}{34}$.

100. What is 7% of 7?

101. Solve the proportion: $\dfrac{x}{18} = \dfrac{12.5}{45}$.

102. Convert 120 yards to feet.

9.2 Parallel and Perpendicular Lines

- Parallel and perpendicular lines • Transversals and angles • Properties of parallel lines

In this section, we consider *parallel* and *perpendicular* lines. Since parallel lines are always the same distance apart, the railroad tracks shown in Figure 9-15(a) illustrate one application of parallel lines. Figure 9-15(b) shows one of the events of men's gymnastics, the parallel bars. Since perpendicular lines meet and form right angles, the monument and the ground shown in Figure 9-15(c) illustrate one application of perpendicular lines.

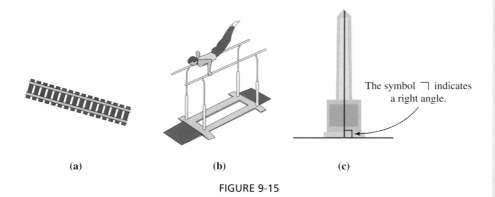

The symbol ⌐ indicates a right angle.

(a)　　　　　　　(b)　　　　　　　(c)

FIGURE 9-15

Parallel and perpendicular lines

If two lines lie in the same plane, they are called **coplanar.** Two coplanar lines that do not intersect are called **parallel lines.** See Figure 9-16(a).

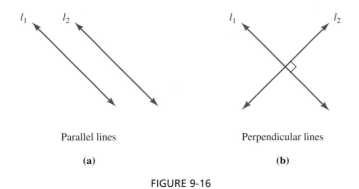

Parallel lines　　　　　　　Perpendicular lines

(a)　　　　　　　(b)

FIGURE 9-16

Parallel lines

Parallel lines are coplanar lines that do not intersect.

If lines l_1 (read as "l sub 1") and l_2 (read as "l sub 2") are parallel, we can write $l_1 \parallel l_2$, where the symbol $\parallel$ is read as "is parallel to."

Perpendicular lines

Perpendicular lines are lines that intersect and form right angles.

In Figure 9-16(b), $l_1 \perp l_2$, where the symbol $\perp$ is read as "is perpendicular to."

Transversals and angles

A line that intersects two or more coplanar lines is called a **transversal.** For example, line l_1 in Figure 9-17 is a transversal intersecting lines l_2, l_3, and l_4.

When two lines are cut by a transversal, the following types of angles are formed.

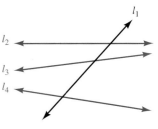

FIGURE 9-17

Alternate interior angles:

∠4 and ∠5

∠3 and ∠6

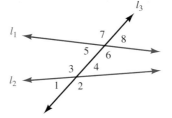

Corresponding angles:

∠1 and ∠5

∠3 and ∠7

∠2 and ∠6

∠4 and ∠8

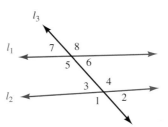

Interior angles:

∠3, ∠4, ∠5, and ∠6

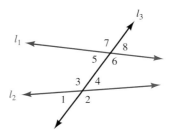

EXAMPLE 1   In Figure 9-18, identify   **a.** all pairs of alternate interior angles,   **b.** all pairs of corresponding angles,   and   **c.** all interior angles.

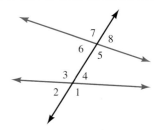

Solution

a. Pairs of alternate interior angles are

∠3 and ∠5, ∠4 and ∠6

b. Pairs of corresponding angles are

FIGURE 9-18

∠1 and ∠5, ∠4 and ∠8, ∠2 and ∠6, ∠3 and ∠7

c. Interior angles are

∠3, ∠4, ∠5, and ∠6

■ Properties of parallel lines

1. If two parallel lines are cut by a transversal, alternate interior angles are congruent. (See Figure 9-19.) If $l_1 \parallel l_2$, then $\angle 2 \cong \angle 4$ and $\angle 1 \cong \angle 3$.

2. If two parallel lines are cut by a transversal, corresponding angles are congruent. (See Figure 9-20.) If $l_1 \parallel l_2$, then $\angle 1 \cong \angle 5$, $\angle 3 \cong \angle 7$, $\angle 2 \cong \angle 6$, and $\angle 4 \cong \angle 8$.

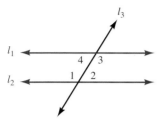

FIGURE 9-19

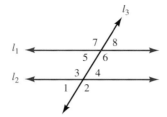

FIGURE 9-20

3. If two parallel lines are cut by a transversal, interior angles on the same side of the transversal are supplementary. (See Figure 9-21.) If $l_1 \parallel l_2$, then $\angle 1$ is supplementary to $\angle 2$ and $\angle 4$ is supplementary to $\angle 3$.

4. If a transversal is perpendicular to one of two parallel lines, it is also perpendicular to the other line. (See Figure 9-22.) If $l_1 \parallel l_2$ and $l_3 \perp l_1$, then $l_3 \perp l_2$.

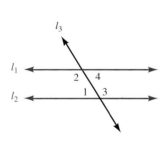

FIGURE 9-21

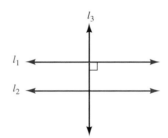

FIGURE 9-22

5. If two lines are parallel to a third line, they are parallel to each other. (See Figure 9-23.) If $l_1 \parallel l_2$ and $l_1 \parallel l_3$, then $l_2 \parallel l_3$.

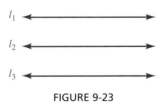

FIGURE 9-23

EXAMPLE 2   See Figure 9-24. If $l_1 \parallel l_2$ and $m(\angle 3) = 120°$, find the measures of the other angles.

Solution

$m(\angle 1) = 60°$ $\angle 3$ and $\angle 1$ are supplementary.

$m(\angle 2) = 120°$ Vertical angles are congruent: $m(\angle 2) = m(\angle 3)$.

$m(\angle 4) = 60°$ Vertical angles are congruent: $m(\angle 4) = m(\angle 1)$.

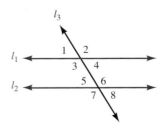

FIGURE 9-24

Self Check 2

If $l_1 \parallel l_2$ and $m(\angle 8) = 50°$, find the measures of the other angles. (See Figure 9-24.)

m($\angle$5) = 60°   If two parallel lines are cut by a transversal, alternate interior angles are congruent: m($\angle$5) = m($\angle$4).

m($\angle$6) = 120°   If two parallel lines are cut by a transversal, alternate interior angles are congruent: m($\angle$6) = m($\angle$3).

m($\angle$7) = 120°   Vertical angles are congruent: m($\angle$7) = m($\angle$6).

m($\angle$8) = 60°   Vertical angles are congruent: m($\angle$8) = m($\angle$5).

EXAMPLE 3   See Figure 9-25. If $\overline{AB} \parallel \overline{DE}$, which pairs of angles are congruent?

Solution   Since $\overline{AB} \parallel \overline{DE}$, corresponding angles are congruent. So we have

$$\angle A \cong \angle 1 \quad \text{and} \quad \angle B \cong \angle 2$$

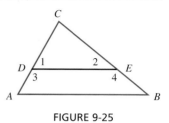

FIGURE 9-25

Self Check 4
In the figure below, $l_1 \parallel l_2$. Find y.

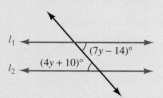

Answer 8

EXAMPLE 4   In Figure 9-26, $l_1 \parallel l_2$. Find x.

Solution   The angles involving x are corresponding angles. Since $l_1 \parallel l_2$, all pairs of corresponding angles are congruent.

$9x - 15 = 6x + 30$ The angle measures are equal.

$3x - 15 = 30$ Subtract $6x$ from both sides.

$\quad\quad 3x = 45$ To undo the subtraction of 15, add 15 to both sides.

$\quad\quad\quad x = 15$ To undo the multiplication by 3, divide both sides by 3.

Thus, x is 15.

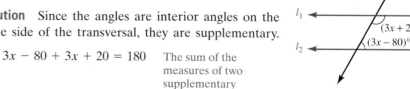

FIGURE 9-26

EXAMPLE 5   In Figure 9-27, $l_1 \parallel l_2$. Find x.

Solution   Since the angles are interior angles on the same side of the transversal, they are supplementary.

$3x - 80 + 3x + 20 = 180$ The sum of the measures of two supplementary angles is 180°.

$\quad\quad\quad 6x - 60 = 180$ Combine like terms.

$\quad\quad\quad\quad\quad 6x = 240$ To undo the subtraction of 60, add 60 to both sides.

$\quad\quad\quad\quad\quad\;\; x = 40$ To undo the multiplication by 6, divide both sides by 6.

Thus, x is 40.

FIGURE 9-27

Section 9.2 STUDY SET

VOCABULARY *Fill in the blanks.*

1. Two lines in the same plane are _____.
2. _____ lines do not intersect.
3. If two lines intersect and form right angles, they are _____.
4. A _____ intersects two or more coplanar lines.
5. In the following illustration, $\angle 4$ and $\angle 6$ are _____ interior angles.
6. In the following illustration, $\angle 2$ and $\angle 6$ are _____ angles.

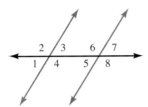

CONCEPTS

7. Which pairs of angles in the illustration above are alternate interior angles?

8. Which pairs of angles in the illustration above are corresponding angles?

9. Which angles shown in the illustration above are interior angles?

10. In the following illustration, $l_1 \parallel l_2$. What can you conclude about l_1 and l_3?

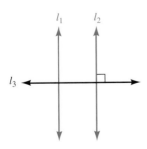

11. In the illustration, $l_1 \parallel l_2$ and $l_2 \parallel l_3$. What can you conclude about l_1 and l_3?

12. In the following illustration, $\overline{AB} \parallel \overline{DE}$. What pairs of angles are congruent?

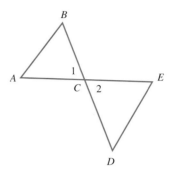

NOTATION *Fill in the blanks.*

13. The symbol ⌐ indicates _____.
14. The symbol $\parallel$ is read as "_____."
15. The symbol $\perp$ is read as "_____."
16. The symbol l_1 is read as "_____."

PRACTICE

17. In the illustration, $l_1 \parallel l_2$ and m($\angle 4$) = 130°. Find the measures of the other angles.

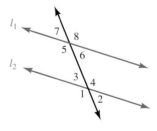

18. In the illustration, $l_1 \parallel l_2$ and m($\angle 2$) = 40°. Find the measures of the other angles.

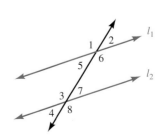

19. In the illustration, $l_1 \parallel \overrightarrow{AB}$. Find the measure of each angle.

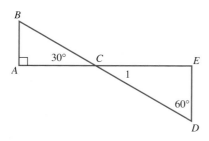

20. In the illustration, $\overline{AB} \parallel \overline{DE}$. Find m($\angle B$), m($\angle E$), and m($\angle 1$).

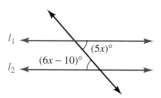

Find x given that $l_1 \parallel l_2$.

21.

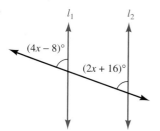

22.

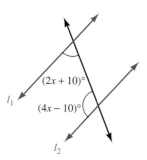

23.

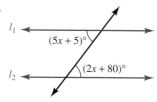

24.

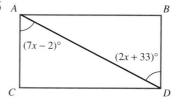

Find x.

25. $l_1 \parallel \overrightarrow{AC}$

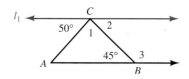

26. $\overline{AB} \parallel \overline{DE}$

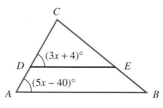

27. $\overline{AB} \parallel \overline{DE}$

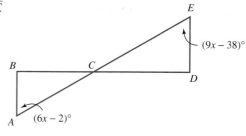

28. $\overline{AC} \parallel \overline{BD}$

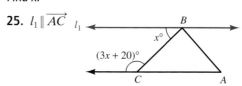

■ APPLICATIONS

29. CONSTRUCTING PYRAMIDS The Egyptians used a device called a **plummet** to determine whether stones were properly leveled. A plummet, shown in the illustration, is made up of an A-frame and a plumb bob suspended from the peak of the frame. How could a builder use a plummet to tell that the stone on the left is not level and that the stones on the right are level?

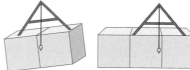

Plummet

Plumb bob

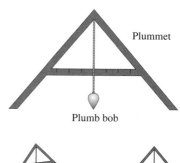

30. DIAGRAMMING SENTENCES English instructors have their students diagram sentences to help teach proper sentence structure. The illustration is a diagram of the sentence *The cave was rather dark and damp.* Point out pairs of parallel and perpendicular lines used in the diagram.

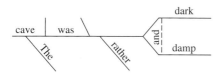

31. LOGOS Point out any perpendicular lines that can be found on the BMW company logo shown below.

32. PAINTING SIGNS For many sign painters, the most difficult letter to paint is a capital E, because of all of the right angles involved. How many right angles are there?

33. HANGING WALLPAPER Explain why the concepts of perpendicular and parallel are both important when hanging wallpaper.

34. TOOLS What geometric concepts are seen in the design of the rake?

WRITING

35. PARKING DESIGNS Using terms from this chapter, write a paragraph describing the parking layout shown below.

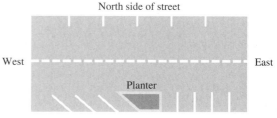

36. In your own words, explain what is meant by each of the following sentences.

  a. The hikers were told that the path *parallels* the river.

  b. John's quick rise to fame and fortune *paralleled* that of his older brother.

  c. The judge stated that the case that was before her court was without *parallel.*

37. Why do you think that $\angle 4$ and $\angle 6$ shown in the illustration for Problem 5 are called alternate interior angles?

38. Why do you think that $\angle 4$ and $\angle 8$ shown in the illustration for Problem 5 are called corresponding angles?

39. Are pairs of alternate interior angles always congruent? Explain.

40. Are pairs of interior angles always supplementary? Explain.

REVIEW

41. Find 60% of 120.

42. 80% of what number is 400?

43. What percent of 500 is 225?

44. Simplify: $3.45 + 7.37 \cdot 2.98$.

45. Is every whole number an integer?

46. Multiply: $2\frac{1}{5} \cdot 4\frac{3}{7}$.

47. Express the phrase as a ratio in lowest terms: 4 ounces to 12 ounces.

48. Convert 5,400 milligrams to kilograms.

9.3 Polygons

- Polygons • Triangles • Properties of isosceles triangles
- The sum of the measures of the angles of a triangle • Quadrilaterals
- Properties of rectangles • The sum of the measures of the angles of a polygon

In this section, we discuss figures called *polygons*. We see these shapes every day. For example, the walls in most buildings are rectangular. We also see rectangular shapes in doors, windows, and sheets of paper.

The gable ends of many houses are triangular, as are the sides of the Great Pyramid in Egypt. Triangular shapes are especially important because triangles are rigid and contribute strength and stability to walls and towers.

The designs in tile or linoleum floors often use the shapes of a pentagon or a hexagon. Stop signs are in the shape of an octagon.

Polygons

> **Polygon**
>
> A **polygon** is a closed geometric figure with at least three line segments for its sides.

The figures in Figure 9-28 are **polygons.** They are classified according to the number of sides they have. The points where the sides intersect are called **vertices.** If a polygon has sides that are all the same length and angles that have the same measure, we call it a **regular polygon.**

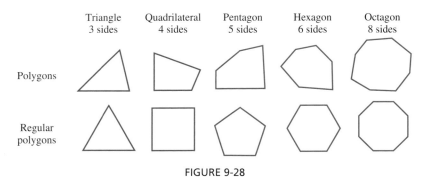

FIGURE 9-28

Self Check 1
Give the number of vertices of
a. a quadrilateral

b. a pentagon

Answers **a.** 4, **b.** 5

EXAMPLE 1 Give the number of vertices of **a.** a triangle and **b.** a hexagon.

Solution
a. From Figure 9-28, we see that a triangle has three angles and therefore three vertices.

b. From Figure 9-28, we see that a hexagon has six angles and therefore six vertices.

From the results of Example 1, we see that the number of vertices of a polygon is equal to a number of its sides.

Triangles

A **triangle** is a polygon with three sides. Figure 9-29 illustrates some common triangles. The slashes on the sides of a triangle indicate which sides are of equal length.

Vertex angle

Base angles

Equilateral triangle
(all sides equal length)

Isosceles triangle
(at least two sides of
equal length)

Scalene triangle
(no sides equal length)

Right triangle
(has a right angle)

FIGURE 9-29

! COMMENT Since equilateral triangles have at least two sides of equal length, they are also isosceles. However, isosceles triangles are not necessarily equilateral.

Since every angle of an equilateral triangle has the same measure, an equilateral triangle is also **equiangular.**

In an isosceles triangle, the angles opposite the sides of equal length are called **base angles,** the sides of equal length form the **vertex angle,** and the third side is called the **base.**

The longest side of a right triangle is called the **hypotenuse,** and the other two sides are called **legs.** The hypotenuse of a right triangle is always opposite the 90° angle.

Properties of isosceles triangles

1. Base angles of an isosceles triangle are congruent.
2. If two angles in a triangle are congruent, the sides opposite the angles have the same length, and the triangle is isosceles.

EXAMPLE 2   Is the triangle in Figure 9-30 an isosceles triangle?

Solution $\angle A$ and $\angle B$ are angles of the triangle. Since $m(\angle A) = m(\angle B)$, we know that $m(\overline{AC}) = m(\overline{BC})$ and that $\triangle ABC$ (read as "triangle ABC") is isosceles.

FIGURE 9-30

Self Check 2
In the figure below, $l_1 \parallel \overline{AB}$. Is the triangle an isosceles triangle?

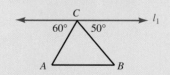

Answer no

The sum of the measures of the angles of a triangle

If you draw several triangles and carefully measure each angle with a protractor, you will find that the sum of the angle measures in each triangle is 180°.

Angles of a triangle

The sum of the angle measures of any triangle is 180°.

Self Check 3

In the figure below, find y.

Answer 90

EXAMPLE 3   See Figure 9-31. Find x.

FIGURE 9-31

Solution   Since the sum of the angle measures of any triangle is 180°, we have

$$x + 40 + 90 = 180$$

$\qquad x + 130 = 180$ Perform the addition: $40 + 90 = 130$.

$\qquad\qquad x = 50$ To undo the addition of 130, subtract 130 from both sides.

Thus, $x = 50$.

EXAMPLE 4   See Figure 9-32. If one base angle of an isosceles triangle measures 70°, how large is the vertex angle?

FIGURE 9-32

Solution   Since one of the base angles measures 70°, so does the other. If we let x represent the measure of the vertex angle, we have

$$x + 70 + 70 = 180$$ The sum of the measures of the angles of a triangle is 180°.

$\qquad x + 140 = 180$ Perform the addition: $70 + 70 = 140$.

$\qquad\qquad x = 40$ To undo the addition of 140, subtract 140 from both sides.

The vertex angle measures 40°.

Quadrilaterals

A **quadrilateral** is a polygon with four sides. Some common quadrilaterals are shown in Figure 9-33.

 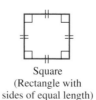

| Parallelogram | Rectangle | Square | Rhombus | Trapezoid |
| (Opposite sides parallel) | (Parallelogram with four right angles) | (Rectangle with sides of equal length) | (Parallelogram with sides of equal length) | (Exactly two sides parallel) |

FIGURE 9-33

Properties of rectangles

1. All angles of a rectangle are right angles.
2. Opposite sides of a rectangle are parallel.
3. Opposite sides of a rectangle are of equal length.
4. The diagonals of a rectangle are of equal length.
5. If the diagonals of a parallelogram are of equal length, the parallelogram is a rectangle.

EXAMPLE 5   Squaring a foundation.   A carpenter intends to build a shed with an 8-by-12-foot base. How can he make sure that the rectangular foundation is square?

Solution See Figure 9-34. The carpenter can use a tape measure to find the lengths of diagonals AC and BD. If these diagonals are of equal length, the figure will be a rectangle and have four right angles. Then the foundation will be square.

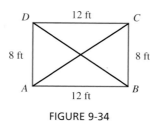

FIGURE 9-34

EXAMPLE 6 In rectangle $ABCD$ (Figure 9-35), the length of $\overline{AC}$ is 20 centimeters. Find each measure: **a.** m($\overline{BD}$), **b.** m($\angle 1$), and **c.** m($\angle 2$).

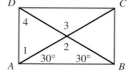

FIGURE 9-35

Solution

a. Since the diagonals of a rectangle are of equal length, m($\overline{BD}$) is also 20 centimeters.

b. We let m($\angle 1$) = x. Then, since the angles of a rectangle are right angles, we have

$$x + 30 = 90$$
$$x = 60 \quad \text{To undo the addition of 30, subtract 30 from both sides.}$$

Thus, m($\angle 1$) = 60°.

c. We let m($\angle 2$) = y. Then, since the sum of the angle measures of a triangle is 180°, we have

$$30 + 30 + y = 180$$
$$60 + y = 180 \quad \text{Simplify: } 30 + 30 = 60.$$
$$y = 120 \quad \text{To undo the addition of 60, subtract 60 from both sides.}$$

Thus, m($\angle 2$) = 120°.

Self Check 6
In rectangle $ABCD$ (Figure 9-35), the length of $\overline{DC}$ is 16 centimeters. Find each measure:

a. m($\overline{AB}$)

b. m($\angle 3$)

c. m($\angle 4$)

Answers **a.** 16 cm, **b.** 120°, **c.** 60°

The parallel sides of a trapezoid are called **bases,** the nonparallel sides are called **legs,** and the angles on either side of a base are called **base angles.** If the nonparallel sides are the same length, the trapezoid is an **isosceles trapezoid.** In an isosceles trapezoid, the base angles are congruent.

EXAMPLE 7 **Cross section of a drainage ditch.** A cross section of a drainage ditch (Figure 9-36) is an isosceles trapezoid with $\overline{AB} \parallel \overline{CD}$. Find x and y.

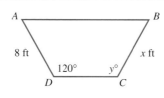

FIGURE 9-36

Solution Since the figure is an isosceles trapezoid, its nonparallel sides have the same length. So m($\overline{AD}$) and m($\overline{BC}$) are equal, and $x = 8$.

Since the base angles of an isosceles trapezoid are congruent, m($\angle D$) = m($\angle C$). Thus, y is 120.

■ The sum of the measures of the angles of a polygon

We have seen that the sum of the angle measures of any triangle is 180°. Since a polygon with n sides can be divided into $n - 2$ triangles, the sum of the angle measures of the polygon in degrees is $(n - 2)180$.

> **Angles of a polygon**
>
> The sum S, in degrees, of the measures of the angles of a polygon with n sides is given by the formula
>
> $$S = (n - 2)180$$

Self Check 8
Find the sum of the angle measures of a quadrilateral.

EXAMPLE 8   Find the sum of the angle measures of a pentagon.

Solution   Since a pentagon has 5 sides, we substitute 5 for n in the formula and simplify.

$$S = (\boldsymbol{n} - 2)180$$
$$= (\boldsymbol{5} - 2)180 \qquad \text{Substitute 5 for } n.$$
$$= (3)180 \qquad \text{Perform the subtraction within the parentheses.}$$
$$= 540$$

Answer 360°

The sum of the angles of a pentagon is 540°.

Self Check 9
The sum of the measures of the angles of a polygon is 720°. Find the number of sides the polygon has.

EXAMPLE 9   The sum of the measures of the angles of a polygon is 1,080°. Find the number of sides the polygon has.

Solution   To find the number of sides the polygon has, we substitute 1,080 for S in the formula and then solve for n.

$$S = (n - 2)180$$
$$\boldsymbol{1,080} = (n - 2)180 \qquad \text{Substitute 1,080 for } S.$$
$$1,080 = 180n - 360 \qquad \text{Distribute the multiplication by 180.}$$
$$1,080 + \boldsymbol{360} = 180n - 360 + \boldsymbol{360} \qquad \text{To undo the subtraction of 360, add 360 to both sides.}$$
$$1,440 = 180n \qquad \text{Simplify.}$$
$$\frac{1,440}{\boldsymbol{180}} = \frac{180n}{\boldsymbol{180}} \qquad \text{To undo the multiplication of 180, divide both sides by 180.}$$
$$8 = n \qquad \text{Perform the division: } \tfrac{1,440}{180} = 8.$$

Answer 6

The polygon has 8 sides. It is an octagon.

Section 9.3   STUDY SET

■ **VOCABULARY**   *Fill in the blanks.*

1. A _____ polygon has sides that are all the same length and angles that all have the same measure.

2. A polygon with four sides is called a _____.
   A _____ is a polygon with three sides.

3. A _____ is a polygon with six sides.

4. A polygon with five sides is called a _____.

5. An eight-sided polygon is an _____.

6. The points where the sides of a polygon intersect are called _____.

7. A triangle with three sides of equal length is called an _____ triangle.

8. An _____ triangle has two sides of equal length.

9. The longest side of a right triangle is the _____.

10. The _____ angles of an isosceles triangle have the same measure.

11. A _____ with a right angle is a rectangle.

12. A rectangle with all sides of equal length is a _____.

13. A _____ is a parallelogram with four sides of equal length.

14. A _____ has two sides that are parallel and two sides that are not parallel.

15. The legs of an _____ trapezoid have the same length.

16. The _____ of a polygon is the distance around it.

CONCEPTS *Give the number of sides each polygon has and classify it as a triangle, quadrilateral, pentagon, hexagon, or octagon. Then give the number of vertices it has.*

17.

18.

19.

20.

21.

22.

23.

24.

Classify each triangle as an equilateral triangle, an isosceles triangle, a scalene triangle, or a right triangle.

25.

26.

$55°$
$55°$

27.

28.

29.

$60°$ $60°$
$60°$

30.

$30°$
$60°$

31.

20 cm 20 cm

32.

$50°$
$70°$
$60°$

Classify each quadrilateral as a rectangle, a square, a rhombus, or a trapezoid. More than one name can be used for some figures.

33.

4 in.
4 in. 4 in.
4 in.

34.

35.

36.

5.5 ft
$90°$
$90°$ 3.7 ft

37.

$2\frac{1}{3}$ yd
$4\frac{2}{3}$ yd

38.

8 cm
8 cm 8 cm
8 cm

39.

40.

NOTATION *Fill in the blanks.*

41. The symbol $\triangle$ means _____.

42. The symbol $m(\angle 1)$ means the _____ of angle 1.

PRACTICE *See the following illustration. The measures of two angles of △ABC are given. Find the measure of the third angle.*

43. $m(\angle A) = 30°$ and $m(\angle B) = 60°$
$m(\angle C) = $ _____.

44. $m(\angle A) = 45°$ and $m(\angle C) = 105°$
$m(\angle B) = $ _____.

45. $m(\angle B) = 100°$ and $m(\angle A) = 35°$
$m(\angle C) = $ _____.

46. $m(\angle B) = 33°$ and $m(\angle C) = 77°$
$m(\angle A) = $ _____.

47. $m(\angle A) = 25.5°$ and $m(\angle B) = 63.8°$
$m(\angle C) = $ _____.

48. $m(\angle B) = 67.25°$ and
$m(\angle C) = 72.5°$
$m(\angle A) = $ _____.

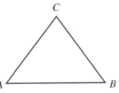

Refer to rectangle ABCD, shown below.

49. $m(\angle 1) = $ _____.
50. $m(\angle 3) = $ _____.
51. $m(\angle 2) = $ _____.
52. If $m(\overline{AC})$ is 8 cm, then $m(\overline{BD}) = $ _____.

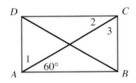

Find the sum of the angle measures of each polygon.

53. A hexagon
54. An octagon
55. A decagon (10 sides)
56. A dodecagon (12 sides)

Find the number of sides a polygon has if the sum of its angle measures is the given number.

57. 900°
58. 1,260°
59. 2,160°
60. 3,600°

APPLICATIONS

61. Give three uses of triangles in everyday life.
62. Give three uses of rectangles in everyday life.
63. Give three uses of squares in everyday life.
64. Give a use of a trapezoid in everyday life.

65. POLYGONS IN NATURE As we see in illustration (a), a starfish has the approximate shape of a pentagon. What approximate polygon shape do you see in each of the other objects? **b.** Lemon
c. Chili pepper   **d.** Apple

(a)   (b)

(c)   (d)

66. FLOWCHARTS A flowchart shows a sequence of steps to be performed by a computer to solve a given problem. When designing a flowchart, the programmer uses a set of standardized symbols to represent various operations to be performed by the computer. Locate a rectangle, a rhombus, and a parallelogram in the flowchart shown.

67. CHEMISTRY Polygons are used to represent the chemical structure of compounds graphically. In the illustration, what types of polygons are used to represent methylprednisolone, the active ingredient in an antiinflammatory medication?

Methylprednisolone

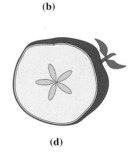

68. PODIUMS What polygon describes the shape of the upper portion of the podium?

69. EASELS Show how two of the legs of the easel form the equal sides of an isosceles triangle.

70. AUTOMOBILE JACKS Refer to the illustration. Show that no matter how high the jack is raised, it always forms two isosceles triangles.

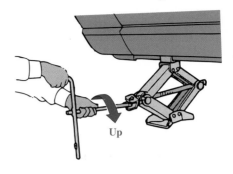

WRITING

71. Explain why a square is a rectangle.

72. Explain why a trapezoid is not a parallelogram.

REVIEW

73. Find 20% of 110.

74. Find 15% of 50.

75. What percent of 200 is 80?

76. 20% of what number is 500?

77. Simplify: $0.85 \div 2(0.25)$.

78. FIRST AID When checking an accident victim's pulse, a paramedic counted 13 beats during a 15-second span. How many beats would be expected in 60 seconds?

9.4 Properties of Triangles

- Congruent triangles • Similar triangles • The Pythagorean theorem

Proportions and triangles are often used to measure distances indirectly. For example, by using a proportion, Eratosthenes (275–195 B.C.) was able to estimate the circumference of the Earth with remarkable accuracy. On a sunny day, we can use properties of similar triangles to calculate the height of a tree while staying safely on the ground. By using a theorem proved by the Greek mathematician Pythagoras (about 500 B.C.), we can calculate the length of the third side of a right triangle whenever we know the lengths of two sides.

Congruent triangles

Triangles that have the same area and the same shape are called **congruent triangles.** In Figure 9-37, triangles ABC and DEF are congruent.

$$\triangle ABC \cong \triangle DEF \quad \text{Read as "Triangle } ABC \text{ is congruent to triangle } DEF.\text{"}$$

Corresponding angles and corresponding sides of congruent triangles are called **corresponding parts.** The notation $\triangle ABC \cong \triangle DEF$ shows which vertices are corresponding parts.

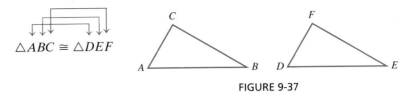

FIGURE 9-37

Corresponding parts of congruent triangles always have the same measure. For the congruent triangles shown in Figure 9-37,

$$m(\angle A) = m(\angle D), \quad m(\angle B) = m(\angle E), \quad m(\angle C) = m(\angle F),$$
$$m(\overline{BC}) = m(\overline{EF}), \quad m(\overline{AC}) = m(\overline{DF}), \quad m(\overline{AB}) = m(\overline{DE})$$

EXAMPLE 1 Name the corresponding parts of the congruent triangles in Figure 9-38.

Solution The corresponding angles are

$\angle A$ and $\angle E, \quad \angle B$ and $\angle D, \quad \angle C$ and $\angle F$

Since corresponding sides are always opposite corresponding angles, the corresponding sides are

$\overline{BC}$ and $\overline{DF}, \quad \overline{AC}$ and $\overline{EF}, \quad \overline{AB}$ and $\overline{ED}$

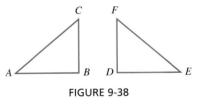

FIGURE 9-38

We will discuss three ways of showing that two triangles are congruent.

SSS property

If three sides of one triangle are congruent to three sides of a second triangle, the triangles are congruent.

The triangles in Figure 9-39 are congruent because of the SSS property.

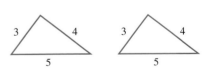

FIGURE 9-39

SAS property

If two sides and the angle between them in one triangle are congruent, respectively, to two sides and the angle between them in a second triangle, the triangles are congruent.

The triangles in Figure 9-40 are congruent because of the SAS property.

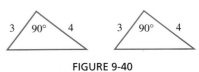

FIGURE 9-40

ASA property

If two angles and the side between them in one triangle are congruent, respectively, to two angles and the side between them in a second triangle, the triangles are congruent.

The triangles in Figure 9-41 are congruent because of the ASA property.

FIGURE 9-41

! COMMENT There is no SSA property. To illustrate this, consider the triangles in Figure 9-42. Two sides and an angle of △*ABC* are congruent to two sides and an angle of △*DEF*. But the congruent angle is not between the congruent sides.

We refer to this situation as SSA. Obviously, the triangles are not congruent, because they have different areas.

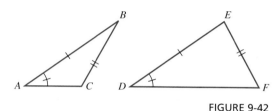

FIGURE 9-42

The slash marks indicate congruent parts. That is, the sides with one slash are the same length, the sides with two slashes are the same length, and the angles with one slash have the same measure.

EXAMPLE 2 Explain why the triangles in Figure 9-43 are congruent.

Solution Since vertical angles are congruent,

$$m(\angle 1) = m(\angle 2)$$

From the figure, we see that

$$m(\overline{AC}) = m(\overline{EC}) \quad \text{and} \quad m(\overline{BC}) = m(\overline{DC})$$

Since two sides and the angle between them in one triangle are congruent, respectively, to two sides and the angle between them in a second triangle, △*ABC* ≅ △*EDC* by the SAS property.

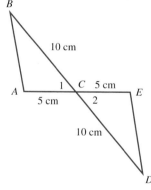

FIGURE 9-43

Similar triangles

If two angles of one triangle are congruent to two angles of a second triangle, the triangles have the same shape. Triangles with the same shape are called **similar triangles.** In Figure 9-44, △*ABC* ~ △*DEF* (read the symbol ~ as "is similar to").

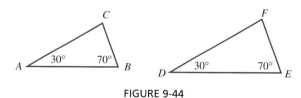

FIGURE 9-44

! COMMENT Note that congruent triangles are always similar, but similar triangles are not always congruent.

> **Property of similar triangles**
>
> If two triangles are similar, all pairs of corresponding sides are in proportion.

In the similar triangles shown in Figure 9-44 on the previous page, the following proportions are true.

$$\frac{AB}{DE} = \frac{BC}{EF}, \qquad \frac{BC}{EF} = \frac{CA}{FD}, \qquad \text{and} \qquad \frac{CA}{FD} = \frac{AB}{DE}$$

EXAMPLE 3   Finding the height of a tree.   A tree casts a shadow 18 feet long at the same time a woman 5 feet tall casts a shadow that is 1.5 feet long. (See Figure 9-45.) Find the height of the tree.

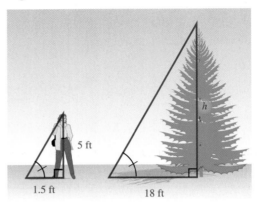

FIGURE 9-45

Solution   The figure shows the triangles determined by the tree and its shadow and by the woman and her shadow. Since the triangles have the same shape, they are similar, and the lengths of their corresponding sides are in proportion. If we let h represent the height of the tree, we can find h by solving the following proportion.

$$\frac{h}{5} = \frac{18}{1.5} \qquad \frac{\text{Height of the tree}}{\text{Height of the woman}} = \frac{\text{shadow of the tree}}{\text{shadow of the woman}}$$

$1.5h = 5(18)$ In a proportion, the product of the extremes is equal to the product of the means.

$1.5h = 90$ Perform the multiplication: $5(18) = 90$.

$h = 60$ To undo the multiplication by 1.5, divide both sides by 1.5.

The tree is 60 feet tall.

▌ The Pythagorean theorem

In the movie *The Wizard of Oz,* the scarecrow was in search of a brain. To prove that he had found one, he tried to recite the Pythagorean theorem. In words, the Pythagorean theorem can be stated as:

> *In a right triangle, the square of the hypotenuse is equal to the sum of squares of the other two sides.*

> **Pythagorean theorem**
>
> If the length of the hypotenuse of a right triangle is c and the lengths of its legs are a and b, then
>
> $a^2 + b^2 = c^2$

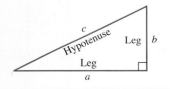

EXAMPLE 4 **Constructing a high-ropes adventure course.** A builder of a high-ropes adventure course wants to secure the pole shown in Figure 9-46 by attaching a cable from the anchor stake 8 feet from its base to a point 6 feet up the pole. How long should the cable be?

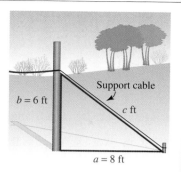

Support cable
$b = 6$ ft
c ft
$a = 8$ ft

FIGURE 9-46

Solution The support cable, the pole, and the ground form a right triangle. If we let c represent the length of the cable (the hypotenuse), then we can use the Pythagorean theorem with $a = 8$ and $b = 6$ to find c.

$c^2 = a^2 + b^2$ This is the Pythagorean theorem.

$c^2 = 8^2 + 6^2$ Substitute 8 for a and 6 for b.

$c^2 = 64 + 36$ Evaluate the exponential expressions.

$c^2 = 100$ Simplify the right-hand side.

To find c, we must find a number that, when squared, is 100. There are two such numbers, one positive and one negative; they are the square roots of 100. Since c represents the length of a support cable, c cannot be negative. For this reason, we need only find the positive square root of 100 to get c.

$c^2 = 100$ This is the equation to solve.

$c = \sqrt{100}$ The symbol $\sqrt{}$ is used to indicate the positive square root of a number.

$c = 10$ $\sqrt{100} = 10$, because $10^2 = 100$.

The support cable should be 10 feet long.

Self Check 4
A 26-foot ladder rests against the side of a building. If the base of the ladder is 10 feet from the wall, how far up the side of the building does the ladder reach?

Answer 24 ft

Finding the width of a television screen   CALCULATOR SNAPSHOT

The size of a television screen is the diagonal measure of its rectangular screen. (See Figure 9-47.) To find the width of a 27-inch screen that is 17 inches high, we use the Pythagorean theorem with $c = 27$ and $b = 17$.

$$c^2 = a^2 + b^2$$
$$27^2 = a^2 + 17^2$$
$$27^2 - 17^2 = a^2$$

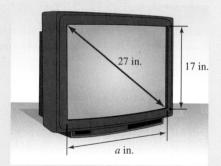

27 in.
17 in.
a in.

FIGURE 9-47

The variable a represents the width of a television screen, so it must be positive. To find a, we find the positive square root of the result when 17^2 is subtracted from 27^2. Using a radical symbol to indicate this, we have

$$\sqrt{27^2 - 17^2} = a$$

We can evaluate the expression on the left-hand side by entering these numbers and pressing these keys.

(27 x^2 − 17 x^2) $\sqrt{}$ 20.97617696

To the nearest inch, the width of the television screen is 21 inches.

It is also true that

If the square of one side of a triangle is equal to the sum of the squares of the other two sides, the triangle is a right triangle.

Self Check 5
Is a triangle with sides of 9, 40, and 41 meters a right triangle?

EXAMPLE 5   Is a triangle with sides of 5, 12, and 13 meters a right triangle?

Solution   We can use the Pythagorean theorem to answer this question. Since the longest side of the triangle is 13 meters, we must substitute 13 for c. It doesn't matter which of the two remaining side lengths we substitute for a and which we substitute for b.

$$c^2 = a^2 + b^2 \qquad \text{This is the Pythagorean theorem.}$$
$$13^2 \stackrel{?}{=} 5^2 + 12^2 \qquad \text{Substitute 13 for } c, \text{ 5 for } a, \text{ and 12 for } b.$$
$$169 \stackrel{?}{=} 25 + 144 \qquad \text{Evaluate the exponential expressions.}$$
$$169 = 169 \qquad \text{Simplify the right-hand side.}$$

Since the square of the longest side is equal to the sum of the squares of the other two sides, the triangle is a right triangle.

Answer yes

Self Check 6
Is a triangle with sides of 4, 5, and 6 inches a right triangle?

EXAMPLE 6   Is a triangle with sides of 1, 2, and 3 feet a right triangle?

Solution   We check to see whether the square of the longest side is equal to the sum of the squares of the other two sides.

$$c^2 = a^2 + b^2 \qquad \text{This is the Pythagorean theorem.}$$
$$3^2 \stackrel{?}{=} 1^2 + 2^2 \qquad \text{Substitute 3 for } c, \text{ 1 for } a, \text{ and 2 for } b.$$
$$9 \stackrel{?}{=} 1 + 4 \qquad \text{Evaluate the exponential expressions.}$$
$$9 \neq 5 \qquad \text{Simplify the right-hand side.}$$

Since the square of the longest side is not equal to the sum of the squares of the other two sides, the triangle is not a right triangle.

Answer no

Section 9.4   STUDY SET

VOCABULARY   *Fill in the blanks.*

1. _____ triangles are the same size and the same shape.

2. All _____ parts of congruent triangles have the same measure.

3. If two triangles are _____, they have the same shape.

4. The _____ is the longest side of a right triangle.

CONCEPTS   *Determine whether each statement is true. If a statement is false, tell why.*

5. If three sides of one triangle are the same length as three sides of a second triangle, the triangles are congruent.

6. If two sides of one triangle are the same length as two sides of a second triangle, the triangles are congruent.

7. If two sides and an angle of one triangle are congruent, respectively, to two sides and an angle of a second triangle, the triangles are congruent.

8. If two angles and the side between them in one triangle are congruent, respectively, to two angles and the side between them in a second triangle, the triangles are congruent.

9. In a proportion, the product of the means is equal to the product of the extremes.

10. If two angles of one triangle are congruent to two angles of a second triangle, the triangles are similar.

11. Are the triangles shown below congruent?

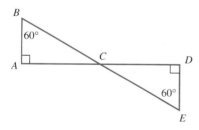

12. Are the triangles shown below congruent?

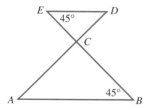

13. Are the triangles shown below similar?

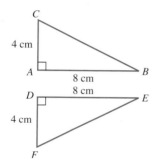

14. Are the triangles shown below similar?

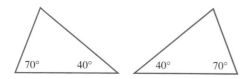

15. The Pythagorean theorem states that for a right triangle, $c^2 = a^2 + b^2$. What do the variables a, b, and c represent?

16. A triangle has sides of length 3, 4, and 5 centimeters. Substitute the lengths into $c^2 = a^2 + b^2$ and show that a true statement results. From the result, what can we conclude about the triangle?

17. Suppose that c represents the length of the hypotenuse of a right triangle and $c^2 = 25$. Fill in the blanks: To find c, we must find a number that, when squared, is ▧. Since c represents a positive number, we need only find the positive _____ _____ of 25 to get c.

$$c^2 = 25$$
$$▨ = \sqrt{25}$$ A radical symbol indicates the positive square root.
$$c = ▧$$

18. Solve: $\dfrac{h}{2.6} = \dfrac{27}{13}$.

◼ NOTATION *Fill in the blanks.*

19. The symbol ≅ is read as "_____."

20. The symbol ~ is read as "_____."

◼ PRACTICE *Name the corresponding parts of the congruent triangles.*

21. Refer to the illustration.

$\overline{AC}$ corresponds to ____.

$\overline{DE}$ corresponds to ____.

$\overline{BC}$ corresponds to ____.

$\angle A$ corresponds to ____.

$\angle E$ corresponds to ____.

$\angle F$ corresponds to ____.

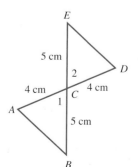

22. Refer to the illustration.

$\overline{AB}$ corresponds to ____.

$\overline{EC}$ corresponds to ____.

$\overline{AC}$ corresponds to ____.

$\angle D$ corresponds to ____.

$\angle B$ corresponds to ____.

$\angle 1$ corresponds to ____.

Determine whether each pair of triangles is congruent.
If they are explain why.

23.
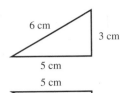

24.
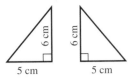

36.
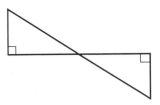

Refer to the illustration below and find the length of the unknown side.

37. $a = 3$ and $b = 4$. Find c.

38. $a = 12$ and $b = 5$. Find c.

39. $a = 15$ and $c = 17$. Find b.

40. $b = 45$ and $c = 53$. Find a.

41. $a = 5$ and $c = 9$. Find b.

42. $a = 1$ and $b = 7$. Find c.

25.

26.
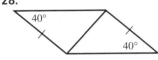

27.

28.
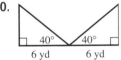

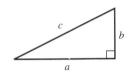

29.
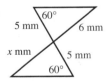

30.
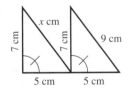

The lengths of the three sides of a triangle are given. Determine whether the triangle is a right triangle.

43. 8, 15, 17   **44.** 6, 8, 10

45. 7, 24, 26   **46.** 9, 39, 40

Find x.

31.
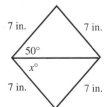

32.
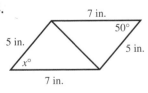

APPLICATIONS   *Solve each problem. If an answer is not exact, give the answer to the nearest tenth.*

47. HEIGHT OF A TREE The tree in the illustration casts a shadow 24 feet long when a man 6 feet tall casts a shadow 4 feet long. Find the height of the tree.

33.
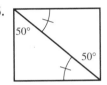

34.

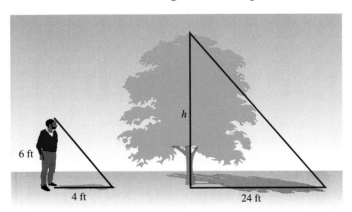

Determine whether the triangles are similar.

35.
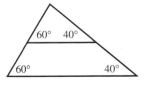

48. HEIGHT OF A BUILDING A man places a mirror on the ground and sees the reflection of the top of a building, as shown on the next page. Find the height of the building.

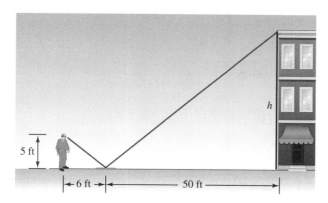

49. WIDTH OF A RIVER Use the dimensions in the illustration to find *w*, the width of the river.

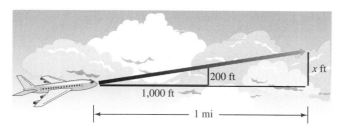

50. FLIGHT PATHS The airplane shown below ascends 200 feet as it flies a horizontal distance of 1,000 feet. How much altitude is gained as it flies a horizontal distance of 1 mile? (*Hint:* 1 mile = 5,280 feet.)

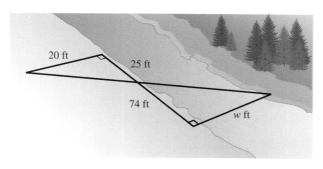

51. FLIGHT PATHS An airplane descends 1,200 feet as it files a horizontal distance of 1.5 miles. How much altitude is lost as it flies a horizontal distance of 5 miles?

52. GEOMETRY If segment *DE* in the following illustration is parallel to segment *AB*, △*ABC* is similar to △*DEC*. Find *x*.

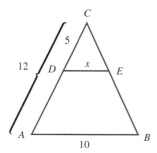

53. ADJUSTING LADDERS A 20-foot ladder reaches a window 16 feet above the ground. How far from the wall is the base of the ladder?

54. LENGTH OF GUY WIRES A 30-foot tower is to be fastened by three guy wires attached to the top of the tower and to the ground at positions 20 feet from its base. How much wire is needed?

55. PICTURE FRAMES After gluing and nailing two pieces of picture frame molding together, a frame maker checks her work by making a diagonal measurement. If the sides of the frame form a right angle, what measurement should the frame maker read on the yardstick?

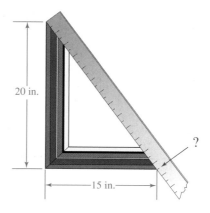

56. CARPENTRY The gable end of the roof shown is divided in half by a vertical brace, 8 feet in height. Find the length of the roof line.

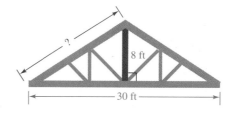

57. BASEBALL A baseball diamond is a square with each side 90 feet long. How far is it from home plate to second base?

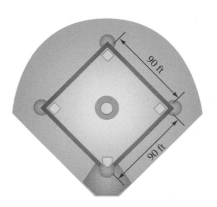

58. TELEVISION What size is the television screen shown?

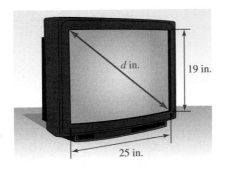

■ **WRITING**

59. Explain the Pythagorean theorem.

60. Explain the procedure used to solve the equation $c^2 = 64$. (Assume that c is positive.)

■ **REVIEW**   *Estimate the answer to each problem.*

61. $\dfrac{0.95 \cdot 3.89}{2.997}$ **62.** 21% of 42

63. 32% of 60   **64.** $\dfrac{4.966 + 5.001}{2.994}$

65. 49.5% of 18.1   **66.** 98.7% of 0.03

9.5 Perimeters and Areas of Polygons

- Perimeters of polygons • Perimeters of figures that are combinations of polygons
- Areas of polygons • Areas of figures that are combinations of polygons

In this section, we discuss how to find perimeters and areas of polygons. Finding perimeters is important when estimating the cost of fencing or the cost of woodwork in a house. Finding areas is important when calculating the cost of carpeting, the cost of painting a house, or the cost of fertilizing a yard.

■ Perimeters of polygons

Recall that the **perimeter** of a polygon is the distance around it. Since a square has four sides of equal length s, its perimeter P is $s + s + s + s$, or $4s$.

> **Perimeter of a square**
>
> If a square has a side of length s, its perimeter P is given by the formula
>
> $$P = 4s$$

Self Check 1

Find the perimeter of a square whose sides are 23.75 centimeters long.

EXAMPLE 1   Find the perimeter of a square whose sides are 7.5 meters long.

Solution   The perimeter of a square is given by the formula $P = 4s$. We substitute 7.5 for s and simplify.

$$P = 4s$$
$$= 4(7.5)$$
$$= 30$$

The perimeter is 30 meters.

Answer 95 cm

Since a rectangle has two lengths *l* and two widths *w*, its perimeter *P* is *l* + *l* + *w* + *w*, or 2*l* + 2*w*.

Perimeter of a rectangle

If a rectangle has length *l* and width *w*, its perimeter *P* is given by the formula

$$P = 2l + 2w$$

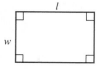

EXAMPLE 2 Find the perimeter of the rectangle in Figure 9-48.

Solution The perimeter is given by the formula $P = 2l + 2w$. We substitute 10 for *l* and 6 for *w* and simplify.

$$P = 2l + 2w$$
$$= 2(\mathbf{10}) + 2(\mathbf{6})$$
$$= 20 + 12$$
$$= 32$$

The perimeter is 32 centimeters.

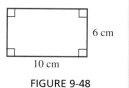

FIGURE 9-48

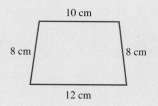

Self Check 2
Find the perimeter of the isosceles trapezoid below.

Answer **38 cm**

EXAMPLE 3 Find the perimeter of the rectangle in Figure 9-49, in meters.

Solution Since 1 meter = 100 centimeters, we can convert 80 centimeters to meters by multiplying 80 centimeters by the unit conversion factor $\frac{1\,m}{100\,cm}$.

$$80 \text{ cm} = 80 \text{ cm} \cdot \frac{1 \text{ m}}{100 \text{ cm}} \qquad \text{Multiply by 1: } \frac{1 \text{ m}}{100 \text{ cm}} = 1.$$

$$= \frac{80}{100} \text{ m} \qquad \text{The units of centimeters divide out.}$$

$$= 0.8 \text{ m} \qquad \text{Divide by 100 by moving the decimal point 2 places to the left.}$$

We can now substitute 3 for *l* and 0.8 for *w* to get

$$P = 2l + 2w$$
$$= 2(\mathbf{3}) + 2(\mathbf{0.8})$$
$$= 6 + 1.6$$
$$= 7.6$$

The perimeter is 7.6 meters.

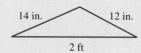

FIGURE 9-49

Self Check 3
Find the perimeter of the triangle below, in inches.

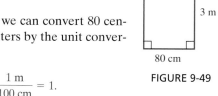

Answer **50 in.**

EXAMPLE 4 The perimeter of the isosceles triangle in Figure 9-50 is 45 meters. Find the length of its base.

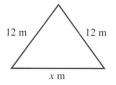

FIGURE 9-50

Self Check 4
The perimeter of an isosceles triangle is 58 meters. If one of its sides of equal length is 15 meters long, how long is its base?

Solution Two sides are 12 meters long, and the perimeter is 45 meters. If x represents the length of the base, we have

$$12 + 12 + x = 45$$
$$24 + x = 45 \qquad \text{Simplify: } 12 + 12 = 24.$$
$$24 + x - 24 = 45 - 24 \qquad \text{To undo the addition of 24, subtract 24 from both sides.}$$
$$x = 21$$

Answer 28 m

The length of the base is 21 meters.

Perimeters of figures that are combinations of polygons

CALCULATOR SNAPSHOT   **Perimeter of a figure**

See Figure 9-51. To find the perimeter, we need to know the values of x and y. Since the figure is a combination of two rectangles, we can use a calculator to see that

$$x = 20.25 - 10.17 \quad \text{and} \quad y = 12.5 - 4.75$$
$$= 10.08 \qquad\qquad\qquad = 7.75$$

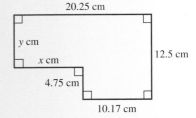

FIGURE 9-51

The perimeter P of the figure is

$$P = 20.25 + 12.5 + 10.17 + 4.75 + \textbf{\textit{x}} + \textbf{\textit{y}}$$
$$= 20.25 + 12.5 + 10.17 + 4.75 + \textbf{10.08} + \textbf{7.75}$$

We can use a calculator to evaluate the expression on the right-hand side by entering these numbers and pressing these keys.

20.25 $\boxed{+}$ 12.5 $\boxed{+}$ 10.17 $\boxed{+}$ 4.75 $\boxed{+}$ 10.08 $\boxed{+}$ 7.75 $\boxed{=}$

$$\boxed{ 65.5}$$

The perimeter is 65.5 centimeters.

Areas of polygons

Recall that the **area** of a polygon is the measure of the amount of surface it encloses. Area is measured in square units, such as square inches or square centimeters. See Figure 9-52.

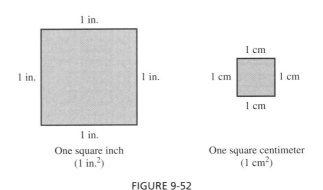

One square inch
(1 in.²)

One square centimeter
(1 cm²)

FIGURE 9-52

In everyday life, we often use areas. For example,

- To carpet a room, we buy square yards.
- A can of paint will cover a certain number of square feet.
- To measure vast amounts of land, we often use square miles.
- We buy house roofing by the "square." One square is 100 square feet.

The rectangle shown in Figure 9-53 has a length of 10 centimeters and a width of 3 centimeters. If we divide the rectangle into squares as shown in the figure, each square represents an area of 1 square centimeter — a surface enclosed by a square measuring 1 centimeter on each side. Because there are 3 rows with 10 squares in each row, there are 30 squares. Since the rectangle encloses a surface area of 30 squares, its area is 30 square centimeters, often written as 30 cm^2.

This example illustrates that to find the area of a rectangle, we multiply its length by its width.

10 cm

3 cm

1 cm^2

FIGURE 9-53

! COMMENT Do not confuse the concepts of perimeter and area. Perimeter is the distance around a polygon. It is measured in linear units, such as centimeters, feet, or miles. Area is a measure of the surface enclosed within a polygon. It is measured in square units, such as square centimeters, square feet, or square miles.

In practice, we do not find areas by counting squares in a figure. Instead, we use formulas for finding areas of geometric figures, as shown in Table 9-1.

Figure	Name	Formula for area
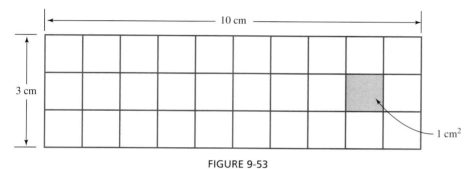	Square	$A = s^2$, where s is the length of one side.
	Rectangle	$A = lw$, where l is the length and w is the width.
	Parallelogram	$A = bh$, where b is the length of the base and h is the height. (A height is always perpendicular to the base.)

(continued)

Figure	Name	Formula for area
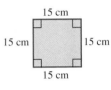	Triangle	$A = \frac{1}{2}bh$, where b is the length of the base and h is the height. The segment perpendicular to the base and representing the height is called an **altitude.**
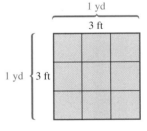	Trapezoid	$A = \frac{1}{2}h(b_1 + b_2)$, where h is the height of the trapezoid and b_1 and b_2 represent the lengths of the bases.

TABLE 9-1

Self Check 5

Find the area of the square shown below.

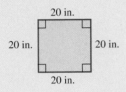

20 in.

20 in.   20 in.

20 in.

Answer 400 in.²

EXAMPLE 5   Find the area of the square in Figure 9-54.

Solution   We can see that the length of one side of the square is 15 centimeters. We can find its area by using the formula $A = s^2$ and substituting 15 for s.

$$A = s^2$$
$$= (\mathbf{15})^2 \quad \text{Substitute 15 for } s.$$
$$= 225 \quad \text{Evaluate the exponential expression: } 15 \cdot 15 = 225.$$

The area of the square is 225 cm².

15 cm

15 cm   15 cm

15 cm

FIGURE 9-54

Self Check 6

Find the number of square centimeters in 1 square meter.

Answer 10,000 cm²

EXAMPLE 6   Find the number of square feet in 1 square yard. (See Figure 9-55.)

Solution   Since 3 feet = 1 yard, each side of 1 square yard is 3 feet long.

$$1 \text{ yd}^2 = (\mathbf{1} \text{ yd})^2$$
$$= (\mathbf{3} \text{ ft})^2 \quad \text{Substitute 3 feet for 1 yard.}$$
$$= 9 \text{ ft}^2 \quad (3 \text{ ft})^2 = (3 \text{ ft})(3 \text{ ft}) = 9 \text{ ft}^2.$$

There are 9 square feet in 1 square yard.

1 yd
3 ft

1 yd { 3 ft

FIGURE 9-55

Self Check 7

Find the area in square inches of a rectangle with dimensions of 6 inches by 2 feet.

EXAMPLE 7 Women's sports.
Field hockey is a team sport in which players use sticks to try to hit a ball into their opponents' goal. Find the area of the rectangular field shown in Figure 9-56. Give the answer in square feet.

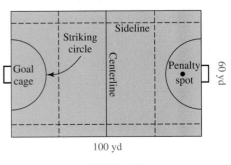

100 yd

FIGURE 9-56

Solution   To find the area in square yards, we substitute 100 for l and 60 for w in the formula for the area of a rectangle, and simplify.

$$A = lw$$
$$= 100(60)$$
$$= 6,000$$

The area is 6,000 square yards. Since there are 9 square feet per square yard, we can convert this number to square feet by multiplying 6,000 square yards by $\frac{9 \text{ ft}^2}{1 \text{ yd}^2}$.

$$6,000 \text{ yd}^2 = 6,000 \text{ yd}^2 \cdot \frac{9 \text{ ft}^2}{1 \text{ yd}^2} \qquad \text{Multiply by the unit conversion factor: } \frac{9 \text{ ft}^2}{1 \text{ yd}^2}.$$
$$= 6,000 \cdot 9 \text{ ft}^2 \qquad \text{The units of square yards divide out.}$$
$$= 54,000 \text{ ft}^2 \qquad \text{Multiply: } 6,000 \cdot 9 = 54,000.$$

The area of the field is 54,000 ft².

Answer 144 in.²

Dorm Rooms   THINK IT THROUGH

"The United States has more than 4,000 colleges and universities, with 2 million students living in college dorms." Washingtonpost.com, 2004

The average dormitory room in a residence hall has about 180 square feet of floor space. The rooms are usually furnished with the following items having the given dimensions:

- 2 extra-long twin beds (39 in. wide × 80 in. long × 24 in. high)
- 2 dressers (18 in. wide × 36 in. long × 48 in. high)
- 2 bookcases (12 in. wide × 24 in. long × 40 in. high)
- 2 desks (24 in. wide × 48 in. long × 28 in. high)

How many square feet of floor space are left?

EXAMPLE 8   Find the area of the parallelogram in Figure 9-57.

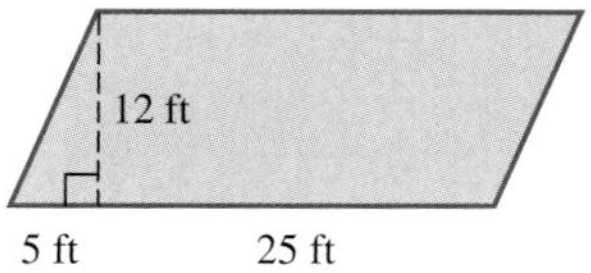

Solution   The length of the base of the parallelogram is

5 feet + 25 feet = 30 feet

The height is 12 feet. To find the area, we substitute 30 for b and 12 for h in the formula for the area of a parallelogram and simplify.

$$A = bh$$
$$= 30(12)$$
$$= 360$$

The area of the parallelogram is 360 ft².

FIGURE 9-57

Self Check 8
Find the area of the parallelogram below.

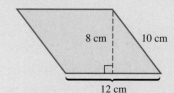

Answer 96 cm²

Self Check 9
Find the area of the triangle
below.

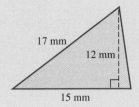

Answer 90 mm²

EXAMPLE 9   Find the area of the triangle in Figure 9-58.

Solution   We substitute 8 for b and 5 for h in the formula for the area of a triangle, and simplify. (The side having length 6 cm is additional information that is not used to find the area.)

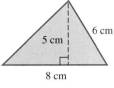

FIGURE 9-58

$$A = \frac{1}{2}bh$$

$$= \frac{1}{2}(8)(5) \quad \text{The length of the base is 8 cm. The height is 5 cm.}$$

$$= 4(5) \quad \text{Perform the multiplication: } \frac{1}{2}(8) = 4.$$

$$= 20$$

The area of the triangle is 20 cm².

EXAMPLE 10   Find the area of the triangle in Figure 9-59.

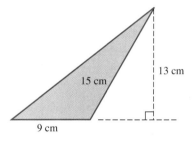

FIGURE 9-59

Solution   In this case, the altitude falls outside the triangle.

$$A = \frac{1}{2}bh$$

$$= \frac{1}{2}(9)(13) \quad \text{Substitute 9 for } b \text{ and 13 for } h.$$

$$= \frac{1}{2}\left(\frac{9}{1}\right)\left(\frac{13}{1}\right) \quad \text{Write 9 as } \frac{9}{1} \text{ and 13 as } \frac{13}{1}.$$

$$= \frac{117}{2} \quad \text{Multiply the fractions.}$$

$$= 58.5 \quad \text{Perform the division.}$$

The area of the triangle is 58.5 cm².

EXAMPLE 11   Find the area of the trapezoid in Figure 9-60.

Solution   In this example, $b_1 = 10$ and $b_2 = 6$. It is incorrect to say that $h = 1$, because the height of 1 foot must be expressed as 12 inches to be consistent with the units of the bases. Thus, we substitute 10 for b_1, 6 for b_2, and 12 for h in the formula for finding the area of a trapezoid and simplify.

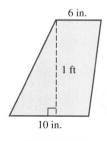

FIGURE 9-60

$$A = \frac{1}{2}h(b_1 + b_2)$$

$$= \frac{1}{2}(12)(10 + 6)$$ The length of the lower base is 10 in. The length of the upper base is 6 in. The height is 12 in.

$$= \frac{1}{2}(12)(16)$$ Perform the addition within the parentheses.

$$= 6(16)$$ Perform the multiplication: $\frac{1}{2}(12) = 6$.

$$= 96$$

The area of the trapezoid is 96 in.^{2}.

Self Check 11
Find the area of the trapezoid below.

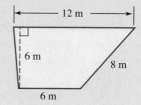

Answer 54 m^2

■ Areas of figures that are combinations of polygons

EXAMPLE 12 **Carpeting a room.** A living room/dining room area has the floor plan shown in Figure 9-61. If carpet costs $29 per square yard, including pad and installation, how much will it cost to carpet the room? (Assume no waste.)

Solution First we must find the total area of the living room and the dining room:

$$A_{total} = A_{living\ room} + A_{dining\ room}$$

Since $\overline{CF}$ divides the space into two rectangles, the areas of the living room and the dining room are found by multiplying their respective lengths and widths.

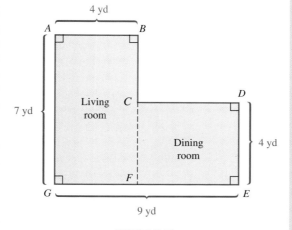

FIGURE 9-61

> Area of living room = lw
> $= 7(4)$
> $= 28$

The area of the living room is 28 yd^2.

To find the area of the dining room, we find its length by subtracting 4 yards from 9 yards to obtain 5 yards. We note that its width is 4 yards.

> Area of dining room = lw
> $= 5(4)$
> $= 20$

The area of the dining room is 20 yd^2.

The total area to be carpeted is the sum of these two areas.

> $A_{total} = A_{living\ room} + A_{dining\ room}$
> $= 28\ yd^2 + 20\ yd^2$
> $= 48\ yd^2$

At $29 per square yard, the cost to carpet the room will be 48 · $29, or $1,392.

EXAMPLE 13 Area of one side of a tent.

Find the area of one side of the tent in Figure 9-62.

Solution Each side is a combination of a trapezoid and a triangle. Since the bases of each trapezoid are 30 feet and 20 feet and the height is 12 feet, we substitute 30 for b_1, 20 for b_2, and 12 for h into the formula for the area of a trapezoid.

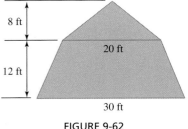

FIGURE 9-62

$$A_{\text{trap.}} = \frac{1}{2}h(b_1 + b_2)$$

$$= \frac{1}{2}(12)(30 + 20)$$

$$= 6(50)$$

$$= 300$$

The area of the trapezoid is 300 ft^2.

Since the triangle has a base of 20 feet and a height of 8 feet, we substitute 20 for b and 8 for h in the formula for the area of a triangle.

$$A_{\text{triangle}} = \frac{1}{2}bh$$

$$= \frac{1}{2}(20)(8)$$

$$= 80$$

The area of the triangle is 80 ft^2.

The total area of one side of the tent is

$$A_{\text{total}} = A_{\text{trap.}} + A_{\text{triangle}}$$

$$= 300 \text{ ft}^2 + 80 \text{ ft}^2$$

$$= 380 \text{ ft}^2$$

The total area is 380 ft^2.

Section 9.5   STUDY SET

VOCABULARY   *Fill in the blanks.*

1. The distance around a polygon is called the _____.

2. The perimeter of a polygon is measured in _____ units.

3. The measure of the surface enclosed by a polygon is called its _____.

4. If each side of a square measures 1 foot, the area enclosed by the square is 1 _____ foot.

5. The area of a polygon is measured in _____ units.

6. The segment that represents the height of a triangle is called an _____.

CONCEPTS   *Sketch and label each of the figures described.*

7. Two different rectangles, each having a perimeter of 40 in.

8. Two different rectangles, each having an area of 40 in.^{2}

9. A square with an area of 25 m²

10. A square with a perimeter of 20 m

11. A parallelogram with an area of 15 yd²

12. A triangle with an area of 20 ft²

13. A figure consisting of a combination of two rectangles whose total area is 80 ft²

14. A figure consisting of a combination of a rectangle and a square whose total area is 164 ft²

NOTATION   *Fill in the blanks*

15. The formula for the perimeter of a square is $P = $ ___ .

16. The formula for the perimeter of a rectangle is $P = $ ___ .

17. The symbol 1 in.² means one _____ _____.

18. One square meter is expressed as 1 m ___ .

19. The formula for the area of a square is $A = $ ___ .

20. The formula for the area of a rectangle is $A = $ ___ .

21. The formula $A = \frac{1}{2}bh$ gives the area of a _____ .

22. The formula $A = \frac{1}{2}h(b_1 + b_2)$ gives the area of a _____ .

PRACTICE   *Find the perimeter of each figure.*

23.
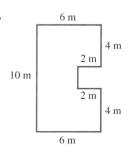

24.
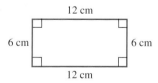

25.
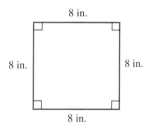

26.
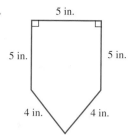

27.
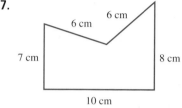

28.
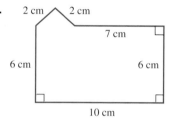

Solve each problem.

29. Find the perimeter of an isosceles triangle with a base of length 21 centimeters and sides of length 32 centimeters.

30. The perimeter of an isosceles triangle is 80 meters. If the length of one side is 22 meters, how long is the base?

31. The perimeter of an equilateral triangle is 85 feet. Find the length of each side.

32. An isosceles triangle with sides of 49.3 inches has a perimeter of 121.7 inches. Find the length of the base.

Find the area of the shaded part of each figure.

33.
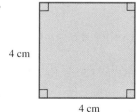

34.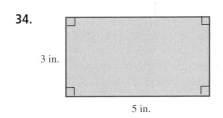
3 in.
5 in.

35.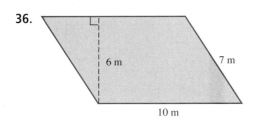
4 cm
6 cm
15 cm

36.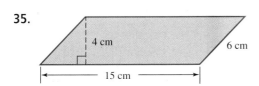
6 m
7 m
10 m

37.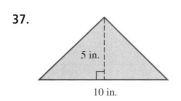
5 in.
10 in.

38.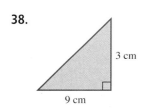
3 cm
9 cm

39.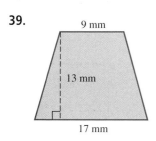
9 mm
13 mm
17 mm

40.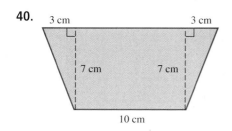
3 cm   3 cm
7 cm   7 cm
10 cm

41.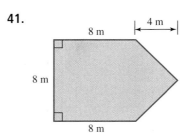
8 m   4 m
8 m
8 m

42.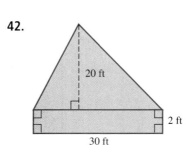
20 ft
2 ft
30 ft

43.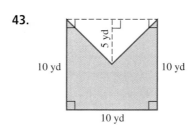
5 yd
10 yd   10 yd
10 yd

44.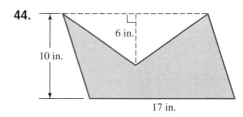
6 in.
10 in.
17 in.

45.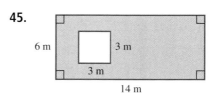
6 m   3 m
3 m
14 m

46.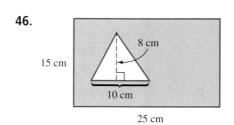
8 cm
15 cm
10 cm
25 cm

47. How many square inches are in 1 square foot?

48. How many square inches are in 1 square yard?

APPLICATIONS

49. FENCING YARDS A man wants to enclose a rectangular yard with fencing that costs $12.50 a foot, including installation. Find the cost of enclosing the yard if its dimensions are 110 ft by 85 ft.

50. FRAMING PICTURES Find the cost of framing a rectangular picture with dimensions of 24 inches by 30 inches if framing material costs $8.46 per foot, including matting.

51. PLANTING SCREENS A woman wants to plant a pine-tree screen around three sides of her backyard. If she plants the trees 3 feet apart, how many trees will she need?

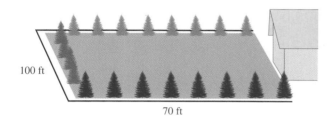

52. PLANTING MARIGOLDS A gardener wants to plant a border of marigolds around the garden shown to keep out rabbits. How many plants will she need if she allows 6 inches between plants?

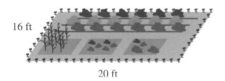

53. BUYING A FLOOR Which is more expensive: a ceramic-tile floor costing $3.75 per square foot or linoleum costing $34.95 per square yard?

54. BUYING A FLOOR Which is cheaper: a hardwood floor costing $5.95 per square foot or a carpeted floor costing $37.50 per square yard?

55. CARPETING A ROOM A rectangular room is 24 feet long and 15 feet wide. At $30 per square yard, how much will it cost to carpet the room? (Assume no waste.)

56. CARPETING A ROOM A rectangular living room measures 30 by 18 feet. At $32 per square yard, how much will it cost to carpet the room? (Assume no waste.)

57. TILING A FLOOR A rectangular basement room measures 14 by 20 feet. Vinyl floor tiles that are 1 ft^2 cost $1.29 each. How much will the tile cost to cover the floor? (Disregard any waste.)

58. PAINTING A BARN The north wall of a barn is a rectangle 23 feet high and 72 feet long. There are five windows in the wall, each 4 by 6 feet. If a gallon of paint will cover 300 ft^2, how many gallons of paint must the painter buy to paint the wall?

59. MAKING A SAIL If nylon is $12 per square yard, how much would the fabric cost to make a triangular sail with a base of 12 feet and a height of 24 feet?

60. PAINTING A GABLE The gable end of a warehouse is an isosceles triangle with a height of 4 yards and a base of 23 yards. It will require one coat of primer and one coat of finish to paint the triangle. Primer costs $17 per gallon, and the finish paint costs $23 per gallon. If one gallon covers 300 square feet, how much will it cost to paint the gable, excluding labor?

61. GEOGRAPHY Use the dimensions of the trapezoid that is superimposed over the state of Nevada to estimate the area of the "Silver State."

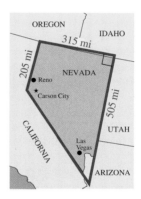

62. SWIMMING POOLS A swimming pool has the shape shown. How many square meters of plastic sheeting will be needed to cover the pool? How much will the sheeting cost if it is $2.95 per square meter? (Assume no waste.)

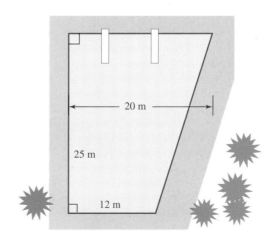

63. CARPENTRY How many sheets of 4-foot-by-8-foot sheetrock are needed to drywall the inside walls on the first floor of the barn shown? (Assume that the carpenters will cover each wall entirely and then cut out areas for the doors and windows.)

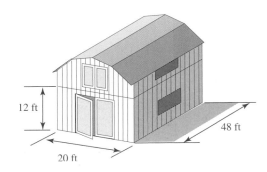

12 ft

48 ft

20 ft

64. CARPENTRY If it costs $90 per square foot to build a one-story home in northern Wisconsin, estimate the cost of building the house with the floor plan shown.

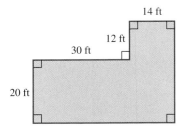

14 ft

12 ft

30 ft

20 ft

65. DRIVING SAFETY The illustration shows the areas on a highway that a truck driver cannot see in the truck's rear view mirrors. Use the scale to determine the approximate dimensions of each blind spot. Then estimate the area of each of them.

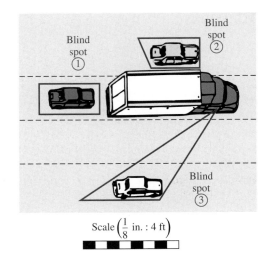

Blind spot ①

Blind spot ②

Blind spot ③

Scale $\left(\frac{1}{8} \text{ in.} : 4 \text{ ft}\right)$

66. ESTIMATING AREA Estimate the area of the sole plate of the iron by thinking of it as a combination of a trapezoid and a triangle.

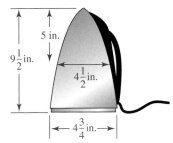

5 in.

$9\frac{1}{2}$ in.

$4\frac{1}{2}$ in.

$4\frac{3}{4}$ in.

WRITING

67. Explain the difference between perimeter and area.

68. Why is it necessary that area be measured in square units?

REVIEW *Perform the calculations. Write all improper fractions as mixed numbers.*

69. $\frac{3}{4} + \frac{2}{3}$

70. $\frac{7}{8} - \frac{2}{3}$

71. $3\frac{3}{4} + 2\frac{1}{3}$

72. $7\frac{5}{8} - 2\frac{5}{6}$

73. $7\frac{1}{2} \div 5\frac{2}{5}$

74. $5\frac{3}{4} \cdot 2\frac{5}{6}$

9.6 Circles

• Circles • Circumference of a circle • Area of a circle

In this section, we discuss circles, one of the most useful geometric figures. In fact, the discovery of fire and the circular wheel were two of the most important events in the history of the human race.

Circles

> ### Circle
>
> A **circle** is the set of all points in a plane that lie a fixed distance from a point called its **center.**

A segment drawn from the center of a circle to a point on the circle is called a **radius.** (The plural of *radius* is *radii.*) From the definition, it follows that all radii of the same circle are the same length.

A **chord** of a circle is a line segment connecting two points on the circle. A **diameter** is a chord that passes through the center of the circle. Since a diameter D of a circle is twice as long as a radius r, we have

$$D = 2r$$

Each of the previous definitions is illustrated in Figure 9-63, in which O is the center of the circle.

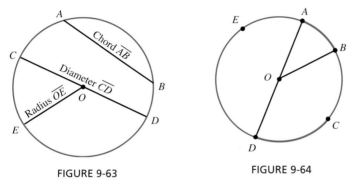

FIGURE 9-63   FIGURE 9-64

Any part of a circle is called an **arc.** In Figure 9-64, the part of the circle from point A to point B is $\overset{\frown}{AB}$, read as "arc AB." $\overset{\frown}{CD}$ is the part of the circle from point C to point D. An arc that is half of a circle is a **semicircle.**

> ### Semicircle
>
> A **semicircle** is an arc of a circle whose endpoints are the endpoints of a diameter.

If point O is the center of the circle in Figure 9-64, $\overline{AD}$ is a diameter and $\overset{\frown}{AED}$ is a semicircle. The middle letter E is used to distinguish semicircle $\overset{\frown}{AED}$ from semicircle $\overset{\frown}{ABCD}$.

An arc that is shorter than a semicircle is a **minor arc.** An arc that is longer than a semicircle is a **major arc.** In Figure 9-64,

$\overset{\frown}{AB}$ is a minor arc   and $\overset{\frown}{ABCDE}$ is a major arc

Circumference of a circle

Since early history, mathematicians have known that the ratio of the distance around a circle (the circumference) divided by the length of its diameter is approximately 3. First Kings, Chapter 7, of the Bible describes a round bronze tank that was 15 feet from brim to brim and 45 feet in circumference, and $\frac{45}{15} = 3$. Today, we have a better value for this ratio, known as π (pi). If C is the circumference of a circle and D is the length of its diameter, then

$$\pi = \frac{C}{D}, \quad \text{where } \pi = 3.141592653589\ldots \qquad \frac{22}{7} \text{ and 3.14 are often used as estimates of } \pi.$$

If we multiply both sides of $\pi = \frac{C}{D}$ by D, we have the following formula.

Circumference of a circle

The circumference of a circle is given by the formula

$$C = \pi D \quad \text{where } C \text{ is the circumference and } D \text{ is the length of the diameter}$$

Since a diameter of a circle is twice as long as a radius r, we can substitute $2r$ for D in the formula $C = \pi D$ to obtain another formula for the circumference C.

$$C = 2\pi r$$

Self Check 1
To the nearest tenth, find the circumference of a circle that has a radius of 12 meters.

EXAMPLE 1   Find the circumference of a circle that has a diameter of 10 centimeters. (See Figure 9-65.)

Solution   We substitute 10 for D in the formula for the circumference of a circle.

$$\begin{aligned} C &= \pi D \\ &= \pi(10) \\ &\approx 3.14(10) \quad \text{Replace } \pi \text{ with an approximation: } \pi \approx 3.14. \\ &\approx 31.4 \end{aligned}$$

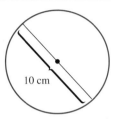

10 cm

FIGURE 9-65

The circumference is approximately 31.4 centimeters.

Answer 75.4 m

CALCULATOR SNAPSHOT   **Calculating revolutions of a tire**

When the $\boxed{\pi}$ key on a scientific calculator is pressed (on some models, the $\boxed{\text{2nd}}$ key must be pressed first), an approximation of π is displayed. To illustrate how to use this key, consider the following problem. How many times does a 15-inch tire revolve when a car makes a 25-mile trip?

We first find the circumference of the tire.

$$\begin{aligned} C &= \pi D \\ &= \pi(15) \quad \text{Substitute 15 for } D \text{, the diameter of the tire.} \\ &= 15\pi \quad \text{Normally, we rewrite a product such as } \pi(15) \text{ so that } \pi \text{ is the} \\ &\qquad\qquad \text{second factor.} \end{aligned}$$

The circumference of the tire is 15π inches.

We then change 25 miles to inches using two unit conversion factors.

$$\frac{25}{1} \text{ miles} \cdot \frac{5{,}280 \text{ feet}}{1 \text{ mile}} \cdot \frac{12 \text{ inches}}{1 \text{ foot}} = 25(5{,}280)(12) \text{ in.}$$

The total distance of the trip is $25(5{,}280)(12)$ inches.

Finally, we divide the total distance of the trip by the circumference of the tire to get

$$\text{The number of revolutions of the tire} = \frac{25(5{,}280)(12)}{15\pi}$$

To approximate the value of $\dfrac{25(5{,}280)(12)}{15\pi}$ using a scientific calculator, we enter

$\boxed{(}\ 25\ \boxed{\times}\ 5280\ \boxed{\times}\ 12\ \boxed{)}\ \boxed{\div}\ \boxed{(}\ 15\ \boxed{\times}\ \boxed{\pi}\ \boxed{)}\ \boxed{=}\ \boxed{33613.52398}$

The tire makes about 33,614 revolutions.

EXAMPLE 2 Architecture.   A Norman window is constructed by adding a semicircular window to the top of a rectangular window. Find the perimeter of the Norman window shown in Figure 9-66.

8 ft   8 ft

6 ft

FIGURE 9-66

Solution   The window is a combination of a rectangle and a semicircle. The perimeter of the rectangular part is

$$P_{\text{rectangular part}} = 8 + 6 + 8 = 22 \quad \text{Add only 3 sides.}$$

The perimeter of the semicircle is one-half the circumference of a circle that has a 6-foot diameter.

$$P_{\text{semicircle}} = \frac{1}{2}\pi D$$

$$= \frac{1}{2}\pi(6) \qquad \text{Substitute 6 for } D.$$

$$\approx 9.424777961 \qquad \text{Use a calculator.}$$

The total perimeter is the sum of the two parts.

$$P_{\text{total}} \approx 22 + 9.424777961$$

$$\approx 31.424777961$$

To the nearest hundredth, the perimeter of the window is 31.42 feet.

▪ Area of a circle

If we divide the circle shown in Figure 9-67(a) into an even number of pie-shaped pieces and then rearrange them as shown in Figure 9-67(b), we have a figure that looks like a parallelogram. The figure has a base that is one-half the circumference of the circle, and its height is about the same length as a radius of the circle.

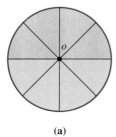
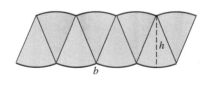

(a)   (b)

FIGURE 9-67

If we divide the circle into more and more pie-shaped pieces, the figure will look more and more like a parallelogram, and we can find its area by using the formula for the area of a parallelogram.

$$A = bh$$

$$= \frac{1}{2}Cr \qquad \text{Substitute } \frac{1}{2} \text{ of the circumference for } b, \text{ and } r \text{ for the height.}$$

$$= \frac{1}{2}(2\pi r)r \qquad \text{Make a substitution: } C = 2\pi r.$$

$$= \pi r^2 \qquad \text{Simplify: } \frac{1}{2} \cdot 2 = 1 \text{ and } r \cdot r = r^2.$$

Area of a circle

The **area of a circle** with radius r is given by the formula

$$A = \pi r^2$$

Self Check 3
To the nearest tenth, find the area of a circle with a diameter of 12 feet.

EXAMPLE 3   To the nearest tenth, find the area of the circle in Figure 9-68.

Solution   Since the length of the diameter is 10 centimeters and the length of a diameter is twice the length of a radius, the length of the radius is 5 centimeters. To find the area of the circle, we substitute 5 for r in the formula for the area of a circle.

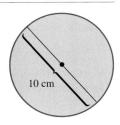

10 cm

FIGURE 9-68

$$A = \pi r^2$$

$$= \pi(5)^2$$

$$= 25\pi$$

$$\approx 78.53981634 \qquad \text{Use a calculator.}$$

Answer 113.1 ft^2.

To the nearest tenth, the area is 78.5 cm^2.

CALCULATOR SNAPSHOT   **Painting a helicopter pad**

Orange paint is available in gallon containers at \$19 each, and each gallon will cover 375 ft^2. To calculate how much the paint will cost to cover a circular helicopter pad 60 feet in diameter, we first calculate the area of the helicopter pad.

$$A = \pi r^2$$

$$= \pi(30)^2 \qquad \text{Substitute one-half of 60 for } r.$$

$$= 30^2\pi$$

The area of the pad is $30^2\pi$ ft^2. Since each gallon of paint will cover 375 ft^2, we can find the number of gallons of paint needed by dividing $30^2\pi$ by 375.

$$\text{Number of gallons needed} = \frac{30^2\pi}{375}$$

To approximate the value of $\dfrac{30^2\pi}{375}$ using a scientific calculator, we enter

30 $\boxed{x^2}$ $\boxed{\times}$ $\boxed{\pi}$ $\boxed{=}$ $\boxed{\div}$ 375 $\boxed{=}$ $\boxed{\text{7.539822369}}$

The result is approximately 7.54. Because paint comes only in full gallons, the painter will need to purchase 8 gallons. The cost of the paint will be 8(\$19), or \$152.

EXAMPLE 4   Find the shaded area in Figure 9-69.

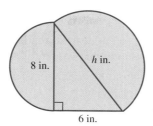

8 in.   h in.

6 in.

FIGURE 9-69

Solution   The figure is a combination of a triangle and two semicircles. By the Pythagorean theorem, the hypotenuse h of the right triangle is

$$h = \sqrt{6^2 + 8^2} = \sqrt{36 + 64} = \sqrt{100} = 10$$

The area of the triangle is

$$A_{\text{right triangle}} = \frac{1}{2}bh = \frac{1}{2}(6)(8) = 3(8) = 24$$

The area enclosed by the smaller semicircle is

$$A_{\text{smaller semicircle}} = \frac{1}{2}\pi r^2 = \frac{1}{2}\pi(4)^2 = \frac{1}{2}\pi(16) = 8\pi$$

The area enclosed by the larger semicircle is

$$A_{\text{larger semicircle}} = \frac{1}{2}\pi r^2 = \frac{1}{2}\pi(5)^2 = \frac{1}{2}\pi(25) = 12.5\pi$$

The total area is

$$A_{\text{total}} = 24 + 8\pi + 12.5\pi \approx 88.4026494 \quad \text{Use a calculator.}$$

To the nearest hundredth, the area is 88.40 in.^{2}.

Section 9.6   STUDY SET

VOCABULARY   *Fill in the blanks.*

1. A segment drawn from the center of a circle to a point on the circle is called a _____.

2. A segment joining two points on a circle is called a _____.

3. A _____ is a chord that passes through the center of a circle.

4. An arc that is one-half of a complete circle is a _____.

5. An arc that is shorter than a semicircle is called a _____ arc.

6. An arc that is longer than a semicircle is called a _____ arc.

7. The distance around a circle is called its _____.

8. The surface enclosed by a circle is called its _____.

CONCEPTS   *Refer to the illustration.*

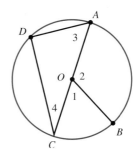

9. Name each radius.

10. Name a diameter.

11. Name each chord.

12. Name each minor arc.

13. Name each semicircle.

14. Name each major arc.

15. If you know the radius of a circle, how can you find its diameter?

16. If you know the diameter of a circle, how can you find its radius?

17. Suppose the two legs of the compass shown on the right are adjusted so that the distance between the pointed ends is 1 inch. Then a circle is drawn.

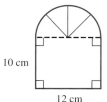

   a. What will the radius of the circle be?

   b. What will the diameter of the circle be?

   c. What will the circumference of the circle be?

   d. What will the area of the circle be?

18. Suppose we find the distance around a can and the distance across the can using a measuring tape, as shown. Then we make a comparison, in the form of a ratio:

$$\frac{\text{the distance around the can}}{\text{the distance across the top of the can}}$$

After we do the indicated division, the result will be close to what number?

19. When evaluating $\pi(6)^2$, what operation should be performed first?

20. Round $\pi = 3.141592653589\ldots$ to the nearest hundredth.

NOTATION   *Fill in the blanks.*

21. The symbol $\overset{\frown}{AB}$ is read as _____ _____.

22. To the nearest hundredth, the value of π is _____.

23. The formula for the circumference of a circle is $C =$ _____ or $C = 2\pi$ _____.

24. The formula $A = \pi r^2$ gives the area of a _____.

25. If C is the circumference of a circle and D is its diameter, then $\frac{C}{D} =$ _____.

26. If D is the diameter of a circle and r is its radius, then $D =$ _____ r.

27. Write $\pi(8)$ in a better form.

28. What does $2\pi r$ mean?

PRACTICE   *Solve each problem. Answers may vary slightly depending on which approximation of π is used.*

29. To the nearest hundredth, find the circumference of a circle that has a diameter of 12 inches.

30. To the nearest hundredth, find the circumference of a circle that has a radius of 20 feet.

31. Find the diameter of a circle that has a circumference of 36π meters.

32. Find the radius of a circle that has a circumference of 50π meters.

Find the perimeter of each figure to the nearest hundredth.

33.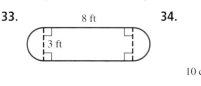

34.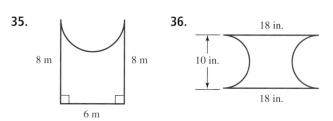

35.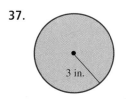

36.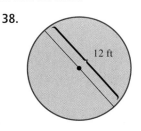

Find the area of each circle to the nearest tenth.

37.   **38.**

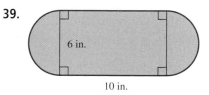

Find the total area of each figure to the nearest tenth.

39.

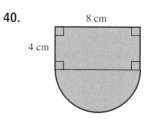

40.

41.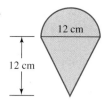

42.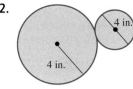

Find the area of each shaded region to the nearest tenth.

43.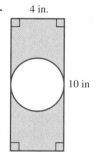

44.

45.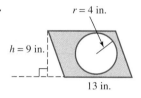

46.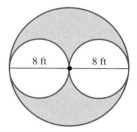

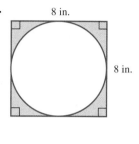

APPLICATIONS *Give each answer to the nearest hundredth. Answers may vary slightly depending on which approximation of π is used.*

47. AREA OF ROUND LAKE Round Lake has a circular shoreline that is 2 miles in diameter. Find the area of the lake.

48. HELICOPTERS How far does a point on the tip of a rotor blade travel when it makes one complete revolution?

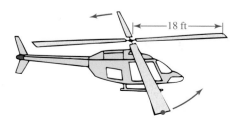

49. GIANT SEQUOIAS The largest sequoia tree is the General Sherman Tree in Sequoia National Park in California. In fact, it is considered to be the largest living thing in the world. According to the *Guinness Book of World Records,* it has a circumference of 102.6 feet, measured $4\frac{1}{2}$ feet above the ground. What is the diameter of the tree at that height?

50. TRAMPOLINES The distance from the center of the trampoline to the edge of its steel frame is 7 feet. The protective padding covering the springs is 15 inches wide. Find the area of the circular jumping surface of the trampoline, in square feet.

51. JOGGING Joan wants to jog 10 miles on a circular track $\frac{1}{4}$ mile in diameter. How many times must she circle the track?

52. FIXING THE ROTUNDA The rotunda at a state capitol is a circular area 100 feet in diameter. The legislature wishes to appropriate money to have the floor of the rotunda tiled. The lowest bid is $83 per square yard, including installation. How much must the legislature spend?

53. BANDING THE EARTH A steel band is drawn tightly about Earth's equator. The band is then loosened by increasing its length by 10 feet, and the resulting slack is distributed evenly along the band's entire length. How far above Earth's surface is the band? (*Hint:* You don't need to know Earth's circumference.)

54. CONCENTRIC CIRCLES Two circles are called **concentric circles** if they have the same center. Find the area of the band between two concentric circles if their diameters are 10 centimeters and 6 centimeters.

55. ARCHERY See the illustration. Find the area of the entire target and the bull's eye. What percent of the area of the target is the bull's eye?

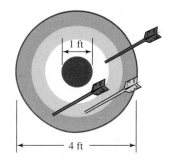

56. LANDSCAPE DESIGNS See the illustration. How much of the lawn does not get watered by the sprinklers at the center of each circle?

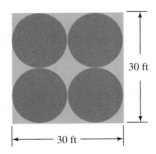

30 ft

30 ft

WRITING

57. Explain what is meant by the circumference of a circle.

58. Explain what is meant by the area of a circle.

59. Explain the meaning of π.

60. Distinguish between a major arc and a minor arc.

61. Explain what it means for a car to have a small turning radius.

62. The word *circumference* means the distance around a circle. In your own words, explain what is meant by each of the following sentences.

   a. A boat owner's dream was to *circumnavigate* the globe.

   b. The teenager's parents felt that he was always trying to *circumvent* the rules.

   c. The class was shown a picture of a circle *circumscribed* about an equilateral triangle.

REVIEW

63. Change $\dfrac{9}{10}$ to a percent.

64. Change $\dfrac{7}{8}$ to a percent.

65. UNIT COSTS A 24-ounce package of green beans sells for $1.29. Give the unit cost in cents per ounce.

66. MILEAGE One car went 1,235 miles on 51.3 gallons of gasoline, and another went 1,456 on 55.78 gallons. Which car got the better gas mileage?

67. How many sides does a pentagon have?

68. What is the sum of the measures of the angles of a triangle?

9.7 Surface Area and Volume

- Volumes of solids   •   Surface areas of rectangular solids
- Volumes and surface areas of spheres   •   Volumes of cylinders
- Volumes of cones   •   Volumes of pyramids

In this section, we discuss a measure of capacity called **volume.** Volumes are measured in cubic units, such as cubic inches, cubic yards, or cubic centimeters. For example,

- We buy gravel or topsoil by the cubic yard.
- We measure the capacity of a refrigerator in cubic feet.
- We often measure amounts of medicine in cubic centimeters.

We also discuss surface area. The ability to compute surface area is necessary to solve problems such as calculating the amount of material necessary to make a cardboard box or a plastic beach ball.

Volumes of solids

A **rectangular solid** and a **cube** are two common geometric solids. (See Figure 9-70 on the next page.)

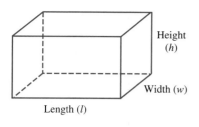

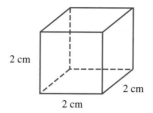

Height (h)

Width (w)

Length (l)

A rectangular solid

2 cm

2 cm

2 cm

A cube

FIGURE 9-70

The **volume** of a rectangular solid is a measure of the space it encloses. Two common units of volume are cubic inches (in.³) and cubic centimeters (cm³). (See Figure 9-71.)

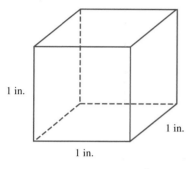

1 in.

1 in.

1 in.

1 cubic inch (1 in.³)

1 cm

1 cm

1 cm

1 cubic centimeter (1 cm³)

FIGURE 9-71

If we divide the rectangular solid shown in Figure 9-72 into cubes, each cube represents a volume of 1 cm³. Because there are 2 levels with 12 cubes on each level, the volume of the rectangular solid is 24 cm³.

In practice, we do not find volumes by counting cubes. Instead, we use the formulas shown in Table 9-2.

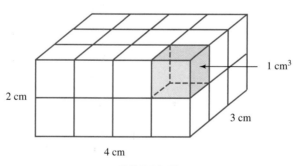

2 cm

3 cm

4 cm

1 cm³

FIGURE 9-72

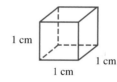

Figure	Name	Volume	Figure	Name	Volume
	Cube	$V = s^3$		Cylinder	$V = \pi r^2 h$ or $V = Bh$*
	Rectangular solid	$V = lwh$		Cone	$V = \frac{1}{3}\pi r^2 h$ or $V = \frac{1}{3}Bh$*

*B represents the area of the base that is shaded in the figure.

(*continued*)

Figure	Name	Volume	Figure	Name	Volume
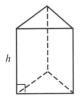	Prism	$V = Bh*$	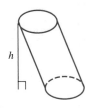 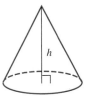	Pyramid	$V = \dfrac{1}{3}Bh*$
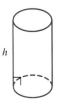	Sphere	$V = \dfrac{4}{3}\pi r^3$			

*B represents the area of the base that is shaded in the figure.

TABLE 9-2

! COMMENT The height of a geometric solid is always measured along a line perpendicular to its base. In each of the solids in Figure 9-73, h is the height.

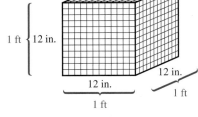

FIGURE 9-73

Self Check 1
How many cubic centimeters are in 1 cubic meter?

EXAMPLE 1 How many cubic inches are there in 1 cubic foot? (See Figure 9-74.)

Solution Since a cubic foot is a cube with each side measuring 1 foot, each side also measures 12 inches. Thus, the volume in cubic inches is

$$V = s^3 \qquad \text{This is the formula for the volume of a cube.}$$

$$= (12)^3 \qquad \text{Substitute 12 for } s.$$

$$= 1{,}728$$

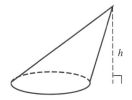

1 ft { 12 in.

12 in.
12 in.
1 ft
1 ft

FIGURE 9-74

Answer 1,000,000 cm³

There are 1,728 cubic inches in 1 cubic foot.

Self Check 2
Find the volume of a rectangular solid with dimensions of 8 by 12 by 20 meters.

EXAMPLE 2 **Volume of an oil storage tank.** An oil storage tank is in the form of a rectangular solid with dimensions of 17 by 10 by 8 feet. (See Figure 9-75.) Find its volume.

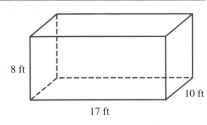

8 ft

10 ft

17 ft

FIGURE 9-75

Solution To find the volume, we substitute 17 for l, 10 for w, and 8 for h in the formula $V = lwh$ and simplify.

$$V = lwh$$
$$= 17(10)(8)$$
$$= 1,360$$

The volume is 1,360 ft³.

Answer 1,920 m³

EXAMPLE 3 Volume of a triangular prism.

Find the volume of the triangular prism in Figure 9-76.

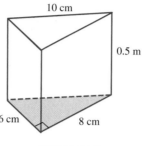

FIGURE 9-76

Solution The volume of the prism is the area of its base multiplied by its height. Since there are 100 centimeters in 1 meter, the height in centimeters is

$$0.5 \text{ m} = 0.5(\mathbf{1\ m})$$
$$= 0.5(\mathbf{100\ cm}) \quad \text{Substitute 100 centimeters for 1 meter.}$$
$$= 50 \text{ cm}$$

The area of the triangular base is $\frac{1}{2}(6)(8) = 24$ square centimeters. The height of the prism is 50 centimeters. Substituting into the formula for the volume of a prism, we have

$$V = Bh$$
$$= 24(50)$$
$$= 1,200$$

The volume of the prism is 1,200 cm³.

Self Check 3
Find the volume of the triangular prism below.

Answer 200 in.³

■ Surface areas of rectangular solids

The **surface area** of a rectangular solid is the sum of the areas of its six faces. Figure 9-77 shows how we can unfold the faces of a cardboard box to derive a formula for its surface area (SA).

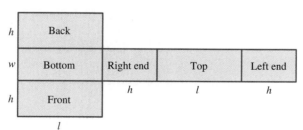

FIGURE 9-77

$$SA = A_{\text{bottom}} + A_{\text{back}} + A_{\text{front}} + A_{\text{right end}} + A_{\text{top}} + A_{\text{left end}}$$
$$= \quad lw \quad + \quad lh \quad + \quad lh \quad + \quad hw \quad + \quad lw \quad + \quad hw$$
$$= 2lw + 2lh + 2hw \quad \text{Combine like terms.}$$

Surface area of a rectangular solid

The surface area of a rectangular solid is given by the formula

$$SA = 2lw + 2lh + 2hw$$

where l is the length, w is the width, and h is the height.

**EXAMPLE 4   Surface area of an oil
tank.**   An oil storage tank is in the form of a
rectangular solid with dimensions of 17 by 10
by 8 feet. (See Figure 9-78.) Find the surface
area of the tank.

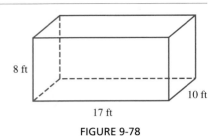

8 ft

10 ft

17 ft

FIGURE 9-78

Solution   To find the surface area, we substitute 17 for *l*, 10 for *w*, and 8 for *h* in the formula for surface area and simplify.

$$SA = 2lw + 2lh + 2hw$$
$$= 2(17)(10) + 2(17)(8) + 2(8)(10)$$
$$= 340 + 272 + 160$$
$$= 772$$

The surface area is 772 ft^2.

▮ Volumes and surface areas of spheres

A **sphere** is a hollow, round ball. (See Figure 9-79.) The points
on a sphere all lie at a fixed distance *r* from a point called its *center*.
A segment drawn from the center of a sphere to a point on the
sphere is called a *radius*.

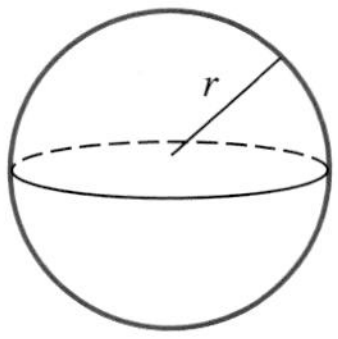

r

FIGURE 9-79

CALCULATOR SNAPSHOT   **Filling a water tank**

See Figure 9-80. To calculate how many cubic feet of water are needed to fill a spherical water tank with a radius of 15 feet, we substitute 15 for *r* in the formula for the volume of a sphere.

$$V = \frac{4}{3}\pi r^3$$

$$= \frac{4}{3}\pi(15)^3$$

15 ft

FIGURE 9-80

To approximate the value of $\frac{4}{3}\pi(15)^3$ using a scientific calculator, we enter

15 $\boxed{y^x}$ 3 $\boxed{=}$ $\boxed{\times}$ 4 $\boxed{\div}$ 3 $\boxed{=}$ $\boxed{\times}$ $\boxed{\pi}$ $\boxed{=}$ $\boxed{\text{14137.16694}}$

To the nearest tenth, 14,137.2 ft^3 of water are needed to fill the tank.

There is a formula to find the surface area of a sphere.

> **Surface area of a sphere**
>
> The surface area of a sphere with radius r is given by the formula
>
> $$SA = 4\pi r^2$$

EXAMPLE 5 Manufacturing beach balls. A beach ball is to have a diameter of 16 inches. (See Figure 9-81.) How many square inches of material will be needed to make the ball? (Disregard any waste.)

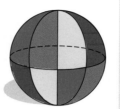

Solution Since a radius r of the ball is one-half the diameter, $r = 8$ inches. We can now substitute 8 for r in the formula for the surface area of a sphere.

FIGURE 9-81

$$SA = 4\pi r^2$$
$$= 4\pi(\mathbf{8})^2$$
$$= 4\pi(64)$$
$$= 256\pi \qquad \text{Simplify: } 4 \cdot 64 = 256.$$
$$\approx 804.2477193 \qquad \text{Use a calculator.}$$

A little more than 804 in.^{2} of material is needed to make the ball.

▌ Volumes of cylinders

A **cylinder** is a hollow figure like a piece of pipe. (See Figure 9-82.)

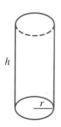

FIGURE 9-82

EXAMPLE 6 Find the volume of the cylinder in Figure 9-83.

Solution Since a radius is one-half the diameter of the circular base, $r = 3$ cm. From the figure, we see that the height of the cylinder is 10 cm. So we can substitute 3 for r and 10 for h in the formula for the volume of a cylinder.

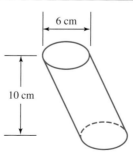

$$V = \pi r^2 h$$
$$= \pi(\mathbf{3})^2(\mathbf{10})$$
$$= 90\pi \qquad \text{Simplify: } (3)^2(10) = 9(10) = 90.$$
$$\approx 282.7433388 \qquad \text{Use a calculator.}$$

FIGURE 9-83

To the nearest hundredth, the volume of the cylinder is 282.74 cm^3.

Volume of a silo

A silo is a structure used for storing grain. The silo in Figure 9-84 is a cylinder 50 feet tall topped with a **hemisphere** (a half-sphere). To find the volume of the silo, we add the volume of the cylinder to the volume of the dome.

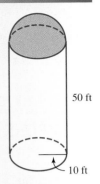

50 ft

10 ft

FIGURE 9-84

$$\text{Volume}_{cylinder} + \text{volume}_{dome} = (\text{area}_{cylinder's\ base})(\text{height}_{cylinder}) + \frac{1}{2}(\text{volume}_{sphere})$$

$$= \pi r^2 h + \frac{1}{2}\left(\frac{4}{3}\pi r^3\right)$$

$$= \pi r^2 h + \frac{2\pi r^3}{3} \qquad \frac{1}{2}\left(\frac{4}{3}\pi r^3\right) = \frac{1}{2}\cdot\frac{4}{3}\pi r^3 = \frac{4}{6}\pi r^3 = \frac{2\pi r^3}{3}$$

$$= \pi(10)^2(50) + \frac{2\pi(10)^3}{3} \qquad \text{Substitute 10 for } r \text{ and 50 for } h.$$

To approximate the value of $\pi(10)^2(50) + \dfrac{2\pi(10)^3}{3}$ using a scientific calculator, we enter

$\boxed{\pi}\ \boxed{\times}\ 10\ \boxed{x^2}\ \boxed{\times}\ 50\ \boxed{=}\ \boxed{+}\ \boxed{(}\ 2\ \boxed{\times}\ \boxed{\pi}\ \boxed{\times}\ 10\ \boxed{y^x}\ 3$
$\boxed{\div}\ 3\ \boxed{)}\ \boxed{=}$

$\boxed{17802.35837}$

The volume of the silo is approximately 17,802 ft³.

EXAMPLE 7 Machining a block of metal.
See Figure 9-85. Find the volume that is left when the hole is drilled through the metal block.

Solution We must find the volume of the rectangular solid and then subtract the volume of the cylinder. We will think of the rectangular solid and the cylinder as lying on their sides. Thus, the height is 18 cm when we find each volume.

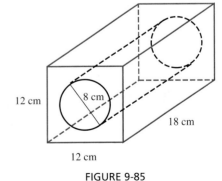

12 cm

8 cm

18 cm

12 cm

FIGURE 9-85

$$V_{rect.\ solid} = lwh$$
$$= 12(12)(18)$$
$$= 2,592$$

$$V_{cylinder} = \pi r^2 h$$
$$= \pi(4)^2(18)$$
$$= 288\pi \qquad \text{Simplify: } (4)^2(18) = 16(18) = 288.$$
$$\approx 904.7786842 \qquad \text{Use a calculator.}$$

$$V_{\text{drilled block}} = V_{\text{rect. solid}} - V_{\text{cylinder}}$$
$$\approx 2{,}592 - 904.7786842$$
$$\approx 1{,}687.221316 \qquad \text{Use a calculator.}$$

To the nearest hundredth, the volume is 1,687.22 cm³.

Volumes of cones

Two **cones** are shown in Figure 9-86. Each cone has a height h and a radius r, which is the radius of the circular base.

 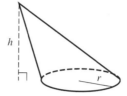

FIGURE 9-86

EXAMPLE 8   To the nearest tenth, find the volume of the cone in Figure 9-87.

Solution   Since the radius is one-half the diameter, $r = 4$ cm. We then substitute 4 for r and 6 for h in the formula for the volume of a cone.

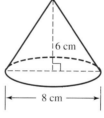

FIGURE 9-87

$$V = \frac{1}{3}\pi r^2 h$$

$$= \frac{1}{3}\pi(4)^2(6)$$

$$= \frac{1}{3}\pi(96) \qquad \text{Simplify: } (4)^2(6) = 16(6) = 96.$$

$$= 32\pi \qquad \text{Multiply: } \frac{1}{3}(96) = 32.$$

$$\approx 100.5309649$$

To the nearest tenth, the volume is 100.5 cubic centimeters.

Volumes of pyramids

Two **pyramids** with a height h are shown in Figure 9-88.

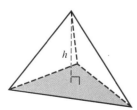   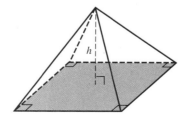

The base is a triangle.   The base is a square.

(a)   (b)

FIGURE 9-88

Self Check 9
Find the volume of the pyramid shown below.

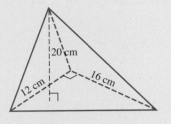

Answer 640 cm³

EXAMPLE 9 Find the volume of a pyramid that has a square base with each side 6 meters long and a height of 9 meters.

Solution Since the base is a square with each side 6 meters long, the area of the base is 6^2 m², or 36 m². We can then substitute 36 for the area of the base and 9 for the height in the formula for the volume of a pyramid.

$$V = \frac{1}{3}Bh$$

$$= \frac{1}{3}(36)(9)$$

$$= 12(9) \qquad \text{Multiply: } \frac{1}{3}(36) = 12.$$

$$= 108$$

The volume of the pyramid is 108 m³.

Section 9.7 STUDY SET

VOCABULARY *Fill in the blanks.*

1. The space contained within a geometric solid is called its _____.

2. A _____ solid is like a hollow shoe box.

3. A _____ is a rectangular solid with all sides of equal length.

4. The volume of a cube with each side 1 inch long is 1 _____ inch.

5. The _____ area of a rectangular solid is the sum of the areas of its faces.

6. The point that is equidistant from every point on a sphere is its _____.

7. A _____ is a hollow figure like a drinking straw.

8. A _____ is one-half of a sphere.

9. A _____ looks like a witch's pointed hat.

10. A figure that has a polygon for its base and that rises to a point is called a _____.

CONCEPTS *Write the formula used for finding the volume of each solid.*

11. A rectangular solid

12. A prism

13. A sphere

14. A cylinder

15. A cone

16. A pyramid

17. Write the formula for finding the surface area of a rectangular solid.

18. Write the formula for finding the surface area of a sphere.

19. How many cubic feet are in 1 cubic yard?

20. How many cubic inches are in 1 cubic yard?

21. How many cubic decimeters are in 1 cubic meter?

22. How many cubic millimeters are in 1 cubic centimeter?

Which geometric concept (perimeter, circumference, area, volume, or surface area) should be applied to find each of the following?

23. a. The size of a room to be air conditioned

   b. The amount of land in a national park

   c. The amount of space in a refrigerator freezer

   d. The amount of cardboard in a shoe box

   e. The distance around a checkerboard

   f. The amount of material used to make a basketball

24. a. The amount of cloth in a car cover

   b. The size of a trunk of a car

   c. The amount of paper used for a postage stamp

   d. The amount of storage in a cedar chest

   e. The amount of beach available for sunbathing

   f. The distance the tip of a propeller travels

25. In the following illustration, the unit of measurement of length that was used to draw the figure was the inch.

   a. What is the volume of the figure?

   b. What is the area of the front of the figure?

   c. What is the area of the base of the figure?

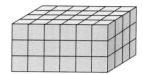

26. The cardboard box shown is a cube. Suppose the six faces were unfolded to lie flat on a table. Draw a picture of what this would look like.

NOTATION *Fill in the blanks.*

27. The notation 1 in.^{3} is read as one _____ _____.

28. One cubic centimeter is represented as 1 cm▪.

PRACTICE *Find the volume of each solid. If an answer is not exact, round to the nearest hundredth. (Answers may vary slightly, depending on which approximation of π is used.)*

29. A rectangular solid with dimensions of 3 by 4 by 5 centimeters

30. A rectangular solid with dimensions of 5 by 8 by 10 meters

31. A prism whose base is a right triangle with legs 3 and 4 meters long and whose height is 8 meters

32. A prism whose base is a right triangle with legs 5 and 12 feet long and whose height is 10 feet

33. A sphere with a radius of 9 inches

34. A sphere with a diameter of 10 feet

35. A cylinder with a height of 12 meters and a circular base with a radius of 6 meters

36. A cylinder with a height of 4 meters and a circular base with a diameter of 18 meters

37. A cone with a height of 12 centimeters and a circular base with a diameter of 10 centimeters

38. A cone with a height of 3 inches and a circular base with a radius of 4 inches

39. A pyramid with a square base 10 meters on each side and a height of 12 meters

40. A pyramid with a square base 6 inches on each side and a height of 4 inches

Find the surface area of each solid. If an answer is not exact, round to the nearest hundredth.

41. A rectangular solid with dimensions of 3 by 4 by 5 centimeters

42. A cube with a side 5 centimeters long

43. A sphere with a radius of 10 inches

44. A sphere with a diameter of 12 meters

Find the volume of each figure. If an answer is not exact, round to the nearest hundredth. (Answers may vary slightly, depending on which approximation of π is used.)

45.

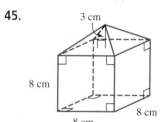

46.

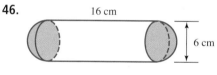

47.

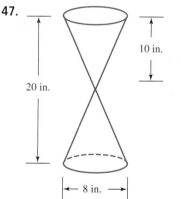

48.

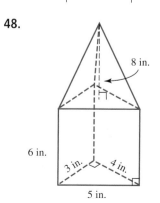

APPLICATIONS *Solve each problem. If an answer is not exact, round to the nearest hundredth.*

49. VOLUME OF A SUGAR CUBE A sugar cube is $\frac{1}{2}$ inch on each edge. How much volume does it occupy?

50. VOLUME OF A CLASSROOM A classroom is 40 feet long, 30 feet wide, and 9 feet high. Find the number of cubic feet of air in the room.

51. WATER HEATERS Complete the ad for the high-efficiency water heater shown.

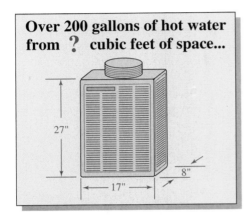

Over 200 gallons of hot water from **?** cubic feet of space...

27"

17"   8"

52. REFRIGERATOR CAPACITY The largest refrigerator advertised in a J. C. Penney catalog has a capacity of 25.2 cubic feet. How many cubic inches is this?

53. VOLUME OF AN OIL TANK A cylindrical oil tank has a diameter of 6 feet and a length of 7 feet. Find the volume of the tank.

54. VOLUME OF A DESSERT A restaurant serves pudding in a conical dish that has a diameter of 3 inches. If the dish is 4 inches deep, how many cubic inches of pudding are in each dish?

55. HOT-AIR BALLOONS The lifting power of a spherical balloon depends on its volume. How many cubic feet of gas will a balloon hold if it is 40 feet in diameter?

56. VOLUME OF A CEREAL BOX A box of cereal measures 3 by 8 by 10 inches. The manufacturer plans to market a smaller box that measures $2\frac{1}{2}$ by 7 by 8 inches. By how much will the volume be reduced?

57. ENGINES The *compression ratio* of an engine is the volume in one cylinder with the piston at bottom-dead-center (B.D.C.), divided by the volume with the piston at top-dead-center (T.D.C.). From the data given in the illustration in the next column, what is the compression ratio of the engine? Use a colon to express your answer as a ratio.

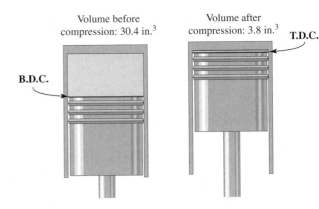

Volume before compression: 30.4 in.^{3}

Volume after compression: 3.8 in.^{3} T.D.C.

B.D.C.

58. LINT REMOVERS The illustration shows a handy gadget; it uses a cylinder of sheets of sticky paper that can be rolled over clothing and furniture to pick up lint and pet hair. After the paper is full, that sheet is peeled away to expose another sheet of sticky paper. Find the area of the first sheet by using the formula $LSA = 2\pi rh$, where LSA represents the lateral surface area of the cylinder.

$2\frac{1}{2}$ in.

4 in.

WRITING

59. What is meant by the *volume* of a cube?

60. What is meant by the *surface area* of a cube?

61. Are the units used to measure area different from the units used to measure volume? Explain.

62. The dimensions (length, width, and height) of one rectangular solid are entirely different numbers from the dimensions of another rectangular solid. Would it be possible for the rectangular solids to have the same volume? Explain.

REVIEW

63. Evaluate: $-5(5 - 2)^2 + 3$.

64. BUYING PENCILS Carlos bought 6 pencils at $0.60 each and a notebook for $1.25. He gave the clerk a $5 bill. How much change did he receive?

65. Solve: $\dfrac{x}{-4} = \dfrac{1}{4}$.

66. 38 is what percent of 40?

67. Express the phrase "3 inches to 15 inches" as a ratio in lowest terms.

68. Convert 40 ounces to pounds.

69. Convert 2.4 meters to millimeters.

70. State the Pythagorean theorem.

KEY CONCEPT

Formulas

A **formula** is a mathematical expression that is used to express a relationship between quantities. We have studied formulas used in mathematics, business, geometry, and science.

Write a formula describing the mathematical relationship between the given quantities.

1. Distance traveled (d), rate traveled (r), time traveling at that rate (t)

2. Sale price (s), original price (p), discount (d)

3. Perimeter of a rectangle (P), length of the rectangle (l), width of the rectangle (w)

4. Amount of simple interest earned (I), principal (P), interest rate (r), time the money is invested (t)

Use a formula to solve each problem.

5. Find the area (A) of the triangular lot.

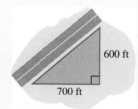

600 ft

700 ft

6. Find the volume (V) of the ice chest.

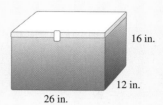

16 in.

12 in.

26 in.

7. Find the retail price (p) of a cookware set that costs the store owner $45.50 and is marked up $35.

8. Find the profit (p) made by a school T-shirt sale if revenue was $14,500 and costs were $10,200.

9. Find the distance (d) that a rock falls in 3 seconds after being dropped from the edge of a cliff.

10. Find the temperature in degrees Celsius (C) if the temperature in degrees Fahrenheit is 59.

Sometimes we use the same formula to answer several related questions. The results can be displayed in a table.

11. Find the interest earned by each account.

Type of account	Principal	Annual rate earned	Time invested	Interest earned
Savings	$5,000	5%	3 yr	
Passbook	$2,250	2%	1 yr	
Trust fund	$10,000	6.25%	10 yr	

12. Complete the table.

Type of coin	Number	Value (¢)	Total value (¢)
Penny	15		
Nickel	n		
Dime	d		
Quarter	q		

ACCENT ON TEAMWORK

SECTION 9.1

WRITING DIGITS In the illustration, the digit 1 is drawn using one angle, and the digit 2 is drawn using two angles. Draw the digit 3 using three angles, the digit 4 using four angles, and so on for all of the digits up to and including 9.

SECTION 9.2

CONSTRUCTIONS

Step 1: See Illustration (a). Using a straightedge, draw $\overline{AB}$. Then place the sharp point of a compass at A and draw an arc.

Step 2: With the same compass setting, place the sharp point at B. As shown in Illustration (b), draw another arc that intersects the arc from step 1 at two points. Label these points C and D.

Step 3: Using a straightedge, draw a line through points C and D. Label the point where line CD intersects $\overline{AB}$ as point E. Does $m(\overline{AE}) = m(\overline{EB})$?

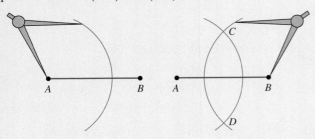

SECTION 9.3

TANGRAM A tangram is a puzzle in which geometric shapes are arranged to form other shapes. Cut out the pieces in the illustration. Assemble them so that they form a square. There should be no gaps, overlaps, or holes.

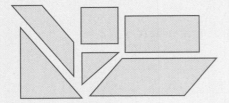

SECTION 9.4

CONGRUENT TRIANGLES Draw a triangle on a piece of paper. Then measure the lengths of its sides (with a ruler) and the angle measures (with a protractor). Choose a combination of any three measurements and tell them to your partner. Are the given facts sufficient for your partner to construct a triangle congruent to yours?

SECTION 9.5

AREA Find the area of the shaded figure on the square grid.

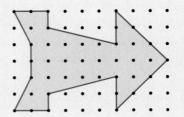

SECTION 9.6

PI Carefully measure the circumference and the diameter of different-size circles. Record the measurements in a table like the one below. Then use a calculator to find $\frac{C}{D}$. The result should be a number close to π.

Object	Circumference	Diameter	$\frac{C}{D}$
Jar	$3\frac{1}{2}$ in.	$1\frac{1}{8}$ in.	3.11

SECTION 9.7

PYRAMIDS Cut out, fold, and glue together the pattern shown. Estimate the volume and surface area of the pyramid.

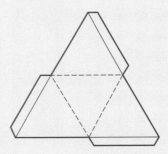

CHAPTER REVIEW

Some Basic Definitions

CONCEPTS

In geometry, we study *points, lines,* and *planes.*

A *line segment* is a part of a line with two endpoints. A *ray* is a part of a line with one endpoint.

An *angle* is a figure formed by two rays with a common endpoint. The common endpoint is called the *vertex* of the angle.

A *protractor* is used to find the measure of an angle.

An *acute angle* is greater than 0° but less than 90°. A *right angle* measures 90°. An *obtuse angle* is greater than 90° but less than 180°. A *straight angle* measures 180°.

Two angles that have the same vertex and are side-by-side are called *adjacent angles.*

REVIEW EXERCISES

1. In the illustration, identify a point, a line, and a plane.

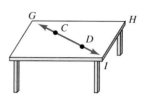

2. In the illustration, find m($\overline{AB}$).

3. In the illustration below, give four ways to name the angle.

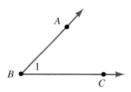

4. In the illustration above, use a protractor to find the measure of the angle.

5. In the illustration below, identify each acute angle, right angle, obtuse angle, and straight angle.

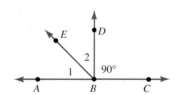

The measures of several angles are given. Identify each angle as an acute angle, a right angle, an obtuse angle, or a straight angle.

6. m($\angle A$) = 150°

7. m($\angle B$) = 90°

8. m($\angle C$) = 180°

9. m($\angle D$) = 25°

10. The two angles shown are adjacent angles. Find *x*.

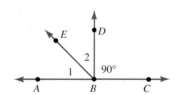

11. Line AB is shown. Find y.

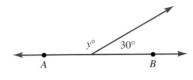

12. Find **a.** m($\angle 1$) and **b.** m($\angle 2$).

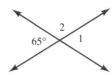

13. Find the complement of an angle that measures 50°.

14. Find the supplement of an angle that measures 140°.

15. Are angles measuring 30°, 60°, and 90° supplementary?

SECTION 9.2   *Parallel and Perpendicular Lines*

16. Which part of the illustration represents parallel lines?

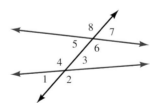

          (a)   **(b)**

17. Identify all pairs of alternate interior angles shown in the illustration below.

18. Identify all pairs of corresponding angles shown in the illustration below.

19. Identify all pairs of vertical angles shown in the illustration below.

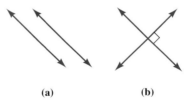

20. In the illustration below, $l_1 \parallel l_2$. Find the measure of each angle.

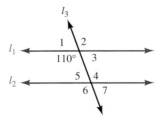

21. In the illustration below, $\overline{DC} \parallel \overline{AB}$. Find the measure of each angle.

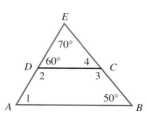

22. In Illustration (a), $l_1 \parallel l_2$. Find x.

23. In Illustration (b), $l_1 \parallel l_2$. Find x.

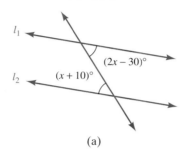

(a)

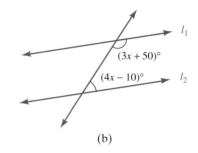

(b)

Polygons

A *polygon* is a closed geometric figure. The points at which the sides intersect are called *vertices*. A *regular polygon* has sides that are all the same length and angles that are all the same measure.

Polygons are classified as follows:

Number of sides	Name
3	triangle
4	quadrilateral
5	pentagon
6	hexagon
8	octagon

Identify each polygon as a triangle, a quadrilateral, a pentagon, a hexagon, or an octagon.

24.

25.

26.

27.

28.

Give the number of vertices in each polygon.

29. Triangle

30. Quadrilateral

31. Octagon

32. Hexagon

An *equilateral triangle* has three sides of equal length.
An *isosceles triangle* has at least two sides of equal length.
A *scalene triangle* has no sides of equal length.
A *right triangle* has one right angle.

Classify each of the triangles as an equilateral triangle, an isosceles triangle, a scalene triangle, or a right triangle.

33.
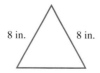

34.
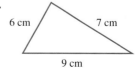

In an isosceles triangle, the angles opposite the sides of equal length are called *base angles*. The third angle is called the *vertex angle*. The third side is called the *base*.

Properties of isosceles triangles:

1. The base angles are congruent.

2. If two angles in a triangle are congruent, the sides opposite the angles are congruent, and the triangle is isosceles.

The sum of the measures of the angles of any triangle is 180°.

35.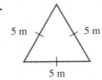

36.

Determine whether each triangle is isosceles.

37.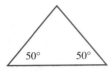

38.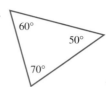

In each triangle, find x.

39.

40.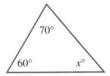

41. If one base angle of an isosceles triangle measures 65°, how large is the vertex angle?

42. If one base angle of an isosceles triangle measures 60°, what can you conclude about the triangle?

Quadrilaterals are classified as follows:

Property	Name
Opposite sides parallel	parallel-ogram
Parallelogram with four right angles	rectangle
Rectangle with all sides equal	square
Parallelogram with sides of equal length	rhombus
Exactly two sides parallel	trapezoid

Classify each quadrilateral as a parallelogram, a rectangle, a square, a rhombus, or a trapezoid.

43.

44.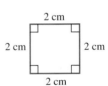

45.

46.

47.

48.

Properties of rectangles:

1. All angles are right angles.
2. Opposite sides are parallel.
3. Opposite sides are of equal length.
4. Diagonals are of equal length.
5. If the diagonals of a parallelogram are of equal length, the parallelogram is a rectangle.

In the illustration below, the length of diagonal $\overline{AC}$ of rectangle ABCD is 15 centimeters. Find each measure.

49. m($\overline{BD}$)

50. m($\angle 1$)

51. m($\angle 2$)

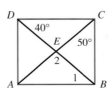

In the illustration below, ABCD is a rectangle. Classify each statement as true or false.

52. $m(\overline{AB}) = m(\overline{DC})$

53. $m(\overline{AD}) = m(\overline{DC})$

54. Triangle ABE is isosceles.

55. $m(\overline{AC}) = m(\overline{BD})$

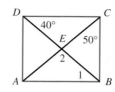

The parallel sides of a trapezoid are called *bases*. The nonparallel sides are called *legs*. If the legs of a trapezoid are of equal length, it is *isosceles*. In an isosceles trapezoid, the angles opposite the sides of equal length are *base angles*, and they are congruent.

In the illustration, ABCD is an isosceles trapezoid. Find each measure.

56. $m(\angle B)$

57. $m(\angle C)$

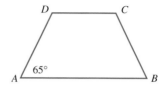

The sum of the measures of the angles of a polygon (in degrees) is given by the formula

$$S = (n - 2)180$$

Find the sum of the angle measures of each polygon.

58. Quadrilateral

59. Hexagon

Properties of Triangles

If two triangles have the same size and the same shape, they are *congruent triangles*.

Corresponding parts of congruent triangles have the same measure.

See the illustration. Complete the list of corresponding parts.

60. $\angle A$ corresponds to _____.

61. $\angle B$ corresponds to _____.

62. $\angle C$ corresponds to _____.

63. $\overline{AC}$ corresponds to _____.

64. $\overline{AB}$ corresponds to _____.

65. $\overline{BC}$ corresponds to _____.

Three ways to show that two triangles are congruent are

1. the SSS property

2. the SAS property

3. the ASA property

Determine whether the triangles in each pair are congruent. If they are, tell why.

66.

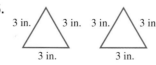

67.

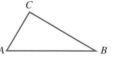

68.

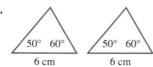

69.

563

If two triangles have the same shape, they are said to be *similar.* If two angles of one triangle have the same measure as two angles of a second triangle, the triangles are similar.

Determine whether the triangles in each pair are similar.

70.

71.

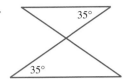

72. If a tree casts a 7-foot shadow at the same time a man 6 feet tall casts a 2-foot shadow, how tall is the tree?

The Pythagorean theorem: If the length of the *hypotenuse* of a right triangle is c, and the lengths of its legs are a and b, then

$$a^2 + b^2 = c^2$$

Refer to the illustration and find the length of the unknown side.

73. If $a = 5$ and $b = 12$, find c.

74. If $a = 8$ and $c = 17$, find b.

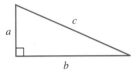

75. 🖩 To the nearest tenth, find the height of the television screen shown.

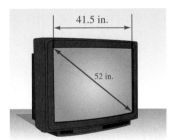

41.5 in.

52 in.

SECTION 9.5 **Perimeters and Areas of Polygons**

The *perimeter* of a polygon is the distance around it.

76. Find the perimeter of a square with sides 18 inches long.

77. Find the perimeter of a rectangle that is 3 meters long and 1.5 meters wide.

Find the perimeter of each polygon.

78.

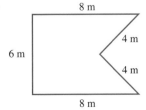

8 m

6 m

4 m

4 m

8 m

79.

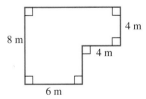

4 m

8 m

4 m

6 m

The *area* of a polygon is the measure of the surface it encloses.

Formulas for area:

Figure	Area
Square	$A = s^2$
Rectangle	$A = lw$
Parallel-ogram	$A = bh$
Triangle	$A = \frac{1}{2}bh$
Trapezoid	$A = \frac{1}{2}h(b_1 + b_2)$

Find the area of each polygon.

80.
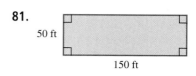
3.1 cm
3.1 cm
3.1 cm
3.1 cm

81.
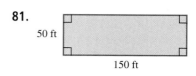
50 ft
150 ft

82.
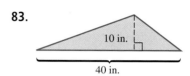
20 ft
15 ft
30 ft

83.
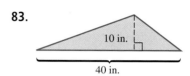
10 in.
40 in.

84.
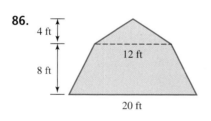
12 cm
8 cm
18 cm

85.
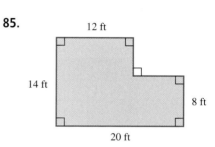
12 ft
14 ft
8 ft
20 ft

86.
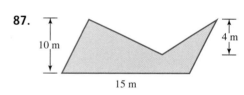
4 ft
12 ft
8 ft
20 ft

87.
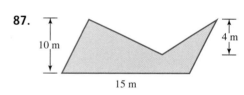
10 m
4 m
15 m

88. How many square feet are there in 1 square yard?

89. How many square inches are in 1 square foot?

SECTION 9.6	*Circles*

A *circle* is the set of all points in a plane that lie a fixed distance from a point called its *center*. The fixed distance is the circle's *radius*.

A *chord* of a circle is a line segment connecting two points on the circle.

Refer to the illustration.

90. Name each chord.

91. Name each diameter.

92. Name each radius.

93. Name the center.

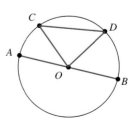

A *diameter* is a chord that passes through the circle's center.

The *circumference* (perimeter) of a circle is given by the formulas

$$C = \pi D \quad \text{or} \quad C = 2\pi r$$

where $\pi = 3.14159\ldots$.

The *area* of a circle is given by the formula

$$A = \pi r^2$$

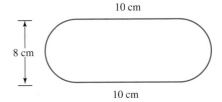

Find each answer to the nearest tenth.

94. Find the circumference of a circle with a diameter of 21 centimeters.

95. Find the perimeter of the figure shown.

10 cm

8 cm

10 cm

96. Find the area of a circle with a diameter of 18 inches.

97. Find the area of the figure shown in above.

SECTION 9.7   *Surface Area and Volume*

The *volume* of a solid is a measure of the space it occupies.

Figure	Volume
Cube	$V = s^3$
Rectangular solid	$V = lwh$
Prism	$V = Bh^*$
Sphere	$V = \frac{4}{3}\pi r^3$
Cylinder	$V = \pi r^2 h$
Cone	$V = \frac{1}{3}\pi r^2 h$
Pyramid	$V = \frac{1}{3}Bh^*$

*B represents the area of the base.

The *surface area* of a rectangular solid is the sum of the areas of its six faces.

The surface area of a sphere is given by the formula

$$SA = 4\pi r^2$$

Find the volume of each solid to the nearest unit.

98.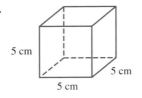

5 cm
5 cm
5 cm

99.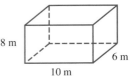

8 m
10 m
6 m

100.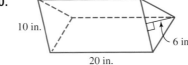

10 in.
20 in.
6 in.

101.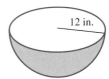

12 in.

102.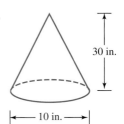

10 ft
16 ft

103.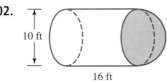

30 in.
10 in.

104.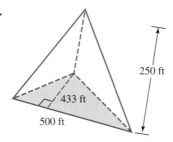

250 ft
433 ft
500 ft

105.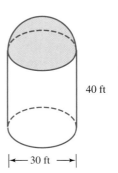

40 ft
30 ft

106. How many cubic inches are there in 1 cubic foot?

107. How many cubic feet are there in 2 cubic yards?

To the nearest tenth, find the surface area of each solid.

108.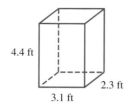

4.4 ft

2.3 ft

3.1 ft

109.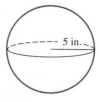

5 in.

1. Find m($\overline{AB}$),

2. Which point is the vertex of $\angle ABC$?

Determine whether each statement is true or false.

3. An angle of 47° is an acute angle.

4. An angle of 90° is a straight angle.

5. An angle of 180° is a right angle.

6. An angle of 132° is an obtuse angle.

7. Find x.

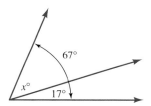

8. Find y.

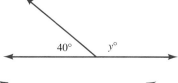

9. Find y.

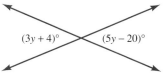

10. CALLIGRAPHY The illustration shows how the tip of the pen should be held at a 45° angle to the horizontal. What is x?

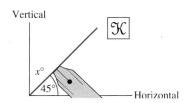

11. Find the complement of an angle measuring 67°.

12. Find the supplement of an angle measuring 117°.

Refer to the illustration below, in which $l_1 \parallel l_2$.

13. m($\angle 1$) = _____.

14. m($\angle 2$) = _____.

15. m($\angle 3$) = _____.

16. Find x.

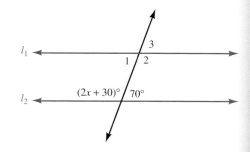

17. Complete the table.

Polygon	Number of sides
Triangle	
Quadrilateral	
Hexagon	
Pentagon	
Octagon	

18. Complete the table about triangles.

Property	Kind of triangle
All sides of equal length	
No sides of equal length	
Two sides of equal length	

Refer to the illustration below.

19. Find m($\angle A$).

20. Find m($\angle C$).

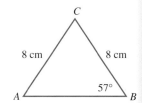

21. If the measures of two angles in a triangle are 65° and 85°, find the measure of the third angle.

22. Find the sum of the measures of the angles in a decagon (a ten-sided polygon).

23. In the illustration, *ABCD* is a rectangle. Name three pairs of segments with equal lengths.

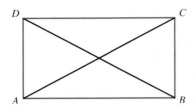

24. In the illustration, *ABCD* is an isosceles trapezoid. Find *x*.

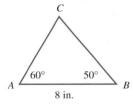

Refer to the illustration, in which △ABC ≅ △DEF.

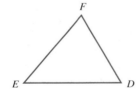

25. Find m($\overline{DE}$).　　　**26.** Find m(∠*E*).

Refer to the illustration in the next column, in which m(∠*A*) = m(∠*D*) *and* m(∠*C*) = m(∠*F*).

27. Find *x*.

28. Find *y*.

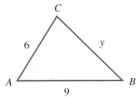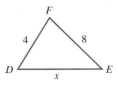

Give each answer to the nearest tenth.

29. A baseball diamond is a square with each side 90 feet long. What is the straight-line distance from third base to first base?

30. Find the area of a triangle with a base 44.5 centimeters long and a height of 17.6 centimeters.

31. Find the area of a trapezoid with a height of 6 feet and bases that are 12.2 feet and 15.7 feet long.

32. THE OLYMPICS Steel rod is to be bent to form the interlocking rings of the Olympic Games symbol. How many feet of steel rod will be needed to make the symbol if the diameter of each ring is to be 6 feet?

33. Find the area of a circle with a diameter that is 6 feet long.

34. Find the volume of a rectangular solid with dimensions 4.3 by 5.7 by 6.5 meters.

35. Find the volume of a sphere that is 8 meters in diameter.

36. Find the volume of a 10-foot-tall pyramid that has a rectangular base 5 feet long and 4 feet wide.

37. Give a real-life example in which the concept of perimeter is used. Do the same for area and for volume. Be sure to discuss the type of units used in each case.

38. Draw a cube. Explain how to find its surface area.

1. AMUSEMENT PARKS Use the data in the table to construct a bar graph on the illustration.

Fatal accidents on amusement park rides										
Year	'93	'94	'95	'96	'97	'98	'99	'00	'01	'02
Number	4	2	4	3	4	7	6	1	3	2

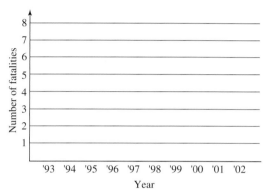

Source: U.S. Product Consumer Safety Commission

2. USED CARS The following ad appeared in *The Car Trader.* (O.B.O. means "or best offer.") If offers of $8,750, $8,875, $8,900, $8,850, $8,800, $7,995, $8,995, and $8,925 were received, what was the selling price of the car?

> 1969 Ford Mustang. New tires
> Must sell!!!! $10,500 O.B.O.

3. Subtract: $35,021 - 23,999$.

4. Divide: $1,353 \div 41$.

5. Round 2,109,567 to the nearest thousand.

6. Prime factor 220.

7. Find all the factors of 24.

8. List the set of integers.

9. Evaluate: $-10(-2) - 2^3 + 1$.

10. Evaluate: $5 - 3[4^2 - (1 + 5 \cdot 2)]$.

11. Evaluate: $|-6 - (-3)|$.

12. Evaluate the expression $\dfrac{2x + 3y}{z - y}$ for $x = 2$, $y = -3$, and $z = -4$.

13. What is the difference between an equation and an expression?

14. Simplify: $4x - 2(3x - 4) - 5(2x)$.

Solve each equation. **Check each result.**

15. $3(p + 15) + 4(11 - p) = 0$

16. $5t - 7 = 7t + 13$

17. $-x + 2 = 13$ **18.** $4x - 40 = -20$

19. SNAILS According to the *Guinness Book of World Records,* in the 1995 World Snail Racing Championships, a snail covered a 13-inch course in 2 minutes. What was the snail's rate in inches per minute?

20. SHOPPING What is the value of x coupons, each of which gives the shopper 50¢ off?

21. Translate to mathematical symbols: Seven squared minus two cubed

22. LUMBER To find the number of board feet (b.f.) in a piece of lumber, use the formula

$$\text{b.f.} = \frac{\text{thickness (in.)} \cdot \text{width (in.)} \cdot \text{length (ft)}}{12}$$

Find the number of board feet in the piece of lumber shown. (*Hint:* the symbol ″ stands for inches and the symbol ′ stands for feet.)

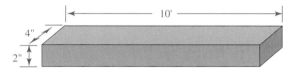

23. Simplify: $\dfrac{35}{28}$.

24. Add: $45\dfrac{2}{3} + 96\dfrac{4}{5}$.

25. Subtract: $\dfrac{3}{4} - \dfrac{3}{5}$.

26. BAKING A 5-pound bag of all-purpose flour contains $17\dfrac{1}{2}$ cups. A baker uses $3\dfrac{3}{4}$ cups. How much flour is left?

27. Multiply: $-\dfrac{6}{25}\left(2\dfrac{7}{24}\right)$.

28. Divide: $\dfrac{15}{8} \div \dfrac{45}{8}$.

29. What is the reciprocal of $\dfrac{9}{8}$?

30. Write $7\dfrac{1}{2}$ as an improper fraction.

31. PET MEDICATION A pet owner was told to use an eye dropper to administer medication to his sick kitten. The cup shown on the next page contains 8 doses of the medication. Determine the size of a single dose.

32. Evaluate: $\dfrac{3}{4} + \left(-\dfrac{1}{3}\right)^2\left(\dfrac{5}{4}\right)$.

33. Simplify: $\dfrac{7 - \dfrac{2}{3}}{4\dfrac{5}{6}}$.

34. GRAVITY Objects on the moon weigh only one-sixth as much as on Earth. If a rock weighs 3 ounces on the moon, how much does it weigh on Earth?

35. GLOBAL WARMING The graph below shows the annual mean global temperature change as measured by satellites orbiting the Earth.

   a. When was the greatest rise in temperature recorded? What was it?

   b. When was the greatest decline in temperature recorded? What was it?

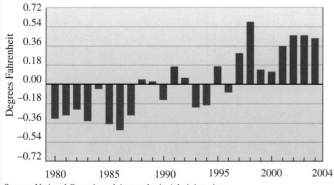

Change In Mean Global Surface Temperature

Source: National Oceanic and Atmospheric Administration

36. Graph each member of the set on the number line: $\left\{-4\dfrac{5}{8}, \sqrt{17}, 2.89, \dfrac{2}{3}, -0.1, -\sqrt{9}, \dfrac{3}{2}\right\}$

37. Round the number pi to the nearest ten thousandth: $\pi = 3.141592654.\ldots$

38. Place the proper symbol ($>$ or $<$) in the blank: 154.34 ▨ 154.33999.

39. Add: $3.4 + 106.78 + 35 + 0.008$.

40. Multiply: $-5.5(-3.1)$.

41. Multiply: $(89.9708)(1,000)$.

42. Divide: $\dfrac{0.0742}{1.4}$.

43. Evaluate: $-8.8 + (-7.3 - 9.5)$.

44. Evaluate: $\dfrac{7}{8}(9.7 + 15.8)$.

45. Change $\dfrac{2}{15}$ to a decimal.

46. Evaluate $\dfrac{(-1.3)^2 + 6.7}{-0.9}$ and round to the nearest hundredth.

47. DECORATIONS A mother has budgeted $20 for decorations for her daughter's birthday party. She decides to buy a tank of helium for $15.15 and some balloons. If the balloons sell for 5 cents apiece, how many balloons can she buy?

48. Find the square root of 100.

49. Evaluate: $2\sqrt{121} - 3\sqrt{64}$.

50. Evaluate: $\sqrt{\dfrac{49}{81}}$.

51. TABLE TENNIS The weights (in ounces) of 8 ping-pong balls that are to be used in a tournament are as follows: 0.85, 0.87, 0.88, 0.88, 0.85, 0.86, 0.84, and 0.85. Find the mean, median, and mode of the weights.

52. a. Consider $(-3)^2$. What is the base and what is the exponent? Evaluate the expression.

   b. Consider -3^2. What is the base and what is the exponent? Evaluate the expression.

Simplify each expression.

53. $s^4 \cdot s^5$

54. $(a^5)^7$

55. $-3h^9(-5h)$

56. $(2b^3c^6)^3$

57. $(y^5)^2(y^4)^3$

58. $x^m \cdot x^n$

59. What percent of the figure is shaded? What percent is not shaded?

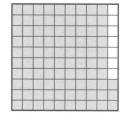

60. What number is 15% of 450?

61. 24.6 is 20.5% of what number?

62. Complete the table.

Percent	Decimal	Fraction
57%		
	0.001	
		$\frac{1}{3}$

63. STUDENT GOVERNMENT In an election for Student Body President, 560 votes were cast. Stan Cisneros received 308 votes, and Amy Huang-Sims received 252 votes. Use a circle graph to show the percent of the vote received by each candidate.

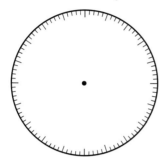

64. SHOPPING What is the regular price of the calculator shown below?

SALE PRICE
$54 $\frac{75}{\text{EA}}$

Save 27%

65. SALES TAX If the sales tax rate is $6\frac{1}{4}$%, how much sales tax will be added to the price of a new car selling for $18,550?

66. COLLECTIBLES A German Hummel porcelain figurine, which was originally purchased for $125, was sold by a collector ten years later for $750. What was the percent increase in the value of the figurine?

67. PAYING OFF LOANS To pay for tuition, a college student borrows $1,500 for two months. If the annual interest rate is 9%, how much will the student have to repay when the loan comes due?

68. RETIREMENT When he got married, a man invested $5,000 in an account that guaranteed to pay 8% interest, compounded monthly, for 50 years. At the end of 50 years, how much will his account be worth?

Write each phrase as a ratio.

69. 3 centimeters to 7 centimeters

70. 13 weeks to 1 year

71. COMPARISON SHOPPING A dry-erase whiteboard with an area of 400 in.² sells for $24. A larger board, with an area of 600 in.², sells for $42. Which board is the better buy?

72. Solve the proportion: $\frac{x}{14} = \frac{13}{28}$.

73. INSURANCE CLAIMS In one year, an auto insurance company had 3 complaints per 1,000 policies. If a total of 375 complaints were filed that year, how many policies did the company have?

74. SCALE DRAWINGS Suppose the house plan shown is drawn on a grid of $\frac{1}{4}$-inch squares. How long is the house?

Scale $\frac{1}{4}$ in. : 3 ft

Make each conversion.

75. 168 inches = ▨ feet

76. 15 yards = ▨ inches

77. 212 ounces = ▨ pounds

78. 30 gallons = ▨ quarts

79. 25 cups = ▨ fluid ounces

80. 738 minutes = ▨ hours

81. 654 milligrams = ▨ centigrams

82. 500 milliliters = ▨ liter

83. 5,890 decimeters = ▨ dekameters

84. 75°C = ▨ F

85. THE AMAZON The Amazon River enters the Atlantic Ocean through a broad estuary, roughly estimated at 240,000 m in width. Convert the width to kilometers.

86. TENNIS A tennis ball weighs between 57 and 58 g. Express this range in centigrams.

87. **OCEAN LINERS** When it was making the transatlantic cruises from England to America, the Queen Mary got 13 feet to the gallon.

  a. How many meters a gallon is this?

  b. The fuel capacity of the ship was 3,000,000 gallons. How many liters is this?

88. **COOKING** What is the weight of a 10-pound ham in kilograms?

89. How many degrees are in a right angle?

90. How many degrees are in an acute angle?

91. Find the supplement of an angle of 105°.

92. Find the complement of an angle of 75°.

Refer to the illustration, in which $l_1 \parallel l_2$. *Find the measure of each angle.*

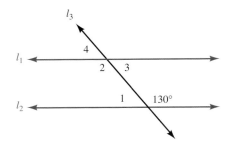

93. m(∠1)
94. m(∠3)
95. m(∠2)
96. m(∠4)

Refer to the illustration, in which $AB \parallel DE$ *and* $m(AC) = m(BC)$. *Find the measure of each angle.*

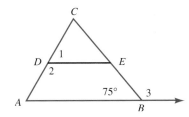

97. m(∠1)
98. m(∠C)
99. m(∠2)
100. m(∠3)

101. **JAVELIN THROW** Determine x and y.

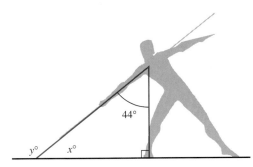

102. Find the sum of the angles of a pentagon.

103. If two sides of a right triangle measure 5 meters and 12 meters, how long is the hypotenuse?

If an answer is not exact, round to the nearest hundredth.

104. Find the perimeter and area of a rectangle with dimensions of 9 meters by 12 meters.

105. Find the area of a triangle with a base that is 14 feet long and an altitude of 18 feet.

106. Find the area of a trapezoid that has bases that are 12 inches and 14 inches long and a height of 7 inches.

107. Find the circumference and area of a circle with a diameter of 14 centimeters.

108. Find the area of the shaded region, which is created using two semicircles.

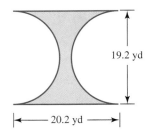

109. Find the volume of a rectangular solid with dimensions of 5 meters by 6 meters by 7 meters.

110. Find the volume of a sphere with a diameter of 10 inches.

111. Find the volume of a cone that has a circular base 8 meters in diameter and a height of 9 meters.

112. Find the volume of a cylindrical pipe that is 20 feet long and 6 inches in diameter.

113. Find the surface area of a block of ice that is in the shape of a rectangular solid with dimensions 15 in. × 24 in. × 18 in.

Polynomials

I.1 Introduction to Polynomials

- Polynomials • Classifying polynomials • Degree of a polynomial
- Evaluating polynomials

Polynomials

Recall that an **algebraic term,** or simply a **term,** is a number or a product of a number and one or more variables, which may be raised to powers. Some examples of terms are

$$17, \quad 5x, \quad 6t^2, \quad \text{and} \quad -8z^3$$

The coefficients of these terms are 17, 5, 6, and -8, respectively.

> **Polynomials**
>
> A **polynomial** is a single term or a sum of terms in which all variables have whole-number exponents. No variable appears in the denominator.

Some examples of polynomials are

$$0, \quad 8y^2, \quad 2x + 1, \quad 4y^2 - 2y + 3, \quad \text{and} \quad 7a^3 + 2a^2 - a - 1$$

The polynomial $8y^2$ has one term. The polynomial $2x + 1$ has two terms, $2x$ and 1. Since $4y^2 - 2y + 3$ can be written as $4y^2 + (-2y) + 3$, it is the sum of three terms, $4y^2$, $-2y$, and 3.

Classifying polynomials

We classify some polynomials by the number of terms they contain. A polynomial with one term is called a **monomial.** A polynomial with two terms is called a **binomial.** A polynomial with three terms is called a **trinomial.** Some examples of these polynomials are shown in Table I-1.

Monomials	Binomials	Trinomials
$5x^2$	$2x - 1$	$5t^2 + 4t + 3$
$-6x$	$18a^2 - 4a$	$27x^3 - 6x + 2$
29	$-27z^4 + 7z^2$	$32r^2 + 7r - 12$

TABLE I-1

Self Check 1
Classify each polynomial as a monomial, a binomial, or a trinomial:

a. $5x$

b. $8x^2 + 7$

c. $x^2 - 2x - 1$

Answers **a.** monomial, **b.** binomial, **c.** trinomial

EXAMPLE 1 Classify each polynomial as a monomial, a binomial, or a trinomial: **a.** $3x + 4$, **b.** $3x^2 + 4x - 12$, and **c.** $25x^3$.

Solution

a. Since $3x + 4$ has two terms, it is a binomial.

b. Since $3x^2 + 4x - 12$ has three terms, it is a trinomial.

c. Since $25x^3$ has one term, it is a monomial.

■ Degree of a polynomial

The monomial $7x^3$ is called a **monomial of third degree** or a **monomial of degree 3,** because the variable occurs three times as a factor.

- $5x^2$ is a monomial of degree 2.   Because the variable occurs two times as a factor: $x^2 = x \cdot x$.

- $-8x^4$ is a monomial of degree 4.   Because the variable occurs four times as a factor: $x^4 = x \cdot x \cdot x \cdot x$.

- $\dfrac{1}{2}x^5$ is a monomial of degree 5.   Because the variable occurs five times as a factor: $x^5 = x \cdot x \cdot x \cdot x \cdot x$.

We define the degree of a polynomial by considering the degrees of each of its terms.

> **Degree of a polynomial**
>
> The **degree of a polynomial** is the same as the degree of its term with largest degree.

For example,

- $x^2 + 5x$ is a binomial of degree 2, because the degree of its term with largest degree (x^2) is 2.
- $4y^3 + 2y - 7$ is a trinomial of degree 3, because the degree of its term with largest degree ($4y^3$) is 3.
- $\frac{1}{2}z + 3z^4 - 2z^2$ is a trinomial of degree 4, because the degree of its term with largest degree ($3z^4$) is 4.

Self Check 2
Find the degree of each polynomial:

a. $3p^3$

b. $17r^4 + 2r^8 - r$

c. $-2g^5 - 7g^6 + 12g^7$

Answers **a.** 3, **b.** 8, **c.** 7

EXAMPLE 2 Find the degree of each polynomial: **a.** $-2x + 4$, **b.** $5t^3 + t^4 - 7$, and **c.** $3 - 9z + 6z^2 - z^3$.

Solution

a. Since $-2x$ can be written as $-2x^1$, the degree of the term with largest degree is 1. Thus, the degree of the polynomial is 1.

b. In $5t^3 + t^4 - 7$, the degree of the term with largest degree (t^4) is 4. Thus, the degree of the polynomial is 4.

c. In $3 - 9z + 6z^2 - z^3$, the degree of the term with largest degree ($-z^3$) is 3. Thus, the degree of the polynomial is 3.

■ Evaluating polynomials

When a number is substituted for the variable in a polynomial, the polynomial takes on a numerical value. Finding this value is called **evaluating the polynomial.**

EXAMPLE 3   Evaluate each polynomial when $x = 3$: **a.** $3x - 2$ and **b.** $-2x^2 + x - 3$.

Solution

a. $3x - 2 = 3(3) - 2$ Substitute 3 for x.

$ = 9 - 2$ Multiply: $3(3) = 9$.

$ = 7$ Subtract: $9 - 2 = 7$.

b. $-2x^2 + x - 3 = -2(3)^2 + 3 - 3$ Substitute 3 for x.

$ = -2(9) + 3 - 3$ Square 3: $3 \cdot 3 = 9$.

$ = -18 + 3 - 3$ Multiply: $-2(9) = -18$.

$ = -15 - 3$ Add: $-18 + 3 = -15$.

$ = -18$ Subtract: $-15 - 3 = -18$.

Self Check 3
Evaluate each polynomial when $x = -1$:
a. $-2x^2 - 4$
b. $3x^2 - 4x + 1$

Answers **a.** -6, **b.** 8

EXAMPLE 4   **Height of an object.**   The polynomial $-16t^2 + 28t + 8$ gives the height (in feet) of an object t seconds after it has been thrown straight up. Find the height of the object in 1 second.

Solution   To find the height at 1 second, we evaluate the polynomial at $t = 1$.

$-16t^2 + 28t + 8 = -16(1)^2 + 28(1) + 8$ Substitute 1 for t.

$ = -16(1) + 28(1) + 8$ Square 1: $1 \cdot 1 = 1$.

$ = -16 + 28 + 8$ Multiply: $-16(1) = -16$ and $28(1) = 28$.

$ = 12 + 8$ Add: $-16 + 28 = 12$.

$ = 20$ Add: $12 + 8 = 20$.

At 1 second, the object is 20 feet above the ground.

Self Check 4
Find the height of the object in 2 seconds.

Answer 0 ft

Section I.1 STUDY SET

VOCABULARY   *Fill in the blanks.*

1. A polynomial with one term is called a _____.

2. A polynomial with three terms is called a _____.

3. A polynomial with two terms is called a _____.

4. The degree of a polynomial is the _____ as the degree of its term with largest degree.

CONCEPTS   *Classify each polynomial as a monomial, a binomial, or a trinomial.*

5. $3x^2 - 4$

6. $5t^2 - t + 1$

7. $17e^4$

8. $x^2 + x + 7$

9. $25u^2$

10. $x^2 - 9$

11. $q^5 + q^2 + 1$

12. $4d^3 - 3d^2$

Find the degree of each polynomial.

13. $5x^3$

14. $3t^5 + 3t^2$

15. $2x^2 - 3x + 2$

16. $\dfrac{1}{2}p^4 - p^2$

17. $2m$

18. $7q - 5$

19. $25w^6 + 5w^7$

20. $p^6 - p^8$

NOTATION *Complete each solution.*

21. Evaluate $3a^2 + 2a - 7$ when $a = 2$.

$$3a^2 + 2a - 7 = 3(\ )^2 + 2(\ ) - 7$$
$$= 3(\ ) + \ - 7$$
$$= 12 + 4 - 7$$
$$= \ - 7$$
$$= 9$$

22. Evaluate $-q^2 - 3q + 2$ when $q = -1$.

$$-q^2 - 3q + 2 = -(\ )^2 - 3(\ ) + 2$$
$$= -(\ ) - 3(-1) + 2$$
$$= -1 + \ + 2$$
$$= \ + 2$$
$$= 4$$

PRACTICE *Evaluate each polynomial for the given value.*

23. $3x + 4$ when $x = 3$

24. $\dfrac{1}{2}x - 3$ when $x = -6$

25. $2x^2 + 4$ when $x = -1$

26. $-\dfrac{1}{2}x^2 - 1$ when $x = 2$

27. $0.5t^3 - 1$ when $t = 4$

28. $0.75a^2 + 2.5a + 2$ when $a = 0$

29. $\dfrac{2}{3}b^2 - b + 1$ when $b = 3$

30. $3n^2 - n + 2$ when $n = 2$

31. $-2s^2 - 2s + 1$ when $s = -1$

32. $-4r^2 - 3r - 1$ when $r = -2$

APPLICATIONS *The height h (in feet) of a ball shot straight up with an initial velocity of 64 feet per second is given by the equation $h = -16t^2 + 64t$. Find the height of the ball after the given number of seconds.*

33. 0 second

34. 1 second

35. 2 seconds

36. 4 seconds

The number of feet that a car travels before stopping depends on the driver's reaction time and the braking distance. For one driver, the stopping distance d is given by the equation $d = 0.04v^2 + 0.9v$, where v is the velocity of the car. Find the stopping distance for each of the following speeds.

37. 30 mph

38. 50 mph

39. 60 mph

40. 70 mph

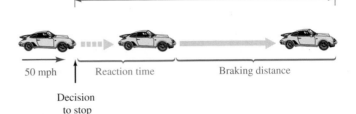

WRITING

41. Explain how to find the degree of the polynomial $2x^3 + 5x^5 - 7x$.

42. Explain how to evaluate the polynomial $-2x^2 - 3$ when $x = 5$.

REVIEW *Perform the operations.*

43. $\dfrac{2}{3} + \dfrac{4}{3}$

44. $\dfrac{1}{2} + \dfrac{2}{3}$

45. $\dfrac{36}{7} - \dfrac{23}{7}$

46. $\dfrac{5}{14} - \dfrac{4}{21}$

47. $\dfrac{5}{12} \cdot \dfrac{18}{5}$

48. $\dfrac{23}{25} \div \dfrac{46}{5}$

Solve each equation.

49. $x - 4 = 12$

50. $4z = 108$

51. $2(x - 3) = 6$

52. $3(a - 5) = 4(a + 9)$

I.2 Adding and Subtracting Polynomials

• Adding polynomials • Subtracting polynomials

Polynomials can be added, subtracted, and multiplied just like numbers in arithmetic. In this section, we show how to find sums and differences of polynomials.

Adding polynomials

Recall that like terms have exactly the same variables and the same exponents. For example, the monomials

$3z^2$ and $-2z^2$ are like terms   Both have the same variable (z) with the same exponent (2).

However, the monomials

$7b^2$ and $8a^2$ are not like terms   They have different variables.

$32p^2$ and $25p^3$ are not like terms   The exponents of p are different.

Also recall that we can combine like terms by adding their coefficients and keeping the same variables and exponents. For example,

$$2y + 5y = (2 + 5)y \qquad \text{and} \qquad -3x^2 + 7x^2 = (-3 + 7)x^2$$
$$= 7y \qquad\qquad\qquad\qquad = 4x^2$$

Thus, *to add monomials that are like terms, we add the coefficients and keep the same variables and exponents.*

EXAMPLE 1   Add: $5x^3 + 7x^3$.

Solution   Since the monomials are like terms, we add the coefficients and keep the variables and exponents.

$$5x^3 + 7x^3 = 12x^3$$

Self Check 1
Add: $7y^3 + 12y^3$.

Answer $19y^3$

EXAMPLE 2   Add: $\dfrac{3}{2}t^2 + \dfrac{5}{2}t^2 + \dfrac{7}{2}t^2$.

Solution   Since the monomials are like terms, we add the coefficients and keep the variables and exponents.

$$\frac{3}{2}t^2 + \frac{5}{2}t^2 + \frac{7}{2}t^2 = \left(\frac{3}{2} + \frac{5}{2} + \frac{7}{2}\right)t^2$$

$$= \frac{15}{2}t^2 \qquad \begin{array}{l}\text{To add the fractions, add the numerators and}\\ \text{keep the denominator: } 3 + 5 + 7 = 15.\end{array}$$

Self Check 2
Add: $3.2m^3 + 4.5m^3 + 7.2m^3$.

Answer $14.9m^3$

To add two polynomials, we write a + sign between them and combine like terms.

EXAMPLE 3   Add: $2x + 3$ and $7x - 1$.

Solution

$$(2x + 3) + (7x - 1) \qquad \text{Write a + sign between the binomials.}$$

$$= (2x + 7x) + (3 - 1) \qquad \begin{array}{l}\text{Use the associative and commutative properties to}\\ \text{group like terms together.}\end{array}$$

$$= 9x + 2 \qquad\qquad\qquad \text{Combine like terms.}$$

Self Check 3
Add: $(5y - 2) + (-3y + 7)$.

Answer $2y + 5$

The binomials in Example 3 can be added by writing the polynomials so that like terms are in columns.

$$\begin{array}{r} 2x + 3 \\ + \underline{7x - 1} \\ 9x + 2 \end{array}$$ Add the like terms, one column at a time.

Self Check 4
Add:
$(2b^2 - 4b) + (b^2 + 3b - 1)$.

EXAMPLE 4 Add: $(5x^2 - 2x + 4) + (3x^2 - 5)$.

Solution

$$
\begin{aligned}
(5x^2 &- 2x + 4) + (3x^2 - 5) \\
&= (5x^2 + 3x^2) + (-2x) + (4 - 5) \quad \text{Use the associative and commutative} \\
&\qquad\qquad\qquad\qquad\qquad\qquad\qquad\quad \text{properties to group like terms together.} \\
&= 8x^2 - 2x - 1 \qquad\qquad\qquad\qquad \text{Combine like terms.}
\end{aligned}
$$

Answer $3b^2 - b - 1$

The polynomials in Example 4 can be added by writing the polynomials so that like terms are in columns.

$$
\begin{array}{r}
5x^2 - 2x + 4 \\
+\,3x^2 \quad\;\; - 5 \\
\hline
8x^2 - 2x - 1
\end{array}
$$
Add the like terms, one column at a time.

Self Check 5
Add:
$(s^2 + 1.2s - 5) + (3s^2 - 2.5s + 4)$.

EXAMPLE 5 Add: $(3.7x^2 + 4x - 2) + (7.4x^2 - 5x + 3)$.

Solution

$$
\begin{aligned}
(3.7x^2 &+ 4x - 2) + (7.4x^2 - 5x + 3) \\
&= (3.7x^2 + 7.4x^2) + (4x - 5x) + (-2 + 3) \quad \text{Use the associative and} \\
&\qquad\qquad\qquad\qquad\qquad\qquad\qquad\qquad\qquad \text{commutative properties to} \\
&\qquad\qquad\qquad\qquad\qquad\qquad\qquad\qquad\qquad \text{group like terms together.} \\
&= 11.1x^2 - x + 1 \qquad\qquad\qquad\qquad\qquad\quad \text{Combine like terms.}
\end{aligned}
$$

Answer $4s^2 - 1.3s - 1$

The trinomials in Example 5 can be added by writing them so that like terms are in columns.

$$
\begin{array}{r}
3.7x^2 + 4x - 2 \\
+\;\; 7.4x^2 - 5x + 3 \\
\hline
11.1x^2 - \;\; x + 1
\end{array}
$$
Add the like terms, one column at a time.

▮ Subtracting polynomials

To subtract one monomial from another, we add the opposite of the monomial that is to be subtracted. In symbols, $x - y = x + (-y)$.

Self Check 6
Subtract: $6y^3 - 9y^3$.

EXAMPLE 6 Subtract: $8x^2 - 3x^2$.

Solution

$$
\begin{aligned}
8x^2 - 3x^2 &= 8x^2 + (-3x^2) \quad \text{Add the opposite of } 3x^2. \\
&= 5x^2 \qquad\qquad\quad\;\; \text{Add the coefficients and keep the same variable and} \\
&\qquad\qquad\qquad\qquad\;\; \text{exponent.}
\end{aligned}
$$

Answer $-3y^3$

To subtract polynomials, we also add the opposite. For example, to subtract $3n^2 - 4n + 2$ from $5n^2 + 2n - 3$, we proceed as follows.

$$(5n^2 + 2n - 3) - (3n^2 - 4n + 2)$$

$$= (5n^2 + 2n - 3) + [-(3n^2 - 4n + 2)] \quad \text{Add the opposite.}$$

$$= (5n^2 + 2n - 3) + (-3n^2 + 4n - 2) \quad \begin{array}{l}\text{Use the distributive property to}\\ \text{change signs.}\end{array}$$

$$= (5n^2 - 3n^2) + (2n + 4n) + (-3 - 2) \quad \begin{array}{l}\text{Use the associative and}\\ \text{commutative properties to}\\ \text{group like terms together.}\end{array}$$

$$= 2n^2 + 6n - 5 \quad \text{Combine like terms.}$$

These polynomials can be subtracted by writing them so that like terms are in columns.

$$
\begin{array}{r}
5n^2 + 2n - 3 \\
-(3n^2 - 4n + 2) \\
\end{array}
\longrightarrow
\begin{array}{r}
5n^2 + 2n - 3 \\
+ \underline{-3n^2 + 4n - 2} \\
2n^2 + 6n - 5
\end{array}
\quad \text{Change signs and add.}
$$

EXAMPLE 7   Subtract: $(3x - 4.2) - (5x + 7.2)$.

Solution

$$(3x - 4.2) - (5x + 7.2)$$

$$= (3x - 4.2) + [-(5x + 7.2)] \quad \text{Add the opposite.}$$

$$= (3x - 4.2) + (-5x - 7.2) \quad \text{Use the distributive property to change signs.}$$

$$= (3x - 5x) + (-4.2 - 7.2) \quad \begin{array}{l}\text{Use the associative and commutative properties}\\ \text{to group like terms together.}\end{array}$$

$$= -2x - 11.4 \quad \text{Combine like terms.}$$

Self Check 7
Subtract:
$(3.3a - 5) - (7.8a + 2)$.

Answer $-4.5a - 7$

The binomials in Example 7 can be subtracted by writing them so that like terms are in columns.

$$
\begin{array}{r}
3x - 4.2 \\
-(5x + 7.2) \\
\end{array}
\longrightarrow
\begin{array}{r}
3x - 4.2 \\
+ \underline{-5x - 7.2} \\
-2x - 11.4
\end{array}
\quad \text{Change signs and add.}
$$

EXAMPLE 8   Subtract: $(3x^2 - 4x - 6) - (2x^2 - 6x + 12)$.

Solution

$$(3x^2 - 4x - 6) - (2x^2 - 6x + 12)$$

$$= (3x^2 - 4x - 6) + [-(2x^2 - 6x + 12)] \quad \text{Add the opposite.}$$

$$= (3x^2 - 4x - 6) + (-2x^2 + 6x - 12) \quad \begin{array}{l}\text{Use the distributive property to}\\ \text{change signs.}\end{array}$$

$$= (3x^2 - 2x^2) + (-4x + 6x) + (-6 - 12) \quad \begin{array}{l}\text{Use the associative and}\\ \text{commutative properties to}\\ \text{group like terms together.}\end{array}$$

$$= x^2 + 2x - 18 \quad \text{Combine like terms.}$$

Self Check 8
Subtract:
$(5y^2 - 4y + 2) - (3y^2 + 2y - 1)$.

Answer $2y^2 - 6y + 3$

The trinomials in Example 8 can be subtracted by writing them so that like terms are in columns.

$$
\begin{array}{r}
3x^2 - 4x - 6 \\
-(2x^2 - 6x + 12) \\
\end{array}
\longrightarrow
\begin{array}{r}
3x^2 - 4x - 6 \\
+ \underline{-2x^2 + 6x - 12} \\
x^2 + 2x - 18
\end{array}
\quad \text{Change signs and add.}
$$

Section I.2 STUDY SET

VOCABULARY   *Fill in the blanks.*

1. If two algebraic terms have exactly the same variables and exponents, they are called _____ terms.

2. $3x^3$ and $3x^2$ are _____ terms.

CONCEPTS   *Fill in the blanks.*

3. To add two monomials, we add the _____ and keep the same _____ and exponents.

4. To subtract one monomial from another, we add the _____ of the monomial that is to be subtracted.

Determine whether the monomials are like terms. If they are, combine them.

5. $3y, 4y$

6. $3x^2, 5x^2$

7. $3x, 3y$

8. $3x^2, 6x$

9. $3x^3, 4x^3, 6x^3$

10. $-2y^4, -6y^4, 10y^4$

11. $-5x^2, 13x^2, 7x^2$

12. $23, 12x, 25x$

NOTATION   *Complete each solution.*

13. $(3x^2 + 2x - 5) + (2x^2 - 7x)$
$$= (3x^2 + \boxed{}) + (2x - \boxed{}) + (-5)$$
$$= \boxed{} + (-5x) - 5$$
$$= 5x^2 - 5x - 5$$

14. $(3x^2 + 2x - 5) - (2x^2 - 7x)$
$$= (3x^2 + 2x - 5) + [-(\boxed{} - 7x)]$$
$$= (3x^2 + 2x - 5) + (\boxed{})$$
$$= (\boxed{}) + (2x + 7x) + (-5)$$
$$= x^2 + 9x - 5$$

PRACTICE   *Add the polynomials.*

15. $4y + 5y$

16. $-2x + 3x$

17. $-8t^2 - 4t^2$

18. $15x^2 + 10x^2$

19. $3s^2 + 4s^2 + 7s^2$

20. $-2a^3 + 7a^3 - 3a^3$

21. $(3x + 7) + (4x - 3)$

22. $(2y - 3) + (4y + 7)$

23. $(2x^2 + 3) + (5x^2 - 10)$

24. $(-4a^2 + 1) + (5a^2 - 1)$

25. $(5x^3 - 4.2x) + (7x^3 - 10.7x)$

26. $(-4.3a^3 + 25a) + (5.8a^3 - 10a)$

27. $(3x^2 + 2x - 4) + (5x^2 - 17)$

28. $(5a^2 - 2a) + (-2a^2 + 3a + 4)$

29. $(7y^2 + 5y) + (y^2 - y - 2)$

30. $(4p^2 - 4p + 5) + (6p - 2)$

31. $(3x^2 - 3x - 2) + (3x^2 + 4x - 3)$

32. $(4c^2 + 3c - 2) + (3c^2 + 4c + 2)$

33. $(3n^2 - 5.8n + 7) + (-n^2 + 5.8n - 2)$

34. $(-3t^2 - t + 3.4) + (3t^2 + 2t - 1.8)$

35. $3x^2 + 4x + 5$
$\underline{2x^2 - 3x + 6}$

36. $2x^2 - 3x + 5$
$\underline{-4x^2 - x - 7}$

37. $-3x^2 - 7$
$\underline{-4x^2 - 5x + 6}$

38. $4x^2 - 4x + 9$
$\underline{9x - 3}$

39. $-3x^2 + 4x + 25.4$
$\underline{5x^2 - 3x - 12.5}$

40. $-6x^3 - 4.2x^2 + 7$
$\underline{-7x^3 + 9.7x^2 - 21}$

Subtract the polynomials.

41. $32u^3 - 16u^3$

42. $25y^2 - 7y^2$

43. $18x^5 - 11x^5$

44. $17x^6 - 22x^6$

45. $(4.5a + 3.7) - (2.9a - 4.3)$

46. $(5.1b - 7.6) - (3.3b + 5.9)$

47. $(-8x^2 - 4) - (11x^2 + 1)$

48. $(5x^3 - 8) - (2x^3 + 5)$

49. $(3x^2 - 2x - 1) - (-4x^2 + 4)$

50. $(7a^2 + 5a) - (5a^2 - 2a + 3)$

51. $(3.7y^2 - 5) - (2y^2 - 3.1y + 4)$

52. $(t^2 - 4.5t + 5) - (2t^2 - 3.1t - 1)$

53. $(2b^2 + 3b - 5) - (2b^2 - 4b - 9)$

54. $(3a^2 - 2a + 4) - (a^2 - 3a + 7)$

55. $(5p^2 - p + 7.1) - (4p^2 + p + 7.1)$

56. $(m^2 - m - 5) - (m^2 + 5.5m - 7.5)$

57. $3x^2 + 4x - 5$
$\underline{- (-2x^2 - 2x + 3)}$

58. $3y^2 - 4y + 7$
$\underline{- (6y^2 - 6y - 13)}$

59. $-2x^2 - 4x + 12$
$\underline{- (10x^2 + 9x - 24)}$

60.
$$25x^3 - 45x^2 + 31x$$
$$- (12x^3 + 27x^2 - 17x)$$

61.
$$4x^3 - 3x + 10$$
$$- (5x^3 - 4x - 4)$$

62.
$$3x^3 + 4x^2 + 12$$
$$- (-4x^3 + 6x^2 - 3)$$

APPLICATIONS *Consider the following information: If a house is purchased for $85,000 and is expected to appreciate $700 per year, its value y after x years is given by the equation $y = 700x + 85,000$.*

63. VALUE OF A HOUSE Find the expected value of the house after 10 years.

64. VALUE OF A HOUSE A second house is purchased for $102,000 and is expected to appreciate $900 per year. Find an equation that will give the value y of the house after x years.

65. VALUE OF A HOUSE Find the value of the house discussed in Exercise 64 after 12 years.

66. VALUE OF TWO HOUSES Find a single polynomial equation that will give the combined value y of both houses after x years.

67. VALUE OF TWO HOUSES In two ways, find the value of the two houses after 15 years

  a. By substituting into the polynomial equations $y = 700x + 85,000$ and $y = 900x + 102,000$ and adding

  b. By substituting into the result of Exercise 66

68. VALUE OF TWO HOUSES In two ways, find the value of the two houses after 25 years.

  a. By substituting into the polynomial equations $y = 700x + 85,000$ and $y = 900x + 102,000$ and adding

  b. By substituting into the result of Exercise 66

Consider the following information. A young couple bought two cars, one for $8,500 and the other for $10,200. The first car is expected to depreciate $800 per year and the second car $1,100 per year.

69. VALUE OF A CAR Write an equation that will give the value y of the first car after x years.

70. VALUE OF A CAR Write an equation that will give the value y of the second car after x years.

71. VALUE OF TWO CARS Find a single equation that will give the value y of both cars after x years.

72. VALUE OF TWO CARS In two ways, find the value of the two cars after 6 years.

WRITING

73. What are *like terms?*

74. Explain how to add two polynomials.

75. Explain how to subtract two polynomials.

76. When two binomials are added, is the result always a binomial? Explain.

REVIEW

77. BASKETBALL SHOES Use the following information to find how much lighter the Kevin Garnett shoe is than the Michael Jordan shoe.

Nike Air Garnett III	Air Jordan XV
Synthetic fade mesh and leather.	Full grain leather upper with woven pattern.
Sizes $6\frac{1}{2}$–18	Sizes $6\frac{1}{2}$–18
Weight: 13.8 oz	Weight: 14.6 oz

78. AEROBICS The number of calories burned when doing step aerobics depends on the step height. How many more calories are burned during a 10-minute workout using an 8-inch step instead of a 4-inch step?

Step height (in.)	Calories burned per minute
4	4.5
6	5.5
8	6.4
10	7.2

Source: *Reebok Instructor News* (Vol. 4, No. 3, 1991)

79. THE PANAMA CANAL A ship entering the Panama Canal from the Atlantic Ocean is lifted up 85 feet to Lake Gatun by the Gatun Lock system. See the illustration. Then the ship is lowered 31 feet by the Pedro Miguel Lock. By how much must the ship be lowered by the Miraflores Lock system for it to reach the Pacific Ocean water level?

80. CANAL LOCKS What is the combined length of the system of locks in the Panama Canal? Express your answer as a mixed number and as a decimal, rounded to the nearest tenth.

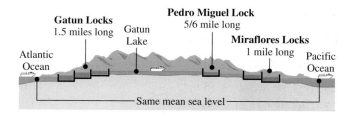

I.3 Multiplying Polynomials

- Multiplying monomials   • Multiplying a polynomial by a monomial
- Multiplying a binomial by a binomial   • Multiplying a polynomial by a binomial

In this section, we discuss how to multiply polynomials.

■ Multiplying monomials

To multiply $4x^2$ by $-2x^3$, we use the commutative and associative properties of multiplication to group the numerical factors and the variable factors and multiply.

$$4x^2(-2x^3) = 4(-2)x^2x^3$$
$$= -8x^5$$

This example suggests the following rule.

> **Multiplying two monomials**
>
> To multiply two monomials, multiply the numerical factors and then multiply the variable factors.

Self Check 1
Multiply: $-7a^3 \cdot 2a^5$.

EXAMPLE 1 Multiply: **a.** $3y \cdot 6y$ and **b.** $-3x^5(2x^5)$.

Solution
a. $3y \cdot 6y = (3 \cdot 6)(y \cdot y)$ Multiply the numerical factors and multiply the variables.
$$= 18y^2 \qquad \text{Multiply: } 3 \cdot 6 = 18 \text{ and } y \cdot y = y^2.$$

b. $(-3x^5)(2x^5) = (-3 \cdot 2)(x^5 \cdot x^5)$ Multiply the numerical factors and multiply the variables.
$$= -6x^{10} \qquad \text{Multiply: } -3 \cdot 2 = -6 \text{ and } x^5 \cdot x^5 = x^{10}.$$

Answer $-14a^8$

■ Multiplying a polynomial by a monomial

To find the product of a polynomial and a monomial, we use the distributive property. To multiply $x + 4$ by $3x$, for example, we proceed as follows:

$$3x(x + 4) = 3x(x) + 3x(4) \quad \text{Use the distributive property.}$$
$$= 3x^2 + 12x \qquad \text{Multiply the monomials: } 3x(x) = 3x^2 \text{ and } 3x(4) = 12x.$$

The results of this example suggest the following rule.

> **Multiplying polynomials by monomials**
>
> To multiply a polynomial by a monomial, use the distributive property to remove parentheses and simplify.

Self Check 2
Multiply:
a. $3y(5y^3 - 4y)$
b. $5x(3x^2 - 2x + 3)$

EXAMPLE 2 Multiply: **a.** $2a^2(3a^2 - 4a)$ and **b.** $2x(3x^2 + 2x - 3)$.

Solution
a. $2a^2(3a^2 - 4a)$
$$= 2a^2(3a^2) - 2a^2(4a) \quad \text{Use the distributive property.}$$
$$= 6a^4 - 8a^3 \qquad \text{Multiply: } 2a^2(3a^2) = 6a^4 \text{ and } 2a^2(4a) = 8a^3.$$

b. $2x(3x^2 + 2x - 3)$

$= 2x(3x^2) + 2x(2x) - 2x(3)$ Use the distributive property.

$= 6x^3 + 4x^2 - 6x$ Multiply: $2x(3x^2) = 6x^3$, $2x(2x) = 4x^2$, and $2x(3) = 6x$.

Answers a. $15y^4 - 12y^2$, **b.** $15x^3 - 10x^2 + 15x$

Multiplying a binomial by a binomial

To multiply two binomials, we must use the distributive property more than once. For example, to multiply $2x + 3$ by $3x - 5$, we proceed as follows.

$(3x - 5)(2x + 3) = (3x - 5)2x + (3x - 5)3$ Distribute the factor of $3x - 5$ over the two terms within $(2x + 3)$.

$= 2x(3x - 5) + 3(3x - 5)$ Use the commutative property of multiplication.

$= 2x(3x) - 2x(5) + 3(3x) - 3(5)$ Use the distributive property twice.

$= 6x^2 - 10x + 9x - 15$ Perform the multiplications.

$= 6x^2 - x - 15$ Combine like terms: $-10x + 9x = -x$.

The results of this example suggest the following rule.

Multiplying a binomial by a binomial

To multiply two binomials, multiply each term of one binomial by each term of the other binomial and combine like terms.

EXAMPLE 3   Multiply: $(2x - 4)(3x + 5)$.

Solution

$(2x - 4)(3x + 5)$

$= (2x - 4)3x + (2x - 4)5$ Each term within $(3x + 5)$ is multiplied by $2x - 4$.

$= 3x(2x - 4) + 5(2x - 4)$ Use the commutative property of multiplication.

$= 3x(2x) - 3x(4) + 5(2x) - 5(4)$ Use the distributive property twice.

$= 6x^2 - 12x + 10x - 20$ Perform the multiplications.

$= 6x^2 - 2x - 20$ Combine like terms: $-12x + 10x = -2x$.

Self Check 3
Multiply: $(3x - 2)(2x + 3)$.

Answer $6x^2 + 5x - 6$

EXAMPLE 4   Find: $(5x - 4)^2$.

Solution   In the expression $(5x - 4)^2$, the binomial $5x - 4$ is the base and 2 is the exponent.

$(5x - 4)^2 = (5x - 4)(5x - 4)$ Write $5x - 4$ as a factor two times.

$= (5x - 4)5x - (5x - 4)4$ Distribute the factor of $5x - 4$ over each term within $(5x - 4)$.

$= 5x(5x - 4) - 4(5x - 4)$ Change the order of the factors.

$= 5x(5x) - 5x(4) - 4(5x) - 4(-4)$ Distribute the multiplication by $5x$. Distribute the multiplication by 4.

$= 25x^2 - 20x - 20x + 16$ Perform the multiplications.

$= 25x^2 - 40x + 16$ Simplify: $-20x - 20x = -40x$.

Self Check 4
Find: $(5x + 4)^2$.

Answer $25x^2 + 40x + 16$

! **COMMENT** A common error when squaring a binomial is to square only its first and second terms. For example, it is incorrect to write

$$(5x - 4)^2 = (5x)^2 - (4)^2$$
$$= 25x^2 - 16$$

The correct answer is $25x^2 - 40x + 16$.

■ Multiplying a polynomial by a binomial

We must use the distributive property more than once to multiply a polynomial by a binomial. For example, to multiply $3x^2 + 3x - 5$ by $2x + 3$, we proceed as follows.

$$(2x + 3)(3x^2 + 3x - 5) = (2x + 3)3x^2 + (2x + 3)3x - (2x + 3)5$$
$$= 3x^2(2x + 3) + 3x(2x + 3) - 5(2x + 3)$$
$$= 6x^3 + 9x^2 + 6x^2 + 9x - 10x - 15$$
$$= 6x^3 + 15x^2 - x - 15$$

Self Check 5

Multiply: $(x - 2)(3x^2 + 4x + 1)$.

EXAMPLE 5 Multiply: $(3a + 1)(3a^2 + 2a + 2)$.

Solution

$$(3a + 1)(3a^2 + 2a + 2)$$
$$= (3a + 1)3a^2 + (3a + 1)2a + (3a + 1)2$$
$$= 3a^2(3a + 1) + 2a(3a + 1) + 2(3a + 1)$$
$$= 3a^2(3a) + 3a^2(1) + 2a(3a) + 2a(1) + 2(3a) + 2(1)$$
$$= 9a^3 + 3a^2 + 6a^2 + 2a + 6a + 2$$
$$= 9a^3 + 9a^2 + 8a + 2$$

Answer $3x^3 - 2x^2 - 7x - 2$

We can use a column format to multiply polynomials. To do so, we multiply each term in the top polynomial by each term in the bottom polynomial. To make the addition easy, we will keep like terms in columns. As an example, we can multiply $2x - 4$ by $3x + 2$ as follows.

$$
\begin{array}{r}
2x - 4 \\
\times\ \underline{3x + 2} \\
\end{array}
$$

$3x(2x - 4) \longrightarrow \qquad 6x^2 - 12x$

$2(2x - 4) \longrightarrow \qquad \underline{+\ 4x - 8}$

$\qquad\qquad\qquad 6x^2 - 8x - 8$

Self Check 6

Multiply $3y^2 - 5y + 4$ by $4y - 3$.

EXAMPLE 6 Multiply $3a^2 - 4a + 7$ by $2a + 5$.

Solution

$$
\begin{array}{r}
3a^2 - 4a + 7 \\
\times\ \underline{\qquad 2a + 5} \\
6a^3 - 8a^2 + 14a \\
\underline{+\ 15a^2 - 20a + 35} \\
6a^3 + 7a^2 - 6a + 35
\end{array}
$$

Answer $12y^3 - 29y^2 + 31y - 12$

Section I.3 STUDY SET

VOCABULARY   *Fill in the blanks.*

1. A polynomial with one term is called a _____.
2. A polynomial with two terms is called a _____.
3. A polynomial with ____ terms is called a trinomial.
4. $a(b + c) = ab + ac$ illustrates the _____ property.

CONCEPTS   *Fill in the blanks.*

5. To multiply two monomials, multiply the _____ factors and then multiply the variable _____.
6. To multiply a polynomial by a monomial, use the _____ property to remove parentheses and simplify.
7. To multiply two binomials, multiply each _____ of one binomial by each term of the other binomial and combine _____ terms.
8. To multiply a polynomial by a binomial, we must use the distributive _____ more than once.

NOTATION   *Complete each solution.*

9. $3x(2x - 5) = 3x(\quad) - 3x(\quad)$
   $$= 6x^2 - 15x$$

10. $(3x + 1)(2x^2 - 3x - 2)$

   $$= (3x + 1)\;\rule{1cm}{0.4pt}\; - (3x + 1)\;\rule{1cm}{0.4pt}\; - (3x + 1)\;\rule{1cm}{0.4pt}$$

   $$= 2x^2(3x + 1) - 3x(3x + 1) - 2(3x + 1)$$

   $$= 2x^2(\quad) + 2x^2(1) - 3x(3x) - 3x(1)$$
   $$\quad - 2(3x) - 2(\quad)$$

   $$= 6x^3 + 2x^2 - \rule{0.6cm}{0.4pt} - 3x - \rule{0.6cm}{0.4pt} - 2$$

   $$= 6x^3 - 7x^2 - 9x - 2$$

PRACTICE   *Find each product.*

11. $(3x^2)(4x^3)$
12. $(-2a^3)(3a^2)$
13. $(3b^2)(-2b)$
14. $(3y)(-y^4)$
15. $(-2x^2)(3x^3)$
16. $(-7x^3)(-3x^3)$
17. $\left(-\dfrac{2}{3}y^5\right)\left(\dfrac{3}{4}y^2\right)$
18. $\left(\dfrac{2}{5}r^4\right)\left(\dfrac{3}{5}r^2\right)$
19. $3(x + 4)$
20. $-3(a - 2)$
21. $-4(t + 7)$
22. $6(s^2 - 3)$
23. $3x(x - 2)$
24. $4y(y + 5)$
25. $-2x^2(3x^2 - x)$
26. $4b^3(2b^2 - 2b)$

27. $2x(3x^2 + 4x - 7)$
28. $3y(2y^2 - 7y - 8)$
29. $-p(2p^2 - 3p + 2)$
30. $-2t(t^2 - t + 1)$
31. $3q^2(q^2 - 2q + 7)$
32. $4v^3(-2v^2 + 3v - 1)$
33. $(a + 4)(a + 5)$
34. $(y - 3)(y + 5)$
35. $(3x - 2)(x + 4)$
36. $(t + 4)(2t - 3)$
37. $(2a + 4)(3a - 5)$
38. $(2b - 1)(3b + 4)$

Square each binomial.

39. $(2x + 3)^2$
40. $(2y + 5)^2$
41. $(2x - 3)^2$
42. $(2y - 5)^2$
43. $(5t + 1)^2$
44. $(5t - 1)^2$

Multiply the polynomials.

45. $(2x + 1)(3x^2 - 2x + 1)$
46. $(x + 2)(2x^2 + x - 3)$
47. $(x - 1)(x^2 + x + 1)$
48. $(x + 2)(x^2 - 2x + 4)$
49. $(x + 2)(x^2 - 3x + 1)$
50. $(x + 3)(x^2 + 3x + 2)$

Find each product.

51. $4x + 3$
     $\underline{\;x + 2}$

52. $5r + 6$
     $\underline{\;2r - 1}$

53. $4x - 2$
     $\underline{\;3x + 5}$

54. $6r + 5$
     $\underline{\;2r - 3}$

55. $x^2 - x + 1$
     $\underline{\;\;\;\;\;x + 1}$

56. $4x^2 - 2x + 1$
     $\underline{\;\;\;\;\;2x + 1}$

APPLICATIONS

57. **GEOMETRY** Express the area of the rectangle.

$(x + 2)$ ft

$(x - 2)$ ft

58. SAILING The height h of the triangular sail is $4x$ feet, and the base b is $3x - 2$ feet. Express the area of the sail. (*Hint:* The area of a triangle is given by the formula $A = \frac{1}{2}bh$.)

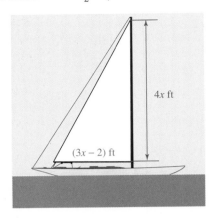

4x ft

(3x − 2) ft

59. ECONOMICS The revenue R received from selling clock radios is the product of their price and the number that are sold. If the price of each radio is given by the formula $-\frac{x}{100} + 30$ and x is the number sold, find a formula that gives the amount of revenue received.

60. ECONOMICS If the pricing formula given in Exercise 59 changes to $-\frac{x}{100} + 40$, find the formula for revenue received.

WRITING

61. Explain how to multiply two binomials.

62. Explain how to find $(2x + 1)^2$.

63. Explain why $(x + 1)^2 \neq x^2 + 1^2$. (Read $\neq$ as "is not equal to.")

64. If two terms are to be added, they have to be like terms. If two terms are to be multiplied, must they be like terms? Explain.

REVIEW

65. THE EARTH It takes 23 hours, 56 minutes, and 4.091 seconds for the Earth to rotate around its axis once. Write 4.091 in words.

66. TAKE-OUT FOOD The sticker shows the amount and the price per pound of some spaghetti salad that was purchased at a delicatessen. Find the total price of the salad.

> *Joan's Spaghetti Salad*
> 303 Foothill Plaza
> **Plaza Deli**
> 0.78   3.95   00.00
> NET WT. LB.   PRICE/ LB. $ (TOTAL PRICE $)

67. What is $\frac{7}{64}$ in decimal form?

68. Write $-\frac{6}{10}$ as a decimal.

69. Evaluate: $56.09 + 78 + 0.567$.

70. Evaluate: $-679.4 - (-599.89)$.

71. Evaluate: $\sqrt{16} + \sqrt{36}$.

72. Find: $103.6 \div 0.56$.

APPENDIX II

Inductive and Deductive Reasoning

- Inductive reasoning • Deductive reasoning

To reason means to think logically. The objective of this appendix is to develop your problem-solving ability by improving your reasoning skills. We introduce two funda-mental types of reasoning that can be applied in a wide variety of settings. They are known as *inductive reasoning* and *deductive reasoning*.

■ Inductive reasoning

In a laboratory, scientists conduct experiments and observe outcomes. After several repetitions with similar outcomes, the scientist generalizes the results into a statement that appears to be true.

- If I heat water to 212°F, it will boil.
- If I drop a weight, it will fall.
- If I combine an acid with a base, a chemical reaction occurs.

When we draw general conclusions from specific observations, we are using **inductive reasoning.** The next examples show how inductive reasoning can be used in mathematical thinking. Given a list of numbers or symbols, called a *sequence,* we can often find a missing term of the sequence by looking for patterns and applying induc-tive reasoning.

EXAMPLE 1   An increasing pattern.   Find the next number in the sequence $5, 8, 11, 14, \ldots$.

Solution   The terms of the sequence are increasing. To discover the pattern, we find the *difference* between each pair of successive terms.

$8 - 5 = 3$ Subtract the first term from the second term.

$11 - 8 = 3$ Subtract the second term from the third term.

$14 - 11 = 3$ Subtract the third term from the fourth term.

The difference between each pair of numbers is 3. This means that each successive number is 3 greater than the previous one. Thus, the next number in the sequence is $14 + 3$, or 17.

Self Check 1
Find the next number in the sequence $-3, -1, 1, 3, \ldots$.

Answer 5

Self Check 2

Find the next number in the sequence −0.1, −0.3, −0.5, −0.7 . . . .

Answer −0.9

EXAMPLE 2 A decreasing pattern.   Find the next number in the sequence −2, −4, −6, −8, ...

Solution   The terms of the sequence are decreasing. Since each successive term is 2 less than the previous one, the next number in the pattern is −8 − 2, or −10.

Self Check 3

Find the next entry in the sequence Z, A, Y, B, X, C, ...

EXAMPLE 3 An alternating pattern.   Find the next letter in the sequence A, D, B, E, C, F, D, ...

Solution   The letter A is the first letter of the alphabet, D is the fourth letter, B is the second letter, and so on. We can create the following letter–number correspondence.

$A \rightarrow 1$ ⎞ Add 3.
$D \rightarrow 4$ ⎠ Subtract 2.
$B \rightarrow 2$ ⎞ Add 3.
$E \rightarrow 5$ ⎠ Subtract 2.
$C \rightarrow 3$ ⎞ Add 3.
$F \rightarrow 6$ ⎠ Subtract 2.
$D \rightarrow 4$

The numbers in the sequence 1, 4, 2, 5, 3, 6, 4, . . . alternate in size. They change from smaller to larger, to smaller, to larger, and so on.

We see that 3 is added to the first number to get the second number. Then 2 is subtracted from the second number to get the third number. To get successive terms in the sequence, we alternately add 3 to one number and then subtract 2 from that result to get the next number.

If we apply this pattern, the next number in the numerical sequence would be 4 + 3, or 7. The next letter in the original sequence would be G, because it is the seventh letter of the alphabet.

Answer W

Self Check 4

Find the next geometric shape in the sequence below.

 , , . . .

Answer

EXAMPLE 4 Two patterns.   Find the next geometric shape in the sequence below.

Solution   This sequence has two patterns occurring at the same time. The first figure has three sides and one dot, the second figure has four sides and two dots, and the third figure has five sides and three dots. Thus, we would expect the next figure to have six sides and four dots, as shown in Figure A-1.

FIGURE A-1

EXAMPLE 5   A circular pattern. Find the next geometric shape in the sequence below.

Self Check 5
Find the next geometric shape in the sequence below.

Solution   From figure to figure, we see that each dot moves from one point of the star to the next, in a counterclockwise direction. This is a circular pattern. The next shape in the sequence will be the one shown in Figure A-2.

FIGURE A-2

Answer

Deductive reasoning

As opposed to inductive reasoning, **deductive reasoning** moves from the general case to the specific. For example, if we know that the sum of the angles in any triangle is 180°, we know that the sum of the angles of $\triangle ABC$ is 180°. Whenever we apply a general principle to a particular instance, we are using deductive reasoning.

A deductive reasoning system is built on four elements.

1. **Undefined terms:** terms that we accept without giving them formal meaning
2. **Defined terms:** terms that we define in a formal way
3. **Axioms** or **postulates:** statements that we accept without proof
4. **Theorems:** statements that we can prove with formal reasoning

Many problems can be solved by deductive reasoning. For example, suppose that we plan to enroll in an early-morning algebra class, and we know that Professors Perry, Miller, and Tveten are scheduled to teach algebra next semester. After some investigating, we find out that Professor Perry teaches only in the afternoon and Professor Tveten teaches only in the evenings. Without knowing anything about Professor Miller, we can conclude that he will be our teacher, since he is the only remaining possibility.

The following examples show how to use deductive reasoning to solve problems.

EXAMPLE 6   Scheduling classes. Four professors are scheduled to teach mathematics next semester, with the following course preferences.

1. Professors A and B don't want to teach calculus.
2. Professor C wants to teach statistics.
3. Professor B wants to teach algebra.

Who will teach trigonometry?

Solution   The following chart shows each course, with each possible instructor.

Calculus	Algebra	Statistics	Trigonometry
A	A	A	A
B	B	B	B
C	C	C	C
D	D	D	D

Since Professors A and B don't want to teach calculus, we can cross them off the calculus list. Since Professor C wants to teach statistics, we can cross her off every other list. This leaves Professor D as the only person to teach calculus, so we can cross her off every other list. Since Professor B wants to teach algebra, we can cross him off every other list. Thus, the only remaining person left to teach trigonometry is Professor A.

Calculus	Algebra	Statistics	Trigonometry
A̶	A	A	A
B̶	B	B̶	B̶
C̶	C̶	C	C̶
D	D̶	D̶	D̶

Self Check 7
Of the 50 cars on a used-car lot, 9 are red, 31 are foreign models, and 6 are red foreign models. If a customer wants to buy an American model that is not red, how many cars does she have to choose from?

EXAMPLE 7   State flags.   The graph in Figure A-3 gives the number of state flags that feature an eagle, a star, or both. How many state flags have neither an eagle nor a star?

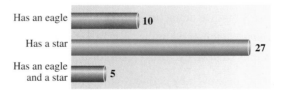

FIGURE A-3

Solution   In Figure A-4(a), the intersection (overlap) of the circles is a way to show that there are 5 state flags that have both an eagle and a star. If an eagle appears on a total of 10 flags, then the left circle must contain 5 more flags outside of the intersection. See Figure A-4(b). If a total of 27 flags have a star, the right circle must contain 22 more flags outside the intersection.

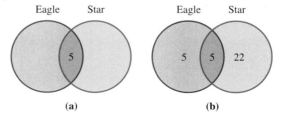

FIGURE A-4

From Figure A-4, we see that $5 + 5 + 22$, or 32 flags have an eagle, a star, or both. To find how many flags have neither an eagle nor a star, we subtract this total from the number of state flags, which is 50.

$$50 - 32 = 18$$

There are 18 state flags that have neither an eagle nor a star.

Answer 16

Appendix II STUDY SET

VOCABULARY *Fill in the blanks.*

1. _____ reasoning draws general conclusions from specific observations.

2. _____ reasoning moves from the general case to the specific.

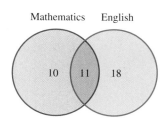

Mathematics   English

CONCEPTS *Determine whether the pattern shown is increasing, decreasing, alternating, or circular.*

3. 2, 3, 4, 2, 3, 4, 2, 3, 4, . . .

4. 8, 5, 2, −1, . . .

5. −2, −4, 2, 0, 6, . . .

6. 0.1, 0.5, 0.9, 1.3, . . .

7. a, c, b, d, c, e, . . .

8.

9. ROOM SCHEDULING From the chart, determine what time(s) on a Wednesday morning a practice room in a music building is available. The symbol X indicates that the room has already been reserved.

	M	T	W	Th	F
9 A.M.	X		X	X	
10 A.M.	X	X			X
11 A.M.			X		X

10. QUESTIONNAIRES A group of college students were asked if they were taking a mathematics course and if they were taking an English course. The results are displayed in the illustration.

   a. How many students were taking a mathematics course and an English course?

   b. How many students were taking an English course but not a mathematics course?

   c. How many students were taking a mathematics course?

PRACTICE *Find the number that comes next in each sequence.*

11. 1, 5, 9, 13, . . .

12. 15, 12, 9, 6, . . .

13. −3, −5, −8, −12, . . .

14. 5, 9, 14, 20, . . .

15. −7, 9, −6, 8, −5, 7, −4, . . .

16. 2, 5, 3, 6, 4, 7, 5, . . .

17. 9, 5, 7, 3, 5, 1, . . .

18. 1.3, 1.6, 1.4, 1.7, 1.5, 1.8, . . .

19. −2, −3, −5, −6, −8, −9, . . .

20. 8, 11, 9, 12, 10, 13, . . .

21. 6, 8, 9, 7, 9, 10, 8, 10, 11, . . .

22. 10, 8, 7, 11, 9, 8, 12, 10, 9, . . .

Find the figure that comes next in each sequence.

23.

24.

Find the missing figure in each sequence.

25. , , , ? ,

26. , , ? , ,

Find the next letter or letters in the sequence.

27. A, c, E, g, . . .   **28.** R, SS, TTT, . . .

29. d, h, g, k, j, n, . . .   **30.** B, N, C, N, D, . . .

What conclusion(s) can be drawn from each set of information?

31. Four people named John, Luis, Maria, and Paula have occupations as teacher, butcher, baker, and candlestick maker.

   1. John and Paula are married.

   2. The teacher plans to marry the baker in December.

   3. Luis is the baker.

   Who is the teacher?

32. In a zoo, a zebra, a tiger, a lion, and a monkey are to be placed in four cages numbered from 1 to 4, from left to right. The following decisions have been made:

   1. The lion and the tiger should not be side by side.

   2. The monkey should be in one of the end cages.

   3. The tiger is to be in cage 4.

   In which cage is the zebra?

33. A Ford, a Buick, a Dodge, and a Mercedes are parked side by side.

   1. The Ford is between the Mercedes and the Dodge.

   2. The Mercedes is not next to the Buick.

   3. The Buick is parked on the left end.

   Which car is parked on the right end?

34. Four divers at the Olympics finished first, second, third, and fourth.

   1. Diver A beat diver B.

   2. Diver C placed between divers B and D.

   3. Diver B beat diver D.

   In which order did they finish?

35. A green, a blue, a red, and a yellow flag are hanging on a flagpole.

   1. The blue flag is between the green and yellow flags.

   2. The red flag is next to the yellow flag.

   3. The green flag is above the red flag.

   What is the order of the flags from top to bottom?

36. Andres, Barry, and Carl each have two occupations: bootlegger, musician, painter, chauffeur, barber, and gardener. From the following facts, find the occupations of each man.

   1. The painter bought a quart of spirits from the boot-legger.

   2. The chauffeur offended the musician by laughing at his mustache.

   3. The chauffeur dated the painter's sister.

   4. Both the musician and the gardener used to go hunting with Andres.

   5. Carl beat both Barry and the painter at monopoly.

   6. Barry owes the gardener $100.

APPLICATIONS

37. JURY DUTY The results of a jury service questionnaire are shown. Determine how many of the 20,000 respondents have served on neither a criminal court nor a civil court jury.

Jury Service Questionnaire

997	Served on a criminal court jury
103	Served on a civil court jury
35	Served on both

38. ELECTRONIC POLLS In the illustration, the Internet poll shows that 124 people voted for the first choice, 27 people voted for the second choice, and 19 people voted for both the first and the second choices. How many people clicked the third choice, "Neither"?

Internet Poll	You may vote for more than one.
What would you do if gasoline reached $2.50 a gallon?	⦿ Cut down on driving ⦿ Buy a more fuel-efficient car ⦿ Neither
	Number of people voting 178

39. THE SOLAR SYSTEM The graph shows some important characteristics of the nine planets in our solar system. How many planets are neither rocky nor have moons?

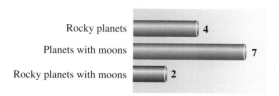

Rocky planets 4

Planets with moons 7

Rocky planets with moons 2

40. Write a problem in such a way that the following diagram can be used to solve it.

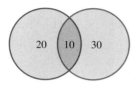

20 | 10 | 30

WRITING

41. Describe deductive reasoning.

42. Describe a real-life situation in which you might use deductive reasoning.

43. Describe inductive reasoning.

44. Describe a real-life situation in which you might use inductive reasoning.

Roots and Powers

n	n^2	$\sqrt{n}$	n^3	$\sqrt[3]{n}$	n	n^2	$\sqrt{n}$	n^3	$\sqrt[3]{n}$
1	1	1.000	1	1.000	51	2,601	7.141	132,651	3.708
2	4	1.414	8	1.260	52	2,704	7.211	140,608	3.733
3	9	1.732	27	1.442	53	2,809	7.280	148,877	3.756
4	16	2.000	64	1.587	54	2,916	7.348	157,464	3.780
5	25	2.236	125	1.710	55	3,025	7.416	166,375	3.803
6	36	2.449	216	1.817	56	3,136	7.483	175,616	3.826
7	49	2.646	343	1.913	57	3,249	7.550	185,193	3.849
8	64	2.828	512	2.000	58	3,364	7.616	195,112	3.871
9	81	3.000	729	2.080	59	3,481	7.681	205,379	3.893
10	100	3.162	1,000	2.154	60	3,600	7.746	216,000	3.915
11	121	3.317	1,331	2.224	61	3,721	7.810	226,981	3.936
12	144	3.464	1,728	2.289	62	3,844	7.874	238,328	3.958
13	169	3.606	2,197	2.351	63	3,969	7.937	250,047	3.979
14	196	3.742	2,744	2.410	64	4,096	8.000	262,144	4.000
15	225	3.873	3,375	2.466	65	4,225	8.062	274,625	4.021
16	256	4.000	4,096	2.520	66	4,356	8.124	287,496	4.041
17	289	4.123	4,913	2.571	67	4,489	8.185	300,763	4.062
18	324	4.243	5,832	2.621	68	4,624	8.246	314,432	4.082
19	361	4.359	6,859	2.668	69	4,761	8.307	328,509	4.102
20	400	4.472	8,000	2.714	70	4,900	8.367	343,000	4.121
21	441	4.583	9,261	2.759	71	5,041	8.426	357,911	4.141
22	484	4.690	10,648	2.802	72	5,184	8.485	373,248	4.160
23	529	4.796	12,167	2.844	73	5,329	8.544	389,017	4.179
24	576	4.899	13,824	2.884	74	5,476	8.602	405,224	4.198
25	625	5.000	15,625	2.924	75	5,625	8.660	421,875	4.217
26	676	5.099	17,576	2.962	76	5,776	8.718	438,976	4.236
27	729	5.196	19,683	3.000	77	5,929	8.775	456,533	4.254
28	784	5.292	21,952	3.037	78	6,084	8.832	474,552	4.273
29	841	5.385	24,389	3.072	79	6,241	8.888	493,039	4.291
30	900	5.477	27,000	3.107	80	6,400	8.944	512,000	4.309
31	961	5.568	29,791	3.141	81	6,561	9.000	531,441	4.327
32	1,024	5.657	32,768	3.175	82	6,724	9.055	551,368	4.344
33	1,089	5.745	35,937	3.208	83	6,889	9.110	571,787	4.362
34	1,156	5.831	39,304	3.240	84	7,056	9.165	592,704	4.380
35	1,225	5.916	42,875	3.271	85	7,225	9.220	614,125	4.397
36	1,296	6.000	46,656	3.302	86	7,396	9.274	636,056	4.414
37	1,369	6.083	50,653	3.332	87	7,569	9.327	658,503	4.431
38	1,444	6.164	54,872	3.362	88	7,744	9.381	681,472	4.448
39	1,521	6.245	59,319	3.391	89	7,921	9.434	704,969	4.465
40	1,600	6.325	64,000	3.420	90	8,100	9.487	729,000	4.481
41	1,681	6.403	68,921	3.448	91	8,281	9.539	753,571	4.498
42	1,764	6.481	74,088	3.476	92	8,464	9.592	778,688	4.514
43	1,849	6.557	79,507	3.503	93	8,649	9.644	804,357	4.531
44	1,936	6.633	85,184	3.530	94	8,836	9.695	830,584	4.547
45	2,025	6.708	91,125	3.557	95	9,025	9.747	857,375	4.563
46	2,116	6.782	97,336	3.583	96	9,216	9.798	884,736	4.579
47	2,209	6.856	103,823	3.609	97	9,409	9.849	912,673	4.595
48	2,304	6.928	110,592	3.634	98	9,604	9.899	941,192	4.610
49	2,401	7.000	117,649	3.659	99	9,801	9.950	970,299	4.626
50	2,500	7.071	125,000	3.684	100	10,000	10.000	1,000,000	4.642

Answers to Selected Exercises

Chapter 1   Check Your Knowledge   (page 2)

1. natural, whole   **2.** perimeter   **3.** commutative, associative
4. factors, product, quotient   **5.** prime
6. 3 thousands + 7 hundreds + 3 tens + 7 ones   **7.** 186,300
8.

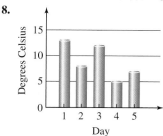

9.

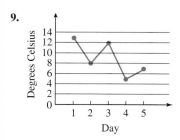

10. >   **11.** 5,121   **12.** 58
13. 2,061,497   **14.** 24,624
15. 57 R 34   **16.** 64 ft,
247 sq. ft   **17.** $2 \cdot 5^2 \cdot 19$
18. 5   **19.** 103   **20.** 3
21. 19   **22.** 63

Think It Through   (page 9)

1. c   **2.** b   **3.** e   **4.** d   **5.** a

Study Set   Section 1.1   (page 10)

1. set   **3.** expanded   **5.** number   **7.** 3   **9.** 6
11. whole numbers
13.

15.

17. >   **19.** >   **21.** <   **23.** >   **25.** braces
27. 2 hundreds + 4 tens + 5 ones; two hundred forty-five
29. 3 thousands + 6 hundreds + 9 ones; three thousand
six hundred nine   **31.** 3 ten thousands + 2 thousands +
5 hundreds; thirty-two thousand five hundred   **33.** 1 hundred
thousand + 4 thousands + 4 hundreds + 1 one; one hundred
four thousand four hundred one   **35.** 425   **37.** 2,736   **39.** 456
41. 27,598   **43.** 9,113   **45.** 10,700,506   **47.** 79,590   **49.** 80,000
51. 5,926,000   **53.** 5,900,000   **55.** $419,160   **57.** $419,000

59. Aisha   **61. a.** the 70s, 7   **b.** the 60s, 9   **c.** the 60s, 12
63.

65.

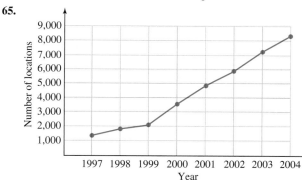

67.
a.

No. 201		March 9 , 20 05
Payable to Davis Chevrolet		$ 15,601.00
Fifteen thousand six hundred one and $\frac{00}{100}$		DOLLARS
		Don Smith
45-365-02		

b.

No. 7890		Aug. 12 , 20 05
Payable to Dr. Anderson		$ 3,433.00
Three thousand four hundred thirty three and $\frac{00}{100}$		DOLLARS
		Juan Decito
45-828-02		

69. 1,865,593; 482,880; 1,503; 269; 43,449
71. a. 299,800,000 m/sec   **b.** 300,000,000 m/sec

Study Set   Section 1.2   (page 19)

1. addends **3.** rectangle, square **5.** dimensions
7. commutative **9.** perimeter **11.** commutative
property of addition **13.** associative property of addition
15. commutative property of addition **17.** 0 **19.** parentheses
21. 33 plus 12 equals 45 **23.** 47 **25.** 38 **27.** 461 **29.** 150
31. 363 **33.** 121 **35.** 140 **37.** 979 **39.** 1,985 **41.** 10,000
43. 15,907 **45.** 1,861 **47.** 5,312 **49.** 88 ft **51.** 68 in.
53. 91 ft **55.** $767,496,832 **57.** 5,843,737 **59.** 10,057 mi
61. $6,233,000,000 **63.** 196 in. **67.** 3 thousands +
1 hundred + 2 tens + 5 ones **69.** 6,354,780 **71.** 6,350,000

Study Set   Section 1.3   (page 26)

1. difference **3.** addition **5.** left, right **7.** subtraction,
addition **9.** $83 - 30$ **11.** correct **13.** incorrect **15.** 3
17. 25 **19.** 103 **21.** 59 **23.** 24 **25.** 118 **27.** 958 **29.** 1,689
31. 10,457 **33.** 303 **35.** 43,559 **37.** 19,299 **39.** 102
41. 49,760 **43.** 5 **45.** 197 **47.** 110 **49.** 6 **51.** $18
53. 33 points **55.** $213 **57.** 1,263,030 **59.** $1,322
61. 39,732 **67.** 5,370,650 **69.** 5,371,000 **71.** 5,400,000
73. 52 in. **75.** 1,530

Study Set   Section 1.4   (page 36)

1. addition **3.** product **5.** associative **7.** area **9. a.** $4 \cdot 8$
b. $7 \cdot 15$ **11.** $5 \cdot 8 = 8 \cdot 5$ **13.** 3 **15. a.** area **b.** area
c. perimeter **17.** $\times, \cdot, (\ )$ **19.** 28 **21.** 56 **23.** 84 **25.** 98
27. 1,491 **29.** 6,232 **31.** 3,700 **33.** 750 **35.** 1,070,000
37. 512,000 **39.** 11,200 **41.** 6,300 **43.** 8,250 **45.** 101,000
47. 324 **49.** 5,829 **51.** 7,623 **53.** 1,060 **55.** 2,576
57. 20,079 **59.** 155,832 **61.** 408,758 **63.** 16,969,380
65. 8,945,912 **67.** 2,400 **69.** 45,696 **71.** 84 in.^{2} **73.** 144 in.^{2}
75. $132 **77.** 204 grams **79.** 72 **81.** 125,800 **83.** 312
85. yes **87.** 288 mi **89.** 54 ft^2 **91.** 1,260 mi, 97,200 mi^2
93. 64 **95.** 388 ft^2 **99.** 8 **101.** 872

Study Set   Section 1.5   (page 45)

1. quotient, dividend **3.** divisible **5.** 0 **7.** 0 **9.** sum
11. 0, 5 **13.** sum **15.** 1 **17.** undefined **19.** two
21. $\div,)\overline{}, \overline{}$ **23.** 8 **25.** 6 **27.** 9 **29.** 7 **31.** $3 \cdot 8 = 24$
33. $8 \cdot 7 = 56$ **35.** 72 **37.** 35 **39.** 237 **41.** 617 **43.** 5
45. 3 **47.** 12 **49.** 13 **51.** 73 **53.** 41 **55.** 205 **57.** 210
59. 8 R 25 **61.** 20 R 3 **63.** 30 R 13 **65.** 31 R 28 **67.** yes
69. no **71.** yes **73.** yes **75.** 350 **77.** 89 **79.** 124 **81.** 160
83. 4 **85.** 59,375 gal **87.** 14,500 lb **89.** 13 dozen **91.** 9 girls,
24 teams **93.** $73 **95.** AZ: 50; IN: 170; RI: 1,100; SC: 140 (all
answers given in persons per square mile) **99.** 2 **101.** 23

Study Set   Estimation   (page 48)

1. no **3.** no **5.** no **7.** approx. 8,900 mi **9.** approx. 30 bags
11. 1, 800,000,000

Study Set   Section 1.6   (page 54)

1. factors **3.** factor **5.** composite **7.** prime **9.** base,
exponent **11.** $1 \cdot 27$ or $3 \cdot 9$ **13. a.** 44 **b.** 100 **15. a.** 1 and 11
b. 1 and 23 **c.** 1 and 37 **d.** They are prime numbers.
17. yes **19.** 90 **21.** 605 **23.** no **25.** 2 and 5 **27.** 2
29. $3 \cdot 5 \cdot 2 \cdot 5; 5 \cdot 3 \cdot 5 \cdot 2$; they are the same. **31.** 13, 8, 7

33. 2 **35.** $7 \cdot 7 \cdot 7$ **37.** $3 \cdot 3 \cdot 3 \cdot 3 \cdot 3$ **39.** $5 \cdot 5 \cdot 11$ **41.** 10
43. 2^5 **45.** 5^4 **47.** $4^2(5^2)$ **49.** 1, 2, 5, 10 **51.** 1, 2, 4, 5, 8, 10,
20, 40 **53.** 1, 2, 3, 6, 9, 18 **55.** 1, 2, 4, 11, 22, 44 **57.** 1, 7, 11,
77 **59.** 1, 2, 4, 5, 10, 20, 25, 50, 100 **61.** $3 \cdot 13$ **63.** $3^2 \cdot 11$
65. $2 \cdot 3^4$ **67.** $2^2 \cdot 5 \cdot 11$ **69.** 2^6 **71.** $3 \cdot 7^2$ **73.** 81 **75.** 32
77. 144 **79.** 4,096 **81.** 72 **83.** 3,456 **85.** 12,812,904
87. 1,162,213 **89.** 1, 2, 4, 7, 14, 28; $1 + 2 + 4 + 7 + 14 = 28$
91. 2^2 square units; 3^2 square units; 4^2 square units **97.** 231,000
99. 0 **101.** $A = lw$

Think It Through   (page 62)

10, 5, 12, 3, 4

Study Set   Section 1.7   (page 62)

1. parentheses, brackets **3.** evaluate **5.** 3; square, multiply,
subtract **7.** multiply, subtract **9.** $2 \cdot 3^2 = 2 \cdot 9; (2 \cdot 3)^2 = 6^2$
11. 4, 20, 8 **13.** 9, 36, 30 **15.** 27 **17.** 2 **19.** 15 **21.** 25
23. 5 **25.** 25 **27.** 18 **29.** 813 **31.** 5,239 **33.** 16 **35.** 5
37. 49 **39.** 24 **41.** 13 **43.** 10 **45.** 198 **47.** 18 **49.** 216
51. 17 **53.** 191 **55.** 3 **57.** 29 **59.** 14 **61.** 64 **63.** 192
65. 74 **67.** 137 **69.** 3 **71.** 21 **73.** 11 **75.** 1 **77.** 10,496
79. 2,845 **81.** $2(6) + 4(2) + 2(1)$; $22 **83.** $24 + 6(5) +$
$10(10) + 12(20) + 2(50) + 100$; $594 **85.** brick: $3(3) + 1 + 1 +$
$3 + 3(5)$; 29; aphid: $3[1 + 2(3) + 4 + 1 + 2]$; 42 **87.** 79° **89.** 5
91. 298 **93. a.** 135 **b.** $(1 + 4 + 4)(1 \cdot 4 \cdot 4) = 144$ **99.** 7,300
101. 9,591

Key Concept   (page 66)

1. We can obtain different answers for the same problem;
Method 1 **3.** power, addition, subtraction **5.** multiplication,
division, multiplication, subtraction **7.** 5 **9.** 0 **11.** 206

Chapter Review   (page 68)

1. 2, 5, 9 **2.** 0, 2, 5, 9
3.

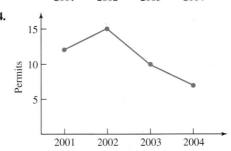

4.

5. 6 **6.** 7 **7.** 5 hundred thousands + 7 ten thousands +
3 hundreds + 2 ones **8.** 3 ten millions + 7 millions + 3 hundred
thousands + 9 thousands + 5 tens + 4 ones **9.** 3,207
10. 23,253,412 **11.** > **12.** < **13.** 2,507,300 **14.** 2,510,000
15. 2,507,350 **16.** 2,500,000 **17.** 13 **18.** 13 **19.** 14 **20.** 14

21. 20 **22.** 18 **23.** 348 **24.** 11,925 **25.** 518 **26.** 6,000 **27. a.** comm. prop. of add. **b.** asso. prop. of add. **28.** 2,746 ft **29.** 208,909,566 **30.** 14,461 **31.** 3 **32.** 4 **33.** 17 **34.** 54 **35.** 74 **36.** 2,075 **37.** $45 **38.** $785 **39.** 147 **40.** 2,737,654 **41.** 56 **42.** 56 **43.** 0 **44.** 7 **45.** 210 **46.** 210 **47.** 3,297 **48.** 178,704 **49.** 31,684 **50.** 455,544 **51.** 330 **52. a.** assoc. prop. of mult. **b.** comm. prop. of mult. **53.** 32 cm^2 **54.** 6,084 in.^{2} **55.** Santiago **56.** 14,400 **57.** 21 **58.** 37 **59.** 19 R 6 **60.** 23 R 27 **61.** 3 **62.** 17 **63.** 0 **64.** undefined **65.** 16, 25 **66.** 34 **67.** 1, 2, 3, 6, 9, 18 **68.** 1, 5, 25 **69.** prime **70.** composite **71.** neither **72.** neither **73.** composite **74.** prime **75.** odd **76.** even **77.** even **78.** odd **79.** $2 \cdot 3 \cdot 7$ **80.** $3 \cdot 5^2$ **81.** 6^4 **82.** $5^3(13^2)$ **83.** 125 **84.** 121 **85.** 200 **86.** 2,700 **87.** 49 **88.** 23 **89.** 75 **90.** 4 **91.** 32 **92.** 72 **93.** 8 **94.** 24 **95.** 1 **96.** 3 **97.** 19 **98.** 7 **99.** 77 **100.** 60

Chapter 1   Test   (page 73)

1. 0, 1, 2, 3, 4 **2.** 5 thousands + 2 hundreds + 6 tens + 6 ones **3.** 7,507 **4.** 35,000,000 **5.**

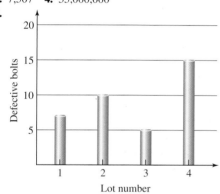

6.

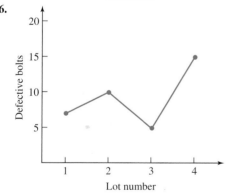

7. > **8.** < **9.** 762 **10.** 248 **11.** 8,100 **12.** 942 **13.** 2,168 in. **14.** $76 **15.** 424 **16.** 26,791 **17.** 72 **18.** 114 R 57 **19.** 92 cm, 529 cm^2 **20. a.** asso. prop. of mult. **b.** Comm. prop. of add. **21. a.** 0 **b.** 0 **22.** 96 **23.** 47 **24.** $2^2 \cdot 3^2 \cdot 5 \cdot 7$ **25.** 44 **26.** 29 **27.** 26 **28.** 1 **29.** 39

Chapter 2   Check Your Knowledge   (page 76)

1. absolute **2.** identity **3.** opposites (or negatives) **4.** unlike **5.** > **6.** 1 **7. a.** −14 **b.** 0 **c.** −8 **d.** 2 **8. a.** −6 **b.** 6 **c.** −12 **9. a.** −9 **b.** 100 **c.** −24 **10.** 3(−6) = −18 **11. a.** −4 **b.** 30 **c.** undefined **12. a.** 7 **b.** −7 **c.** 12 **13. a.** −9 **b.** 9 **c.** 1 **14. a.** 16 **b.** −13 **c.** 29 **d.** 1 **15.** $20 **16.** 15

Think It Through   (page 81)

$4,621, $1,073, $3,325

Study Set   Section 2.1   (page 84)

1. line **3.** graph **5.** inequality **7.** absolute value **9.** integers **11. a.** < **b.** > **c.** <, > **13.** yes **15.** 15 − 8 **17.** 15 > 12 **19. a.** −225 **b.** −10 **c.** −3 **d.** −12,000 **e.** −2 **21.** negative **23.** −4 **25.** −8 and 2 **27.** −7 **29.** 6 − 4, −6, −(−6) (answers may vary) **31. a.** −(−8) **b.** |−8| **c.** 8 − 8 **d.** −|−8| **33.** 9 **35.** 8 **37.** 14 **39.** −20 **41.** −6 **43.** 203 **45.** 0 **47.** 11 **49.** 4 **51.** 12 **53.**

```
←─┼──┼──┼──┼──┼──┼──┼──┼──┼──┼──→
 −5 −4 −3 −2 −1  0  1  2  3  4  5
```

55.

```
←─┼──┼──┼──┼──┼──┼──┼──┼──┼──┼──→
 −5 −4 −3 −2 −1  0  1  2  3  4  5
```

57. < **59.** < **61.** > **63.** > **65.** ≥ **67.** ≥ or ≤ **69.** ≤ **71.** ≤ **73.** 2, 3, 2, 0, −3, −7 **75.** peaks: 2, 4, 0; valleys: −3, −5, −2 **77. a.** −1 (1 below par) **b.** −3 (3 below par) **c.** Most scores are below par. **79. a.** −10° to −20° **b.** 40° **c.** 10° **81. a.** 200 years **b.** A.D. **c.** B.C. **d.** the birth of Christ **83.**

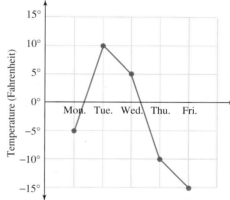

91. 23,500 **93.** 761 **95.** associative property of multiplication

Think It Through   (page 92)

Decrease expenses; increase income; decrease expenses; increase income; increase income; increase income; decrease expenses; decrease expenses; increase income; decrease expenses

Study Set   Section 2.2   (page 95)

1. identity **3.** 3 **5.** −2 **7. a.** yes **b.** yes **9. a.** 7 **b.** 10 **11.** subtract, larger **13.** −18, −19 **15.** 5, 2 **17.** −5 should be within parentheses: −6 + (−5) **19.** 11 **21.** 23 **23.** 0 **25.** −99 **27.** −9 **29.** −10 **31.** 1 **33.** −7 **35.** −20 **37.** 15 **39.** 8 **41.** 2 **43.** −10 **45.** 9 **47.** 8 **49.** −21 **51.** 3 **53.** −10 **55.** −4 **57.** 7 **59.** −21 **61.** −7 **63.** 9 **65.** 0 **67.** 0 **69.** 5 **71.** 0 **73.** −3 **75.** −10 **77.** −1 **79.** −17 **81.** −8,346 **83.** −1,032 **85.** 3G, −3G **87.** no; $70 shortfall each month **89.** 5; 4% risk **91.** −1, 0 **93.** 7 ft over flood stage **95.** about −$2,500 million **101.** 15 ft^2 **103.** 16 ft **105.** 5^3

Study Set   Section 2.3   (page 103)

1. difference **3.** adding, opposite **5.** 6 **7.** + **9.** brackets **11.** −8 − (−4) **13.** 7 **15.** no; 8 − 3 = 5, 3 − 8 = −5 **17.** −3, 2, 0 **19.** −2, −10, 6, −4 **21.** 9 **23.** −13 **25.** −10

27. −1 **29.** 0 **31.** 8 **33.** 5 **35.** −4 **37.** −4 **39.** −20
41. 0 **43.** 0 **45.** −15 **47.** −9 **49.** 3 **51.** 9 **53.** −2
55. −10 **57.** −14 **59.** 3 **61.** −8 **63.** −18 **65.** −6 **67.** 10
69. −4 **71.** −2,447 **73.** 20,503 **75.** −1,676 **77.** −120 ft
79. 16 points **81.** −8 **83.** 1,066 ft **85.** −4 yd
87. a.

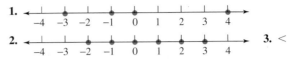
Water
Bottom level   Platform
−12   0   25

b. 37 ft

89. No; he will be $244 overdrawn (−244). **95.** 5,990
97. 1, 2, 4, 5, 10, 20 **99.** 143

Study Set   Section 2.4 (page 111)

1. factors, product **3.** 3, exponent **5.** unlike
7. commutative **9.** −9, the opposite of that number
11. pos · pos, pos · neg, neg · pos, neg · neg **13. a.** negative
b. positive **15. a.** 3 **b.** 12 **c.** 5 **d.** 9 **e.** 10 **f.** 25
17. a. 2, 4, 4, 16, 6, 64 **b.** even **19.** 6, −24 **21.** −5 should be
in parentheses: −6(−5) **23.** 54 **25.** −15 **27.** −36 **29.** 56
31. −20 **33.** −120 **35.** 0 **37.** 6 **39.** 7 **41.** −23 **43.** −48
45. 40 **47.** −30 **49.** −60 **51.** −1 **53.** −18 **55.** 0 **57.** 0
59. 60 **61.** 16 **63.** −125 **65.** −8 **67.** 81 **69.** −1 **71.** 1
73. 49, −49 **75.** −144, 144 **77.** −59,812 **79.** 43,046,721
81. −25,728 **83.** 390,625 **85. a.** plan #1: −30 lb, plan #2:
−28 lb **b.** plan #1; the workout time is double that of plan #2
87. a. high 2, low −3 **b.** high 4, low −6 **89.** −20° **91.** −20 ft
93. −$24,330 **99.** 45 **101.** 2,100 **103.** is less than

Study Set   Section 2.5 (page 117)

1. quotient, divisor **3.** absolute value **5.** positive
7. 5(−5) = −25 **9.** 0(?) = −6 **11.** $\frac{-20}{5} = -4$ **13. a.** always
true **b.** sometimes true **c.** always true **15.** −7 **17.** 2
19. 5 **21.** 3 **23.** −20 **25.** −2 **27.** 0 **29.** undefined
31. −5 **33.** 1 **35.** −1 **37.** 10 **39.** −4 **41.** −3 **43.** 5
45. −4 **47.** −5 **49.** −4 **51.** −542 **53.** −16 **55.** −4° per
hour **57.** −1,010 ft **59.** −6 (6 games behind) **61.** −$15
63. −$1,740 **69.** 104 **71.** 2 · 3 · 5 · 7 **73.** yes **75.** 81

Study Set   Section 2.6 (page 123)

1. order **3.** grouping **5.** 3; power, multiplication, subtraction
7. multiplication, subtraction **9.** The base of the first
exponential expression is 3; the base of the second is −3.
11. 4, 20, −20, −28 **13.** 9, −36, −42 **15.** −7 **17.** 1 **19.** −21
21. −14 **23.** −7 **25.** −5 **27.** 12 **29.** −14 **31.** 30 **33.** 2
35. 15 **37.** −42 **39.** −5 **41.** −3 **43.** 4 **45.** 0 **47.** −14
49. 19 **51.** 4 **53.** −3 **55.** 25 **57.** −48 **59.** 44 **61.** 91
63. 3 **65.** −5 **67.** 17 **69.** 11 **71.** 8 **73.** 112 **75.** −1,707
77. −15 **79.** −200 **81.** −320 **83.** −9,000 **85.** −1,200
87. 19 **89.** 11 yd **91.** 60-cent gain **97.** 5,000 **99.** Add the
lengths of all its sides. **101.** no

Key Concept   (page 126)

1. −5 **3.** −30 **5.** +10 or 10 **7.** −205
9.

−4 −3 −2 −1 0 1 2 3 4
Negatives   Positives

11. −6 < −4

13. Like signs: Add their absolute values and attach their
common sign to the sum. Unlike signs: Subtract their absolute

values, the smaller from the larger, and attach the sign of
the number with the larger absolute value to that result.
15. Divide their absolute values. Like signs: The quotient is
positive. Unlike signs: The quotient is negative.

Chapter Review   (page 128)

1.

−4 −3 −2 −1 0 1 2 3 4

2.

−4 −3 −2 −1 0 1 2 3 4

3. <

4. < **5.** ≥ **6.** ≤ or ≥ **7.** −33 ft **8.** −$1,200 **9.** −10 sec
10. 4 **11.** 0 **12.** 43 **13.** −12 **14.** negative **15.** the
opposite **16.** negative **17.** minus **18.** 12 **19.** −8 **20.** 8
21. 0 **22.** 2

4
−2
−4 −3 −2 −1 0 1 2 3 4

23. −4

−3
−1
−5 −4 −3 −2 −1 0 1 2

24. −10 **25.** −83 **26.** −8 **27.** −1 **28.** 112 **29.** −11
30. −3 **31.** −2 **32.** −4 **33.** −20 **34.** 0 **35.** 0 **36.** 11
37. −4 **38.** 65 ft **39.** −3 **40.** −21 **41.** 4 **42.** −112
43. −6 **44.** 6 **45.** −37 **46.** 30 **47.** adding, opposite
48. −4 **49.** 15 **50.** 6 **51.** −8 **52.** −77 **53.** −1
54. −225 ft **55.** Alaska: 180°; Virginia: 140° **56.** −45 **57.** 18
58. −14 **59.** 376 **60.** −100 **61.** 1 **62.** −25 **63.** −150
64. −36 **65.** −36 **66.** 0 **67.** 1 **68.** −3, −6, −9 **69.** 25
70. −32 **71.** 64 **72.** −64 **73.** negative **74.** first expression:
base of 2; second: base of −2; −4, 4 **75.** 5, −3, −15 **76.** −2
77. −5 **78.** −8 **79.** 101 **80.** 0 **81.** undefined **82.** 1
83. 10 **84.** −2 min **85.** −22 **86.** 4 **87.** −43 **88.** 8 **89.** 41
90. 0 **91.** −13 **92.** 32 **93.** 12 **94.** −16 **95.** −4 **96.** 1
97. −1 **98.** −4 **99.** −70 **100.** 20 **101.** −7,000 **102.** 1,100

Chapter 2 Test   (page 133)

1. a. > **b.** < **c.** < **2.** {. . . , −3, −2, −1, 0, 1, 2, 3, . . .}
3. Monroe
4. −5

−2
−3
−6 −5 −4 −3 −2 −1 0 1 2 3 4 5 6

5. a. −34 **b.** −34 **c.** −8 **6. a.** −13 **b.** −1 **c.** −15
d. −150 **7. a.** −70 **b.** −48 **c.** 16 **d.** 0 **8.** −4(5) = −20
9. a. −8 **b.** undefined **c.** −5 **d.** 0 **10.** $3 million
11. 154 ft **12. a.** 6 **b.** 7 **c.** −6 **d.** 132 **13. a.** 16 **b.** −16
c. −1 **14.** −27 **15.** 1 **16.** −34 **17.** 42 **18.** 4 **19.** −72°
20. left, −7 **21.** −15 **22.** −4 + (−4) + (−4) + (−4) +
(−4) = −20 **23.** The absolute value of a number is the
distance from the number to 0 on the number line. Distance is
either positive or 0, but never negative. **24.** It is true because
12 = 12.

Cumulative Review Exercises   (page 135)

1. 1, 2, 5, 9 **2.** 0, 1, 2, 5, 9 **3.** −2, −1 **4.** −2, −1, 0, 1, 2, 5, 9
5. 6 **6.** 3 **7.** 7,326,500 **8.** 7,330,000 **9.** CRF Cable

10.

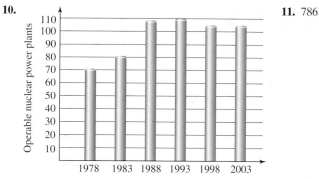

Source: *The World Almanac, 2005*

11. 786

12. 3,806 **13.** 4,684 **14.** 13,136 **15.** 104 ft, 595 ft² **16.** 65
17. 11,745 **18.** 13 **19.** 307,329 **20.** 467 **21.** 1,728
22. 1, 2, 3, 6, 9, 18 **23.** prime, odd **24.** composite, even
25. even **26.** odd **27.** $2^3 \cdot 3^2 \cdot 7$ **28.** 11^4 **29.** 175 **30.** 38
31. 50 **32.** 2 **33.** no **34.** 21
35.

36. **37.** −5 **38.** −5

39. −1 **40.** 8 **41.** 216 **42.** −35 **43.** 2 **44.** −5 **45.** 26
46. 70 **47.** −3 **48.** 4 **49.** −81 **50.** 81 **51.** $126,037
52. −$79 **53.** −100 ft **54.** −83,600,000

Chapter 3 Check Your Knowledge (page 138)

1. numerator, denominator **2.** base, height **3.** reciprocals
4. least **5.** equivalent **6.** $\frac{1}{3}, \frac{2}{3}$ **7. a.** $\frac{4}{5}$ **b.** $\frac{3}{4}$ **8. a.** $\frac{3}{8}$ **b.** $-\frac{1}{3}$
9. $22\frac{1}{2}$ **10. a.** 2 **b.** $\frac{1}{6}$ **11. a.** $\frac{51}{40}$ **b.** $\frac{2}{5}$ **12. a.** $\frac{1}{12}$ **b.** $\frac{5}{8}$

13. $\frac{30}{36}$ **14.**

15. $\frac{5}{8}$ cm² **16.** $313\frac{7}{15}$
17. $4\frac{11}{24}$ **18.** $9\frac{33}{40}$ **19.** $-\frac{2}{3}$ **20.** $\frac{9}{5}$ **21.** $-\frac{2}{5}$ **22.** 60
23. $1\frac{3}{4}$ million dollars

Study Set Section 3.1 (page 144)

1. numerator, denominator **3.** proper, improper
5. equivalent **7.** higher, building **9. a.** 2 **b.** 3 **c.** 5
d. 7 **11.** equivalent fractions: $\frac{2}{6} = \frac{1}{3}$ **13. a.** In the first
case, 20 and 28 were factored. In the second case, they were
prime factored. **b.** yes **15.** The 2's in the numerator and
denominator aren't common factors. **17. a.** $\frac{8}{1}$ **b.** $-\frac{25}{1}$
19. 3, 2, 2, 3, 2, 3 **21.** $\frac{1}{3}$ **23.** $\frac{1}{3}$ **25.** $\frac{2}{3}$ **27.** $\frac{5}{2}$ **29.** $-\frac{1}{2}$
31. $-\frac{6}{7}$ **33.** $\frac{5}{9}$ **35.** $\frac{6}{7}$ **37.** in lowest terms **39.** $\frac{3}{8}$ **41.** $\frac{5}{7}$
43. $\frac{4}{5}$ **45.** in lowest terms **47.** $\frac{7}{8}$ **49.** $-\frac{1}{3}$ **51.** $\frac{3}{5}$ **53.** $\frac{5}{4}$
55. 2 **57.** $\frac{35}{40}$ **59.** $\frac{28}{35}$ **61.** $\frac{45}{54}$ **63.** $\frac{15}{30}$ **65.** $\frac{4}{14}$ **67.** $\frac{54}{60}$ **69.** $-\frac{25}{20}$
71. $-\frac{6}{45}$ **73.** $\frac{15}{5}$ **75.** $\frac{48}{8}$ **77.** $\frac{36}{9}$ **79.** $-\frac{4}{2}$ **81.** $\frac{3}{5}$ **83.** $-\frac{15}{16}$ in.
85. $\frac{7}{10}, \frac{1}{8}$

87.

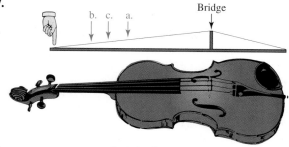

89. one-quarter turn to the left; three-quarters of a turn to the
right **91.**

SNACKS
Potato chips
Peanuts
Pretzels
Tortilla chips

93. $\frac{1}{250}$ **99.** 11 **101.** 564,000
103. −6

Study Set Section 3.2 (page 153)

1. multiply **3.** product **5.** base, height **7.** $\frac{1}{6}$
9.

a. $\frac{1}{4}$ **b.** 12, 1, $\frac{1}{12}$

11. a. negative **b.** positive **13. a.** true **b.** false **17.** $\frac{1}{8}$
19. $\frac{21}{128}$ **21.** $\frac{4}{7}$ **23.** $\frac{77}{60}$ **25.** $-\frac{1}{5}$ **27.** $\frac{2}{9}$ **29.** $\frac{2}{3}$ **31.** 1 **33.** $\frac{1}{20}$
35. $\frac{1}{3}$ **37.** 15 **39.** −12 **41.** $\frac{1}{2}$ **43.** $\frac{8}{3}$ **45.** $-\frac{16}{3}$ **47.** $-\frac{3}{2}$
49. $\frac{5}{6}$ **51.** $\frac{4}{9}$ **53.** $\frac{25}{81}$ **55.** $\frac{16}{9}$ **57.** $-\frac{27}{64}$
59.

·	$\frac{1}{2}$	$\frac{1}{3}$	$\frac{1}{4}$	$\frac{1}{5}$	$\frac{1}{6}$
$\frac{1}{2}$	$\frac{1}{4}$	$\frac{1}{6}$	$\frac{1}{8}$	$\frac{1}{10}$	$\frac{1}{12}$
$\frac{1}{3}$	$\frac{1}{6}$	$\frac{1}{9}$	$\frac{1}{12}$	$\frac{1}{15}$	$\frac{1}{18}$
$\frac{1}{4}$	$\frac{1}{8}$	$\frac{1}{12}$	$\frac{1}{16}$	$\frac{1}{20}$	$\frac{1}{24}$
$\frac{1}{5}$	$\frac{1}{10}$	$\frac{1}{15}$	$\frac{1}{20}$	$\frac{1}{25}$	$\frac{1}{30}$
$\frac{1}{6}$	$\frac{1}{12}$	$\frac{1}{18}$	$\frac{1}{24}$	$\frac{1}{30}$	$\frac{1}{36}$

61. 15 ft² **63.** $\frac{15}{2}$ yd² **65.** 15 ft² **67.** 290 **69.** 18, 6, and 2 in.
71. $\frac{3}{8}$ cup sugar, $\frac{1}{6}$ cup molasses
73.

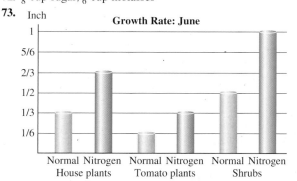

Growth Rate: June

75. 121 in.² **77.** 18 in.² **83.** 6,800 **85.** 26 **87.** 5^3

Study Set   Section 3.3   (page 160)

1. reciprocals 3. $\frac{1}{2}, \frac{3}{2}$ 5.  $4 \div \frac{1}{3}$, 12

7. 1 9. a. 5 b. 5 c. $\frac{1}{3}$ 13. $\frac{5}{6}$ 15. $\frac{27}{16}$ 17. 1 19. $\frac{2}{3}$ 21. 36
23. 50 25. $\frac{2}{15}$ 27. $\frac{1}{192}$ 29. $-\frac{27}{8}$ 31. $-\frac{15}{2}$ 33. $-\frac{1}{64}$ 35. 1
37. $\frac{8}{15}$ 39. $\frac{1}{6}$ 41. $\frac{13}{8}$ 43. $-\frac{5}{8}$ 45. -6 47. $-\frac{5}{2}$ 49. 104
51. 56 53. route 1 55. a. 16 b. $\frac{3}{4}$ in. c. $\frac{1}{120}$ in. 57. 7,855
63. is less than 65. positive 67. false 69. -29

Think It Through   (page 169)

$\frac{7}{20}$

Study Set   Section 3.4   (page 170)

1. least 3. higher 5. numerators, common
7. The denominators are unlike. 9. a. 4 b. 5
11. a. once b. twice c. three times 13. 60
15. a.  $\frac{1}{3}$
b. $\frac{1}{4}$; $\frac{1}{3}$; $\frac{1}{4} = \frac{3}{12}$, $\frac{1}{3} = \frac{4}{12}$; since $\frac{4}{12} > \frac{3}{12}$, $\frac{1}{3} > \frac{1}{4}$. 19. 18 21. 24
23. 40 25. 60 27. $\frac{4}{7}$ 29. $\frac{20}{103}$ 31. $\frac{2}{5}$ 33. $\frac{5}{8}$ 35. $\frac{9}{20}$ 37. $\frac{22}{15}$
39. $\frac{23}{56}$ 41. $\frac{1}{12}$ 43. $\frac{1}{12}$ 45. $\frac{47}{50}$ 47. $-\frac{3}{16}$ 49. $-\frac{2}{3}$ 51. $-\frac{23}{24}$
53. $-\frac{13}{5}$ 55. $-\frac{23}{4}$ 57. $\frac{19}{48}$ 59. $-\frac{43}{45}$ 61. $\frac{26}{75}$ 63. $\frac{17}{54}$ 65. $\frac{341}{400}$
67. $-\frac{1}{50}$ 69. $\frac{47}{60}$ 71. $\frac{3}{4}$ 73. $\frac{23}{10}$ 75. $\frac{5}{36}$ 77. $-\frac{17}{60}$ 79. a $\frac{7}{32}$ in.
b. $\frac{3}{32}$ in. 81. $\frac{17}{24}$; no 83. $\frac{1}{16}$ lb, undercharge 85. $\frac{4}{5}, \frac{3}{4}, \frac{5}{8}$ 87. $\frac{7}{10}$
89. $\frac{1}{6}$ hp 95. 4 97. 576 99. 30 in.

Study Set   The LCM and the GCF   (page 175)

1. 15 3. 56 5. 42 7. 18 9. 660 11. 600 13. 72 15. 378
17. 3 19. 11 21. 4 23. 25 25. 20 27. 12 29. 6 31. 9
33. 360 min (6 hr)

Think It Through   (page 177)

$1\frac{1}{3}$ hr

Study Set   Section 3.5   (page 180)

1. mixed 3. graph 5. a. $-5\frac{1}{2}°$ b. $-6\frac{7}{8}$ in. 7. a. $2\frac{2}{3}$
b. $1\frac{1}{3}$ 9. $-\frac{4}{5}, -\frac{2}{5}, \frac{1}{5}$ 11. $2\frac{1}{2}$
13.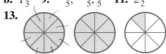

17. $3\frac{3}{4}$ 19. $5\frac{4}{5}$ 21. $-3\frac{1}{3}$
23. $10\frac{7}{12}$ 25. $\frac{13}{2}$ 27. $\frac{104}{5}$
29. $-\frac{56}{9}$ 31. $\frac{602}{3}$

33. [number line showing $-2\frac{8}{9}$, $1\frac{2}{3}$, $\frac{16}{5}$]

35. [number line showing $-\frac{10}{3}$, $-\frac{98}{99}$, $3\frac{1}{7}$]

37. $3\frac{4}{7}$ 39. $10\frac{1}{2}$ 41. 14 43. $-13\frac{3}{4}$ 45. $-8\frac{1}{3}$ 47. $\frac{35}{72}$
49. $-1\frac{1}{4}$ 51. $\frac{25}{9} = 2\frac{7}{9}$ 53. $-\frac{64}{27} = -2\frac{10}{27}$ 55. $1\frac{9}{11}$ 57. $-\frac{9}{10}$
59. 12 61. $\frac{5}{16}$ 63. $-\frac{2}{3}$ 65. $2\frac{1}{2}$ 67. -2 69. 64 calories

71. 357¢ = $3.57 73. 675 75. $2\frac{3}{4}$ in., $1\frac{1}{4}$ in. 77. $42\frac{5}{8}$ in.²
79. 602 81. size 14, slim cut 87. 72 89. 4(8) 91. $\frac{23}{5} = 4\frac{3}{5}$

Study Set   Section 3.6   (page 188)

1. commutative 3. borrow 5. a. $76, \frac{3}{4}$ b. $76 + \frac{3}{4}$
7. the fundamental property of fractions 9. a. $10\frac{1}{16}$
b. $1,290\frac{1}{3}$ c. $17\frac{1}{2}$ d. $46\frac{1}{2}$ 13. $4\frac{2}{5}$ 15. $5\frac{1}{7}$ 17. $7\frac{1}{2}$
19. $5\frac{11}{30}$ 21. $1\frac{1}{4}$ 23. $1\frac{11}{24}$ 25. $9\frac{3}{10}$ 27. $3\frac{5}{14}$ 29. $129\frac{11}{15}$
31. $397\frac{5}{12}$ 33. $273\frac{2}{9}$ 35. $623\frac{8}{21}$ 37. $11\frac{5}{30}$ 39. $101\frac{7}{16}$ 41. $2\frac{1}{2}$
43. $26\frac{7}{24}$ 45. $10\frac{7}{16}$ 47. $320\frac{5}{18}$ 49. $6\frac{1}{3}$ 51. $\frac{1}{4}$ 53. $3\frac{12}{35}$
55. $3\frac{5}{8}$ 57. $4\frac{1}{3}$ 59. $3\frac{7}{8}$ 61. $53\frac{5}{12}$ 63. $460\frac{1}{8}$ 65. $1\frac{37}{70}$
67. $14\frac{5}{24}$ 69. $-5\frac{1}{4}$ 71. $-5\frac{7}{8}$ 73. $2\frac{3}{4}$ mi 75. $7\frac{2}{3}$ cups
77. $48\frac{1}{2}$ ft 79. a. $16\frac{1}{2}, 16\frac{1}{2}; 5\frac{1}{5}, 5\frac{1}{5}$ b. $21\frac{7}{10}$ mi
81. a. 20¢ b. 30¢ 83. $191\frac{2}{3}$ ft 89. 52 91. $\frac{13}{20}$ 93. 6
95. the amount of surface a figure encloses

Study Set   Section 3.7   (page 197)

1. complex 3. $\frac{2}{3} \div \frac{1}{5}$ 5. 15 7. negative 9. subtraction
13. $\frac{1}{3}$ 15. $\frac{31}{45}$ 17. $\frac{37}{40}$ 19. $\frac{3}{10}$ 21. $-1\frac{27}{40}$ 23. $\frac{3}{4}$ 25. $-\frac{3}{64}$
27. $-1\frac{1}{6}$ 29. $8\frac{1}{2}$ 31. $\frac{49}{4} = -15\frac{5}{16}$ 33. $\frac{121}{16}$ 35. 102 37. 36
39. $8\frac{1}{4}$ in. 41. $\frac{5}{6}$ 43. $-1\frac{1}{3}$ 45. $10\frac{1}{2}$ 47. $\frac{4}{9}$ 49. 3 51. 5
53. -20 55. 11 57. $\frac{3}{7}$ 59. $-\frac{3}{8}$ 61. $8\frac{1}{2}$ 63. $14\frac{11}{20}$ mi 65. yes
67. $10\frac{1}{2}$ mi 69. 6 sec 75. 144 77. 37 79. 8 81. $2^5 \cdot 3^2$

Key Concept   (page 201)

1. 5, 5, 5, 5, 5 2. 5, 5, 5, 5, 5, 3 3. 7, 7, 7, 7, 7

Chapter Review   (page 203)

1. $\frac{7}{24}$ 2. The figure is not divided into equal parts. 3. $-\frac{2}{3}, \frac{-2}{3}$
4. equivalent fractions: $\frac{6}{8} = \frac{3}{4}$ 5. The numerator and
denominator of the fraction are being divided by 2.
6. The numerator and denominator of the fraction are being
divided by 2. The answer to each division is 1. 7. $\frac{1}{3}$ 8. $\frac{5}{12}$
9. $-\frac{3}{4}$ 10. $\frac{11}{18}$ 11. The numerator and denominator of
the original fraction are being multiplied by 2 to obtain an
equivalent fraction in higher terms. 12. $\frac{12}{18}$ 13. $-\frac{6}{16}$ 14. $\frac{21}{45}$
15. $\frac{36}{9}$ 16. $\frac{1}{6}$ 17. $-\frac{14}{45}$ 18. $\frac{5}{12}$ 19. $\frac{1}{5}$ 20. $\frac{21}{5}$ 21. $\frac{9}{4}$ 22. 1
23. 1 24. true 25. false 26. $\frac{2}{9}$ 27. $-\frac{8}{21}$ 28. $\frac{1}{21}$ 29. $-\frac{5}{9}$
30. $\frac{9}{16}$ 31. $-\frac{125}{8} = -15\frac{5}{8}$ 32. $\frac{4}{9}$ 33. $-\frac{8}{125}$ 34. 30 lb
35. 60 in.² 36. 8 37. $-\frac{12}{11}$ 38. $\frac{1}{5}$ 39. $\frac{7}{8}$ 40. $\frac{25}{66}$ 41. $-\frac{7}{8}$
42. $\frac{3}{32}$ 43. $\frac{5}{2}$ 44. $-\frac{3}{2}$ 45. $\frac{8}{5}$ 46. $-\frac{4}{9}$ 47. -6 48. 12 49. $\frac{5}{7}$
50. $-\frac{6}{5}$ 51. $\frac{1}{2}$ 52. $\frac{5}{4}$ 53. The denominators are not the same.
54. 90 55. $\frac{5}{6}$ 56. $\frac{1}{40}$ 57. $-\frac{29}{24}$ 58. $\frac{20}{7}$ 59. $-\frac{11}{50}$ 60. $\frac{25}{12}$
61. $-\frac{23}{6}$ 62. $\frac{47}{60}$ 63. $\frac{7}{32}$ in. 64. the second hour 65. $2\frac{1}{6}$
66. $\frac{13}{6}$ 67. $3\frac{3}{5}$ 68. $-3\frac{11}{12}$ 69. 1 70. $2\frac{1}{3}$ 71. $\frac{75}{8}$ 72. $-\frac{11}{5}$
73. $\frac{201}{2}$ 74. $\frac{199}{100}$
75. [number line showing $-2\frac{2}{3}$, $\frac{8}{9}$, $\frac{59}{24}$]
76. $-\frac{3}{10}$ 77. $\frac{21}{22}$ 78. 40 79. $-2\frac{1}{2}$ 80. $48\frac{1}{8}$ in. 81. $3\frac{23}{40}$
82. $6\frac{1}{6}$ 83. $1\frac{1}{12}$ 84. $1\frac{5}{16}$ 85. $39\frac{11}{12}$ gal 86. $182\frac{5}{18}$ 87. $113\frac{3}{20}$

88. $31\frac{11}{24}$ **89.** $316\frac{3}{4}$ **90.** $20\frac{1}{2}$ **91.** $34\frac{3}{8}$ **92.** $\frac{8}{9}$ **93.** $\frac{19}{72}$ **94.** $-\frac{12}{17}$
95. $-\frac{2}{5}$ **96.** $\frac{23}{10} = 2\frac{3}{10}$ **97.** $11\frac{1}{6}$

Chapter 3   Test   (page 209)

1. a. $\frac{4}{5}$ **b.** $\frac{1}{5}$ **2. a.** $\frac{3}{4}$ **b.** $\frac{2}{5}$ **3.** $-\frac{3}{20}$ **4.** 40 **5.** 6 **6.** $\frac{1}{30}$
7. $\frac{21}{24}$ **8.**   **9.** $\$1\frac{1}{2}$ million

$$-1\frac{1}{7} \qquad \frac{7}{6} \qquad 2\frac{4}{5}$$
$$\begin{array}{ccccccc} & & & & & & \\ \hline -2 & -1 & 0 & 1 & 2 & 3 \end{array}$$

10. $\frac{35}{8} = 4\frac{3}{8}$ **11.** $261\frac{11}{36}$ **12.** $37\frac{5}{12}$ **13. a.** 0 lb **b.** $2\frac{3}{4}$ in.
c. $3\frac{3}{4}$ in. **14.** $\frac{11}{7}$ **15.** $11\frac{3}{4}$ in. **16.** perimeter: $53\frac{1}{3}$ in., area:
$106\frac{2}{3}$ in.^{2} **17.** 60 **18.** 12 **19.** $\frac{13}{24}$ **20.** $-\frac{20}{21}$ **21.** $-\frac{5}{3}$ **22.** 144
23. numerator, fraction bar, denominator; equal parts of a
whole or a division **24.** When we multiply a number, such as $\frac{3}{4}$,
and its reciprocal, $\frac{4}{3}$, the result is 1. **25. a.** simplifying a
fraction; dividing the numerator and denominator of a fraction
by the same number **b.** building a fraction; equivalent
fractions: $\frac{1}{2} = \frac{2}{4}$ **c.** multiplying the numerator and denominator
of a fraction by the same number **26.** The denominators are
not the same.

Cumulative Review Exercises   (page 211)

1. 5,434,700 **2.** 5,430,000 **3.** 11,555, 10:30 A.M. **4.** hundred
billions **5.** 8,136 **6.** 3,519 **7.** 299,320 **8.** 991 **9.** 450 ft
10. 11,250 ft^2 **11.** $2^2 \cdot 3 \cdot 7$ **12.** $2 \cdot 3^2 \cdot 5^2$ **13.** $2^3 \cdot 3^2 \cdot 5$
14. $2^4 \cdot 3^2 \cdot 5^2$ **15.** 16 **16.** -35 **17.** 2 **18.** 2 **19.** $\frac{3}{4}$ **20.** $\frac{5}{2}$
21. $-\frac{4}{5}$ **22.** $\frac{1}{2}$ **23.** $1\frac{5}{12}$ **24.** $-\frac{1}{35}$ **25.** $\frac{23}{6}$ **26.** $-\frac{53}{8}$ **27.** $9\frac{11}{12}$
28. $5\frac{11}{15}$ **29.** $\frac{11}{16}$ in. **30.** 30 sec, 60 sec **31.** $\frac{2}{7}$ **32.** $-1\frac{9}{29}$

Chapter 4   Check Your Knowledge   (page 214)

1. sum **2.** right **3.** numerator **4.** root **5.** $\frac{21}{25}$ **6.** 885.81
7. a. 354.2782 **b.** 20,004.78 **8. a.** 7.875 ft^2 **b.** 11.5 ft
9. 3.1 **10. a.** 0.15 **b.** 0.625 **11.** 1.6
12.   **13.** $\frac{5}{12}$
$$-0.375 \quad 0.25 \qquad\qquad \textbf{14. } \$4.75$$
$$\begin{array}{ccccc} \hline -2 & -1 & 0 & 1 & 2 \end{array}$$

15.   **16. a.** 19 **b.** $\frac{19}{20}$
$$-1.73 \qquad\qquad 1.41 \qquad\qquad \textbf{c. } -0.3$$
$$\begin{array}{ccccc} \hline -2 & -1 & 0 & 1 & 2 \end{array} \qquad \textbf{17. a. } >\ \textbf{ b. } <\ \textbf{ c. } >$$

18. 52.6 mph

Study Set   Section 4.1   (page 221)

1. 4 7 8 9 . 0 2 6 5
(thousands, hundreds, tens, ones, tenths, hundredths, thousandths, ten-thousandths)
3. rounding **5. a.** thirty-two and four hundred fifteen
thousandths **b.** 32 **c.** $\frac{415}{1,000}$ **d.** $30 + 2 + \frac{4}{10} + \frac{1}{100} + \frac{5}{1,000}$
7.

$$-3\frac{1}{100} \qquad -0.7 \quad \frac{7}{10} \qquad 3.01$$
$$\begin{array}{ccccccccccc} \hline -5 & -4 & -3 & -2 & -1 & 0 & 1 & 2 & 3 & 4 & 5 \end{array}$$

9. a. true **b.** false **c.** true **d.** true **11.** $\frac{47}{100}$, 0.47
13. _____|_____   **15.** 9,816.0245
$\qquad\qquad$ 0.3

17. fifty and one tenth; $50\frac{1}{10}$ **19.** negative one hundred
thirty-seven ten-thousandths; $-\frac{137}{10,000}$ **21.** three hundred four
and three ten-thousandths; $304\frac{3}{10,000}$ **23.** negative seventy-two
and four hundred ninety-three thousandths; $-72\frac{493}{1,000}$ **25.** -0.39
27. 6.187 **29.** 506.1 **31.** 2.7 **33.** -0.14 **35.** 33.00
37. 3.142 **39.** 1.414 **41.** 39 **43.** 2,988 **45. a.** $3,090
b. $3,090.30 **47.** < **49.** > **51.** 132.64, 132.6401, 132.6499
53. $1,025.78
55.

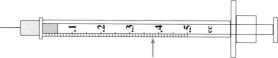

57. a. 0.30 **b.** 1,609.34 **c.** 453.59 **d.** 3.79 **59.** sand, silt,
granule, clay **61.** Texas City, Houston, Westport, Galveston,
White Plains, Crestline **63.** gold: Patterson; silver: Khorkina;
bronze: Zhang **65. a.** Q1, 2004; $0.25 **b.** Q2, 2002; $-$0.25
73. $164\frac{11}{20}$ **75.** $\frac{32}{243}$ **77.** 72 in.^{2} **79.** -1

Study Set   Section 4.2   (page 229)

1. sum **3.** opposite **5.** point **7.** 39.9 **9.** 54.72 **11.** 15.9
13. 0.23064 **15.** 288.46 **17.** 58.04 **19.** 9.53 **21.** 70.29
23. 4.977 **25.** 0.19 **27.** -10.9 **29.** 38.29 **31.** -14.3
33. -0.0355 **35.** -16.6 **37.** 47.91 **39.** 2.598 **41.** 11.01
43. 4.1 **45.** 35.85 **47.** -57.47 **49.** 6.2 **51.** 15.2 **53.** 8.03
55. a. 53.044 sec **b.** 102.38 **57.** 103.4 in. **59.** 1.8, Texas
61. 1.74 mi, 2.32 mi, 4.06 mi, 2.90 mi, 0 mi, 2.90 mi
63. 43.03 sec **65.** $765.69, $740.69 **67. a.** $101.94 **b.** $55.80
69. 8,156.9343 **71.** 1,932.645 **73.** 2,529.0582 **79.** $110\frac{23}{40}$
81. $-\frac{5}{6}$

Study Set   Section 4.3   (page 238)

1. factors, product **3.** whole, sum **5. a.** $\frac{21}{1,000}$
b. $\frac{21}{1,000} = 0.021$. They are the same. **7.** 2.3 **9.** 0.08 **11.** -0.15
13. 0.98 **15.** 0.072 **17.** 12.32 **19.** -0.0049 **21.** -0.084
23. -8.6265 **25.** 9.6 **27.** -56.7 **29.** 12.24 **31.** -18.183
33. 0.024 **35.** -16.5 **37.** 42 **39.** 6,716.4 **41.** -0.56
43. 8,050 **45.** 980 **47.** -200 **49.** 0.01, 0.04, 0.09, 0.16, 0.25,
0.36, 0.49, 0.64, 0.81 **51.** 1.44 **53.** 1.69 **55.** -17.48
57. 14.24 **59.** 0.84 **61.** -3.872 **63.** 18.72 **65.** 86.49
67. 38.16 **69.** 14.6 **71. a.** $12.50, $12,500, $15.75, $1,575
b. $14,075 **73.** 0.75 in. **75.** 136.4 lb **77.** $95.20, $123.75
79. 160.6 m **81.** 0.000000136 in., 0.0000000136 in., 0.00000004 in.
83. 15.29694 **85.** 631.2722 **87.** $102.65 **93.** 7,300
95. the absolute value of negative three **97.** -1

Think It Through   (page 246)

2.86

Study Set   Section 4.4   (page 247)

1. dividend, divisor, quotient **3.** whole, right, above **5.** true
7. 10 **9.** Use multiplication to see whether $2.13 \cdot 0.9 = 1.917$.

11. yes **13.** moving the decimal points in the divisor and dividend 2 places to the right **15.** 4.5 **17.** −9.75 **19.** 6.2 **21.** 32.1 **23.** 2.46 **25.** −7.86 **27.** 2.66 **29.** 7.17 **31.** 130 **33.** 1,050 **35.** 0.6 **37.** 0.6 **39.** 5.3 **41.** −2.4 **43.** 13.60 **45.** 0.79 **47.** 0.07895 **49.** −0.00064 **51.** 0.0348 **53.** 4.504 **55.** −0.96 **57.** 1,027.19 **59.** 3.5 **61.** 58.5 **63.** 280 **65.** 11 hr later: 6 P.M. **67.** 567 **69.** 1998: $13.00; 2003: $15.35 **71.** 0.37 mi **73.** 7.24 **75.** −3.96 **81.** $\frac{7}{6} = 1\frac{1}{6}$ **83.** $\{\ldots, -3, -2, -1, 0, 1, 2, 3, \ldots\}$ **85.** $38\frac{1}{2}$ **87.** 42.05

Study Set   Estimation   (page 251)

1. approx. $240 **3.** approx. 2 cubic feet less **5.** approx. 30 **7.** approx. $330 **9.** approx. $520 **11.** not reasonable **13.** reasonable **15.** reasonable **17.** not reasonable

Study Set   Section 4.5   (page 257)

1. repeating **3.** decimal **5. a.** 7 ÷ 8 **b.** numerator **7.** smaller

9.

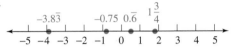

11. a. false **b.** true **c.** true **d.** false **13. a.** no **b.** It is a repeating decimal. **15.** 0.5 **17.** −0.625 **19.** 0.5625 **21.** −0.53125 **23.** 0.55 **25.** 0.775 **27.** −0.015 **29.** 0.002 **31.** $0.\overline{6}$ **33.** $0.\overline{45}$ **35.** $-0.58\overline{3}$ **37.** $0.0\overline{3}$ **39.** 0.23 **41.** 0.38 **43.** 0.152 **45.** 0.370 **47.** 1.33 **49.** −3.09 **51.** 3.75 **53.** −8.67 **55.** 12.6875 **57.** 203.73 **59.** < **61.** < **63.** $\frac{37}{90}$ **65.** $\frac{19}{60}$ **67.** $\frac{3}{22}$ **69.** $-\frac{1}{90}$ **71.** 1 **73.** 0.57 **75.** 5.27 **77.** 0.24 **79.** −2.55 **81.** 0.068 **83.** 7.11 **85.** −1.7 **87.** 4.25 **89.** 18.1 **91.** $0.\overline{2277}$ **93.** 0.03472 **95.** 0.0625, 0.375, 0.5625, 0.9375 **97.** $\frac{3}{40}$ in. **99.** 23.4 sec, 23.8 sec, 24.2 sec, 32.6 sec **101.** 93.6 in.^{2} **107.** −1 **109.** {0, 1, 2, 3, 4, 5, 6, 7} **111.** 15

Study Set   Section 4.6   (page 263)

1. root **3.** radical, positive **5.** radicand **7.** 25, 25 **9.** 7^2 **11.** $\frac{3}{4}$ **13.** $\sqrt{6}, \sqrt{11}, \sqrt{23}, \sqrt{27}$ **15. a.** 1 **b.** 0 **17. a.** 2.4 **b.** 5.76 **c.** 0.24

19.

21. a. 4, 5 **b.** 9, 10 **23.** −7, 8 **25.** 4 **27.** −11 **29.** −0.7 **31.** 0.5 **33.** 0.3 **35.** $-\frac{1}{9}$ **37.** $-\frac{4}{3}$ **39.** $\frac{2}{5}$ **41.** 31 **43.** −20 **45.** $-\frac{7}{20}$ **47.** −70 **49.** 2.56 **51.** −3.6 **53.** 1, 1.414, 1.732, 2, 2.236, 2.449, 2.646, 2.828, 3, 3.162 **55.** 37 **57.** 61 **59.** 3.87 **61.** 8.12 **63.** 4.904 **65.** −3.332 **67.** 4,899 **69.** −0.0333 **71. a.** 5 ft **b.** 10 ft **73.** 127.3 ft **75.** 42-inch **83.** 82.35 **85.** 16 **87.** {0, 1, 2, 3, 4, 5, 6, ...} **89.** $-\frac{17}{144}$

Key Concept   (page 266)

1. {1, 2, 3, 4, 5, ...} **3.** $\{\ldots, -3, -2, -1, 0, 1, 2, 3, \ldots\}$ **5.** nonterminating, nonrepeating decimals; a number that can't be written as a fraction of two integers **7.** false **9.** false **11.** true **13.** false **15.** true

Chapter Review   (page 268)

1. 0.67, $\frac{67}{100}$ **2.**

3. $10 + 6 + \frac{4}{10} + \frac{5}{100} + \frac{2}{1,000} + \frac{3}{10,000}$ **4.** two and three tenths, $2\frac{3}{10}$ **5.** negative fifteen and fifty-nine hundredths, $-15\frac{59}{100}$ **6.** six hundred one ten-thousandths, $\frac{601}{10,000}$ **7.** one hundred-thousandth, $\frac{1}{100,000}$

8.

9. Washington, Diaz, Chou, Singh, Gerbac **10.** true **11.** < **12.** > **13.** = **14.** < **15.** 4.58 **16.** 3,706.090 **17.** −0.1 **18.** 88.1 **19.** 66.7 **20.** 45.188 **21.** 15.17 **22.** 27.71 **23.** −7.7 **24.** 3.1 **25.** −4.8 **26.** −29.09 **27.** −25.6 **28.** 4.939 **29.** $48.21 **30.** 8.15 in. **31.** −0.24 **32.** 2.07 **33.** −17.05 **34.** 197.945 **35.** 0.00006 **36.** 4.2 **37.** 90,145.2 **38.** 2,897 **39.** 0.04 **40.** 0.0225 **41.** 10.89 **42.** 0.001 **43.** −10.61 **44.** 25.82 **45.** 692.25 **46.** 68.62 in.^{2} **47.** 0.07 in. **48.** 1.25 **49.** −10.45 **50.** 1.29 **51.** 4.103 **52.** −2.9 **53.** 0.053 **54.** 63 **55.** 0.81 **56.** 12.9 **57.** −667.3 **58.** 20.22 **59.** $8.34 **60.** 0.8976 **61.** −0.00112 **62.** 13.95 **63.** 14 **64.** 9.5 **65.** 0.875 **66.** −0.4 **67.** 0.5625 **68.** 0.06 **69.** $0.\overline{54}$ **70.** $-0.\overline{6}$ **71.** 0.58 **72.** 1.03 **73.** > **74.** >

75.

76. $\frac{11}{15}$

77. −6.24 **78.** 93 **79.** 39.564 **80.** 33.49 **81.** 34.88 in.^{2} **82. a.** radical **b.** 8^2 **83.** 7 **84.** −4 **85.** 10 **86.** 0.3 **87.** $\frac{8}{5}$ **88.** 0.9 **89.** $-\frac{1}{6}$ **90.** 0 **91.** 9 and 10 **92.** It differs by 0.11.

93.

94. −30 **95.** 2.5 **96.** −27 **97.** 1.5 **98.** 4.36 **99.** 7.68

Chapter 4   Test   (page 273)

1. $\frac{79}{100}$, 0.79 **2.** Selway, Monroe, Paston, Covington, Cadia **3. a.** sixty-two and fifty-five hundredths; $62\frac{55}{100}$ **b.** eight thousand thirteen one hundred thousandths; $\frac{8,013}{100,000}$ **4.** 33.050 **5.** $208.75 **6. a.** 0.567909 **b.** 0.458 **7.** 1.02 in. **8. a.** 10.75 **b.** 6.121 **c.** 0.1024 **d.** 14.07 **9.** 1.25 mi^2 **10.** 0.004 in. **11.** 3.588 **12. a.** 0.34 **b.** $0.41\overline{6}$ **13.** −2.29 **14.** $1.\overline{18}$

15.

16. $\frac{41}{30}$

17. 0.42 g **18.** 80

19.

20. radical **21.** 12^2 **22. a.** 11 **b.** $-\frac{1}{30}$ **23. a.** > **b.** > **c.** > **d.** > **24. a.** −0.2 **b.** 13

Cumulative Review Exercises   (page 275)

1. $788,000 **2.** (3 + 4) + 5 = 3 + (4 + 5) **3.** 27 R 42 **4.** 1,000 **5.** 11 · 5 · 2^2 **6.** 1, 2, 4, 5, 10, 20

7. {0, 1, 2, 3, 4, 5, . . .} **8.** −13 **9.** adding **11.** −5(3) = −15
12. −1 **13.** 35 **14.** 102 **15.** −$1,100 **16.** $\frac{6}{13}$
17. equivalent fractions **18.** $\frac{5}{7}$ **19.** $\frac{21}{128}$ **20.** −$\frac{3}{16}$ **21.** $\frac{34}{21}$
22. $19\frac{1}{8}$ **23.** $26\frac{7}{24}$ **24.** −$\frac{1}{3}$ **25.** 157.5 in.²
26.

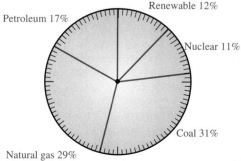

27. 0.001 in. **28.** < **29.** −8.136 **30.** 5.6 **31.** 5,601.2
32. 0.0000897 **33.** 47.95 **34.** 33.6 hr **35.** 232.8°C **36.** $0.41\overline{6}$
37. $50 **38.** 7 **39.** $\frac{25}{4}$ **40.** −6

Chapter 5   Check Your Knowledge (page 278)

1. one hundred **2.** amount, percent, base **3.** discount
4. principal **5. a.** 0.75, 75% **b.** 0.625, 62.5% **c.** 1.45, 145%
6. a. 35%, $\frac{7}{20}$ **b.** 398%, $\frac{199}{50}$ or $3\frac{49}{50}$ **c.** 10.5%, $\frac{21}{200}$
7. a. 0.25, $\frac{1}{4}$ **b.** 2 or 2.0, 2 or $\frac{2}{1}$ **c.** 0.005, $\frac{1}{200}$ **8.** 35%
9. 66.7% **10. a.** 37.5% or $37\frac{1}{2}$% **b.** $33\frac{1}{3}$% **11.** 325
12. 17% **13.** 52 **14.** 250% **15.** $2.99, $11.96 **16.** $29.94
17. $46.00 **18.** $1,676.47 **19.** 93% **20.** $1,045.00
21. $40.71 **22.** 13

Study Set   Section 5.1 (page 285)

1. percent **3.** 100, simplify **5.** right **7. a.** 0.84, 84%, $\frac{21}{25}$
b. 16% **9.** $\frac{17}{100}$ **11.** $\frac{1}{20}$ **13.** $\frac{3}{5}$ **15.** $\frac{5}{4}$ **17.** $\frac{1}{150}$ **19.** $\frac{21}{400}$ **21.** $\frac{3}{500}$
23. $\frac{19}{1,000}$ **25.** 0.19 **27.** 0.06 **29.** 0.408 **31.** 2.5 **33.** 0.0079
35. 0.0025 **37.** 93% **39.** 61.2% **41.** 3.14% **43.** 843%
45. 5,000% **47.** 910% **49.** 17% **51.** 16% **53.** 40%
55. 105% **57.** 62.5% **59.** 18.75% **61.** $66\frac{2}{3}$% **63.** $8\frac{1}{3}$%
65. 11.11% **67.** 55.56% **69. a.** $\frac{15}{191}$ **b.** 8% **71. a.** $\frac{9}{22}$
b. 41% **73. a.** $\frac{5}{29}$ **b.** 17% **c.** 24% **75.** 5 ft **77.** 0.9944
79. as a decimal; 89.6% **81.** torso: 27.5% **83.** 92%
85. 0.27% **93.** −$\frac{1}{12}$ **95.** −1 **97.** 68.25 cm²

Think It Through (page 294)

36% are enrolled in college full time. 69% of full time students read 5 or more assigned text books, manuals, or books during the current school year. 38% occasionally.

Study Set   Section 5.2 (page 295)

1. graph **3.** A = 0.10 · 50 **5.** 48 = p · 47 **7. a.** 0.12
b. 0.056 **c.** 1.25 **d.** 0.0025 **9.** more **11.** 44% **13.** 25
15. 50 **17.** 90 **19.** 80% **21.** 65 **23.** 0.096 **25.** 1.25%
27. 44 **29.** 43.5 **31.** 107.1 **33.** 99 **35.** 60 **37.** 31.25%
39.

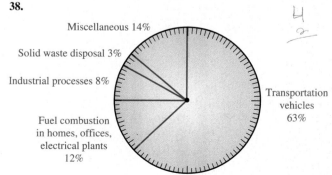

Source: Energy Information Administration

41. 120 **43.** $666 billion **45.** 38,000 = 38K **47.** 24 oz
49. yes **51.** 30, 12 **53.** 2.7 in. **55.** 5% **57.** yes **63.** 18.17
65. 5.001 **67.** 0.008

Think It Through (page 302)

1. 1970–1975, about a 75% increase **2.** 1995–2000, about a 9% decrease

Study Set   Section 5.3 (page 304)

1. commission **3.** discount **5.** subtract, original
7. $42.75 **9.** 8% **11.** $47.34, $2.84, $50.18 **13.** $150
15. 8%, 1.2%, 1.45%, 6.2% **17.** 360 hr **19.** 96 calories
21. 2000–2001; 12% **23.** 10% **25.** 31% **27. a.** 25%
b. 36% **29.** $2,955 **31.** 1.5% **33.** $12,000 **35.** $39.95, 25%
37. $187.49 **39.** $349.97, 13% **41.** $3.60, 23%, $11.88
43. $76.50 **49.** −50 **51.** $\frac{13}{36}$ **53.** 173.4 **55.** 13

Study Set   Estimation (page 309)

1. 164 **3.** $60 **5.** $54,000 **7.** 320 lb **9.** 130 **11.** 21
13. 18,000 **15.** 3,100

Study Set   Section 5.4 (page 314)

1. principal **3.** interest **5.** simple **7. a.** 0.07 **b.** 0.098
c. 0.0625 **9.** $1,800 **11. a.** compound interest **b.** $1,000
c. 4 **d.** $50 **e.** 1 year **13.** multiplication **15.** $5,300
17. $1,472 **19.** $4,262.14 **21.** $10,000, 0.0725, 2 yr, $1,450
23. $192, $1,392, $58 **25.** $18.828 million **27.** $755.83
29. $1,271.22 **31.** $570.65 **33.** $30,915.66 **39.** $\frac{1}{2}$ **41.** $\frac{29}{35}$
43. $8\frac{1}{3}$ **45.** 12

Key Concept (page 317)

67, 100, 0.05, 2,000; 0.67, 0.56, 0.0005; 75%, 0.8, 0.625, 625%
1. 198.4 **2.** 60% **3.** 62.5 **4.** 1,062.5 **5.** 17% **6.** 512
7. $3,000 **8.** $3,468.55

Chapter Review (page 319)

1. 39%, 0.39, $\frac{39}{100}$ **2.** 111%, 1.11, $1\frac{11}{100}$ **3.** 61% **4.** $\frac{3}{20}$ **5.** $\frac{6}{5}$
6. $\frac{37}{400}$ **7.** $\frac{1}{1,000}$ **8.** 0.27 **9.** 0.08 **10.** 1.55 **11.** 0.018
12. 83% **13.** 62.5% **14.** 5.1% **15.** 600% **16.** 50%
17. 80% **18.** 87.5% **19.** 6.25% **20.** $33\frac{1}{3}$% **21.** $83\frac{1}{3}$%
22. 55.56% **23.** 266.67% **24.** 63% **25.** 0.1% = $\frac{1}{1,000}$
26. amount: 15, base: 45, percent: $33\frac{1}{3}$% **27.** A = 32% · 96
28. 200 **29.** 125 **30.** 1.75% **31.** 2,100 **32.** 121 **33.** 30
34. 14.4 gal nitro, 0.6 gal methane **35.** 68 **36.** 87% **37.** $5.43
38.

39. 139,531,200 mi² **40.** $3.30, $63.29 **41.** 4% **42.** $40.20
43. original **44.** 25% **45.** 9.6% **46.** $50, $189.99, 26%
47. $6,000, 8%, 2 years, $960 **48.** $10,308.22 **49.** $134.69
50. $2,142.45 **51.** $6,076.45 **52.** $43,265.78

Chapter 5   Test   (page 323)

1. 61%, $\frac{61}{100}$, 0.61 **2.** 199%, $\frac{199}{100}$, 1.99 **3. a.** 0.67 **b.** 0.123
c. 0.0975 **4. a.** 25% **b.** 62.5% **c.** 12% **5. a.** 19%
b. 347% **c.** 0.5% **6. a.** $\frac{11}{20}$ **b.** $\frac{1}{10,000}$ **c.** $\frac{5}{4}$ **7.** 23.33%
8. 60% **9.** $66\frac{2}{3}$% **10.** 25% **11. a.** 1.02 in. **b.** 32.98 in.
12. 6.5% **13.** $3.81 **14.** 93.9% **15.** 90 **16.** 21 **17.** 144
18. 27% **19.** $35.92 **20.** $41,440 **21.** $11.95, $3, 20%
22. 22% **23.** $150 **24.** $5,079.60 **25.** The phrase "bringing
crime down to 37%" is unclear. The question that arises is: 37%
of what? **26.** Interest is money that is paid for the use of money.

Cumulative Review Exercises   (page 325)

1.

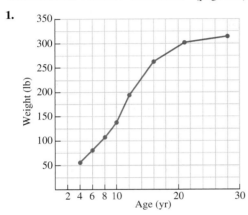

2. $6 \cdot 8 = 8 \cdot 6$ **3. a.** 1, 2, 4, 5, 8, 10, 20, 40 **b.** $5 \cdot 2^3$
4. $2,106 **5.** 64 ft² **6.** −7 **7.** −4 **8.** 55 **9.** 9 **10.** −30
11. 15°C **12.** −10 **13.** $\frac{4}{11}$ **14.** $\frac{2}{3}$ **15.** $-\frac{5}{21}$ **16.** $\frac{2}{5}$ **17.** $\frac{34}{21}$
18. $20\frac{5}{18}$ **19.** $\frac{4}{9}$ **20.** $-\frac{8}{27}$ **21.** 70.29 **22.** −8.6265 **23.** 752
24. 83.4 **25.** 452.03 **26.** 452.030 **27.** $0.7\overline{3}$ **28.** $\frac{2}{3}$ **29.** −29
30. 3.5 hr **31.** 29%, $\frac{29}{100}$, 0.473, $\frac{473}{1,000}$, 87.5%, 0.875 **32.** 125
33.

STEAK STAMPEDE
Bloomington, MN
Server #12\ AT
VISA   67463777288
NAME   DALTON/ LIZ
AMOUNT $75.18
GRATUITY $ 12.00
TOTAL $ 87.18

34. 0.0018% **35.** $1,450 **36.** 50

Chapter 6   Check Your Knowledge   (page 328)

1. ratio **2.** proportion **3.** length, metric **4.** Fahrenheit,
Celsius **5.** 7/5 **6.** 2/9 **7.** 47 mpg **8.** the 12-oz. bag
9. 4/1 **10. a.** yes **b.** no **11.** no **12.** 68 **13.** −40
14. 25.74 **15.** −0.2 **16.** 500 cm **17.** $9\frac{2}{3}$ yd **18.** 26,400 ft
19. 252 in. **20.** 109.4 yds **21.** 175.26 cm **22.** 2,500
23. $3\frac{1}{2}$ in. **24.** −40°C

Think It Through   (page 332)

22 : 1, 24 : 1, 26 : 1; Basic mathematics has the lowest student-to-
instructor ratio.

Study Set   Section 6.1   (page 335)

1. ratio **3.** cost **5.** 3 **7.** 10 **9.** $\frac{11 \text{ minutes}}{60 \text{ minutes}} = \frac{11}{60}$ **11.** $\frac{13}{9}$, 13 to
9, 13 : 9 **13.** $\frac{5}{7}$ **15.** $\frac{1}{2}$ **17.** $\frac{2}{3}$ **19.** $\frac{5}{8}$ **21.** $\frac{2}{7}$ **23.** $\frac{1}{3}$ **25.** $\frac{1}{5}$
27. $\frac{3}{7}$ **29.** $\frac{3}{4}$ **31.** $1,800 **33.** $\frac{1}{3}$ **35.** $8,750 **37.** $\frac{1}{5}$ **39.** $\frac{32 \text{ ft}}{3 \text{ sec}}$
41. $\frac{21 \text{ made}}{25 \text{ attempts}}$ **43.** $\frac{375 \text{ students}}{2 \text{ yr}}$ **45.** $\frac{3 \text{ beats}}{2 \text{ measures}}$ **47.** 12 revolutions
per min **49.** 1.5 errors per hr **51.** 7 presents per child
53. 320 people per square mile **55.** $0.07 per foot **57.** 1.2
cents per ounce **59.** $68 per person **61.** $0.8 billion per
month **63.** $\frac{1}{1}$ **65.** $\frac{3}{2}$ **67.** $\frac{\frac{2}{3}}{3\frac{1}{2}}$ **69.** $\frac{12 \text{ hits}}{22 \text{ at-bats}} = \frac{6 \text{ hits}}{11 \text{ at-bats}}$
71. $\frac{5 \text{ compressions}}{2 \text{ breaths}}$ **73.** $\frac{329 \text{ complaints}}{100,000 \text{ passengers}}$ **75.** $\frac{1 \text{ faculty member}}{16 \text{ students}}$
77. $1.89 per gal **79.** 7¢ per oz **81.** the 6-oz can
83. the 50-tablet boxes **85.** the truck **87.** 440 gal per min
89. 325 mi, 65 mph **91.** the second car **97.** 45.537
99. 192.7012

Study Set   Section 6.2   (page 344)

1. proportion **3.** cross **5.** $2 \cdot 10, 5 \cdot 4$ **7. a.** $\frac{5}{8} = \frac{15}{24}$
b. $\frac{3 \text{ teacher's aides}}{25 \text{ children}} = \frac{12 \text{ teacher's aides}}{100 \text{ children}}$ **9.** i, iv **13.** no **15.** yes
17. no **19.** yes **21.** no **23.** yes **25.** 4 **27.** 6 **29.** 3 **31.** 36
33. 1 **35.** −2 **37.** 18 **39.** −3.1 **41.** 3,500 **43.** 5.625
45. $218.75 **47.** $11.76 **49.** the same **51.** 24 **53.** 975
55. about $4\frac{1}{4}$ **57.** 19 sec **59.** 221 mi **61.** $309 **63.** 10 ft
65. 65.25 ft = 65 ft 3 in. **67.** 2.625 in. = $2\frac{5}{8}$ in. **73.** 90%
75. $\frac{1}{3}$ **77.** 2.6

Study Set   Section 6.3   (page 355)

1. length **3.** 1 **5.** capacity **7.** 1 **9.** 5,280 **11.** 16 **13.** 8
15. 1 **17.** 24 **19.** $\frac{5}{8}$ in., $1\frac{3}{4}$ in., $2\frac{5}{16}$ in. **21. a.** $\frac{1 \text{ ton}}{2,000 \text{ lb}}$ **b.** $\frac{2 \text{ pt}}{1 \text{ qt}}$
23. a. iv **b.** i **c.** ii **d.** iii **25. a.** iii **b.** iv **c.** i **d.** ii
31. $2\frac{5}{8}$ in. **33.** $10\frac{3}{4}$ in. **35.** 48 in. **37.** 42 in. **39.** 2 ft
41. 288 in. **43.** 2.5 yd **45.** $4\frac{2}{3}$ ft **47.** 15 ft **49.** $2\frac{1}{3}$ yd
51. 3 mi **53.** 2,640 ft **55.** 5 lb **57.** 3.5 tons **59.** 24,800 lb
61. 6 pt **63.** 2 gal **65.** 2 pt **67.** 4 hr **69.** 5 days **71.** 150 yd
73. 2,880 in. **75.** 0.28 mi **77.** 61,600 yd **79.** 128 oz **81.** 4.95
tons **83.** 68 **85.** $71\frac{7}{8}$ gal = 71.875 gal **87.** 320 oz **89.** $6\frac{1}{8}$
days = 6.125 days **93.** 3,700 **95.** 3,673.26 **97.** 0.101 **99.** 0.1

Study Set   Section 6.4   (page 365)

1. tens **3.** thousands **5.** hundredths **7.** metric
9. 1 cm, 3 cm, 6 cm **11. a.** $\frac{1 \text{ km}}{1,000 \text{ m}}$ **b.** $\frac{100 \text{ cg}}{1 \text{ g}}$ **c.** $\frac{1,000 \text{ milliliters}}{1 \text{ liter}}$
13. a. iii **b.** i **c.** ii **15. a.** ii **b.** iii **c.** i **17.** 10 **19.** $\frac{1}{100}$
21. $\frac{1}{1,000}$ **23.** 1,000 **25.** 1,000 **27.** 1,000 **29.** $\frac{1}{100}$ **31.** 1
37. 156 mm **39.** 28 cm **41.** 300 **43.** 570 **45.** 3.1
47. 7,680,000 **49.** 0.472 **51.** 4.532 **53.** 0.0325 **55.** 37.5
57. 125 **59.** 675,000 **61.** 6.383 **63.** 0.63 **65.** 69.5
67. 5.689 **69.** 5.762 **71.** 0.000645 **73.** 0.65823
75. 3,000 **77.** 2,000 **79.** 1,000,000 **81.** 0.5 **83.** 3,000
85. 5,000 **87.** 10 **89.** 0.5 km, 1 km, 1.5 km, 5 km, 10 km
91. 3.43 hm **93.** 12 cm, 8 cm **95.** 40 dL **97.** 4 **99.** 3 g
105. $23.99 **107.** $1\frac{9}{35}$

Think It Through (page 371)

1. 216 mm × 279 mm   **2.** 9 kilograms   **3.** 22.5 milliliters

Study Set   Section 6.5   (page 373)

1. Fahrenheit, Celsius   **3. a.** meter   **b.** meter   **c.** inch   **d.** mile
5. a. liter   **b.** liter   **c.** gallon   **11.** 91.4   **13.** 147.6   **15.** 39,372
17. 127   **19.** 1   **21.** 11,350   **23.** 17.5   **25.** 0.6   **27.** 0.1
29. 243.4   **31.** 710   **33.** 0.5   **35.** 10°   **37.** 122°   **39.** 14°
41. −20.6°   **43.** 5 mi   **45.** 70 mph   **47.** 1.9 km   **49.** 1.9 cm
51. 411 lb; 744 lb   **53. a.** 226.8 g   **b.** 0.24 L   **55.** no
57. the 3 quarts   **59.** 62°C   **61.** 28°C   **63.** −5°C and 0°C
69. $\frac{29}{15}$ **71.** $\frac{4}{5}$ **73.** 8.05   **75.** 15.6

Key Concept   (page 376)

1. teacher's aides needed to supervise 75 children; 2, 75, x, 2, x;
15, 75, 150, 15, 15, 10   **2.** 375   **3.** 10,800 ft   **4.** $1,152

Chapter Review   (page 378)

1. $\frac{1}{3}$ **2.** $\frac{1}{4}$ **3.** $\frac{3}{2}$ **4.** $\frac{2}{3}$ **5.** $\frac{37}{32}$ **6.** $7.75   **7.** the 8-oz can   **8.** 75
9. 15   **10.** no   **11.** yes   **12.** yes   **13.** no   **14.** 4.5   **15.** 16
16. 7.2   **17.** −0.12   **18.** 192.5 mi   **19.** 300   **20.** 12 ft
21. $1\frac{1}{2}$ in.   **22.** $\frac{1\ \text{mi}}{5{,}280\ \text{ft}} = 1, \frac{5{,}280\ \text{ft}}{1\ \text{mi}} = 1$ **23.** 15 ft   **24.** 216 in.
25. 5.5 ft   **26.** 306 in.   **27.** 1.75 mi   **28.** 1,760 yd   **29.** 2 lb
30. 275.2 oz   **31.** 96,000 oz   **32.** 2.25 tons   **33.** 80 fl oz
34. 0.5 gal   **35.** 68 c   **36.** 5.5 qt   **37.** 40 pt   **38.** 56 c
39. 1,200 sec   **40.** 15 min   **41.** $8\frac{1}{3}$ days   **42.** 360 min
43. 108 hr   **44.** 86,400 sec   **45.** $484\frac{2}{3}$ yd   **46.** 100   **47.** 4 cm
48. $\frac{1\ \text{km}}{1{,}000\ \text{m}} = 1, \frac{1{,}000\ \text{m}}{1\ \text{km}} = 1$ **49.** 4.75 m   **50.** 8,000 mm
51. 0.03 km   **52.** 2,000 dm   **53.** 50 hm   **54.** 25 hm   **55.** 70 mg
56. 8 g   **57.** 5.425 kg   **58.** 5,425,000 mg   **59.** 7.5 g
60. 0.05 kg   **61.** 50   **62.** 1.5 L   **63.** 3.25 kL   **64.** 1,000 dL
65. 40 cL   **66.** 20 hL   **67.** 400 mL   **68.** 1,000 mL
69. 164.04 ft   **70.** the Sears Tower   **71.** 3,106 km
72. 198.12 cm   **73.** 850.5 g   **74.** 33 lb   **75.** 11,000 g
76. 910 kg   **77.** about 2 lb   **78.** LaCroix   **79.** the 5-liter bottle
80. 25°C   **81.** 30°C

Chapter 6   Test   (page 383)

1. $\frac{3}{4}$ **2.** $\frac{1}{6}$ **3.** the 2-pound can   **4.** 22.5 kwh per day   **5.** $\frac{1}{1}$, 1:1,
1 to 1   **6.** no   **7.** yes   **8.** yes   **9.** 15   **10.** 63.24   **11.** −0.21
12. 0.2   **13.** $3.43   **14.** $1\frac{2}{3}$ c   **15.** 15 ft   **16.** $8\frac{1}{3}$ yd   **17.** 160 oz
18. 3,200 lb   **19.** 128 fl oz   **20.** 115,200 min   **21.** the one on
the left   **22.** the blue one   **23.** the right side   **24.** 0.5 km
25. 500 cm   **26.** 0.08 kg   **27.** 70,000 mL   **28.** 7.5 g
29. the 100-yd race   **30.** Jim   **31.** the 1-liter bottle   **32.** 182°F
33. A scale is a ratio (or rate) comparing the size of a drawing
and the size of an actual object. For example, 1 inch to 6 feet
(1 in. : 6 ft).   **34.** It is easier to convert from one unit to another
in the metric system, because it is based on the number 10.

Cumulative Review Exercises   (page 385)

1. 6 ten thousands + 4 thousands + 5 hundreds + 2 ones
2. 20 R 3   **3.** 587, 278, −6,790   **4. a.** −8   **b.** undefined
c. −8   **d.** 0   **e.** 8   **f.** 0   **5.** 15 shots   **6.** −9, 9   **7.** −61
8. 2,500   **9.** 0.25   **10.** 3.51   **11.** 25.1   **12.** 0.19   **13.** 79

14. $A = \frac{1}{2}bh$ **15.** $\frac{4}{5}$ **16.** $\frac{54}{60}$ **17.** −1   **18.** $\frac{1}{4}$ **19.** $\frac{3}{4}$ hp   **20.** no
21. yes   **22.** 6   **23.** yes   **24.** 15   **25.** 55.1°F   **26.** 23
27. $0.08\overline{3}$ **28.** $66\frac{2}{3}$%   **29.** $I = Prt$ **30.** 99%, $\frac{99}{100}$, 0.013, $\frac{13}{1{,}000}$,
31.25%, 0.3125   **31.** $427.99; about 30%   **32.** $\frac{1}{5}$
33. a. 960 hr   **b.** 4,320 min   **c.** 480 sec   **34.** 2.5 lb
35. 2,400 mm   **36.** 0.32 kg   **37. a.** 1 gal   **b.** a meterstick
38. the 45-kilogram bag

Chapter 7   Check Your Knowledge   (page 388)

1. bar   **2.** pictograph   **3.** frequency   **4.** mean   **5.** median
6. mode   **7.** circle graph   **8.** 4   **9.** 15   **10.** 20%   **11.** O, AB
12. histogram   **13.** 3   **14.** 25   **15.** 40%   **16.** 10   **17.** 20%
18. 7   **19.** 7   **20.** 9   **21.** mean   **22.** mean, median

Study Set   Section 7.1   (page 396)

1. a   **3.** c   **5.** d   **7.** bars, equal   **9.** $7.35   **11.** $4.01
13. $7,895   **15.** $3,110   **17.** nuclear energy   **19.** 29%
21. coal   **23.** 1980   **25.** 1970   **27.** 320 thousand
metric tons   **29.** reckless driving and failure to yield
31. reckless driving   **33.** seniors   **35.** $50   **37.** French and
German   **39.** English   **41.** 51.4%   **43.** about 11%
45. about 49%   **47.** 175%   **49.** $190   **51.** miners   **53.** miners
55. 1   **57.** 1   **59.** Runner 1 was running; runner 2 was stopped.
61. 27   **63.** 90   **71.** −7   **73.** $\frac{25}{36}$ **75.** 11, 13, 17, 19, 23, 29   **77.** 4

Think It Through   (page 405)

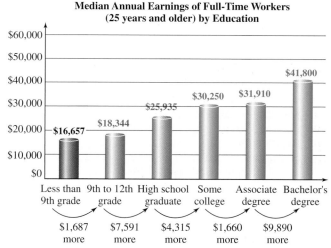

Median Annual Earnings of Full-Time Workers
(25 years and older) by Education

Source: U.S. Census Bureau, June 2004

Study Set   Section 7.2   (page 405)

1. mean   **3.** median   **5.** the number of values   **7.** 8   **9.** 35
11. 19   **13.** 9   **15.** 6   **17.** 17.5   **19.** 3   **21.** none   **23.** 22.7
25. about 63¢   **27.** 60¢   **29.** 50¢   **31.** about 61°   **33.** 64°
35. 2,670 mi   **37.** 89 mi   **39.** Median and mode are 85.
41. same average (56); sister's scores are more consistent
43. 22.525 oz, 25 oz   **45.** $4.15, $4.19, $4.29   **47.** city: mean 43,
median 42, mode 42; hwy: mean 48.8, median 49, mode 49
49. a. 5.5   **b.** 5.6   **c.** 5.6   **53.** 3^4 **55.** $\frac{1}{6}$ **57.** 6   **59.** $\frac{19}{10} = 1\frac{9}{10}$

Key Concept   (page 409)

1. 13.7   **3.** 15   **5.** 6   **7.** 6

Chapter Review   (page 411)

1. $-18°$ **2.** 30 mph   **3.** about 4.9 billion   **4.** 1997 and 1998
5. 1999 and 2000   **6.** 1996 and 1997   **7.** about 830 million
8. about 865 million   **9.** 1987   **10.** about 1,770 million
11. 180   **12.** 160   **13.** yes   **14.** median   **15.** 1.2 oz
16. 1.138 oz   **17.** 7.3 microns, 7.2 microns, 6.9 microns
18. $1.45 billion

Chapter 7   Test   (page 415)

1. about $1,659   **2.** about $11   **3.** about 4.1%   **4.** about 1.2%
5. about 19%   **6.** about 6%   **7.** about 270,000   **8.** 1985
9. about 7,400   **10.** 65.5%   **11.** A   **12.** C   **13.** E
14. bicyclist 1   **15.** 7.5   **16.** 7.5   **17.** 5   **18.** mean
19. 3.6, 3.6, 3.1   **20.** Half the families had more debt and half
had less debt.

Cumulative Review Exercises   (page 417)

1. 358,600,000 gal   **2.** 50,000   **3.** 54,604   **4.** 4,209
5. 23,115   **6.** 87   **7.** $683 + 459 = 1,142$ **8.** 2011
9. $4 \cdot 5 = 5 + 5 + 5 + 5 = 20$ **10.** $10,912 \text{ in.}^2$
11. a. 1, 2, 3, 6, 9, 18   **b.** $3^2 \cdot 2$ **12.** 2, 3, 5, 7, 11, 13, 17,
19, 23, 29   **13.** It has factors other than 1 and itself.
For example, $27 = 3 \cdot 9$.   **14.** 22   **15.** 315   **16.** 6
17.

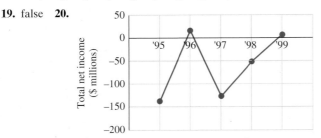

  18. 5
19. false   **20.**

21. -20 **22.** 30   **23.** 125   **24.** 5   **25.** -5 **26.** 429
27. $-3^2 = -(3 \cdot 3) = -9$; $(-3)^2 = (-3)(-3) = 9$ **28.** 1,100°F
29. 800   **30.** 15%   **31.** $\frac{5}{0}; \frac{0}{5}$; division by 0   **32.** $\frac{7}{6} = 1\frac{1}{6}$
33. $-\frac{1}{6}$ **34.** $6\frac{3}{4}$ in.   **35.** $\frac{8}{35}$ **36.** $\frac{21}{20} = 1\frac{1}{20}$ **37.** 220   **38.** 345
39. 0.744   **40.** 745   **41.** $0.\overline{72}$ **42.** 160 min   **43.** 3.1 hr
44. 3.02, 3.005, 2.75

Chapter 8   Check Your Knowledge   (page 420)

1. equal, $=$ **2.** variable, constant   **3.** multiplication
4. formula   **5.** expression, expressions, equations
6. coefficient, variable   **7.** like   **8.** 3   **9.** 2   **10.** 16   **11.** 48
12. 120   **13.** 140   **14.** 17   **15.** 18   **16.** 6   **17.** 45 mi
18. $6x + 21$ **19.** $-6x + 9$ **20. a.** $3x$ **b.** $-x$ **21. a.** $3x$
b. $12ab$ **c.** $x + 1$ **22. a.** 6   **b.** 1   **23. a.** $25q$ cents or $0.25q$
b. $5f$ **24. a.** x^5 **b.** $6x^6$ **c.** $9x^2y^6$ **d.** x^{18}

Study Set   Section 8.1   (page 428)

1. equal, $=$ **3.** right-hand   **5.** equivalent   **7.** y, c
9. addition of 6; subtract 6 from both sides   **13.** yes   **15.** no
17. yes   **19.** yes   **21.** yes   **23.** yes   **25.** no   **27.** no   **29.** no
31. yes   **33.** 10   **35.** 7   **37.** 3   **39.** 4   **41.** 13   **43.** 75
45. 740   **47.** 339   **49.** 3   **51.** 5   **53.** 9   **55.** 10   **57.** 1   **59.** 56
61. 84   **63.** 105   **65.** 4   **67.** 12   **69.** 0.8   **71.** 4.7   **73.** $\frac{1}{6}$

75. $\frac{11}{24}$ **79.** 94,683,948   **81.** 62   **83.** $218,500   **85.** $180 million
87. 25 units   **89.** $190   **97.** 325,780   **99.** 90   **101.** 3

Study Set   Section 8.2   (page 436)

1. division   **3.** x **5.** $\frac{y}{z}$ **7.** It is being multiplied by 4. Divide
by 4.   **9. a.** Subtract 5 from both sides.   **b.** Add 5 to both
sides.   **c.** Divide both sides by 5.   **d.** Multiply both sides by 5.
11. 3, 3, 4, 12, 4   **13.** 1   **15.** 96   **17.** 3   **19.** 6   **21.** 1
23. 2   **25.** 16   **27.** $\frac{2}{15}$ **29.** 14   **31.** 42   **33.** 75   **35.** 39
37. 50   **39.** 49   **41.** 10   **43.** 3   **45.** 2   **47.** 1   **49.** 40
51. 1,200   **53.** 0.36   **55.** $\frac{1}{4}$ **59.** 390 wpm   **61.** 14   **63.** 96
65. 32 calls   **67.** 55 lb   **73.** 48 cm   **75.** $2^3 \cdot 3 \cdot 5$ **77.** 72
79. 26 mpg

Think It Through   (page 443)

Suggested: 4, 6, 8, 10; expanded: 6, 9, 12, 15

Study Set   Section 8.3   (page 446)

1. formula   **3.** substitute   **5.** $2 - 8 + 10$; it looks like
subtraction   **7. a.** x = length part 1; $x - 40$ = length part 2;
$x + 16$ = length part 3 (answers may vary)   **b.** 20 in. and 76 in.
9. $22, $27, $(p + 2)$, $(10p + 2)$ **11.** 50 mi, 48 mi, $3t$ mi, 5 mi,
$3x$ mi   **13.** speedometer: rate; odometer: distance; clock: time;
$d = rt$ **15.** The rate is expressed in miles per hour, and the
time in minutes.   **17. a.** $d = rt$ **b.** $C = \frac{5}{9}(F - 32)$ **c.** $d = 16t^2$
19. 17   **21.** 4   **23.** 40   **25.** -6 **27.** -6 **29.** 23   **31.** -8
33. 100   **35.** -28 **37.** 3   **39.** -3 **41.** -7 **43.** -18 **45.** 25
47. 21   **49.** -5 **51.** -29 **53.** -45 **55.** 70¢   **57.** $8,200
59. $23   **61.** 300 mi   **63.** $-10°$ C   **65.** 240 m^2 **67.** 64 ft
69. 5,213, 5,079, 4,814, 2,053, 2,051, 1,921, 3,160, 3,028, 2,893
71. $D = B - C$; $20, $12, $42   **73.** 5.2 yards per carry   **75.** 40
77. 4   **83.** 17, 37, 41   **85.** 7   **87.** division by 3   **89.** 3

Study Set   Section 8.4   (page 455)

1. distributive, removed   **3.** equivalent   **5.** $x(y + z) = xy + xz$
7. $(w + 7)5$ **9.** 5, 6, 6, 2, 3   **11.** $-y - 9$ **13.** -5 **15. a.** $+$
b. $-$ **c.** $-$ **d.** $+$ **17. a.** x **b.** $x + 5$ **c.** $5x - 10y - 15$
d. $5x$ **19.** $12x$ **21.** $-30y$ **23.** $100t$ **25.** $12s$ **27.** $14c$
29. $-40h$ **31.** $-42xy$ **33.** $16rs$ **35.** $30xy$ **37.** $-30br$
39. $80c$ **41.** $-8e$ **43.** $4x + 4$ **45.** $16 - 4x$ **47.** $-6e - 6$
49. $-16q + 48$ **51.** $12 + 20s$ **53.** $42 + 24d$ **55.** $-25r + 30$
57. $-24 - 18d$ **59.** $9x - 21y + 6$ **61.** $9z + 9x + 15y$
63. $-x - 3$ **65.** $-4t - 5$ **67.** $3w + 4$ **69.** $-5x + 4y - 1$
71. $2(4x + 5)$ **73.** $(-4 - 3x)5$ **75.** $-3(4y - 2)$
77. $3(4 - 7t - 5s)$ **83.** 5   **85.** multiplication, division,
subtraction, addition   **87.** $>$ **89.** carpeting, painting

Study Set   Section 8.5   (page 463)

1. term   **3.** perimeter   **5.** distributive   **7.** sum
9. a. term   **b.** factor   **c.** factor   **d.** factor   **11. a.** 11   **b.** 8
c. -4 **d.** 1   **e.** -1 **f.** 102   **13.** 6, m; -75, t; 1, w; 4, bh
15. It helps identify the like terms.   **17.** $(2d + 15)$ mi
19. To add the like terms, add 9 and 5 and keep the variable.
21. 7   **23.** $2, 5x$ **25. a.** the perimeter of a rectangle   **b.** 2 times
the length   **c.** 2 times the width   **27.** $3x^2, 5x, 4$ **29.** 5, $5t$,
$-8t$, 4   **31.** 2   **33.** 5   **35.** $15t$ **37.** $4s$ **39.** x **41.** $4d$
43. $-4e$ **45.** cannot be simplified   **47.** $-6z$ **49.** $-7x$ **51.** 0
53. 0   **55.** $4x$ **57.** can't be simplified   **59.** $-2y$ **61.** $3a$

63. $11t + 12$ **65.** $2w - 5$ **67.** $-7r + 11R$ **69.** $-50d$
71. $8x - 4y - 9$ **73.** $9x + 34$ **75.** $-22s + 23$ **77.** $19e - 21$
79. $2t + 8$ **81.** $3x + 8$ **83.** $10y - 32$ **85.** \$288 **87.** 36 ft,
48 ft, 60 ft, 72 ft, 84 ft **93.** 2 **95.** $2^2 \cdot 5^2$ **97.** absolute value

Think It Through (page 468)

1. $x + x + 2.5 = 43.1$; men: 22.8 yr; women: 20.3 yr.
2. men: 26.9 yr; women: 25.3 yr

Study Set Section 8.6 (page 470)

1. solve **3.** distributive **5.** combine **7.** When we substitute
-5 for x, the result is a false statement: $-10 = -9$. **9.** $5k$
11. a. $4x$ **b.** $2x$ **13. a.** $2t - 8$ **b.** -4 **c.** -16
15. $2x, 2, 2, -10$ **17.** $9, 45, 45, 5x, 5, 5, 10$ **19.** yes
21. no **23.** 2 **25.** -4 **27.** -30 **29.** -28 **31.** 42 **33.** 37
35. 306 **37.** 735 **39.** 2 **41.** -14 **43.** -8 **45.** 5 **47.** -12
49. 4 **51.** 8 **53.** 10 **55.** 6 **57.** 0 **59.** -4 **61.** -10
63. 0 **65.** 1 **67.** 2 **69.** -11 **71.** 26 **73.** -3 **75.** 7 **77.** 3
83. -16 **85.** -3 **87.** 5 **89.** positive

Study Set Section 8.7 (page 477)

1. base, exponent **3.** like **5.** product **7. a.** x^7 **b.** x^2y^3
c. $3^4a^2b^3$ **9.** $x^2 \cdot x^6 = x^8$ (answers may vary) **11.** $(c^5)^2 = c^{10}$
(answers may vary) **13. a.** x^{m+n} **b.** x^{mn} **c.** x^ny^n **15. a.** 2
b. -10 **c.** x **17. a.** $x^2; 2x$ **b.** $x^3; x + x^2$ **c.** $x^4; 2x^2$
19. a. $4x^2; 5x$ **b.** $12x^2; 7x$ **c.** $12x^3; 4x^2 + 3x$ **21.** 27 **27.** x^5
29. x^{10} **31.** f^{13} **33.** n^{32} **35.** l^{10} **37.** x^{11} **39.** 2^{12} **41.** 5^8
43. $8x^3$ **45.** $5t^{10}$ **47.** $-24x^5$ **49.** $-x^4$ **51.** $36y^8$ **53.** $-40t^{10}$
55. x^3y^3 **57.** b^8c^8 **59.** x^5y^2 **61.** a^4b^4 **63.** x^5y^7 **65.** $18x^3y^4$
67. $16x^4y^2$ **69.** $-24f^6t^4$ **71.** a^4b^3 **73.** $12x^4y^3$ **75.** x^8
77. m^{500} **79.** $8a^3$ **81.** x^4y^4 **83.** $27s^6$ **85.** $4s^4t^6$ **87.** x^{14}
89. c^{30} **91.** $36a^{14}$ **93.** $216a^{15}$ **95.** x^{60} **97.** $32b^{25}$ **103.** $\frac{3}{4}$
105. 5 **107.** 7 **109.** 12

Key Concept (page 479)

1. Let x = the monthly cost to lease the van. **3.** Let x = the
width of the filed. **5.** Let x = the distance traveled by the
motorist. **7.** $a + b = b + a$ **9.** $\frac{b}{1} = b$ **11.** $n - 1 < n$
13. $(r + s) + t = r + (s + t)$

Chapter Review (page 481)

1. no **2.** yes **3.** y **4.** t **5.** 9 **6.** 31 **7.** 340 **8.** 133 **9.** 9
10. 14 **11.** 120 **12.** 5 **13.** \$97,250 **14.** 185 **15.** 4 **16.** 3
17. 21 **18.** 14 **19.** 21 **20.** 36 **21.** 315 **22.** 425 **23.** 6 ft
24. \$128 **25.** h = height of wall, length of upper base = $h - 5$,
length of lower base = $2h - 3$ **26.** 5 ft, 17 ft **27.** 12 **28.** -8
29. 100 **30.** -4 **31.** 130 mi, 114 mi, $6x$ mi, $55t$ mi **32.** \$278
33. \$15,230 **34.** 2002 **35.** 2004 **36.** They decreased
37. The pool is 2° C warmer. **38.** 144 ft **39.** $-10x$
40. $42xy$ **41.** $60de$ **42.** $32s$ **43.** $2e$ **44.** $49xy$ **45.** $84k$
46. $100t$ **47.** $4y + 20$ **48.** $-30t - 45$ **49.** $-21 - 21x$
50. $-12e + 24x + 3$ **51.** $-6t + 4$ **52.** $-5 - x$
53. $-6t + 3s - 1$ **54.** $5a + 3$ **55.** $-4x, 8$ **56.** $-3y, 1$
57. factor **58.** term **59.** factor **60.** term **61.** yes **62.** no
63. yes **64.** no **65.** $7x$ **66.** $-3r$ **67.** $-9t$ **68.** $-3z$
69. $5x$ **70.** $-12y$ **71.** $w - 5$ **72.** $-6x + 2y$ **73.** $-46d + 2a$
74. $10y + 15h - 1$ **75.** $13y + 48$ **76.** $-5t + 22$ **77.** $3x + 12$

78. $-50f + 84$ **79.** 194 ft **80.** yes **81.** -4 **82.** -18 **83.** -3
84. 15 **85.** 4 **86.** 85 **87.** -3 **88.** -2 **90. a.** $4h \cdot 4h \cdot 4h$
b. $5^2d^3m^4$ **91.** h^{10} **92.** t^8 **93.** w^7 **94.** 4^{12} **95.** $8b^7$
96. $-24x^4$ **97.** $24f^7$ **98.** $-a^2b^2$ **99.** x^2y^6 **100.** m^2n^2
101. $27m^3z^7$ **102.** $-20c^3d^6$ **103.** v^{12} **104.** $27y^3$ **105.** $25t^8$
106. $8a^{12}b^{15}$ **107.** c^{26} **108.** $108s^{12}$ **109.** c^{14} **110.** $8x^9$

Chapter 8 Test (page 487)

1. 4 **2.** 30 **3.** 11 **4.** 81 **5.** 200 **6.** 24 **7.** 3,100 **8.** 194 yr
9. -3 **10.** 165 mi **11.** $25x + 5$ **12.** $-42 + 6x$ **13.** $-6y - 4$
14. $6a + 9b - 21$ **15. a.** factor **b.** term **16. a.** $-28y + 10x$
b. $-3t$ **17.** $8x^2, -4x, -6$ **18. a.** $11x$ **b.** $24ce$ **c.** $5x$
d. $30y$ **19.** $-6y - 3$ **20.** 2 **21.** -9 **22.** -3 **23.** 1
24. a. $10k¢$ **b.** \$20$(p + 2)$ **25. a.** h^6 **b.** $-28x^5$ **c.** b^8
d. $24g^5k^{13}$ **26. a.** f^{15} **b.** $4a^4b^2$ **c.** x^{15} **d.** x^{15}

Cumulative Review Exercises (page 489)

1. 7,535,700 **2.** 7,540,000 **3.** 9,137 **4.** 3,322 **5.** 245,870
6. 875 **7.** 260 ft **8.** 4,000 ft² **9.** $2^3 \cdot 3 \cdot 7$ **10.** $3^2 \cdot 5^2$
11. $2^2 \cdot 3^2 \cdot 5$ **12.** $2^4 \cdot 3^2 \cdot 5$ **13.** 18 **14.** -11 **15.** 1
16. -10 **17.** $\frac{1}{7}$ **18.** $\frac{2}{5}$ **19.** $\frac{19}{12} = 1\frac{7}{12}$ **20.** $\frac{1}{4}$ **21.** 7 **22.** 14
23. -52 **24.** -2 **25.** $-15x$ **26.** $28x^2$ **27.** $-6x + 8$
28. $-15x + 10y - 20$ **29.** $5x$ **30.** $7a^2$ **31.** $-x - y$
32. $-4x + 8$ **33.** -5 **34.** -5 **35.** -16 **36.** -60 **37.** 4
38. 4 **39.** 21 **40.** 21 ft by 84 ft **41.** p^8 **42.** $-15t^9$
43. $-x^5y^7$ **44.** $81a^8$ **45.** $-n^6$ **46.** $9x^6y^2$ **47.** $108p^{12}$
48. $-8x^9$

Chapter 9 Check Your Knowledge (page 492)

1. parallel, perpendicular **2.** quadrilateral, triangle **3.** right,
hypotenuse **4.** congruent, similar **5.** perimeter, area
6. radius, circumference, diameter **7.** volume, surface
8. a. III **b.** I **c.** IV **d.** II **9.** B **10.** 14° **11.** 36 **12.** 61°
13. a. 50° **b.** 130° **c.** 130° **d.** 50° **e.** 95 **14.** 80°
15. a. 13 ft **b.** 60 ft² **16.** 17.55 in.² **17.** 28π ft² ≈ 78.5 ft³
18. $10\frac{2}{3}\pi$ ft³ ≈ 33.5 ft³ **19.** 200 ft³ **20.** 160π ft³ ≈ 502.7 ft³

Study Set Section 9.1 (page 499)

1. segment **3.** midpoint **5.** protractor **7.** right **9.** 180°
11. supplementary **13.** true **15.** false **17.** true **19.** true
21. acute **23.** obtuse **25.** right **27.** straight **29.** true
31. false **33.** yes **35.** yes **37.** no **39.** true **41.** true
43. true **45.** true **47.** angle **49.** ray **51.** 3 **53.** 3 **55.** 1
57. B **59.** 40° **61.** 135° **63.** 10 **65.** 27.5 **67.** 30 **69.** 25
71. 60° **73.** 75° **75.** 130° **77.** 230° **79.** 100° **81.** 40°
83.

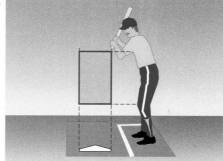

85. 65,115
87. 30°
95. 16

97. $\frac{7}{24}$ **99.** 6 **101.** 5

Study Set Section 9.2 (page 507)

1. coplanar **3.** perpendicular **5.** alternate **7.** $\angle 4$ and $\angle 6$, $\angle 3$ and $\angle 5$ **9.** $\angle 3, \angle 4, \angle 5, \angle 6$ **11.** They are parallel.
13. a right angle **15.** is perpendicular to **17.** m($\angle 1$) = 130°, m($\angle 2$) = 50°, m($\angle 3$) = 50°, m($\angle 5$) = 130°, m($\angle 6$) = 50°, m($\angle 7$) = 50°, m($\angle 8$) = 130° **19.** m($\angle A$) = 50°, m($\angle 1$) = 85°, m($\angle 2$) = 45°, m($\angle 3$) = 135° **21.** 10 **23.** 30 **25.** 40 **27.** 12
29. If the stones are level, the plum bob string should pass through the midpoint of the crossbar of the A-frame.
41. 72 **43.** 45% **45.** yes **47.** $\frac{1}{3}$

Study Set Section 9.3 (page 514)

1. regular **3.** hexagon **5.** octagon **7.** equilateral
9. hypotenuse **11.** parallelogram **13.** rhombus **15.** isosceles
17. 4, quadrilateral, 4 **19.** 3, triangle, 3 **21.** 5, pentagon, 5
23. 6, hexagon, 6 **25.** scalene triangle **27.** right triangle
29. equilateral triangle **31.** isosceles triangle **33.** square, rhombus, rectangle **35.** rhombus **37.** rectangle
39. trapezoid **41.** triangle **43.** 90° **45.** 45° **47.** 90.7°
49. 30° **51.** 60° **53.** 720° **55.** 1,440° **57.** 7 sides
59. 14 sides **65. b.** octagon **c.** triangle **d.** pentagon
67. pentagon, hexagon **73.** 22 **75.** 40% **77.** 0.10625

Study Set Section 9.4 (page 522)

1. congruent **3.** similar **5.** true **7.** false **9.** true **11.** yes
13. yes **15.** a and b represent the lengths of the legs; c represents the length of the hypotenuse. **17.** 25, square root, c, 5 **19.** is congruent to **21.** $\overline{DF}, \overline{AB}, \overline{EF}, \angle D, \angle B, \angle C$
23. yes, SSS **25.** not necessarily **27.** yes, SSS **29.** yes, SAS
31. 6 mm **33.** 50° **35.** yes **37.** 5 **39.** 8 **41.** $\sqrt{56}$
43. yes **45.** no **47.** 36 ft **49.** 59.2 ft **51.** 4,000 ft **53.** 12 ft
55. 25 in. **57.** 127.3 ft **61.** $1\frac{1}{3}$ **63.** 20 **65.** 9

Study Set Section 9.5 (page 534)

1. perimeter **3.** area **5.** square **7.** length 15 in. and width 5 in.; length 16 in. and width 4 in. (answers may vary) **9.** sides of length 5 m **11.** base 5 yd and height 3 yd (answers may vary) **13.** length 5 ft and width 4 ft; length 20 ft and width 3 ft (answers may vary) **15.** $4s$ **17.** square inch **19.** s^2
21. triangle **23.** 32 in. **25.** 36 m **27.** 37 cm **29.** 85 cm
31. $28\frac{1}{3}$ ft **33.** 16 cm² **35.** 60 cm² **37.** 25 in.² **39.** 169 mm²
41. 80 m² **43.** 75 yd² **45.** 75 m² **47.** 144 **49.** $4,875
51. 81 **53.** linoleum **55.** $1,200 **57.** $361.20 **59.** $192
61. 111,825 mi² **63.** 51 **65.** spot 1: l = 20 ft, w = 10 ft, 200 ft²; spot 2: b_1 = 20 ft, b_2 = 16 ft, h = 10 ft, 180 ft²; spot 3: b = 28 ft, h = 28 ft, 392 ft² **69.** $1\frac{5}{12}$ **71.** $6\frac{1}{12}$ **73.** $1\frac{7}{18}$

Study Set Section 9.6 (page 543)

1. radius **3.** diameter **5.** minor **7.** circumference **9.** $\overline{OA}$, $\overline{OC}$, and $\overline{OB}$ **11.** $\overline{DA}, \overline{DC}$ and $\overline{AC}$ **13.** $\overparen{ABC}$ and $\overparen{ADC}$
15. Double the radius. **17. a.** 1 in. **b.** 2 in.
c. 2π in. $\approx$ 6.28 in. **d.** π in.² $\approx$ 3.14 in.² **19.** Square 6.
21. arc AB **23.** $\pi D, 2\pi r$ **25.** π **27.** 8π
29. 37.70 in. **31.** 36 m **33.** 25.42 ft **35.** 31.42 m
37. A = 28.3 in.² **39.** 88.3 in.² **41.** 128.5 cm² **43.** 27.4 in.²
45. 66.7 in.² **47.** 3.14 mi² **49.** 32.66 ft **51.** 12.73 times

53. 1.59 ft **55.** 12.57 ft²; 0.79 ft²; 6.28% **63.** 90%
65. 5.375¢ per oz **67.** five

Study Set Section 9.7 (page 554)

1. volume **3.** cube **5.** surface **7.** cylinder **9.** cone
11. $V = lwh$ **13.** $V = \frac{4}{3}\pi r^3$ **15.** $V = \frac{1}{3}Bh$ or $V = \frac{1}{3}\pi r^2 h$
17. $SA = 2lw + 2lh + 2hw$ **19.** 27 ft³ **21.** 1,000 dm³
23. a. volume **b.** area **c.** volume **d.** surface area
e. perimeter **f.** surface area **25. a.** 72 in.³ **b.** 18 in.²
c. 24 in.² **27.** cubic inch **29.** 60 cm³ **31.** 48 m³
33. 3,053.63 in.³ **35.** 1,357.17 m³ **37.** 314.16 cm³ **39.** 400 m³
41. 94 cm² **43.** 1,256.64 in.² **45.** 576 cm³ **47.** 335.10 in.³
49. $\frac{1}{8}$ in.³ = 0.125 in.³ **51.** 2.125 **53.** 197.92 ft³
55. 33,510.32 ft³ **57.** 8:1 **63.** −42 **65.** −1 **67.** $\frac{1}{5}$
69. 2,400 mm

Key Concept (page 557)

1. $d = rt$ **3.** $P = 2l + 2w$ **5.** 210,000 ft² **7.** $80.50 **9.** 144 ft
11. $750, $45, $6,250

Chapter Review (page 559)

1. points C and D, line CD, plane GHI **2.** 5 units **3.** $\angle ABC$, $\angle CBA, \angle B, \angle 1$ **4.** 48° **5.** $\angle 1$ and $\angle 2$ are acute, $\angle ABD$ and $\angle CBD$ are right angles, $\angle CBE$ is obtuse, and $\angle ABC$ is a straight angle. **6.** obtuse angle **7.** right angle **8.** straight angle **9.** acute angle **10.** 15 **11.** 150 **12. a.** 65° **b.** 115°
13. 40° **14.** 40° **15.** no **16.** part a **17.** $\angle 4$ and $\angle 6$, $\angle 3$ and $\angle 5$ **18.** $\angle 1$ and $\angle 5$, $\angle 4$ and $\angle 8$, $\angle 2$ and $\angle 6$, $\angle 3$ and $\angle 7$
19. $\angle 1$ and $\angle 3$, $\angle 2$ and $\angle 4$, $\angle 5$ and $\angle 7$, $\angle 6$ and $\angle 8$
20. m($\angle 1$) = 70°, m($\angle 2$) = 110°, m($\angle 3$) = 70°, m($\angle 4$) = 110°, m($\angle 5$) = 70°, m($\angle 6$) = 110°, m($\angle 7$) = 70° **21.** m($\angle 1$) = 60°, m($\angle 2$) = 120°, m($\angle 3$) = 130°, m($\angle 4$) = 50° **22.** 40 **23.** 20
24. octagon **25.** pentagon **26.** triangle **27.** hexagon
28. quadrilateral **29.** 3 **30.** 4 **31.** 8 **32.** 6 **33.** isosceles
34. scalene **35.** equilateral **36.** right triangle **37.** yes
38. no **39.** 90 **40.** 50 **41.** 50° **42.** It is equilateral.
43. trapezoid **44.** square **45.** parallelogram **46.** rectangle
47. rhombus **48.** rectangle **49.** 15 cm **50.** 40° **51.** 100°
52. true **53.** false **54.** true **55.** true **56.** 65° **57.** 115°
58. 360° **59.** 720° **60.** $\angle D$ **61.** $\angle E$ **62.** $\angle F$ **63.** $\overline{DF}$
64. $\overline{DE}$ **65.** $\overline{EF}$ **66.** congruent, SSS **67.** congruent, SAS
68. congruent, ASA **69.** not necessarily congruent **70.** yes
71. yes **72.** 21 ft **73.** 13 **74.** 15 **75.** 31.3 in. **76.** 72 in.
77. 9 m **78.** 30 m **79.** 36 m **80.** 9.61 cm² **81.** 7,500 ft²
82. 450 ft² **83.** 200 in.² **84.** 120 cm² **85.** 232 ft² **86.** 152 ft²
87. 120 m² **88.** 9 ft² **89.** 144 in.² **90.** $\overline{CD}, \overline{AB}$ **91.** $\overline{AB}$
92. $\overline{OA}, \overline{OC}, \overline{OD}, \overline{OB}$ **93.** O **94.** 66.0 cm **95.** 45.1 cm
96. 254.5 in.² **97.** 130.3 cm² **98.** 125 in.³ **99.** 480 m³
100. 600 in.³ **101.** 3,619 in.³ **102.** 1,518 ft³ **103.** 785 in.³
104. 9,020,833 ft³ **105.** 35,343 ft³ **106.** 1,728 in.³ **107.** 54 ft³
108. 61.8 ft² **109.** 314.2 in.²

Chapter 9 Test (page 569)

1. 4 units **2.** B **3.** true **4.** false **5.** false **6.** true **7.** 50
8. 140 **9.** 12 **10.** 45 **11.** 23° **12.** 63° **13.** 70° **14.** 110°
15. 70° **16.** 40 **17.** 3, 4, 6, 5, 8 **18.** equilateral triangle, scalene triangle, isosceles triangle **19.** 57° **20.** 66° **21.** 30°
22. 1,440° **23.** m($\overline{AB}$) = m($\overline{DC}$), m($\overline{AD}$) = m($\overline{BC}$), and m($\overline{AC}$) = m($\overline{BD}$) **24.** 130° **25.** 8 in. **26.** 50° **27.** 6

28. 12   **29.** 127.3 ft   **30.** 391.6 cm^2 **31.** 83.7 ft^2 **32.** 94.2 ft
33. 28.3 ft^2 **34.** 159.3 m^3 **35.** 268.1 m^3 **36.** 66.7 ft^3
38. The surface area is 6 times the area of one face of the cube.

Cumulative Review Exercises   (page 571)

1.

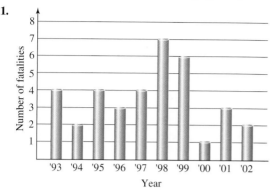

2. $8,995   **3.** 11,022   **4.** 33   **5.** 2,110,000   **6.** $11 \cdot 5 \cdot 2^2$
7. 1, 2, 3, 4, 6, 8, 12, 24   **8.** $\{\ldots -3, -2, -1, 0, 1, 2, 3, \ldots\}$ **9.** 13
10. -10 **11.** 3   **12.** 5   **13.** An equation contains an = sign;
an expression does not.   **14.** $-12x + 8$ **15.** 89   **16.** -10
17. -11 **18.** 5   **19.** 6.5 in./min   **20.** $50x¢$ **21.** $7^2 - 2^3$
22. $6\frac{2}{3}$ b.f.   **23.** $\frac{5}{4}$ **24.** $142\frac{7}{15}$ **25.** $\frac{3}{20}$ **26.** $13\frac{3}{4}$ cups   **27.** $-\frac{11}{20}$
28. $\frac{1}{3}$ **29.** $\frac{8}{9}$ **30.** $\frac{15}{2}$ **31.** $\frac{3}{32}$ fluid oz   **32.** $\frac{8}{9}$ **33.** $1\frac{9}{29}$
34. 18 oz.   **35. a.** 1998; about 0.6° F   **b.** 1986; about $-0.4°$ F
36.

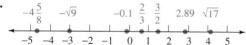

37. 3.1416   **38.** $>$ **39.** 145.188   **40.** 17.05   **41.** 89,970.8
42. 0.053   **43.** -25.6 **44.** 22.3125   **45.** $0.1\overline{3}$ **46.** -9.32
47. 97   **48.** 10   **49.** -2 **50.** $\frac{7}{9}$ **51.** 0.86 oz, 0.855 oz, 0.85 oz
52. a. $-3, 2, 9$ **b.** $3, 2, -9$ **53.** s^9 **54.** a^{35}
55. $15h^{10}$ **56.** $8b^9c^{18}$ **57.** y^{22} **58.** x^{m+n} **59.** 93%, 7%
60. 67.5   **61.** 120   **62.** $0.57, \frac{57}{100}, 0.1\%, \frac{1}{1,000}, 33\frac{1}{3}\%, 0.\overline{3}$
63.

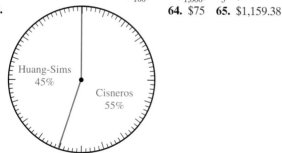

64. $75   **65.** $1,159.38

66. 500%   **67.** $1,522.50   **68.** $269,390.92   **69.** $\frac{3}{7}$ **70.** $\frac{1}{4}$
71. the smaller board   **72.** $6\frac{1}{2}$ **73.** 125,000   **74.** 75 ft
75. 14   **76.** 540   **77.** 13.25   **78.** 120   **79.** 200   **80.** 12.3
81. 65.4   **82.** 0.5   **83.** 58.9   **84.** 167°   **85.** 240 km
86. between 5,700 and 5,800 cg   **87. a.** about 4 m/gal

b. 11,355,000 L   **88.** about 4.5 kg   **89.** 90   **90.** more than 0
but less than 90   **91.** 75°   **92.** 15°   **93.** 50°   **94.** 50°   **95.** 130°
96. 50°   **97.** 75°   **98.** 30°   **99.** 105°   **100.** 105°   **101.** 46, 134
102. 540°   **103.** 13 m   **104.** 42 m, 108 m^2 **105.** 126 ft^2
106. 91 in.^{2} **107.** 43.98 cm, 153.94 cm^2 **108.** 98.31 yd^2
109. 210 m^3 **110.** 523.60 in.^{3} **111.** 150.80 m^3 **112.** 3.93 ft^3
113. 2,124 in.^{2}

Study Set   Section I.1   (page A-3)

1. monomial   **3.** binomial   **5.** binomial   **7.** monomial
9. monomial   **11.** trinomial   **13.** 3   **15.** 2   **17.** 1   **19.** 7
23. 13   **25.** 6   **27.** 31   **29.** 4   **31.** 1   **33.** 0 ft   **35.** 64 ft
37. 63 ft   **39.** 198 ft   **43.** 2   **45.** $\frac{13}{7} = 1\frac{6}{7}$ **47.** $\frac{3}{2} = 1\frac{1}{2}$
49. 16   **51.** 6

Study Set   Section I.2   (page A-8)

1. like   **3.** coefficients, variables   **5.** yes, $7y$ **7.** no   **9.** yes,
$13x^3$ **11.** yes, $15x^2$ **13.** $2x^2, 7x, 5x^2$ **15.** $9y$ **17.** $-12t^2$
19. $14s^2$ **21.** $7x + 4$ **23.** $7x^2 - 7$ **25.** $12x^3 - 14.9x$
27. $8x^2 + 2x - 21$ **29.** $8y^2 + 4y - 2$ **31.** $6x^2 + x - 5$
33. $2n^2 + 5$ **35.** $5x^2 + x + 11$ **37.** $-7x^2 - 5x - 1$
39. $2x^2 + x + 12.9$ **41.** $16u^3$ **43.** $7x^5$ **45.** $1.6a + 8$
47. $-19x^2 - 5$ **49.** $7x^2 - 2x - 5$ **51.** $1.7y^2 + 3.1y - 9$
53. $7b + 4$ **55.** $p^2 - 2p$ **57.** $5x^2 + 6x - 8$
59. $-12x^2 - 13x + 36$ **61.** $-x^3 + x + 14$ **63.** $92,000
65. $112,800   **67.** $211,000   **69.** $y = -800x + 8,500$
71. $y = -1,900x + 18,700$ **77.** 0.8 oz   **79.** 54 ft

Study Set   Section I.3   (page A-13)

1. monomial   **3.** 3   **5.** numerical, factors   **7.** term, like
9. $2x, 5$ **11.** $12x^5$ **13.** $-6b^3$ **15.** $-6x^5$ **17.** $-\frac{1}{2}y^7$
19. $3x + 12$ **21.** $-4t - 28$ **23.** $3x^2 - 6x$ **25.** $-6x^4 + 2x^3$
27. $6x^3 + 8x^2 - 14x$ **29.** $-2p^3 + 3p^2 - 2p$
31. $3q^4 - 6q^3 + 21q^2$ **33.** $a^2 + 9a + 20$ **35.** $3x^2 + 10x - 8$
37. $6a^2 + 2a - 20$ **39.** $4x^2 + 12x + 9$ **41.** $4x^2 - 12x + 9$
43. $25t^2 + 10t + 1$ **45.** $6x^3 - x^2 + 1$ **47.** $x^3 - 1$
49. $x^3 - x^2 - 5x + 2$ **51.** $4x^2 + 11x + 6$ **53.** $12x^2 + 14x - 10$
55. $x^3 + 1$ **57.** $(x^2 - 4)$ ft^2 **59.** $R = -\frac{x^2}{100} + 30x$
65. four and ninety-one thousandths   **67.** 0.109375
69. 134.657   **71.** 10

Study Set   Appendix II   (page A-19)

1. inductive   **3.** circular   **5.** alternating   **7.** alternating
9. 10 A.M.   **11.** 17   **13.** -17 **15.** 6   **17.** 3   **19.** -11 **21.** 9
23.   **25.**   **27.** I   **29.** m   **31.** Maria

33. the Mercedes   **35.** green, blue, yellow, red   **37.** 18,935
39. 0

INDEX